T0213286

Texts in Computational Science and Engineering

20

Editors

Timothy J. Barth
Michael Griebel
David E. Keyes
Risto M. Nieminen
Dirk Roose
Tamar Schlick

More information about this series at http://www.springer.com/series/5151

John A. Trangenstein

Scientific Computing

Vol. III – Approximation and Integration

 Springer

John A. Trangenstein
Professor Emeritus of Mathematics
Department of Mathematics
Duke University
Durham
North Carolina, USA

Additional material to this book can be downloaded from http://extras.springer.com.

ISSN 1611-0994   ISSN 2197-179X   (electronic)
Texts in Computational Science and Engineering
ISBN 978-3-030-09872-8   ISBN 978-3-319-69110-7   (eBook)
https://doi.org/10.1007/978-3-319-69110-7

Mathematics Subject Classification (2010): 34, 41, 65

Printed on acid-free paper

This Springer imprint is published by the registered company Springer International Publishing AG part of Springer Nature.
The registered company address is: Gewerbestrasse 11, 6330 Cham, Switzerland

To my grandsons Jack and Luke

Preface

This is the third volume in a three-volume book about scientific computing. The primary goal in these volumes is to present many of the important computational topics and algorithms used in applications such as engineering and physics, together with the theory needed to understand their proper operation. However, a secondary goal in the design of this book is to allow readers to experiment with a number of interactive programs *within the book*, so that readers can improve their understanding of the problems and algorithms. This interactivity is available in the HTML form of the book, through JavaScript programs.

The intended audience for this book are upper level undergraduate students and beginning graduate students. Due to the self-contained and comprehensive treatment of the topics, this book should also serve as a useful reference for practicing numerical scientists. Instructors could use this book for multisemester courses on numerical methods. They could also use individual chapters for specialized courses such as numerical linear algebra, constrained optimization, or numerical solution of ordinary differential equations. In order to read all volumes of this book, readers should have a basic understanding of both linear algebra and multivariable calculus. However, for this volume it will suffice to be familiar with linear algebra and single variable calculus. Some of the basic ideas for both of these prerequisites are reviewed in this text, but at a level that would be very hard to follow without prior familiarity with those topics. Some experience with computer programming would also be helpful, but not essential. Students should understand the purpose of a computer program, and roughly how it operates on computer memory to generate output.

Many of the computer programming examples will describe the use of a Linux operating system. This is the only publicly available option in our mathematics department, and it is freely available to all. Students who are using proprietary operating systems, such as Microsoft and Apple systems, will need to replace statements specific to Linux with the corresponding statements that are appropriate to their environment.

This book also references a large number of programs available in several programming languages, such as C, C^{++}, Fortran and JavaScript, as well as

MATLAB modules. These programs should provide examples that can train readers to develop their own programs, from existing software whenever possible or from scratch whenever necessary.

Chapters begin with an overview of topics and goals, followed by recommended books and relevant software. Some chapters also contain a case study, in which the techniques of the chapter are used to solve an important scientific computing problem in depth.

Chapters typically begin with a summary of topics and goals, followed by recommended books and relevant software. Many chapters also contain a case study, in which the techniques of the chapter are used to solve an important scientific computing problem in depth.

Chapter 1 discusses interpolation and approximation. These topics are often introduced in introductory calculus, but need much greater elaboration for scientific computing. The techniques in this chapter will be fundamental to the development of numerical methods in the remaining chapters of this volume.

Chapter 2 presents numerical methods for differentiation and integration. Numerical integration can employ important ideas from probability, or useful techniques from polynomial approximation, together with the skills used to overcome rounding errors. The chapter ends with a discussion of multidimensional integration methods, which are not commonly discussed in scientific computing texts. This chapter depends on material in Chap. 1.

Chapter 3 discusses the numerical solution of initial value problems in ordinary differential equations. The mathematical analysis of these problems is fairly straightforward and generally easy for readers to understand. Fortunately, some very sophisticated software is available for just this purpose and represents a level of achievement that should be the goal of software developers working in other problem areas. This chapter depends on material in Chaps. 1 and 2.

The final Chap. 4 examines ordinary differential equations with specified boundary values. The mathematical analysis of these problems is more difficult and is often approached either by eigenfunction expansions (a topic that builds on the discussion in Chap. 1 of Volume II) or by functional analysis (a topic that few students in typical scientific computing classes have studied). This chapter depends on material in the three preceding chapters.

In summary, this volume covers mathematical and numerical analysis, algorithm selection, and software development. The goal is to prepare readers to build programs for solving important problems in their chosen discipline. Furthermore, they should develop enough mathematical sophistication to know the limitations of the pieces of their algorithm and to recognize when numerical features are due to programming bugs rather than the correct response of their problem.

I am indebted to many teachers and colleagues who have shaped my professional experience. I thank Jim Douglas Jr. for introducing me to numerical analysis as an undergrad. (Indeed, I could also thank a class in category theory for motivating me to look for an alternative field of mathematical study.) John Dennis, James Bunch, and Jorge Moré all provided a firm foundation for my training in numerical analysis, while Todd Dupont, Jim Bramble, and Al Schatz gave me important training in finite element analysis for my PhD thesis. But I did not really learn to program until I met Bill Gragg, who also emphasized the importance of classical analysis in the development of fundamental algorithms. I also learned from my students, particularly Randy LeVeque, who was in the first numerical analysis class I ever taught. Finally, I want to thank Bill Allard for many conversations about the deficiencies in numerical analysis texts. I hope that this book moves the field a bit in the direction that Bill envisions.

Most of all, I want to thank my family for their love and support.

Durham, NC, USA   John A. Trangenstein
July 7, 2017

Contents

Contents for Volume 1

Contents for Volume 2

Chapter 1
Interpolation and Approximation

Far better an approximate answer to the right question, which is often vague, than the exact answer to the wrong question, which can always be made precise.

John Tukey *[178, p. 13]*

Abstract This chapter begins with a discussion of interpolation. Polynomial interpolation for a function of a single variable is analyzed, and implemented through Newton, Lagrange and Hermite forms. Intelligent selection of interpolation points is discussed, and extensions to multi-dimensional polynomial interpolation are presented. Rational polynomial interpolation is studied next, and connected to quadric surfaces. Then the discussion turns to piecewise polynomial interpolation and splines. The study of interpolation concludes with a presentation of parametric curves. Afterwards, the chapter moves on to least squares approximation, orthogonal polynomials and trigonometric polynomials. Trigonometric polynomial interpolation or approximation is implemented by the fast Fourier transform. The chapter concludes with wavelets, as well as their application to discrete data and continuous functions.

1.1 Overview

In our earlier work, we have found it useful to replace a complicated function with another that is easier to evaluate. For example, in order to find a zero of a nonlinear function $f(x)$, Newton's method in Sect. 5.4 of Chap. 5 in Volume I found the zero of a linear function that *interpolates* f and its derivative at a specified point. The secant method in Sect. 5.5 of Chap. 5 in Volume I replaced a general function f by a linear function that interpolates f at two points. In order to minimize a function, Newton's method in Sect. 5.7.3 of Chap. 5 in Volume I used a quadratic function that interpolates the function value, first and second derivatives at a given point. Hermite

Additional Material: The details of the computer programs referred in the text are available in the Springer website (http://extras.springer.com/2018/978-3-319-69110-7) for authorized users.

cubic minimization in Sect. 5.7.5 of Chap. 5 in Volume I interpolated function values and slopes at given two points. These ideas could be generalized to incorporate more derivatives and more interpolation points. Rather than continue to analyze each interpolation case separately, we should find a way to analyze the interpolation process in general.

In other circumstances, we replaced complicated functions by simpler functions without using interpolation. In Example 6.2.1 of Chap. 6 in Volume I we discussed *approximating* data points by a straight line. The resulting line may not have reproduced any of the data points exactly, but it did minimize the sum of squares of the errors in the data representation. Also, our analysis of rounding error accumulation in Sect. 3.8 of Chap. 3 in Volume I and similar sections was an attempt to approximate complicated and uncertain numerical results with simpler and more easily understood functions of machine precision. We have also used other kinds of approximations without explicitly examining them. In particular, computer evaluation of functions such as the logarithm, exponential or trigonometric functions all involve clever approximations.

However, the next example will show that some approximation ideas are not successful.

Example 1.1.1 Suppose that we want to approximate

$$\mathrm{erf}(x) \equiv \frac{2}{\sqrt{\pi}} \int_0^x e^{-t^2} dt \text{ for } |x| \leq 1 .$$

We could use the well-known Taylor series for the exponential function to obtain

$$\mathrm{erf}(x) \approx \int_0^x \sum_{j=0}^n \frac{(-1)^j t^{2j}}{j!} \, dt = \sum_{j=0}^n \frac{(-1)^j x^{2j+1}}{(2j+1)!} .$$

Such an approximation might be very accurate for small values of x, and wildly inaccurate for large values of x, even for large values of n. To examine this approximation for various values of n, readers may view Fig. 1.1, which was generated by the C^{++} program erf.C. They may also execute the JavaScript program for **approximating the error function**.

Computer algorithms for evaluating transcendental functions are typically designed to be fast (usually involving approximations by polynomials) and accurate (by minimizing the maximum error over some interval of the function argument). The precise manner in which these approximations are performed varies with the computer hardware manufacturer and each of its chip designs, but some early ideas regarding how to perform these approximations can be found in Fike's "Computer Evaluation of Mathematical Functions" [78].

The goal of this chapter is to analyze the error in replacing a function $f(x)$, that is difficult to evaluate, either because it is complicated or known only through an equation it satisfies, by a linear combination of nice functions $\phi_0(x), \ldots, \phi_n(x)$ that are easy to evaluate. We will want to find scalars $\gamma_0, \ldots, \gamma_n$ so that $f - \sum_{i=0}^n \phi_i \gamma_i$ is

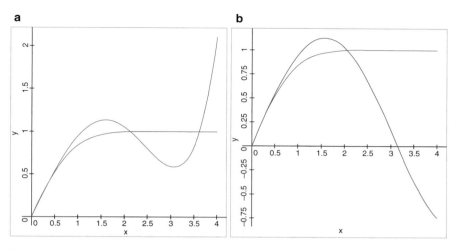

Fig. 1.1 Error function approximation. (**a**) $n = 2$. (**b**) $n = 4$

small. In general, there are two main ways in which we could measure this error. If f is known at m discrete points, we could require zero error at each of the m points. This requirement is equivalent to imposing the linear system

$$\begin{bmatrix} \phi_0(x_0) & \cdots & \phi_n(x_0) \\ \vdots & & \vdots \\ \phi_0(x_m) & \cdots & \phi_n(x_m) \end{bmatrix} \begin{bmatrix} \gamma_0 \\ \vdots \\ \gamma_n \end{bmatrix} = \begin{bmatrix} f(x_0) \\ \vdots \\ f(x_m) \end{bmatrix}. \tag{1.1}$$

When $m = n$ and the matrix $[\phi_j(x_i)]$ is nonsingular, this process is called **interpolation**.

In other cases, we might prefer to overdetermine the linear system by taking $m > n$. This might be done to reduce the effect of random noise in the values of $f(x_i)$, or to give the curve a smoother shape. However, when $m > n$ we must allow the overdetermined linear system for the γ_j to be solved inexactly. More generally, if f is known as a function and we are given a norm on functions, we could choose the coefficients γ_j to minimize $\| f - \sum_{j=0}^{n} \phi_j \gamma_j \|$. This process is call **approximation**.

Of course, the optimization problem in approximation depends on the choice of the norm. There are several popular **function norms** used for approximation of a function f defined on a finite interval:

- **max norm:** $\| f \|_\infty \equiv \max_{x \in [a,b]} |f(x)|$,
- **2-norm:** $\| f \|_2 \equiv \sqrt{\int_a^b f(x)^2 dx}$,
- **weighted 2-norm:** $\| f \|_{2,w} \equiv \sqrt{\int_a^b f(x)^2 w(x) dx}$, where $w(x) > 0$, and
- **p-norm:** $\| f \|_p \equiv \left[\int_a^b |f(x)|^p dx \right]^{1/p}$, where $1 \le p < \infty$.

All of these norms satisfy the requirements in the Definition 3.5.1 of Chap. 3 in Volume I of a norm. We will see other examples of norms as we progress.

We will begin this chapter by studying polynomial interpolation in Sects. 1.2 and 1.3. Polynomials are easy to compute, and Taylor's theorem shows that they have good approximation properties. However, there are problems in using high-order polynomials to approximate some functions over large intervals. Sometimes, good results can be obtained with rational polynomials, especially in graphical display of quadric surfaces; these ideas will be discussed in Sects. 1.4 and 1.5. Another way to overcome this problems with high-order polynomial interpolation is to use different polynomials on sub-intervals of the interpolation domain; such piecewise polynomials are called splines; these will be discussed in Sect. 1.6.

Another way to overcome problems with interpolation is to use approximation. Least-squares approximation is the most common approach, and this idea is examined in Sect. 1.7. There are effective ways to perform least squares polynomial approximation via orthogonal polynomials, and very fast ways to perform least squares trigonometric polynomial approximation. We will conclude the chapter with a discussion of wavelet approximation in Sect. 1.8. Wavelets can be used to approximate functions on a hierarchy of scales, with very fast computations for changing scales.

We should also remark that the ideas in this chapter will be essential to the developments in later chapters. When we study integration and differentiation in Chap. 2, we will make heavy use of both polynomial interpolation and polynomial approximation to develop accurate and efficient methods. Polynomial interpolation will play a crucial role in the development of methods for solving initial value problems in Chap. 3, and polynomial approximation will allow us to develop high-order finite element methods for boundary-value problems in Chap. 4.

For more information about the material in this chapter, we recommend the numerical analysis books by Dahlquist and Björck [57], Greenbaum [91], Henrici [102], Kincaid and Cheney [119], Ralston and Rabinowitz [149] and Stoer [162]. For more information about approximation theory, we recommend the books by Brenner and Scott [23, Chapter 4], Cheney [40], Chui [42], Ciarlet [43], Cohen [48], Daubechies [58], Davis [59], de Boor [61], Foley et al. [80, Chapters 11 and 12], Strang and Nguyen [164] and Trefethen [176].

For interpolation software, readers may be interested in GSL (GNU Scientific Library) Interpolation. The GSL library contains routines to perform linear, polynomial and spline interpolations for functions of a single variable. MATLAB users should become familiar with commands interp1, interp2, interp3 and interpn. These commands can perform various kinds of piecewise polynomial interpolations for 1, 2, 3 or n function arguments. Least squares polynomial approximations can be computed by the MATLAB command polyfit. Similar routines are available in the GSL library routines for least squares fitting. For fast Fourier transforms, we recommend fftpack or fftw, which are publicly available through netlib. Other fast Fourier transform software is available through the GSL fast Fourier transform routines, or through MATLAB commands fft, fft2 or fftn. For discrete wavelets, we recommend either the GSL wavelet transforms or the MATLAB wavelet toolbox.

1.2 Polynomial Interpolation

In this section, we will describe the general polynomial interpolation problem and study its well-posedness. Afterward, we will develop some common representations for polynomial interpolation, including the Newton and Lagrange forms. We will also study Hermite interpolation, in which both function values and derivatives are interpolated. We will see that high-order polynomial interpolation can produce substantial errors between the interpolation points, particularly near and beyond the endpoints of interpolation. In order to reduce this so-called Runge phenomenon, we will re-examine the error estimates to learn how to make better choices for the interpolation points. We will end the section with a brief discussion of Bernstein polynomials, which are the basic tool for proving the famous Weierstrass approximation theorem.

1.2.1 Well-Posedness

The **polynomial interpolation problem** takes the following form:

- **Given** real scalars y_i for $0 \le i \le n$,
- **Find** a polynomial p of degree at most n
- **So that** for $0 \le i \le n$, $p(x_i) = y_i$.

For example, the scalars y_i could be the values $f(x_i)$ for some real-valued function f.

As we stated in Sect. 1.3 of Chap. 1 in Volume I, after specifying a mathematical model for a scientific computing problem, we need to make sure that the problem is well-posed. The following lemma will show that a solution to the polynomial interpolation problem exists and is unique.

Lemma 1.2.1 *Suppose that n is a non-negative integer and $x_0, \ldots, x_n$ are distinct real numbers. Then for any $n + 1$ real numbers $y_0, \ldots, y_n$ there is a unique polynomial p of degree at most n so that for $0 \le i \le n$ we have $p(x_i) = y_i$.*

Proof First, we will prove uniqueness of the interpolant. If p and $\widetilde{p}$ both interpolate $y_0, \ldots, y_n$ then $p - \widetilde{p}$ has $n + 1$ distinct zeros $x_0, \ldots, x_n$. The fundamental theorem of Algebra 1.2.1 of Chap. 1 in Volume II shows that we can use the zeros x_1 through x_n to factor

$$p(x) - \widetilde{p}(x) = c \prod_{i=1}^{n} (x - x_i) .$$

Since we also have $p(x_0) - \widetilde{p}(x_0) = 0$, we conclude that $c = 0$, and thus that $p - \widetilde{p} = 0$.

Next, we will prove existence of the interpolant by induction on the degree of the polynomial. For degree zero, the obvious interpolant is the constant polynomial

$p_0(x) \equiv y_0$. Inductively, we assume that for any $k \geq 1$ there exists a polynomial p_{k-1} of degree at most $k - 1$ so that for all $0 \leq i < k$ we have

$$p_{k-1}(x_i) = y_i \; . '$$

In order to interpolate at the additional point x_k, let

$$p_k(x) = p_{k-1}(x) + c_k(x - x_0) \cdot \ldots \cdot (x - x_{k-1}) \; .$$

It is obvious that for $0 \leq i < k$ we have

$$p_k(x_i) = y_i + c_k(x_i - x_0) \cdot \ldots \cdot (x_i - x_{k-1}) = y_i \; ,$$

no matter what the value of c_k might be. We can easily choose c_k to interpolate at x_k:

$$c_k = \frac{y_k - p_{k-1}(x_k)}{(x_k - x_0) \ldots (x_k - x_{k-1})} \; .$$

Here, the denominator is nonzero because the interpolation points are assumed to be distinct. Thus the induction can be continued for arbitrary degree (up to degree n), and the proof is complete.

The next lemma gives us a representation for the error in polynomial interpolation.

Lemma 1.2.2 *Suppose that the function f is $n + 1$ times continuously differentiable in $[a, b]$, and p is a polynomial of degree at most n that interpolates f at distinct interpolation points $x_0 < \ldots < x_n$ in $[a, b]$. Then for all $x \in [a, b]$ there exists $\xi_x \in (a, b)$ so that*

$$f(x) - p(x) = \frac{f^{(n+1)}(\xi_x)}{(n + 1)!} \prod_{i=0}^{n} (x - x_i) \; . \tag{1.2}$$

Proof The conclusion is obvious if $x \in \{x_0, \ldots, x_n\}$. Otherwise, let

$$w(t) = \prod_{i=0}^{n} (t - x_i) \text{ and } \psi_x(t) = f(t) - p(t) - w(t)\lambda_x \; ,$$

where λ_x is chosen so that $\psi_x(x) = 0$:

$$\lambda_x = \frac{f(x) - p(x)}{w(x)} \; .$$

Note that ψ_x is $n+1$ times continuously differentiable in $[a,b]$, and that $\psi_x = 0$ at $n+2$ points $x, x_0, \ldots, x_n$. By Rolle's theorem, ψ_x' has at least $n+1$ distinct zeros in (a,b), ψ_x'' has at least n distinct zeros in (a,b). In general, the k-th derivative $\psi_x^{(k)}$ has at least $n+2-k$ distinct zeros in (a,b). We conclude that $\psi_x^{(n+1)}$ has at least one zero $\xi_x \in (a,b)$. Then

$$0 = \psi_x^{(n+1)}(\xi_x) = f^{(n+1)}(\xi_x) - p^{(n+1)}(\xi_x) - \lambda_x w^{(n+1)}(\xi_x) = f^{(n+1)}(\xi_x) - \lambda_x(n+1)! \, .$$

This implies that

$$\lambda_x = \frac{f^{(n+1)}(\xi_x)}{(n+1)!} \, ,$$

and the result is proved.

Although this lemma gives us an exact formula for the error in polynomial interpolation, its use is limited by the need to evaluate the $(n+1)$-th order derivative of f, and by the lack of an explicit value for ξ_x. Typically, this error formula is useful only if we can find a convenient upper bound for $|f^{(n+1)}(\xi)|$ for any ξ between the interpolation points.

Exercise 1.2.1 Suppose that p is a polynomial of degree n that interpolates an $(n+1)$-times continuously differentiable function f at points $x_0 < \ldots < x_n$, and that $\widetilde{p}$ interpolates $\widetilde{f}$ at the same points. Find an expression for the error $p(x) - \widetilde{p}(x)$ where $x \in (x_0, x_n)$.

Exercise 1.2.2 If $a \le x_0 < \ldots < x_n \le b$ and $a \le x \le b$, show that

$$\left| \prod_{i=0}^{n} (x - x_i) \right| \le (b-a)^{n+1} \, .$$

Use this inequality to bound the maximum error in polynomial interpolation, in terms of the maximum value of the $n+1$-th derivative of the interpolated function.

Exercise 1.2.3 Suppose that the interpolation points $x_0 < \ldots x_n$ are **equidistant**, meaning that there is a positive number h so that

$$x_i = x_0 + ih \, .$$

In order to simplify the calculations to follow, we will assume that $x_0 = -1$ and $x_n = 1$, and that n is even. Define

$$w(x) = \prod_{i=0}^{n} (x - x_i) \, .$$

1. If $x \in (x_{n-1}, x_n)$, show that $x = rh$ where $n - 2 < 2r < n$, and show that

$$(n-1)! \leq \frac{|w(x)|}{h^{n-1}(x - x_{n-1})(x_n - x)} \leq n! \,.$$

2. Show that for $x \in (x_{n-1}, x_n)$ the maximum value of $(x - x_{n-1})(x_n - x)$ is $h^2/4$, and use this fact to show that

$$|w(x)| \leq \frac{2^{n-1} n!}{n^{n+1}} \,.$$

3. Show that $w(x)$ is an odd function (i.e., $w(-x) = -w(x)$), and conclude that the maximum of $|w(x)|$ for $x \in [-1, 1]$ is equal to the maximum of $|w(x)|$ for $x \in [0, 1]$.
4. Show that if $x \notin \{x_0, \ldots x_n\}$ and $x \in (0, x_{n-1})$, then

$$\left| \frac{w(x + h)}{w(x)} \right| > 1 \,.$$

Conclude that the maximum of $|w(x)|$ occurs in (x_{n-1}, x_n), and that interpolation at equidistant points is usually more accurate for x near the center of the interpolation region.

Exercise 1.2.4 Suppose that we interpolate $f(x) = \cos(x)$ at the points $x_0 = 0.3$, $x_1 = 0.4$, $x_2 = 0.5$ and $x_3 = 0.6$. First, determine the error at $x = 0.44$ in the polynomial interpolation of f at these interpolation points. Then use Eq. (1.2) to find a bound on the error at x. How does your error bound compare with the true error?

1.2.2 Newton Interpolation

In Sect. 1.2.1, we showed that polynomial interpolants exist and are unique. We also found a formula for the error in polynomial interpolation. Now we are ready to develop strategies for computing polynomial interpolants.

Some choices for representing the interpolating polynomial are clearly better than others. For example, if we choose the monomials as basis polynomials, then we have to solve the linear system

$$\begin{bmatrix} x_0^0 & x_0^1 & \cdots & x_0^n \\ x_1^0 & x_1^1 & \cdots & x_1^n \\ \vdots & \vdots & \cdots & \vdots \\ x_n^0 & x_n^1 & \cdots & x_n^n \end{bmatrix} \begin{bmatrix} \gamma_0 \\ \gamma_1 \\ \vdots \\ \gamma_n \end{bmatrix} = \begin{bmatrix} y_0 \\ y_1 \\ \vdots \\ y_n \end{bmatrix} \,.$$

The matrix in this linear system is called a **Vandermonde matrix**. It is a full matrix that would require some kind of matrix factorization so that we could solve for the γ_j. Further, if we want to increase the order of the interpolation, we have to change all of the previous unknown γ_j. Worst of all, Vandermonde matrices tend to be poorly conditioned [84].

However, it is possible to choose a different basis for polynomials, so that we obtain a left-triangular system of equations for the coefficients γ_j. This form has several advantages. First, no factorization of the matrix is required. Second, if we increase the order of the interpolation, then the lower-order coefficients remain unchanged. Best of all, our work in Sect. 3.8.6 of Chap. 3 in Volume I shows that the conditioning of triangular linear systems seldom becomes a significant numerical issue. Let us make these claims more concrete.

In **Newton interpolation** we choose the basis polynomials to be

$$v_j(x) = \prod_{i=0}^{j-1}(x - x_i) \ .$$

Then Newton interpolation takes the form

$$p(x) = \sum_{j=0}^{n}\gamma_j v_j(x) = \gamma_0 + \gamma_1(x - x_0) + \ldots + \gamma_n(x - x_0)(x - x_1)\ldots(x - x_{n-1}) \ .$$

Furthermore, the basis polynomials can be computed easily by recursion:

$$v_0(x) = 1$$
$$\text{for } 0 < j \le n \ , \ v_j(x) = v_{j-1}(x)(x - x_{j-1}) \ .$$

All that remains is to develop techniques for evaluating the coefficients γ_j in Newton interpolation. Consider the linear system for the coefficients in Newton interpolation:

$$\begin{bmatrix} 1 \\ 1 & x_1 - x_0 \\ 1 & x_2 - x_0 & (x_2 - x_0)(x_2 - x_1) \\ \vdots & \vdots & \vdots & \ddots \\ 1 & x_n - x_0 & (x_n - x_0)(x_n - x_1) & \ldots & \prod_{i=0}^{n-1}(x_n - x_i) \end{bmatrix} \begin{bmatrix} \gamma_0 \\ \gamma_1 \\ \gamma_2 \\ \vdots \\ \gamma_n \end{bmatrix} = \begin{bmatrix} f(x_0) \\ f(x_1) \\ f(x_2) \\ \vdots \\ f(x_n) \end{bmatrix} \quad (1.3)$$

Obviously, $\gamma_0 = f(x_0)$. Let us subtract the first equation from the other equations, then divide the ith equation by $x_i - x_0$ for $1 \le i \le n$:

$$\begin{bmatrix} 1 \\ 0 & 1 \\ 0 & 1 & x_2 - x_1 \\ \vdots & \vdots & \vdots & \ddots \\ 0 & 1 & x_n - x_1 & \cdots & \prod_{i=1}^{n-1}(x_n - x_i) \end{bmatrix} \begin{bmatrix} \gamma_0 \\ \gamma_1 \\ \gamma_2 \\ \vdots \\ \gamma_n \end{bmatrix} = \begin{bmatrix} f(x_0) \\ [f(x_1) - f(x_0)]/[x_1 - x_0] \\ [f(x_2) - f(x_0)]/[x_2 - x_0] \\ \vdots \\ [f(x_n) - f(x_0)]/[x_n - x_0] \end{bmatrix}$$

$$\equiv \begin{bmatrix} f[x_0] \\ f[x_0, x_1] \\ f[x_0, x_2] \\ \vdots \\ f[x_0, x_n] \end{bmatrix}$$

Thus

$$\gamma_1 = \frac{f(x_1) - f(x_0)}{x_1 - x_0} = f[x_0, x_1] \ .$$

Repeating the same process on the last n equations and unknowns in this system leads

$$\begin{bmatrix} 1 \\ 0 & 1 \\ 0 & 0 & 1 \\ 0 & 0 & 1 & x_3 - x_2 \\ \vdots & \vdots & \vdots & \ddots \\ 0 & 0 & 1 & x_n - x_2 & \cdots & \prod_{i=2}^{n-1}(x_n - x_i) \end{bmatrix} \begin{bmatrix} \gamma_0 \\ \gamma_1 \\ \gamma_2 \\ \vdots \\ \gamma_n \end{bmatrix} = \begin{bmatrix} f(x_0) \\ f[x_0, x_1] \\ \{f[x_0, x_2] - f[x_0, x_1]\}/\{x_2 - x_1\} \\ \vdots \\ \{f[x_0, x_n] - f[x_0, x_1]\}/\{x_n - x_1\} \end{bmatrix}$$

$$\equiv \begin{bmatrix} f[x_0] \\ f[x_0, x_1] \\ f[x_0, x_1, x_2] \\ \vdots \\ f[x_0, x_1, x_n] \end{bmatrix}$$

This process can be continued until all of the coefficients γ_j have been computed.

The solution process in the previous paragraph leads to the following definition.

Definition 1.2.1 Given a function $f(x)$ and distinct points $x_0, \ldots, x_n$, a **divided difference** $f[x_0, \ldots, x_\ell]$ is defined recursively for all $\ell \le n$ by

$$f[x_0] = f(x_0) \, , \tag{1.4}$$

for $0 \le \ell < n$ and $k > \ell$ $\tag{1.5}$

$$f[x_0, \ldots, x_{\ell-1}, x_\ell] = \frac{f[x_0, \ldots, x_{\ell-2}, x_\ell] - f[x_0, \ldots, x_{\ell-2}, x_{\ell-1}]}{x_\ell - x_{\ell-1}} \, . \tag{1.6}$$

We note that an alternative form of this recursion, which will be more useful for computation, can be found later in Eq. (1.10).

Definition (1.6) allows us to prove the next lemma.

Lemma 1.2.3 *Given an integer $n \ge 0$ and a function f, suppose that the polynomial*

$$p(x) = \sum_{j=0}^{n} \gamma_j \prod_{i=0}^{j-1} (x - x_i)$$

interpolates $f(x)$ at $x = x_0, \ldots, x_n$. Then for all $0 \le \ell \le n$

$$\gamma_\ell = f[x_0, \ldots, x_\ell] \tag{1.7}$$

and for all $\ell \le k \le n$

$$f[x_0, \ldots, x_{\ell-1}, x_k] = \sum_{j=\ell}^{k} \gamma_j \prod_{i=\ell}^{j-1} (x_k - x_i) \, . \tag{1.8}$$

Proof We will prove the result by induction on ℓ.

First, we consider the case $\ell = 0$. Since p interpolates f at $x_0, \ldots, x_n$, it follows that for all $0 \le k \le n$

$$f[x_k] = f(x_k) = p(x_k) = \sum_{j=0}^{n} \gamma_j \prod_{i=0}^{j-1} (x_k - x_i) = \sum_{j=0}^{k} \gamma_j \prod_{i=0}^{j-1} (x_k - x_i)$$

In the sum on the far right, we have eliminated the terms involving zero factors in the product. This proves that conclusion (1.8) is true for $\ell = 0$. In particular, taking $k = 0$ in this expression leads to a sum with only one term, namely

$$f[x_0] == \sum_{j=0}^{0} \gamma_j \prod_{i=0}^{j-1} (x_k - x_i) = \gamma_0 \, .$$

This proves conclusion (1.7) for $\ell = 0$.

Inductively, assume that the results are true for divided differences of order $\ell > 0$; in other words, assume that (1.7) and (1.8) hold as written. We will prove that the results hold for divided differences of order $\ell + 1$. Note that for all $\ell < k \leq n$ the inductive hypotheses imply that

$$
\begin{aligned}
f[x_0, \ldots, x_\ell, x_k] &= \frac{f[x_0, \ldots, x_{\ell-1}, x_k] - f[x_0, \ldots, x_{\ell-1}, x_\ell]}{x_k - x_\ell} \\
&= \frac{\sum_{j=\ell}^{k} \gamma_j \prod_{i=\ell}^{j-1}(x_k - x_i) - \sum_{j=\ell}^{\ell} \gamma_j \prod_{i=\ell}^{j-1}(x_\ell - x_i)}{x_k - x_\ell} \\
&= \frac{\sum_{j=\ell+1}^{k} \gamma_j \prod_{i=\ell}^{j-1}(x_k - x_i)}{x_k - x_\ell} = \sum_{j=\ell+1}^{k} \gamma_j \prod_{i=\ell+1}^{j-1} (x_k - x_i) .
\end{aligned}
$$

This proves (1.8) for $\ell + 1$. Taking $k = \ell + 1$ in this equation leads to

$$
f[x_0, \ldots, x_\ell, x_{\ell+1}] = \sum_{j=\ell+1}^{\ell+1} \gamma_j \prod_{i=\ell+1}^{j-1} (x_{\ell+1} - x_i) = \gamma_{\ell+1} .
$$

This proves (1.7) for $\ell + 1$, and completes the induction.
The previous lemma has the following easy consequence, which allows us to provide an explicit formula for the interpolating polynomial.

Corollary 1.2.1 *Suppose that the function f and data points $x_0, \ldots, x_n$ are given. Define the polynomial $p_f(x)$ by*

$$
p_f(x) \equiv \sum_{j=0}^{n} f[x_0, \ldots, x_j] \prod_{i=0}^{j-1}(x - x_i) .
$$

Then p_f interpolates f at $x_0, \ldots, x_n$.

Proof For all $0 \leq \ell \leq k \leq n$, we can use the definition of $p_f(x_\ell)$, remove zero terms from the sum, and then apply Eq. (1.8) to get

$$
p_f(x_\ell) = \sum_{j=0}^{n} f[x_0, \ldots, x_j] \prod_{i=0}^{j-1}(x_\ell - x_i) = \sum_{j=0}^{\ell} f[x_0, \ldots, x_j] \prod_{i=0}^{j-1}(x_\ell - x_i)
$$

$$
= f[x_\ell] = f(x_\ell) .
$$

Our next goal is to find an efficient algorithm for evaluating the divided differences. The next lemma will make this goal easier.

Lemma 1.2.4 *The divided difference $f[x_0, \ldots, x_\ell]$ is independent of the ordering of the points $x_0, \ldots, x_\ell$.*

Proof We will prove this result by induction. The claim is obviously true for $\ell = 0$, since there is only one ordering of a set with a single element. For $\ell = 1$, we note that

$$f[x_1, x_0] \equiv \frac{f(x_0) - f(x_1)}{x_0 - x_1} = \frac{f(x_1) - f(x_0)}{x_1 - x_0} = f[x_0, x_1] .$$

Let us assume inductively that the claim is true for $\ell \geq 1$. For divided differences with fewer than $\ell + 1$ interpolation points, the divided differences depend only on the set of interpolation points and not the ordering. We will denote these divided differences by putting the set of points inside the square brackets for the divided differences. For example, we will write the full set minus the member x_ℓ in the form $\{x_0, \ldots, x_{\ell+1}\} \setminus \{x_\ell\}$.

Since any permutation is a composition of interchanges, it will suffice to show that an interchange of two interpolation points does not change the divided difference. There are four cases.

First, suppose that we interchange x_i and x_j, where $0 \leq i < j < \ell$. Then

$$f[x_0, \ldots, x_{i-1}, x_j, x_{i+1}, \ldots, x_{j-1}, x_i, x_{j+1}, \ldots, x_\ell, x_{\ell+1}]$$
$$= \frac{f[\{x_0, \ldots, x_{\ell+1}\} \setminus \{x_\ell\}] - f[\{x_0, \ldots, x_{\ell+1}\} \setminus \{x_{\ell+1}\}]}{x_{\ell+1} - x_\ell}$$
$$= f[x_0, \ldots, x_\ell, x_{\ell+1}] .$$

Next, suppose that we interchange x_ℓ and $x_{\ell+1}$. Then

$$f[x_0, \ldots, x_{\ell-1}, x_{\ell+1}, x_\ell]$$
$$= \frac{f[\{x_0, \ldots, x_{\ell+1}\} \setminus \{x_\ell\}] - f[\{x_0, \ldots, x_{\ell+1}\} \setminus \{x_\ell\}]}{x_\ell - x_{\ell+1}}$$
$$= \frac{f[\{x_0, \ldots, x_{\ell+1}\} \setminus \{x_\ell\}] - f[\{x_0, \ldots, x_{\ell+1}\} \setminus \{x_\ell\}]}{x_{\ell+1} - x_\ell}$$
$$= f[x_0, \ldots, x_{\ell-1}, x_\ell, x_{\ell+1}] .$$

In the third case, we consider an interchange of $x_{\ell+1}$ and x_i where $0 \leq i < \ell$. We compute

$$f[x_0, \ldots, x_{i-1}, x_{\ell+1}, x_{i+1}, \ldots, x_{ell}, x_i]$$
$$= \frac{f[\{x_0, \ldots, x_{\ell+1}\} \setminus \{x_\ell\}] - f[\{x_0, \ldots, x_{\ell+1}\} \setminus \{x_i\}]}{x_i - x_\ell}$$

$$= \frac{1}{x_i - x_\ell} \left(\frac{f[\{x_0,\dots,x_{\ell+1}\} \setminus \{x_\ell,x_i\}] - f[\{x_0,\dots,x_{\ell+1}\} \setminus \{x_\ell,x_{\ell+1}\}]}{x_{\ell+1} - x_i} \right.$$

$$\left. - \frac{f[\{x_0,\dots,x_{\ell+1}\} \setminus \{x_i,x_\ell\}] - f[\{x_0,\dots,x_{\ell+1}\} \setminus \{x_i,x_{\ell+1}\}]}{x_{\ell+1} - x_\ell} \right)$$

$$= \frac{f[\{x_0,\dots,x_{\ell+1}\} \setminus \{x_i,x_\ell\}]}{x_i - x_\ell} \left(\frac{1}{x_{\ell+1} - x_i} - \frac{1}{x_{\ell+1} - x_\ell} \right)$$

$$+ \frac{f[\{x_0,\dots,x_{\ell+1}\} \setminus \{x_\ell,x_{\ell+1}\}]}{(x_{\ell+1} - x_i)(x_\ell - x_i)} + \frac{f[\{x_0,\dots,x_{\ell+1}\} \setminus \{x_i,x_{\ell+1}\}]}{(x_i - x_\ell)(x_{\ell+1} - x_\ell)}$$

$$= \frac{f[\{x_0,\dots,x_{\ell+1}\} \setminus \{x_i,x_\ell\}]}{(x_i - x_{\ell+1})(x_\ell - x_{\ell+1})} + \frac{f[\{x_0,\dots,x_{\ell+1}\} \setminus \{x_\ell,x_{\ell+1}\}]}{x_{\ell+1} - x_\ell} \left(\frac{1}{x_\ell - x_i} - \frac{1}{x_{\ell+1} - x_i} \right)$$

$$+ \frac{f[\{x_0,\dots,x_{\ell+1}\} \setminus \{x_i,x_{\ell+1}\}]}{(x_i - x_\ell)(x_{\ell+1} - x_\ell)}$$

$$= \frac{1}{x_{\ell+1} - x_\ell} \left(\frac{f[\{x_0,\dots,x_{\ell+1}\} \setminus \{x_i,x_\ell\}] - f[\{x_0,\dots,x_{\ell+1}\} \setminus \{x_\ell,x_{\ell+1}\}]}{x_{\ell+1} - x_i} \right.$$

$$\left. - \frac{f[\{x_0,\dots,x_{\ell+1}\} \setminus \{x_i,x_{\ell+1}\}] - f[\{x_0,\dots,x_{\ell+1}\} \setminus \{x_\ell,x_{\ell+1}\}]}{x_\ell - x_i} \right)$$

$$= \frac{f[\{x_0,\dots,x_{\ell+1}\} \setminus \{x_\ell\}] - f[\{x_0,\dots,x_{\ell+1}\} \setminus \{x_{\ell+1}\}]}{x_{\ell+1} - x_\ell} = f[x_0,\dots,x_{\ell+1}] \, .$$

In the final case, we consider an interchange of x_ℓ and x_i where $0 \leq i < \ell$. Using case two, then case three, and finally case two of this proof, we compute

$$f[x_0,\dots,x_{i-1},x_\ell,x_{i+1},\dots,x_{\ell-1},x_i,x_{\ell+1}]$$

$$= f[x_0,\dots,x_{i-1},x_\ell,x_{i+1},\dots,x_{\ell-1},x_{\ell+1},x_i]$$

$$= f[x_0,\dots,x_{i-1},x_i,x_{i+1},\dots,x_{\ell-1},x_{\ell+1},x_\ell]$$

$$= f[x_0,\dots,x_{i-1},x_i,x_{i+1},\dots,x_{\ell-1},x_\ell,x_{\ell+1}] \, .$$

Before we present our algorithm for divided differences, we will note that the divided differences are related to derivatives at intermediate points.

Lemma 1.2.5 *If f is n times continuously differentiable in $[a,b]$ and $x_0 < \dots < x_n$ are distinct points in $[a,b]$, then there exists $\xi \in (x_0, x_n)$ so that*

$$f[x_0,\dots,x_n] = \frac{f^{(n)}(\xi)}{n!} \, . \tag{1.9}$$

Proof Using Lemma 1.2.3 we see that

$$p_f(x) = \sum_{j=0}^{n} f[x_0, \ldots, x_j] \prod_{i=0}^{j-1} (x - x_i)$$

interpolates f at $x_0, \ldots, x_n$. It follows that $f(x) - p_f(x)$ has at least $n+1$ distinct zeros in $[x_0, x_n]$. By Rolle's theorem, $f'(x) - p'_f(x)$ has at least n distinct zeros in (x_0, x_n). In general, for all $0 < k \leq n$, we see that $f^{(k)} - p_f^{(k)}$ has at least $n + 1 - k$ distinct zeros in (x_0, x_n). In particular, $f^{(n)}(x) - p_f^{(n)}(x)$ has at least one zero $\xi \in (a, b)$. Since

$$0 = f^{(n)}(\xi) - p_f^{(n)}(\xi_x) = f^{(n)}(\xi) - f[x_0, \ldots, x_n] n!$$

the claimed results follows easily.

The results above can be used to develop a more computationally useful form for the divided difference recursion. If we use Lemma 1.2.4 to reorder the points in the divided difference, and the definition of the divided difference (1.6), we can obtain

$$
\begin{aligned}
f[x_0, \ldots, x_{n-1}, x_n] &= f[x_1, \ldots, x_n, x_0] \equiv \frac{f[x_1, \ldots, x_{n-1}, x_0] - f[x_1, \ldots, x_{n-1}, x_n]}{x_0 - x_n} \\
&= \frac{f[x_0, \ldots, x_{n-1}] - f[x_1, \ldots, x_n]}{x_0 - x_n} = \frac{f[x_1, \ldots, x_n] - f[x_0, \ldots, x_{n-1}]}{x_n - x_0}.
\end{aligned}
$$

$$(1.10)$$

To find all of the divided differences $\gamma_j = f[x_0, \ldots, x_j]$ at once, we can execute the following

Algorithm 1.2.1 (Divided Difference Evaluation)

> for $j = 0, \ldots, n$, $\gamma_j = f_j$
>
> for $j = 1, \ldots, n$
>
> > for $i = n, \ldots, j$, $\gamma_i = (\gamma_i - \gamma_{i-1})/(x_i - x_{i-j})$

To evaluate the interpolation polynomial $p(x)$, we perform the following

Algorithm 1.2.2 (Newton Interpolation Polynomial Evaluation)

> $p = \gamma_n$
>
> for $j = n - 1, \ldots, 0$, $p = \gamma_j + p * (x - x_j)$;

Example 1.2.1 Suppose that we want to find a cubic polynomial p that interpolates some function f at four interpolation points. In general, we would compute the divided difference table

$$
\begin{array}{ll}
x_0 & f(x_0) \\
& \quad\quad f[x_0, x_1] \\
x_1 & f(x_1) \quad\quad\quad\quad f[x_0, x_1, x_2] \\
& \quad\quad f[x_1, x_2] \quad\quad\quad\quad\quad f[x_0, x_1, x_2, x_3] \\
x_2 & f(x_2) \quad\quad\quad\quad f[x_1, x_2, x_3] \\
& \quad\quad f[x_2, x_3] \\
x_3 & f(x_3)
\end{array}
$$

Now suppose that we are given the following data:

x	1	2	4	5
f(x)	0	2	12	20

For these data, the divided difference table is

$$
\begin{array}{ll}
1 \quad 0 \\
& (2-0)/(2-1) = 2 \\
2 \quad 2 \quad\quad\quad\quad\quad\quad\quad (5-2)/(4-1) = 1 \\
& (12-2)/(4-2) = 5 \quad\quad\quad\quad\quad\quad (1-1)/(5-1) = 0 \\
4 \quad 12 \quad\quad\quad\quad\quad\quad\quad (8-5)/(5-2) = 1 \\
& (20-12)/(5-4) = 8 \\
5 \quad 20
\end{array}
$$

It follows that the interpolating polynomial is

$$
p(x) = 0 + (x-1)\{2 + (x-2)[1 + (x-4) \cdot 0]\} = (x-1)x \, .
$$

To be able to add another point to the interpolation without recomputing the coefficients γ_j, we need to store the top and bottom of the interpolation table. If this feature is important, then we need to modify the algorithm that generates the divided differences in the following form:

Algorithm 1.2.3 (Divided Differences for New Interpolation Point)

$$
\begin{aligned}
&\text{for } 0 \leq j \leq n \,, \ \gamma_j = f_j \\
&\delta_0 = \gamma_n \\
&\text{for } 0 \leq j \leq n \\
&\quad \text{for } n \geq i \geq j \ \gamma_i = (\gamma_i - \gamma_{i-1})/(x_i - x_{i-j}) \\
&\quad \delta_j = \gamma_n
\end{aligned}
$$

Afterward, we can add a new interpolation point as follows:

Algorithm 1.2.4 (Add New Point for Newton Interpolation)

$$\tau = \delta_0;$$

$$\delta_0 = f_{n+1};$$

$$\text{for } 1 \leq j \leq n+1$$

$$\sigma = \delta_j$$

$$\delta_j = (\tau - \delta_{j-1})/(x_{n+1-j} - x_{n+1})$$

$$\tau = \sigma$$

$$\gamma_{n+1} = \delta_{n+1}$$

Figure 1.2 shows polynomial interpolants to the sine function. In this case, low-order polynomials do an excellent job of representing the sine function over a large interval.

A Fortran program to implement Newton divided differences can be found in divided_difference.f, and a C^{++} program to perform polynomial interpolation can be found in interpolate.C. This C^{++} program also plots the interpolation and performs a numerical error analysis by measuring the interpolation errors for a range of orders of interpolating polynomials. Alternatively, readers can execute a JavaScript program for interpolating the sine function.

Exercise 1.2.5 Count the number of arithmetic operations in Algorithm 1.2.1 for computing the divided differences of a function f at $n+1$ interpolation points.

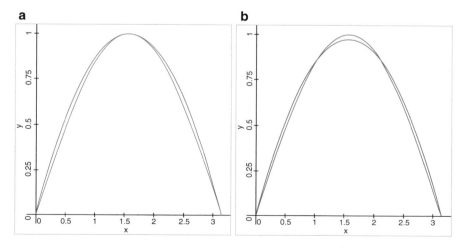

Fig. 1.2 Polynomial interpolation for sine. (**a**) 3 points. (**b**) 4 points

Table 1.1 Values for
polynomial of unknown
degree

x	$p(x)$
-2	-5
-1	1
0	1
1	1
2	7
3	25

Exercise 1.2.6 Count the number of arithmetic operations in Algorithm 1.2.2 for evaluating the Newton interpolation polynomial at a point x, given the divided differences γ_j.

Exercise 1.2.7 Write a program to interpolate $f(x) = e^x$ at $n + 1$ equidistant points in $[0, 1]$, and experimentally determine the number of interpolation points needed so that the maximum error in polynomial interpolation over this interval is at most 10^{-8}.

Exercise 1.2.8 What is the minimum degree of a polynomial $p(x)$ that has the specified values at the points given in Table 1.1?

Exercise 1.2.9 Prove that if p is a polynomial of degree k, then for all $n > k$ and for all distinct points $x_0 < \ldots < x_n$ we have $p[x_0, \ldots, x_n] = 0$.

Exercise 1.2.10 Let $\phi_n(x) = (x - x_0) \ldots (x - x_n)$. Show that

$$f[x_0, \ldots, x_n] = f(x_0))/\phi'(x_0) + \ldots + f(x_n)/\phi'(x_n) \ .$$

Exercise 1.2.11 Prove the Leibniz formula

$$(fg)[x_0, \ldots, x_n] = \sum_{k=0}^{n} f[x_0, \ldots, x_k] g[x_k, \ldots, x_n] \ .$$

1.2.3 Lagrange Interpolation

Our second interpolation technique chooses the basis polynomials $\phi_j(x)$ so that the matrix $[\phi_j(x_i)]$ in the interpolation linear system (1.1) is the identity matrix. Given interpolation points $x_0, \ldots, x_n$, it is not hard to see that the basis polynomials

$$\lambda_{j,n}(x) = \prod_{\substack{i=0 \\ i \neq j}}^{n} \frac{x - x_i}{x_j - x_i} \tag{1.11}$$

will accomplish this goal. The resulting **Lagrange interpolation** formula is

$$f(x) \approx p_f(x) \equiv \sum_{j=0}^{n} f(x_j) \prod_{i \neq j} \frac{x - x_i}{x_j - x_i} . \tag{1.12}$$

The Lagrange interpolation polynomial can be computed as follows

$$
\begin{aligned}
&p = 0 \\
&\text{for } j = 0 , \ldots , n \\
&\quad t = f(x_j) \\
&\quad \text{for } i = 0 , \ldots , j - 1 \\
&\qquad t = t * (x - x_i)/(x_j - x_i) \\
&\quad \text{for } i = j + 1 , \ldots , n \\
&\qquad t = t * (x - x_i)/(x_j - x_i) \\
&\quad p = p + t
\end{aligned} \tag{1.13}
$$

However, the following lemma will give us an alternative way to evaluate the interpolation polynomial without computing divided differences.

Lemma 1.2.6 *Let $\{x_j\}_{j=0}^{n}$ be a set of $n + 1$ distinct real numbers, and let $\{y_j\}_{j=0}^{n}$ be a set of $n + 1$ real numbers. Suppose that $p_{j,k}(x)$ is the polynomial of degree at most j such that*

$$p_{j,k}(x_i) = y_i \text{ for } 0 \leq i < j \text{ and } i = k . \tag{1.14}$$

Then for $0 \leq j < n$

$$p_{j+1,k}(x) = \frac{(x - x_k)p_{j,j}(x) - (x - x_j)p_{j,k}(x)}{x_j - x_k} . \tag{1.15}$$

Proof Since $p_{0,k}(x)$ is a polynomial of degree at most zero that takes value y_k at x_k, Eq. (1.14) implies that we must have

$$p_{0,k}(x) = y_k .$$

Next, note that Eq. (1.15) implies that $p_{1,k}(x)$ is a polynomial of degree at most one, and that for $1 \leq k \leq n$ we have

$$p_{1,k}(x_0) = \frac{(x_0 - x_k)p_{0,0}(x_0) - (x_0 - x_0)p_{0,k}(x_0)}{x_0 - x_k} = \frac{(x_0 - x_k)y_0}{x_0 - x_k} = y_0 \text{ and}$$

$$p_{1,k}(x_k) = \frac{(x_k - x_k)p_{0,0}(x_k) - (x_k - x_0)p_{0,k}(x_k)}{x_0 - x_k} = \frac{-(x_k - x_0)y_k}{x_0 - x_k} = y_k .$$

Thus $p_{1,k}(x)$ is the polynomial of degree at most 1 that passes through (x_0, y_0) and (x_k, y_k). Inductively, assume that $p_{j,k}(x)$ satisfies the interpolation conditions (1.14) for $j \leq k \leq n$. Let $p_{j+1,k}(x)$ be defined by (1.15). Then for $0 \leq i < j$ we get

$$p_{j+1,k}(x_i) = \frac{(x_i - x_k)p_{j,j}(x_i) - (x_i - x_j)p_{j,k}(x_i)}{x_j - x_k} = \frac{(x_k - x_k)y_i - (x_i - x_j)y_i}{x_j - x_k} = y_0 .$$

We also have

$$p_{j+1,k}(x_j) = \frac{(x_j - x_k)p_{j,j}(x_j) - (x_j - x_j)p_{j,k}(x_j)}{x_j - x_k} = \frac{(x_j - x_k)y_j}{x_j - x_k} = y_j \text{ and}$$

$$p_{j+11,k}(x_k) = \frac{(x_k - x_k)p_{j,j}(x_k) - (x_k - x_j)p_{j,k}(x_k)}{x_j - x_k} = \frac{-(x_k - x_j)y_k}{x_j - x_k} = y_k .$$

Thus $p_{j+11,k}(x)$ is the polynomial of degree at most $j + 1$ that passes through (x_i, y_i) for $0 \leq i \leq j$ and $i = k$. This completes the induction and the proof.
Lemma 1.2.6 leads to the following algorithm for polynomial interpolation:

Algorithm 1.2.5 (Neville-Aitken Polynomial Interpolation)

$$\text{for } 0 \leq k \leq n , \ p_k = y_k$$
$$\text{for } j = 0, \ldots, n - 1$$
$$t = x - x_j$$
$$\text{for } k = j + 1, \ldots, n$$
$$q_k = ((x - x_k)p_j - t * p_k)/(x_j - x_k)$$
$$\text{for } k = j + 1, \ldots, n$$
$$p_k = q_k$$
$$p = p_n$$

Some Lagrange interpolation basis polynomials of degree six can be viewed in Fig. 1.3.

The advantage of Lagrange interpolation is that the interpolation polynomial is easy to find. Compared to Newton interpolation however, it is harder with Lagrange interpolation to add a new point to the interpolation, since all the basis functions will change. Further, it is harder to evaluate the Lagrange interpolation polynomial than to evaluate the Newton interpolation polynomial.

Numerical experiments with the C^{++} program GUIInterpolate.C show that Newton interpolation can be significantly faster than Lagrange interpolation. For example, interpolation by a polynomial of degree 30 is performed more than 25

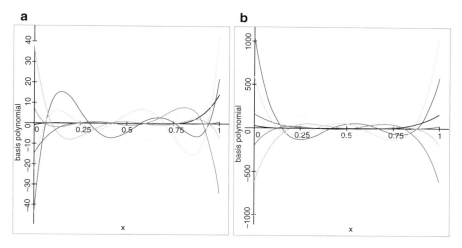

Fig. 1.3 Degree 6 Lagrange polynomial values and slopes. (**a**) Polynomial values. (**b**) Polynomial slopes

times faster by Newton interpolation than by Lagrange interpolation. Neville-Aitken interpolation is much more competitive with Newton interpolation.

The scientific computing literature contains few discussions of rounding errors in polynomial interpolation. One discussion is due to Brügner [24], who summarized his Ph.D. thesis regarding equidistant polynomial interpolation. He found that the Lagrange and Neville-Aitken interpolation algorithms were similar to each other in sensitivity to rounding errors, and significantly less sensitive than Newton interpolation.

Readers can view a C^{++} class to implement various polynomial representations in files Polynomial.H and Polynomial.C. There is an abstract class called `Polynomial`, with specific implementations such as `Monomial`, `C0LagrangePolynomial`, `C1LagrangePolynomial` and `C2Lagrange Polynomial`. Class `C0LagrangePolynomial` corresponds to the Lagrange polynomials we have studied in this section, and class `C1LagrangePolynomial` corresponds to the polynomials we are about to discuss in the next section.

Alternatively, readers may use the GSL (GNU Scientific Library) to perform certain kinds of polynomial interpolation. This software library can perform two types of polynomial interpolation, namely `gsl_interp_linear` and `gsl_interp_polynomial`. On the other hand, MATLAB apparently does not offer any commands to perform polynomial interpolation, although it does provide routines to perform certain kinds of spline interpolations.

Exercise 1.2.12 Count the number of arithmetic operations in Algorithm 1.13. How does this compare with Newton interpolation?

Exercise 1.2.13 Count the number of arithmetic operations in Algorithm 1.2.5. How does this compare with Newton interpolation and Lagrange interpolation?

Exercise 1.2.14 Suppose you are to make a table of values of $\log(x)$, $1/2 \le x \le 3/2$, on an equidistant mesh of width $1/(N+1)$, for some positive integer N. Assume that polynomial interpolation will be applied to the table, and that the maximum error in interpolation is to be at most 10^{-6}. What value of N should you choose? To answer this question, write a computer program to

- determine the interpolating polynomial $p(x)$ for arbitrary values of the number of interpolation points
- evaluate the error in the interpolation at arbitrary points $x \in [1/2, 3/2]$
- plot the error $\log(x) - p(x)$ as a function of x
- determine the maximum error at points $x_j = 1/2 + j/100(N+1)$ for $0 \le j \le 100(N+1)$, and
- plot the logarithm of the maximum error as a function of N.

1.2.4   Hermite Interpolation

For smooth functions, we may find it advantageous to interpolate both f and f' at each interpolation point. In this case, we want to find a polynomial p of degree at most $2n + 1$ so that for $0 \le i \le n$

$$p(x_i) = f(x_i) \text{ and } p'(x_i) = f'(x_i) .$$

We can symbolically modify Newton interpolation to perform Hermite interpolation as follows. We begin by forming a divided difference table in which the interpolation points are duplicated:

$$
\begin{array}{lll}
x_0 & f_0 & \\
 & & f'_0 \\
x_0 & f_0 & \\
 & & f[x_0, x_1] \\
x_1 & f_1 & \\
 & & f'_1 \\
x_1 & f_1 & \\
\vdots & \vdots & \vdots \\
x_n & f_n & \\
 & & f'_n \\
x_n & f_n &
\end{array}
$$

Additional columns of the Newton divided difference table are computed in the usual way.

Example 1.2.2 Suppose we want to find a cubic polynomial that interpolates f and f' at x_0 and x_1. We would form the table

$$
\begin{array}{llll}
x_0 & f_0 & & \\
& f'_0 & & \\
x_0 & f_0 & f[x_0,x_0,x_1] & \\
& f[x_0,x_1] & & f[x_0,x_0,x_1,x_1] \\
x_1 & f_1 & f[x_0,x_1,x_1] & \\
& f'_1 & & \\
x_1 & f_1 & &
\end{array}
$$

In particular, if $f(0) = 1$ and $f'(0) = 0 = f(1) = f'(1)$, then the divided difference table is

$$
\begin{array}{llll}
0 & 1 & & \\
& & 0 & \\
0 & 1 & & -1 \\
& & -1 & 2 \\
1 & 0 & & 1 \\
& & 0 & \\
1 & 0 & &
\end{array}
$$

Thus the Hermite interpolant is

$$
\begin{aligned}
p_f(x) &= f[x_0] + f[x_0,x_0](x-x_0) + f[x_0,x_0,x_1](x-x_0)^2 \\
&\quad + f[x_0,x_0,x_1,x_1](x-x_0)^2(x-x_1) \\
&= 1 - x^2 + 2x^2(x-1) = 1 - x^2(3-2x)\ .
\end{aligned}
$$

Exercise 1.2.15 Use divided differences and Newton interpolation to construct the cubic interpolant to f and f' at two points, thus verifying the formula (5.53) in Chap. 5 of Volume I, which we used to compute the minimizer of a functional.

Exercise 1.2.16 Suppose that we would like to generalize Hermite interpolation to interpolate zeroth, first and second derivatives at the interpolation points. Use Eq. (1.9) to determine the values that should be entered into the divided difference table, and verify your choice for interpolation of the zeroth, first and second derivative of a quadratic function at a single point x_0.

Exercise 1.2.17 Revise the proof of Lemma 1.2.5 to apply to divided differences for Hermite interpolation.

Exercise 1.2.18 Revise Lemma 1.2.2 to apply to interpolation of both f and f' at $n + 1$ distinct interpolation points.

Exercise 1.2.19 Let

$$\lambda_{j,n}(x) = \prod_{\substack{i=0 \\ i \neq j}}^{n} \frac{x - x_i}{x_j - x_i}$$

be the Lagrange interpolation polynomials. Show that the Lagrange form of Hermite interpolation is

$$f(x) \approx \sum_{j=0}^{n} f(x_j) \left[1 - 2(x - x_j)\lambda'_{j,n}(x_j) \right] \lambda_{j,n}(x)^2 + \sum_{j=0}^{n} f'(x_j) \left[x - x_j \right] \lambda_{j,n}(x)^2 ,$$

and describe how you would compute $\lambda'_{j,n}(x_j)$.

Exercise 1.2.20 Can you generalize the Neville-Aitken Algorithm 1.2.5 to Hermite interpolation? For assistance, see Mühlbach [136].

1.2.5 Runge Phenomenon

The third exercise in 1.2.1 was designed to show that the largest errors in polynomial interpolation at equidistant points are likely to occur near the ends of the interpolation interval. The following example illustrates this observation.

Example 1.2.3 (Runge)   If we interpolate

$$f(x) = \frac{1}{25x^2 + 1}$$

at equidistant points in $[-1, 1]$, then for $n \geq 10$ we find large oscillations in the interpolating polynomial $p(x)$ near the endpoints $x = \pm 1$. In fact, the maximum error in the interpolation at points in $[-1, 1]$ approaches infinity as the order n of the polynomial becomes infinite. Figure 1.4 shows polynomial interpolants to Runge's function. Note that the higher-order polynomials develop large oscillations near the ends of the interpolation interval. Alternatively, readers may execute the following JavaScript program for the **Runge phenomenon.**

By changing variables, it is possible to see that given any set of interpolation points $x_0^{(n)}, \ldots, x_n^{(n)} \in [a, b]$, there is a continuous function f defined on $[a, b]$ so that $\|f - p_n(x)\|_\infty \to \infty$ as $n \to \infty$. This pessimistic result is due to Faber [73].

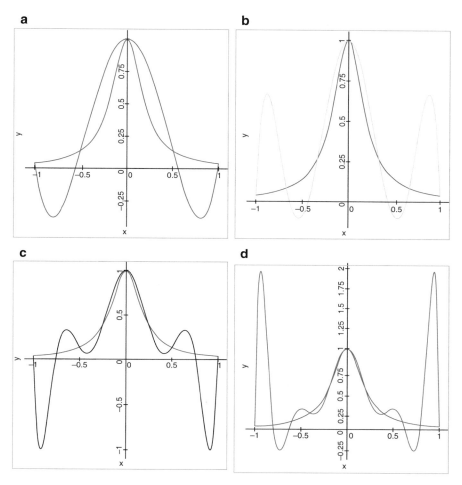

Fig. 1.4 Polynomial interpolation for Runge's function. (**a**) 5 points. (**b**) 7 points. (**c**) 9 points. (**d**) 11 points

1.2.6 *Chebyshev Interpolation Points*

Equally-spaced interpolation of Runge's function makes it obvious that we cannot always improve the accuracy of interpolation by increasing the number of inter-polation points. In order to understand how we might improve the accuracy of polynomial interpolation, let us re-examine the error estimate (1.2). This equation shows that the interpolation error depends on the order n of the interpolation polynomial, the value of the n-th derivative of f in the interpolation interval, and the interpolation points $x_0, \ldots, x_n$ through the Newton polynomial $v_{n+1}(x) = \prod_{i=0}^{n}(x - x_i)$.

Before we are given a function f to be interpolated by a polynomial on some interval $[a, b]$, we can try to control the interpolation error by choosing the interpolation points in order to solve

$$\min_{a \leq x_0 < \ldots < x_n \leq b} \max_{x \in [a,b]} \left| \prod_{i=0}^{n} (x - x_i) \right| . \tag{1.16}$$

The solution of this **minimax problem** can be determined from the **equal alternation theorem** of minimax approximation theory. This theorem says that, under certain conditions on the set of approximating functions $\phi_0, \ldots, \phi_n$, the linear combination $\sum_{j=0}^{n} \gamma_j \phi_j$ that minimizes the max-norm of the error $f - \sum_{j=0}^{n} \gamma_j \phi_j$ is such that the error has $n + 2$ equal maximum alternations on the approximating interval, with successive alternations taking alternating signs. Interested readers may read Atkinson [8, p. 190ff], Cheney [40, p. 75] or Ralston and Rabinowitz [149, p. 301] for more details about this theorem.

For our purposes, it will suffice to provide a solution to the minimax problem (1.16). First, we claim that if $[a, b] = [-1, 1]$, the solution is given by the Chebyshev polynomial

$$T_{n+1}(\xi) = 2^{-n} \cos\left([n + 1] \cos^{-1} \xi\right) .$$

This is because T_{n+1} is a polynomial of degree $n + 1$ with leading coefficient 1, and T_{n+1} has $n + 2$ equal alternations at the points $\cos(j\pi/[n + 1])$ for $0 \leq j \leq n + 1$. The zeros of T_{n+1} are the points

$$\xi_{n-j} = \cos\left(\frac{2j + 1}{2n + 2}\pi\right)$$

for $0 \leq j \leq n$. It follows that we can write T_{n+1} in factored form as

$$T_{n+1}(\xi) = \prod_{j=0}^{n} (\xi - \xi_j) .$$

In order to choose interpolation points on $[a, b]$ to minimize the maximum value of $\prod_{j=0}^{n}(x - x_j)$, we will map the Chebyshev zeros from $[-1, 1]$ to $[a, b]$. This can be done by the change of variables

$$x = \xi \frac{b - a}{2} + \frac{b + a}{2} .$$

In other words, the mapped Chebyshev interpolation points are

$$x_{n-j} = \frac{b - a}{2} \cos\left(\frac{2j + 1}{2n + 2}\pi\right) + \frac{b + a}{2}$$

for $0 \le j \le n$. With this choice,

$$\prod_{j=0}^{n}(x - x_j) = \left(\frac{b-a}{2}\right)^{n+1} T_{n+1}(\xi) \, ,$$

and

$$\left|\prod_{j=0}^{n}(x - x_j)\right| = \left(\frac{b-a}{2}\right)^{n+1} |T_{n+1}(\xi)| \le \left(\frac{b-a}{2}\right)^{n+1} 2^{-n} \, .$$

Thus the error in polynomial interpolation using Chebyshev interpolation points can be estimated as follows:

$$|f(x) - p_f(x)| \le (b-a)^{n+1} \frac{\max_{\xi \in (a,b)} |f^{(n+1)}(\xi)|}{2^{2n+1}(n+1)!} \text{ for all } x \in (a, b) \, .$$

The computational difficulty with the Chebyshev nodes is that whenever we increase the number n of interpolation points, we have to change all of the interpolation points.

Figure 1.5 shows the polynomial interpolants to Runge's function at several choices of numbers of Chebyshev points. This figure should be compared to Fig. 1.4, which showed polynomial interpolation at equidistant points. Unlike the latter figure, interpolation at the Chebyshev points is demonstrating convergence as the number of interpolation points increases. Both figures were generated by the C^{++} program GUIInterpolate.C.

Interested readers may also execute the JavaScript program for interpolating **Runge's function at the Chebyshev points** (see /home/faculty/johnt/ book/code/approximation/runge_chebyshev.html). The use of Chebyshev interpolation points eliminates the oscillations produced by equidistant interpolation points.

Exercise 1.2.21 Experimentally determine the minimum number of equidistant interpolation points needed to produce a maximum error of at most 10^{-4} in polynomial interpolation of the exponential function on $[0, 1]$. Also determine the minimum number of Chebyshev interpolation points needed for the same maximum error for the exponential function on $[0, 1]$. How do these numbers of interpolation points compare as the maximum error is decreased to 10^{-6}, 10^{-8} and 10^{-10}?

Exercise 1.2.22 Use the formula (1.12) for the Lagrange interpolation polynomial to derive a perturbation estimate. In other words, bound the error in the Lagrange interpolation polynomial at a point x due to errors in the function values at the interpolation points. In particular, how does the perturbation bound depend on the order of the interpolation and the width of the interpolation interval if we use Chebyshev interpolation points?

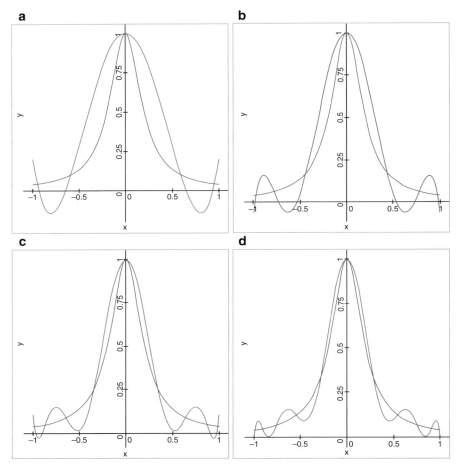

Fig. 1.5 Interpolation of Runge's function at Chebyshev points. (**a**) 5 points. (**b**) 7 points. (**c**) 9 points. (**d**) 11 points

1.2.7   Bernstein Polynomials

The following important result in analysis shows that polynomials can be used to approximate continuous functions with arbitrarily high uniform precision.

Theorem 1.2.1 (Weierstrass Approximation)   *Given a continuous function f on the closed interval $[a, b]$ and a real scalar $\varepsilon > 0$ there is a degree n and a polynomial p of degree n so that $\|f - p\|_\infty < \varepsilon$.*

Proof See, for example, Cheney [40, p. 68], Feller [77, vol. II, p. 222], Kincaid and Cheney [119, p. 289] or Yosida [187, p. 8].

The proofs all proceed by using the **Bernstein polynomials**

$$b_{j,n}(x) \equiv \binom{n}{j} x^j (1-x)^{n-j}$$

to show that the linear combinations

$$p_n(x) = \sum_{j=0}^{n} f\left(\frac{j}{n}\right) b_{j,n}(x)$$

converge uniformly to functions f that are continuous on $[0,1]$. The linear combinations p_n of the Bernstein polynomials do not necessarily interpolate f at any point. However, their approximation errors provide an upper bound for the errors associated with minimax approximation polynomials, which do interpolate f between the equal alternations.

The reader may execute a JavaScript program for **approximating Runge's function by Bernstein polynomials.** Interpolation of Runge's function at the Chebyshev points produces better approximations of Runge's function than the Bernstein polynomials. However, we will find other uses for Bernstein polynomials in Sect. 1.6.8.2.

Exercise 1.2.23 Show that for all $0 \le j \le n$ we have

$$\int_0^1 b_{j,n}(x)\, \mathrm{d}x = \frac{1}{n+1} \ .$$

1.3   Multidimensional Interpolation

In order to plot a data in multiple dimensions, it is common to plot a polynomial interpolant of the data on a grid. The grid could be a union of simplices (triangles in two dimensions or tetrahedra in three dimensions) or blocks (quadrilaterals in two dimensions or hexahedra in three dimensions). Either choice allows us to make a **tiling** or **tessellation** of a d-dimensional region. In three dimensions, it is sometimes useful to use pentahedra, which are more commonly chosen to be wedges (*a.k.a.* prisms) rather than pyramids. We will discuss simplices and blocks below; for a discussion of wedges, see, for example, Trangenstein [174]. Our discussion in this section is intended as a preparation for later study of finite element methods, and not for interpolation of more general data, particularly in high-dimensional spaces.

Multidimensional polynomial interpolation and approximation is also very important in finite element methods for solving partial differential equations. We will not discuss the numerical solution of partial differential equations in this text; instead, we recommend reading finite element books by Aziz [9], Babuška and Strouboulis [10], Bathe and Wilson [14], Braess [20], Brenner and Scott [23],

Chen [37], Ciarlet [43], Hughes [106], Johnson [116], Strang and Fix [163], Szabó and Babuška [169], Trangenstein [174], Wait and Mitchell [180] and Zienkiewicz [189]. Each of these texts deals with multidimensional polynomial interpolation and approximation within the context of solving partial differential equations.

1.3.1 Multi-Indices

In order to describe polynomials in multiple dimensions, we will begin with the following useful definition.

Definition 1.3.1 Given any positive integer d, a **multi-index** is a vector $\boldsymbol{\alpha}$ with d nonnegative integer components. If $\boldsymbol{\alpha}$ is a multi-index, then

$$|\boldsymbol{\alpha}| \equiv \sum_{i=1}^{d} \alpha_i \text{ and } \boldsymbol{\alpha}! \equiv \prod_{i=1}^{d} \alpha_i! \ .$$

If $\mathbf{x}$ is a d-vector and α is a multi-index, then

$$\mathbf{x}^{\boldsymbol{\alpha}} \equiv \prod_{i=1}^{d} \mathbf{x}_i^{\alpha_i} \ .$$

If $\boldsymbol{\alpha}$ and $\boldsymbol{\beta}$ are multi-indices of the same size then

$$\boldsymbol{\alpha}\boldsymbol{\beta} \equiv \prod_{i=1}^{d} \alpha_i \beta_i \ ,$$

and

$$\boldsymbol{\beta} \leq \boldsymbol{\alpha} \text{ if and only if for all } i \text{ we have } \beta_i \leq \alpha_i \ .$$

If $\boldsymbol{\beta} \leq \boldsymbol{\alpha}$, then we define the **binomial coefficient** to be

$$\binom{\boldsymbol{\alpha}}{\boldsymbol{\beta}} \equiv \frac{\boldsymbol{\alpha}!}{\boldsymbol{\beta}! \, (\boldsymbol{\alpha} - \boldsymbol{\beta})!} \ .$$

Using multi-indices, we can prove the following generalization of the binomial expansion (2.1) in Chap. 2 of Volume I.

Theorem 1.3.1 (Multinomial Expansion) *Suppose that n is a nonnegative integer, $\boldsymbol{\alpha}$ is a multi-index of size d and $\mathbf{x}$ is a d-vector. Then*

$$\left(\sum_{|\boldsymbol{\alpha}|=1} \mathbf{x}^{\boldsymbol{\alpha}} \right)^n = \sum_{|\boldsymbol{\alpha}|=n} \frac{n!}{\boldsymbol{\alpha}!} \mathbf{x}^{\boldsymbol{\alpha}} \ . \tag{1.17}$$

Proof Since

$$\left(\sum_{|\alpha|=1} \mathbf{x}^\alpha \right)^0 = 1 = \frac{0!}{0!}\mathbf{x}^0 \, ,$$

the theorem is obviously true for $n = 0$. Assume inductively that the theorem is true for $n - 1 \geq 0$. Then

$$\left(\sum_{|\alpha|=1} \mathbf{x}^\alpha \right)^n = \sum_{|\alpha|=1} \mathbf{x}^\alpha \sum_{|\beta|=n-1} \frac{(n-1)!}{\beta!}\mathbf{x}^\beta = \sum_{i=1}^d \sum_{|\beta|=n-1} \frac{(n-1)!}{\beta!}\mathbf{x}^{\beta+\mathbf{e}_i}$$

$$= \sum_{|\gamma|=n} (n-1)!\mathbf{x}^\gamma \sum_{i=1}^d \frac{1}{(\gamma - \mathbf{e}_i)!} \, .$$

Note that for all $|\gamma| = n$ we have

$$\sum_{i=1}^d \frac{1}{(\gamma - \mathbf{e}_i)!} = \frac{1}{\gamma!} \sum_{i=1}^d \gamma_i = \frac{n}{\gamma!} \, . \tag{1.18}$$

As usual, $\mathbf{e}_i$ is the i-th **axis** d-**vector**. By combining the last two equations, we can prove that the inductive hypothesis is true for n.

Example 1.3.1 As an example of a multinomial expansion, we note that

$$(\mathbf{x}_0 + \mathbf{x}_1 + \mathbf{x}_2)^3 = \frac{3!}{3!\, 0!\, 0!}\mathbf{x}_0^3 + \frac{3!}{0!\, 3!\, 0!}\mathbf{x}_1^3 + \frac{3!}{0!\, 0!\, 3!}\mathbf{x}_2^3$$

$$+ \frac{3!}{2!\, 1!\, 0!}\mathbf{x}_0^2\mathbf{x}_1^1 + \frac{3!}{2!\, 0!\, 1!}\mathbf{x}_0^2\mathbf{x}_2^1 + \frac{3!}{1!\, 2!\, 0!}\mathbf{x}_0^1\mathbf{x}_1^2$$

$$+ \frac{3!}{0!\, 2!\, 1!}\mathbf{x}_1^2\mathbf{x}_2^1 + \frac{3!}{1!\, 0!\, 2!}\mathbf{x}_0^1\mathbf{x}_2^2 + \frac{3!}{0!\, 1!\, 2!}\mathbf{x}_1^1\mathbf{x}_2^2 + \frac{3!}{1!\, 1!\, 1!}\mathbf{x}_0^1\mathbf{x}_1^1\mathbf{x}_2^1 \, .$$

Next, let us define polynomials in multiple dimensions.

Definition 1.3.2 Suppose that n is a nonnegative integer n and $\boldsymbol{\alpha}$ is a multi-index of size d. Let $\mathbf{x}$ be a d-vector, and for each multi-index $\boldsymbol{\alpha}$ with $|\boldsymbol{\alpha}| \leq n$ let a_α be a scalar. Then

$$p(\mathbf{x}) \equiv \sum_{|\alpha|\leq n} a_\alpha \mathbf{x}^\alpha$$

is a d-**dimensional polynomial** of degree at most n.

Example 1.3.2 The general polynomial of degree 3 in two dimensions is

$$p(\mathbf{x}) = a_{0,0}\mathbf{x}_0^0\mathbf{x}_1^0 + a_{1,0}\mathbf{x}_0^1\mathbf{x}_1^0 + a_{0,1}\mathbf{x}_0^0\mathbf{x}_1^1 + a_{2,0}\mathbf{x}_0^2\mathbf{x}_1^0 + a_{1,1}\mathbf{x}_0^1\mathbf{x}_1^1 + a_{0,2}\mathbf{x}_0^0\mathbf{x}_1^2$$
$$+ a_{3,0}\mathbf{x}_0^3\mathbf{x}_1^0 + a_{2,1}\mathbf{x}_0^2\mathbf{x}_1^1 + a_{1,2}\mathbf{x}_0^1\mathbf{x}_1^2 + a_{0,3}\mathbf{x}_0^0\mathbf{x}_1^3 \ .$$

In order to determine the dimension of the linear space of all polynomials of degree at most n in d dimensions, we will need some information about binomial coefficients. The following result was previously proved by a combinatorial argument in Sect. 2.3.1.10 of Chap. 2 in Volume I, and used to compute powers of matrices in Sect. 1.6.2 of Chap. 1 in Volume II.

Lemma 1.3.1 *For all* $1 \le k \le n$ *we have the* **Pascal's triangle** *formula*

$$\binom{n}{k} = \binom{n-1}{k} + \binom{n-1}{k-1} . \tag{1.19}$$

Proof Using the definitions of the binomial coefficients, we can compute

$$\binom{n-1}{k} + \binom{n-1}{k-1} = \frac{(n-1)!}{k!(n-k-1)!} + \frac{(n-1)!}{(k-1)!(n-k)!}$$
$$= \frac{(n-1)!}{(k-1)!(n-k-1)!}\left[\frac{1}{k} + \frac{1}{n-k}\right] = \frac{(n-1)!}{(k-1)!(n-k-1)!}\frac{n}{k(n-k)}$$
$$= \frac{n!}{k!(n-k)!} = \binom{n}{k} .$$

The following result can be found in Feller [77, vol. 1, p. 64]. We will use it to determine the dimension of the linear space of multi-dimensional polynomials in Lemma 1.3.3.

Lemma 1.3.2 *For all nonnegative integers* k *and* n,

$$\sum_{i=0}^{n}\binom{i+k}{k} = \binom{n+k+1}{k+1} . \tag{1.20}$$

Proof This result is obviously true for $n = 0$ and all k, since

$$\sum_{i=0}^{0}\binom{i+k}{k} = \binom{k}{k} = 1 = \binom{0+k+1}{k+1} .$$

Inductively, assume that the claim is true for $n-1$ and for all k. We will prove inductively that the claim is true for n and all k.

If $k = 0$, then

$$\sum_{i=0}^{n} \binom{i+0}{0} = \sum_{i=0}^{n} 1 = n+1 = \binom{n+0+1}{0+1}.$$

Assume inductively that the claim is true for n whenever $0 \le k < j$. Then Eq. (1.19) gives us

$$\binom{n+j+1}{j+1} = \binom{n+j}{j+1} + \binom{n+j}{j} = \binom{(n-1)+j+1}{j+1} + \binom{n+(j-1)+1}{(j-1)+1}$$

then the inductive hypotheses for $n-1$, and for n with $k = j-1$ give us

$$= \sum_{i=0}^{n-1} \binom{i+j}{j} + \sum_{i=0}^{n} \binom{i+j-1}{j-1}$$

then a change of summation variables in the first sum produces

$$= \sum_{i=1}^{n} \binom{i-1+j}{j} + \sum_{i=1}^{n} \binom{i+j-1}{j-1} + 1 = \sum_{i=1}^{n} \left[\binom{i+j-1}{j-1} + \binom{i-1+j}{j} \right] + 1$$

and finally Eq. (1.19) gives us

$$= \sum_{i=1}^{n} \binom{i+j}{j} + 1 = \sum_{i=0}^{n} \binom{i+j}{j}.$$

Finally, we are able to determine the dimension of the linear space of multi-dimensional polynomials. This result will help us to determine the number of basis functions needed to represent them.

Lemma 1.3.3 *The dimension of the linear space of polynomials of degree at most n in d dimensions is* $\binom{n+d}{d}$.

Proof In any number of dimensions d, the linear space of polynomials of degree at most n has a basis given by the monomials $\{x^\alpha : |\alpha| \le n\}$. It follows that the dimension of this space of polynomials is equal to the number of multi-indices of modulus at most n. In one dimension, this is

$$\sum_{|\alpha| \le n} 1 = \sum_{\alpha=0}^{n} 1 = n+1 = \binom{n+1}{1}.$$

Assume inductive that the claim is true in $d - 1$ dimensions. Given a multi-index $\boldsymbol{\alpha}$ in d dimensions, partition

$$\boldsymbol{\alpha} = \begin{bmatrix} \boldsymbol{\beta} \\ \alpha_d \end{bmatrix}$$

where $\boldsymbol{\beta}$ is a multi-index in $d - 1$ dimensions. Then our inductive hypothesis and Eq. (1.20) imply that

$$\sum_{|\boldsymbol{\alpha}| \le n} 1 = \sum_{\alpha_d=0}^{n} \sum_{|\boldsymbol{\beta}| \le n - \alpha_d} 1 = \sum_{\alpha_d=0}^{n} \binom{n - \alpha_d + d - 1}{d - 1} = \binom{n + d}{d} .$$

Exercise 1.3.1 Use Pascal's triangle to show that the Bernstein polynomials satisfy the recursion

$$b_{j,n}(t) = (1 - t)b_{j,n-1}(t) + t b_{j-1,n-1}(t) .$$

1.3.2 Simplices

Let us begin with the definition of our first geometric region for interpolation.

Definition 1.3.3 Let d be a positive integer, and let

$$\mathbf{X} = \{\mathbf{x}_0, \dots, \mathbf{x}_d\}$$

be a real $d \times (d + 1)$ matrix. A **simplex** in d dimensions with **vertices** given by the columns of $\mathbf{X}$ is the set of points

$$\{\mathbf{Xb} \ : \ \mathbf{b} \ge \mathbf{0} \text{ and } \mathbf{e}^\top \mathbf{b} = 1\} .$$

A simplex is with vertices $\mathbf{x}_0, \dots, \mathbf{x}_d$ is **positively oriented** if and only if

$$\det[\mathbf{x}_1 - \mathbf{x}_0, \dots, \mathbf{x}_d - \mathbf{x}_0] > 0 .$$

The **reference simplex** in d dimensions is the simplex with vertices given by the columns of $\mathbf{X} = \begin{bmatrix} \mathbf{0}, & \mathbf{I} \end{bmatrix}$. Here $\mathbf{I}$ represents the **identity matrix**.

Thus, the reference simplex is the set $\{\boldsymbol{\xi} \ : \ \boldsymbol{\xi} \ge \mathbf{0} \text{ and } \mathbf{e}^\top \boldsymbol{\xi} \le 1\}$. In two dimensions, a simplex is called a triangle, and in three dimensions it is called a tetrahedron.

Next, let us describe a useful coordinate system on reference simplices.

Definition 1.3.4 A real vector $\mathbf{b}$ is a vector of **barycentric coordinates** if and only if $\mathbf{e}^\top \mathbf{b} = 1$.

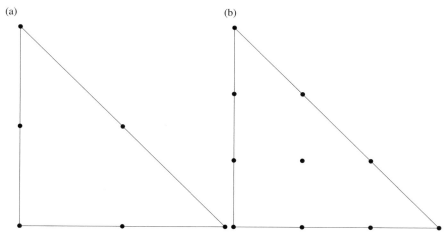

Fig. 1.6 Triangle lattice points. (**a**) $n = 2$. (**b**) $n = 3$

It should be obvious that the reference simplex in d dimensions is precisely the set of those points $\boldsymbol{\xi}$ whose barycentric coordinates

$$\mathbf{b} = \begin{bmatrix} \boldsymbol{\xi} \\ 1 - \mathbf{e}^\top \boldsymbol{\xi} \end{bmatrix}$$

in $d + 1$ dimensions are nonnegative. Let us also define some useful points in reference simplices.

Definition 1.3.5 Given a positive integer n, the set of all **simplex lattice points** with spacing $1/n$ is

$$\mathscr{L}_{d,n} \equiv \left\{ \frac{\boldsymbol{\alpha}}{n} \; : \; \boldsymbol{\alpha} \text{ is a multi-index in } d \text{ dimensions and } |\boldsymbol{\alpha}| \leq n \right\} .$$

For example, some triangle lattice points are shown in Fig. 1.6.

Definition 1.3.5 has the following immediate consequence.

Lemma 1.3.4 *For any positive integers d and n, the set $\mathscr{L}_{d,n}$ of all simplex lattice points in d dimensions is contained in the reference simplex in d dimensions, and $\mathscr{L}_{d,n}$ has $\binom{n+d}{d}$ members.*

Proof If $\boldsymbol{\alpha}/n \in \mathscr{L}_{d,n}$, then $\boldsymbol{\alpha}/n \geq \mathbf{0}$ and $\mathbf{e}^\top (\boldsymbol{\alpha}/n) = |\boldsymbol{\alpha}|/n \leq 1$, so $\boldsymbol{\alpha}/n$ is in the reference simplex. Furthermore, the number of members of $\mathscr{L}_{d,n}$ is the same as the number of multi-indices $\boldsymbol{\alpha}$ such that $|\boldsymbol{\alpha}| \leq n$, and the proof of Lemma 1.3.3 shows that the latter is equal to $\binom{n+d}{d}$.

We will now generalize Lagrange interpolation to multiple dimensions.

Definition 1.3.6 Assume that d and n are positive integers, and that $\boldsymbol{\alpha}$ is a multi-index in d dimensions such that $|\boldsymbol{\alpha}| \leq n$. Let $\boldsymbol{\xi}$ be a point in the reference simplex in d dimensions. Define the multi-index

$$\mathbf{a} = \begin{bmatrix} \boldsymbol{\alpha} \\ n - |\boldsymbol{\alpha}| \end{bmatrix},$$

in $d + 1$ dimensions, and define the barycentric coordinates

$$\mathbf{b}(\boldsymbol{\xi}) = \begin{bmatrix} \boldsymbol{\xi} \\ 1 - \mathbf{e}^\top \boldsymbol{\xi} \end{bmatrix} \geq \mathbf{0}$$

for $\boldsymbol{\xi}$. Then

$$\lambda_{\boldsymbol{\alpha},n}(\boldsymbol{\xi}) \equiv \prod_{j=0}^{d} \prod_{i=0}^{\mathbf{a}_j-1} \frac{n\mathbf{b}_j(\boldsymbol{\xi}) - i}{\mathbf{a}_j - i} \tag{1.21}$$

is a Lagrange polynomial on the reference simplex in d dimensions.
This definition has the following easy implication.

Lemma 1.3.5 *Assume that d and n are positive integers, and that $\boldsymbol{\alpha}$ is a multi-index in d dimensions such that $|\boldsymbol{\alpha}| \leq n$. Define the functions $\lambda_{\boldsymbol{\alpha},n}$ by Eq. (1.21). Then the set $\{\lambda_{\boldsymbol{\alpha},n} : |\boldsymbol{\alpha}| \leq n\}$ is a basis for the set of all polynomials of degree at most n in d dimensions.*

Proof Since each component of $\mathbf{b}(\boldsymbol{\xi})$ is a polynomial of degree one in one or more components of $\boldsymbol{\xi}$, we see that for each multi-index $\boldsymbol{\alpha}$ in d dimensions with $|\boldsymbol{\alpha}| \leq n$, $\lambda_{\boldsymbol{\alpha},n}$ is a polynomial of degree

$$\sum_{j=0}^{d} \mathbf{a}_i = |\mathbf{a}| = n .$$

Next, we note that if $\boldsymbol{\zeta}$ is a multi-index in d dimensions satisfying $|\boldsymbol{\zeta}| \leq n$, then we can define the multi-index

$$\mathbf{z} = \begin{bmatrix} \boldsymbol{\zeta} \\ n - |\boldsymbol{\zeta}| \end{bmatrix}$$

in $d + 1$ dimensions. Note that if $\boldsymbol{\alpha} \neq \boldsymbol{\zeta}$ then there is a component index j such that $\mathbf{a}_j > \mathbf{z}_j$; for that component index j, $\mathbf{z}_j$ is an integer less than $\mathbf{a}_j$, so

$$\prod_{i=0}^{\mathbf{a}_j-1} \frac{\mathbf{z}_j - i}{\mathbf{a}_j - i} = 0 .$$

It follows that

$$\lambda_{\alpha,n}\left(\frac{\zeta}{n}\right) = \prod_{j=0}^{d}\prod_{i=0}^{\mathbf{a}_j-1}\frac{\mathbf{z}_j - i}{\mathbf{a}_j - i} = 0 \, .$$

It is also easy to see that

$$\lambda_{\alpha,n}\left(\frac{\alpha}{n}\right) = \prod_{j=0}^{d}\prod_{i=0}^{\mathbf{a}_j-1}\frac{\mathbf{a}_j - i}{\mathbf{a}_j - i} = 1 \, .$$

In order to show that the Lagrange polynomials on the reference simplex form a basis for all polynomials, Corollary 3.2.2 of Chap. 3 in Volume I implies that we need only show that the Lagrange polynomials are linearly independent. Suppose that there is a collection of scalars γ_α such that

$$\sum_{|\alpha|\le n} \gamma_\alpha \lambda_{\alpha,n}(\xi) = 0$$

for all ξ in the reference simplex. Then for any multi-index ζ with $|\zeta| \le n$ we can define the multi-index

$$\mathbf{z} = \begin{bmatrix} \zeta \\ n - |\zeta| \end{bmatrix}$$

and see that

$$0 = \sum_{|\alpha|\le n} \gamma_\alpha \lambda_{\alpha,n}\left(\frac{\zeta}{n}\right) = \gamma_\zeta \lambda_{\zeta,n}\left(\frac{\zeta}{n}\right) = \gamma_\zeta \, .$$

This shows that the Lagrange polynomials on the reference simplex are linearly independent, and completes the proof of the lemma.

Example 1.3.3 A table of the Lagrange polynomials of degree 2 on the reference triangle can be found in Table 1.2.

Lemma 1.3.5 can now be used to develop Lagrange interpolation of functions defined on simplices.

Lemma 1.3.6 *Let d and n be positive integers, and let* $\mathbf{X}$ *be a real* $d \times (d + 1)$ *matrix. Assume that* $\mathscr{S}_\mathbf{X}$ *is the simplex in d dimensions with vertices given by the columns of* $\mathbf{X}$, *and that* $\mathscr{S}_\mathbf{X}$ *is positively oriented. For any* ξ *in the reference simplex in d dimensions, define*

$$\mathbf{b}(\xi) = \begin{bmatrix} \xi \\ 1 - \mathbf{e}^\top \xi \end{bmatrix} \, .$$

Table 1.2 Lagrange polynomials of degree 2 on reference triangle

$\boldsymbol{\alpha}$	$\lambda_{\alpha,2}(\boldsymbol{\xi})$
$\begin{bmatrix} 0 \\ 0 \end{bmatrix}$	$\{1\}\{1\}\left\{\frac{2(1-\xi_0-\xi_1)-0}{2-0}\ \frac{2(1-\xi_0-\xi_1)-1}{2-1}\right\}$
$\begin{bmatrix} 1 \\ 0 \end{bmatrix}$	$\left\{\frac{2\xi_0-0}{1-0}\right\}\{1\}\left\{\frac{2(1-\xi_0-\xi_1)-0}{2-0}\right\}$
$\begin{bmatrix} 2 \\ 0 \end{bmatrix}$	$\left\{\frac{2\xi_0-0}{2-0}\ \frac{2\xi_0-1}{2-1}\right\}\{1\}\{1\}$
$\begin{bmatrix} 0 \\ 1 \end{bmatrix}$	$\{1\}\left\{\frac{2\xi_1-0}{1-0}\right\}\left\{\frac{2(1-\xi_0-\xi_1)-0}{2-0}\right\}$
$\begin{bmatrix} 1 \\ 1 \end{bmatrix}$	$\left\{\frac{2\xi_0-0}{1-0}\right\}\left\{\frac{2\xi_1-0}{1-0}\right\}\{1\}$
$\begin{bmatrix} 0 \\ 2 \end{bmatrix}$	$\{1\}\left\{\frac{2\xi_1-0}{2-0}\ \frac{2\xi_1-1}{2-1}\right\}\{1\}$

Then for any function f defined on $\mathscr{S}_{\mathbf{X}}$, the function

$$p_f\left(\mathbf{Xb}(\boldsymbol{\xi})\right) = \sum_{|\boldsymbol{\alpha}|\le n} f\left(\mathbf{Xb}(\boldsymbol{\alpha}/n)\right)\lambda_{\alpha,n}(\boldsymbol{\xi})\ . \tag{1.22}$$

is also defined on the simplex $\mathscr{S}_{\mathbf{X}}$, and $p_f(\mathbf{Xb}(\boldsymbol{\xi}))$ is a polynomial of degree at most n in $\boldsymbol{\xi}$. Furthermore, p_f interpolates f at the mapped lattice points in $\mathscr{S}_{\mathbf{X}}$.

Proof For any multi-index $\boldsymbol{\alpha}$ in d dimensions with $|\boldsymbol{\alpha}| \le n$, $\boldsymbol{\alpha}/n$ is a lattice point in the reference simplex, and $\mathbf{Xb}(\boldsymbol{\alpha}/n)$ is a mapped lattice point in $\mathscr{S}_{\mathbf{X}}$. Thus $f(\mathbf{Xb}(\boldsymbol{\alpha}/n))$ is defined for each lattice point in the reference simplex. It follows that the right-hand side of (1.22) is defined for all $\boldsymbol{\xi}$ in the reference simplex. Lemma 1.3.5 also shows that the right-hand side of this equation is a polynomial of degree at most n in $\boldsymbol{\xi}$.

If $\mathbf{x} \in \mathscr{S}_{\mathbf{X}}$, then there exists $\mathbf{c} \ge \mathbf{0}$ with $\mathbf{e}^\top\mathbf{c} = 1$ such that

$$\mathbf{x} = \mathbf{Xc}\ .$$

If $\mathbf{c} = [\gamma_0, \ldots, \gamma_d]$, then this equation can be rewritten

$$\sum_{j=1}^d (\mathbf{x}_j - \mathbf{x}_0)\gamma_j = \mathbf{x} - \mathbf{x}_0\ .$$

Since $\mathscr{S}_{\mathbf{X}}$ is positively-oriented, this linear system is nonsingular. In other words, for any $\mathbf{x} \in \mathscr{S}_{\mathbf{X}}$ we can find barycentric coordinates $\mathbf{c}$ for a point in the reference simplex such that $\mathbf{x} = \mathbf{Xc}$. Thus p_f is defined on the simplex $\mathscr{S}_{\mathbf{X}}$.

The proof of Lemma 1.3.5 also shows that each Lagrange polynomial is zero at all simplex lattice points other than its associated simplex lattice point. It follows that for each lattice point α/n in the reference simplex, we have

$$p_f\left(\mathbf{Xb}(\alpha/n)\right) = \sum_{|\beta|\le n} f\left(\mathbf{Xb}(\beta/n)\right) \lambda_{\beta,n}(\alpha/n) = f\left(\mathbf{Xb}(\alpha/n)\right) \ .$$

This proves that p_f interpolates f at the mapped simplex lattice points.

Readers who are interested in modifying Lagrange interpolation on triangles to use Chebyshev interpolation points instead of lattice points should read Chen and Babuška [38]. For similar ideas on tetrahedra, read Chen and Babuška [39].

Exercise 1.3.2 Write an algorithm for evaluating the Lagrange interpolation polynomial p_f at a point $\mathbf{Xb}(\xi)$ mapped into the simplex $\mathscr{S}_\mathbf{X}$ with vertices given by the columns of the $d \times (d+1)$ real matrix $\mathbf{X}$. Determine the number of arithmetic operations involved in your algorithm.

Exercise 1.3.3 Experimentally determine the number of lattice points needed to interpolate $f(\mathbf{x}) = e^{x_1 x_2}$ on the reference triangle so that the error in interpolation is at most 10^{-8}.

1.3.3 Blocks

In $d > 1$ dimensions, it is easy to describe interpolation on polygonal regions bounded by $2d$ faces. Let us begin by describing the simplest of these regions.

Definition 1.3.7 The **reference hypercube** in d dimensions is the set

$$\{\xi \ : \ \xi \text{ is a real } d\text{-vector}, \ \xi \ge \mathbf{0} \text{ and } \|\xi\|_\infty \le 1\} \ .$$

In two dimensions, hypercubes are called squares, and in three dimensions they are called cubes. We will also identify some useful points within reference hypercubes.

Definition 1.3.8 Given a positive integer n, the set of all **hypercube lattice points** with spacing $1/n$ is

$$\mathscr{L}_{d,n} = \left\{\frac{\alpha}{n} \ : \ \alpha \text{ is a multi-index in } d \text{ dimensions and } \|\alpha\|_\infty \le n\right\} \ .$$

Definition 1.3.8 has the following easy consequence.

Lemma 1.3.7 *Given positive integers d and n, the set $\mathscr{L}_{d,n}$ of hypercube lattice points has $(n+1)^d$ members.*

Proof The lattice point $\alpha/n \in \mathscr{L}_{d,n}$ if and only if for each $0 \le i < d$ we have $0 \le \alpha_i \le n$. Thus there are $n+1$ possible values for each of the d components of α. The claim follows easily.

We would like to develop polynomial interpolation in multiple dimensions. Let us begin with the following definition.

Definition 1.3.9 Assume that d and n are positive integers, and that $\boldsymbol{\alpha}$ is a multi-index in d dimensions with $\|\boldsymbol{\alpha}\|_\infty \leq n$. Then

$$\lambda_{\boldsymbol{\alpha},n}(\boldsymbol{\xi}) = \prod_{j=0}^{d-1} \prod_{\substack{i=0 \\ i \neq \alpha_j}}^{n} \frac{n\xi_j - i}{\alpha_j - i} \tag{1.23}$$

is a Lagrange polynomial on the reference hypercube in d dimensions.
Using Eq. (1.11), it should be easy to see that each Lagrange polynomial is a product of one-dimensional Lagrange basis polynomials in individual components of $\boldsymbol{\xi}$. Definition 1.3.9 also has the following easy implication.

Lemma 1.3.8 *Assume that d and n are positive integers, and that $\boldsymbol{\alpha}$ is a multi-index in d dimensions with $\|\boldsymbol{\alpha}\|_\infty \leq n$. Define the functions $\lambda_{\boldsymbol{\alpha},n}$ by (1.23). Then the set $\{\lambda_{\boldsymbol{\alpha},n} : \|\boldsymbol{\alpha}\|_\infty \leq n\}$ is a basis for the set of all polynomials of degree at most n in each of the d dimensions.*

Proof It is easy to see that for each $j \in [0, d)$ the function

$$\prod_{\substack{i=0 \\ i \neq \alpha_j}}^{n} \frac{n\xi_j - i}{\alpha_j - i}$$

is a polynomial of degree n. Thus $\lambda_{\boldsymbol{\alpha},n}$ is a polynomial of degree n in each component of $\boldsymbol{\xi}$.

Suppose that $\boldsymbol{\zeta}$ is a multi-index with $\|\boldsymbol{\zeta}\|_\infty \leq n$. If $\boldsymbol{\zeta} \neq \boldsymbol{\alpha}$, then there is a component index $j \in [0, d)$ so that $\zeta_j \neq \alpha_j$. It follows that

$$\prod_{\substack{i=0 \\ i \neq \alpha_j}}^{n} \frac{\zeta_j - i}{\alpha_j - i} = 0 \, ,$$

so $\lambda_{\boldsymbol{\alpha},n}(\boldsymbol{\zeta}/n) = 0$. It is also easy to show that $\lambda_{\boldsymbol{\alpha},n}(\boldsymbol{\alpha}/n) = 1$.

The dimension of the linear space of polynomials that have order at most n in each component is n^d. This dimension is the same as the number of Lagrange polynomials $\lambda_{\boldsymbol{\alpha},n}$ for $\|\boldsymbol{\alpha}\|_\infty \leq n$. In order to show that the Lagrange polynomials form a basis for all such polynomials, Corollary 3.2.2 of Chap. 3 in Volume I implies that we need only show that they are linearly independent. Suppose that there is a collection of scalars γ_α such that

$$\sum_{\|\boldsymbol{\alpha}\|_\infty \leq n} \gamma_\alpha \lambda_{\boldsymbol{\alpha},n}(\boldsymbol{\xi}) = 0$$

for all $\boldsymbol{\xi}$ in the reference hypercube. Then for any multi-index $\boldsymbol{\zeta}$ with $\|\boldsymbol{\zeta}\|_\infty \le n$ we have

$$0 = \sum_{\|\boldsymbol{\alpha}\|_\infty \le n} \gamma_\alpha \lambda_{\alpha,n}\left(\frac{\boldsymbol{\zeta}}{n}\right) = \gamma_\zeta$$

This shows that the Lagrange polynomials on the reference hypercube are linearly independent, and completes the proof.

In order to handle interpolation on blocks in multiple dimensions, let us first describe the interpolation region.

Definition 1.3.10 Let d be a positive integer, and for each multi-index $\boldsymbol{\alpha}$ in d-dimensions with $\|\boldsymbol{\alpha}\|_\infty \le 1$ let $\mathbf{x}_\alpha$ be a real d-vector. Recall the one-dimensional first-order Lagrange basis polynomials

$$\lambda_{0,1}(\xi) = 1 - \xi \text{ and } \lambda_{1,1}(\xi) = \xi .$$

For any d-vector $\boldsymbol{\xi}$ in the reference hypercube, define the mapping

$$\boldsymbol{\mu}(\boldsymbol{\xi}) = \sum_{\|\boldsymbol{\alpha}\|_\infty \le 1} \mathbf{x}_\alpha \prod_{j=0}^{d-1} \lambda_{\alpha_j,1}(\xi_j) . \tag{1.24}$$

Assume that the vectors $\mathbf{x}_\alpha$ have been chosen so that

$$\det \frac{\partial \boldsymbol{\mu}(\boldsymbol{\xi})}{\partial \boldsymbol{\xi}} > 0$$

for all $\boldsymbol{\xi}$ in the reference hypercube. Then a **block** in d dimensions with **vertices** given by the vectors $\mathbf{x}_\alpha$ is the set of points

$$\mathbf{B} = \{\boldsymbol{\mu}(\boldsymbol{\xi}) : \boldsymbol{\xi} \ge \mathbf{0} \text{ and } \|\boldsymbol{\xi}\|_\infty \le 1\} .$$

Now we are ready to describe polynomial interpolation on a block.

Lemma 1.3.9 *Let d and n be positive integers, and let $\mathbf{B}$ be a block in d dimensions with vertices given by the vectors $\mathbf{x}_\alpha$ for each multi-index $\boldsymbol{\alpha}$ with $\|\boldsymbol{\alpha}\|_\infty \le 1$. If the Lagrange polynomials $\lambda_{\alpha,n}$ are defined by (1.23) and the mapping $\boldsymbol{\mu}$ from the reference hypercube to $\mathbf{B}$ is given by (1.24), then for any function f defined on $\mathbf{B}$ the function*

$$p_f(\boldsymbol{\mu}(\boldsymbol{\xi})) = \sum_{\|\boldsymbol{\alpha}\|_\infty \le n} f\left(\boldsymbol{\mu}(\boldsymbol{\alpha}/n)\right) \lambda_{\alpha,n}(\boldsymbol{\xi}) \tag{1.25}$$

is also defined on $\mathbf{B}$, and $p_f \circ \boldsymbol{\mu}$ is a polynomial of degree at most n in each dimension. Furthermore, p_f interpolates f at the mapped lattice points in $\mathbf{B}$.

Note that we are claiming that $p_f \circ \boldsymbol{\mu}$ is a polynomial; since $\boldsymbol{\mu}$ is generally nonlinear in multiple dimensions, this does not imply that p_f is a polynomial.

Proof Lemma 1.3.8 shows that the right-hand side of (1.25) is a polynomial of degree at most n in each component of $\boldsymbol{\xi}$. By the definition of $\mathbf{B}$, p_f is defined on $\mathbf{B}$.

The proof of Lemma 1.3.8 also shows that each Lagrange polynomial is zero at all hypercube lattice points other than its associated lattice point. It follows that for each lattice point $\boldsymbol{\alpha}/n$ in the reference hypercube, we have

$$p_f(\boldsymbol{\mu}(\boldsymbol{\alpha}/n)) = \sum_{\|\boldsymbol{\beta}\|_\infty \leq n} f\left(\boldsymbol{\mu}(\boldsymbol{\beta}/n)\right) \lambda_{\boldsymbol{\beta},n}(\boldsymbol{\alpha}/n) = f\left(\boldsymbol{\mu}(\boldsymbol{\alpha}/n)\right) .$$

This proves that p_f interpolates f at the mapped hypercube lattice points.

The reader should note that it is fairly easy to use Chebyshev interpolation points instead of lattice points on the reference hypercube in each coordinate direction. For high-order interpolation, the Chebyshev points are preferable to the equidistant lattice points. For generalizations of the Neville-Aitken algorithm for evaluating polynomial interpolants in multiple dimension, see Sauer and Xu [156].

Readers can view C^{++} classes to implement various multi-dimensional polynomial representations in files ShapeFunction.H and ShapeFunction.C. There is an abstract class called `ShapeFunction`, with derived abstract classes such as `TensorProductPolynomials`, `TrianglePolynomials` and `TetrahedronPolynomials`. Non-abstract classes include

- `C0LagrangeIntervalPolynomials`,
- `C0LagrangeTrianglePolynomials`,
- `C0LagrangeQuadrilateralPolynomials`,
- `C0LagrangeTetrahedronPolynomials` and
- `C0LagrangeHexahedronPolynomials`.

These files are also available online from Cambridge University Press.

The GSL (GNU Scientific Library) does not contain any routines for multidimensional interpolation. MATLAB offers commands interp2 and interp3 to perform either linear or cubic *spline* interpolation on a rectangular mesh.

Exercise 1.3.4 Write an algorithm for evaluating the Lagrange interpolation polynomial p_f at a point $\boldsymbol{\mu}(\boldsymbol{\xi})$ mapped from the reference hypercube to the region with vertices given by the columns of the $d \times 2^d$ real matrix $\mathbf{X}$. Determine the number of arithmetic operations involved in your algorithm.

Exercise 1.3.5 Modify the definition (1.23) of Lagrange polynomials on a square to use Chebyshev interpolation point locations in each coordinate direction.

Exercise 1.3.6 Read Sauer and Xu [156]. to learn how to generalize Neville-Aitken interpolation to hypercubes. Compare the number of arithmetic operations for the Neville-Aitken algorithm on a square to straightforward evaluation of the Lagrange polynomials in evaluating p_f.

1.3.4 Error Estimate

Now that we have described interpolation on simplices and hypercubes in multiple dimensions, we are ready to estimate the errors in multi-dimensional interpolation. Our discussion will follow Ciarlet and Raviart [44].

First, we note that multi-indices make it easy to develop a the Taylor expansion in multiple dimensions.

Lemma 1.3.10 (Taylor Expansion) *Suppose that k is a positive integer, $\mathbf{x}$ is a d-vector, and the function f is k-times continuously differentiable in a convex neighborhood of $\mathbf{x}$. Let*

$$
\mathbf{D} = \begin{bmatrix} \partial_0 \\ \vdots \\ \partial_{d-1} \end{bmatrix}
$$

be the vector of partial differentiation operators. Then for all d-vectors $\mathbf{y}$ so that $\mathbf{x} + \mathbf{y}$ is in that neighborhood,

$$
f(\mathbf{x}+\mathbf{y}) = \sum_{|\alpha|<k} \frac{1}{\alpha!} \mathbf{y}^\alpha \mathbf{D}^\alpha f(\mathbf{x}) + \sum_{|\alpha|=k} \frac{k}{\alpha!} \mathbf{y}^\alpha \int_0^1 s^{k-1} \mathbf{D}^\alpha f(\mathbf{x}+\mathbf{y}[1-s]) \, ds . \quad (1.26)
$$

Proof Let

$$
\phi(s) = f(\mathbf{x}+\mathbf{y}[1-s]) .
$$

Then $\phi(0) = f(\mathbf{x}+\mathbf{y})$ and $\phi(1) = f(\mathbf{x})$. The fundamental theorem of calculus shows that

$$
f(\mathbf{x}) - f(\mathbf{x}+\mathbf{y}) = \phi(1) - \phi(0) = \int_0^1 \phi'(s) \, ds = -\int_0^1 \sum_{i=1}^d \partial_i f(\mathbf{x}+\mathbf{y}[1-s]) \mathbf{y}_i \, ds
$$

This is equivalent to the claim for $k = 1$.

Next, we assume inductively that the claim is true for some integer $k - 1 \geq 0$. In other words, our inductive hypothesis is that

$$
f(\mathbf{x}+\mathbf{y}) - \sum_{|\alpha|<k-1} \frac{1}{\alpha!} \mathbf{y}^\alpha \mathbf{D}^\alpha f(\mathbf{x}) = \sum_{|\alpha|=k-1} \frac{k-1}{\alpha!} \mathbf{y}^\alpha \int_0^1 s^{k-2} \mathbf{D}^\alpha f(\mathbf{x}+\mathbf{y}[1-s]) \, ds
$$

from which integration by parts gives us

$$
= \sum_{|\alpha|=k-1} \frac{k-1}{\alpha!} \mathbf{y}^\alpha \left\{ \frac{1}{k-1} s^{k-1} \mathbf{D}^\alpha f(\mathbf{x} + \mathbf{y}[1-s]) \right\} \Big|_{s=0}^{1}
$$

$$
+ \int_0^1 \frac{1}{k-1} s^{k-1} \sum_{i=1}^d \partial_i \mathbf{D}^\alpha f(\mathbf{x} + \mathbf{y}[1-s]) y_i \, ds \Bigg\}
$$

$$
= \sum_{|\alpha|=k-1} \frac{1}{\alpha!} \mathbf{y}^\alpha \mathbf{D}^\alpha f(\mathbf{x}) + \sum_{|\alpha|=k-1} \sum_{i=1}^d \frac{1}{\alpha!} \mathbf{y}^{\alpha+\mathbf{e}_i} \int_0^1 s^{k-1} \mathbf{D}^{\alpha+\mathbf{e}_i} f(\mathbf{x} + \mathbf{y}[1-s]) \, ds
$$

$$
= \sum_{|\alpha|=k-1} \frac{1}{\alpha!} \mathbf{y}^\alpha \mathbf{D}^\alpha f(\mathbf{x}) + \sum_{|\beta|=k} \sum_{i=1}^d \frac{1}{(\beta-\mathbf{e}_i)!} \mathbf{y}^\beta \int_0^1 s^{k-1} \mathbf{D}^\beta f(\mathbf{x} + \mathbf{y}[1-s]) \, ds .
$$

Note that for all $|\beta| = k$,

$$
\sum_{i=1}^d \frac{1}{(\beta-\mathbf{e}_i)!} = \frac{1}{\beta!} \sum_{i=1}^d \beta_i = \frac{k}{\beta!} .
$$

By combining this equation and (1.18), we see that we have proved that the inductive hypothesis is true for k.

Example 1.3.4 Let us expand the Taylor series approximation of order two for a function of three variables:

$$
f(\mathbf{x} + \mathbf{y}) \approx \frac{1}{0!0!0!} y_0^0 y_1^0 y_2^0 f(\mathbf{x})
$$

$$
+ \frac{1}{1!0!0!} y_0^1 y_1^0 y_2^0 \partial_0^1 f(\mathbf{x}) + \frac{1}{0!1!0!} y_0^0 y_1^1 y_2^0 \partial_1^1 f(\mathbf{x}) + \frac{1}{0!0!1!} y_0^0 y_1^0 y_2^1 \partial_2^1 f(\mathbf{x})
$$

$$
+ \frac{1}{2!0!0!} y_0^2 y_1^0 y_2^0 \partial_0^2 f(\mathbf{x}) + \frac{1}{0!2!0!} y_0^0 y_1^2 y_2^0 \partial_1^2 f(\mathbf{x}) + \frac{1}{0!0!2!} y_0^0 y_1^0 y_2^2 \partial_2^2 f(\mathbf{x})
$$

$$
+ \frac{1}{1!1!0!} y_0^1 y_1^1 y_2^0 \partial_0^1 \partial_1^1 f(\mathbf{x}) + \frac{1}{1!0!1!} y_0^1 y_1^0 y_2^1 \partial_0^1 \partial_2^1 f(\mathbf{x}) + \frac{1}{0!1!1!} y_0^0 y_1^1 y_2^1 \partial_1^1 \partial_2^1 f(\mathbf{x}) .
$$

Next, we will prove a curious fact about Lagrange interpolation polynomials.

Lemma 1.3.11 ([44, p. 181]) *Suppose that k and d are positive integers, and that $n \geq \binom{k+d}{d}$ in d dimensions. Let $\{\mathbf{a}_i\}_{i=0}^{n-1}$ be a set of distinct real d-vectors, and let the corresponding set $\{\lambda_i(\mathbf{x})\}_{i=0}^{n-1}$ of Lagrange interpolation polynomials be such that for all polynomials of degree at most $k-1$ and for all d-vectors $\mathbf{x}$*

$$
\sum_{i=0}^{n-1} q(\mathbf{a}_i) \lambda_i(\mathbf{x}) = q(\mathbf{x}) . \tag{1.27}
$$

Then for all multi-indices α and β of modulus less than k and for all d-vectors $\mathbf{x}$,

$$\frac{1}{\alpha!} \sum_{i=0}^{n-1} (\mathbf{a}_i - \mathbf{x})^\alpha \mathbf{D}^\beta \lambda_i(\mathbf{x}) = \begin{cases} 1, \alpha = \beta \\ 0, \alpha \neq \beta \end{cases}. \tag{1.28}$$

Proof The multinomial expansion (1.17) and the interpolation assumption (1.27) imply that

$$\frac{1}{\alpha!} \sum_{i=0}^{n-1} \mathbf{D}^\beta \lambda_i(\mathbf{x})(\mathbf{a}_i - \mathbf{x})^\alpha = \frac{1}{\alpha!} \sum_{i=0}^{n-1} \mathbf{D}^\beta \lambda_i(\mathbf{x}) \sum_{\gamma \leq \alpha} \frac{\alpha!}{\gamma!(\alpha-\gamma)!} (-\mathbf{x})^\gamma \mathbf{a}_i^{\alpha-\gamma}$$

$$= \frac{1}{\alpha!} \sum_{\gamma \leq \alpha} \frac{\alpha!}{\gamma!(\alpha-\gamma)!} (-\mathbf{x})^\gamma \mathbf{D}^\beta \left\{ \sum_{i=0}^{n-1} \mathbf{a}_i^{\alpha-\gamma} \lambda_i(\mathbf{x}) \right\} = \sum_{\gamma \leq \alpha} \frac{1}{\gamma!(\alpha-\gamma)!} (-\mathbf{x})^\gamma \mathbf{D}^\beta \mathbf{x}^{\alpha-\gamma}. \tag{1.29}$$

If $\alpha \not\geq \beta$, then each term in the last sum is zero and the claim is satisfied. Otherwise, we have $\alpha \geq \beta$. The (1.29) and the multinomial expansion (1.17) imply that

$$\frac{1}{\alpha!} \sum_{i=0}^{n-1} \mathbf{D}^\beta \lambda_i(\mathbf{x})(\mathbf{a}_i - \mathbf{x})^\alpha = \sum_{\gamma \leq \alpha} \frac{1}{\gamma!(\alpha-\gamma)!} (-\mathbf{x})^\gamma \mathbf{D}^\beta \mathbf{x}^{\alpha-\gamma}$$

$$= \sum_{\gamma \leq \alpha - \beta} \frac{1}{\gamma!(\alpha-\gamma-\beta)!} (-1)^{|\gamma|} \mathbf{x}^{\alpha-\beta} = \frac{\mathbf{x}^{\alpha-\beta}}{(\alpha-\beta)!} \sum_{\gamma \leq \alpha - \beta} \frac{(\alpha-\beta)!}{\gamma!(\alpha-\beta-\gamma)!} (-1)^{|\gamma|}$$

$$= \frac{\mathbf{x}^{\alpha-\beta}}{(\alpha-\beta)!} (\mathbf{e} - \mathbf{e})^{\alpha-\beta}.$$

The final expression is nonzero if and only if $\alpha = \beta$, in which case it simplifies to one.

Lemma 1.3.11 allows us to prove the following error estimate for Lagrange interpolation.

Theorem 1.3.2 (Multidimensional Interpolation Error Estimate [44, p. 181])
Suppose that Ω is a convex set in d dimensions, k is a positive integer and $n \geq \binom{k+d}{d}$. Let $\{\mathbf{a}_i\}_{i=0}^{n-1} \subset \overline{\Omega}$ be a set of distinct points, and let the corresponding set $\{\lambda_i(\mathbf{x})\}_{i=0}^{n-1}$ of Lagrange interpolation polynomials be such that for all polynomials q of degree less than k and for all d-vectors $\mathbf{x}$ we have

$$\sum_{i=1}^{n} q(\mathbf{a}_i) \lambda_i(\mathbf{x}) = q(\mathbf{x}).$$

Suppose that Ω has diameter h, meaning that for all $|\alpha| \leq k$ and all $0 \leq i < n$ we have

$$\max_{\mathbf{x} \in \Omega} |(\mathbf{a}_i - \mathbf{x})^\alpha| \leq h^{|\alpha|}. \tag{1.30}$$

If the function f is defined in Ω, define the Lagrange interpolation polynomial p_f for f by

$$p_f(\mathbf{x}) = \sum_{i=0}^{n-1} f(\mathbf{a}_i)\lambda_i(\mathbf{x}) .$$

Then for all functions f with bounded derivatives of order at most k on Ω, and for all $|\beta| < k$ we have

$$\sup_{\mathbf{x}\in\Omega} \left|\mathbf{D}^\beta f(\mathbf{x}) - \mathbf{D}^\beta p_f(\mathbf{x})\right| \le h^k \left(\sum_{|\alpha|=k} \frac{1}{\alpha!} \sup_{\mathbf{x}\in\Omega} |\mathbf{D}^\alpha f(\mathbf{x})|\right) \left(\sum_{i=0}^{n-1} \sup_{\mathbf{x}\in\Omega} \left|\mathbf{D}^\beta \lambda_i(\mathbf{x})\right|\right) .$$

Proof If $\mathbf{x} \in \Omega$ and f is k-times continuously differentiable in Ω, we can use the Taylor expansion (1.26) to get

$$f(\mathbf{a}_i) = \sum_{|\alpha|<k} \frac{1}{\alpha!}(\mathbf{a}_i-\mathbf{x})^\alpha \mathbf{D}^\alpha f(\mathbf{x}) + k \sum_{|\alpha|=k} \frac{1}{\alpha!}(\mathbf{a}_i-\mathbf{x})^\alpha \int_0^1 (1-t)^{k-1}\mathbf{D}^\alpha f(\mathbf{x}+[\mathbf{a}_i-\mathbf{x}]t) \, dt .$$

For all $|\beta| < k$, Lagrange interpolation and the Taylor expansion imply

$$\mathbf{D}^\beta p_f(\mathbf{x}) = \sum_{i=0}^{n-1} f(\mathbf{a}_i)\mathbf{D}^\beta\lambda_i(\mathbf{x})$$

$$= \sum_{|\alpha|<k} \frac{1}{\alpha!} \sum_{i=0}^{n-1} \{(\mathbf{a}_i - \mathbf{x})^\alpha \mathbf{D}^\alpha f(\mathbf{x})\} \, \mathbf{D}^\beta\lambda_i(\mathbf{x})$$

$$+ k \sum_{|\alpha|=k} \frac{1}{\alpha!} \sum_{i=0}^{n-1} \left\{(\mathbf{a}_i - \mathbf{x})^\alpha \int_0^1 (1-t)^{k-1}\mathbf{D}^\alpha f(\mathbf{x}+[\mathbf{a}_i-\mathbf{x}]t) \, dt\right\} \mathbf{D}^\beta\lambda_i(\mathbf{x})$$

then Eq. (1.28) produces

$$= \mathbf{D}^\beta f(\mathbf{x}) + k \sum_{|\alpha|=k} \frac{1}{\alpha!} \sum_{i=0}^{n-1} \left\{(\mathbf{a}_i - \mathbf{x})^\alpha \int_0^1 (1-t)^{k-1}\mathbf{D}^\alpha f(\mathbf{x}+[\mathbf{a}_i-\mathbf{x}]t) \, dt\right\} \mathbf{D}^\beta\lambda_i(\mathbf{x}) .$$

It follows that

$$\sup_{\mathbf{x}\in\Omega} \left|\mathbf{D}^\beta f(\mathbf{x}) - \mathbf{D}^\beta p_f(\mathbf{x})\right|$$

$$\le k \sum_{|\alpha|=k} \frac{1}{\alpha!} \sum_{i=0}^{n-1} \sup_{\mathbf{x}\in\Omega} \left\{|(\mathbf{a}_i - \mathbf{x})^\alpha| \int_0^1 (1-t)^{k-1} |\mathbf{D}^\alpha f(\mathbf{x}+[\mathbf{a}_i-\mathbf{x}]t)| \, dt \left|\mathbf{D}^\beta\lambda_i(\mathbf{x})\right|\right\} .$$

We can bound

$$\sup_{\mathbf{x}\in\Omega} \int_0^1 (1-t)^{k-1}\,\left|\mathbf{D}^\alpha f(\mathbf{x}+[\mathbf{a}_i-\mathbf{x}]t)\right|\,dt \leq \sup_{\mathbf{x}\in\Omega}\left|\mathbf{D}^\alpha f(\mathbf{x})\right| \int_0^1 (1-t)^{k-1}\,dt = \frac{1}{k}\sup_{\mathbf{x}\in\Omega}\left|\mathbf{D}^\alpha f(\mathbf{x})\right|.$$

As a result,

$$\sup_{\mathbf{x}\in\Omega}\left|\mathbf{D}^\beta f(\mathbf{x}) - \mathbf{D}^\beta\lambda\{f\}(\mathbf{x})\right| \leq h^k \left(\sum_{|\alpha|=k}\frac{1}{\alpha!}\sup_{\mathbf{x}\in\Omega}\left|\mathbf{D}^\alpha f(\mathbf{x})\right|\right)\left(\sum_{i=0}^{n-1}\sup_{\mathbf{x}\in\Omega}\left|\mathbf{D}^\beta\lambda_i(\mathbf{x})\right|\right).$$

Exercise 1.3.7 Bound the maximum error in interpolating a twice continuously differentiable function f by a linear function on the reference triangle, with interpolation points at the vertices.

Exercise 1.3.8 Bound the maximum error in interpolating a thrice continuously differentiable function f by a bilinear function on the reference square, with interpolation points at the vertices.

1.4   Rational Polynomials

Polynomial interpolation is the most common form of interpolation. However, there are some advantages to approximating functions by quotients of two polynomials. In this section, we will briefly examine some ways to find accurate rational polynomial approximations to functions. We will also see that rational polynomials can be evaluated very efficiently.

1.4.1   Padé Approximation

Let us begin by defining our new class of approximations.

Definition 1.4.1 The function $p(x)/q(x)$ is a **rational polynomial** of order (n, m) if and only if p is a polynomial of degree n, q is a polynomial of degree m, and there is no scalar z such that $p(z) = 0 = q(z)$.

The following theorem shows that rational polynomials can have very good approximation properties near the origin.

Theorem 1.4.1 (Padé Approximation)   *Given nonnegative integers n and m, if f has $n + m + 1$ continuous derivatives in a neighborhood of $z = 0$, then there exists a polynomial $p_{n/m}$ of degree n, a polynomial $q_{n/m}$ of degree m with $q_{n/m}(0) \neq 0$, a scalar $\delta > 0$, an order $\nu > n$ and a scalar $\mu > 0$ so that for all $|z| < \delta$*

$$\left|f(z) - \frac{p_{n/m}(z)}{q_{n/m}(z)}\right| \leq \mu\,|z|^\nu.$$

Proof We will follow the presentation in Baker and Graves-Morris [12, p. 6f]. Define the coefficients

$$\gamma_k = \begin{cases} 0, & k < 0 \\ \frac{f^{(k)}(0)}{k!}, & 0 \leq k \leq n + m \end{cases}$$

as well as the $(m + 1) \times (m + 1)$ matrices

$$\mathbf{B}_{n/m}(z) = \begin{bmatrix} \gamma_{n-m+1} & \gamma_{n-m+2} & \cdots & \gamma_n & \gamma_{n+1} \\ \gamma_{n-m+2} & \gamma_{n-m+3} & \cdots & \gamma_{n+1} & \gamma_{n+2} \\ \vdots & \vdots & \ddots & \vdots & \vdots \\ \gamma_n & \gamma_{n+1} & \cdots & \gamma_{n+m-1} & \gamma_{n+m} \\ z^m & z^{m-1} & \cdots & z & 1 \end{bmatrix} \quad (1.31)$$

and

$$\mathbf{A}_{n/m}(z) = \begin{bmatrix} \gamma_{n-m+1} & \gamma_{n-m+2} & \cdots & \gamma_n & \gamma_{n+1} \\ \gamma_{n-m+2} & \gamma_{n-m+3} & \cdots & \gamma_{n+1} & \gamma_{n+2} \\ \vdots & \vdots & \ddots & \vdots & \vdots \\ \gamma_n & \gamma_{n+1} & \cdots & \gamma_{n+m-1} & \gamma_{n+m} \\ \sum_{k=-m}^{n-m} \gamma_k z^{m+k} & \sum_{k=-m+1}^{n-m+1} z^{m+k-1} & \cdots & \sum_{k=-1}^{n-1} \gamma_k z^{k+1} & \sum_{k=0}^{n} \gamma_k z^k \end{bmatrix}. \quad (1.32)$$

Also define the polynomial

$$b_{n/m}(z) = \det \mathbf{B}_{n/m}(z) \quad (1.33)$$

of degree at most m, and the polynomial

$$a_{n/m}(z) = \det \mathbf{A}_{n/m}(z) \quad (1.34)$$

of degree at most n. Then

$$b_{n/m}(z)f(z) = \begin{bmatrix} \gamma_{n-m+1} & \gamma_{n-m+2} & \cdots & \gamma_n & \gamma_{n+1} \\ \gamma_{n-m+2} & \gamma_{n-m+3} & \cdots & \gamma_{n+1} & \gamma_{n+2} \\ \vdots & \vdots & \ddots & \vdots & \vdots \\ \gamma_n & \gamma_{n+1} & \cdots & \gamma_{n+m-1} & \gamma_{n+m} \\ \sum_{k=0}^{\infty} \gamma_k z^{m+k} & \sum_{k=0}^{\infty} \gamma_k z^{m+k-1} & \cdots & \sum_{k=0}^{\infty} \gamma_k z^{k+1} & \sum_{k=0}^{\infty} \gamma_k z^k \end{bmatrix}$$

then we subtract z^{n+i} times row i from the last row for $1 \le i \le m$ to get

$$a_{n/m}(z) + \begin{bmatrix} \gamma_{n-m+1} & \gamma_{n-m+2} & \cdots & \gamma_n & \gamma_{n+1} \\ \gamma_{n-m+2} & \gamma_{n-m+3} & \cdots & \gamma_{n+1} & \gamma_{n+2} \\ \vdots & \vdots & \ddots & \vdots & \vdots \\ \gamma_n & \gamma_{n+1} & \cdots & \gamma_{n+m-1} & \gamma_{n+m} \\ \sum_{k=n+1}^{\infty} \gamma_k z^{m+k} & \sum_{k=n+2}^{\infty} \gamma_k z^{m+k-1} & \cdots & \sum_{k=n+m}^{\infty} \gamma_k z^{k+1} & \sum_{k=n+m+1}^{\infty} \gamma_k z^k \end{bmatrix} .$$

The matrix on the right in this expression is $O(z^{n+m+1})$, so we conclude that

$$b_{n/m}(z)f(z) = a_{n/m}(z) + O\left(z^{n+m+1}\right) .$$

We can eliminate all lowest-order coefficients in $b_{n/m}(z)$ that are zero, and divide both $a_{n/m}(z)$ and $b_{n/m}(z)$ by a common power of z to obtain the polynomials $p_{n/m}(z)$ and $q_{(n/m}z)$, respectively.

We should note that if we write

$$p_{n/m}(x) = \sum_{j=0}^{n} p_j x^j \text{ and } q_{n/m}(x) = \sum_{j=0}^{m} q_j x^j ,$$

and take $p_{n+i} = 0 = q_{m+i}$ for all $i \ge 0$, then the coefficients for the Padé approximation polynomials satisfy the linear system

$$\sum_{j=0}^{k} \frac{f^{(j)}(0)}{j!} q_{k-j} = p_k$$

for $0 \le k < \nu$.

Example 1.4.1 Suppose we want to find a Padé approximation of order $(1, 1)$ to e^x. Since the numerator $p_{1/1}$ and denominator $q_{1/1}$ are both first-order polynomials, we have

$$p_0 + p_1 x \approx \left[\sum_{j=0}^{\infty} \frac{x^j}{j!} \right] [q_0 + q_1 x] = \sum_{j=0}^{\infty} \frac{q_0}{j!} x^j + \sum_{j=0}^{\infty} \frac{q_1}{j!} x^{j+1} = q_0 + \sum_{j=1}^{\infty} \left[\frac{q_0}{j!} + \frac{q_1}{(j-1)!} \right] x^j .$$

We will choose $q_0 = 1$, and equate coefficients of powers of x to get

$$p_0 = 1$$

$$p_1 = \frac{q_0}{1!} + \frac{q_1}{0!} = 1 + q_1$$

$$0 = \frac{q_0}{2!} + \frac{q_1}{1!} = \frac{1}{2} + q_1 .$$

After we solve this linear system, we see that the Padé approximation is

$$e^x \approx \frac{1 + x/2}{1 - x/2} .$$

The determination of Padé approximations requires symbolic computation. As a result, they are not available in MATLAB *per se*. However, Padé approximations are available in Maple, Mathematica and the MATLAB Symbolic Math Toolbox.

Before leaving this section, we would like to present an important identity involving the polynomials in Padé approximations.

Theorem 1.4.2 (Wynn's Identity) *Suppose that n and m are nonnegative integers, and that the function f has $n + m + 2$ continuous derivatives in a neighborhood of $z = 0$. Define the scalars*

$$
\gamma_k = \begin{cases} 0, & k < 0 \\ \frac{f^{(k)}(0)}{k!}, & 0 \le k \le n + m + 1 \end{cases}
$$

the $m \times m$ matrix

$$
\mathbf{C}_{n/m} = \begin{bmatrix} \gamma_{n-m+1} & \gamma_{n-m+2} & \cdots & \gamma_n \\ \gamma_{n-m+2} & \gamma_{n-m+3} & \cdots & \gamma_{n+1} \\ \vdots & \vdots & \ddots & \vdots \\ \gamma_n & \gamma_{n+1} & \cdots & \gamma_{n+m-1} \end{bmatrix}
$$

and its determinant

$$
c_{n/m} = \det \mathbf{C}_{n/m} \ .
$$

If $c_{(n+1)/m} c_{n/(m+1)} \ne 0$, then the Padé approximations $p_{n/m}(z)/q_{n/m}(z)$ to $f(z)$ satisfy

$$
\left[\frac{p_{n/(m+1)}(z)}{q_{n/(m+1)}(z)} - \frac{p_{n/m}(z)}{q_{n/m}(z)} \right]^{-1} + \left[\frac{p_{n/(m-1)}(z)}{q_{n/(m-1)}(z)} - \frac{p_{n/m}(z)}{q_{n/m}(z)} \right]^{-1}
$$
$$
= \left[\frac{p_{(n+1)/m}(z)}{q_{(n+1)/m}(z)} - \frac{p_{n/m}(z)}{q_{n/m}(z)} \right]^{-1} + \left[\frac{p_{(n-1)/m}(z)}{q_{(n-1)/m}(z)} - \frac{p_{n/m}(z)}{q_{n/m}(z)} \right]^{-1} . \tag{1.35}
$$

Proof Our proof will follow that in Baker and Graves-Morris [12, p. 81ff].

The matrix $\mathbf{B}_{n/(m+1)}(z)$ defined in Eq. (1.31) is $(m + 2) \times (m + 2)$, and has determinant given by $b_{n/m}(z)$. Sylvester's determinant identity (3.9) in Chap. 3 of Volume I involving the first and last rows and columns of this matrix says that

$$
b_{n/(m+1)}(z) c_{(n+1)/m} = b_{(n+1)/m}(z) c_{n/(m+1)} - z b_{n/m}(z) c_{(n+1)/(m+1)} \ .
$$

We can divide this equation by $c_{(n+1)/m} c_{n/(m+1)}$ to obtain

$$
\frac{b_{(n+1)/m}(z)}{c_{(n+1)/m}} - \frac{b_{n/(m+1)}(z)}{c_{n/(m+1)}} = \frac{z b_{n/m}(z) c_{(n+1)/(m+1)}}{c_{(n+1)/m} c_{n/(m+1)}} \ . \tag{1.36}
$$

Next, we define

$$
\mathbf{M}_{n/m}(z) =
\begin{bmatrix}
\gamma_{n-m} & \gamma_{n-m+1} & \cdots & \gamma_{n-1} & \gamma_n & 0 \\
\gamma_{n-m+1} & \gamma_{n-m+2} & \cdots & \gamma_n & \gamma_{n+1} & 0 \\
\vdots & \vdots & \ddots & \vdots & \vdots & \vdots \\
\gamma_{n-1} & \gamma_n & \cdots & \gamma_{n+m-2} & \gamma_{n+m-1} & 0 \\
\gamma_n & \gamma_{n+1} & \cdots & \gamma_{n+m-1} & \gamma_{n+m} & 1 \\
z^m & z^{m-1} & \cdots & z & 1 & 0
\end{bmatrix},
$$

and note that expansion by minors (3.8) in Chap. 3 of Volume I in the last column gives us

$$
\det \mathbf{M}_{n/m}(z) = (-1)^{(m+1)+(m+2)} \det
\begin{bmatrix}
\gamma_{n-m} & \gamma_{n-m+1} & \cdots & \gamma_{n-1} & \gamma_n \\
\gamma_{n-m+1} & \gamma_{n-m+2} & \cdots & \gamma_n & \gamma_{n+1} \\
\vdots & \vdots & \ddots & \vdots & \vdots \\
\gamma_{n-1} & \gamma_n & \cdots & \gamma_{n+m-2} & \gamma_{n+m-1} \\
z^m & z^{m-1} & \cdots & z & 1
\end{bmatrix}
$$

$$
= -b_{(n-1)/m}(z) .
$$

Then Sylvester's determinant identity (3.9) in Chap. 3 of Volume I involving the first and last rows and columns of $\mathbf{M}_{n/m}(z)$ produces

$$
-b_{(n-1)/m}(z)c_{(n+1)/m} = -b_{n/(m-1)}(z)c_{n/(m+1)} - b_{n/m}(z)c_{n/m} .
$$

We can divide this equation by $c_{(n+1)/m}c_{n/(m+1)}$ to obtain

$$
\frac{b_{(n-1)/m}(z)}{c_{n/(m+1)}} - \frac{b_{n/(m-1)}(z)}{c_{(n+1)/m}} = \frac{b_{n/m}(z)c_{n/m}}{c_{(n+1)/m}c_{n/(m+1)}} . \tag{1.37}
$$

Recall the definition (1.32) of the $(m+1) \times (m+1)$ matrix $\mathbf{A}_{n/m}(z)$ and its determinant $a_{n/m}(z)$ Since Theorem 1.4.1 proved that

$$
f(z) - \frac{a_{n/m}(z)}{b_{n/m}(z)} = O\left(z^{n+m+1}\right) \text{ as } z \to 0
$$

and since $b_{n/m}(z) = O(1)$ as $z \to 0$, it follows that

$$
\frac{a_{(n+1)/m}(z)b_{n/m}(z) - a_{n/m}(z)b_{(n+1)/m}(z)}{b_{n/m}(z)b_{(n+1)/m}(z)} = \frac{a_{(n+1)/m}(z)}{b_{(n+1)/m}(z)} - \frac{a_{n/m}(z)}{b_{n/m}(z)}
$$

$$
= \frac{O\left(z^{n+m+1}\right)}{b_{n/m}(z)b_{(n+1)/m}(z)}
$$

as $z \to 0$. The only term of order z^{n+m+1} in the numerator on the left is the leading coefficient of $a_{(n+1)/m}(z) b_{n/m}(z)$. Note that

$$
b_{n/m}(z) = \det \mathbf{B}_{n/m}(z) =
\begin{bmatrix}
\gamma_{n-m+1} & \gamma_{n-m+2} & \cdots & \gamma_{n+1} \\
\vdots & \vdots & \ddots & \vdots \\
\gamma_n & \gamma_{n+1} & \cdots & \gamma_{n+m} \\
z^m & z^{m-1} & \cdots & 1
\end{bmatrix}
$$

then expansion by minors in the last row shows that

$$
= (-1)^{(m+1)+1} z^m \det
\begin{bmatrix}
\gamma_{n-m+2} & \cdots & \gamma_{n+1} \\
\vdots & \ddots & \vdots \\
\gamma_{n+1} & \cdots & \gamma_{n+m}
\end{bmatrix}
+ \text{ lower order terms}
$$

$$
= (-1)^m z^m c_{(n+1)/m} + \text{ lower order terms} .
$$

In order to find the leading coefficient of $a_{n/m}(z)$, we use the definition

$$
a_{n/m}(z) = \det
\begin{bmatrix}
\gamma_{n-m+1} & \gamma_{n-m+2} & \cdots & \gamma_{n+1} \\
\vdots & \vdots & \ddots & \vdots \\
\gamma_n & \gamma_{n+1} & \cdots & \gamma_{n+m} \\
\sum_{k=-m}^{n-m} \gamma_k z^{m+k} & \sum_{k=-m+1}^{n-m+1} \gamma_k z^{m+k-1} & \cdots & \sum_{k=0}^{n} \gamma_k z^k
\end{bmatrix}
$$

then the multilinearity of the determinant in its last row shows that

$$
= z^n \det
\begin{bmatrix}
\gamma_{n-m+1} & \gamma_{n-m+2} & \cdots & \gamma_{n+1} \\
\vdots & \vdots & \ddots & \vdots \\
\gamma_n & \gamma_{n+1} & \cdots & \gamma_{n+m} \\
\gamma_{n-m} & \gamma_{n-m+1} & \cdots & \gamma_n
\end{bmatrix}
+ \text{ lower order terms}
$$

then we can reorder the rows to get

$$
= z^n (-1)^m \det
\begin{bmatrix}
\gamma_{n-m} & \gamma_{n-m+1} & \cdots & \gamma_n \\
\gamma_{n-m+1} & \gamma_{n-m+2} & \cdots & \gamma_{n+1} \\
\vdots & \vdots & \ddots & \vdots \\
\gamma_n & \gamma_{n+1} & \cdots & \gamma_{n+m}
\end{bmatrix}
+ \text{ lower order terms}
$$

$$
= z^n (-1)^m c_{n/(m+1)} + \text{ lower order terms} .
$$

Thus

$$\frac{a_{(n+1)/m}(z)}{b_{(n+1)/m}(z)} - \frac{a_{n/m}(z)}{b_{n/m}(z)} = \frac{z^{n+m+1}c_{(n+1)/(m+1)}c_{(n+1)/m}}{b_{(n+1)/m}(z)b_{n/m}(z)} . \qquad (1.38)$$

Similar arguments give us

$$\frac{a_{n/m}(z)}{b_{n/m}(z)} - \frac{a_{(n-1)/m}(z)}{b_{(n-1)/m}(z)} = \frac{z^{n+m}c_{n/(m+1)}c_{n/m}}{b_{n/m}(z)b_{(n-1)/m}(z)} , \qquad (1.39)$$

$$\frac{a_{n/(m+1)}(z)}{b_{n/(m+1)}(z)} - \frac{a_{n/m}(z)}{b_{n/m}(z)} = \frac{z^{n+m+1}c_{(n+1)/(m+1)}c_{n/(m+1)}}{b_{n/(m+1)}(z)b_{n/m}(z)} \qquad (1.40)$$

and

$$\frac{a_{n/m}(z)}{b_{n/m}(z)} - \frac{a_{n/(m-1)}(z)}{b_{n/(m-1)}(z)} = \frac{z^{n+m}c_{(n+1)/m}c_{n/m}}{b_{n/m}(z)b_{n/(m-1)}(z)} \qquad (1.41)$$

We are now ready to prove Wynn's identity (1.35). First, we use Eqs. (1.38) and (1.40) to see that

$$\left[\frac{a_{(n+1)/m}(z)}{b_{(n+1)/m}(z)} - \frac{a_{n/m}(z)}{b_{n/m}(z)}\right]^{-1} - \left[\frac{a_{n/(m+1)}(z)}{b_{n/(m+1)}(z)} - \frac{a_{n/m}(z)}{b_{n/m}(z)}\right]^{-1}$$

$$= \frac{b_{(n+1)/m}(z)b_{n/m}(z)}{z^{n+m+1}c_{(n+1)/(m+1)}c_{(n+1)/m}} - \frac{b_{n/(m+1)}(z)b_{n/m}(z)}{z^{n+m+1}c_{(n+1)/(m+1)}c_{n/(m+1)}}$$

$$= \frac{b_{n/m}(z)}{z^{n+m+1}c_{(n+1)/(m+1)}}\left[\frac{b_{(n+1)/m}(z)}{c_{(n+1)/m}} - \frac{b_{n/(m+1)}(z)}{c_{n/(m+1)}}\right]$$

then we use Eq. (1.36) to obtain

$$= \frac{b_{n/m}(z)}{z^{n+m+1}c_{(n+1)/(m+1)}}\left[\frac{zb_{n/m}(z)c_{(n+1)/(m+1)}}{c_{(n+1)/m}c_{n/(m+1)}}\right] = \frac{b_{n/m}(z)^2}{z^{n+m}c_{(n+1)/m}c_{n/(m+1)}}$$

$$= \frac{b_{n/m}(z)}{z^{n+m}c_{n/m}}\left[\frac{b_{n/m}(z)c_{n/m}}{c_{(n+1)/m}c_{n/(m+1)}}\right]$$

then Eq. (1.37) gives us

$$= \frac{b_{n/m}(z)}{z^{n+m}c_{n/m}}\left[\frac{b_{(n-1)/m}(z)}{c_{n/(m+1)}} - \frac{b_{n/(m-1)}(z)}{c_{(n+1)/m}}\right] = \frac{b_{n/m}(z)b_{(n-1)/m}(z)}{z^{n+m}c_{n/m}c_{n/(m+1)}}$$

$$= \frac{b_{n/m}(z)b_{n/(m-1)}(z)}{z^{n+m}c_{n/m}c_{(n+1)/m}}$$

and finally Eqs. (1.39) and (1.41) produce

$$= \left[\frac{a_{n/m}(z)}{b_{n/m}(z)} - \frac{a_{(n-1)/m}(z)}{b_{(n-1)/m}(z)}\right]^{-1} - \left[\frac{a_{n/m}(z)}{b_{n/m}(z)} - \frac{a_{n/(m-1)}(z)}{b_{n/(m-1)}(z)}\right]^{-1}.$$

This is equivalent to the claimed result (1.35).

Exercise 1.4.1 Find the Padé approximation of order $(2, 2)$ to e^x.

Exercise 1.4.2 Some programming languages, such as JavaScript, do not provide an internal function for computing a hyperbolic sine. Develop Padé approximations to $\sinh(x)$ of order (n, n) for $n = 1, 2$ and 3. For what ranges of x are each of these Padé approximations accurate to a relative error of 10^{-15}?

1.4.2 Continued Fractions

Our next goal is to show that every rational polynomial can be evaluated efficiently as a **continued fraction**. We will use the following notation.

Definition 1.4.2 If p is a polynomial, then ∂p is the degree of p.
Suppose that $r(x) = p_0(x)/p_1(x)$ is a rational polynomial of order (n, m) with $n \geq m$, and assume that the highest-order term in $p_1(x)$ is x^m (i.e., the coefficient of the highest-order term is one). Then we can perform the following algorithm:

Algorithm 1.4.1 (Continued Fraction Determination)

$k = 1$

while $\partial p_k > 0$

    find the quotient polynomial q_k in dividing p_{k-1} by p_k with remainder $\widetilde{p}_{k+1}$

    if $\widetilde{p}_{k+1} \equiv 0$ break

    $c_{k+1} = $ coefficient of highest-order term in $\widetilde{p}_{k+1}$

    $p_{k+1} = \widetilde{p}_{k+1}/c_{k+1}$

    $k = k + 1$

$\ell = k$

Then for each k we have

$$\frac{p_{k-1}(x)}{p_k(x)} = q_k(x) + \frac{c_{k+1}p_{k+1}(x)}{p_k(x)} = q_k(x) + \frac{c_{k+1}}{p_k(x)/p_{k+1}(x)}.$$

This algorithm also shows that for $k \geq 1$ we have

$$\partial p_{k-1} = \partial p_k + \partial q_k \text{ and } \partial p_{k+1} < \partial p_k .$$

Note that when $n \geq m$, we stop at $\ell \leq m \leq n$, since the degrees of the polynomials p_k decrease as k increases in Algorithm 1.4.1.

Example 1.4.2 Algorithm 1.4.1 produces

$$\frac{p_0(x)}{p_1(x)} \equiv \frac{2x^4 - 4x^3 - 2x^2 + 12x - 4}{x^3 - 2x^2 - x + 5} = 2x + \frac{2x - 4}{x^3 - 2x^2 - x + 5} \equiv q_1(x) + 2\frac{p_2(x)}{p_1(x)} ,$$

$$\frac{p_1(x)}{p_2(x)} \equiv \frac{x^3 - 2x^2 - x + 5}{x - 2} = x^2 - 1 + \frac{3}{x - 2} \equiv q_2(x) + 3\frac{p_3(x)}{p_2(x)} ,$$

$$\frac{p_2(x)}{p_3(x)} \equiv \frac{x - 2}{1} = x - 2 \equiv q_3(x) .$$

After we perform Algorithm 1.4.1 to determine the quotient polynomials $q_k(x)$ and scalars c_k, we can evaluate $p_0(x)/p_1(x)$ as a continued fraction by performing the following.

Algorithm 1.4.2 (Continued Fraction Evaluation)

$$r = q_\ell(x)$$

$$\text{for } k = \ell - 1, \ldots, 1$$

$$r = q_k(x) + \frac{c_{k+1}}{r}$$

This algorithm involves $\ell - 1$ divisions, and the evaluation of the polynomials $q_\ell, \ldots, q_1$, presumably by Horner's rule (i.e., Algorithm 2.3.2 of Chap. 2 in Volume I). The first polynomial q_1 may have an arbitrary nonzero leading coefficient, but each of $q_2, \ldots, q_\ell$ have leading coefficient equal to one. Evaluation of q_1 involves as many multiplications as its degree, and the evaluation each of $q_2, \ldots, q_\ell$ involves as many multiplications as the degree minus one. The total number of multiplications in Algorithm 1.4.2 is

$$\partial q_1 + \sum_{k=2}^{\ell} (\partial q_k - 1) = \sum_{k=1}^{\ell} \partial q_k + 1 - \ell = \sum_{k=1}^{\ell} [\partial p_{k-1} - \partial p_k] + 1 - \ell$$

$$= \partial o_0 - \partial p_\ell + 1 - \ell = \partial p_0 + 1 - \ell .$$

The total number of multiplications and divisions is thus $\partial p_0 = n$, where we have assumed that $n = \partial p_0 > \partial p_1 = m$.

On the other hand, if $r(x) = p_1(x)/p_0(x)$ is a rational polynomial of order (n, m) with $n < m$, then we can take $q_1(x) = 0$ in Algorithm 1.4.1. In this case, we have $\ell \leq n$, and the total number of multiplications in computing $q_2, \ldots, q_\ell$ is $m - n$. The total work in computing r by continued fractions is $\ell + (m - n) \leq m$. Thus, we can always evaluate a rational polynomial of order (n, m) in at most $\max\{m, n\}$ multiplications and divisions.

Example 1.4.3 Suppose that we want to evaluate the rational polynomial

$$\frac{p_0(x)}{p_1(x)} = \frac{2x^4 - 4x^3 - 2x^2 + 12x - 4}{x^3 - 2x^2 - x + 5} .$$

If we use Horner's rule (Algorithm 2.3.2 of Chap. 2 in Volume I) to evaluate p_0 and p_1, we would use 4 multiplications and additions to compute $p_0(x)$, and 3 multiplications and additions to compute $p_1(x)$, for a total of 7 multiplications and additions. Using the continued fractions expansion of this rational polynomial from Example 1.4.2 and then evaluating the rational polynomial by Algorithm 1.4.2, we would compute

$$r = x - 2$$

$$r = x^2 - 1 + \frac{3}{r}$$

$$r = 2x + \frac{2}{r}$$

Even if we do not take advantage of zero coefficients in the quotient polynomials, we would require no multiplications and one addition for q_3, one addition for q_1, and one multiplications and two additions for q_2. The total work would be 2 multiplications, 4 additions, and 2 divisions.

The determination of continued fraction expansions for rational polynomials requires symbolic computation. As a result, such computations are not available in MATLAB. However, continued fraction expansions of rational polynomials are available in Maple and Mathematica.

Exercise 1.4.3 For $n \geq 1$ define the continued fractions

$$f_n(z) = \frac{z}{1-} \Big/ \frac{z^2}{3-} \Big/ \cdots \Big/ \frac{z^2}{2n - 1} .$$

Gauss [83, pp. 134–138] proved that $\lim_{n \to \infty} f_n(z) = \tan z$ for all complex numbers z. Verify that $f_1, \ldots, f_3$ are the Padé approximations to $\tan z$.

Exercise 1.4.4 Find the Padé approximations of order $(1, 1)$, $(2, 2)$ and $(3, 3)$ to $\sin z$, and describe how to evaluate each of these as continued fractions.

1.4.3 Rational Interpolation

It is tempting to try to interpolate functions with rational polynomials. The following example, adapted from Stoer [162, p. 45] shows that this is not always possible.

Example 1.4.4 Suppose that we want to find a rational polynomial of the form

$$r(x) = \frac{a + bx}{1 + cx}$$

so that $r(0) = 1$, $r(1) = 2$ and $r(2) = 2$. Since there are three unknown coefficients in the expression for $r(x)$, these interpolation conditions seem reasonable. Note that $r(0) = 1$ implies that $a = 1$. Next, $r(1) = 2$ implies that $b = 1 + 2c$. Finally, $r(2) = 2$ implies that $2 + 4c = 1 + 2(1 + 2c)$. Since the last expression is equal to $3 + 4c$, we get a contradiction. Thus there is no rational polynomial of this form that interpolates these data.

Nevertheless, rational polynomials are very useful in minimax approximation. Since this topic is beyond the scope of this book, we recommend that interested readers examine, for example, Cheney's "Introduction to Approximation Theory" [40, p. 153ff].

1.5 Quadric Surfaces

In computer graphics, it is sometime advantageous to represent certain common-place surfaces implicitly, as solutions of a quadratic equation.

Definition 1.5.1 Suppose that we are given ten values for the real coefficients c_α, where α is a multi-index in 3 dimensions satisfying $|\alpha| \leq 2$. Then the corresponding **quadric surface** is the set

$$\left\{ \mathbf{x} \ : \ \mathbf{x} \text{ is a 3-vector and } \sum_{|\alpha| \leq 2} c_\alpha \mathbf{x}^\alpha = 0 \right\} .$$

Lemma 1.3.3 shows that the number of coefficients c_α with $|\alpha| \leq 2$ for α a multi-index in 3 dimensions is $\binom{3+2}{2} = 10$. Definition 1.5.1 leads to the following easy result.

Lemma 1.5.1 *For any quadric surface with coefficients c_α there is a 4×4 symmetric matrix $\mathbf{Q}$ so that for all 3-vectors $\mathbf{x}$ we have*

$$\sum_{|\alpha| \leq 2} c_\alpha \mathbf{x}^\alpha = \begin{bmatrix} \mathbf{x}^\top & 1 \end{bmatrix} \mathbf{Q} \begin{bmatrix} \mathbf{x} \\ 1 \end{bmatrix} = 0 .$$

Proof It is easy to see that

$$2\mathbf{Q} = \begin{bmatrix} 2c_{\mathbf{e}_0+\mathbf{e}_0} & c_{\mathbf{e}_0+\mathbf{e}_1} & c_{\mathbf{e}_0+\mathbf{e}_2} & c_{\mathbf{e}_0} \\ c_{\mathbf{e}_1+\mathbf{e}_0} & 2c_{\mathbf{e}_1+\mathbf{e}_1} & c_{\mathbf{e}_1+\mathbf{e}_2} & c_{\mathbf{e}_1} \\ c_{\mathbf{e}_2+\mathbf{e}_0} & c_{\mathbf{e}_2+\mathbf{e}_1} & 2c_{\mathbf{e}_2+\mathbf{e}_2} & c_{\mathbf{e}_2} \\ c_{\mathbf{e}_0} & c_{\mathbf{e}_1} & c_{\mathbf{e}_2} & 2c_0 \end{bmatrix}.$$

The number of distinct entries of a 4×4 symmetric matrix is $\sum_{i=1}^4 i = (4)(4 + 1)/2 = 10$. Careful readers will also recall from our discussion in Sect. 4.3.3 of Chap. 4 in Volume I that the use of 4-vectors to represent three-dimensional coordinates in computer graphics has many advantages.

Quadric surfaces are often available in three-dimensional graphics packages. Given two components of any point on the quadric surface, the third component can be found by solving a quadratic equation. Often, one of the two possible solutions of the quadratic equation corresponds to a part of the surface hidden from view.

Suppose that we partition

$$\mathbf{Q} = \begin{bmatrix} \mathbf{A} & \mathbf{a} \\ \mathbf{a}^\top & \alpha \end{bmatrix}.$$

We will assume that $\mathbf{A}$ is nonzero, and has at least as many positive eigenvalues as negative eigenvalues; otherwise, we replace $\mathbf{Q}$ with $-\mathbf{Q}$ in the description of the quadric surface. We can use the spectral Theorem 1.3.1 of Chap. 1 in Volume II to write $\mathbf{A} = \mathbf{X}\mathbf{\Lambda}\mathbf{X}^\top$, where the columns of $\mathbf{X}$ are the eigenvectors of $\mathbf{A}$ and $\mathbf{\Lambda}$ is the diagonal matrix of eigenvalues of $\mathbf{A}$ in decreasing order.

If $\mathbf{A}$ has three positive eigenvalues (i.e. $\mathbf{A}$ is positive-definite), then the quadric surface is an **ellipsoid**. If $\mathbf{A}$ has two positive eigenvalues and one zero eigenvalue, then we have an **elliptic paraboloid** if the third component of $\mathbf{X}^\top \mathbf{a}$ is nonzero, else we have an **elliptic cylinder**. If $\mathbf{A}$ has two positive eigenvalues and one negative eigenvalue, then we examine the value of

$$\gamma = \alpha - \mathbf{a}^\top \mathbf{A}^{-1} \mathbf{a} \; ;$$

we have a **hyperboloid of one sheet** when $\gamma < 0$, a **hyperboloid of two sheets** when $\gamma > 0$, and a **cone** when $\gamma = 0$. Here $\mathbf{A}^{-1}$ represents the **inverse** of the matrix $\mathbf{A}$. If $\mathbf{A}$ has one positive eigenvalue, one zero eigenvalue and one negative eigenvalue, then we have a **hyperbolic paraboloid** when the third entry of $\mathbf{X}^\top \mathbf{a}$ is nonzero, otherwise we have a **hyperbolic cylinder**. Finally, if $\mathbf{A}$ has one positive eigenvalue and two zero eigenvalues, we have a **parabolic cylinder**.

For more information about quadric surfaces and computer graphics in general, we recommend Foley et al. [80], or the OpenGL Programming Guide. Quadric surfaces are not provided by any MATLAB command.

Exercise 1.5.1 Describe a method to determine the normal to a quadric surface at some given point on the surface.

Exercise 1.5.2 Write a routine to compute points on a quadric surface. Describe how you choose which spatial variable to be dependent on the other two.

Exercise 1.5.3 Describe a plane in three dimensions in terms of the coefficients of a quadric surface, as in Definition 1.5.1.

Exercise 1.5.4 Describe a sphere in three dimensions in terms of the matrix $\mathbf{Q}$ in Lemma 1.5.1.

Exercise 1.5.5 Given three points that are not co-linear, determine a plane that passes through those three points. This is interpolation by a linear surface.

Exercise 1.5.6 Given four points that are not co-planar, determine a sphere that passes through those four points. This is interpolation by a particular quadric surface.

1.6 Splines

In Sect. 1.2 we saw that it is difficult to get a uniformly good polynomial interpolation over a large domain. To avoid this problem, we will break the interpolation domain into a number of pieces and use different polynomials in the different subdomains. The resulting **piecewise polynomial** functions are commonly used to represent general curves and surfaces in computer graphics (see Sect. 1.6.8), and are commonly used to represent solutions of differential equations solved by finite element methods (see Sect. 4.6). Some splines are also used as wavelets (see Sect. 1.8.2.9).

We will begin with a general definition of the idea.

Definition 1.6.1 Suppose that we are given a problem domain $[a, b]$ and an integer $n > 0$ Then the set

$$\mathscr{M}_n = \{x_0, \ldots, x_n\}$$

is a **mesh** if and only if its elements, called the **mesh nodes**, satisfy $x_0 < x_1 < \ldots < x_n$ with $x_0 = a$ and $x_n = b$. The closed intervals $[x_i, x_{i+1}]$ for all $0 \leq i < n$ are called the **mesh elements**. Next, if m and k are positive integers then $s_{m,k}$ is a **spline** on the **mesh** $\mathscr{M}_n$ if and only if

- for all $0 \leq i < n$ $s_{m,k}$ is a polynomial of degree at most m in the mesh element $[x_i, x_{i+1}]$, and
- $s_{m,k}$ is $k - 1$ times continuously differentiable on $[a, b]$.

Our goal is to choose a spline so that it is close in some sense to a given function f. The tactics for achieving this goal will depend on the polynomial degree m and the continuity index k.

In Sects. 1.6.1, 1.6.2 and 1.6.3, we will develop splines in groups, depending on the continuity index k. Within each group, the development may vary due to the element-wise polynomial order m. For $0 < k = m - 1$, we will see that it is difficult to develop local basis functions on non-uniform meshes. Such cases usually require the solution of a system of linear equations to determine the spline coefficients. However, on uniform meshes it is possible to develop *b-splines*, which are also important in wavelet analysis. We will present a general discussion of errors with spline approximation in Sect. 1.6.6.

We will generalize splines in two ways. In Sect. 1.6.7, we will show how to introduce a parameter to vary the approximation from piecewise linear to twice continuously differentiable piecewise cubic. Afterward, in Sect. 1.6.8 we will show how to use splines to generate curves in higher-dimensional spaces, using some old and a couple of new ideas.

1.6.1 Continuous

We will begin by studying splines that are continuous piecewise polynomials of degree at most m on each mesh element. With n mesh elements, such a spline has a total of $(m+1)n$ unknowns, namely the $m+1$ coefficients of each of the polynomials on each of the n mesh elements. Continuity of the piecewise polynomials at the interior nodes $x_1, \ldots, x_{n-1}$ gives us $n-1$ constraints. This leaves us with $(m+1)n - (n-1) = mn + 1$ degrees of freedom.

We can determine the remaining degrees of freedom by requiring the piecewise polynomial to interpolate the function f at each of the $n + 1$ mesh nodes, and at $m - 1$ additional points in the interior of each of the mesh elements. Interpolation at the mesh nodes provides $n + 1$ additional equations, and interpolation at $m - 1$ interior points in each mesh element gives us $(m - 1)n$ additional equations, for a total of $n + 1 + (m - 1)n = mn + 1$ interpolation conditions.

Normally, we will choose the interpolation points in the interior of each mesh element as follows. First, we choose **canonical interpolation points**

$$0 = \xi_0 < \xi_1 < \ldots < \xi_{m-1} < \xi_m = 1 \, ,$$

We will assume that the canonical interpolation points ξ_j are **symmetrically placed**, meaning that for all $0 \leq j \leq m$ we have

$$\xi_{m-j} = 1 - \xi_j \, .$$

For example, we could choose $\xi_j = j/m$, so that the interpolation points are equidistant. Optimal locations for the interior interpolation points have been determined by Angelos et al. [5], and have been discussed by Chen and Babuška [38].

On each mesh element $[x_{i-1}, x_i]$, we will define the **element width**

$$h_i = x_i - x_{i-1} \, ,$$

and then interpolate the given function at the points $x_{i-1} + \xi_j h_i$ for $0 \leq j \leq m$ and $1 \leq i \leq n$.

In some applications, it is useful to rewrite the spline $s_{m,1}$ in terms of basis functions associated with the nodes, as well as basis functions associated with any interpolation points interior to the mesh elements. First, we define the **canonical nodal basis function**

$$\beta_{0;m,1}(\xi) = \begin{cases} \prod_{k=1}^m (1 - |\xi/\xi_k|), & |\xi| \leq 1 \\ 0, & |\xi| \geq 1 \end{cases} . \tag{1.42}$$

Note that since $\xi_m = 1$, the function $\beta_{0;m,1}$ is continuous at $\xi = \pm 1$. Next, for $1 \leq j < m$ we define the **canonical interior basis functions**

$$\beta_{j;m,1} = \begin{cases} \prod_{\substack{0 \leq k \leq m \\ k \neq j}} (\xi - \xi_k)/(\xi_j - \xi_k), & 0 \leq \xi \leq 1 \\ 0, & \xi < 0 \text{ or } \xi > 1 \end{cases} .$$

Note that since $\xi_0 = 0$ and $\xi_m = 1$, the function $\beta_{j;m,1}$ is continuous at $\xi = 0$ and $\xi = 1$.

For node indices $0 \leq i \leq n$, we can perform a change of variables to get the **nodal basis functions**

$$b_{i;0;m,1}(x) = \begin{cases} \beta_{0;m,1}\left((x - x_i)/h_{i+1}\right), & x \geq x_i \\ \beta_{0;m,1}\left((x - x_i)/h_i\right), & x \leq x_i \end{cases} .$$

In addition, for $1 \leq i \leq n$ and $1 \leq j < m$ we define the **interior basis functions** by

$$b_{i;j;m,1}(x) = \beta_{j;m,1}\left((x - x_{i-1})/h_i\right) \, .$$

Then for all $x \in [a, b]$, the spline can be written in terms of the two types of basis functions as follows:

$$s_{m,1}(x) = \sum_{i=0}^n f(x_i) b_{i;0;m,1}(x) + \sum_{i=1}^n \sum_{j=1}^{m-1} f(x_{i-1} + \xi_j h_i) b_{i;j;m,1}(x) \, .$$

This formula clearly describes how the spline relates to function values at the interpolation points, and how the spline uses the various basis functions. However, for any individual point $x \in [a, b]$, most of the basis functions in this formula have zero value.

In other applications, it is useful to write the spline $s_{m,1}$ in terms of basis functions associated with the elements. In this case, we can use the basis polynomials for

Lagrange interpolation by polynomials of degree m on $[0, 1]$, namely, for $0 \leq j \leq m$

$$\lambda_{j,m}(\xi) = \prod_{\substack{0 \leq k \leq m \\ k \neq j}} \frac{\xi - \xi_k}{\xi_j - \xi_k} . \tag{1.43}$$

Then for all mesh element numbers $1 \leq i \leq n$ and all $x \in [x_{i-1}, x_i]$ we can evaluate the spline by the following

Algorithm 1.6.1 (C^0 **Spline Evaluation**)

$$h_i = x_i - x_{i-1}$$

$$\xi = \frac{x - x_{i-1}}{h_i}$$

$$s_{m,1}(x) = f(x_{i-1})\lambda_{0,m}(\xi) + f(x_i)\lambda_{m,m}(\xi) + \sum_{j=1}^{m-1} f\left(x_{i-1} + \xi_j h_i\right)\lambda_{j,m}(\xi) .$$

With either representation of the continuous spline, at any fixed value of $x \in [a, b]$ there are at most $m + 1$ basis functions that are nonzero in the evaluation of $s_{m,1}(x)$.

Please note that we will discuss error estimates for continuous spline interpolation in Sect. 1.6.6.

Example 1.6.1 Suppose that $m = 1$, meaning that the spline $s_{1,1}(x)$ is piecewise linear. Then for all $1 \leq i \leq n$ and for all $x \in [x_{i-1}, x_i]$ we let $\xi = (x - x_{i-1})/h_i$ and compute

$$s_{1,1}(x) = f(x_{i-1})(1 - \xi) + f(x_i)\xi .$$

Example 1.6.2 Suppose that $m = 2$, meaning that the spline $s_{2,1}(x)$ is piecewise quadratic. Then for all $1 \leq i \leq n$ and for all $x \in [x_i, x_{i+1}]$ we let $\xi = (x - x_{i-1})/h_i$ and compute

$$s_{2,1}(x) = f(x_{i-1})(1 - \xi)(1 - 2\xi) + f\left(\frac{x_{i-1} + x_i}{2}\right)4\xi(1 - \xi) + f(x_i)\xi(2\xi - 1) .$$

A Fortran program to implement continuous piecewise polynomial interpolation is available as routine `c0_spline_eval` in piecewise_poly.f. This is called from the C^{++} program piecewisePolynomial.C. The C^{++} program plots the interpolation and performs a numerical error analysis by measuring the interpolation errors for a range of numbers of interpolation points. Readers may also execute a JavaScript program for approximating Runge's function by **continuous splines.** Some results for piecewise linear interpolation to Runge's function are shown in Fig. 1.7.

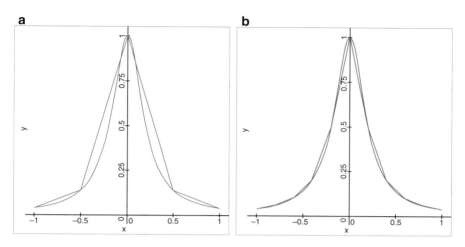

Fig. 1.7 Piecewise linear interpolation for Runge's function. (**a**) 5 points. (**b**) 11 points

MATLAB users should use command interp1 with method = 'linear' to perform piecewise linear interpolation. GSL users should examine interpolation type gsl_interp_linear.

Exercise 1.6.1 Use the definition (1.43) to prove that $\lambda_{m-j,m}(1 - \xi) = \lambda_{j,m}(\xi)$ whenever the canonical interpolation points are symmetrically placed.

Exercise 1.6.2 Write a program to perform continuous piecewise linear interpolation to a given function f on a uniform mesh. Then perform a **mesh refinement study** with Runge's function, to determine how the maximum error in the spline approximation over $x \in [-1, 1]$ varies as a function of the mesh width $h = \max_{1 \le i \le n} x_i - x_{i-1}$. Plot the logarithm of the maximum error versus minus the logarithm of h, and determine the power of h evidenced by the graph.

Exercise 1.6.3 Repeat the previous exercise with $f(x) = \sqrt{(x)}$ for $x \in [0, 1]$.

Exercise 1.6.4 Perform a mesh refinement study for continuous piecewise quadratic interpolation to Runge's function. Plot the logarithm of the maximum error versus minus the logarithm of the mesh width h, and determine the apparent power of h for the error function in this graph.

1.6.2 Continuously Differentiable

Next, we will consider splines $s_{m,2}$ (see Definition 1.6.1) that are continuously differentiable piecewise polynomials of degree at most m on the mesh elements. On a mesh with n elements, such a spline has a total of $(m + 1)n$ undetermined coefficients: $m + 1$ coefficients for each polynomial on each of the n individual elements. Continuity of the function values and derivatives at the interior mesh nodes gives us $2(n - 1)$ constraints on these coefficients. This leaves us with $(m + 1)n - 2(n - 1) = (m - 1)n + 2$ degrees of freedom. How we specify these degrees of freedom will depend on the order m of the piecewise polynomials.

1.6.2.1   Piecewise Quadratics

If $m = 2$, then there are $n + 2$ degrees of freedom in a continuously differentiable spline on a mesh with n elements. We will determine all but one of the remaining degrees of freedom by requiring the spline to interpolate f at the $n + 1$ mesh nodes. We can specify the remaining degree of freedom by interpolating f' at either the first or last mesh node.

Let

$$z_i = s'_{2,2}(x_i)$$

be the derivative of the spline at the i-th node. On each mesh element $[x_{i-1}, x_i]$ the spline derivative $s'_{2,2}$ is linear, so we can write

$$s'_{2,2}(x) = z_{i-1} \frac{x_i - x}{x_i - x_{i-1}} + z_i \frac{x - x_{i-1}}{x_i - x_{i-1}} .$$

For $x \in [x_{i-1}, x_i]$, we can integrate this equation to get

$$s_{2,2}(x) = -\frac{z_{i-1}}{2} \frac{(x_i - x)^2}{x_i - x_{i-1}} + \frac{z_i}{2} \frac{(x - x_{i-1})^2}{x_i - x_{i-1}} + c_{i-1} ,$$

where c_{i-1} is a constant. Since $s_{2,2}$ interpolates f at x_{i-1}, we must have

$$f(x_{i-1}) = s_{2,2}(x_{i-1}) = -\frac{z_{i-1}}{2}(x - x_{i-1}) + c_{i-1}$$

$$\implies c_{i-1} = f(x_{i-1}) + z_{i-1} \frac{x_i - x_{i-1}}{2} .$$

Since $s_{2,2}$ also interpolates f at x_i, we must have

$$f(x_i) = s_{2,2}(x_i) = \frac{z_i}{2}(x_i - x_{i-1}) + c_{i-1} = \frac{z_i}{2}(x_i - x_{i-1}) + f(x_{i-1}) + \frac{z_{i-1}}{2}(x_i - x_{i-1})$$

$$\implies z_{i-1} + z_i = 2 \frac{f(x_i) - f(x_{i-1})}{x_i - x_{i-1}} = 2f[x_{i-1}, x_i] .$$

This gives us a bi-diagonal linear system for the unknown slopes z_i. If we interpolate f' at x_0, then we can forward-solve the linear system as follows:

Algorithm 1.6.2 (C^1 Quadratic Spline Coefficients)

$$z_0 = f'(x_0)$$
$$\text{for } 1 \leq i \leq n \,, \; z_i = 2f[x_{i-1}, x_i] - z_{i-1} \,.$$

There is a similar algorithm, involving back-solving, if we specify f' at x_n. Once we have determined the slopes z_i, we can evaluate $s_{2,2}$ at a point x in the mesh element (x_{i-1}, x_i) as follows:

Algorithm 1.6.3 (C^1 Quadratic Spline Evaluation)

$$h_i = x_i - x_{i-1}$$
$$\xi = \frac{x - x_{i-1}}{h_i}$$
$$s_{2,2}(x) = f_{i-1} + [z_{i-1}(2 - \xi) + z_i \xi] \xi h_i / 2$$

However, we should remark that continuously differentiable quadratic splines are seldom used in practice. First, the interpolation of f' at one of the endpoints of the mesh induces an asymmetric treatment of the boundaries. Second, the linear system for the slopes z_i is somewhat of a nuisance.

A Fortran program to implement continuously differentiable piecewise quadratic interpolation is available as routines `c1_quadratic_coefs` and `c1_quadratic_eval` in piecewise_poly.f. This is called from the C^{++} program piecewisePolynomial.C. The C^{++} program plots the interpolation and performs a numerical error analysis by measuring the interpolation errors for a range of numbers of interpolation points.

Neither MATLAB nor the GSL provide continuously differentiable quadratic splines.

1.6.2.2 Piecewise Cubics (or Higher)

Recall that there are $(m-1)n + 2$ degrees of freedom in a continuously differentiable spline $s_{m,2}$ on a mesh with n elements. For $m \geq 3$, we will determine some of these degrees of freedom by requiring the piecewise polynomial to interpolate both f and f' at the $n + 1$ mesh nodes. This leaves us with $(m - 1)n + 2 - 2(n + 1) = (m - 3)n$ remaining degrees of freedom. The remaining degrees of freedom can be specified by interpolating f at $m - 3$ additional points in the interior of each mesh element. In a fashion analogous to our work in Sect. 1.6.1, we will assume that the interpolation points interior to a particular mesh element (x_{i-1}, x_i) are of the form

$x_{i-1}(1 - \xi_j) + x_i \xi_j$, where the canonical interpolation points ξ_j for $1 \leq j \leq m - 3$ are symmetrically placed in $(0, 1)$.

Let us begin by determining the basis functions associated with mesh nodes $x_0, \ldots, x_n$. First, we will find a canonical nodal basis function that has value one and slope zero at $\xi = 0$, value and slope zero at $\xi = \pm 1$, and value zero at $\pm \xi_j$ for $1 \leq j \leq m - 3$. The canonical basis function is easily seen to be

$$
\beta_{0;m,2}(\xi) = \begin{cases} \left(1 + \left[2 + \sum_{j=1}^{m-3} 1/\xi_j\right] |\xi|\right) (1 - |\xi|)^2 \prod_{j=1}^{m-3}(1 - |\xi|/\xi_j), & |\xi| \leq 1 \\ 0, & |\xi| \geq 1 \end{cases}.
$$

Next, we will find a canonical nodal basis function that has value zero and slope one at $\xi = 0$, value and slope zero at $\xi = 1$, and value zero at ξ_j for $1 \leq j \leq m - 3$. This canonical basis function is

$$
\gamma_{m,2}(\xi) = \begin{cases} \xi(1 - |\xi|)^2 \prod_{j=1}^{m-3}(1 - |\xi|/\xi_j), & |\xi| \leq 1 \\ 0, & |\xi| \geq 1 \end{cases}.
$$

Finally, we will find a canonical nodal basis function that has value and slope zero at both $\xi = 0$ and $\xi = 1$, value zero at ξ_k for $k \neq j$, and value one at ξ_j. This gives us

$$
\beta_{j;m,2}(\xi) = \begin{cases} \dfrac{\xi^2}{\xi_j^2} \dfrac{(1-\xi)^2}{(1-\xi_j)^2} \prod_{\substack{0 \leq k \leq m-3 \\ k \neq j}} \dfrac{\xi - \xi_k}{\xi_j - \xi_k}, & 0 \leq \xi \leq 1 \\ 0, & \xi \leq 0 \text{ or } \xi \geq 1 \end{cases}.
$$

Next, for $1 \leq i \leq n$ we define the element width

$$
h_i = x_i - x_{i-1}
$$

For $0 \leq i \leq n$, these can be used with a change of variables to produce the nodal basis functions

$$
b_{i;0;m,2}(x) = \begin{cases} \beta_{0;m,2}([x - x_i]/h_{i+1}), & x \geq x_i \\ \beta_{0;m,2}([x - x_i]/h_i), & x \leq x_i \end{cases} \text{ and}
$$

$$
c_{i;0;m,2}(x) = \begin{cases} h_{i+1}\gamma_{m,2}([x - x_i]/h_{i+1}), & x \geq x_i \\ h_i\gamma_{m,2}([x - x_i]/h_i), & x \leq x_i \end{cases}.
$$

In addition, for $1 \leq i \leq n$ and $1 \leq j \leq m-3$ we define the interior basis functions by

$$
b_{i-1;j;m,2}(x) = \beta_{j;m,2}((x - x_{i-1})/h_i) .
$$

Then for all $x \in [a, b]$, the spline can be written in terms of the two types of basis functions as follows:

$$s_{m,2}(x) = \sum_{i=0}^{n} f(x_i) b_{i;0;m,2}(x) + \sum_{i=0}^{n} f'(x_i) h_i c_{i;0;m,2}(x)$$

$$+ \sum_{i=1}^{n} \sum_{j=1}^{m-3} f\left(x_{i-1} + \xi_j h_i\right) b_{i-1;j;m,2}(x) .$$

This expression for the spline function clearly displays the dependence on the interpolation points for f, but involves many zero terms. Alternatively, for any $x \in (x_{i-1}, x_i)$ we can evaluate the spline by performing the

Algorithm 1.6.4 (C^1 Higher-Order Spline Evaluation)

$$h_i = x_i - x_{i-1}$$

$$\xi = \frac{x - x_{i-1}}{h_i}$$

$$s_{m,2}(x) = f(x_{i-1}) \beta_{0,m,2}(\xi) + f(x_i) \beta_{0,m,2}(1 - \xi) + f'(x_{i-1}) \gamma_{m,2}(\xi) h_i$$

$$- f'(x_i) \gamma_{m,2}(1 - \xi) h_i + \sum_{j=1}^{m-3} f(x_{i-1} + \xi_j h_i) \beta_{j,m,2}(\xi)$$

Example 1.6.3 The spline $s_{3,2}$ is continuously differentiable and piecewise cubic. For all $x \in [x_{i-1}, x_i]$ we compute

$$h_i = x_i - x_{i-1}$$

$$\xi = (x - x_{i-1})/(x_i - x_{i-1})$$

$$s_{3,2}(x) = f(x_{i-1})(1 + 2\xi)(1 - \xi)^2 + f(x_i)\xi^2(3 - 2\xi) + h_i \xi(1 - \xi)$$

$$\{f'(x_{i-1})(1 - \xi) - f'(x_i)\xi\} .$$

This spline $s_{3,2}$ is commonly called a **Hermite cubic spline**.

Some results with continuously differentiable piecewise cubic interpolation to Runge's function are shown in Fig. 1.8. In this case, the continuous cubic splines appear to be more accurate than the continuously differentiable cubic spline for the same number of mesh nodes. This figure was generated by the C^{++} program GUI-PiecewisePolynomial.C. Readers can see another implementation of continuously differentiable splines in Fortran routine c1_spline_eval, which is available in piecewise_poly.f. This is called from the C^{++} program piecewisePolynomial.C.

Readers may also execute a JavaScript program for approximating Runge's function by **continuously differentiable splines**.

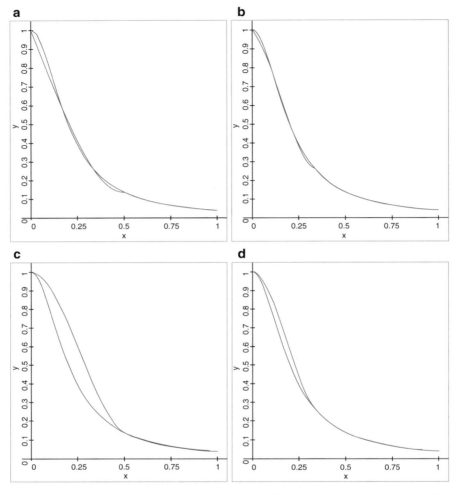

Fig. 1.8 Cubic interpolation for Runge's function. (**a**) C^0, 3 nodes. (**b**) C^0, 4 nodes. (**c**) C^1, 3 nodes. (**d**) C^1, 4 nodes

   MATLAB users could choose method = 'cubic' in command interp1 to produce "shape-preserving" piecewise cubic interpolants. These are continuously differentiable, but do not require derivatives of the interpolated function at the interpolation points. For more information about this interpolation process, see the paper by Costantini and Morandi [51].

Exercise 1.6.5 Write a program to compute the piecewise Hermite cubic interpolant to

$$f(x) = \frac{1}{x^2 + 25}$$

at equidistant nodes in the interval $[-1, 1]$. Plot f and the interpolant for 2 and 3 nodes. By sampling the error at 100 points between each node, determine a (computer generated) estimate for the interpolation error, and plot the log of the error versus the log of the mesh width.

Exercise 1.6.6 Describe an algorithm to evaluate the continuously differentiable quartic spline $s_{4,2}$ on a mesh element $[x_{i-1}, x_i]$.

1.6.3 Twice Continuously Differentiable

Next, we will develop splines $s_{m,3}$ (recall Definition 1.6.1) that are twice continuously differentiable piecewise polynomials of degree at most m on each of the mesh elements. With n elements, such a spline has a total of $(m + 1)n$ undetermined coefficients: $m + 1$ coefficients for each polynomial on each of the n mesh elements. Continuity of the function values and first two derivatives at the interior mesh nodes gives us $3(n - 1)$ constraints on these coefficients. This leaves us with $(m + 1)n - 3(n - 1) = (m - 2)n + 3$ degrees of freedom. We will specify these remaining degrees of freedom in different ways, depending on the order m of the piecewise polynomials.

1.6.3.1 Piecewise Cubics

There are $n + 3$ degrees of freedom in a twice continuously differentiable cubic spline on a mesh with n elements. We will determine $n + 1$ of these degrees of freedom by interpolating the given function f at the mesh nodes. This will leave us with 2 degrees of freedom, which will be specified at the initial and final mesh nodes. One common choice for the endpoint conditions is

$$s'_{3,3}(x_0) = f'(x_0), \quad s'_{3,3}(x_n) = f'(x_n) ;$$

another common choice for the endpoint conditions is

$$s''_{3,3}(x_0) = 0 = s''_{3,3}(x_n) .$$

With either choice, both endpoints are treated symmetrically.

Note that the second derivative $s''_{3,3}$ is linear on the mesh element $[x_{i-1}, x_i]$. This suggests that we define

$$z_{i-1} = s''_{3,3}(x_{i-1}) \text{ and } h_i = x_i - x_{i-1} .$$

Since $s_{3,3}''$ is continuous, we have

$$s_{3,3}''(x) = z_{i-1}\frac{x_i - x}{h_i} + z_i\frac{x - x_{i-1}}{h_i} \ .$$

Thus there are constants c_i and d_i such that for $x_{i-1} \le x \le x_i$

$$s_{3,3}(x) = \frac{z_{i-1}}{6}\frac{(x_i - x)^3}{h_i} + \frac{z_i}{6}\frac{(x - x_{i-1})^3}{h_i} + c_i(x - x_{i-1}) + d_i(x_i - x) \ .$$

Since $s_{3,3}$ interpolates f at the nodes,

$$f_{i-1} = s_{3,3}(x_{i-1}) = \frac{z_{i-1}}{6}h_i^2 + d_i h_i \implies d_i = \frac{f_{i-1}}{h_i} - \frac{z_{i-1}h_i}{6} \text{ and}$$

$$f_i = s_{3,3}(x_i) = \frac{z_i}{6}h_i^2 + c_i h_i \implies c_i = \frac{f_i}{h_i} - \frac{z_i h_i}{6} \ .$$

This implies that for $x_{i-1} \le x \le x_i$,

$$s_{3,3}(x) = f_{i-1} + \frac{f_i - f_{i-1}}{h_i}(x - x_{i-1}) + z_{i-1}\frac{(x_i - x)[(x_i - x)^2 - h_i^2]}{6h_i}$$

$$+ z_i\frac{(x - x_{i-1})[(x - x_{i-1})^2 - h_i^2]}{6h_i} \ .$$

If we define $\xi = (x - x_{i-1})/h_i$, then we can write

$$s_{3,3}(x) = f_{i-1} + (f_i - f_{i-1})\xi + \frac{z_{i-1}h_i^2}{6}(1 - \xi)[(1 - \xi)^2 - 1] + \frac{z_i h_i^2}{6}\xi[\xi^2 - 1]$$

$$= f_{i-1}(1 - \xi) + f_i\xi - \frac{\xi(1 - \xi)h_i^2}{6}[(z_{i-1}(2 - \xi) + z_i(1 + \xi)] \ .$$

The unknown second derivatives of the spline are determined by the conditions imposed at the endpoints, and by continuity of s' at the interior nodes. Note that

$$\lim_{x \downarrow x_i} s_{3,3}'(x) = -\frac{1}{2}z_i h_{i+1} + \frac{f_{i+1} - f_i}{h_{i+1}} - \frac{1}{6}(z_{i+1} - z_i)h_{i+1}$$

$$= -\frac{h_{i+1}z_i}{3} - \frac{h_{i+1}z_{i+1}}{6} + \frac{f_{i+1} - f_i}{h_{i+1}} \ ,$$

and

$$\lim_{x \uparrow x_i} s_{3,3}'(x) = -\frac{1}{2}z_i h_i + \frac{f_i - f_{i-1}}{h_i} - \frac{1}{6}(z_i - z_{i-1})h_i$$

$$= \frac{h_i z_i}{3} + \frac{h_i z_{i-1}}{6} + \frac{f_i - f_{i-1}}{h_i} \ .$$

Thus continuity of $s'_{3,3}$ can be written as the linear equation

$$\frac{h_i}{6}z_{i-1} + \frac{h_{i+1} + h_i}{3}z_i + \frac{h_{i+1}}{6}z_{i+1} = \frac{f_{i+1} - f_i}{h_{i+1}} - \frac{f_i - f_{i-1}}{h_i}.$$

If we choose our endpoint conditions to be $s''_{3,3}(x_0) = 0 = s''_{3,3}(x_n)$, then we must have $z_0 = 0 = z_n$ and we are left with the symmetric positive tridiagonal linear system

$$\begin{bmatrix} 2(h_2 + h_1) & h_2 & & & & \\ h_2 & 2(h_3 + h_2) & \ddots & & & \\ & \ddots & \ddots & \ddots & & \\ & & \ddots & 2(h_{n-1} + h_{n-2}) & h_{n-1} \\ & & & h_{n-1} & 2(h_n + h_{n-1}) \end{bmatrix} \begin{bmatrix} z_1 \\ z_2 \\ \vdots \\ \vdots \\ z_{n-1} \end{bmatrix}$$

$$= \begin{bmatrix} 6\left(f[x_2, x_1] - f[x_1, x_0]\right) \\ 6\left(f[x_3, x_2] - f[x_2, x_1]\right) \\ \vdots \\ 6\left(f[x_{n-1}, x_{n-2}] - f[x_{n-2}, x_{n-3}]\right) \\ 6\left(f[x_n, x_{n-1}] - f[x_{n-1}, x_{n-2}]\right) \end{bmatrix}.$$

If, on the other hand, we choose our endpoint conditions to be $s'_{3,3}(x_0) = f'_0, s'_{3,3}(x_n) = f'_n$, then we are left with the symmetric positive tridiagonal linear system

$$\begin{bmatrix} 2h_1 & h_1 & & & & \\ h_1 & 2(h_2 + h_1) & h_2 & & & \\ & h_2 & 2(h_3 + h_2) & \ddots & & \\ & & \ddots & \ddots & \ddots & \\ & & & \ddots & 2(h_n + h_{n-1}) & h_n \\ & & & & h_n & 2h_n \end{bmatrix} \begin{bmatrix} z_0 \\ z_1 \\ z_2 \\ \vdots \\ z_{n-1} \\ z_n \end{bmatrix}$$

$$= \begin{bmatrix} 6(f[x_1, x_0] - f'(x_0)) \\ 6(f[x_2, x_1] - f[x_1, x_0]) \\ 6(f[x_3, x_2] - f[x_2, x_1]) \\ \vdots \\ 6(f[x_n, x_{n-1}] - f[x_{n-1}, x_{n-2}]) \\ 6(f'(x_n) - f[x_n, x_{n-1}]) \end{bmatrix}.$$

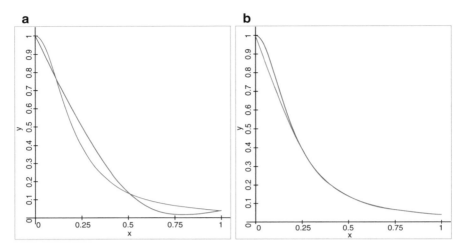

Fig. 1.9 C^2 piecewise cubic interpolation for Runge's function. (**a**) 2 nodes. (**b**) 4 nodes

Some computational results for twice continuously differentiable piecewise cubic interpolation to Runge's function are shown in Fig. 1.9. Readers can see an implementation of twice continuously differentiable cubic splines in Fortran routines `c2_cubic_coefs` and `c2_cubic_eval`, which are available in piecewise_poly.f. This is called from the C^{++} program piecewisePolynomial.C.

Alternative twice continuously differentiable cubic spline routines are available as Forsythe, Malcolm and Moler routines spline seval, or as SLATEC routines polint, bint4, bintk, bvalu and bspev. MATLAB users can use command interp1 with `method = 'spline'` to perform twice continuously differentiable cubic spline interpolation. GSL programmers can should examine gsl_interp_spline.

1.6.3.2 Piecewise Quartics

Our next goal is to determine a twice continuously differentiable piecewise polynomial of degree at most 4 on each mesh element. Since there are 5 undetermined coefficients for each quartic on each of the n mesh elements, there are a total of $5n$ undetermined coefficients. Since there are 3 continuity conditions at each of the interior mesh nodes, we need to specify $5n - 3(n - 1) = 2n + 3$ additional conditions. We will interpolate f and f' at each of the $n + 1$ mesh nodes, leaving $(2n + 3) - 2(n + 1) = 1$ remaining degree of freedom. This final degree of freedom can be determined by providing the value of f'' at either x_0 or x_n.

For any $x \in [x_{i-1}, x_i]$, we can write

$$x = x_{i-1} + \xi h_i \text{ where } h_i = x_i - x_{i-1} \text{ and } \xi \in [0, 1] .$$

Since $s_{4,3}$ interpolates f and f' at x_{i-1} and x_i, we must have

$$
\begin{aligned}
s_{4,3}(x) =\ & f(x_{i-1})(1+4\xi)(1-2\xi)(1-\xi)^2 + f'(x_{i-1})h_i\xi(1-2\xi)(1-\xi)^2 \\
& + f(x_i)\xi^2(2\xi-1)(5-4\xi) - f'(x_i)h_i\xi^2(2\xi-1)(1-\xi) \\
& + c_i 16\xi^2(1-\xi)^2 ,
\end{aligned}
$$

where c_i is the value of $s_{4,3}$ at the midpoint of the mesh element:

$$
s_{4,3}((x_{i-1}+x_i)/2) = c_i .
$$

Since $d\xi/dx = 1/h_i$, we can take two derivatives with respect to x to get

$$
\begin{aligned}
s_{4,3}''(x) =\ & \frac{f(x_{i-1})}{h_i^2}\left[-96\xi^2 + 108\xi - 22\right] + \frac{f'(x_{i-1})}{h_i}\left[-24\xi^2 + 30\xi - 8\right] \\
& + \frac{f(x_i)}{h_i^2}\left[-96\xi^2 + 84\xi - 10\right] - \frac{f'(x_i)}{h_i}\left[-24\xi^2 + 18\xi - 2\right] \\
& + \frac{c_i}{h_i^2}\left[192\xi^2 - 192\xi + 32\right] .
\end{aligned}
$$

In particular,

$$
s_{4,3}''(x_{i-1}) = -22\frac{f(x_{i-1})}{h_i^2} - 8\frac{f'(x_{i-1})}{h_i} - 10\frac{f(x_i)}{h_i^2} + 2\frac{f'(x_i)}{h_i} + 32\frac{c_i}{h_i^2}
$$

and

$$
s_{4,3}''(x_i) = -10\frac{f(x_{i-1})}{h_i^2} - 2\frac{f'(x_{i-1})}{h_i} - 22\frac{f(x_i)}{h_i^2} + 8\frac{f'(x_i)}{h_i} + 32\frac{c_i}{h_i^2} .
$$

In order for $s_{4,3}''$ to be continuous at x_i, we require that

$$
\begin{aligned}
\lim_{x\uparrow x_i} s_{4,3}(x) &= -10\frac{f(x_{i-1})}{h_i^2} - 2\frac{f'(x_{i-1})}{h_i} - 22\frac{f(x_i)}{h_i^2} + 8\frac{f'(x_i)}{h_i} + 32\frac{c_i}{h_i^2} \\
&= \lim_{x\downarrow x_i} s_{4,3}(x) = -22\frac{f(x_i)}{h_{i+1}^2} - 8\frac{f'(x_i)}{h_{i+1}} - 10\frac{f(x_{i+1})}{h_{i+1}^2} + 2\frac{f'(x_{i+1})}{h_{i+1}} + 32\frac{c_{i+1}}{h_{i+1}^2} .
\end{aligned}
$$

This gives us a bi-diagonal system of linear equations for the unknowns c_i.
    If we provide a value for $f''(x_0)$, then we can compute

$$
c_1 = \frac{11}{16}f(x_0) + \frac{5}{16}f(x_1) + \frac{1}{4}f'(x_0)h_1 - \frac{1}{16}f'(x_1)h_1 + \frac{1}{32}f''(x_0)h_1^2 .
$$

Afterward, we can forward-solve to get

$$
c_{i+1} = -\frac{5}{16}f(x_{i-1})\left(\frac{h_{i+1}}{h_i}\right)^2 - \frac{1}{16}f'(x_{i-1})\frac{h_{i+1}^2}{h_i} + \frac{11}{16}f(x_i)\left[1 - \frac{h_{i+1}^2}{h_i^2}\right]
$$

$$
+ \frac{1}{4}f'(x_i)\left[h_{i+1} + \frac{h_{i+1}^2}{h_i}\right] + \frac{5}{16}f(x_{i+1}) - \frac{1}{16}f'(x_{i+1})h_{i+1} + c_i\frac{h_{i+1}^2}{h_i^2} \; .
$$

A Fortran program to implement twice continuously differentiable piecewise quartic interpolation is available as routines c2_quartic_coefs and c2_quartic_eval in piecewise_poly.f. This is called from the C++ program piecewisePolynomial.C. The C++ program plots the interpolation and performs a numerical error analysis by measuring the interpolation errors for a range of numbers of interpolation points.

1.6.3.3 Piecewise Quintics (or Higher)

Finally, we will determine a twice continuously differentiable polynomial of degree $m \geq 5$ on each of the mesh elements. Such a spline has a total of $(m + 1)n$ undetermined coefficients. Specifying 3 continuity conditions at each of the interior mesh nodes leaves us with $(m + 1)n - 3(n - 1) = (m - 2)n + 3$ remaining degrees of freedom. We will specify f, f' and f'' at each of the $n + 1$ mesh nodes, after which we have $[(m - 2)n + 3] - 3(n + 1) = (m - 5)n$ remaining degrees of freedom. These can be determined by interpolating f at $m - 5$ additional points interior to each mesh element.

We choose $m - 5$ canonical points $0 < \xi_1 < \ldots < \xi_{m-5} < 1$. These points should be symmetrically placed, so that $\xi_{m-4-j} = 1 - \xi_j$. We can use these scalars to compute

$$
\sigma = \sum_{j=1}^{m-5} \frac{1}{\xi_j} \text{ and } \tau = \sum_{j=1}^{m-5}\sum_{\substack{k=1 \\ k \neq j}}^{m-5} \frac{1}{\xi_j \xi_k} \; .
$$

If $x \in [x_{i-1}, x_i]$, we define

$$
h_i = x_i - x_{i-1} \text{ and } \xi = \frac{x - x_{i-1}}{h_i} \; .
$$

Then we can verify that

$$
s_{m,3}(x) = f(x_{i-1})\left[1 + \{3 + \sigma\}\xi + \left\{6 + 3\sigma + \sigma^2 - \frac{\tau}{2}\right\}\xi^2\right](1 - \xi)^3 \prod_{j=1}^{m-5}(1 - \xi/\xi_j)
$$

$$
+ f'(x_{i-1})h_i\xi \left[1 + \{3 + \sigma\}\xi\right](1 - \xi)^3 \prod_{j=1}^{m-5}(1 - \xi/\xi_j)
$$

$$+ f''(x_{i-1}) h_i^2 \frac{1}{2} \xi^2 (1 - \xi)^3 \prod_{j=1}^{m-5} (1 - \xi/\xi_j)$$

$$+ f(x_i) \left[1 + \{3 + \sigma\} [1 - \xi] + \left\{ 6 + 3\sigma + \sigma^2 - \frac{\tau}{2} \right\} [1 - \xi]^2 \right] \xi^3 \prod_{j=1}^{m-5} \frac{\xi - \xi_j}{1 - \xi_j}$$

$$- f'(x_i) h_i (1 - \xi) [1 + \{3 + \sigma\} [1 - \xi]] \xi^3 \prod_{j=1}^{m-5} \frac{\xi - \xi_j}{1 - \xi_j}$$

$$+ f''(x_i) h_i^2 \frac{1}{2} (1 - \xi)^2 \xi^3 \prod_{j=1}^{m-5} \frac{\xi - \xi_j}{1 - \xi_j}$$

$$+ \sum_{j=1}^{m-5} f\left(x_{i-1} + \xi_j h_i \right) \left(\frac{\xi}{\xi_j} \frac{1 - \xi}{1 - \xi_j} \right)^3 \prod_{\substack{k=1 \\ k \neq j}}^{m-5} \frac{\xi - \xi_k}{\xi_j - \xi_k} .$$

A Fortran program to implement twice continuously differentiable spline interpolation is available as routine c2_spline_eval in piecewise_poly.f. This is
called from the C++ program piecewisePolynomial.C. The C++ program plots the
interpolation and performs a numerical error analysis by measuring the interpolation
errors for a range of numbers of interpolation points.

Exercise 1.6.7 Show how to find the twice continuously differentiable cubic spline
that interpolates the function f at mesh nodes $a = x_0 < x_1 < \ldots < x_{n-1} < b$
and uses periodicity for the endpoint conditions. In other words, we assume that f
is extended periodically outside the interval $[a, b)$, so that $f(b) = f(x_n) = f(a)$,
etc. Determine how many interpolation points are needed for the periodic extension
outside $[a, b)$ in order to determine $s_{3,3}$ uniquely, where the interpolation points for
the extension should lie, and what the values of f should be at those points. Then
describe an algorithm for computing the coefficients for $s_{3,3}$ and how to evaluate $s_{3,3}$
at a point $x \in (a, b)$.

Exercise 1.6.8 Write a program to compute the twice continuously differentiable
cubic spline interpolant to

$$f(x) = \frac{1}{x^2 + 25}$$

at equidistant nodes in the interval $[-1, 1]$. Plot f and the interpolant for 2 and
3 nodes. By sampling the error at 100 points between each node, determine a
(computer generated) estimate for the interpolation error, and plot the log of the
error versus the log of the mesh width.

1.6.4 Case Study: Electro-Cardiology

The Luo-Rudy I model [130] for electrical wave propagation in the heart involves several ordinary differential equations for gating variables, which control the transmission of various ions across the cell membrane. The reactions have the form

$$\frac{dg}{dt} = a - (a + b)g$$

where g is the gating variable and a and b are coefficients that depend on the membrane potential. For example, the coefficients for one of the potassium currents have the forms

$$a(V) = \begin{cases} 1.02\exp(-\delta(V))/[1 + \exp(-\delta(V))], & \delta(V) \geq 0 \\ 1.02/[1 + \exp(\delta(V))], & \delta(V) < 0 \end{cases}$$

where $\delta(V) \equiv 0.2385(V + 87.26 - 59.215)$;

$$b(V) = \begin{cases} \frac{0.49124\exp(v_{,1}(V)-\delta(V))+\exp(v_2(V)-\delta(V))}{1+\exp(-\delta(V))}, & \varepsilon(V) \geq 0 \\ \frac{0.49124\exp(v_1(V))+\exp(v_2(V))}{1+\exp(\varepsilon(V))}, & \varepsilon(V) < 0 \end{cases}$$

where $v_1(V) \equiv 0.08032(V + 87.26 + 5.476)$,

$v_2(V) \equiv 0.06175(V + 87.26 - 594.31)$,

$\varepsilon(V) \equiv -0.5143(V + 87.26 + 4.753)$.

These ordinary differential equations are solved on a grid across the heart tissue, at times running over possibly several heart beats. This could mean that these ordinary differential equations are solved many millions of times. Thus the speed of evaluation of these coefficients is important.

The exponentials in these expressions are somewhat expensive to evaluate. It is more efficient to replace these analytical expressions by moderate order spline approximations. For example, Cherry et al. [41] used Hermite cubic splines to evaluate these coefficients. The details of the interpolations are missing from that paper, but readers may find the discussion of the problem and other numerical methods interesting.

A program to compute the Hermite cubic spline interpolant to these Luo-Rudy functions can be found in the C++ file hermiteCubicSpline.C and the Fortran file lr.f.

1.6.5 Cardinal B-Splines

In Sects. 1.6.2 and 1.6.3, we found it necessary to solve linear systems in order to construct piecewise polynomial splines of degree k with $k-1$ continuous derivatives.

In this section, we will show how to construct basis functions for such splines, provided that the mesh is uniform. In order to do this, we will provide the following definition.

Definition 1.6.2 (Schoenberg [157])   The **cardinal B-splines** B_k are defined for $k \geq 1$ by the recursion

$$B_1(t) \equiv \begin{cases} 1, \, 0 \leq t < 1 \\ 0, \, \text{otherwise} \end{cases},$$

$$B_k(t) = (B_{k-1} * B_1)(t) \equiv \int_{-\infty}^{\infty} B_{k-1}(t - \tau) B_1(\tau) \, d\tau \, . \tag{1.44}$$

Definition 1.6.2 has many useful consequences. We will begin with the following.

Lemma 1.6.1   *For all $k \geq 1$, $B_k(t)$ is a nonnegative piecewise polynomial of degree $k - 1$, and is zero outside the interval $[0, k)$. Further, for $k \geq 2$ we have*

$$B_k(t) = \int_0^1 B_{k-1}(t - \tau) \, d\tau = \int_{t-1}^t B_{k-1}(\sigma) \, d\sigma \, . \tag{1.45}$$

Finally, for all $k \geq 2$ and all t

$$B_k(k - t) = B_k(t) \, . \tag{1.45}$$

Proof First, we note that the definition (1.44) implies that for all $k \geq 2$ we have

$$B_k(t) = B_{k-1} * B_1(t) = \int_{-\infty}^{\infty} B_{k-1}(t - \tau) B_1(\tau) \, d\tau = \int_0^1 B_{k-1}(t - \tau) \, d\tau$$

$$= \int_{t-1}^t B_{k-1}(\sigma) \, d\sigma \, .$$

Note that $B_1 = \chi_{[0,1)}$ is nonnegative. Assume inductively for $k - 1 \geq 1$ that B_{k-1} is nonnegative. Then $B_k = B_{k-1} * B_1$ is a convolution of nonnegative functions, and is therefore nonnegative.

For $k = 1$, $B_1 = \chi_{[0,1)}$ is a piecewise polynomial of degree 0 with support equal to $[0, 1]$. Assume inductively for $k - 1 \geq 1$ that B_{k-1} is a piecewise polynomial of degree $k - 2$ with support equal to $[0, k - 1]$. Equation (1.45) now shows that B_k is a piecewise polynomial of degree $k - 1$, and that its support is $[0, k]$.

Finally, we turn to the proof of (1.6.1). Note that

$$B_2(t) = \max\{0, 1 - |1 - t|\} = \max\{0, 1 - |t - 1|\} = \max\{0, 1 - |1 - (2 - t)|\} = B_2(2 - t) \, .$$

Assume inductively that (1.6.1) is true for $k - 1 \geq 2$. Then

$$B_k(k - t) = \int_0^1 B_{k-1}(k - t - \tau)\, d\tau = \int_0^1 B_{k-1}(k - 1 - [t + \tau - 1])\, d\tau$$

$$= \int_0^1 B_{k-1}(t + \tau - 1)\, d\tau = \int_0^1 B_{k-1}(t - \sigma)\, d\sigma = B_k(t) .$$

Next, we will examine derivatives of the cardinal B-spline.

Lemma 1.6.2 *Let the cardinal B-splines B_k be defined for $k \geq 1$ by (1.44). Then for $k \geq 2$, B_k is continuous and has $k - 2$ continuous derivatives. For $k \geq 3$ and $1 \leq n \leq k - 2$ we have*

$$\frac{d^n B_k}{dt^n}(t) = \sum_{j=0}^n (-1)^j \binom{n}{j} B_{k-n}(t - j) . \tag{1.46}$$

In particular, for $k \geq 2$ we have

$$\frac{d^{k-2} B_k}{dt^{k-2}}(0) = 0 = \frac{d^{k-2} B_k}{dt^{k-2}}(k) .$$

Proof Note that

$$B_2(t) \equiv (B_1 * B_1)(t) = \int_{-\infty}^\infty B_1(s) B_1(t - s)\, ds = \int_0^1 B_1(t - s)\, ds = \int_{\max\{0, t-1\}}^{\min\{1, t\}} ds$$

$$= \begin{cases} \int_0^t ds, & 0 < t < 1 \\ \int_{t-1}^1 ds, & 1 < t < 2 \\ 0, & \text{otherwise} \end{cases} = \begin{cases} t, & 0 < t < 1 \\ 2 - t, & 1 < t < 2 \\ 0, & \text{otherwise} \end{cases} = \max\{0, 1 - |1 - t|\} .$$

Thus B_2 is a continuous piecewise polynomial of degree 1, with support given by the interval $[0, 2]$, and $B_2(0) = 0 = B_2(2)$. Thus B_2 satisfies the claims in the lemma.

For $k \geq 3$, we can differentiate Eq. (1.45) to get

$$B_k'(t) = \int_0^1 B_{k-1}'(t - \tau)\, d\tau = \int_{t-1}^t B_{k-1}'(\sigma)\, d\sigma = B_{k-1}(t) - B_{k-1}(t - 1) .$$

This verifies (1.46) for $n = 1$. Inductively, suppose that (1.46) is true for $n - 1$. Then

$$\frac{d^n B_k}{dt^n}(t) = \frac{d}{dt}\left[\frac{d^{n-1} B_k}{dt^{n-1}}\right](t) = \frac{d}{dt}\left[\sum_{j=0}^{n-1}(-1)^j \binom{n-1}{j} B_{k+1-n}(t - j)\right]$$

$$= \sum_{j=0}^{n-1}(-1)^j \binom{n-1}{j}[B_{k-n}(t - j) - B_{k-n}(t - j - 1)]$$

$$= \sum_{j=0}^{n-1} (-1)^j \binom{n-1}{j} B_{k-n}(t-j) - \sum_{J=0}^{n-1} (-1)^J \binom{n-1}{J} B_{k-n}(t-J-1)$$

$$= \sum_{j=0}^{n-1} (-1)^j \binom{n-1}{j} B_{k-n}(t-j) - \sum_{j=1}^{n} (-1)^{j-1} \binom{n-1}{j-1} B_{k-n}(t-j)$$

$$= (-1)^0 \binom{n-1}{0} B_{k-n}(t) + \sum_{j=1}^{n-1} (-1)^j \left[\binom{n-1}{j} + \binom{n-1}{j-1} \right] B_{k-n}(t-j)$$

$$+ (-1)^n \binom{n-1}{n-1} B_{k-n}(t-n)$$

then Pascal's triangle (1.19) leads to

$$= (-1)^0 \binom{n}{0} B_{k-n}(t) + \sum_{j=1}^{n-1} (-1)^j \binom{n}{j} B_{k-n}(t-j) + (-1)^n \binom{n}{n} B_{k-n}(t-n)$$

$$= \sum_{j=0}^{n} \binom{n}{j} B_{k-n}(t-j) \ .$$

This completes the inductive proof of (1.46).

From this equation, we can see that B_k is a $(k-2)$-times continuously differentiable piecewise polynomial of degree $k-1$. Also, for all $k \geq 2$ we have

$$\frac{d^{k-2}B_k}{dt^{k-2}}(0) = \sum_{j=0}^{k-2} (-1)^j \binom{k-2}{j} B_2(-j) = B_2(0) = 0 \ ,$$

and

$$\frac{d^{k-2}B_k}{dt^{k-2}}(k) = \sum_{j=0}^{k-2} (-1)^j \binom{k-2}{j} B_2(k-j) = B_2(2) = 0 \ .$$

Next, we will develop a recursion for evaluating cardinal B-splines.

Lemma 1.6.3 *If the cardinal B-splines B_k are defined for $k \geq 1$ by (1.44), then for all $k \geq 2$ and all t we have*

$$B_k(t) = \frac{t}{k-1} B_{k-1}(t) + \frac{k-t}{k-1} B_{k-1}(t-1) \ . \tag{1.47}$$

Proof Note that

$$B_2(t) = \int_0^1 B_1(t - \tau)\, d\tau = \int_{\max\{0, t-1\}}^{\min\{1, t\}} d\tau = \min\{1, t\} - \max\{0, t - 1\}$$

$$= \begin{cases} t, & 0 \le t < 1 \\ 2 - t, & 1 \le t < 2 \\ 0, & \text{otherwise} \end{cases} = tB_1(t) + (1 - t)B_1(t - 1)\,.$$

Thus (1.47) is true for $k = 2$. Assume inductively that (1.47) is true for $k - 1 \ge 2$. Then definition (1.44) says that

$$B_k(t) = \int_0^1 B_{k-1}(t - \tau)\, d\tau$$

and the inductive hypothesis gives us

$$= \int_0^1 \frac{t - \tau}{k - 2} B_{k-2}(t - \tau) + \frac{k - 1 - t + \tau}{k - 2} B_{k-2}(t - \tau - 1)\, d\tau$$

then we can collect terms to get

$$= \frac{k - 1}{k - 2} \int_0^1 B_{k-2}(t - \tau - 1)\, d\tau + \int_0^1 \frac{t - \tau}{k - 2} \left[B_{k-2}(t - \tau) - B_{k-2}(t - \tau - 1) \right]\, d\tau$$

then definition (1.44) and Eq. (1.46) imply that

$$= \frac{k - 1}{k - 2} B_{k-1}(t - 1) + \int_0^1 \frac{t - \tau}{k - 2} B'_{k-1}(t - \tau)\, d\tau$$

then we change variables of integration to produce

$$= \frac{k - 1}{k - 2} B_{k-1}(t - 1) + \int_{t-1}^t \frac{\sigma}{k - 2} B'_{k-1}(\sigma)\, d\sigma$$

find that integration by parts gives us

$$= \frac{k - 1}{k - 2} B_{k-1}(t - 1) + \frac{\sigma}{k - 2} B_{k-1}(\sigma) \Big|_{t-1}^t - \int_{t-1}^t \frac{1}{k - 2} B_{k-1}(\sigma)\, d\sigma$$

and finally use Eq. (1.45) to arrive at

$$= \frac{k - 1}{k - 2} B_{k-1}(t - 1) + \frac{t}{k - 2} B_{k-1}(t) - \frac{t - 1}{k - 2} B_{k-1}(t - 1) - \frac{1}{k - 2} B_k(t)\,.$$

We can collect terms involving cardinal B-splines of the same order to get

$$\frac{k-1}{k-2}B_k(t) = \frac{t}{k-2}B_{k-1}(t) + \frac{k-t}{k-2}B_{k-1}(t-1) .$$

This is equivalent to (1.47).

Next, we will compute the Fourier transform of cardinal B-splines, and examine some consequences of its form.

Lemma 1.6.4 *If the cardinal B-splines B_k are defined for $k \geq 1$ by (1.44), and if the Fourier transform of a function f is defined by*

$$\mathcal{F}\{f\}(\omega) = \int_{-\infty}^{\infty} f(x)e^{-i\omega x} \, dx ,$$

then for all $k \geq 1$ we have

$$\mathcal{F}\{B_k\}(\omega) = \left[\frac{1-e^{-i\omega}}{i\omega}\right]^k . \tag{1.48}$$

As a result,

$$\int_{-\infty}^{\infty} B_k(t) \, dt = \int_0^k B_k(t) \, dt = 1 . \tag{1.49}$$

Furthermore, for all $k \geq 1$ and all real t we have the **partition of unity** *equation*

$$\sum_{n=-\infty}^{\infty} B_k(t-n) = 1 . \tag{1.50}$$

Proof Note that

$$\mathcal{F}\{B_1\}(\omega) = \int_0^1 e^{-i\omega t} \, dt = \frac{1-e^{-i\omega}}{i\omega} .$$

Assume inductively that the claim is true for $k - 1 \geq 1$. Then

$$\mathcal{F}\{B_k\}(\omega) = \mathcal{F}\{B_{k-1} * B_1\}(\omega) = \mathcal{F}\{B_{k-1}\}(\omega)\mathcal{F}\{B_1\}(\omega)$$

$$= \mathcal{F}\{B_1\}(\omega)^{k-1}\mathcal{F}\{B_1\}(\omega) = \mathcal{F}\{B_1\}(\omega)^k = \left[\frac{1-e^{-i\omega}}{i\omega}\right]^k .$$

It follows that for all $k \geq 1$ we have

$$\int_{\infty}^{\infty} B_k(t) \, dt = \mathcal{F}\{B_k\}(0) = \mathcal{F}\{B_1\}(0)^k = 1 .$$

Finally, we will prove (1.50) by induction. If $k = 1$, we have

$$\sum_{n=-\infty}^{\infty} B_1(t-n) = \sum_{n=-\infty}^{\infty} \chi_{[0,1)}(t-n) = \chi_{[0,1)}(t - \lfloor t \rfloor) = 1 \ .$$

Inductively, assume that (1.50) is true for $k - 1 \geq 1$. Then

$$\sum_{n=-\infty}^{\infty} B_k(t-n) = \sum_{n=-\infty}^{\infty} \int_{t-n-1}^{t-n} B_{k-1}(\sigma) \, d\sigma = \int_{-\infty}^{\infty} B_{k-1}(\sigma) \, d\sigma = 1 \ .$$

Lemma 1.6.4 has the following useful corollary.

Corollary 1.6.1 *If the cardinal B-splines B_k are defined for $k \geq 1$ by (1.44), then for all $k \geq 2$ we have*

$$\sum_{n=-\infty}^{\infty} B_k(n) = \sum_{n=1}^{k-1} B_k(n) = 1 \ . \tag{1.51}$$

Proof Equations (1.49) and (1.45) imply that for $k \geq 2$

$$\sum_{n=1}^{k-1} B_k(n) = \sum_{n=1}^{k-1} \int_{n-1}^{n} B_{k-1}(t) \, dt = \int_0^{k-1} B_{k-1}(t) \, dt = 1 \ .$$

This proves (1.51).

Next, we will prove a **refinement equation** for the cardinal B-splines. We hope that the reader will compare Eq. (1.52) with the refinement Eq. (1.124) for a general wavelet scaling function in Definition 1.8.9.

Lemma 1.6.5 *If the cardinal B-splines B_k are defined for $k \geq 1$ by (1.44), then for all $k \geq 1$ we have*

$$B_k(t) = \sum_{n=0}^{k} 2^{1-k} \binom{k}{n} B_k(2t - n) \ . \tag{1.52}$$

Proof Using the Eq. (1.48) for the Fourier transform of B_k, we can compute

$$\mathscr{F}\left\{ \sum_{n=0}^{k} 2^{1-k} \binom{k}{n} B_k(2t - n) \right\}(\omega) = \int_{-\infty}^{\infty} \sum_{n=0}^{k} 2^{1-k} \binom{k}{n} B_k(2t - n) e^{-i\omega t} \, dt$$

$$= \sum_{n=0}^{k} 2^{1-k} \binom{k}{n} \int_{-\infty}^{\infty} B_k(s) e^{-i\omega[s+n]/2} \, ds/2 = \sum_{n=0}^{k} 2^{-k} \binom{k}{n} e^{-i\omega n/2} \mathscr{F}\{B_k\}\left(\frac{\omega}{2}\right)$$

$$= \left[\frac{1 + e^{-i\omega/2}}{2} \right]^k \left[\frac{1 - e^{-i\omega/2}}{i\omega/2} \right]^k = \left[\frac{1 - e^{-i\omega}}{i\omega} \right]^k = \mathscr{F}\{B_k\}(\omega) \ .$$

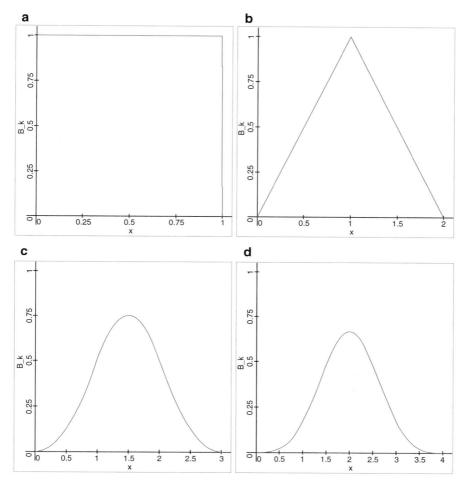

Fig. 1.10 Cardinal B-splines. (**a**) $k = 1$. (**b**) $k = 2$. (**c**) $k = 3$. (**d**) $k = 4$

Figure 1.10 shows the cardinal B-splines for $1 \leq k \leq 4$. This figure was generated by the C^{++} program bSpline.C. Alternatively, readers may execute the JavaScript program **bspline** (see `/home/faculty/johnt/book/code/` `approximation/bspline.html`).

MATLAB users can use command bspline to plot a B-spline for interpolation at a given sequence of points. GSL users should examine the documentation on B-splines.

We are not really interested in using cardinal B-splines for interpolation. Instead, we will use these functions to develop maxflat biorthogonal filters for wavelet approximations in Sect. 1.8.1.9.

Exercise 1.6.9 Show that the maximum value of B_k occurs at $k/2$, and find its value.

Exercise 1.6.10 Write a program to evaluate the cardinal B-spline of degree k at a point t, using the recursion (1.47).

1.6.6   Error Estimate

We can use Theorem 1.3.2 to provide a useful error estimate for continuous splines. In the following theorem, the Lagrange interpolation polynomials $\lambda_{j,n}$ are defined in Eq. (1.43).

Theorem 1.6.1 (Continuous Spline Error Estimate)   *Suppose that we are given a polynomial degree m, a problem domain $[a, b]$, a number n of mesh elements, mesh nodes $a = x_0 < \ldots x_n = b$, a degree of smoothness σ with $1 \leq \sigma \leq m + 1$, a function f with σ bounded derivatives in $[a, b]$, and symmetrically placed canonical interpolation points $\xi_1 < \ldots < \xi_{m-1} \in (0, 1)$. Let $s_{m,1}$ be the continuous spline that interpolates f at the mesh nodes $x_0, \ldots, x_n$ and at the element interior points $x_{i-1} + \xi_j [x_i - x_{i-1}]$ for $1 \leq j < m$ and for $1 \leq i \leq n$, and let $\lambda_{0,m}, \ldots, \lambda_{m,m}$ be the Lagrange interpolation polynomials associated with the interpolation points $0 = \xi_0 < \xi_1 < \ldots < \xi_{m-1} < \xi_m = 1$. Define the mesh width by*

$$h = \max_{1 \leq i \leq n} x_i - x_{i-1} .$$

Then the error in the spline approximation satisfies

$$\sup_{x \in [a,b]} |f(x) - s_{m,1}(x)| \leq \frac{h^\sigma}{\sigma !} \left[\sum_{j=0}^{m} \sup_{\xi \in [0,1]} |\lambda_{j,m}(\xi)| \right] \sup_{x \in [a,b]} \left| f^{(\sigma)}(x) \right| .$$

Proof Since $s_{m,1}(x)$ interpolates f at $m + 1$ distinct points in each mesh element $[x_{i-1}, x_i]$, Theorem 1.3.2 implies that

$$\sup_{x \in [a,b]} |f(x) - s_{m,1}(x)| = \max_{1 \leq i \leq n} \sup_{x \in [x_{i-1}, x_i]} |f(x) - s_{m,1}(x)|$$

$$\leq \max_{1 \leq i \leq n} \left\{ \frac{[x_i - x_{i-1}]^\sigma}{\sigma !} \sup_{x \in [x_{i-1}, x_i]} \left| f^{(\sigma)}(x) \right| \sum_{j=0}^{m} \sup_{x \in [x_{i-1}, x_i]} \left| \lambda_{j,m} \left(\frac{x - x_{i-1}}{x_i - x_{i-1}} \right) \right| \right\}$$

$$\leq \frac{h^\sigma}{\sigma !} \sup_{x \in [a,b]} \left| f^{(\sigma)}(x) \right| \sum_{j=0}^{m} \sup_{\xi \in [0,1]} |\lambda_{j,m}(\xi)| .$$

Readers may execute a JavaScript refinement study program with continuous splines. This program allows the reader to observe the effective computational rate of convergence in continuous spline approximation, by determining the slope of the graph of the log of the maximum interpolation error versus the log of the number of elements.

For splines with one or more continuous derivatives, theorem (1.6.1) does not apply. Instead, we may use results due either to Deny and Lions [65] or Bramble and Hilbert [21]. The former shows that the error in approximating a function f with σ derivatives by piecewise polynomials of degree m is of order $h^{\min\{\sigma, m+1\}}$. The latter is more general, but draws the same conclusion for functions of a single variable that are approximated by piecewise polynomials.

To verify the order of the error in continuously differentiable spline interpolation, readers may execute the JavaScript refinement study programs for either **continuously differentiable splines**, or **twice continuously differentiable splines** (see /home/faculty/johnt/book/code/approximation/ c2spline_ error.html). These programs allow the reader to observe the effective computational rate of convergence in continuously differentiable spline approximations.

Exercise 1.6.11 Suppose that continuous piecewise linear interpolation is performed on noisy data $f_i = f(x_i) + \varepsilon_i$. (For example, the numbers f_i may be the floating-point numbers closest to the function values.) Find an estimate for the $\|\cdot\|_\infty$ error in continuous piecewise linear interpolation with this noisy data: how does the error depend on the noise ε_i?

Exercise 1.6.12 Explain why the error in piecewise Hermite cubic splines should be at best $O(h^4)$ in general.

Exercise 1.6.13 What should be the order the error in approximating functions by twice continuously differentiable splines $s_{3,3}$ that are provided the values of the first derivative at the endpoints of the mesh?

Exercise 1.6.14 What should be the order the error in approximating functions by natural splines, namely twice continuously differentiable splines $s_{3,3}$ that have zero second derivative at the endpoints of the mesh?

Exercise 1.6.15 Perform a mesh refinement study to determine the observed order of convergence in the mesh width for natural splines. Use the Runge function for the mesh refinement study.

Exercise 1.6.16 Perform a mesh refinement study to determine the observed order of convergence for continuous splines in approximating $f(x) = \sqrt{x}$ on the interval $[0, 1]$. Try continuous piecewise polynomials of degree 1, 2 and 3.

1.6.7   Tension Splines

Sometimes we want to approximate functions that may have large derivatives. Often, the derivative of the function is either unknown or expensive. For such functions, continuous cubic splines $s_{3,1}$ may appear too "rough", continuously differentiable (Hermite) cubic splines $s_{3,2}$ would require evaluation of derivatives at the mesh nodes, and twice continuously differentiable cubic splines $s_{3,3}$ may be too smooth to handle the large derivatives. It might be nice in such circumstances to have an adjustable parameter to control the smoothness (or curvature) of the spline. Of course, some users just appreciate having another knob to turn.

The following is a common approach for controlling the smoothness of a piecewise interpolant. We assume that we are given a mesh $x_0 < x_1 < \ldots < x_n$, as well as function values f_i for $0 \le i \le n$ at the mesh points, and a tension parameter $\tau > 0$. We want to find a twice continuously differentiable function $s_\tau(x)$ that interpolates the data on the mesh,

$$s_\tau(x_i) = f_i \text{ for } 0 \le i \le n$$

and that solves the differential equation

$$s_\tau''''(x) - \tau^2 s_\tau''(x) = 0 \text{ for all } x \in (x_{i-1}, x_i)$$

for each $1 \le i \le n$.

To determine $s_\tau(x)$, we will make use of the values

$$z_i = s_\tau''(x_i) \text{ for } 0 \le i \le n .$$

On the mesh element $[x_{i-1}, x_i]$, the solution of the two-point boundary-value problem

$$s_\tau''''(x) - \tau^2 s_\tau''(x) = 0 \text{ for } x_{i-1} < x < x_i$$
$$s_\tau(x_{i-1}) = f_{i-1} , \quad s_\tau(x_i) = f_i ,$$
$$s_\tau''(x_{i-1}) = z_{i-1} , \quad s_\tau''(x_i) = z_i$$

is

$$s_\tau(x) = \frac{z_{i-1} \sinh[\tau h_i(1 - \xi)] + z_i \sinh[\tau h_i \xi]}{\tau^2 \sinh(\tau h_i)}$$
$$+ \left(f_{i-1} - \frac{z_{i-1}}{\tau^2} \right)(1 - \xi) + \left(f_i - \frac{z_i}{\tau^2} \right)\xi$$

where

$$h_i = x_i - x_{i-1} \text{ and } \xi = \frac{x - x_{i-1}}{h_i} \, .$$

Note that s_τ is not a polynomial.

We still need to determine the unknown second derivatives z_i at the mesh nodes. It is straightforward to compute

$$s'_\tau(x) = \frac{-z_{i-1}\cosh[\tau h_i(1 - \xi)] + z_i \cosh[\tau(x - x_{i-1})]}{\tau \sinh[\tau(x_i - x_{i-1})]} - \frac{f_{i-1} - z_{i-1}/\tau^2}{h_i} + \frac{f_i - z_i/\tau^2}{h_i} \, ,$$

from which it follows that

$$s'_\tau(x_{i-1}) = \frac{-z_{i-1}\cosh[\tau h_i] + z_i}{\tau \sinh[\tau h_i]} - \frac{f_{i-1} - z_{i-1}/\tau^2}{h_i} + \frac{f_i - z_i/\tau^2}{h_i} \text{ and}$$

$$s'_\tau(x_i) = \frac{-z_{i-1} + z_i\cosh[\tau h_i]}{\tau \sinh[\tau h_i]} - \frac{f_{i-1} - z_{i-1}/\tau^2}{h_i} + \frac{f_i - z_i/\tau^2}{h_i} \, .$$

Continuity of s'_τ at the interior mesh points gives us the equations

$$\alpha_{i-1/2}z_{i-1} + (\beta_{i-1/2} + \beta_{i+1/2})z_i + \alpha_{i+1/2}z_{i+1} = \gamma_{i+1/2} - \gamma_{i-1/2}$$

where for $1 \leq i \leq n$ we have

$$\alpha_{i-1/2} = \frac{1}{h_i} - \frac{\tau}{\sinh(\tau h_i)} \, ,$$

$$\beta_{i-1/2} = \frac{\tau \cosh(\tau h_i)}{\sinh(\tau h_i)} - \frac{1}{h_i} \, ,$$

$$\gamma_{i-1/2} = \tau^2 \frac{f_i - f_{i-1}}{h_i} \, .$$

Note that $\alpha_{i-1/2} > 0$ and $\beta_{i-1/2} > 0$.

Typically we choose $z_0 = 0 = z_n$ and obtain the tridiagonal system for the **natural tension spline**

$$\begin{bmatrix} \beta_{1/2} + \beta_{3/2} & \alpha_{3/2} & & \\ \alpha_{3/2} & \beta_{3/2} + \beta_{5/2} & \ddots & \\ & \ddots & \ddots & \alpha_{n-3/2} \\ & & \alpha_{n-3/2} & \beta_{n-3/2} + \beta_{n-1/2} \end{bmatrix} \begin{bmatrix} z_1 \\ z_2 \\ \vdots \\ z_{n-1} \end{bmatrix} = \begin{bmatrix} \gamma_{3/2} - \gamma_{1/2} \\ \gamma_{5/2} - \gamma_{3/2} \\ \vdots \\ \gamma_{n-1/2} - \gamma_{n-3/2} \end{bmatrix} \, .$$

Alternatively, we might assume periodicity, i.e. $s'_\tau(x_0) = s'_\tau(x_n)$ and $z_0 = s''_\tau(x_0) = s''_\tau(x_n) = z_n$. This leads to the linear system

$$
\begin{bmatrix}
\beta_{1/2} + \beta_{3/2} & \alpha_{3/2} & 0 & \cdots & 0 & \alpha_{1/2} \\
\alpha_{3/2} & \beta_{3/2} + \alpha_{5/2} & \alpha_{5/2} & \cdots & 0 & 0 \\
0 & \alpha_{5/2} & \beta_{5/2} + \beta_{7/2} & \ddots & 0 & 0 \\
\vdots & \vdots & & \ddots & \ddots & \vdots \\
0 & 0 & 0 & \ddots & \beta_{n-3/2} + \beta_{n-1/2} & \alpha_{n-1/2} \\
\alpha_{1/2} & 0 & 0 & \cdots & \alpha_{n-1/2} & \beta_{n-1/2} + \beta_{1/2}
\end{bmatrix}
\begin{bmatrix}
z_1 \\ z_2 \\ z_3 \\ \vdots \\ z_{n-1} \\ z_n
\end{bmatrix}
$$

$$
=
\begin{bmatrix}
\gamma_{3/2} - \gamma_{1/2} \\
\gamma_{5/2} - \gamma_{3/2} \\
\gamma_{7/2} - \gamma_{5/2} \\
\vdots \\
\gamma_{n-1/2} - \gamma_{n-3/2} \\
\gamma_{1/2} - \gamma_{n-1/2}
\end{bmatrix}.
$$

Note that as $\tau \to \infty$, the tension spline $s_\tau(x)$ tends to the continuous piecewise linear spline $s_{1,1}$; this is because the differential equation solved by $s_\tau(x)$ tends to the equation $s''_\tau = 0$. Similarly, as $\tau \to 0$, the tension spline $s_\tau(x)$ tends to the twice continuously differentiable cubic spline $s_{3,3}$.

On the practical side, it is important to note that evaluation of the tension spline $s_\tau(x)$ involves hyperbolic sines, which are significantly more expensive to evaluate than cubic polynomials.

Readers can view a C++ program for computing a tension spline in tension-Spline.C. The user can choose the number of interpolation points, and whether the points are equidistant or random. Readers can also choose to interpolate piecewise constant, piecewise linear, piecewise quadratic or smooth functions, and see the behavior of the errors as the mesh is refined. And, of course, readers can choose the tension. Alternatively, readers may execute a JavaScript program for **tension splines.**

Exercise 1.6.17 Write a program to compute the tension spline interpolant to

$$
f(x) = \frac{1}{x^2 + 25}
$$

at equidistant nodes in the interval $[-1, 1]$. For tension $= 10^{-2}, 10^{-1}, 10^0, 10^1, 10^2$, plot f and the interpolant for 5 nodes.

Exercise 1.6.18 Discuss how the linear systems for the second derivatives of the tension spline should behave as the tension τ becomes large, and as τ approaches zero. Then discuss what will happened in the evaluation of $s_\tau(x)$ for large τ, and for

τ near zero. Is the tension spline sensitive to rounding errors? If we want the tension spline s_τ to be close to the continuous piecewise linear spline $s_{1,1}$, how large should we choose τ? If we want the tension spline s_τ to be close to the twice continuously differentiable spline $s_{3,3}$, how small should we choose τ?

Exercise 1.6.19 Perform a mesh refinement study to determine the order of accuracy for tension splines at various values of τ.

Exercise 1.6.20 Digitize the boundary of some interesting figure and enter the points (x_i, y_i) into the computer. Interpolate the points with a tension spline, and adjust the tension to produce a reasonable reproduction of your original figure.

1.6.8 Parametric Curves

Suppose that we want to use splines to draw the outline of some two-dimensional object, such as the side view of an enclosed object. Since there are (at least) two values of the y coordinate for each x coordinate in the image (and *vice versa*), we cannot use a spline to fit the values y_i to mesh values x_i.

Instead, we can let t be a parameter that measures distance around the outside of the image. Given the three sets $\{t_0, \ldots, t_n\}$, $\{x_0, \ldots, x_n\}$ and $\{y_0, \ldots, y_n\}$ we can fit splines for $x(t)$ and $y(t)$. More generally, given d-vectors $\mathbf{x}_0, \ldots, \mathbf{x}_n$ and scalars $t_0 < \ldots t_n$ we can use the ideas in Sect. 1.6.1 to construct a piecewise linear spline $\mathbf{s}_{1,1}(t)$, which is defined for $1 \le i \le n$ and $t_{i-1} \le t \le t_i$ by

$$\mathbf{s}_{1,1}(t) = \mathbf{x}_{i-1} \frac{t_i - t}{t_i - t_{i-1}} + \mathbf{x}_i \frac{t - t_{i-1}}{t_i - t_{i-1}} \ .$$

We could also generalize this approach to use higher-order piecewise polynomials.

In computer graphics, it is common to use cubic polynomials to draw parametric curves. We will discuss two ideas for describing these cubics. The first is based on Hermite cubic interpolation ideas in Sect. 1.6.2, and the second is related to the Bernstein polynomials we discussed in Sect. 1.2.7. Afterward, we will examine the use of rational polynomials for parametric curves.

1.6.8.1 Hermite Curves

Given scalars $t_0 < \ldots < t_n$ and two sets of d-vectors $\{\mathbf{x}_0, \ldots, \mathbf{x}_n\}$ and $\{\mathbf{x}'_0, \ldots, \mathbf{x}'_n\}$, we can generate a continuously differentiable parametric curve $\mathbf{s}_{3,3}(t)$ for $1 \le i \le n$ and $t_{i-1} \le t \le t_i$ by

$$\mathbf{s}_{3,3}(t) = \left[\mathbf{x}_{i-1}(1 + 2\tau) + \mathbf{x}'_{i-1} h_i \tau \right] (1 - \tau)^2$$
$$+ \left[\mathbf{x}_i(3 - 2\tau)) - \mathbf{x}'_i h_i(1 - \tau) \right] \tau^2 \ ,$$

where

$$h_i = t_i - t_{i-1} \text{ and } \tau = \frac{t - t_{i-1}}{h_i} ,$$

We can rewrite this Hermite curve $s_{3,3}(t)$ in the form

$$s_{3,3}(t) = \begin{bmatrix} \mathbf{x}_{i-1} & \mathbf{x}'_{i-1}h_i & \mathbf{x}'_i h_i & \mathbf{x}_i \end{bmatrix} \begin{bmatrix} (1+2\tau)(1-\tau)^2 \\ \tau(1-\tau)^2 \\ -\tau^2(1-\tau) \\ \tau^2(3-2\tau) \end{bmatrix}$$

$$= \begin{bmatrix} \mathbf{x}_{i-1} & \mathbf{x}'_{i-1}h_i & \mathbf{x}'_i h_i & \mathbf{x}_i \end{bmatrix} \begin{bmatrix} 2 & -3 & 0 & 1 \\ 1 & -2 & 1 & 0 \\ 1 & -1 & 0 & 0 \\ -2 & 3 & 0 & 0 \end{bmatrix} \begin{bmatrix} \tau^3 \\ \tau^2 \\ \tau \\ 1 \end{bmatrix} \equiv \mathbf{G}_H \mathbf{M}_H \mathbf{t}$$

In computer graphics, $\mathbf{G}_H$ is called the Hermite **geometry matrix**, and $\mathbf{M}_H$ is called the Hermite **blending basis matrix**. The vector $\mathbf{M}_H \mathbf{t}$ is called the Hermite **blending function**. Some computer graphics software packages provide routines to plot parametric curves described by a geometry matrix and a blending basis matrix. See, for example, OpenGL Evaluators and NURBS.

1.6.8.2 Bézier Curves

In many cases it is difficult to determine values for the derivatives $\mathbf{x}'$ of a parametric curve at the mesh nodes t_i. Instead, we can develop a **Bézier curve**, which has the form

$$\mathbf{x}(\tau) = \begin{bmatrix} \mathbf{x}_{i-1} & \mathbf{x}_{i-2/3} & \mathbf{x}_{i-1/3} & \mathbf{x}_i \end{bmatrix} \begin{bmatrix} (1-\tau)^3 \\ 3\tau(1-\tau)^2 \\ 3\tau^2(1-\tau) \\ \tau^3 \end{bmatrix}$$

$$= \begin{bmatrix} \mathbf{x}_{i-1} & \mathbf{x}_{i-2/3} & \mathbf{x}_{i-1/3} & \mathbf{x}_i \end{bmatrix} \begin{bmatrix} -1 & 3 & -3 & 1 \\ 3 & -6 & 3 & 0 \\ -3 & 3 & 0 & 0 \\ 1 & 0 & 0 & 0 \end{bmatrix} \begin{bmatrix} \tau^3 \\ \tau^2 \\ \tau \\ 1 \end{bmatrix} \equiv \mathbf{G}_B \mathbf{M}_B \mathbf{t} .$$

Here, the blending functions are the Bernstein polynomials

$$b_{j,3}(\tau) = \binom{3}{j} \tau^j (1-\tau)^{3-j} ,$$

which are obviously nonnegative for $\tau \in [0, 1]$. Also, the multinomial expansion (1.17) shows that the Bernstein polynomials sum to one for all τ:

$$1 = (\tau + [1 - \tau])^n = \sum_{j=0}^{n} \binom{n}{j} \tau^j [1 - \tau]^{n-j} .$$

It follows that

$$\mathbf{x}(\tau) = \sum_{j=0}^{3} \mathbf{x}_{i-1+j/3} b_{j,n}(\tau)$$

is a **convex combination** of the vectors $\mathbf{x}_{i-1}$, $\mathbf{x}_{i-2/3}$, $\mathbf{x}_{i-1/3}$ and $\mathbf{x}_i$. This implies that the parametric curve $\mathbf{x}(\tau)$ lies in the **convex hull** of these four points.

Let us compute the derivative of this parametric curve at the endpoints of the interval in t. By using the definitions

$$h_i = t_i - t_{i-1} \text{ and } \tau = (t - t_{i-1})/h_i ,$$

we get

$$\mathbf{x}'(t) = \sum_{j=0}^{3} \mathbf{x}_{i-1+j/3} b'_{j,3}(\tau) h_i .$$

In particular,

$$\mathbf{x}'(t_{i-1}) = (-3\mathbf{x}_{i-1} + 3\mathbf{x}_{i-2/3}) h_i \text{ and}$$
$$\mathbf{x}'(t_i) = (-3\mathbf{x}_{i-1/3} + 3\mathbf{x}_i) h_i .$$

Thus a parametric curve using the Bernstein polynomials as the blending functions will be continuously differentiable if and only if at each t_i we have

$$(\mathbf{x}_i - \mathbf{x}_{i-1/3}) h_i = (\mathbf{x}_{i+1/3} - \mathbf{x}_i) h_{i+1} .$$

The point $\mathbf{x}_{i+1/3}$ is called the **control point** for the Bézier curve at the point $\mathbf{x}_i$ on (t_i, t_{i+1}), and $\mathbf{x}_{i-1/3}$ is the control point for $\mathbf{x}_i$ on (t_{i-1}, t_i).

In computer graphics, we generally require $\mathcal{G}^1$ **continuity** at the nodes, which means equal tangent directions:

$$\lim_{t \uparrow t_i} \frac{\mathbf{x}'(t)}{\|\mathbf{x}'(t)\|} = \lim_{t \downarrow t_i} \frac{\mathbf{x}'(t)}{\|\mathbf{x}'(t)\|} .$$

It is easy to see that a Bézier curve is $\mathcal{G}^1$ continuous at t_i if and only if $\mathbf{x}_i - \mathbf{x}_{i-1/3} = \mathbf{x}_{i+1/3} - \mathbf{x}_i$.

If the limiting tangent vectors on either side of t_i are nonzero and the parametric curve is continuously differentiable, then the curve is $\mathcal{G}^1$ continuous. However, $\mathcal{G}^1$ continuity does not imply that the curve is continuously differentiable; the tangent directions may be equal even though the magnitudes are not.

1.6.8.3   Rational Splines

As we saw in Sect. 4.3.3 of Chap. 4 in Volume I, computer graphics often employs **homogeneous coordinates**. A three-vector $\mathbf{x}$ has homogeneous coordinates $(\mathbf{x}w, w)$ for any scalar $w > 0$. The homogeneous coordinates $(\mathbf{x}, 0)$ correspond to the "point at infinity" on the line through the origin passing through $\mathbf{x}$. Homogeneous coordinates allow us to rescale a curve by merely changing the fourth coordinate.

Rational spline curves are splines fit to homogeneous coordinates. Physical coordinates are then ratios of piecewise polynomials

$$\mathbf{r}(t) = \frac{\mathbf{x}(t)}{w(t)} .$$

Rational splines of degree at least 2 can represent conic sections exactly, as the following lemma will show.

Lemma 1.6.6 *Suppose that we are given a unit three-vector vector* $\mathbf{a}$ *(representing the axis of a cone), a three-vector* $\mathbf{c}$ *(representing the vertex of a cone), and an angle* $\phi \in (0, \pi/2)$. *Let the cone* C *be given by*

$$C = \{\mathbf{x} : |(\mathbf{x} - \mathbf{c}) \cdot \mathbf{a}| = \|\mathbf{x} - \mathbf{c}\| \cos \phi\} .$$

Here $\mathbf{x} \cdot \mathbf{y}$ *denotes the* **inner product** *of* $\mathbf{x}$ *and* $\mathbf{y}$. *Suppose that* $\mathbf{x}_0$ *and* $\mathbf{x}_2$ *are two distinct three-vectors on the cone* C:

$$(\mathbf{x}_0 - \mathbf{c}) \cdot \mathbf{a} = \sigma_0 \|\mathbf{x}_0 - \mathbf{c}\| \cos \phi \ \text{where} \ |\sigma_0| = 1 \ \text{and}$$

$$(\mathbf{x}_2 - \mathbf{c}) \cdot \mathbf{a} = \sigma_2 \|\mathbf{x}_2 - \mathbf{c}\| \cos \phi \ \text{where} \ |\sigma_2| = 1 . \tag{1.53}$$

Also suppose that $\mathbf{x}_1$ *is outside of the cone*

$$|(\mathbf{x}_1 - \mathbf{c}) \cdot \mathbf{a}| < \|\mathbf{x}_1 - \mathbf{c}\| \cos \phi ,$$

that $\mathbf{x}_1$ *is not co-linear with* $\mathbf{x}_0$ *and* $\mathbf{x}_2$, *and that*

$$\sigma_0 (\mathbf{x}_1 - \mathbf{c}) \cdot a \|\mathbf{x}_0 - \mathbf{c}\| = (\mathbf{x}_0 - \mathbf{c}) \cdot (\mathbf{x}_1 - \mathbf{c}) \cos \phi \ \text{and}$$

$$\sigma_2 (\mathbf{x}_1 - \mathbf{c}) \cdot a \|\mathbf{x}_2 - \mathbf{c}\| = (\mathbf{x}_2 - \mathbf{c}) \cdot (\mathbf{x}_1 - \mathbf{c}) \cos \phi . \tag{1.54}$$

Define

$$w_1 = \sqrt{\frac{\sigma_0\sigma_2\|\mathbf{x}_0 - \mathbf{c}\|\|\mathbf{x}_2 - \mathbf{c}\| - (\mathbf{x}_0 - \mathbf{c})\cdot(\mathbf{x}_2 - \mathbf{c})}{2\|\mathbf{x}_1 - \mathbf{c}\|^2 \cos^2\phi - [(\mathbf{x}_1 - \mathbf{c})\cdot\mathbf{a}]^2}} \cos\phi . \tag{1.55}$$

Let

$$b_{0,2}(\tau) = (1 - \tau)^2 , \ b_{1,2}(\tau) = 2\tau(1 - \tau)^2 \text{ and } b_{2,2}(\tau) = \tau^2$$

be the Bernstein polynomials of degree two. Then the rational curve

$$\mathbf{r}(\tau) = \frac{\mathbf{x}_0 b_{0,2}(\tau) + \mathbf{x}_1 w_1 b_{1,2}(\tau) + \mathbf{x}_{2,2} b_2(\tau)}{b_{0,2}(\tau) + w_1 b_{1,2}(\tau) + b_{2,2}(\tau)} \tag{1.56}$$

lies in the intersection of the cone C and the plane determined by the three vectors $\mathbf{x}_0$, $\mathbf{x}_1$ *and* $\mathbf{x}_2$.

Proof First, we use the definition (1.56) of $\mathbf{r}(\tau)$ to obtain

$$\left\{([\mathbf{r}(\tau) - \mathbf{c}]\cdot\mathbf{a})^2 - \|\mathbf{r}(\tau) - \mathbf{c}\|^2 \cos^2\phi\right\}[b_{0,2}(\tau) + w_1 b_{1,2}(\tau) + b_{2,2}(\tau)]^2$$

$$= [(\mathbf{x}_0 - \mathbf{c})\cdot\mathbf{a}b_{0,2}(\tau) + (\mathbf{x}_1 - \mathbf{c})\cdot\mathbf{a}w_1 b_{1,2}(\tau) + (\mathbf{x}_2 - \mathbf{c})\cdot\mathbf{a}b_{2,2}(\tau)]^2$$

$$- \|(\mathbf{x}_0 - \mathbf{c})b_{0,2}(\tau) + (\mathbf{x}_1 - \mathbf{c})w_1 b_{1,2}(\tau) + (\mathbf{x}_2 - \mathbf{c})b_{2,2}(\tau)\|^2 \cos^2\phi$$

then we expand the quadratic terms to get

$$= [(\mathbf{x}_0 - \mathbf{c})\cdot\mathbf{a}]^2 b_{0,2}(\tau)^2 + [(\mathbf{x}_1 - \mathbf{c})\cdot\mathbf{a}]^2 w_1^2 b_{1,2}(\tau)^2 + (\mathbf{x}_2 - \mathbf{c})\cdot\mathbf{a}]^2 b_{2,2}(\tau)^2$$

$$+ 2(\mathbf{x}_0 - \mathbf{c})\cdot\mathbf{a}(\mathbf{x}_1 - \mathbf{c})\cdot\mathbf{a}w_1 b_{0,2}(\tau)b_{1,2}(\tau)$$

$$+ 2(\mathbf{x}_1 - \mathbf{c})\cdot\mathbf{a}(\mathbf{x}_2 - \mathbf{c})\cdot\mathbf{a}w_1 b_{1,2}(\tau)b_{2,2}(\tau)$$

$$+ 2(\mathbf{x}_0 - \mathbf{c})\cdot\mathbf{a}(\mathbf{x}_2 - \mathbf{c})\cdot\mathbf{a}b_{0,2}(\tau)b_{2,2}(\tau)$$

$$- \|\mathbf{x}_0 - \mathbf{c}\|^2 b_{0,2}(\tau)^2 \cos^2\phi - \|\mathbf{x}_1 - \mathbf{c}\|^2 w_1^2 b_{1,2}(\tau)^2 \cos^2\phi$$

$$- \|\mathbf{x}_2 - \mathbf{c}\|^2 b_{2,2}(\tau)^2 \cos^2\phi$$

$$- 2(\mathbf{x}_0 - \mathbf{c})\cdot(\mathbf{x}_1 - \mathbf{c})w_1 b_{0,2}(\tau)b_{1,2}(\tau) \cos^2\phi$$

$$- 2(\mathbf{x}_1 - \mathbf{c})\cdot(\mathbf{x}_2 - \mathbf{c})w_1 b_{1,2}(\tau)b_{2,2}(\tau) \cos^2\phi$$

$$- 2(\mathbf{x}_0 - \mathbf{c})\cdot(\mathbf{x}_2 - \mathbf{c})b_{0,2}(\tau)b_{2,2}(\tau) \cos^2\phi$$

then we use assumption (1.53) to eliminate the terms that do not vanish at $\tau = 0$ or 1:

$$= [(\mathbf{x}_1 - \mathbf{c}) \cdot \mathbf{a}]^2 w_1^2 b_{1,2}(\tau)^2 + 2(\mathbf{x}_0 - \mathbf{c}) \cdot a(\mathbf{x}_1 - \mathbf{c}) \cdot \mathbf{a} w_1 b_{0,2}(\tau) b_{1,2}(\tau)$$
$$+ 2(\mathbf{x}_1 - \mathbf{c}) \cdot a(\mathbf{x}_2 - \mathbf{c}) \cdot \mathbf{a} w_1 b_{1,2}(\tau) b_{2,2}(\tau)$$
$$+ 2(\mathbf{x}_0 - \mathbf{c}) \cdot a(\mathbf{x}_2 - \mathbf{c}) \cdot \mathbf{a} b_{0,2}(\tau) b_{2,2}(\tau)$$
$$- \|\mathbf{x}_1 - \mathbf{c}\|^2 w_1^2 b_{1,2}(\tau)^2 \cos^2 \phi - 2(\mathbf{x}_0 - \mathbf{c}) \cdot (\mathbf{x}_1 - \mathbf{c}) w_1 b_{0,2}(\tau) b_{1,2}(\tau) \cos^2 \phi$$
$$- 2(\mathbf{x}_1 - \mathbf{c}) \cdot (\mathbf{x}_2 - \mathbf{c}) w_1 b_{1,2}(\tau) b_{2,2}(\tau) \cos^2 \phi$$
$$- 2(\mathbf{x}_0 - \mathbf{c}) \cdot (\mathbf{x}_2 - \mathbf{c}) b_{0,2}(\tau) b_{2,2}(\tau) \cos^2 \phi$$

then we use assumption (1.54) to eliminate the terms that do not vanish to second order at $\tau = 0$ or 1:

$$= [(\mathbf{x}_1 - \mathbf{c}) \cdot \mathbf{a}]^2 w_1^2 b_{1,2}(\tau)^2 + 2(\mathbf{x}_0 - \mathbf{c}) \cdot a(\mathbf{x}_2 - \mathbf{c}) \cdot \mathbf{a} b_{0,2}(\tau) b_{2,2}(\tau)$$
$$- \|\mathbf{x}_1 - \mathbf{c}\|^2 w_1^2 b_{1,2}(\tau)^2 \cos^2 \phi - 2(\mathbf{x}_0 - \mathbf{c}) \cdot (\mathbf{x}_2 - \mathbf{c}) b_{0,2}(\tau) b_{2,2}(\tau) \cos^2 \phi$$

then we use the identity $4b_{0,2}(\tau) b_{2,2}(\tau) = b_{1,2}(\tau)^2$ to get

$$= \left\{ [(\mathbf{x}_1 - \mathbf{c}) \cdot \mathbf{a}]^2 w_1^2 - \|\mathbf{x}_1 - \mathbf{c}\|^2 \cos^2 \phi \right\} w_1^2 b_{1,2}(\tau)^2$$
$$+ \frac{1}{2} \left\{ (\mathbf{x}_0 - \mathbf{c}) \cdot a(\mathbf{x}_2 - \mathbf{c}) \cdot \mathbf{a} - (\mathbf{x}_0 - \mathbf{c}) \cdot (\mathbf{x}_2 - \mathbf{c}) \cos^2 \phi \right\} b_{1,2}(\tau)^2 .$$

The choice (1.55) of w_1 now shows that this final expression vanishes for all τ.

Lemma 1.6.6 shows us that only one rational polynomial is needed to draw hyperbolas, parabolas and ellipses. For non-rational splines, a large number of nodes is needed to obtain a reasonable approximation to conic sections. More information regarding rational splines and conic sections is available in Böhm et al. [18], Foley et al. [80, p. 501ff] or Tiller [173].

1.7 Least Squares Approximation

The previous sections in this chapter have discussed interpolation of functions, mostly by polynomials. In this section, we will adopt a different approach. Given some complicated, expensive or sparsely defined function f, we will find a simple, inexpensive and continuous function that comes as close to f as possible. Closeness will be determined by a function norm, which we will discuss in Sect. 1.7.1. In the remainder of our study of approximation, we will confine our discussion to Euclidean norms. We will discover in Sect. 1.7.2 that the best approximation with respect to an Euclidean norm is determined by the normal equations, in a fashion similar to the normal equations we developed for overdetermined linear systems in Corollary 6.3.1 of Chap. 6 in Volume I. As a careful reader might expect, the normal equations for function approximation have serious numerical difficulties, which can be overcome by the selection of a mutually orthogonal basis for the approximating functions. In particular, we will develop orthogonal polynomials in Sect. 1.7.3, and trigonometric polynomial approximations in Sect. 1.7.4. The latter section will introduce the very-important **fast Fourier transform**, which achieves its speed by a recursion over a hierarchy of scales. This use of a hierarchy of scales will help us to study wavelets in Sect. 1.8.

1.7.1 Norms and Inner Products

Recall that we presented several function norms in the introduction to this chapter. Any function norm could be used to formulate the following important problem.

Definition 1.7.1 Given a function f, a norm $\|\cdot\|$ and linearly independent functions $\phi_0, \ldots, \phi_n$, the **approximation problem** involves finding scalars $c_0, \ldots, c_n$ so that $\left\| f(x) - \sum_{j=0}^{n} \phi_j(x)c_j \right\|_2$ is as small as possible.

The **least squares approximation problem** uses either the continuous Euclidean norm

$$\|f\|_2 = \left[\int_a^b |f(x)|^2 \, w(x) dx \right]^{1/2}$$

or the discrete Euclidean norm

$$\|f\|_2^2 = \sum_{i=0}^{m} |f(x_i)|^2 \, w_i \,.$$

In the continuous Euclidean norm, we assume that $w(x) > 0$ for $x \in (a, b)$, and in the discrete Euclidean norm we assume that $w_i > 0$ for all $0 \le i \le n$.

Both Euclidean norms have associated inner products, namely

$$(f, g) = \int_a^b f(x)\overline{g(x)}w(x)dx \tag{1.57}$$

and

$$(f, g) = \sum_{i=0}^{m} f(x_i)\overline{g(x_i)}w_i \,,$$

respectively. In these definitions, the overline indicates complex conjugation, if appropriate. These inner products satisfy the general properties of an abstract inner product on a linear space, as described by Halmos [95, p. 121]. The reader is asked to verify these properties in the exercises below.

We also note the following important inequality.

Theorem 1.7.1 (Cauchy-Schwarz Inequality)   *Suppose that $(\cdot,\cdot)$ is an inner product, and define* $\|f\|_2 = \sqrt{(f,f)}$. *If* $\|f\|_2$ *and* $\|g\|_2$ *are finite, then*

$$|(f,g)| \le \|f\|_2\,\|g\|_2 \;. \tag{1.58}$$

Furthermore, $\|\cdot\|_2$ *is a norm.*

Proof The Cauchy-Schwarz inequality is obvious if $\|g\|_2 = 0$. If $\|g\|_2 > 0$, then

$$0 \le \left\| f - g\frac{(f,g)}{\|g\|_2^2} \right\|_2^2 = \left(f - g\frac{(f,g)}{\|g\|_2^2} \,,\, f - g\frac{(f,g)}{\|g\|_2^2} \right)$$

$$= (f,f) - (f,g)\frac{\overline{(f,g)}}{\|g\|_2^2} - (g,f)\frac{(f,g)}{\|g\|_2^2} + (g,g)\frac{(f,g)}{\|g\|_2^2}\frac{\overline{(f,g)}}{\|g\|_2^2}$$

and since $(g,f) = \overline{(f,g)}$ we get

$$= \|f\|_2^2 - \frac{|(f,g)|^2}{\|g\|_2^2} \;.$$

This final inequality is equivalent to the Cauchy-Schwarz inequality

Next, let us prove that $\|f\|_2 = \sqrt{(f,f)}$ is a norm. Since an inner product is nonnegative and definite, $\|f\|_2$ is nonnegative and definite. Since an inner product is linear and self-adjoint, for any scalar α we have

$$\|f\alpha\|_2 = \sqrt{(f\alpha,f\alpha)} = \sqrt{(f\alpha,f)\alpha} = \sqrt{(f,f)\overline{\alpha}\alpha} = \sqrt{\|f\|_2^2|\alpha|^2} = \|f\|_2\,|\alpha| \;.$$

All that remains is to prove the triangle inequality:

$$\|f+g\|_2^2 = (f+g,f+g) = (f,f) + (f,g) + (g,f) + (g,g)$$

$$= \|f\|_2^2 + 2\Re\{(f,g)\} + \|g\|_2^2 \le \|f\|_2^2 + 2|(f,g)| + \|g\|_2^2$$

$$\le \|f\|_2^2 + 2\|f\|_2\|g\|_2 + \|g\|_2^2 = (\|f\|_2 + \|g\|_2)^2 \;.$$

Next, we will use the notion of an inner product to define orthogonality for functions.

Definition 1.7.2 Two functions f and g are **orthogonal** with respect to the inner product $(\cdot,\cdot)$ if and only if $(f,g) = 0$. The functions $\phi_0,\ldots,\phi_n$ are an **orthogonal**

system if and only if $(\phi_i, \phi_i) > 0$ for all i and $(\phi_i, \phi_j) = 0$ for all $i \neq j$. Finally, an orthogonal system is **orthonormal** if and only if $(\phi_j, \phi_j) = 1$ for all j.
This definition leads to the following famous theorem.

Theorem 1.7.2 (Pythagorean) *If $(f, g) = 0$, then $\|f + g\|_2^2 = \|f\|_2^2 + \|g\|_2^2$.*

Proof $\|f + g\|_2^2 = (f + g, f + g) = (f, f) + (g, f) + (f, g) + (g, g) = \|f\|_2^2 + \|g\|_2^2$.
Next, we will define linear independence for functions.

Definition 1.7.3 The functions $\phi_0, \ldots, \phi_n$ are **linearly independent** with respect to the norm $\| \cdot \|_2$ if and only if $\| \sum_{j=0}^n \phi_j c_j \|_2 = 0$ implies that all of the coefficients c_j are zero.
This definition leads to the following useful result.

Corollary 1.7.1 *If the functions $\phi_0, \ldots, \phi_n$ form an orthogonal system, then they are linearly independent with respect to the Euclidean norm.*

Proof Since the functions $\phi_0, \ldots, \phi_n$ are mutually orthogonal, repeated application of the Pythagorean theorem leads to

$$\left\| \sum_{j=0}^n \phi_j(x)c_j \right\|_2^2 = \sum_{j=0}^n \|\phi_j\|_2^2 c_j^2 \ .$$

If the linear combination $\sum_{j=0}^n \phi_j(x)c_j$ is zero, then the right hand side of the previous equation is zero. Since the functions ϕ_j all have positive norms, we conclude that all the c_j are zero.

Exercise 1.7.1 Show that the inner product (1.57) satisfies the following conditions.

1. For all square-integrable functions f and g,

$$(f, g) = \overline{(g, f)} \ .$$

2. For all square-integrable functions f_1, f_2 and g, and all scalars α_1 and α_2,

$$(f_1\alpha_1 + f_2\alpha_2, g) = (f_1, g)\alpha_1 + (f_2, g)\alpha_2 \ .$$

3. For all square-integrable functions f,

$$(f, f) \geq 0 \ .$$

4. For all square-integrable functions f, $(f, f) = 0$ implies that $f = 0$.

Exercise 1.7.2 Show that the functions

$$\phi_j(x) = \cos(jx) \ , \ j = 0, \ldots, m$$

form an orthogonal system with respect to the continuous inner product

$$(f, g) = \int_0^\pi f(x)g(x)dx$$

and have norms given by

$$\|\phi_j(x)\|_2^2 = \pi/2 \ .$$

Exercise 1.7.3 Show that the functions

$$\phi_j(x) = \cos(jx) \ , \ j = 0, \ldots, m$$

form an orthogonal system with respect to the discrete inner product

$$(f, g) = \sum_{k=0}^m f(x_k)g(x_k) \ , \ \text{where } x_k = \frac{\pi}{2} \frac{2k+1}{m+1} \ .$$

and

$$\|\phi_j(x)\|_2^2 = \begin{cases} \frac{m+1}{2}, & j \geq 1 \\ m+1, j = 0 \end{cases} \ .$$

Exercise 1.7.4 Show that $\|f\|_1 \leq \sqrt{b-a}\|f\|_2$ and that $\|f\|_2 \leq \sqrt{b-a}\|f\|_\infty$.

1.7.2   Normal Equations

At this point in our discussion of approximation, we have presented the least squares problem, and described some general properties of inner products and their associated norms. Our next goal is to determine circumstances under which the least squares problem has a unique solution.

Theorem 1.7.3 (Normal Equations)   *Suppose that f is a function, and that the functions $\phi_0, \ldots, \phi_n$ are linearly independent with respect to the norm $\| \cdot \|_2$. Then the solution to the least squares problem*

$$\min \left\| f - \sum_{j=0}^n \phi_j c_j \right\|_2$$

*exists and is unique; furthermore, the optimal coefficients c_j satisfy the **normal equations**: for $0 \le j \le n$,*

$$\sum_{i=0}^{n} c_i(\phi_i, \phi_j) = (f, \phi_j) \,. \tag{1.59}$$

Proof First, note that if the coefficients c_i satisfy the normal Eq. (1.59), then for all $0 \le j \le n$

$$\left(f - \sum_{i=0}^{n} \phi_i c_i, \phi_j \right) = 0 \,.$$

It follows that for all coefficients γ_j we have

$$\left\| f - \sum_{i=0}^{n} \phi_i \gamma_i \right\|_2^2 = \left(f - \sum_{i=0}^{n} \phi_i \gamma_i \,, f - \sum_{k=0}^{n} \phi_k \gamma_k \right)$$

$$= \left(f - \sum_{i=0}^{n} \phi_i c_i + \sum_{i=0}^{n} \phi_i [c_i - \gamma_i] \,, f - \sum_{k=0}^{n} \phi_k c_k + \sum_{k=0}^{n} \phi_k [c_k - \gamma_k] \right)$$

$$= \left(f - \sum_{i=0}^{n} \phi_i c_i \,, f - \sum_{k=0}^{n} \phi_k c_k \right) - \left(f - \sum_{i=0}^{n} \phi_i c_j \,, \sum_{k=0}^{n} \phi_k [c_k - \gamma_k] \right)$$

$$- \left(\sum_{i=0}^{n} \phi_i [c_i - \gamma_i] \,, f - \sum_{k=0}^{n} \phi_k c_k \right) + \left(\sum_{i=0}^{n} \phi_i [c_i - \gamma_i] \,, \sum_{k=0}^{n} \phi_k [c_k - \gamma_k] \right)$$

and finally the normal equations show that

$$= \left\| f - \sum_{i=0}^{n} \phi_i c_i \right\|_2^2 + \left\| \sum_{i=0}^{n} \phi_i [c_i - \gamma_i] \right\|_2^2 \,. \tag{1.60}$$

Thus the norm of the error is minimized when $\gamma_i = c_j$ for all $0 \le i \le n$.

Next, we will show that the normal equations are nonsingular whenever the set of functions $\{\phi_0, \dots, \phi_n\}$ is linearly independent. Suppose that the complex conjugate of the vector $[\gamma_0, \dots, \gamma_n]$ lies in the null space of the matrix in the normal equations:

$$0 = \sum_{j=0}^{n} (\phi_i, \phi_j) \overline{\gamma_j} \,.$$

Then we can multiply this equation by γ_i and sum over i to get

$$0 = \sum_{i=0}^{n} \gamma_i \left\{ \sum_{j=0}^{n} (\phi_i, \phi_j) \overline{\gamma_j} \right\} i = \left(\sum_{i=0}^{n} \phi_i \gamma_i , \sum_{j=0}^{n} \phi_j \gamma_j \right) = \left\| \sum_{j=0}^{n} \phi_j \gamma_j \right\|_2^2 .$$

It follows that $\sum_{j=0}^{n} \phi_j \gamma_j = 0$ with respect to the norm $\| \cdot \|_2$. Since the functions $\phi_0, \ldots, \phi_n$ are linearly independent with respect to this norm, we conclude that the γ_j are all zero. Then Lemma 3.2.9 of Chap. 3 in Volume I shows that the matrix in the normal Eq. (1.59) is nonsingular.

Example 1.7.1 Suppose that we want to find the least squares approximation to a function f by a constant. In this case, we take $n = 0$ and $\phi_0(x) = 1$. If we use the discrete weighted inner product

$$(f, g) = \sum_{i=0}^{m} f(x_i) g(x_i) w_i ,$$

then the normal Eq. (1.59) can be rewritten in the form

$$\left(\sum_{i=0}^{m} w_i \right) c_0 = \sum_{i=0}^{m} f(x_i) w_i .$$

Thus the optimal coefficient is the weighted average

$$c_0 = \frac{\sum_{i=0}^{m} f(x_i) w_i}{\sum_{i=0}^{m} w_i} .$$

Theorem 1.7.3 provides us with reasonable circumstances under which the least squares problem has a solution, and the solution is unique. Next, we will discuss circumstances under which the solution of the normal equations is easy.

Corollary 1.7.2 *Suppose that* $(\cdot , \cdot)$ *is an inner product,* $\| \cdot \|_2$ *is the associated norm, this norm of the function f is finite, and* $\phi_0, \ldots, \phi_n$ *is an orthogonal system with respect to the given inner product. Then the normal equation matrix* $[(\phi_i, \phi_j)]$ *is diagonal, and the best least squares approximation to f by a linear combination of the functions* ϕ_j *uses coefficients*

$$c_j = \frac{(f, \phi_j)}{(\phi_j, \phi_j)} . \tag{1.61}$$

Furthermore, the error in the best least squares approximation satisfies

$$\left\| f - \sum_{j=0}^{n} \frac{(f, \phi_j)}{(\phi_j, \phi_j)} \phi_j \right\|_2^2 = \|f\|_2^2 - \sum_{j=0}^{n} \frac{|(f, \phi_j)|^2}{\|\phi_j\|_2^2} .$$

Proof Since the functions $\phi_0, \ldots, \phi_n$ form an orthogonal system, the normal equations involve a diagonal matrix, and the solution of the normal equations leads to (1.61).

If we choose $\gamma_j = 0$ for all j in (1.60), we get

$$\|f\|_2^2 = \left\| f - \sum_{j=0}^{n} \phi_j c_j \right\|_2^2 + \left\| \sum_{j=0}^{n} \phi_j c_j \right\|_2^2 .$$

then the orthogonality of the ϕ_j implies that

$$= \left\| f - \sum_{j=0}^{n} \phi_j c_j \right\|_2^2 + \sum_{j=0}^{n} |c_j|^2 \|\phi_j\|_2^2 . \tag{1.62}$$

This result is equivalent to the claim in the corollary.
We also obtain the following famous result.

Corollary 1.7.3 (Bessel's Inequality) *Suppose that $(\cdot\,,\,\cdot)$ is an inner product, $\|\cdot\|_2$ is the associated norm, this norm of the function f is finite, and $\phi_0, \ldots, \phi_n$ is an orthogonal system with respect to the given inner product. Then*

$$\sum_{j=0}^{n} \frac{|(f\,,\,\phi_j)|^2}{\|\phi_j\|_2^2} \leq \|f\|_2^2 .$$

Proof We can rewrite (1.62) in the form

$$\sum_{j=0}^{n} |c_j|^2 \|\phi_j\|_2^2 = \|f\|_2^2 - \left\| f - \sum_{j=0}^{n} \phi_j c_j \right\|_2^2 \leq \|f\|_2^2 .$$

The claim follows by using Eq. (1.61) for the optimal coefficients c_j^*.

In many cases, it is possible to show that the norm of the error in the least squares approximation tends to zero as n tends to infinity. Under such circumstances, Bessel's inequality will imply that

$$\|f\|^2 = \sum_{j=0}^{\infty} \frac{|(f\,,\,\phi_j)|^2}{\|\phi_j\|_2^2} ,$$

and Corollary 1.7.2 will imply that the error in the least squares approximation satisfies

$$\left\| f - \sum_{j=0}^{n} \frac{(f\,,\,\phi_j)}{(\phi_j\,,\,\phi_j)} \phi_j \right\|_2^2 = \sum_{j=n+1}^{\infty} \frac{|(f\,,\,\phi_j)|^2}{\|\phi_j\|_2^2} .$$

If we have some way to estimate the term on the right, then we can obtain an error estimate for the least squares approximation.

Finally, we will end our discussion of normal equations with the following cautionary example.

Example 1.7.2 (Hilbert Matrix) Suppose that we use the monomials to represent the least-squares polynomial approximation to a function f on the interval $[0, 1]$. In other words, our basis functions are $\phi_j(x) = x^j$ for $0 \leq j \leq m$. Then the normal equations involve the matrix with entries

$$(\phi_i, \phi_j) = \int_0^1 x^{i+j} dx = \frac{1}{i+j+1} .$$

This matrix is called the **Hilbert matrix**, and it is extremely ill-conditioned. For example, the condition number of the 12×12 Hilbert matrix is slightly more than 10^{16}. This implies that on many personal computers, double-precision solutions of linear systems involving this matrix cannot be guaranteed to have any significant digits.

Exercise 1.7.5 Find the continuous least squares approximation to $\sin(x)$ for $x \in [-\pi, \pi]$ by a straight line.

Exercise 1.7.6 Find the continuous least squares approximation to e^x for $x \in [0, 1]$ by a straight line.

Exercise 1.7.7 Show that the discrete least squares approximation to a function f using the discrete points $x_0 < \ldots < x_n$ by a polynomial of degree n is the polynomial interpolant at those points.

1.7.3 Orthogonal Polynomials

In order to develop a computationally reliable method for least-squares polynomial approximation, we could use the Gram-Schmidt process (previously described in Sect. 6.8 of Chap. 6 in Volume I) to generate an orthogonal set of polynomials. To begin the Gram-Schmidt process, we could use the monomials as the initial linearly independent set of functions from which the Gram-Schmidt process would compute the orthogonal polynomials. In such a scenario, we would begin with

$$\phi_0(x) = x^0 / \|x^0\|_2 = 1 / \|1\|_2 .$$

After we have determined $\phi_0, \ldots, \phi_{k-1}$, we could compute

$$p_k(x) = x^k - \sum_{j=0}^{k-1} \phi_j(x)(x^k, \phi_j) ,$$

and define the next orthonormal polynomial by

$$\phi_k = p_k / \| p_k \|_2 \ .$$

The resulting functions $\phi_0, \ldots, \phi_k$ would be an orthonormal system. However, such a Gram-Schmidt process would be expensive, because it would require k inner products (x^k, ϕ_j) for the calculation of each ϕ_k.

1.7.3.1   Recurrences

It is more efficient to generate orthogonal polynomials by means of a recurrence relation. This process is described in the following lemma.

Lemma 1.7.1 *Suppose that $(\cdot \ , \ \cdot)$ is an inner product on functions. Define the sequence of polynomials $\{\phi_k\}$ by the*

Algorithm 1.7.1 **(Three-Term Recurrence)**

$$\phi_0(x) = 1 \ ,$$

$$\phi_1(x) = x - \alpha_1 \ ,$$

$$\text{for } k \geq 2 \ , \ \ \phi_k(x) = (x - \alpha_k)\phi_{k-1}(x) - \beta_k \phi_{k-2}(x) \ . \tag{1.63}$$

Then $\{\phi_k\}_{k=0}^{\infty}$ is orthogonal if and only if

$$\alpha_k = \frac{(x\phi_{k-1}, \phi_{k-1})}{(\phi_{k-1}, \phi_{k-1})} \text{ for } k \geq 1 \text{ and} \tag{1.64a}$$

$$\beta_k = \frac{(x\phi_{k-1}, \phi_{k-2})}{(\phi_{k-2}, \phi_{k-2})} \text{ for } k \geq 2 \ . \tag{1.64b}$$

With these choices of the coefficients α_k and β_k, for all $k \geq 0$ the leading coefficient of $\phi_k(x)$ is one. Finally, for all $k \geq 2$, we have

$$\beta_k = \frac{(\phi_{k-1} \ , \ \phi_{k-1})}{(\phi_{k-2} \ , \ \phi_{k-2})} > 0 \ . \tag{1.65}$$

Proof We will prove our claims by induction. When $k = 0$, we need only observe that the leading coefficient of $\phi_0(x) = 1$ is x^0.

Next, we note that for $k = 1$ we have $\phi_1(x) = x - \alpha_1$. Thus the leading coefficient of ϕ_1 is x^1, and the definition (1.64a) of α_1 implies that

$$(\phi_1 \ , \ \phi_0) = (x - \alpha_1 \ , \ 1) = (x\phi_0 \ , \ \phi_0) - (\phi_1 \ , \ \phi_0)\alpha_1 = 0 \ .$$

For polynomials of degree at most $k - 1$ with $k \geq 2$, our inductive hypothesis is that the leading coefficient of ϕ_{k-1} is one, and that for all $0 \leq j < k - 1$ we have $(\phi_{k-1}, \phi_j) = 0$. Then it is obvious that the coefficient of x^k in

$$\phi_k(x) = (x - \alpha_k)\phi_{k-1}(x) - \beta_k\phi_{k-2}(x)$$

is also one. For $j < k - 2$ the orthogonality of ϕ_{k-1} and ϕ_{k-2} to ϕ_j implies that

$$(\phi_k, \phi_j) = (x\phi_{k-1}, \phi_j) - \alpha_k(\phi_{k-1}, \phi_j) - \beta_k(\phi_{k-2}, \phi_j) = (\phi_{k-1}, x\phi_j) = 0 .$$

The final inner product in this equation is zero because $x\phi_j$ is a polynomial of degree less than $k - 1$ and the inductive hypothesis guarantees that ϕ_{k-1} is orthogonal to all polynomials of degree less than $k - 1$. Next, we use the definition (1.64a) of α_k to show that ϕ_k is orthogonal to ϕ_{k-1}:

$$(\phi_k , \phi_{k-1}) = (x\phi_{k-1}, \phi_{k-1}) - \alpha_k(\phi_{k-1}, \phi_{k-1}) - \beta_k(\phi_{k-2}, \phi_{k-1})$$
$$= (x\phi_{k-1}, \phi_{k-1}) - \alpha_k(\phi_{k-1}, \phi_{k-1}) = 0 .$$

We can use the definition (1.64b) of β_k to show that ϕ_k is orthogonal to ϕ_{k-2}:

$$(\phi_k, \phi_{k-2}) = (x\phi_{k-1}, \phi_{k-2}) - \alpha_k(\phi_{k-1}, \phi_{k-2}) - \beta_k(\phi_{k-2}, \phi_{k-2})$$
$$= (x\phi_{k-1}, \phi_{k-2}) - \beta_k(\phi_{k-2}, \phi_{k-2}) = 0 .$$

At this point, we have completed the induction.

All that remains is to prove the final claim (1.65). First, for $k = 2$ we note that

$$\beta_2(\phi_0 , \phi_0) = (x\phi_1 , \phi_0) = (\phi_1 , x\phi_0) = (\phi_1 , \phi_1 + \alpha_1\phi_0) = (\phi_1 , \phi_1) .$$

Assume inductively that (1.65). is true for $k - 1$ with $k > 2$. Then

$$\beta_k(\phi_{k-2}, \phi_{k-2}) = (\phi_{k-1}, x\phi_{k-2}) = (\phi_{k-1}, \phi_{k-1} + \alpha_{k-1}\phi_{k-2} + \beta_{k-1}\phi_{k-3})$$
$$= (\phi_{k-1}, \phi_{k-1}) .$$

This completes the inductive proof of (1.65).

In summary, we can implement the three-term recurrence for computing orthogonal polynomials by means of the following algorithm:

Algorithm 1.7.2 (Generate Orthogonal Polynomials)

$$\phi_{-1}(x) \equiv 0$$
$$\phi_0(x) \equiv 1$$
$$\delta_{-1} = 1$$

for $1 \leq j \leq n$

$$v_{j-1} = (x\phi_{j-1}, \phi_{j-1})$$
$$\delta_{j-1} = (\phi_{j-1}, \phi_{j-1})$$
$$\alpha_j = v_{j-1}/\delta_{j-1}$$
$$\beta_j = b_{j-1}/\delta_{j-2}$$
$$\phi_j(x) = (x - \alpha_j)\phi_{j-1}(x) - \beta_j\phi_{j-2}(x) .$$

From this algorithm, it is easy to see that calculation of orthogonal polynomials requires at most two inner products per polynomial. Once the coefficients α_j and β_j have been computed, it is easy to evaluate the polynomials at a point by means of the following algorithm:

Algorithm 1.7.3 (Evaluate Orthogonal Polynomials)

$$\phi_{-1}(x) = 0$$
$$\phi_0(x) = 1$$
$$\text{for } 1 \leq j \leq n , \ \phi_j(x) = (x - \alpha_j)\phi_{j-1}(x) - \beta_j\phi_{j-2}(x) .$$

In order to determine the least-squares polynomial approximation $\sum_{j=0}^{n} c_j\phi_j$ to a function f, we would compute the coefficients c_j by

$$c_j = \frac{(f, \phi_j)}{(\phi_j, \phi_j)} .$$

Note that the denominator in this expression was computed as the scalar δ_j in Algorithm 1.7.2. We conclude that the solution of least squares polynomial approximation via orthogonal polynomials is not only more numerically stable, but also less computationally expensive than solution by solving the normal equations.

Often, orthogonal polynomials are described by a three-term recurrence in the form

$$p_0(x) = c_0$$
$$p_1(x) = (c_1x - a_1)p_0(x)$$
$$\text{for} k \geq 2 , \ p_k(x) = (c_kx - a_k)p_{k-1}(x) - b_kp_{k-2}(x) , \tag{1.66}$$

where the additional coefficient c_k is chosen to simplify the other coefficients or to scale the polynomials appropriately. The connection between this form of the three-term recurrence and our original form (1.63) can be found in the next lemma.

Lemma 1.7.2 *Suppose that sequence $\{\phi_k\}$ of polynomials is generated by the three-term recurrence (1.63), and the sequence $\{p_k\}$ of polynomials is generated by the* *three-term recurrence* (1.66). *If we have*

$$a_k = c_k \alpha_k \text{ for all } k \geq 1 \text{ and}$$

$$b_k = c_k c_{k-1} \beta_k \text{ for all } k \geq 2,$$

then for all $k \geq 0$ we have

$$p_k(x) = \phi_k(x) \prod_{j=0}^{k} c_j.$$

Proof We will prove this claim by induction. First, we note that algorithm (1.66) begins with $p_0(x) = c_0 = \phi_0(x)c_0$. Next, it computes

$$p_1(x) = (c_1 x - a_1)p_0(x) = (c_1 x - a_1)\phi_0(x)c_0 = (x - \alpha_1)\phi_0(x)c_1 c_0 = \phi_1(x)c_1 c_0.$$

Assume inductively that $p_n(x) = \phi_n(x) \prod_{j=0}^{n} c_j$ for $n - 1 \geq 2$. At this point, algorithm (1.66) implies that

$$p_k(x) = (c_k x - a_k)p_{k-1}(x) - b_k p_{k-2}(x)$$

then the definitions of a_k and b_k give us

$$= (c_k x - c_k \alpha_k)\phi_{k-1}(x) \prod_{j=0}^{k-1} c_j - c_k c_{k-1} \beta_k \phi_{k-2}(x) \prod_{j=0}^{k-2} c_j$$

and finally Algorithm 1.7.3 implies that

$$= [(x - \alpha_k)\phi_{k-1}(x) - \beta_k \phi_{k-2}(x)] \prod_{j=0}^{k} c_j = \phi_k(x) \prod_{j=0}^{k} c_j.$$

Example 1.7.3 **Legendre polynomials** are orthogonal with respect to the continuous inner product

$$(f, g) = \int_{-1}^{1} f(x)\overline{g(x)}dx.$$

It is often convenient to generate the Legendre polynomials so that

$$p_k(1) = 1$$

for all $k \geq 0$. The recurrence that produces these values takes the form

$$p_0(x) = 1$$
$$p_1(x) = x$$
$$p_k(x) = \frac{2k-1}{k} x p_{k-1}(x) - \frac{k-1}{k} p_{k-2}(x) . \tag{1.67}$$

It is also possible to prove that this form of the Legendre polynomials satisfies

$$\| p_k \|^2 = \frac{2}{2k+1} . \tag{1.68}$$

Readers may view several different orthogonal polynomial families by running the JavaScript program for **orthogonal polynomials** (see `/home/faculty/johnt/book/code/approximation/orthogonal_polys.html`). This program plots the Legendre, Laguerre, Chebyshev and Hermite polynomials. For the definitions of the last three of these orthogonal polynomial families, please see the exercises below.

Let us summarize how we would use orthogonal polynomials to determine a least-square polynomial approximation to some function f. As we showed in Corollary 1.7.2, for $0 \leq j \leq n$ we would compute the least-squares coefficients

$$c_j = \frac{(f , \phi_j)}{(\phi_j , \phi_j)} .$$

This would typically involve numerical quadrature to compute the integrals; see Sect. 2.3 for various numerical integration methods. Afterward, we can modify Algorithm 1.7.3 to compute the least-square polynomial approximation $p(x) = \sum_{j=0}^{n} c_j \phi_j(x)$:

Algorithm 1.7.4 (Evaluate Least Squares Polynomial Approximant)

$$\phi_{.-2} = 0$$
$$\phi_{.-1} = 1$$
$$p = c_0$$
$$\text{for } 1 \leq j \leq n$$
$$\quad \phi = (x - \alpha_j)\phi_{.-1}(x) - \beta_j \phi_{.-2}(x)$$
$$\quad p = p + c_j \phi$$
$$\quad \phi_{.-2} = \phi_{.-1}$$
$$\quad \phi_{.-1} = \phi .$$

We would like to remark that there are some important error estimates for least squares polynomial approximation, but the most useful ones are actually derived from estimates for minimax polynomial approximation. For example, the Weierstrass approximation Theorem 1.2.1 implies that the maximum error in the uniformly best polynomial approximation to a continuous function can be made arbitrarily small by choosing the polynomial order sufficiently large. This theorem implies that the least squares polynomial approximation to a continuous function over a bounded interval converges as well. For functions f with $k \leq n$ higher-order continuous derivatives on $[a, b]$, there is a **Jackson theorem** [40, p. 147] proving that

$$\min_{q_n \text{ a polynomial of degree } n} \max_{-1 \leq x \leq 1} |f(x) - q_n(x)| \leq \frac{(n - k + 1)!}{(n + 1)!} \frac{\pi (b - a)^k}{4} \|f^{(k)}\|_2 .$$

(1.69)

If $\{\phi_k\}_{k=0}^{\infty}$ is a sequence of orthogonal polynomials for some inner product, then the sequence of least squares polynomial approximations to a function f is $\{p_n\}_{n=0}^{\infty}$ where

$$p_n = \sum_{k=0}^{n} \frac{(f, \phi_k)}{(\phi_k, \phi_k)} \phi_k ; .$$

Let q_n be the polynomial of degree n that minimizes the maximum norm of the error in approximating f on $[a, b]$. Then

$$\|f - p_n\|^2 \leq \|f - q_n\|^2 = \int_a^b |f(x) - q_n(x)|^2 w(x) \, dx \leq \max_{x \in [a,b]} |f(x) - q_n(x)| \int_a^b w(x) \, dx .$$

Since $\max_{x \in [a,b]} |f(x) - q_n(x)| \to 0$ as $n \to \infty$, it follows that $\|f - p_n\| \to 0$ as $n \to \infty$. In other words, the error in least squares polynomial approximation tends to zero as the order of the polynomial tends to infinity. If f has $k \leq n$ continuous derivatives, we can use a change of variables and the Jackson theorem estimate (1.69) to provide an upper bound for the error in least squares polynomial approximation. For more details, we suggest that the reader consult Cheney [40, p. 101ff].

Before we end this section, we would like to present one more result that will be used in developing Gaussian quadrature formulas in Sect. 2.3.10.

Theorem 1.7.4 (Christoffel-Darboux) *Suppose that $(\cdot, \cdot)$ is an inner product on functions such that*

$$(xf, g) = (f, xg) .$$

Let the sequence of polynomials $\{p_k\}_{k=0}^\infty$ be orthogonal in this inner product, and satisfy the three-term recurrence (1.66). Then for all $x \neq y$ we have

$$\frac{p_{m+1}(x)p_m(y) - p_m(x)p_{m+1}(y)}{c_{m+1}(x-y)(p_m,\, p_m)} = \sum_{k=0}^m \frac{p_k(x)p_k(y)}{(p_k,\, p_k)} . \qquad (1.70)$$

Proof For $m \geq 1$ we can use the three-term recurrence to obtain

$$p_{m+1}(x)p_m(y) = (c_{m+1}x - a_{m+1})p_m(x)p_m(y) - b_{m+1}p_{m-1}(x)p_m(y) \text{ and}$$

$$p_m(x)p_{m+1}(y) = (c_{m+1}y - a_{m+1})p_m(x)p_m(y) - b_{m+1}p_m(x)p_{m-1}(y) .$$

We can subtract the second of these equations from the first to get

$$p_{m+1}(x)p_m(y) - p_m(x)p_{m+1}(y)$$
$$= c_{m+1}(x-y)p_m(x)p_m(y) + b_{m+1}[p_m(x)p_{m-1}(y) - p_{m-1}(x)p_m(y)] . \qquad (1.71)$$

Next, we use the recurrence (1.66) and, if necessary, the understanding that $p_{-1}(x) = 0$ to get

$$b_{m+1}(p_{m-1},\, p_{m-1}) = ((c_{m+1}x - a_{m+1})p_m - p_{m+1},\, p_{m-1})$$
$$= c_{m+1}(xp_m,\, p_{m-1}) = c_{m+1}(p_m,\, xp_{m-1})$$
$$= \frac{c_{m+1}}{c_m}(p_m + a_m p_{m-1} + b_m p_{m-2},\, p_m) = \frac{c_{m+1}}{c_m}(p_m,\, p_m) .$$

Thus Eq. (1.71) is equivalent to

$$\frac{p_{m+1}(x)p_m(y) - p_m(x)p_{m+1}(y)}{c_{m+1}(x-y)(p_m,\, p_m)} = \frac{p_m(x)p_m(y)}{(p_m,\, p_m)} + \frac{p_m(x)p_{m-1}(y) - p_{m-1}(x)p_m(y)}{c_m(x-y)(p_{m-1},\, p_{m-1})} .$$

We can continue to apply this equation until we obtain

$$\frac{p_{m+1}(x)p_m(y) - p_m(x)p_{m+1}(y)}{c_{m+1}(x-y)(p_m,\, p_m)} = \sum_{k=1}^m \frac{p_k(x)p_k(y)}{(p_k,\, p_k)} + \frac{p_1(x)p_0(y) - p_0(x)p_1(y)}{c_1(x-y)(p_0,\, p_0)}$$

$$= \sum_{k=1}^m \frac{p_k(x)p_k(y)}{(p_k,\, p_k)} + \frac{(c_1 x - a_1)c_0^2 - c_0^2(c_1 y - a_1)}{c_1(x-y)(p_0,\, p_0)} = \sum_{k=0}^m \frac{p_k(x)p_k(y)}{(p_k,\, p_k)} .$$

Exercise 1.7.8 Laguerre polynomials are orthogonal with respect to the continuous inner product

$$(f, g) = \int_0^\infty f(x)g(x)e^{-x}dx .$$

Show that Laguerre polynomials satisfy the recurrence relation

$$\ell_{-1}(x) = 0$$

$$\ell_0(x) = 1$$

$$\text{for all } n \geq 0 \; \ell_{n+1}(x) = \frac{2n+1-x}{n+1}\ell_n(x) - \frac{n}{n+1}\ell_{n-1}(x) \;.$$

Exercise 1.7.9 Chebyshev polynomials are orthogonal with respect to the continuous inner product

$$(f, g) = \int_{-1}^{1} \frac{f(x)g(x)}{\sqrt{1-x^2}} dx \;.$$

Show that Chebyshev polynomials satisfy the recurrence relation

$$t_0(x) = 1$$

$$t_1(x) = x$$

$$\text{for all } n \geq 1 \; t_{n+1}(x) = 2xt_n(x) - t_{n-1}(x) \;.$$

Also show that $t_n(x) = \cos(n \cos^{-1}(x))$.

Exercise 1.7.10 Hermite polynomials are orthogonal with respect to the continuous inner product

$$(f, g) = \int_{-\infty}^{\infty} f(x)g(x)e^{-x^2} dx \;.$$

Show that Hermite polynomials satisfy the recurrence relation

$$h_{-1}(x) = 0$$

$$h_0(x) = 1$$

$$\text{for all } n \geq 0 \; h_{n+1}(x) = 2xh_n(x) - 2nh_{n-1}(x) \;.$$

Exercise 1.7.11 Gram polynomials are orthogonal with respect to the discrete inner product

$$(f, g) = \sum_{i=0}^{m} f(x_i)g(x_i) \;, x_i = -1 + \frac{2i}{m} \;.$$

This is a discrete inner product with equidistant nodes in $[-1, 1]$. If we define

$$p_0(x) \equiv \frac{1}{\sqrt{m+1}} \;, \; p_{-1}(x) \equiv 0 \;,$$

show that the higher-order polynomials satisfy the recurrence relation

$$p_k(x) = C_k x p_{k-1}(x) - \frac{C_k}{C_{k-1}} p_{k-2}(x) \,,$$

where

$$C_k = \frac{m}{k} \sqrt{\frac{4k^2 - 1}{(m+1)^2 - k^2}} \,.$$

Also show that with this modified choice of $p_0(x)$ we have

$$\|p_k\|^2 = 1 \text{ for } 0 \le k \,.$$

These polynomials are very similar to Legendre polynomials for $n \ll \sqrt{m}$; for $n \gg \sqrt{m}$ they have very large oscillations between the nodes x_i and a large maximum norm in $[-1, 1]$. Thus, when fitting a polynomial of degree n to m equidistant data points, we should never choose $n > 2\sqrt{m}$.

Exercise 1.7.12 Discrete Chebyshev polynomials are orthogonal with respect to the discrete inner product

$$(f, g) = \sum_{i=0}^{m} f(x_i) g(x_i) \,, x_i = \cos\left(\frac{2i+1}{m+1} \frac{\pi}{2}\right) \,.$$

Show that discrete Chebyshev polynomials satisfy

$$\|p_k\|^2 = \begin{cases} 1, & k = 0 \\ \frac{k+1}{2}, & k > 0 \end{cases} \,.$$

Exercise 1.7.13 Write a program to evaluate all Legendre polynomials of degree at most n at a given point x. Use the three-term recurrence (1.67) to generate the polynomials.

Exercise 1.7.14 Write a program to compute the least squares cubic polynomial approximation to $f(x) = e^{-x}$ on the interval $[0, 1]$. Plot the function f and its cubic polynomial approximation on $[0, 1]$.

1.7.3.2 Legendre Polynomial Identities

We will use a number of identities for Legendre polynomials to derive numerical integration techniques in Sects. 2.3.10, 2.3.11 and 2.3.12. Proof of these identities are seldom included or referenced in numerical analysis texts, so we will provide them here.

Theorem 1.7.5 (Generating Function for Legendre Polynomials) *The Legendre polynomials, defined by the three-term recurrence* (1.67), *are the expansion coefficients for the* **generating function**

$$(1 + t^2 - 2xt)^{-1/2} = \sum_{k=0}^{\infty} p_k(x) t^k \ . \tag{1.72}$$

Furthermore,

$$p_k(-x) = (-1)^k p_k(x) \ for \ k \geq 0 \ , \tag{1.73a}$$

$$p_k(1) = 1 \ for \ k \geq 0 \ , \tag{1.73b}$$

$$p_k'(1) = k(k+1)/2 \ for \ k \geq 0 \ , \tag{1.73c}$$

$$(2k+1) p_k(x) = p_{k+1}'(x) - p_{k-1}'(x) \ for \ k \geq 1 \ , \tag{1.73d}$$

$$k p_k(x) = x p_k'(x) - p_{k-1}'(x) \ for \ k \geq 1 \ and \tag{1.73e}$$

$$(x^2 - 1) p_k'(x) = k x p_k(x) - k p_{k-1}(x) \ for \ k \geq 1 \ . \tag{1.73f}$$

Proof Let us verify the generating function Eq. (1.72). Since

$$g(x, t) \equiv (1 + t^2 - 2xt)^{-1/2}$$

is analytic in t, we can write

$$g(x, t) = \sum_{k=0}^{\infty} q_k(x) t^k$$

for some functions $q_k(x)$. Since

$$\frac{\partial g}{\partial t}(x, t) = (x - t)(1 + t^2 - 2xt)^{-3/2} \ ,$$

we see that

$$q_0(x) = g(x, 0) = 1 \ \text{and} \ q_1(x) = \frac{\partial g}{\partial t}(x, 0) = (x - 0)[1 - 0^2 - 2(x)(0)]^{-3/2} = x \ .$$

Since

$$\frac{x - t}{1 + t^2 - 2xt} g(x, t) = \frac{\partial g}{\partial t}(x, t) = \sum_{k=0}^{\infty} k q_k(x) t^{k-1} \ ,$$

we can multiply this equation by $1 + t^2 - 2xt$ to obtain

$$(x - t) \sum_{k=0}^{\infty} q_k(x)t^k = \sum_{k=0}^{\infty} xq_k(x)t^k - \sum_{k=0}^{\infty} q_k(x)t^{k+1}$$

$$= (1 + t^2 - 2xt) \sum_{k=0}^{\infty} kq_k(x)t^{k-1}$$

$$= \sum_{k=1}^{\infty} kq_k(x)t^{k-1} + \sum_{k=0}^{\infty} kq_k(x)t^{k+1} - 2 \sum_{k=0}^{\infty} kxq_k(x)t^k .$$

This can be simplified to

$$\sum_{k=0}^{\infty} (2k + 1)xq_k(x)t^k = \sum_{k=0}^{\infty} (k + 1)q_k(x)t^{k+1} + \sum_{k=1}^{\infty} kq_k(x)t^{k-1}$$

$$= \sum_{\ell=0}^{\infty} \ell q_{\ell-1}(x)t^{\ell} + \sum_{m=0}^{\infty} (m + 1)q_{m+1}(x)t^m .$$

Equating coefficients of equal powers of t gives us

$$(2k + 1)xq_k(x) = (k + 1)q_{k+1}(x) + kq_{k-1}(x) .$$

Since the polynomials $q_k(x)$ satisfy the recurrence relation for the Legendre polynomials, and since $q_0(x) = p_0(x)$ and $q_1(x) = p_1(x)$, we conclude that $q_k(x) = p_k(x)$ for all $k \geq 0$.

Equation (1.73a) can be proved by equating expansion coefficients in

$$g(x, t) = g(-x, -t) .$$

Also, Eq. (1.73b) can be verified by equating the expansion coefficients of

$$g(1, t) = (1 - 2t + t^2)^{-1/2} = (1 - t)^{-1} = \sum_{k=0}^{\infty} t^k .$$

Furthermore, Eq. (1.73c) is the result of equating the expansion coefficients in

$$\frac{\partial g}{\partial x}(1, t) = t(1 - 2t + t^2)^{-3/2} = t(1 - t)^{-3} = \sum_{k=1}^{\infty} \frac{(k + 1)k}{2} t^k .$$

Equation (1.73d). can be obtained by equating expansion coefficients in the equation

$$2t\frac{\partial g}{\partial t}(x, t) + g(x, t) = \frac{1}{t}\frac{\partial g}{\partial x}(x, t) - t\frac{\partial g}{\partial x}(x, t) .$$

Also, Eq. (1.73e) follows from equating the expansion coefficients in the equation

$$t\frac{\partial g}{\partial t}(x, t) = x\frac{\partial g}{\partial x}(x, t) - t\frac{\partial g}{\partial x}(x, t) .$$

Finally, Eq. (1.73f). is the result of equating expansion coefficients in the equation

$$(x^2 - 1)\frac{\partial g}{\partial x}(x, t) = t\frac{\partial (x - t)g}{\partial t}(x, t) .$$

Here is another very useful set of identities.

Theorem 1.7.6 (Rodriques' Formula) *The Legendre polynomials, defined by the three-term recurrence (1.67), also satisfy*

$$p_n(x) = \frac{1}{2^n n!}\frac{d^n}{dx^n}(x^2 - 1)^n \text{ for } n \geq 0 .$$

Furthermore,

$$\int_{-1}^{1} p_n(x)^2 \, dx = \frac{2}{2n + 1} \text{ for } n \geq 0 , \tag{1.74}$$

and the Legendre polynomial p_n satisfies the **Legendre differential equation**

$$\frac{d}{dx}\left[(1 - x^2)p_n'(x)\right] + n(n + 1)p_n(x) = 0 . \tag{1.75}$$

Proof Define

$$q_n(x) = \frac{1}{2^n n!}\frac{d^n}{dx^n}(x^2 - 1)^n .$$

Since $(x^2 - 1)^n$ is a polynomial of degree $2n$, it follows that q_n is a polynomial of degree n. The Leibniz rule for differentiation implies that

$$2^n n! q_n(x) = \frac{d^n}{dx^n}\left[(x - 1)^n(x + 1)^n\right] = \sum_{k=0}^{n}\binom{n}{k}\frac{d^{n-k}}{dx^{n-k}}(x - 1)^n\frac{d^k}{dx^k}(x + 1)^n$$

$$= \sum_{k=0}^{n}\binom{n}{k}\frac{n!}{k!}(x - 1)^k\frac{n!}{(n - k)!}(x + 1)^{n-k} .$$

It follows that

$$q_n(1) = \frac{1}{2^n n!}\binom{n}{0}\frac{n!\, n!}{0!\, n!}2^n = 1 \,.$$

Thus q_n has the same degree as p_n and the same value at $x = 1$. If we show that the polynomials q_n are mutually orthogonal, then we must have $q_n = p_n$ for all $n \geq 0$.
For $0 \leq m < n$ the Leibniz rule for differentiation implies that

$$\frac{d^m}{dx^m}(x^2 - 1)^n = \frac{d^m}{dx^m}[(x-1)^n(x+1)^n] = \sum_{k=0}^{m}\binom{m}{k}\frac{d^{m-k}}{dx^{m-k}}(x-1)^n\frac{d^k}{dx^k}(x+1)^n$$

$$= \sum_{k=0}^{m}\binom{m}{k}\frac{n!}{(n+k-m)!}(x-1)^{n+k-m}\frac{n!}{(n-k)!}(x+1)^{n-k} \,.$$

We conclude that

$$\frac{d^m}{dx^m}(x^2 - 1)^n \big|_{x=1} = 0 \text{ and } \frac{d^m}{dx^m}(x^2 - 1)^n \big|_{x=-1} = 0 \tag{1.76}$$

for all $0 \leq m < n$. Consequently, integration by parts gives us

$$\int_{-1}^{1}\frac{d^m}{dx^m}(x^2 - 1)^m\frac{d^n}{dx^n}(x^2 - 1)^n\, dx$$

$$= \left[\frac{d^m}{dx^m}(x^2-1)^m\frac{d^{n-1}}{dx^{n-1}}(x^2-1)^n\right]_{-1}^{1} - \int_{-1}^{1}\frac{d^{m+1}}{dx^{m+1}}(x^2-1)^m\frac{d^{n-1}}{dx^{n-1}}(x^2-1)^n\, dx$$

$$= -\int_{-1}^{1}\frac{d^{m+1}}{dx^{m+1}}(x^2-1)^m\frac{d^{n-1}}{dx^{n-1}}(x^2-1)^n\, dx$$

then repeated integration by parts produces

$$= \ldots = (-1)^{m-1}\left[\frac{d^{2m-1}}{dx^{2m-1}}(x^2-1)^m\frac{d^{n-m}}{dx^{n-m}}(x^2-1)^n\right]_{-1}^{1}$$

$$+ (-1)^m\int_{-1}^{1}\frac{d^{2m}}{dx^{2m}}(x^2-1)^m\frac{d^{n-m}}{dx^{n-m}}(x^2-1)^n\, dx$$

$$= (-1)^m\int_{-1}^{1}\frac{d^{2m}}{dx^{2m}}(x^2-1)^m\frac{d^{n-m}}{dx^{n-m}}(x^2-1)^n\, dx$$

and since $d^{2m}/dx^{2m}(x^2-1)^m$ is a constant, we get

$$= (-1)^m \frac{d^{2m}}{dx^{2m}}(x^2-1)^m \left[\frac{d^{n-m-1}}{dx^{n-m-1}}(x^2-1)^n \right]_{-1}^{1} = 0 .$$

Since this shows that the polynomials $q_n(x)$ are mutually orthogonal, and since $q_n(1) = p_n(1)$ for all n, we conclude that $q_n(x) = p_n(x)$ for all $n \geq 0$.

On the other hand, if $m = n$ the same chain of repeated integrations by parts gives us

$$\int_{-1}^{1} \frac{d^n}{dx^n}(x^2-1)^n \frac{d^n}{dx^n}(x^2-1)^n \, dx = (-1)^n \frac{d^{2n}}{dx^{2n}}(x^2-1)^n \int_{-1}^{1}(x^2-1)^n \, dx$$

The Leibniz rule shows that

$$\frac{d^{2n}}{dx^{2n}}(x^2-1)^n = \frac{d^{2n}}{dx^{2n}}[(x-1)^n(x+1)^n] = \sum_{k=0}^{2n} \binom{2n}{k} \frac{d^{2n-k}}{dx^{2n-k}}(x-1)^n \frac{d^k}{dx^k}(x+1)^n$$

$$= \binom{2n}{n} \frac{d^n}{dx^n}(x-1)^n \frac{d^n}{dx^n}(x+1)^n = \binom{2n}{n}(n!)^2 = (2n)! .$$

A trigonometric substitution shows that

$$(-1)^n \int_{-1}^{1}(x^2-1)^n \, dx = 2 \int_{0}^{1}(1-x^2)^n \, dx = 2 \int_{0}^{\pi/2} \sin^{2n+1}\theta \, d\theta = 2 \frac{(2^n n!)^2}{(2n+1)!} .$$

Thus

$$\int_{-1}^{1} p_n(x)^2 \, dx = \frac{1}{(2^n n!)^2} \int_{-1}^{1} \left[\frac{d^n}{dx^n}(x^2-1)^n \right]^2 dx = \frac{1}{(2^n n!)^2}(2n)! 2 \frac{(2^n n!)^2}{(2n+1)!}$$

$$= \frac{2}{2n+1} .$$

Finally, let us prove that p_n satisfies (1.75). Define

$$v_n(x) = (x^2 01)^n .$$

Then

$$(x^2-1)v_n'(x) - 2nxv_n(x) = 0 .$$

We can use the Leibniz rule for differentiation to take $n + 1$ derivatives of this equation:

$$0 = \sum_{j=0}^{n+1} \binom{n+1}{j} \frac{d^{n+1-j} v_n'}{dx^{n+1-j}}(x) \frac{d^j(x^2-1)}{dx^j} - \sum_{j=0}^{n+1} \binom{n+1}{j} \frac{d^{n+1-j} v_n}{dx^{n+1-j}}(x) \frac{d^j 2nx}{dx^j}$$

$$= \binom{n+1}{0}(x^2-1)\frac{d^{n+2} v_n}{dx^{n+2}}(x) + \binom{n+1}{1}2x\frac{d^{n+1} v_n}{dx^{n+1}}(x) + \binom{n+1}{2}2\frac{d^n v_n}{dx^n}(x)$$

$$- \binom{n+1}{0}2nx\frac{d^{n+1} v_n}{dx^{n+1}}(x) - \binom{n+1}{1}2n\frac{d^n v_n}{dx^n}(x)$$

then we use Rodrigues' formula to write

$$= (x^2-1)p_n''(x) + 2(n+1)xp_n'(x) + n(n+1)p_n(x) - 2nxp_n'(x) - 2n(n+1)p_n(x)$$
$$= (x^2-1)p_n''(x) + 2xp_n'(x) - n(n+1)p_n(x) \ .$$

This is equivalent to the claimed differential Eq. (1.75).

Exercise 1.7.15 Laguerre polynomials can be defined by the recurrence relation

$$\ell_{-1}(x) = 0$$
$$\ell_0(x) = 1$$
$$\text{for all } n \geq 0 \ \ell_{n+1}(x) = \frac{2n+1-x}{n+1}\ell_n(x) - \frac{n}{n+1}\ell_{n-1}(x) \ .$$

In the exercises of Sect. 1.7.3.1, readers were given that these polynomials are orthogonal with respect to the continuous inner product

$$(f, g) = \int_0^\infty f(x)g(x)e^{-x}dx \ .$$

1. Show that Laguerre polynomials are the expansion coefficients in the generating function

$$\frac{1}{1-t}e^{-tx/(1-t)} = \sum_{n=0}^\infty \ell_n(x)t^n \ .$$

2. Show that Laguerre polynomials are given by Rodrigues' formula

$$\ell_n(x) = \frac{1}{n!}\left(\frac{d}{dx} - 1\right)^n x^n$$

for $n \geq 0$.

3. Show that the n-th Laguerre polynomial ℓ_n satisfies the **Laguerre differential equation**

$$xu''(x) + (1 - x)u'(x) + nu(x) = 0 .$$

Exercise 1.7.16 Chebyshev polynomials can be defined by the recurrence relation

$$t_0(x) = 1$$
$$t_1(x) = x$$
$$\text{for all } n \geq 1 \ t_{n+1}(x) = 2xt_n(x) - t_{n-1}(x) .$$

In the exercises of Sect. 1.7.3.1, readers were given that these polynomials are orthogonal with respect to the continuous inner product

$$(f, g) = \int_{-1}^{1} \frac{f(x)g(x)}{\sqrt{1 - x^2}} dx .$$

1. Show that Chebyshev polynomials are the expansion coefficients in the generating function

$$\frac{1 - tx}{1 - 2tx + t^2} = \sum_{n=0}^{\infty} t_n(x)t^n .$$

2. Show that the n-th Chebyshev polynomial t_n satisfies the **Chebyshev differential equation**

$$(1 - x^2)u''(x) - xu'(x) + n^2 u(x) = 0 .$$

Exercise 1.7.17 Hermite polynomials can be defined by the recurrence relation

$$h_{-1}(x) = 0$$
$$h_0(x) = 1$$
$$\text{for all } n \geq 0 \ h_{n+1}(x) = 2xh_n(x) - 2nh_{n-1}(x) .$$

In the exercises of Sect. 1.7.3.1, readers were given that these polynomials are orthogonal with respect to the continuous inner product

$$(f, g) = \int_{-\infty}^{\infty} f(x)g(x)e^{-x^2} dx .$$

1. Show that Hermite polynomials are the expansion coefficients in the generating function

$$e^{2xt-t^2} = \sum_{n=0}^{\infty} h_n(x) \frac{t^n}{n!} \ .$$

2. Show that Hermite polynomials are given by Rodrigues' formula

$$h_n(x) = (-1)^n e^{x^2} \frac{d^n}{dx^n} e^{-x^2} \ .$$

for $n \geq 0$.

3. Show that the n-th Hermite polynomial h_n satisfies the **Hermite differential equation**

$$u''(x) - 2xu'(x) + 2nu(x) = 0 \ .$$

1.7.4 Trigonometric Polynomials

In Sect. 1.7.3, we described a general method for approximating square-integrable functions f by orthogonal polynomials. We showed that the orthogonal polynomials can be generated very efficiently by a three-term recurrence. In this section, we will examine the special case in which the function f to be approximated is periodic. We will see that such functions can be approximated very even more efficiently by trigonometric polynomials.

1.7.4.1 Orthogonality

Let us begin by defining the class of functions we would like to approximate.

Definition 1.7.4 Suppose that f is a complex-valued function of a real variable t. Then f is **periodic** if and only if there is a smallest positive scalar p, called the **period**, so that for all t we have $f(t + p) = f(t)$. Note that if $f(t)$ is periodic with period p, then

$$g(t) \equiv f\left(\frac{tp}{2\pi}\right)$$

is periodic with period 2π. In the remainder of this section, we will assume that periodic functions all have period 2π.

The next lemma describes some important properties of trigonometric polynomials.

Lemma 1.7.3 *Suppose that k is a nonzero integer and t is a real scalar. Then the* **trigonometric polynomial**

$$\phi_k(t) = \mathrm{e}^{\mathrm{i}kt}$$

is periodic with period 2π. Also, the set $\{\phi_k\}_{k=-\infty}^{\infty}$ of all trigonometric polynomials is orthonormal with respect to the continuous inner product

$$(f, g) = \frac{1}{2\pi} \int_0^{2\pi} f(t)\overline{g(t)} \, \mathrm{d}t \; . \tag{1.77}$$

Furthermore, the set $\{\phi_k\}_{k=0}^{m-1}$ is orthonormal with respect to the discrete inner product

$$(f, g) = \frac{1}{m} \sum_{j=0}^{m-1} f(t_j)\overline{g(t_j)} \text{ where } t_j = \frac{2\pi j}{m} \; . \tag{1.78}$$

Finally, the set $\{\phi_k\}_{k=-n}^{n}$ is orthonormal with respect to the discrete inner product

$$(f, g) = \frac{1}{2n+1} \sum_{j=-n}^{n} f(t_j)\overline{g(t_j)} \text{ where } t_j = \frac{2\pi j}{2n+1} \; . \tag{1.79}$$

Proof To prove periodicity, we note that **Euler's identity**

$$\mathrm{e}^{\mathrm{i}\theta} = \cos\theta + \mathrm{i}\sin\theta$$

implies that

$$f(t + 2\pi) = \mathrm{e}^{\mathrm{i}k(t+2\pi)} = \mathrm{e}^{\mathrm{i}kt}\mathrm{e}^{\mathrm{i}2k\pi} = f(t)[\cos(2k\pi) + \mathrm{i}\sin(2k\pi)] = f(t) \; .$$

It is also easy to prove orthonormality in the continuous inner product (1.77):

$$(\phi_k, \phi_\ell) = \frac{1}{2\pi} \int_0^{2\pi} \mathrm{e}^{\mathrm{i}(k-\ell)t} \, \mathrm{d}t = \begin{cases} \frac{2\pi}{2\pi}, & k = \ell \\ \frac{1}{2\pi\mathrm{i}(k-\ell)} \left[\mathrm{e}^{2\pi\mathrm{i}(k-\ell)} - 1\right], & k \neq \ell \end{cases} = \begin{cases} 1, k = \ell \\ 0, k \neq \ell \end{cases} \; .$$

Next, let us consider the first discrete inner product (1.78). If $0 \leq \ell < k < m$, let

$$z = \mathrm{e}^{2\mathrm{i}(k-\ell)\pi/m} \; .$$

Since $0 < k - \ell < m$, it follows that $z \neq 1$. Nevertheless, $z^m = 1$, and

$$m(\phi_k, \phi_\ell) = \sum_{j=0}^{m-1} \mathrm{e}^{2\mathrm{i}(k-\ell)j\pi/m} = \sum_{j=0}^{m-1} z^j = \frac{z^m - 1}{z - 1} = 0 \; .$$

This proves orthogonality. To prove orthonormality, we note that

$$(\phi_k, \phi_k) = \frac{1}{m} \sum_{j=0}^{m-1} 1 = 1 \ .$$

Finally, let us examine the second discrete inner product (1.79). If $-n \le \ell < k \le n$, define

$$z = e^{2i(k-\ell)\pi/(2n+1)} \ .$$

Since $0 \le k - \ell \le 2n$, it follows that $z \ne 1$. But we also note that $z^{2n+1} = 1$, and that

$$(2n+1)(\phi_k, \phi_\ell) = \sum_{j=-n}^{n} e^{2i(k-\ell)j\pi/(2n+1)} = \sum_{j=-n}^{n} z^j = \frac{z^{-n} - z^{n+1}}{1 - z}$$

$$= z^{-n} \frac{1 - z^{2n+1}}{1 - z} = 0 \ .$$

This proves orthogonality. To prove orthonormality, we note that

$$(\phi_k, \phi_k) = \frac{1}{2n+1} \sum_{j=-n}^{n} 1 = 1 \ .$$

In order to find the best approximation to f by a trigonometric polynomial, we should use the orthogonality conditions in Lemma 1.7.3. If we use the discrete inner product in Eq. (1.78), the best discrete least squares approximation to f is

$$f(t) \approx \sum_{k=0}^{m-1} c_k e^{ikt}$$

where

$$c_k = \frac{(f, \phi_k)}{(\phi_k, \phi_k)} = \frac{1}{2\pi m} \sum_{j=0}^{m-1} f(t_j) e^{-ikt_j} \ . \tag{1.80}$$

Note that c_k is a **trapezoidal rule** approximation on the mesh $t_0 < t_1 < \ldots < t_{m-1}$ to the integral

$$c(k) = \frac{1}{2\pi} \int_0^{2\pi} f(t) e^{-ikt} \, dt = \frac{1}{2\pi} \int_{-\pi}^{\pi} f(t) e^{-ikt} \, dt \ . \tag{1.81}$$

This integral $c(k)$ is the coefficient of ϕ_k in the least squares approximation of f with respect to the continuous inner product (1.77). The function $c(k)$ is called the **Fourier transform** of the 2π-periodic function f.

1.7.4.2   Theory

Since the trigonometric polynomials e^{ikt} are orthonormal, Theorem 1.7.3 guarantees that any square-integrable function f has a best least-squares trigonometric polynomial approximation, and Corollary 1.7.2 shows that the optimal coefficient for $\phi_k(t) = e^{ikt}$ is $c_k = (f, \phi_k)/(\phi_k, \phi_k)$. In this section, we will present a number of theoretical results regarding least-squares trigonometric polynomial approximations.

Our first theorem provides minimal conditions under which a trigonometric series will converge to f, except at possible discontinuities. For a proof of the following theorem, see either Gottlieb and Orszag [89, p. 22ff] or Zygmund [190, p. 57ff].

Theorem 1.7.7 (Dirichlet-Jordan Test)   *Suppose that f is 2π-periodic and has* **bounded variation**, *meaning that*

$$\text{for all } N > 0 \quad \sup_{0=t_0<t_1<...<t_N=2\pi} \sum_{i=1}^{N} |f(t_i) - f(t_{i-1})| < \infty .$$

Then, with respect to the continuous inner product (1.77),

1. *for all $t \in [0, 2\pi]$, the least squares trigonometric polynomial approximation to f at t converges to $\frac{1}{2}[f(t+0) + f(t-0)]$*
2. *if f is continuous at every point of a closed interval $I \subset [0, 2\pi]$, then the Fourier series converges uniformly in I.*

As a result, if f is both continuous and periodic, then the least squares trigonometric polynomial approximation to f **converges uniformly** *to f, meaning that for every $\varepsilon > 0$ there exists $M > 0$ so that for all $N \geq M$*

$$\max_{0 \leq t \leq 2\pi} \left| f(t) - \sum_{i=0}^{N} c_k e^{2\pi ikt} \right| < \varepsilon .$$

The next theorem shows that the convergence of trigonometric polynomials can involve significant oscillations near jump discontinuities in f. A derivation of the following famous result can be found in Gottlieb and Orszag [89, p. 26].

Theorem 1.7.8 (Gibbs Phenomenon)   *Suppose that f is periodic with bounded variation, d is a discontinuity of f,*

$$p_n(t) = \sum_{k=-n}^{n} c_k e^{ikt}$$

is the least squares trigonometric polynomial approximation to f with respect to the continuous inner product (1.77), and

$$Si(z) \equiv \int_0^z \frac{\sin s}{s} \, ds \ .$$

Then

$$\lim_{n\to\infty} p_n\left(d + \frac{\pi}{n+1/2}\right) - f(d+) = \left\{\frac{Si(\pi)}{\pi} - \frac{1}{2}\right\} [f(d+) - f(d-)] \ and$$

$$\lim_{n\to\infty} p_n\left(d - \frac{\pi}{n+1/2}\right) - f(d-) = -\left\{\frac{Si(\pi)}{\pi} - \frac{1}{2}\right\} [f(d+) - f(d-)] \ .$$

Since $Si(\pi)/\pi \approx 0.58949$, this means that the least squares trigonometric approximation overshoots by about $Si(\pi)/\pi - 1/2 \approx 9\%$ on either side of a discontinuity in f, and this overshoot does not decrease as n becomes infinitely large.

If f has several continuous derivatives, then the next theorem shows that its best trigonometric polynomial approximation converges to f very rapidly. This result is proved in Gottlieb and Orszag [89, p. 26].

Lemma 1.7.4   *Suppose that f is 2π periodic and $m - 1$ times continuously differentiable in $[0, 2\pi]$. Let*

$$p_n(t) = \sum_{k=-n}^{n} c_k e^{ikt}$$

be the least squares trigonometric polynomial approximation to f with respect to the continuous inner product (1.77). Then

$$|c_k| \le k^{-m} \frac{1}{2\pi} \int_0^{2\pi} \left|f^{(m)}(t)\right| \, dt \ .$$

If $f^{(m)}$ is piecewise continuous, then $p_n(t) - f(t) = O(n^{-m})$ away from discontinuities of $f^{(m)}$, and $p_n(t) - f(t) = O(n^{1-m})$ near discontinuities of $f^{(m)}$.

This result implies that if f is infinitely-many times continuously differentiable, then for all t the least squares trigonometric polynomial approximation p_n converges to f more rapidly than any finite power of $1/n$ for all t.

We will present one more theoretical result, which is not directly related to convergence of trigonometric polynomials.

Theorem 1.7.9 (Discrete Fourier Inversion) *Let m be a positive integer, and let $\{f_j\}_{j=0}^{m-1}$ be a set of m scalars. For $0 \leq j < m$, define*

$$c_k = \frac{1}{m} \sum_{j=0}^{m-1} f_j e^{-2\pi i k j / m} .$$

Then for all $0 \leq j < m$ we have

$$f_j = \sum_{k=0}^{m-1} c_k e^{2\pi i k j / m} . \tag{1.82}$$

Proof The right-hand side of (1.82) is

$$\sum_{k=0}^{m-1} c_k e^{2\pi i k j / m} = \frac{1}{m} \sum_{k=0}^{m-1} \left[\sum_{\ell=0}^{m-1} f_\ell e^{-2\pi i k \ell / m} \right] e^{2\pi i k j / m} = \frac{1}{m} \sum_{\ell=0}^{m-1} f_\ell \left[\sum_{k=0}^{m-1} e^{2\pi i k (\ell - j) / m} \right]$$

then we evaluate the inner sum, which is a geometric series

$$= \frac{1}{m} \sum_{\ell=0}^{m-1} f_\ell \left\{ \begin{array}{l} m, \ \ell = j \\ \frac{e^{2\pi i (\ell - j)} - 1}{e^{2\pi i (\ell - j)/m}}, \ \ell \neq j \end{array} \right. = f_j .$$

1.7.4.3 Fast Fourier Transform

So far, we have shown that least squares trigonometric polynomial approximations exist, are unique, and converge to some given function under reasonable circumstances. In this section, we would like to develop an important algorithm for computing the optimal coefficients, and for evaluating the least squares trigonometric polynomial at certain points. Our basic computational tool, due originally to Cooley and Tukey [50], is very powerful, and very fundamental. It has lead to seemingly unrelated fast techniques for multiplying matrices [165], multiplying integers [158], and multipole expansions [151].

Given a trigonometric polynomial

$$p_m(t) = \sum_{k=0}^{m-1} c_k e^{ikt} ,$$

suppose that we would like to compute the array of values of p_m at the equidistant mesh points

$$t_j = \frac{2\pi j}{m}$$

for $0 \leq j < m$. We will define

$$z \equiv e^{2\pi i/m} \tag{1.83}$$

and write the interpolation conditions in the form

$$
\begin{bmatrix} f_0 \\ f_1 \\ \vdots \\ f_{m-1} \end{bmatrix}
=
\begin{bmatrix} p_m(t_0) \\ p_m(t_1) \\ \vdots \\ p_m(t_{m-1}) \end{bmatrix}
=
\begin{bmatrix}
z^0 & z^0 & \cdots & z^0 \\
z^0 & z^1 & \cdots & z^{(m-1)} \\
\vdots & \vdots & \ddots & \vdots \\
z^0 & z^{(m-1)} & \cdots & z^{(m-1)^2}
\end{bmatrix}
\begin{bmatrix} c_0 \\ c_1 \\ \vdots \\ c_{m-1} \end{bmatrix}.
$$

Thus each f_j is a polynomial in z^j, and each power of z is an m-th **root of unity**, meaning that $(z^j)^m = 1$.

If we use **Horner's rule** to evaluate each of these polynomials, we will expend $O(m)$ operations for each $p_m(t_j)$, or $O(m^2)$ operations to find all of $f_0, \ldots, f_{m-1}$. Instead, we will find a way to evaluate the array of trigonometric polynomial values on the equidistant mesh in an order of $m \log_2 m$ operations.

Because of the Fourier inversion formula (1.82), it is useful to consider the related problem of evaluating the coefficients c_k from the trigonometric polynomial values f_j. In other words, given $f_0, \ldots, f_{m-1}$, we might want to compute

$$c_k = \frac{1}{m} \sum_{j=0}^{m-1} f_j z^{-jk}.$$

Because z, which was defined in (1.83), is an m'th root of unity, we see that $z^{mk} = 1$ for all $0 \leq k < m$. This implies that $z^{-jk} = z^{(m-j)k}$ for all $0 \leq j < m$ as well. In other words,

$$
\begin{bmatrix} c_0 \\ c_1 \\ \vdots \\ c_{m-1} \end{bmatrix}
=
\begin{bmatrix}
z^0 & z^0 & \cdots & z^0 \\
z^0 & z^{-1} & \cdots & z^{-(m-1)} \\
\vdots & \vdots & \ddots & \vdots \\
z^0 & z^{-(m-1)} & \cdots & z^{-(m-1)^2}
\end{bmatrix}
\begin{bmatrix} f_0 \\ f_1 \\ \vdots \\ f_{m-1} \end{bmatrix} \frac{1}{m}
$$

$$
=
\begin{bmatrix}
1 & 0 & \cdots & 0 \\
0 & 0 & \cdots & 1 \\
\vdots & \vdots & & \vdots \\
0 & 1 & \cdots & 0
\end{bmatrix}
\begin{bmatrix}
z^0 & z^0 & \cdots & z^0 \\
z^0 & z^1 & \cdots & z^{m-1} \\
\vdots & \vdots & \ddots & \vdots \\
z^0 & z^{m-1} & \cdots & z^{(m-1)^2}
\end{bmatrix}
\begin{bmatrix} f_0 \\ f_1 \\ \vdots \\ f_{m-1} \end{bmatrix} \frac{1}{m}.
$$

Thus the computation of the coefficients c_k from the trigonometric polynomial values f_j involves the same matrix of powers of an m-th root of unity, followed by a permutation and a scalar multiplication. As a result, we can use the same algorithm that evaluates the f_j from the c_k to evaluate the c_k from the f_j; afterward, we just reorder and scale the results.

Next, we will present the basic step in the **fast Fourier transform**, or **FFT**. If m is even, then we can group the even- and odd-indexed terms together, and factor out z^j in the sum over odd indices:

$$f_j = \sum_{k=0}^{m-1} c_k z^{jk} = \sum_{\substack{k=0 \\ k \text{ even}}}^{m-2} c_k z^{jk} + \sum_{\substack{k=1 \\ k \text{ odd}}}^{m-1} c_k z^{jk}$$

$$= \sum_{\ell=0}^{m/2-1} c_{2\ell}(z^2)^{j\ell} + z^j \sum_{\ell=0}^{m/2-1} c_{2\ell+1}(z^2)^{j\ell} \ .$$

Note that since z is an m'th root of unity, z^2 is an $(m/2)$'th root of unity. Thus if $m/2$ is also even, this same basic step could be repeated. If $m = 2^n$ is a power of two, then the basic step can be repeated $n - 1 = \log_2 m - 1$ times.

When we compute all of the f_j for $0 \leq j < m$, it is important to note that some of the even or odd sums occur for more than one value of j. Since

$$z^{2(j+m/2)} = z^{2j} \ ,$$

there are at most $m/2$ distinct sums of the form

$$\sum_{\ell=0}^{m/2-1} c_{2\ell} z^{2j\ell} \ , \ 0 \leq j < m/2$$

and at most $m/2$ distinct sums of the form

$$\sum_{\ell=0}^{m/2-1} c_{2\ell+1} z^{2j\ell} \ , \ 0 \leq j < m/2 \ .$$

Before we develop the algorithm in more detail, let us estimate the work. Suppose that $M(\ell)$ multiplications and $A(\ell)$ additions are needed to compute all of the desired sums for $0 \leq j \leq 2^\ell - 1$. If we apply the scheme recursively, then the computations at the n'th stage require at most

$$2M(n-1) \text{ multiplications}$$
$$2A(n-1) \text{ additions}$$

to compute the sums, and an additional

$$2^n \text{ multiplications}$$

to multiply the odd-index sums by z^j. This gives us the recursive inequalities

$$M(0) = 0, \; M(n) \le 2M(n-1) + 2^n \implies M(n) \le 2^n(n-1) = m(\log_2 m - 1)$$
$$A(0) = 0, \; A(n) \le 2A(n-1) + 2^n \implies A(n) \le 2^n(n-1) = m(\log_2 m - 1)$$

These inequalities are easily proved by induction.

Let us begin our development of the algorithmic details by considering the special case when $m = 4 = 2^2$. Then $z = e^{2\pi i/4} = i$, and $z^4 = 1$. We want to compute

$$\begin{bmatrix} f_0 \\ f_1 \\ f_2 \\ f_3 \end{bmatrix} = \begin{bmatrix} \sum_{k=0}^{3} z^0 c_k \\ \sum_{k=0}^{3} z^k c_k \\ \sum_{k=0}^{3} z^{2k} c_k \\ \sum_{k=0}^{3} z^{3k} c_k \end{bmatrix} = \begin{bmatrix} z^0 c_0 + z^0 c_2 + z^0 c_1 + z^0 c_3 \\ z^0 c_0 + z^2 c_2 + z^1 c_1 + z^3 c_3 \\ z^0 c_0 + z^4 c_2 + z^2 c_1 + z^6 c_3 \\ z^0 c_0 + z^6 c_2 + z^3 c_1 + z^9 c_3 \end{bmatrix}$$

$$= \begin{bmatrix} \sum_{\ell=0}^{1} z^0 c_{2\ell} + z^0 \sum_{\ell=0}^{1} z^0 c_{2\ell+1} \\ \sum_{\ell=0}^{1} z^{2\ell} c_{2\ell} + z^1 \sum_{\ell=0}^{1} z^{2\ell} c_{2\ell+1} \\ \sum_{\ell=0}^{1} z^0 c_{2\ell} + z^2 \sum_{\ell=0}^{1} z^0 c_{2\ell+1} \\ \sum_{\ell=0}^{1} z^{2\ell} c_{2\ell} + z^3 \sum_{\ell=0}^{1} z^{2\ell} c_{2\ell+1} \end{bmatrix} = \begin{bmatrix} d_0 + d_2 \\ d_1 + d_3 \\ d_0 - d_2 \\ d_1 - d_3 \end{bmatrix}.$$

We can reorganize the needed sums into the calculation

$$\begin{bmatrix} d_0 \\ d_1 \\ d_2 \\ d_3 \end{bmatrix} \equiv \begin{bmatrix} \sum_{\ell=0}^{1} z^0 c_{2\ell} \\ \sum_{\ell=0}^{1} (z^2)^\ell c_{2\ell} \\ \sum_{\ell=0}^{1} z^0 c_{2\ell+1} \\ z \sum_{\ell=0}^{1} (z^2)^\ell c_{2\ell+1} \end{bmatrix} = \begin{bmatrix} z^0 c_0 + z^0 c_2 \\ z^0 c_0 + z^2 c_2 \\ z^0 c_1 + z^0 c_3 \\ z(z^0 c_1 + z^2 c_3) \end{bmatrix} = \begin{bmatrix} c_0 + c_2 \\ c_0 - c_2 \\ c_1 + c_3 \\ z(c_1 - c_3) \end{bmatrix}.$$

To summarize, first we compute

$$\begin{bmatrix} d_0 \\ d_1 \\ d_2 \\ d_3 \end{bmatrix} = \begin{bmatrix} c_0 + c_2 \\ c_0 - c_2 \\ c_1 + c_3 \\ z(c_1 - c_3) \end{bmatrix},$$

and then we compute

$$\begin{bmatrix} f_0 \\ f_1 \\ f_2 \\ f_3 \end{bmatrix} = \begin{bmatrix} d_0 + d_2 \\ d_1 + d_3 \\ d_0 - d_2 \\ d_1 - d_3 \end{bmatrix}.$$

These computations involve only one multiplication, and eight additions or subtractions.

Now we will interpret these same FFT computations in terms of matrices. We want to compute

$$
\begin{bmatrix} f_0 \\ f_1 \\ f_2 \\ f_3 \end{bmatrix} = \begin{bmatrix} z^0 & z^0 & z^0 & z^0 \\ z^0 & z^1 & z^2 & z^3 \\ z^0 & z^2 & z^4 & z^6 \\ z^0 & z^3 & z^6 & z^9 \end{bmatrix} \begin{bmatrix} c_0 \\ c_1 \\ c_2 \\ c_3 \end{bmatrix} \equiv \mathbf{Zc} .
$$

When we break the sums into even and odd terms, we rewrite the system as two matrix-vector products

$$
\begin{bmatrix} f_0 \\ f_1 \\ f_2 \\ f_3 \end{bmatrix} = \begin{bmatrix} 1 & 0 & z^0 & 0 \\ 0 & 1 & 0 & z^0 \\ 1 & 0 & z^2 & 0 \\ 0 & 1 & 0 & z^2 \end{bmatrix} \begin{bmatrix} d_0 \\ d_1 \\ d_2 \\ d_3 \end{bmatrix}
$$

and

$$
\begin{bmatrix} d_0 \\ d_1 \\ d_2 \\ d_3 \end{bmatrix} = \begin{bmatrix} 1 & 0 & z^0 & 0 \\ 1 & 0 & z^2 & 0 \\ 0 & z^0 & 0 & z^0 \\ 0 & z^1 & 0 & z^3 \end{bmatrix} \begin{bmatrix} c_0 \\ c_1 \\ c_2 \\ c_3 \end{bmatrix} .
$$

Next, we note that

$$
\begin{bmatrix} 1 & 0 & z^0 & 0 \\ 0 & 1 & 0 & z^0 \\ 1 & 0 & z^2 & 0 \\ 0 & 1 & 0 & z^2 \end{bmatrix} = \begin{bmatrix} 1 & 0 & 0 & 0 \\ 0 & 0 & 1 & 0 \\ 0 & 1 & 0 & 0 \\ 0 & 0 & 0 & 1 \end{bmatrix} \begin{bmatrix} 1 & z^0 & & \\ 1 & z^2 & & \\ & & 1 & z^0 \\ & & 1 & z^2 \end{bmatrix} \begin{bmatrix} 1 & 0 & 0 & 0 \\ 0 & 0 & 1 & 0 \\ 0 & 1 & 0 & 0 \\ 0 & 0 & 0 & 1 \end{bmatrix} \equiv \mathbf{QSP} .
$$

Here $\mathbf{P}$ is the permutation that re-orders the even indices first, followed by odd indices. Also note that

$$
\begin{bmatrix} 1 & 0 & z^0 & 0 \\ 1 & 0 & z^2 & 0 \\ 0 & z^0 & 0 & z^0 \\ 0 & z^1 & 0 & z^3 \end{bmatrix} = \begin{bmatrix} 1 & & & \\ & 1 & & \\ & & 1 & \\ & & & z \end{bmatrix} \begin{bmatrix} 1 & z^0 & & \\ 1 & z^2 & & \\ & & 1 & z^0 \\ & & 1 & z^2 \end{bmatrix} \begin{bmatrix} 1 & 0 & 0 & 0 \\ 0 & 0 & 1 & 0 \\ 0 & 1 & 0 & 0 \\ 0 & 0 & 0 & 1 \end{bmatrix} \equiv \mathbf{D_1 SP} .
$$

In other words, we have factored

$$
\mathbf{Z} = (\mathbf{QSP})(\mathbf{D_1 SP}) .
$$

We can use this matrix factorization to provide another useful view of the calculations in tabular form:

$$\mathbf{PcP}^\top\mathbf{d} = \mathbf{P}^\top\mathbf{D}_1\mathbf{SPc} \qquad\qquad \mathbf{Q}^\top\mathbf{f} = \mathbf{SPd}$$

$$c_0 d_0 = c_0 + c_2 \qquad\qquad f_0 = d_0 + d_2$$

$$c_2 d_2 = c_0 - c_2 \qquad\qquad f_2 = d_0 - d_2$$

$$c_{11} = c_1 + c_3 \qquad\qquad f_1 = d_1 + d_3$$

$$c_3 d_3 = z(c_1 - c_3) \qquad\qquad f_3 = d_1 - d_3$$

For general $m = 2^n$, Stoer [162, p. 66f] proves that we can factor

$$\mathbf{Z} = (\mathbf{QSP})(\mathbf{D}_{n-1}\mathbf{SP})\dots(\mathbf{D}_1\mathbf{SP}) ,$$

where $\mathbf{S}$ is the block diagonal matrix

$$\mathbf{S} = \begin{bmatrix} 1 & 1 & & & & & \\ 1 & -1 & & & & & \\ & & 1 & 1 & & & \\ & & 1 & -1 & & & \\ & & & & \ddots & & \\ & & & & & 1 & 1 \\ & & & & & 1 & -1 \end{bmatrix} ,$$

and $\mathbf{D}_j$ is the diagonal matrix

$$\mathbf{D}_j = \begin{bmatrix} \delta_0^{(j)} & & & \\ & \delta_1^{(j)} & & \\ & & \ddots & \\ & & & \delta_{m-1}^{(j)} \end{bmatrix} .$$

The scalars $\delta_\ell^{(j)}$ are defined in terms of the binary representation of the integer ℓ. If ℓ has the binary representation

$$\ell = \sum_{k=0}^{j-1} b_k 2^k + 2^j \ell_j^*$$

where $b_k = 0$ or $1, 0 \le k < j$ and $0 \le \ell_j^* < 2^{n-1}$, then

$$\delta_\ell^{(j)} = \exp\left(2\pi i \frac{b_0 \ell_j^*}{2^{n-j+1}}\right) = z^{b_0 \ell_j^* 2^{l-1}} .$$

The permutation matrix $\mathbf{P}$ is defined by

$$(\mathbf{P}\mathbf{x})_{k+2j} = x_{j+k2^{n-1}} \,,$$

or

$$\mathbf{P} \begin{bmatrix} x_0 \\ x_1 \\ x_2 \\ x_3 \\ \vdots \\ x_{m-1} \end{bmatrix} = \begin{bmatrix} x_0 \\ x_{m/2} \\ x_1 \\ x_{m/2+1} \\ \vdots \\ x_{m/2-1} \\ x_{m-1} \end{bmatrix} .$$

The permutation matrix $\mathbf{Q}$ is defined by

$$\mathbf{Q} \begin{bmatrix} x_0 \\ x_1 \\ x_2 \\ x_3 \\ x_4 \\ x_5 \\ x_6 \\ x_7 \\ \vdots \end{bmatrix} = \begin{bmatrix} x_0 \\ x_{m/2} \\ x_{m/4} \\ x_{3m/4} \\ x_{m/8} \\ x_{5m/8} \\ x_{3m/8} \\ x_{7m/8} \\ \vdots \end{bmatrix} .$$

Thus the case $m = 8 = 2^3$ leads to the following table of computations:

$\mathbf{Pcd} = \mathbf{D}_1\mathbf{SPc}$	$\mathbf{e} = \mathbf{D}_2\mathbf{SPdQ}^\top\mathbf{f} = \mathbf{SPe}$
$c_0d_0 = c_0 + c_4$	$e_0 = d_0 + d_4f_0 = e_0 + e_4$
$c_4d_1 = c_0 - c_4$	$e_1 = d_0 - d_4f_4 = e_0 - e_4$
$c_1d_2 = c_1 + c_5$	$e_2 = d_1 + d_5f_2 = e_1 + e_5$
$c_5d_3 = z(c_1 - c_5)$	$e_3 = d_1 - d_5f_6 = e_1 - e_5$
$c_2d_4 = c_2 + c_6$	$e_4 = d_2 + d_6f_1 = e_2 + e_6$
$c_6d_5 = z^2(c_2 - c_6)$	$e_5 = z^2(d_2 - d_6)f_5 = e_2 - e_6$
$c_3d_6 = c_3 + c_7$	$e_6 = d_3 + d_7f_3 = e_3 + e_7$
$c_7d_7 = z^3(c_3 - c_7)$	$e_7 = z^2(d_3 - d_7)f_7 = e_3 - e_7$

There are several programs available to the public for computing fast Fourier transforms. To begin, readers should examine the very short Fortran program Cooley, Lewis and Welch FFT, which appears in Dahlquist and Björck [57, p. 416]. The more robust C program FFTW won the 1999 Wilkinson prize for numerical software. The Fortran program FFTPACK is also available from netlib. There are also fast Fourier transform algorithms within the GSL. MATLAB users should become familiar with the command fft.

Readers can view a program for performing fast trigonometric polynomial interpolation in the C^{++} program GUIFourier.C. This program demonstrates how to call either FFTW or FFTPACK to compute the fast Fourier transform.

Readers may also execute a JavaScript program to show the **Gibbs phenomenon.** This program plots the trigonometric polynomial approximation to

$$f(t) = \begin{cases} t, \ t < \pi \\ 0, \ t = \pi \\ t - 2\pi, \ t > \pi \end{cases} .$$

Finally, readers may execute the JavaScript program for **smooth function FFT error.** This program plots the logarithm of the error in the trigonometric polynomial approximation to

$$f(t) = e^{1/\pi^2 - 1/[t(2\pi - t)]} .$$

This function is periodic with infinitely many continuous derivatives, and the computed results show that the error does not appear to be converging to zero proportional to any power of the number of mesh points.

1.7.4.4   Real FFT

Although it is convenient to write trigonometric polynomials in the form $\sum_{k=0}^{m-1} c_k e^{ikt}$, this gives a misleading view of the range of frequencies available in the approximation. Assuming that m is even, we note that if t is a lattice point (i.e. $e^{itm} = 1$), then

$$\sum_{k=0}^{m-1} c_k e^{ikt} = \sum_{k=0}^{m/2} c_k e^{ikt} + \sum_{k=m/2+1}^{m-1} c_k e^{ikt} = \sum_{k=0}^{m/2} c_k e^{ikt} + \sum_{k=m/2+1}^{m-1} c_k e^{i(k-m)t}$$

$$= \sum_{k=0}^{m/2} c_k e^{ikt} + \sum_{\ell=m/2+1-m}^{-1} c_{\ell+m} e^{i\ell t} = \sum_{k=1-m/2}^{m/2} d_j e^{ijt}$$

where

$$d_j = \begin{cases} c_j & , 0 \le j \le m/2 \\ c_{j+m} & , -m/2 < j < 0 \end{cases} .$$

Next, we note that since $e^{imt/2} = \pm 1$ is real,

$$\sum_{k=1-m/2}^{m/2} d_j e^{ijt} = d_0 + d_{m/2} e^{imt/2} + \sum_{j=1}^{m/2-1} [d_j e^{ijt} + d_{-j} e^{-ijt}]$$

$$= \Re(d_0) + \Re(d_{m/2}) e^{imt/2} + \sum_{j=1}^{m/2-1}$$

$$\{[\Re(d_j) + \Re(d_{-j})] \cos(jt) + [\Im(d_{-j}) - \Im(d_j)] \sin(jt)\}$$

$$+ i \left[\Im(d_0) + \Im(d_{m/2}) e^{imt/2} \right] + i \sum_{j=1}^{m/2-1}$$

$$\{[\Im(d_j) + \Im(d_{-j})] \cos(jt) + [\Re(d_j) - \Re(d_{-j})] \sin(jt)\}$$

When we are working with real functions, it is possible to avoid complex arithmetic by working with Fourier sine and cosine series of the form $a_0 + \sum_{j=1}^{m/2-1} [a_j \cos(jt) + b_j \sin(jt)] + a_{m/2} \cos(mt/2)$. It should also be clear that the use of trigonometric polynomials through order $m - 1$ leads to the representation of frequencies from 0 to $m/2$ only.

The FFTW package has the ability to compute sine and cosine series approximations, as does the GSL.

1.7.4.5   Fast Evaluation

Suppose that we are given a trigonometric polynomial $\sum_{k=0}^{m-1} c_k e^{ikt}$, and we want to evaluate it on a lattice of points $2\pi J/M$ where $M > m$ and both m and M are even. We recall that if $e^{imt} = 1$, then

$$\sum_{k=0}^{m-1} c_k e^{ikt} = \sum_{j=-m/2+1}^{m/2} d_j e^{ijt}$$

where

$$d_j = \begin{cases} c_j & , 0 \le j \le m/2 \\ c_{j+m} & , -m/2 < j < 0 \end{cases}$$

We can write the same trigonometric polynomial in the form $\sum_{J=-M/2+1}^{M/2} D_J e^{iJt}$ provided that we take

$$D_J = \begin{cases} d_J, & -m/2 < J \le m/2 \\ 0 & , \text{ otherwise} \end{cases}$$

Then we can rewrite

$$\sum_{J=-M/2+1}^{M/2} D_J e^{iJt} = \sum_{K=0}^{M-1} C_K e^{iKt}$$

where

$$C_K = \begin{cases} D_K & ,0 \le K \le M/2 \\ D_{K-M} & ,M/2 < K < M \end{cases}$$

$$= \begin{cases} d_K & ,0 \le K \le m/2 \\ d_{K-M} & ,M-m/2 < K < M \\ 0 & , \text{ otherwise} \end{cases}$$

$$= \begin{cases} c_K & ,0 \le K \le m/2 \\ c_{K-M+m} & ,M-m/2 < K < M \\ 0 & , \text{ otherwise} \end{cases}$$

In other words, given the coefficients c_k for $0 \le k < m$ of the coarse trigonometric polynomial, we can compute the coefficients C_K for $0 \le K < M$ of the fine trigonometric polynomial, and then use the fast Fourier transform to compute the values of the original coarse trigonometric polynomial on the fine mesh.

Readers can view a program for fast evaluation of a trigonometric polynomial on a mesh finer than the corresponding interpolation in the C++ program fourier.C. This program also demonstrates how to call either FFTW or FFTPACK to compute the fast Fourier transform.

Exercise 1.7.18 Write down the steps in the calculation of the fast Fourier transformation for $m = 2^4$.

Exercise 1.7.19 If you are a C or C++ programmer, read the documentation for fftw at FFTW Home Page. If you are a Fortran programmer, read the user guide for FFTPack at FFTPack web page. Use this information to program a low-pass filter, described as follows. Suppose that we have a trigonometric polynomial $p_m(x) = \sum_{k=-m}^{m-1} e^{ikx}$ where m is even (for simplicity, assume that m is a power of 2). We want to drop the highest frequencies from this function to form the lower-order trigonometric polynomial $p_{m/2}(x) = \sum_{k=-m/2}^{m/2-1} e^{ikx}$. Write a program to do this using either fftw or FFTPACK. Test your low-pass filter on the trigonometric interpolation to the periodic step function $f(x) = \text{sign}(x)$.

Exercise 1.7.20 Get a copy of an FFT routine, and use it to compute the trigonometric polynomial interpolant to the following functions:

1. $f_1(x) = \exp(-2(x - \pi)^2), 0 < x < 2\pi$,
2.

$$
f_2(x) = \begin{cases} 0,\ 0 < x < \pi/3 \text{ or } 5\pi/3 < x < 2\pi\ , \\ 1,\ \pi/3 < x < \pi \text{ and} \\ -1,\ \pi < x < 5\pi/3 \end{cases}.
$$

Plot each function and its trigonometric polynomial interpolant for $n = 4, 16, 64$ and 256 equidistant interpolation points. Also plot the log of the max error in the interpolation versus the log of the number of interpolation points.

Exercise 1.7.21 Compare execution time for FFTW for trigonometric polynomial interpolation to the smooth gaussian in the previous problem for the following sequences of numbers of interpolation points:

1. n=31,211,2311,30031,510511,9699691,223092871,6469693231
2. n=32,256,4096,32768,524288,16777216,268435456,8589934592

The former sequence involves only prime numbers, and the second sequence involves powers of 2 just greater than the corresponding prime in the former sequence.

1.8   Wavelets

So far, we have studied interpolation and approximation by polynomials, rational polynomials and trigonometric polynomials. Each of these approximating functions localize the approximation in either polynomial order or frequency, but not in terms of the function argument. We have also studied approximation by splines, which involve localization in both polynomial order and in the function argument. However, the connection between splines of the same family on different spatial scales is not always simple, especially when compared to the connection between frequencies in the fast Fourier transform.

In this section, we would like to develop a different kind of approximation, involving localization in the function argument, but designed to make the connection between scales or locales especially simple. The resulting constructions result in fast evaluation methods for approximations on some finite range of scales, and ways to filter out information on other scales.

Wavelet discussions are sometimes simple, such as the description of the Haar wavelet applied to a finite discrete dataset in Jensen and la Cour-Harbo [115]. Other wavelet discussions can be tedious, as in Daubechies [58] or Radunovic [147]. We have found some useful introductory presentations of filter banks in Strang and Nguyen [164]. Other useful discussions of biorthogonal wavelets can be found in Chui [42]. Cohen [48] has some helpful discussions of error estimates, and applications such as the numerical solution of differential equations. Our goal in this chapter is to introduce the reader to some fundamental concepts involving wavelets, and to relate those concepts to our previous work in approximation and interpolation.

Here is an overview of our presentation of wavelets. First, we will examine discrete *signals*, which are just sequences of numbers. We will discuss some important operators on signals, such as shifts, reversals, and convolutions. We will see that Fourier transforms map convolutions into products, so they can be used to simplify some important computations. We will also see that the Shannon Sampling Theorem allows us to reconstruct a band-limited continuous signal from values of the original signal at an infinite sequence of discrete points. This observation implies that we can apply filters to signals in order to limit the band width of the output signal.

In simple cases, we will combine a lowpass filter with a highpass filter to form a filter bank. The filter bank can be used to analyze a signal, to examine its lowpass and highpass content. Under certain conditions on the two filters, we can synthesize the output from these two filters to reconstruct the original signal perfectly. Such a process is analogous to the action of a sound system amplifier, with separate output channels for low frequency sound components (sent to the bass speakers, sometimes called "woofers") and high frequency sound components (sent to the treble speakers, sometimes called "tweeters"). In other cases of signal transmission (such as television signals), the analysis step may indicate that there is little information in the highpass signal output, and this information will not need to be transmitted.

The conditions for perfect signal reconstruction provide guidance for the design of filters in a filter bank. We will examine two important filter bank constructions. The first will show how to develop the Daubechies orthogonal filters, and the second construction will develop spline filters. Both examples of filter banks will use filters with finite responses to a unit impulse.

Following our discussion of discrete signals, we extend the ideas to construct approximations to functions on a continuum. First, we will use the lowpass filters to construct scaling functions. The values of these scaling functions are often known only at the integers, and are evaluated at binary numbers by means of a refinement equation. Wavelets are companion functions that are constructed from a corresponding highpass filter. Wavelets may also be known only at integers, and evaluated at binary numbers by another appropriate refinement equation.

Both scaling functions and wavelets have associated dual functions. The dual functions may be used to project a given function onto the space of linear combinations of scaling functions on some scale; this is the function approximation we seek. The form of the approximation allows us to determine the corresponding function approximations on nearby scales, using the same computations that were developed for analysis and synthesis of discrete signals.

Under certain circumstances, scaling functions and wavelets can be used to construct useful approximations to arbitrary square-integrable functions. Furthermore, if the Fourier transform of the scaling function vanishes to order $k - 1$ at nonzero integer multiples of π, then linear combinations of the scaling function can be used to reproduce exactly all polynomials of degree at most $k - 1$. It will follow that such scaling functions will approximate sufficiently smooth functions with error that is proportional to the kth power of the scale.

1.8.1   Discrete Time Signals

We will begin our discussion of wavelets by developing some basic ideas regarding discrete data. First, let us present a simple example that illustrates the general ideas.

1.8.1.1   Haar Wavelet

Suppose that we are given a $(2n)$-vector $\mathbf{f}$. We can use this vector to define two n-vectors $\mathbf{s}$ and $\mathbf{d}$ by

Algorithm 1.8.1 (Haar Cascade)

$$\text{for } 0 \leq i < n$$
$$\mathbf{s}_i = (\mathbf{f}_{2i} + \mathbf{f}_{2i+1}) * 0.5$$
$$\mathbf{d}_i = \mathbf{f}_{2i} - \mathbf{s}_i .$$

This is an example of one step of what is often called the **cascade algorithm**. A cascade algorithm takes data on a fine scale and computes data on a coarser scale. The total cost of the Haar cascade Algorithm 1.8.1 is $2n$ additions or subtractions, and n multiplications. (In this case, the multiplications could be performed simply as exponent shifts in the floating point number representations.) Note that the data on the finer scale can subsequently be recovered by the

Algorithm 1.8.2 (Haar Pyramid)

$$\text{for } 0 \leq i < n$$
$$\mathbf{f}_{2i} = \mathbf{s}_i + \mathbf{d}_i$$
$$\mathbf{f}_{2i+1} = \mathbf{s}_i - \mathbf{d}_i .$$

This is an example of a one step of a **pyramid algorithm**. A pyramid algorithm takes data on a coarse scale and computes data on a finer scale. The total cost of the Haar pyramid Algorithm 1.8.2 is $2n$ additions or subtractions. Since this pyramid algorithm can recover the input to the cascade algorithm from its output, no information on the fine scale was lost in the cascade algorithm. Consequently, the vectors $\mathbf{s}$ and $\mathbf{d}$ can be stored in the same computer memory as $\mathbf{f}$.

If n is even, the cascade algorithm can be repeated on the n-vector $\mathbf{s}$. If $n = 2^m$ is a power of two, the cascade algorithm can be repeated m times, leading to a full cascade algorithm that might look like

Algorithm 1.8.3 (Full Haar Cascade)

$$n = 2^m$$
$$p = 1$$
$$\text{for } 1 \leq j < m$$
$$\quad n = n/2$$
$$\quad t = 2 * p$$
$$\quad k = 0$$
$$\quad \text{for } 0 \leq i < n$$
$$\quad\quad s = (f[k] + f[k + p]) * 0.5$$
$$\quad\quad f[k + p] = f[k] - s$$
$$\quad\quad f[k] = s$$
$$\quad\quad k = k + t$$
$$\quad p = t$$

The corresponding full pyramid algorithm might look like

Algorithm 1.8.4 (Full Haar Pyramid)

$$p = 2^m$$
$$n = 1$$
$$\text{for } m > j \geq 1$$
$$\quad k = 0$$
$$\quad h = p/2$$
$$\quad \text{for } 0 \leq i < n$$
$$\quad\quad \phi = f[k] + f[k + h]$$
$$\quad\quad f[k + h] = f[k] - f[k + h]$$
$$\quad\quad f[k] = \phi$$
$$\quad\quad k = k + p$$
$$\quad p = h$$
$$\quad n = n * 2$$

The total cost of the full cascade algorithm is

$$\sum_{j=1}^{m-1} 2^{j+1} = 2^{m+1} - 4 = 2n - 4 \text{ additions or subtractions and}$$

$$\sum_{j=1}^{m-1} 2^j = 2^m - 2 = n - 2 \text{ multiplications .}$$

Similarly, the total cost of the full pyramid algorithm is $n - 2$ additions or subtractions. Thus the total work in either of these algorithms is at most a small multiple of the total number of entries in the original data vector. In particular, the work estimate does not involve any $\log n$ terms, so the cost to operate the cascade and pyramid algorithms is of a lower order of work than that involved with the fast Fourier transform.

The sum vectors **s** contain information about averages in various locales on increasing larger scales, while the difference vectors **d** contain information about variations in locales on various scales. The separation of scales and locales can aid our understanding of the behavior of the data, in applications such as stock markets or seismic waves.

In Fig. 1.11, we show the results of the cascade and pyramid algorithms for the Haar wavelet applied to discrete values of the exponential function on the interval [0, 1]. The cascade algorithm corresponds to descending within this figure, and the pyramid algorithm corresponds to ascending. The entries for the signals **s** are labeled "lowpass", and the entries for the signals **d** are labeled "highpass." The justification for these labels will be provided in Sect. 1.8.1.3.

Readers may execute the JavaScript program for the **Haar wavelet** (see /home/ faculty/johnt/book/code/approximation/haar_wavelet.html). This program begins with pointwise data taken from the sine function at 512 equidistant points in $[0, 2\pi)$. The program executes the cascade algorithm to decrease the resolution, or the pyramid algorithm to increase the resolution. Note that the data is stored in a single array, no matter which scale of resolution is selected.

1.8.1.2 Signals

Almost all of us are familiar with home entertainment systems. Often, these use filters to separate the musical signal into low frequency components (for reproduction by the "woofer") and high frequency component (for reproduction by the "tweeter"). Modern music itself is also distributed as a digital signal. We will begin by providing a simple mathematical description of this ideal.

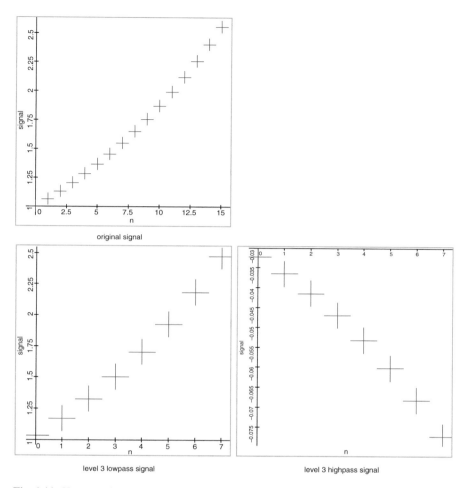

Fig. 1.11 Haar wavelet analysis (going down) and synthesis (going up)

Definition 1.8.1 A discrete time signal is a sequence $\mathbf{x} = \{\mathbf{x}_n\}_{n=-\infty}^{\infty}$ of scalars that is **absolutely summable**, meaning that

$$\sum_{n=-\infty}^{\infty} |\mathbf{x}_n| < \infty \, .$$

If δ_{ij} is the **Kronecker delta**, defined by

$$\delta_{ij} = \begin{cases} 1, \, i = j \, , \\ 0, \, i \neq j \end{cases} ,$$

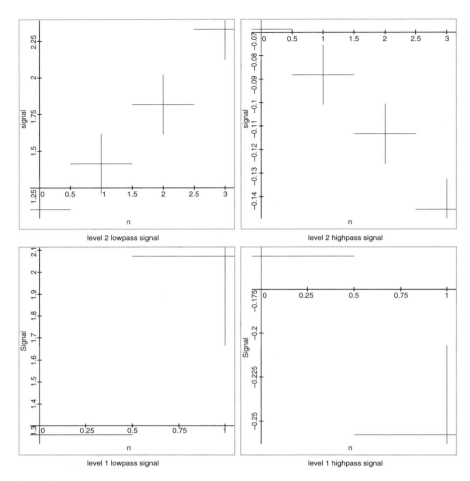

Fig. 1.11 (continued)

then the **unit impulse signal** δ is the discrete time signal $\delta = \{\delta_{0n}\}_{n=-\infty}^{\infty}$.

Next, we will define some operators on signals.

Definition 1.8.2 The **shift operator** (or translation operator) $\mathbf{T}$ is the linear operator on discrete time signals $\mathbf{x}$ that returns the discrete time signal with components

$$(\mathbf{T}\mathbf{x})_n = \mathbf{x}_{n-1} .\tag{1.84}$$

The **reversal operator $\mathbf{R}$** operates on a discrete time signal $\mathbf{x}$ as follows:

$$(\mathbf{R}\mathbf{x})_n = (-1)^n \mathbf{x}_n .\tag{1.85}$$

If σ is a positive integer, then the **downsampling operator** $\downarrow_\sigma$ performs the following operation on a discrete time signal $\mathbf{x}$:

$$[(\downarrow_\sigma)\mathbf{x}]_n = \mathbf{x}_{\sigma n} , \tag{1.86}$$

and the **upsampling operator** $\uparrow_\sigma$ operates on $\mathbf{x}$ to produce

$$[(\uparrow_\sigma)\mathbf{x}]_n = \begin{cases} \mathbf{x}_{n/\sigma}, & \text{if } n \text{ is divisible by } \sigma \\ 0, & \text{otherwise} \end{cases} . \tag{1.87}$$

The previous definitions lead to some simple but useful observations regarding the invertibility of these operations.

Lemma 1.8.1 *The shift operator and the reversal operators are invertible:*

$$(\mathbf{T}^{-1}\mathbf{x})_n = \mathbf{x}_{n+1} \ and$$

$$(\mathbf{R}^{-1}\mathbf{x})_n = (-1)^n\mathbf{x}_n = (\mathbf{R}\mathbf{x})_n .$$

Also, the downsampling operator is a left inverse for the upsampling operator:

$$(\downarrow_\sigma)(\uparrow_\sigma) = \mathbf{I} . \tag{1.88}$$

Proof Since

$$\left[\mathbf{T}\left(\mathbf{T}^{-1}\mathbf{x}\right)\right]_n = \left(\mathbf{T}^{-1}\mathbf{x}\right)_{n-1} = \mathbf{x}_{(n-1)+1} = \mathbf{x}_n ,$$

we conclude that we have found the inverse of the shift operator. Also note that

$$[\mathbf{R}(\mathbf{R}\mathbf{x})]_n = (-1)^n(\mathbf{R}\mathbf{x})_n = (-1)^n(-1)^n\mathbf{x}_n = \mathbf{x}_n ,$$

from which we conclude that $\mathbf{R}^{-1} = \mathbf{R}$. To prove (1.88), we compute

$$[(\downarrow_\sigma)\{(\uparrow_\sigma)\mathbf{x}\}]_n = [(\uparrow_\sigma)\mathbf{x}]_{n\sigma} = \mathbf{x}_{(n\sigma)/\sigma} = \mathbf{x}_n .$$

We also have the following identities concerning products of these operations.

Lemma 1.8.2 *The shift operator* $\mathbf{T}$, *reversal operator* $\mathbf{R}$, *downsampling operator* $(\downarrow_2)$ *and upsampling operator* $(\uparrow_2)$ *satisfy*

$$\mathbf{T}\mathbf{R} = -\mathbf{R}\mathbf{T} , \tag{1.89}$$

$$\mathbf{T}(\downarrow_2) = (\downarrow_2)\mathbf{T}^2 , \tag{1.90}$$

$$\mathbf{R}(\uparrow_2) = (\uparrow_2) \,, \tag{1.91}$$

$$(\downarrow_2)\mathbf{R} = (\downarrow_2) \,, \tag{1.92}$$

$$\mathbf{I} + \mathbf{R} = 2(\uparrow_2)(\downarrow_2) \ and \tag{1.93}$$

$$\mathbf{I} - \mathbf{R} = 2\mathbf{T}^{-1}(\uparrow_2)(\downarrow_2)\mathbf{T} = 2\mathbf{T}(\uparrow_2)(\downarrow_2)\mathbf{T}^{-1} \,. \tag{1.94}$$

Proof To prove (1.89), note that

$$(\mathbf{TRx})_n = (\mathbf{Rx})_{n-1} = (-1)^{n-1}\mathbf{x}_{n-1} = -(-1)^n\mathbf{x}_{n-1} = -(-1)^n(\mathbf{Tx})_n = -(\mathbf{RTx})_n \,.$$

Next, we will prove (1.90):

$$[\mathbf{T}(\downarrow_2)\mathbf{x}]_n = [(\downarrow_2)\mathbf{x}]_{n-1} = \mathbf{x}_{2(n-1)} = \mathbf{x}_{2n-2} = (\mathbf{T}^2\mathbf{x})_n \,.$$

Since

$$[\mathbf{R}(\uparrow_2)\mathbf{x}]_n = (-1)^n\,[(\uparrow_2)\mathbf{x}]_n = (-1)^n \begin{cases} \mathbf{x}_n,\ n\ \text{even} \\ 0,\ \ n\ \text{odd} \end{cases} = \begin{cases} \mathbf{x}_n,\ n\ \text{even} \\ 0,\ \ n\ \text{odd} \end{cases} = [(\uparrow_2)\mathbf{x}]_n \,,$$

equation (1.91) is satisfied.

To prove (1.92), we compute

$$[(\downarrow_2)\mathbf{Rx}]_n = (\mathbf{Rx})_{2n} = (-1)^{2n}\mathbf{x}_{2n} = \mathbf{x}_{2n} = [(\downarrow_2)\mathbf{x}]_n \,.$$

Equation (1.93) can be proved as follows:

$$\mathbf{x}_n + (\mathbf{Rx})_n = [1 + (-1)^n]\mathbf{x}_n = \begin{cases} 2\mathbf{x}_n,\ n\ \text{even} \\ 0,\ \ n\ \text{odd} \end{cases} = 2 \begin{cases} \mathbf{x}_{2(n/2)},\ n\ \text{even} \\ 0,\ \ n\ \text{odd} \end{cases}$$

$$= 2 \begin{cases} [(\uparrow_2)\mathbf{x}]_{n/2}\,,\ n\ \text{even} \\ 0,\ \ n\ \text{odd} \end{cases} = [2(\uparrow_2)(\downarrow_2)\mathbf{x}]_n \,,$$

Finally, we will prove (1.94):

$$\mathbf{x}_n - (\mathbf{Rx})_n = [1 - (-1)^n]\mathbf{x}_n = \begin{cases} 2\mathbf{x}_n,\ \ n\ \text{odd} \\ 0,\ n\ \text{even} \end{cases} = 2\mathbf{T}^{-1} \begin{cases} \mathbf{x}_{2(n-1)/2},\ n-1\ \text{even} \\ 0,\ \ n-1\ \text{odd} \end{cases}$$

$$= \left[2\mathbf{T}^{-1}(\uparrow_2)(\downarrow_2)\mathbf{Tx}\right]_n \,.$$

The alternative form of this expression is proved in a similar fashion.

Readers may view a program to implement signals in the C^{++} files Signal.H and Signal.C. This C^{++} class implements the shift and reverse operators.

Exercise 1.8.1 Prove that $\mathbf{T} + \mathbf{T}^{-1}$ is not invertible, by finding a nonzero discrete time signal $\mathbf{x}$ such that $\mathbf{Tx} + \mathbf{T}^{-1}\mathbf{x} = \mathbf{0}$.

Exercise 1.8.2 Prove that $\mathbf{I} + \mathbf{R}$ is not invertible.

Exercise 1.8.3 Prove that $\mathbf{I} + \mathbf{T}$ is not invertible.

1.8.1.3 Filters

Next, let us introduce a common operation on a pair of arrays.

Definition 1.8.3 If $\mathbf{x}$ and $\mathbf{y}$ are two discrete time signals, their **convolution** is the discrete time signal $\mathbf{y} * \mathbf{x}$, with entries

$$(\mathbf{y} * \mathbf{x})_n = \sum_{k=-\infty}^{\infty} \mathbf{y}_{n-k}\mathbf{x}_k .$$

This definition leads to the following useful results.

Lemma 1.8.3 *Convolution of discrete time signals is* **commutative***, meaning that for all discrete time signals* $\mathbf{x}$ *and* $\mathbf{y}$

$$\mathbf{x} * \mathbf{y} = \mathbf{y} * \mathbf{x} .$$

Convolution of discrete time signals is also **associative***, meaning that for all discrete time signals* $\mathbf{x}$, $\mathbf{y}$ *and* $\mathbf{z}$

$$\mathbf{x} * (\mathbf{y} * \mathbf{z}) = (\mathbf{x} * \mathbf{y}) * \mathbf{z} .$$

Finally, the unit impulse signal is the identity element for convolution, since

$$\mathbf{x} * \delta = \mathbf{x} .$$

Proof For any two discrete time signals $\mathbf{x}$ and $\mathbf{y}$, we have

$$(\mathbf{x} * \mathbf{y})_n = \sum_{k=-\infty}^{\infty} \mathbf{x}_{n-k}\mathbf{y}_k = \sum_{j=-\infty}^{\infty} \mathbf{x}_j\mathbf{y}_{n-j} = (\mathbf{y} * \mathbf{x})_n .$$

This proves commutativity.

For any three discrete time signals $\mathbf{x}$, $\mathbf{y}$ and $\mathbf{z}$, we have

$$[\mathbf{x} * (\mathbf{y} * \mathbf{z})]_n = \sum_{j=-\infty}^{\infty} \mathbf{x}_{n-j} \sum_{k=-\infty}^{\infty} \mathbf{y}_{j-k}\mathbf{z}_k = \sum_{k=-\infty}^{\infty} \left[\sum_{j=-\infty}^{\infty} \mathbf{x}_{n-j}\mathbf{y}_{j-k} \right] \mathbf{z}_k$$

$$= \sum_{k=-\infty}^{\infty} \left[\sum_{\ell=-\infty}^{\infty} \mathbf{x}_\ell \mathbf{y}_{n-\ell-k} \right] \mathbf{z}_k = [(\mathbf{x} * \mathbf{y}) * \mathbf{z}]_n .$$

Thus, convolution is associative.

Finally, note that

$$(\boldsymbol{\delta} * \mathbf{x})_n = \sum_{k=-\infty}^{\infty} \delta_{0,n-k} \mathbf{x}_k = \mathbf{x}_n \, ,$$

so $\boldsymbol{\delta}$ is the convolution identity element.

Here are some useful identities involving convolution and familiar operators.

Lemma 1.8.4 *Let the shift operator* $\mathbf{T}$ *and reversal operator* $\mathbf{R}$ *be defined by Eqs.* (1.84) *and* (1.85)*, respectively. If* $\mathbf{x}$ *and* $\mathbf{y}$ *are two discrete time signals, then*

$$(\mathbf{Ty}) * \mathbf{x} = \mathbf{T}(\mathbf{y} * \mathbf{x}) = \mathbf{y} * (\mathbf{Tx}) \, , \tag{1.95}$$

$$(\mathbf{Ry}) * \mathbf{x} = \mathbf{R}[\mathbf{y} * (\mathbf{Rx})] \, , \ and \tag{1.96}$$

$$\mathbf{R}(\mathbf{x} * \mathbf{y}) = (\mathbf{Rx}) * (\mathbf{Ry}) \, . \tag{1.97}$$

Proof To prove the first claim, we note that

$$[(\mathbf{Ty}) * \mathbf{x}]_n = \sum_{m=-\infty}^{\infty} (\mathbf{Ty})_m \mathbf{x}_{n-m} = \sum_{m=-\infty}^{\infty} \mathbf{y}_{m-1} \mathbf{x}_{n-m} = \sum_{k=-\infty}^{\infty} \mathbf{y}_k \mathbf{x}_{n-1-k}$$

$$= (\mathbf{y} * \mathbf{x})_{n-1} = [\mathbf{T}(\mathbf{y} * \mathbf{x})]_n = \sum_{k=-\infty}^{\infty} \mathbf{y}_k \mathbf{x}_{n-1-k} = [\mathbf{y} * (\mathbf{Tx})]_n \, .$$

The second claim is proved as follows:

$$[(\mathbf{Ry}) * \mathbf{x}]_n = \sum_{m=-\infty}^{\infty} (-1)^m \mathbf{y}_m \mathbf{x}_{n-m} = \sum_{k=-\infty}^{\infty} (-1)^{n-k} \mathbf{y}_{n-k} \mathbf{x}_k$$

$$= (-1)^n \sum_{k=-\infty}^{\infty} \mathbf{y}_{n-k} (\mathbf{Rx})_k = (-1)^n [\mathbf{y} * (\mathbf{Rx})]_n = \{\mathbf{R}[\mathbf{y} * (\mathbf{Rx})]\}_n \, .$$

Here is a proof of the final claim:

$$(\mathbf{R}[\mathbf{x} * \mathbf{y}])_n = (-1)^n \sum_{m=-\infty}^{\infty} \mathbf{x}_m \mathbf{y}_{n-m} = \sum_{m=-\infty}^{\infty} [(-1)^m \mathbf{x}_m] [(-1)^{n-m} \mathbf{y}_{n-m}]$$

$$= [(\mathbf{Rx}) * (\mathbf{R} * \mathbf{y})]_n \, .$$

The next definition shows that convolution is the basic idea behind digital filters.

Definition 1.8.4 If $\mathbf{f}$ is a discrete time signal, then the linear operator $\mathbf{Fx} \equiv \mathbf{f} * \mathbf{x}$ acting on discrete time signals $\mathbf{x}$ is called a **digital filter**, and $\mathbf{f}$ is called its **impulse response**. A digital filter $\mathbf{F}$ is a **finite impulse response (FIR) filter** if and only if there exists an integer $N \geq 0$ so that for all indices n with $|n| > N$ the impulse response satisfies $\mathbf{f}_n = 0$. An FIR filter is **causal** if and only if for all $n < 0$ its impulse response satisfies $\mathbf{f}_n = 0$.

Note that if $\mathbf{F}$ is a digital filter with impulse response $\mathbf{f}$, then

$$\mathbf{F}\delta = \mathbf{f} * \delta = \mathbf{f} .$$

This equation explains why $\mathbf{f}$ is called the impulse response for $\mathbf{F}$. Also note that digital filters can be applied in any order. If $\mathbf{F}$ and $\mathbf{G}$ are two digital filters with impulse response $\mathbf{f}$ and $\mathbf{g}$, respectively, then for any digital signal $\mathbf{x}$ we have

$$(\mathbf{GFx})_n = (\mathbf{g} * [\mathbf{f} * \mathbf{x}])_n = ([\mathbf{g} * \mathbf{f}] * \mathbf{x})_n = ([\mathbf{f} * \mathbf{g}] * \mathbf{x})_n = (\mathbf{f} * [\mathbf{g} * \mathbf{x}])_n = (\mathbf{FGx})_n .$$

It should be easy to see that neither the downsampling operator nor the upsampling operator is a digital filter. However, here is one useful example of a digital filter.

Example 1.8.1 The shift operator $\mathbf{T}$ is a digital filter with impulse response $\mathbf{t}_n = \delta_{1,n}$. This is because

$$\mathbf{t}_n = (\mathbf{T}\delta)_n = (\delta)_{n-1} = \delta_{0,n-1} = \delta_{1,n} .$$

The shift operator is also a causal FIR filter.

Note that this example implies that any digital filter $\mathbf{F}$ is **shift invariant**, meaning that $\mathbf{FT} = \mathbf{TF}$.

Here are two more examples of digital filters.

Example 1.8.2 The **Haar filter** has impulse response $\mathbf{f}$ where

$$\mathbf{f}_n = \begin{cases} 1/2, & n = 0, 1 \\ 0, & \text{otherwise} \end{cases} .$$

This is also a causal FIR filter.

Example 1.8.3 The **Daubechies D_4 filter** has impulse response $\mathbf{f}$ where

$$\mathbf{f}_n = \begin{cases} (1 + \sqrt{3})/8, & n = 0 \\ (3 + \sqrt{3})/8, & n = 1 \\ (3 - \sqrt{3})/8, & n = 2 \\ (1 - \sqrt{3})/8, & n = 3 \\ 0, \text{ otherwise} \end{cases} .$$

This is also a causal FIR filter.

Readers may view a program to implement filters in the C^{++} files Filter.H and Filter.C. This C^{++} class involves several functions, including one to operate the given filter on a signal.

Exercise 1.8.4 Show that downsampling is not a digital filter, by showing that it is not shift invariant.

Exercise 1.8.5 Show that upsampling is not a digital filter.

Exercise 1.8.6 Show that reversal is not a digital filter.

1.8.1.4   Discrete Fourier Transform

In Sect. 1.7.4.3, we studied one form of a Fourier transform, involving trigonometric polynomials of finite order. In this section, we will study another discrete Fourier transform, but this time the order of the trigonometric polynomial may be infinite.

Definition 1.8.5 If $\mathbf{x}$ is a discrete time signal, then its **discrete time Fourier transform** is the linear operator

$$\mathscr{F}^d\{\mathbf{x}\}(\omega) = \sum_{n=-\infty}^{\infty} \mathbf{x}_n e^{-in\omega} \ .$$

There are several useful results regarding the discrete time Fourier transform. We will begin by showing the discrete time Fourier transform is a bounded linear operator.

Lemma 1.8.5 *If $\mathbf{x}$ is a discrete time signal, then its discrete time Fourier transform $\mathscr{F}^d\{\mathbf{x}\}$ is a continuous 2π-periodic function, and*

$$\left|\mathscr{F}^d\{\mathbf{x}\}(\omega)\right| \le \sum_{n=-\infty}^{\infty} |\mathbf{x}_n|$$

for all $\omega \in [-\pi, \pi]$.

Proof Continuity and periodicity are obvious from the definition of the discrete time Fourier transform. To complete the proof, we note that for all $\omega \in [-\pi, \pi]$ we have

$$\left|\mathscr{F}^d\{\mathbf{x}\}(\omega)\right| = \left|\sum_{n=-\infty}^{\infty} \mathbf{x}_n e^{-in\omega}\right| \le \sum_{n=-\infty}^{\infty} \left|\mathbf{x}_n e^{-in\omega}\right| = \sum_{n=-\infty}^{\infty} |\mathbf{x}_n| \ .$$

Next, we will find the inverse operator for the discrete time Fourier transform.

Theorem 1.8.1 (Discrete Time Fourier Inversion) *If $\mathbf{x}$ is a discrete time signal with discrete time Fourier transform $\mathscr{F}^d\{\mathbf{x}\}$, then for all indices n we have*

$$\mathbf{x}_n = \frac{1}{2\pi} \int_{-\pi}^{\pi} \mathscr{F}^d\{\mathbf{x}\}(\omega) e^{i\omega n} \ d\omega \ .$$

Proof We compute

$$\frac{1}{2\pi}\int_{-\pi}^{\pi}\mathscr{F}^d\{\mathbf{x}\}(\omega)e^{i\omega n}\,d\omega = \frac{1}{2\pi}\int_{-\pi}^{\pi}\sum_{k=-\infty}^{\infty}\mathbf{x}_k e^{i\omega(n-k)}\,d\omega$$

$$= \sum_{k=-\infty}^{\infty}\mathbf{x}_k\frac{1}{2\pi}\int_{-\pi}^{\pi}e^{i\omega(n-k)}\,d\omega = \sum_{k=-\infty}^{\infty}\mathbf{x}_k\delta_{n,k} = \mathbf{x}_n\;.$$

Note that the interchange of integration and summation is allowable, because any discrete time signal $\mathbf{x}$ is absolutely summable.

The next two theorems are useful for working with inner products of discrete time signals.

Theorem 1.8.2 (Plancherel)   *If $\mathbf{x}$ is a square-summable discrete time signal, then $\mathscr{F}^d\{\mathbf{x}\}$ is square-integrable on $[-\pi, \pi]$ and*

$$\sum_{n=-\infty}^{\infty}|\mathbf{x}_n|^2 = \frac{1}{2\pi}\int_{-\pi}^{\pi}\left|\mathscr{F}^d\{\mathbf{x}\}(\omega)\right|^2\,d\omega\;.$$

Proof We compute

$$\frac{1}{2\pi}\int_{-\pi}^{\pi}\left|\mathscr{F}^d\{\mathbf{x}\}(\omega)\right|^2\,d\omega = \frac{1}{2\pi}\int_{-\pi}^{\pi}\left|\sum_{n=-\infty}^{\infty}\mathbf{x}_n e^{-in\omega}\right|^2\,d\omega$$

$$= \frac{1}{2\pi}\int_{-\pi}^{\pi}\sum_{n=-\infty}^{\infty}\sum_{m=-\infty}^{\infty}\mathbf{x}_n\overline{\mathbf{x}_m}e^{i(m-n)\omega}\,d\omega = \sum_{n=-\infty}^{\infty}\sum_{m=-\infty}^{\infty}\mathbf{x}_n\overline{\mathbf{x}_m}\frac{1}{2\pi}\int_{-\pi}^{\pi}e^{i(m-n)\omega}\,d\omega$$

$$= \sum_{n=-\infty}^{\infty}\sum_{m=-\infty}^{\infty}\mathbf{x}_n\overline{\mathbf{x}_m}\delta_{n,m} = \sum_{n=-\infty}^{\infty}|\mathbf{x}_n|^2\;.$$

Theorem 1.8.3 (Parseval Identity)   *If $\mathbf{x}$ and $\mathbf{y}$ are square-summable discrete time signals, then*

$$\sum_{n=-\infty}^{\infty}\mathbf{x}_n\overline{\mathbf{y}_n} = \frac{1}{2\pi}\int_{-\pi}^{\pi}\mathscr{F}^d\{\mathbf{x}\}(\omega).\overline{\mathscr{F}^d\{\mathbf{y}\}(\omega)}\,d\omega\;.$$

Proof We have

$$\frac{1}{2\pi}\int_{-\pi}^{\pi}\mathscr{F}^d\{\mathbf{x}\}(\omega)\overline{\mathscr{F}^d\{\mathbf{y}\}(\omega)}\,d\omega = \frac{1}{2\pi}\int_{-\pi}^{\pi}\sum_{n=-\infty}^{\infty}\sum_{m=-\infty}^{\infty}\mathbf{x}_n\overline{\mathbf{y}_m}e^{i(m-n)\omega}\,d\omega$$

$$= \sum_{n=-\infty}^{\infty}\sum_{m=-\infty}^{\infty}\mathbf{x}_n\overline{\mathbf{y}_m}\frac{1}{2\pi}\int_{-\pi}^{\pi}e^{i(m-n)\omega}\,d\omega = \sum_{n=-\infty}^{\infty}\sum_{m=-\infty}^{\infty}\mathbf{x}_n\overline{\mathbf{y}_m}\delta_{n,m} = \sum_{n=-\infty}^{\infty}\mathbf{x}_n\overline{\mathbf{y}_n}\;.$$

The following lemma is extremely important to our work with filters.

Lemma 1.8.6 *If* **f** *and* **g** *are two discrete time signals, then* $\mathscr{F}^d\{\mathbf{g} * \mathbf{f}\} = \mathscr{F}^d\{\mathbf{g}\}\mathscr{F}^d\{\mathbf{f}\}.$

Proof We can compute

$$
\mathscr{F}^d\{\mathbf{g} * \mathbf{f}\}(\omega) = \sum_{n=-\infty}^{\infty} (\mathbf{g} * \mathbf{f})_n e^{-in\omega} = \sum_{n=-\infty}^{\infty} \sum_{k=-\infty}^{\infty} \mathbf{g}_{n-k} \mathbf{f}_k e^{-in\omega}
$$

$$
= \sum_{n=-\infty}^{\infty} \sum_{k=-\infty}^{\infty} \left[\mathbf{g}_{n-k} e^{-i(n-k)\omega}\right]\left[\mathbf{f}_k e^{-ik\omega}\right] = \sum_{\ell=-\infty}^{\infty} \sum_{k=-\infty}^{\infty} \left[\mathbf{g}_\ell e^{-i\ell\omega}\right]\left[\mathbf{f}_k e^{-ik\omega}\right]
$$

$$
= \mathscr{F}^d\{\mathbf{g}\}(\omega)\mathscr{F}^d\{\mathbf{f}\}(\omega) .
$$

The next lemma determines circumstances under which it is possible to find the inverse of a digital filter.

Lemma 1.8.7 *Suppose that* **F** *is a digital filter with impulse response* **f***. Then* **F** *is invertible if and only if* $1/\mathscr{F}^d\{\mathbf{f}\}(\omega)$ *is* **absolutely integrable** *in* $(-\pi, \pi)$*, meaning that*

$$
\int_{-\pi}^{\pi} |f(t)| \ dt < \infty .
$$

If **F** *is invertible, then the impulse response for the inverse is*

$$
\mathbf{f}_n^{-1} = \frac{1}{2\pi} \int_{-\pi}^{\pi} \frac{e^{in\omega}}{\mathscr{F}^d\{\mathbf{f}\}(\omega)} \ d\omega . \tag{1.98}
$$

Proof If **F** is invertible and its inverse has impulse response $\mathbf{f}^{-1}$, then for all discrete time signals **x** we have

$$
\mathscr{F}^d\{\mathbf{x}\} = \mathscr{F}^d\left\{\mathbf{f}^{-1} * (\mathbf{f} * \mathbf{x})\right\} = \mathscr{F}^d\left\{\mathbf{f}^{-1}\right\} \mathscr{F}^d\{\mathbf{f}\}\mathscr{F}^d\{\mathbf{x}\}
$$

so for all $\omega \in [-\pi, \pi]$ we have

$$
\mathscr{F}^d\left\{\mathbf{f}^{-1}\right\}(\omega) = \frac{1}{\mathscr{F}^d\{\mathbf{f}\}(\omega)} .
$$

The formula for $\mathbf{f}^{-1}$ follows from the discrete time Fourier inversion formula.

Conversely, if $1/\mathscr{F}^d\{\mathbf{f}\}(\omega)$ is absolutely integrable over $\omega \in [-\pi, \pi]$, let the discrete time signal $\mathbf{f}^{-1}$ be given by (1.98). Then for all discrete time signals **x** we have

$$\left[\mathbf{f}^{-1} * (\mathbf{f} * \mathbf{x})\right]_n = \sum_{m=-\infty}^{\infty} \sum_{k=-\infty}^{\infty} \mathbf{f}_m^{-1} \mathbf{f}_{n-m-k} \mathbf{x}_k$$

$$= \sum_{k=-\infty}^{\infty} \mathbf{x}_k \left[\sum_{m=-\infty}^{\infty} \mathbf{f}_{n-m-k} \frac{1}{2\pi} \int_{-\pi}^{\pi} \frac{e^{im\omega}}{\mathscr{F}^d\{\mathbf{f}\}(\omega)} \, d\omega \right]$$

$$= \sum_{k=-\infty}^{\infty} \mathbf{x}_k \left[\frac{1}{2\pi} \int_{-\pi}^{\pi} \frac{e^{i(n-k)\omega}}{\mathscr{F}^d\{\mathbf{f}\}(\omega)} \left\{ \sum_{m=-\infty}^{\infty} \mathbf{f}_{n-m-k} e^{i(n-m-k)\omega} \right\} d\omega \right]$$

$$= \sum_{k=-\infty}^{\infty} \mathbf{x}_k \left[\frac{1}{2\pi} \int_{-\pi}^{\pi} \frac{e^{i(n-k)\omega}}{\mathscr{F}^d\{\mathbf{f}\}(\omega)} \mathscr{F}^d\{\mathbf{f}\}(\omega) \, d\omega \right]$$

$$= \sum_{k=-\infty}^{\infty} \mathbf{x}_k \left[\frac{1}{2\pi} \int_{-\pi}^{\pi} e^{i(n-k)\omega} \, d\omega \right] = \sum_{k=-\infty}^{\infty} \mathbf{x}_k \delta_{k,n} = \mathbf{x}_n \ .$$

We will complete this section by finding the discrete time Fourier transform of some operators on discrete time signals.

Lemma 1.8.8 *Suppose that $\mathbf{x}$ is a discrete time signal. If $\mathbf{T}$ is the shift operator defined in Eq. (1.84), then*

$$\mathscr{F}^d\{\mathbf{Tx}\}(\omega) = e^{-i\omega} \mathscr{F}^d\{\mathbf{x}\}(\omega) \ . \tag{1.99}$$

If $\mathbf{R}$ is the reversal operator defined in Eq. (1.85), then

$$\mathscr{F}^d\{\mathbf{Rx}\}(\omega) = \mathscr{F}^d\{\mathbf{x}\}(\omega + \pi) \ . \tag{1.100}$$

If σ is a positive integer and $(\uparrow_\sigma)$ is the upsampling operator defined in Eq. (1.87), then

$$\mathscr{F}^d\{(\uparrow_\sigma)\mathbf{x}\}(\omega) = \mathscr{F}^d\{\mathbf{x}\}(\sigma\omega) \ . \tag{1.101}$$

Finally, if $(\downarrow_2)$ is the downsampling operator defined in Eq. (1.86), then

$$\mathscr{F}^d\{(\downarrow_2)\mathbf{x}\}(\omega) = \frac{1}{2}\left[\mathscr{F}^d\{\mathbf{x}\}(\omega/2) + \mathscr{F}^d\{\mathbf{x}\}(\pi + \omega/2)\right] \ . \tag{1.102}$$

Proof First, we compute

$$\mathscr{F}^d\{\mathbf{Tx}\}(\omega) = \sum_{n=-\infty}^{\infty} \mathbf{x}_{n-1} e^{-in\omega} = \sum_{m=-\infty}^{\infty} \mathbf{x}_m e^{-i(m+1)\omega} = e^{-i\omega} \sum_{m=-\infty}^{\infty} \mathbf{x}_m e^{-im\omega}$$

$$= e^{-i\omega} \mathscr{F}^d\{\mathbf{x}\}(\omega) \ .$$

This proves (1.99).

Next, we note that

$$\mathscr{F}^d\{\mathbf{x}\}(\omega + \pi) = \sum_{n=-\infty}^{\infty} \mathbf{x}_n e^{-in(\omega+\pi)} = \sum_{n=-\infty}^{\infty} \mathbf{x}_n(-1)^n e^{-in\omega} = \sum_{n=-\infty}^{\infty} (\mathbf{Rx})_n e^{-in\omega}$$

$$= \mathscr{F}^d\{\mathbf{Rx}\}(\omega) .$$

This proves (1.100).

To prove (1.101), we evaluate

$$\mathscr{F}^d\left\{(\uparrow_\sigma)\mathbf{x}\right\}(\omega) = \sum_{n=-\infty}^{\infty} [(\uparrow_\sigma)\mathbf{x}]_n\, e^{-in\omega} = \sum_{\substack{n=-\infty \\ n \text{ divisible by } \sigma}}^{\infty} \mathbf{x}_{n/\sigma} e^{-in\omega}$$

$$= \sum_{m=-\infty}^{\infty} \mathbf{x}_m e^{-im\sigma\omega} = \mathscr{F}^d\{\mathbf{x}\}(\sigma\omega) .$$

Finally, note that

$$\mathscr{F}^d\left\{(\downarrow_2)\mathbf{x}\right\}(\omega) = \sum_{n=-\infty}^{\infty} (\downarrow_2 \mathbf{x})_n e^{-in\omega} = \sum_{n=-\infty}^{\infty} \mathbf{x}_{2n} e^{-in\omega} = \sum_{\substack{m=-\infty \\ m \text{ even}}}^{\infty} \mathbf{x}_m e^{-im\omega/2}$$

$$= \sum_{m=-\infty}^{\infty} \mathbf{x}_m \frac{1 + (-1)^m}{2} e^{-im\omega/2} = \frac{1}{2} \sum_{m=-\infty}^{\infty} \mathbf{x}_m \left[e^{-im\omega/2} + e^{-im\pi - im\omega/2}\right]$$

$$= \frac{1}{2} \left[\mathscr{F}^d\{\mathbf{x}\}(\omega/2) + \mathscr{F}^d\{\mathbf{x}\}(\pi + \omega/2)\right] .$$

This proves (1.102)

Exercise 1.8.7 Prove that if $\mathbf{f}$ is a discrete time signal, then for all discrete time signals $\mathbf{x}$ and for all positive integers σ

$$\mathbf{f} * [(\downarrow_\sigma)\mathbf{x}] = (\downarrow_\sigma)\{[(\uparrow_\sigma)\mathbf{f}] * \mathbf{x}\} .$$

Show that the left-hand side of this equation involves less work. This is called the **first Noble identity**.

Exercise 1.8.8 Prove that if $\mathbf{f}$ is a discrete time signal, then for all discrete time signals $\mathbf{x}$ and all positive integers σ we have

$$\mathbf{f} * \mathbf{x} = (\downarrow_\sigma)\{[(\uparrow_\sigma)\mathbf{f}] * [(\uparrow_\sigma)\mathbf{x}]\} .$$

and

$$(\uparrow_\sigma)[\mathbf{f} * \mathbf{x}] = (\uparrow_\sigma)(\downarrow_\sigma)\{[(\uparrow_\sigma)\mathbf{f}] * [(\uparrow_\sigma)\mathbf{x}]\} .$$

Show that the left-hand side of either identity involves less work. These are called the **second Noble identities**.

1.8.1.5 Adjoints

In Sect. 1.8.1.6, we will combine downsampling and upsampling operators with filters to form filter banks. These operations will have continuous analogues in Sect. 1.8.2.7 below. The continuous operations will involve inner products with functions that will be identified as "adjoint" to some given function. Let us explore the basic idea, which is applicable to both the discrete and continuous cases.

Definition 1.8.6 Suppose that L is a **linear operator** on an inner product space $\mathscr{H}$, meaning that for all $h_1, h_2 \in \mathscr{H}$ and all scalars α_1 and α_2 we have

$$L(h_1\alpha_1 + h_2\alpha_2) = (Lh_1)\alpha_1 + (Lh_2)\alpha_2 .$$

Then the **adjoint** L^* of L is such that for all $h_1, h_2 \in \mathscr{H}$

$$(Lh_1, h_2) = (h_1, L^*h_2) .$$

It is easy to see that the adjoint is also a linear operator, and that for any linear operator L we have $(L^*)^* = L$. In linear algebra, using the inner product $(\mathbf{x}, \mathbf{y}) = \mathbf{y}^\top \mathbf{x}$ for real vectors, the adjoint of a real matrix $\mathbf{A}$ is its transpose $\mathbf{A}^* = \mathbf{A}^\top$. The adjoint of a complex matrix is its **Hermitian** $\mathbf{A}^* = \mathbf{A}^H = \overline{\mathbf{A}^\top} = \overline{\mathbf{A}}^\top$.

Lemma 1.8.9 *Let the inner product of two discrete time signals* $\mathbf{x}$ *and* $\mathbf{y}$ *be*

$$(\mathbf{x}, \mathbf{y}) = \sum_{n=-\infty}^{\infty} \mathbf{x}_n \overline{\mathbf{y}_n} .$$

Then the adjoint of the downsampling operator $(\downarrow_2)$ *is the upsampling operator:*

$$(\downarrow_2)^* = (\uparrow_2) .$$

Also the adjoint of the shift operator $\mathbf{T}$ *is the backward shift operator* $\mathbf{T}^{-1}$:

$$\mathbf{T}^* = \mathbf{T}^{-1} .$$

The adjoint of the reversal operator $\mathbf{R}$ *is* $\mathbf{R}$:

$$\mathbf{R}^* = \mathbf{R} .$$

Proof If $\mathbf{x}$ and $\mathbf{y}$ are two discrete time signals and σ is a positive integer, then

$$\sum_{n=-\infty}^{\infty} [(\downarrow_\sigma)\mathbf{x}]_n \,\overline{\mathbf{y}_n} = \sum_{n=-\infty}^{\infty} \mathbf{x}_{\sigma n}\overline{\mathbf{y}_n} = \sum_{\substack{m=-\infty \\ m \text{ divisible by } \sigma}}^{\infty} \mathbf{x}_m \overline{\mathbf{y}_{m/\sigma}}$$

$$= \sum_{n=-\infty}^{\infty} \mathbf{x}_m \overline{[(\uparrow_\sigma)\mathbf{y}]_m} \;.$$

Similarly,

$$\sum_{n=-\infty}^{\infty} (\mathbf{T}\mathbf{x})_n \,\overline{\mathbf{y}_n} = \sum_{n=-\infty}^{\infty} \mathbf{x}_{n-1}\overline{\mathbf{y}_n} = \sum_{m=-\infty}^{\infty} \mathbf{x}_m \overline{\mathbf{y}_{m+1}} = \sum_{m=-\infty}^{\infty} \mathbf{x}_m \overline{(\mathbf{T}^{-1}\mathbf{y})_m} \;.$$

Finally,

$$\sum_{n=-\infty}^{\infty} (\mathbf{R}\mathbf{x})_n \,\overline{\mathbf{y}_n} = \sum_{n=-\infty}^{\infty} (-1)^n \mathbf{x}_n \overline{\mathbf{y}_n} = \sum_{n=-\infty}^{\infty} \mathbf{x}_n \overline{(\mathbf{R}\mathbf{y})_n} \;.$$

Lemma 1.8.10 *Let the inner product of two discrete time signals $\mathbf{x}$ and $\mathbf{y}$ be*

$$(\mathbf{x}, \mathbf{y}) = \sum_{n=-\infty}^{\infty} \mathbf{x}_n \overline{\mathbf{y}_n} \;.$$

Suppose that $\mathbf{f}$ is a discrete time signal. Then the adjoint of convolution with $\mathbf{f}$ is convolution with the discrete time signal $\mathbf{f}^*$, where for all n

$$\mathbf{f}_n^* = \overline{\mathbf{f}_{-n}} \;. \tag{1.103}$$

Furthermore,

$$\mathscr{F}^d \{\mathbf{f}^*\} (\omega) = \overline{\mathscr{F}^d\{\mathbf{f}\}(\omega)} \;. \tag{1.104}$$

Proof If $\mathbf{x}$ and $\mathbf{y}$ are two discrete time signals, then

$$\sum_{n=-\infty}^{\infty} (\mathbf{f} * \mathbf{x})_n \overline{\mathbf{y}_n} = \sum_{n=-\infty}^{\infty} \left(\sum_{m=-\infty}^{\infty} \mathbf{f}_{n-m}\mathbf{x}_m \right) \overline{\mathbf{y}_n} = \sum_{m=-\infty}^{\infty} \mathbf{x}_m \left(\sum_{n=-\infty}^{\infty} \mathbf{f}_{n-m}\overline{\mathbf{y}_n} \right)$$

$$= \sum_{m=-\infty}^{\infty} \mathbf{x}_m \overline{\left(\sum_{n=-\infty}^{\infty} \mathbf{f}_{m-n}^* \mathbf{y}_n \right)} = \sum_{m=-\infty}^{\infty} \mathbf{x}_m \overline{(\mathbf{f}^* * \mathbf{y})_m} \;.$$

Also note that

$$\mathscr{F}^d\{\mathbf{f^*}\}(\omega) = \sum_{n=-\infty}^{\infty} \mathbf{f}_n^* e^{-i\omega n} = \sum_{n=-\infty}^{\infty} \overline{\mathbf{f}_{-n}} e^{-i\omega n} = \overline{\sum_{n=-\infty}^{\infty} f_{-n} e^{i\omega n}}$$

$$= \overline{\sum_{m=-\infty}^{\infty} f_m e^{-i\omega m}} = \overline{\mathscr{F}^d\{\mathbf{f}\}(\omega)} \ .$$

1.8.1.6 Filter Banks

One of the purposes of filters is to analyze signals. Typically, the analysis will use filters of at least two types. The most common types are described in the following definition

Definition 1.8.7 A **lowpass filter** has impulse response ℓ such that its time discrete Fourier transform satisfies

$$\mathscr{F}^d\{\boldsymbol{\ell}\}(0) = 1 \text{ and } \mathscr{F}^d\{\boldsymbol{\ell}\}(\pm\pi) = 0 \ .$$

A **highpass filter** has impulse response $\mathbf{h}$ such that its time discrete Fourier transform satisfies

$$\mathscr{F}^d\{\mathbf{h}\}(0) = 0 \text{ and } \left|\mathscr{F}^d\{\mathbf{h}\}(\pm\pi)\right| = 1 \ .$$

If $\mathbf{f}$ is the impulse response for a filter, then the **passband** is an interval of frequencies ω for which $\mathscr{F}^d\{\mathbf{f}\}(\omega) \approx 1$, and the **stopband** is an interval of frequencies for which $\mathscr{F}^d\{\mathbf{f}\}(\omega) \approx 0$.

In order to analyze a discrete time signal, it is common to apply both a lowpass filter and a highpass filter to gain information about the frequency content of the signal. Information about successively lower frequencies can be obtained by downsampling, although this operator can mislead the analysis because it involves **aliasing**, meaning that different frequencies may be made indistinguishable by downsampling.

After analysis, it may be important to reconstruct the signal from the output of the various filters. This synthesis would be impossible if we used a single filter, but may be accomplished by using multiple filters in what is typically called a **filter bank**. Many forms of filter banks are used in signal processing. We will limit our discussion to two-channel filter banks, typically using a lowpass filter in one channel and a highpass filter in the other. Ideally, the synthesis will reproduce the original signal perfectly. In some cases, the goal of analysis will be to discard unimportant information, thereby reducing the memory required to represent the signal without introducing significant error in the synthesis.

Example 1.8.4 Suppose that we separate the signal $\mathbf{x}$ into two channels, with the first channel containing the components of $\mathbf{x}$ with even indices, and the second channel containing the components with odd indices:

$$\epsilon_n = \mathbf{x}_{2n} = [(\downarrow_2)\mathbf{x}]_n$$

$$\omega_n = \mathbf{x}_{2n+1} = [(\downarrow_2)\mathbf{T}^{-1}\mathbf{x}]_n \ .$$

Here $\mathbf{T}$ is the shift operator, defined in Eq. (1.84), and $(\downarrow_2)$ is the downsampling operator, defined in Eq. (1.86). Note that both of these channels involve downsampling, but only the second channel involves a (reverse) shift. In order to store these channels efficiently as we process the discrete time signal $\mathbf{x}$, we will define the analysis bank to be

$$\mathbf{A}\{\mathbf{x}\}_n = \begin{cases} \epsilon_{n/2}, \ n \text{ even} \\ \omega_{(n-1)/2}, \ n \text{ odd} \end{cases} = \begin{cases} [(\uparrow_2)\epsilon]_n, \ n \text{ even} \\ [\mathbf{T}(\uparrow_2)\omega]_n, \ n \text{ odd} \end{cases} .$$

In order to reconstruct the original signal, our synthesis bank will use upsampling from the two channels. If the channels were stored separately, the synthesis would be $[(\uparrow_2)\epsilon]_n + [\mathbf{T}(\uparrow_2)\omega]_n$. Since we have stored the two channels in the even and odd entries of the output from the analysis bank, we will take

$$\mathbf{S}\{\mathbf{a}\}_n = [(\uparrow_2)(\downarrow_2)\mathbf{a}]_n + [\mathbf{T}(\uparrow_2)(\downarrow_2)\mathbf{T}^{-1}\mathbf{a}]_n \ .$$

Then for all n

$$\mathbf{S}\{\mathbf{A}\{\mathbf{x}\}\}_n = [(\uparrow_2)(\downarrow_2)\mathbf{A}\{\mathbf{x}\}]_n + [\mathbf{T}(\uparrow_2)(\downarrow_2)\mathbf{T}^{-1}\mathbf{A}\{\mathbf{x}\}]_n$$

$$= \begin{cases} [(\downarrow_2)\mathbf{x}]_{n/2} + [(\uparrow_2)(\downarrow_2)\mathbf{T}^{-1}\mathbf{x}]_{n-1}, \ n \text{ even} \\ 0 + [(\uparrow_2)(\downarrow_2)\mathbf{T}^{-1}\mathbf{x}]_{n-1}, \ n \text{ odd} \end{cases} = \begin{cases} \mathbf{x}_n + 0, \ n \text{ even} \\ [(\downarrow_2)\mathbf{T}^{-1}\mathbf{x}]_{(n-1)/2}, \ n \text{ odd} \end{cases}$$

$$= \begin{cases} \mathbf{x}_n, \ n \text{ even} \\ [\mathbf{T}^{-1}\mathbf{x}]_{n-1}, \ n \text{ odd} \end{cases} = \mathbf{x}_n \ .$$

In this case, we achieve perfect synthesis.

Example 1.8.5 Let us define two impulse responses:

$$\ell_n = \begin{cases} 1/2, \ n = -1 \text{ or } 0 \\ 0, \quad \text{otherwise} \end{cases} \text{ and } \mathbf{h}_n = \begin{cases} 1/2, \quad n = -1 \\ -1/2, \quad n = 0 \\ 0, \text{ otherwise} \end{cases} .$$

Then

$$\mathscr{F}^d\{\boldsymbol{\ell}\}(\omega) = \frac{1}{2}\left[e^{i\omega} + 1\right] = e^{i\omega/2}\cos(\omega/2) \text{ and}$$

$$\mathscr{F}^d\{\mathbf{h}\}(\omega) = \frac{1}{2}\left[e^{i\omega} - 1\right] = ie^{i\omega/2}\sin(\omega/2)\,.$$

Thus $\boldsymbol{\ell}$ is the impulse response for a lowpass filter, and $\mathbf{h}$ is the impulse response for a highpass filter. For any discrete time signal $\mathbf{x}$, we have

$$\mathbf{x}_{2n} = \frac{1}{2}\left[\mathbf{x}_{2n} + \mathbf{x}_{2n+1}\right] - \frac{1}{2}\left[-\mathbf{x}_{2n} + \mathbf{x}_{2n+1}\right] = [\boldsymbol{\ell} * \mathbf{x}]_{2n} - [\mathbf{h} * \mathbf{x}]_{2n} \text{ and}$$

$$\mathbf{x}_{2n+1} = \frac{1}{2}\left[\mathbf{x}_{2n} + \mathbf{x}_{2n+1}\right] + \frac{1}{2}\left[-\mathbf{x}_{2n} + \mathbf{x}_{2n+1}\right] = [\boldsymbol{\ell} * \mathbf{x}]_{2n} + [\mathbf{h} * \mathbf{x}]_{2n}\,.$$

Thus it is possible to recover the original signal from the output of the filters.

Next, let us define the **analysis bank** to be the linear operator $\mathbf{A}$ on discrete time signals with entries

$$\mathbf{A}\{\mathbf{x}\}_n = \begin{cases} (\boldsymbol{\ell} * \mathbf{x})_n, & n \text{ even} \\ (\mathbf{h} * \mathbf{x})_{n-1}, & n \text{ odd} \end{cases} = \begin{cases} \frac{1}{2}(\mathbf{x}_n + \mathbf{x}_{n+1}), & n \text{ even} \\ \frac{1}{2}(-\mathbf{x}_{n-1} + \mathbf{x}_n), & n \text{ odd} \end{cases},$$

and define the **synthesis bank** to be the linear operator $\mathbf{S}$ on discrete time signals with entries

$$\mathbf{S}\{\mathbf{a}\}_n = \begin{cases} -2(\mathbf{h} * \mathbf{a})_n, & n \text{ even} \\ 2(\boldsymbol{\ell} * \mathbf{a})_{n-1}, & n \text{ odd} \end{cases} = \begin{cases} \mathbf{a}_n - \mathbf{a}_{n+1}, & n \text{ even} \\ \mathbf{a}_{n-1} + \mathbf{a}_n, & n \text{ odd} \end{cases}.$$

Then for all discrete time signals $\mathbf{x}$ and all n we have

$$\mathbf{S}\{\mathbf{A}\{\mathbf{x}\}\}_n = \begin{cases} \mathbf{A}\{\mathbf{x}\}_n - \mathbf{A}\{\mathbf{x}\}_{n+1}, & n \text{ even} \\ \mathbf{A}\{\mathbf{x}\}_{n-1} + \mathbf{A}\{\mathbf{x}\}_n, & n \text{ odd} \end{cases}$$

$$= \frac{1}{2}\begin{cases} (\mathbf{x}_n + \mathbf{x}_{n+1}) - (-\mathbf{x}_n + \mathbf{x}_{n+1}), & n \text{ even} \\ (\mathbf{x}_{n-1} + \mathbf{x}_n) + (-\mathbf{x}_{n-1} + \mathbf{x}_n), & n \text{ odd} \end{cases} = \mathbf{x}_n\,.$$

Similarly, for all discrete time signals $\mathbf{a}$ and all n we have

$$\mathbf{A}\{\mathbf{S}\{\mathbf{a}\}\}_n = \frac{1}{2}\begin{cases} \mathbf{S}\{\mathbf{a}\}_n + \mathbf{S}\{\mathbf{a}\}_{n+1}, & n \text{ even} \\ -\mathbf{S}\{\mathbf{a}\}_{n-1} + \mathbf{S}\{\mathbf{a}\}_n, & n \text{ odd} \end{cases}$$

$$= \frac{1}{2}\begin{cases} (\mathbf{a}_n - \mathbf{a}_{n+1}) + (\mathbf{a}_n + \mathbf{a}_{n+1}), & n \text{ even} \\ -(\mathbf{a}_{n-1} - \mathbf{a}_n) + (\mathbf{a}_{n-1} + \mathbf{a}_n), & n \text{ odd} \end{cases} = \mathbf{a}_n\,.$$

Thus this synthesis bank is the inverse of this analysis bank.

In signal processing, it is common for synthesis of the analysis bank to return a delayed signal. This gives the lowpass and highpass filters a chance to look a bit into the future in the analysis phase. This suggests that we define the alternative analysis bank

$$\mathbf{A}'\{\mathbf{x}\}_n = \begin{cases} \frac{1}{2}(\mathbf{x}_{n-1} + \mathbf{x}_n), \ n \text{ even} \\ \frac{1}{2}(-\mathbf{x}_{n-2} + \mathbf{x}_{n-1}), \ n \text{ odd} \end{cases},$$

and the alternative synthesis bank

$$\mathbf{S}'\{\mathbf{a}\}_n = \begin{cases} \mathbf{a}_n - \mathbf{a}_{n+1}, \ n \text{ even} \\ \mathbf{a}_{n-1} + \mathbf{a}_n, \ n \text{ odd} \end{cases}.$$

Then for all discrete time signals $\mathbf{x}$ and all n we have

$$\mathbf{S}'\left\{\mathbf{A}'\{\mathbf{x}\}\right\}_n = \begin{cases} \mathbf{A}'\{\mathbf{x}\}_n - \mathbf{A}'\{\mathbf{x}\}_{n+1}, \ n \text{ even} \\ \mathbf{A}'\{\mathbf{x}\}_{n-1} + \mathbf{A}'\{\mathbf{x}\}_n, \ n \text{ odd} \end{cases}$$

$$= \frac{1}{2} \begin{cases} (\mathbf{x}_{n-1} + \mathbf{x}_n) - (-\mathbf{x}_{n-1} + \mathbf{x}_n), \ n \text{ even} \\ (\mathbf{x}_{n-2} + \mathbf{x}_{n-1}) + (-\mathbf{x}_{n-2} + \mathbf{x}_{n-1}), \ n \text{ odd} \end{cases} = \mathbf{x}_{n-1} = \mathbf{T}\{\mathbf{x}\}_n.$$

In this case, synthesis leads to a time delay in the signal. Since we have shown that $\mathbf{S}'\{\mathbf{A}'\{\mathbf{x}\}\} = \mathbf{T}\mathbf{x}$ for all discrete time signals $\mathbf{x}$, readers should be able to show that $\mathbf{A}'\{\mathbf{T}\mathbf{S}'\{\mathbf{a}\}\} = \mathbf{a}$ for all discrete time signals $\mathbf{a}$.

The next theorem provides some general principles for designing a filter bank.

Theorem 1.8.4 (Filter Bank Perfect Synthesis) *Assume that we have a filter bank with two channels using a lowpass and a highpass filter in the following forms:*

$$\mathbf{L}\{\mathbf{x}\} = (\downarrow_2)\left[\boldsymbol{\lambda}^* * \mathbf{x}\right] \ and \ \mathbf{H}\{\mathbf{x}\} = (\downarrow_2)\left[\boldsymbol{\eta}^* * \mathbf{x}\right].$$

Here $\boldsymbol{\lambda}_n^* = \overline{\boldsymbol{\lambda}_{-n}}$ is the impulse response for the adjoint of the digital filter with impulse response $\boldsymbol{\lambda}$ (see Lemma 1.8.10 for more details). Assume that synthesis combines the discrete time signals from the two channels in the following way:

$$\mathbf{S}\{\mathbf{x}\} = 2\boldsymbol{\ell} * \left[(\uparrow_2)\mathbf{L}\{\mathbf{x}\}\right] + 2\mathbf{h} * \left[(\uparrow_2)\mathbf{H}\{\mathbf{x}\}\right].$$

Then the synthesis is perfect, meaning that for all discrete time signals $\mathbf{x}$ we have

$$\mathbf{S}\{\mathbf{x}\} = \mathbf{x},$$

whenever the **modulation matrices** *for the analysis and synthesis filters are* **biorthogonal***, meaning that*

$$\begin{bmatrix} \boldsymbol{\ell} & \mathbf{h} \\ \mathbf{R}\boldsymbol{\ell} & \mathbf{R}\mathbf{h} \end{bmatrix} * \begin{bmatrix} \boldsymbol{\lambda}^* & \mathbf{R}\boldsymbol{\lambda}^* \\ \boldsymbol{\eta}^* & \mathbf{R}\boldsymbol{\eta}^* \end{bmatrix} = \begin{bmatrix} \boldsymbol{\delta} & 0 \\ 0 & \boldsymbol{\delta} \end{bmatrix}. \qquad (1.105)$$

Here $\mathbf{R}$ *is the reversal operator, defined in Eq.* (1.85). *Alternatively, let the impulse response* $\mathbf{p}$ *for the* **product filter** *be defined by*

$$\mathbf{p} = 2\boldsymbol{\ell} * \boldsymbol{\lambda}^* , \qquad (1.106)$$

and let the synthesis filters be related to the analysis filters by

$$\boldsymbol{\eta}^* = \mathbf{RT}^\alpha \boldsymbol{\ell} \ and$$
$$\mathbf{h} = \mathbf{RT}^{-\alpha}\boldsymbol{\lambda}^* ,$$

Here $\mathbf{T}$ *is the shift operator (defined in Eq.* (1.84)) *and* α *is an odd integer. Then the synthesis is perfect if and only if the product filter satisfies*

$$(\uparrow_2)(\downarrow_2)\mathbf{p} = \boldsymbol{\delta} . \qquad (1.107)$$

Another alternative is to choose an impulse response $\mathbf{k}$ *such that for all* ω

$$\mathscr{F}^d\{\mathbf{k}\}(\omega) \neq 0$$

(in other words, $\mathbf{k}$ *has an inverse for convolution), and let the impulse responses* $\mathbf{h}$ *and* $\boldsymbol{\eta}^*$ *be related to* $\boldsymbol{\lambda}^*$ *and* $\boldsymbol{\ell}$ *by*

$$\boldsymbol{\eta}^* = (\mathbf{RT}^\alpha \boldsymbol{\ell}) * [(\uparrow_2)\mathbf{k}] \ and$$
$$\mathbf{h} * [(\uparrow_2)\mathbf{k}] = \mathbf{RT}^{-\alpha}\boldsymbol{\lambda}^* . \qquad (1.108)$$

Then the synthesis is perfect if and only if the product filter again satisfies (1.107). The notation in this theorem uses the following mnemonics: $\boldsymbol{\ell}$ and λ are impulse responses for lowpass filters, while $\mathbf{h}$ and η are impulse responses for highpass filters.

Proof We can use the definitions of the filters $\mathbf{L}$ and $\mathbf{H}$ to compute

$$\mathbf{S}\{\mathbf{x}\} = 2\boldsymbol{\ell} * \left\{(\uparrow_2)(\downarrow_2)\left[\boldsymbol{\lambda}^* * \mathbf{x}\right]\right\} + 2\mathbf{h} * \left\{(\uparrow_2)(\downarrow_2)\left[\boldsymbol{\eta}^* * \mathbf{x}\right]\right\}$$

then Eq. (1.93) gives us

$$= \boldsymbol{\ell} * \left\{\boldsymbol{\lambda}^* * \mathbf{x} + \mathbf{R}\left(\boldsymbol{\lambda}^* * \mathbf{x}\right)\right\} + \mathbf{h} * \left\{\boldsymbol{\eta}^* * \mathbf{x} + \mathbf{R}\left(\boldsymbol{\eta}^* * \mathbf{x}\right)\right\}$$

then we use the fact that convolution is associative to get

$$= \left[\boldsymbol{\ell} * \boldsymbol{\lambda}^* + \mathbf{h} * \boldsymbol{\eta}^*\right] * \mathbf{x} + \left[\boldsymbol{\ell} * \mathbf{R}\boldsymbol{\lambda}^* + \mathbf{h} * \mathbf{R}\boldsymbol{\eta}^*\right] * (\mathbf{Rx}) .$$

We obtain perfect reconstruction for all discrete time signals **x** whenever

$$\begin{bmatrix} \ell & h \end{bmatrix} * \begin{bmatrix} \lambda^* & \mathbf{R}\lambda^* \\ \eta^* & \mathbf{R}\eta^* \end{bmatrix} = \begin{bmatrix} \delta & 0 \end{bmatrix} .$$

This condition is equivalent to

$$\begin{bmatrix} 0 & \delta \end{bmatrix} = \begin{bmatrix} 0 & \mathbf{R}\delta \end{bmatrix} = \begin{bmatrix} \mathbf{R}\left(\ell * \mathbf{R}\lambda^* + h * \mathbf{R}\eta^* \right) & \mathbf{R}\left(\ell * \lambda^* + h * \eta^* \right) \end{bmatrix}$$

which Eq. (1.97) shows is equivalent to

$$= \begin{bmatrix} \mathbf{R}\ell * \lambda^* + \mathbf{R}h * \eta^* & \mathbf{R}\ell * \mathbf{R}\lambda^* + \mathbf{R}h * \mathbf{R}\eta^* \end{bmatrix}$$

$$= \begin{bmatrix} \mathbf{R}\ell & \mathbf{R}h \end{bmatrix} * \begin{bmatrix} \lambda^* & \mathbf{R}\lambda^* \\ \eta^* & \mathbf{R}\eta^* \end{bmatrix} .$$

These two conditions prove the first claim in the theorem.

Suppose that we define the impulse response **p** for the product filter by (1.106), then choose $\eta^* = \mathbf{R}\mathbf{T}^\alpha \ell$ and $h = \mathbf{R}\mathbf{T}^{-\alpha}\lambda^*$. These assumptions imply that

$$\begin{bmatrix} \ell & h \\ \mathbf{R}\ell & \mathbf{R}h \end{bmatrix} * \begin{bmatrix} \lambda^* & \mathbf{R}\lambda^* \\ \eta^* & \mathbf{R}\eta^* \end{bmatrix} = \begin{bmatrix} \ell & \mathbf{R}\mathbf{T}^{-\alpha}\lambda^* \\ \mathbf{R}\ell & \mathbf{T}^{-\alpha}\lambda^* \end{bmatrix} * \begin{bmatrix} \lambda^* & \mathbf{R}\lambda^* \\ \mathbf{R}\mathbf{T}^\alpha \ell & \mathbf{T}^\alpha \ell \end{bmatrix}$$

then Eqs. (1.89) and (1.97) yield

$$= \begin{bmatrix} \ell * \lambda^* + \mathbf{R}\left(\mathbf{T}^{-\alpha}\lambda^* * \mathbf{T}^\alpha \ell \right) & \ell * \mathbf{R}\lambda^* - \mathbf{T}^{-\alpha}\mathbf{R}\lambda^* * \mathbf{T}^\alpha \ell \\ \mathbf{R}\ell * \lambda^* - \mathbf{T}^{-\alpha}\lambda^* * \mathbf{T}^\alpha \mathbf{R}\ell & \mathbf{R}\left(\ell * \lambda^* \right) + \mathbf{T}^{-\alpha}\lambda^* * \mathbf{T}^\alpha \ell \end{bmatrix}$$

then Eq. (1.95) gives us

$$= \begin{bmatrix} \ell * \lambda^* + \mathbf{R}\left(\lambda^* * \ell \right) & \ell * \mathbf{R}\lambda^* - \mathbf{R}\lambda^* * \ell \\ \mathbf{R}\ell * \lambda^* - \lambda^* * \mathbf{R}\ell & \mathbf{R}\left(\ell * \lambda^* \right) + \lambda^* * \ell \end{bmatrix}$$

and finally Eq. (1.93) produces

$$= \begin{bmatrix} 2(\uparrow_2)(\downarrow_2)[\ell * \lambda^*] & 0 \\ 0 & 2(\uparrow_2)(\downarrow_2)[\ell * \lambda^*] \end{bmatrix} = \begin{bmatrix} (\uparrow_2)(\downarrow_2)\mathbf{p} & 0 \\ 0 & (\uparrow_2)(\downarrow_2)\mathbf{p} \end{bmatrix}$$

This proves the claim (1.107).

The final claim in the theorem depends on the fact that the modulation matrices satisfy

$$\begin{bmatrix} \ell & h \\ \mathbf{R}\ell & \mathbf{R}h \end{bmatrix} * \begin{bmatrix} \lambda^* & \mathbf{R}\lambda^* \\ \eta^* & \mathbf{R}\eta^* \end{bmatrix} = \begin{bmatrix} \ell & h \\ \mathbf{R}\ell & \mathbf{R}h \end{bmatrix} * \begin{bmatrix} \lambda^* & \mathbf{R}\lambda^* \\ (\mathbf{R}\mathbf{T}^\alpha \ell) * [(\uparrow_2)\mathbf{k}] & \mathbf{R}\{[\mathbf{R}\mathbf{T}^\alpha \ell] * [(\uparrow_2)\mathbf{k}]\} \end{bmatrix}$$

then Eq. (1.97), Lemma 1.8.1, Equation (1.91) and Lemma 1.8.3 produce

$$= \begin{bmatrix} \ell & h \\ R\ell & Rh \end{bmatrix} * \begin{bmatrix} \lambda^* & R\lambda^* \\ [(\uparrow_2)k] * RT^\alpha \ell & [(\uparrow_2)k] * T^\alpha \ell \end{bmatrix}$$

$$= \begin{bmatrix} \ell & h * [(\uparrow_2)k] \\ R(\ell & h * [(\uparrow_2)k)] \end{bmatrix} * \begin{bmatrix} \lambda^* & R\lambda^* \\ RT^\alpha \ell & T^\alpha \ell \end{bmatrix}$$

then Eq. (1.108) gives us

$$= \begin{bmatrix} \ell & RT^{-\alpha}\lambda^* \\ R\ell & T^{-\alpha}\lambda^* \end{bmatrix} * \begin{bmatrix} \lambda^* & R\lambda^* \\ RT^\alpha \ell & T^\alpha \ell \end{bmatrix}$$

then Eqs. (1.89), (1.97) and (1.95) yield

$$= \begin{bmatrix} \ell * \lambda^* + R\{T^{-\alpha}\lambda^* * T^\alpha \ell\} & \ell * R\lambda^* - T^{-\alpha}R\lambda^* * T^\alpha \ell \\ R\ell * \lambda^* - T^{-\alpha}\lambda^* * T^\alpha R\ell & R\{\lambda^* * \ell\} + \lambda^* * \ell \end{bmatrix}$$

$$= \begin{bmatrix} (I + R)[\ell * \lambda^*] & 0 \\ 0 & (I + R)[\ell * \lambda^*] \end{bmatrix}$$

then the definition (1.106) of the product filter and Eq. (1.93) give us

$$= \begin{bmatrix} \frac{1}{2}\{p + Rp\} & 0 \\ 0 & \frac{1}{2}\{p + Rp\} \end{bmatrix} = \begin{bmatrix} (\uparrow_2)(\downarrow_2)p & 0 \\ 0 & (\uparrow_2)(\downarrow_2)p \end{bmatrix} = \begin{bmatrix} \delta & 0 \\ 0 & \delta \end{bmatrix} .$$

Note that the analysis filter in Theorem 1.8.4 is identical to the cascade algorithm in Eq. (1.134), and the synthesis filter in this theorem is identical to the pyramid algorithm in Eq. (1.135).

Recall that the product filter in Theorem 1.8.4 satisfies (1.107) whenever we have perfect synthesis. The next lemma gives us a bit more information about such product filters.

Lemma 1.8.11 *Suppose that*

$$(\uparrow_2)(\downarrow_2)p = \delta .$$

Then $p_0 = 1$ and $p_{2m} = 0$ for all $m \neq 0$. Further, $(\downarrow_2)p$ is the impulse response for a lowpass filter.

Proof Note that

$$[(\uparrow_2)(\downarrow_2)p]_n = \begin{cases} p_n, & n \text{ even} \\ 0, & n \text{ odd} \end{cases}$$

so $(\uparrow_2)(\downarrow_2)\mathbf{p} = \boldsymbol{\delta}$ if and only if all of the components of $\mathbf{p}$ with nonzero even indices are zero, and $\mathbf{p}_0 = 1$. It follows that

$$\mathscr{F}^d\{(\downarrow_2)\mathbf{T}\mathbf{p}\}(0) = \sum_{n=-\infty}^{\infty} \mathbf{p}_{2n} = 1 \; ,$$

so $(\downarrow_2)\mathbf{T}\mathbf{p}$ is the impulse response for a lowpass filter.

One approach to designing a filter bank with perfect synthesis is the following. First, find a lowpass filter with impulse response $\mathbf{p}$ so that (1.107) is satisfied. Next, factor

$$\mathscr{F}^d\{\mathbf{p}\}(\omega) = 2\mathscr{F}^d\{\boldsymbol{\ell}\}(\omega)\mathscr{F}^d\{\boldsymbol{\lambda}^*\}(\omega) \; .$$

Afterward, given an odd integer α we take

$$\boldsymbol{\eta}^* = \mathbf{R}\mathbf{T}^{\alpha}\boldsymbol{\ell} \text{ and } \mathbf{h} = \mathbf{R}\mathbf{T}^{-\alpha}\boldsymbol{\lambda}^* \; .$$

The integer α can be used to shift the nonzero components of $\mathbf{h}$ or $\boldsymbol{\eta}^*$ into convenient locations.

We would like to make another comment regarding the use of a filter bank. Note that the lowpass filter $\mathbf{L}$ operates on a discrete time signal $\mathbf{x}$ to produce a discrete time signal $\mathbf{L}\{\mathbf{x}\}$. The downsampling operator in $\mathbf{L}$ indicates that the output is associated with a coarser scale of discrete time signal indices. This lowpass filter output could also be decomposed into low and high components on the next coarser scale. This process could be repeated multiple times.

Commonly, both the discrete time signal $\mathbf{x}$ and the lowpass filter response $\boldsymbol{\lambda}^*$ have a finite number of nonzero entries. Suppose that

$$\mathbf{x}_n = 0 \text{ whenever } n \le \xi \text{ or } n > \xi + 2N \text{ , and}$$

$$\boldsymbol{\lambda}_n{}^* = 0 \text{ whenever } n \le \alpha \text{ or } n > \alpha + 2M \; .$$

Then

$$\left((\downarrow_2)[\boldsymbol{\lambda}^*\mathbf{x}]\right)_k = (\boldsymbol{\lambda}^* * \mathbf{x})_{2k} = \sum_{n=-\infty}^{\infty} \boldsymbol{\lambda}^*_{2k-n}\mathbf{x}_n = \sum_{n=\xi+1}^{\xi+2N} \boldsymbol{\lambda}^*_{k-n}\mathbf{x}_n$$

is nonzero only if $\alpha < 2k - n \le \alpha + 2M$ for some $n \in [\alpha + 1, \alpha + 2M]$. This implies that

$$\xi + \alpha + 2 \le 2k \le \xi + \alpha + 2(N + M) \; .$$

The storage required for the lowpass filter output is $N + M$ entries, compared to $2N$ entries for the original signal.

If desired, the lowpass filter output can be stored in the even components of the original filter $\mathbf{x}$, and the highpass filter output can be stored in the odd components of $\mathbf{x}$. On the second pass of the filters, the lowpass filter output can be stored in components that are even multiples of two, and the highpass filter output can be stored in components that are odd multiples of two. Repeated applications of the filters can be stored in components of $\mathbf{x}$ with indices that are successively higher powers of two.

1.8.1.7 Orthogonal Filter Banks

We will begin this section by describing a specialization of a two-channel filter bank with perfect synthesis. This filter bank will allow us to describe the Maxflat (Daubechies) filters in Sect. 1.8.1.8.

Theorem 1.8.5 *Suppose that we have a filter bank with two channels using a lowpass filter and a highpass filter in the following forms*

$$\mathbf{L}^*\{\mathbf{x}\} = \sqrt{2}(\downarrow_2)(\boldsymbol{\ell}^* * \mathbf{x}) \text{ and}$$

$$\mathbf{H}^*\{\mathbf{x}\} = \sqrt{2}(\downarrow_2)(\mathbf{h}^* * \mathbf{x}) .$$

Next, assume that the filter bank uses the synthesis

$$\mathbf{S}\{\mathbf{x}\} = \mathbf{L}\{\mathbf{L}^*\{\mathbf{x}\}\} + \mathbf{H}\{\mathbf{H}^*\{\mathbf{x}\}\} ,$$

where $\mathbf{L}^*$ is the adjoint of $\mathbf{L}$, and $\mathbf{H}^*$ is the adjoint of $\mathbf{H}$. Then the synthesis is perfect whenever

$$\begin{bmatrix} \boldsymbol{\ell} & \mathbf{h} \\ \mathbf{R}\boldsymbol{\ell} & \mathbf{R}\mathbf{h} \end{bmatrix} * \begin{bmatrix} \boldsymbol{\ell}^* & \mathbf{R}\boldsymbol{\ell}^* \\ \mathbf{h}^* & \mathbf{R}\mathbf{h}^* \end{bmatrix} = \begin{bmatrix} \delta & 0 \\ 0 & \delta \end{bmatrix} . \tag{1.109}$$

The Eqs. (1.109) are equivalent to the following set of equations:

$$\frac{1}{2} = \sum_{\substack{n=-\infty \\ n \text{ even}}}^{\infty} \left[|\boldsymbol{\ell}_n|^2 + |\mathbf{h}_n|^2 \right] = \sum_{\substack{n=-\infty \\ n \text{ odd}}}^{\infty} \left[|\boldsymbol{\ell}_n|^2 + |\mathbf{h}_n|^2 \right] \text{ and} \tag{1.110a}$$

$$0 = \sum_{\substack{n=-\infty \\ n \text{ even}}}^{\infty} \left[\boldsymbol{\ell}_n \overline{\boldsymbol{\ell}_{n-m}} + \mathbf{h}_n \overline{\mathbf{h}_{n-m}} \right] = \sum_{\substack{n=-\infty \\ n \text{ odd}}}^{\infty} \left[\boldsymbol{\ell}_n \overline{\boldsymbol{\ell}_{n-m}} + \mathbf{h}_n \overline{\mathbf{h}_{n-m}} \right] \text{ for all } m \neq 0$$

$$\tag{1.110b}$$

Alternatively, suppose that the filter banks are as described above, and that for some odd integer α we have

$$\mathbf{h} = \mathbf{R}\mathbf{T}^{\alpha}\boldsymbol{\ell}^{*} .$$

Then the synthesis is perfect if and only if

$$2(\uparrow_2)(\downarrow_2)\left[\boldsymbol{\ell} * \boldsymbol{\ell}^{*}\right] = \boldsymbol{\delta} , \tag{1.111}$$

which is equivalent to the conditions

$$\frac{1}{2} = \sum_{n=-\infty}^{\infty} |\ell_n|^2 \ \ and$$

$$0 = \sum_{n=-\infty}^{\infty} \ell_n \overline{\ell_{n-2k}} \ for \ all \ k \neq 0 .$$

Proof By taking adjoints of the definitions of $\mathbf{L}^{*}$ and $\mathbf{H}^{*}$, and then applying Lemma 1.8.9, we see that

$$\mathbf{L}\{\mathbf{a}\} = \sqrt{2}\boldsymbol{\ell} * [(\uparrow_2)\mathbf{a}] \ \ and \ \ \mathbf{H}\{\mathbf{a}\} = \sqrt{2}\mathbf{h} * [(\uparrow_2)\mathbf{a}] .$$

Theorem 1.8.4 now implies that the synthesis is perfect whenever (1.109) is satisfied. These equations are equivalent to

$$\boldsymbol{\delta} = \boldsymbol{\ell} * \boldsymbol{\ell}^{*} + \mathbf{h} * \mathbf{h}^{*} \ \ and \ \ 0 = \boldsymbol{\ell} * (\mathbf{R}\boldsymbol{\ell}^{*}) + \mathbf{h} * (\mathbf{R}\mathbf{h}^{*}) ,$$

which are in turn equivalent to

$$\delta_n = \sum_{m=-\infty}^{\infty} \left[\ell_m \overline{\ell_{m-n}} + \mathbf{h}_m \overline{\mathbf{h}_{m-n}} \right] \ \ and$$

$$0 = \sum_{m=-\infty}^{\infty} \left[(-1)^m \ell_m \overline{\ell_{m-n}} + (-1)^m \mathbf{h}_m \overline{\mathbf{h}_{m-n}} \right] .$$

If $n = 0$, these two equations take the form

$$1 = \sum_{m=-\infty}^{\infty} \left[|\ell_m|^2 + |\mathbf{h}_m|^2 \right] \ \ and$$

$$0 = \sum_{m=-\infty}^{\infty} (-1)^m \left[|\ell_m|^2 + |\mathbf{h}_m|^2 \right] .$$

We can sum and difference these two equations to get (1.110a). If $n \neq 0$, we can sum and difference to get (1.110b). Thus we have proved (1.110).

Next, suppose that the highpass filter in our self-adjoint filter bank satisfies

$$\mathbf{h} = \mathbf{R}\mathbf{T}^\alpha \boldsymbol{\ell}^* ,$$

where α is odd. The definition (1.85) of $\mathbf{R}$, the definition (1.84) of $\mathbf{T}$ and Eq. (1.103) show that the components of $\mathbf{h}$ are

$$\mathbf{h}_n = [\mathbf{R}\mathbf{T}^\alpha \boldsymbol{\ell}^*]_n = (-1)^n [\mathbf{T}^\alpha \boldsymbol{\ell}^*]_n = (-1)^n [\boldsymbol{\ell}^*]_{n-\alpha} = (-1)^n \overline{\boldsymbol{\ell}_{\alpha-n}} .$$

Then Eq. (1.103) shows that

$$[\mathbf{h}^*]_n = \overline{\mathbf{h}_{-n}} = \overline{(-1)^n \overline{\boldsymbol{\ell}_{\alpha_n}}} = (-1)^n \boldsymbol{\ell}_{\alpha+n} = (-1)^n [\mathbf{T}^{-\alpha}\boldsymbol{\ell}]_n = [\mathbf{R}\mathbf{T}^{-\alpha}\boldsymbol{\ell}]_n .$$

Thus

$$\mathbf{h}^* = \mathbf{R}\mathbf{T}^{-\alpha}\boldsymbol{\ell} ,$$

and the biorthogonality conditions (1.109) take the form

$$\begin{bmatrix} \delta & 0 \\ 0 & \delta \end{bmatrix} = \begin{bmatrix} \boldsymbol{\ell} & \mathbf{h} \\ \mathbf{R}\boldsymbol{\ell} & \mathbf{R}\mathbf{h} \end{bmatrix} * \begin{bmatrix} \boldsymbol{\ell}^* & \mathbf{R}\boldsymbol{\ell}^* \\ \mathbf{h}^* & \mathbf{R}\mathbf{h}^* \end{bmatrix} = \begin{bmatrix} \boldsymbol{\ell} & \mathbf{R}\mathbf{T}^\alpha \boldsymbol{\ell}^* \\ \mathbf{R}\boldsymbol{\ell} & \mathbf{T}^\alpha \boldsymbol{\ell}^* \end{bmatrix} * \begin{bmatrix} \boldsymbol{\ell}^* & \mathbf{R}\boldsymbol{\ell}^* \\ \mathbf{R}\mathbf{T}^{-\alpha}\boldsymbol{\ell} & \mathbf{T}^{-\alpha}\boldsymbol{\ell} \end{bmatrix}$$

then identities (1.89) and (1.93) produce

$$= \begin{bmatrix} \boldsymbol{\ell}*\boldsymbol{\ell}^* + \mathbf{R}\left(\mathbf{T}^\alpha \boldsymbol{\ell}^* * \mathbf{T}^{-\alpha}\boldsymbol{\ell}\right) & \boldsymbol{\ell}*\mathbf{R}\boldsymbol{\ell}^* - \mathbf{T}^\alpha \mathbf{R}\boldsymbol{\ell}^* * \mathbf{T}^{-\alpha}\boldsymbol{\ell} \\ \mathbf{R}\boldsymbol{\ell}*\boldsymbol{\ell}^* - \mathbf{T}^\alpha \boldsymbol{\ell}^* * \mathbf{T}^{-\alpha}\mathbf{R}\boldsymbol{\ell} & \mathbf{R}\left(\boldsymbol{\ell}*\boldsymbol{\ell}^*\right) + \mathbf{T}^\alpha \boldsymbol{\ell}^* * \mathbf{T}^{-\alpha}\boldsymbol{\ell} \end{bmatrix}$$

then Eq. (1.95) gives us

$$= \begin{bmatrix} \boldsymbol{\ell}*\boldsymbol{\ell}^* + \mathbf{R}\left(\boldsymbol{\ell}^**\boldsymbol{\ell}\right) & \boldsymbol{\ell}*\mathbf{R}\boldsymbol{\ell}^* - \mathbf{R}\boldsymbol{\ell}^**\boldsymbol{\ell} \\ \mathbf{R}\boldsymbol{\ell}*\boldsymbol{\ell}^* - \boldsymbol{\ell}^**\mathbf{R}\boldsymbol{\ell} & \mathbf{R}\left(\boldsymbol{\ell}*\boldsymbol{\ell}^*\right) + \boldsymbol{\ell}^**\boldsymbol{\ell} \end{bmatrix}$$

and finally Eq. (1.93) leads to

$$= \begin{bmatrix} 2(\uparrow_2)(\downarrow_2)\left[\boldsymbol{\ell}*\boldsymbol{\ell}^*\right] & 0 \\ 0 & 2(\uparrow_2)(\downarrow_2)\left[\boldsymbol{\ell}*\boldsymbol{\ell}^*\right] \end{bmatrix} .$$

For an arbitrary integer k, the condition on the diagonal can be written

$$2 \sum_{n=-\infty}^{\infty} \boldsymbol{\ell}_n \boldsymbol{\ell}^*_{2k-n} = \delta_{0,k} .$$

If $k = 0$, we get

$$\sum_{n=-\infty}^{\infty} |\ell_n|^2 = \frac{1}{2} \, ;$$

otherwise, for $k \neq 0$ we get

$$\sum_{n=-\infty}^{\infty} \ell_n \overline{\ell}_{2k-n} = 0 \, .$$

This completes the proof of the final claim.

Let us return to our theory in Theorem 1.8.4. The product filter for our orthogonal filter bank is

$$\mathbf{p} = 2\boldsymbol{\ell} * \boldsymbol{\ell}^* \, ,$$

so

$$\mathscr{F}^d\{\mathbf{p}\}(\omega) = 2\mathscr{F}^d\{\boldsymbol{\ell}\}(\omega).\mathscr{F}^d\{\boldsymbol{\ell}^*\}(\omega) = 2\left|\mathscr{F}^d\{\boldsymbol{\ell}\}(\omega)\right|^2 \, .$$

In order to factor this product filter, we note the following result, due to Fejér and Riesz [150, p. 117].

Lemma 1.8.12 *If* $\mathbf{p}$ *is a finite impulse response and* $\mathscr{F}^d\{\mathbf{p}\}(\omega) \geq 0$ *for all* ω*, then there exists a finite impulse response* $\boldsymbol{\ell}^*$ *so that for all* ω

$$\mathscr{F}^d\{\mathbf{p}\}(\omega) = \left|\mathscr{F}^d\{\boldsymbol{\ell}^*\}(\omega)\right|^2 \, .$$

The Fejér and Riesz lemma implies that we can begin the construction of an orthogonal filter bank by choosing a nonnegative product filter, then factoring it to find the lowpass filter in the filter bank.

The proof of Lemma 1.8.12 contains an algorithm for performing the factorization. We will use this lemma only for certain trigonometric polynomials in Sect. 1.8.1.8. These polynomials will all be symmetric, meaning that they have the form

$$b\left(e^{-i\omega}\right) = \mathscr{F}^d\{\boldsymbol{\beta}\}(\omega) = \sum_{n=1-k}^{k-1} \boldsymbol{\beta}_n e^{-i\omega n} \, ,$$

where $\boldsymbol{\beta}_{-n} = \boldsymbol{\beta}_n$. In other words,

$$b(z) = \sum_{n=1-k}^{k-1} \boldsymbol{\beta}_n z^n \, .$$

It follows that

$$b(1/z) = \sum_{n=1-k}^{k-1} \boldsymbol{\beta}_n z^{-n} = \sum_{m=1-k}^{k-1} \boldsymbol{\beta}_{-m} z^m = \sum_{m=1-k}^{k-1} \boldsymbol{\beta}_m z^m = b(z)$$

for all z. Thus, if $b(z) = 0$, we must also have that $b(1/z) = 0$. If $b(z) \geq 0$ for all $z = e^{-i\omega}$ on the unit circle, then any root of b on the unit circle must have even multiplicity. Since the coefficients of b are real, any complex root of b must be matched by a complex conjugate root. Now, suppose that we use a polynomial root-finder to compute the roots of $z^{k-1} b(z)$:

$$z^{k-1} b(z) = \boldsymbol{\beta}_{k-1} \left[\prod_{i=0}^{k-2} (z - r_i)(z - 1/r_i) \right].$$

Here, we may assume that $|r_i| \leq 1$ for all i. Then

$$b(z) = \boldsymbol{\beta}_{k-1} \prod_{i=0}^{k-2} (z - r_i)(1 - 1/[r_i z]) = \frac{\boldsymbol{\beta}_{k-1}}{\prod_{j=0}^{k-2} \overline{r}_j} \prod_{i=0}^{k-2} (z - r_i)(\overline{r}_i - 1/z)$$

$$= (-1)^{k-1} \frac{\boldsymbol{\beta}_{k-1}}{\prod_{j=0}^{k-2} \overline{r}_j} \prod_{i=0}^{k-2} (z - r_i)(1/z - \overline{r}_i).$$

This suggests that we take

$$q(z) = \sqrt{\frac{(-1)^{k-1} \boldsymbol{\beta}_{k-1}}{\prod_{j=0}^{k-2} \overline{r}_j}} \prod_{i=0}^{k-2} (z - r_i),$$

so that for z on the unit circle, we have

$$b(z) = q(z) \overline{q(z)}.$$

In the product that appears in the denominator in the square root, for each complex root r_j within or on the unit circle its complex conjugate must also be within or on the unit circle, and therefore also in this product. Thus the product in the denominator inside the square root is real. A similar observation shows that the polynomial $\prod_{i=0}^{k-2} (z - r_i)$ must have only real coefficients. The power of -1 will cause $q(z)$ to have only real coefficients.

For another discussion of factorization methods for such trigonometric polynomials, see Strang and Nguyen [164, p. 158ff].

1.8.1.8 Daubechies Filters

If k is a positive integer, then the corresponding **maxflat (Daubechies) product filter** takes its impulse response to have discrete Fourier transform

$$\mathscr{F}^d\{\mathbf{p}\}(\omega) = 2q_k\left(\frac{1-\cos\omega}{2}\right) \text{ where } q_k(y) = (1-y)^k \sum_{n=0}^{k-1}\binom{k+n-1}{n}y^n .$$

$$(1.112)$$

Since

$$(1-y)^{-k} = \sum_{n=0}^{\infty}\binom{k+n-1}{n}y^n$$

it follows that

$$q_k(y) = 1 + O\left(y^k\right) \text{ as } y \to 0 .$$

Thus for $1 \le j < k$ we have

$$\frac{d^j q_k}{dy^j}(0) = 0 .$$

It is also obvious from the form of q_k that for $0 \le j < k$ we have

$$\frac{d^j q_k}{dy^j}(1) = 0 .$$

As a result, q_k is the Hermite interpolation polynomial of degree $2k - 1$ that has value one at $y = 0$, value zero at $y = 1$, and zero derivatives of orders 1 through $k - 1$ at both $y = 0$ and $y = 1$. Consequently, $q_k(y) + q_k(1 - y) = 1$, since this is the Hermite interpolation polynomial to the function that takes the constant value of one. This implies that

$$\mathscr{F}^d\{(\downarrow_2)\mathbf{p}\}(\omega) = \frac{1}{2}\left[\mathscr{F}^d\{\mathbf{p}\}\left(\frac{\omega}{2}\right) + \mathscr{F}^d\{\mathbf{p}\}\left(\pi + \frac{\omega}{2}\right)\right]$$

$$= q_k\left(\frac{1-\cos(\omega/2)}{2}\right) + q_k\left(\frac{1-\cos(\pi+\omega/2)}{2}\right)$$

$$= q_k\left(\frac{1-\cos(\omega/2)}{2}\right) + q_k\left(\frac{1+\cos(\omega/2)}{2}\right)$$

$$= q_k\left(\frac{1-\cos(\omega/2)}{2}\right) + q_k\left(1 - \frac{1-\cos(\omega/2)}{2}\right) = 1 .$$

We conclude that $(\downarrow_2)\mathbf{p} = \boldsymbol{\delta}$; in other words, $\mathbf{p}$ is a **halfband filter**. This equation also implies that

$$1 = \frac{1}{2}\mathscr{F}^d\{\mathbf{p}\}(\omega) + \frac{1}{2}\mathscr{F}^d\{\mathbf{p}\}(\pi + \omega) = \mathscr{F}^d\{(\uparrow_2)(\downarrow_2)\mathbf{p}\}(\omega)$$

for all ω; this condition appeared in Theorem 1.8.4 in order to guarantee a perfect synthesis.

Furthermore, $\mathscr{F}^d\{\mathbf{p}\}(\omega) \geq 0$ for all ω, so Lemma 1.8.12 guarantees that there exists an impulse response $\boldsymbol{\ell}$ such that

$$\mathscr{F}^d\{\mathbf{p}\}(\omega) = \left| \sqrt{2}\mathscr{F}^d\{\boldsymbol{\ell}\}(\omega) \right|^2 .$$

From the factored form of q_k, it is easy to see that there is a trigonometric polynomial $Q_{k-1}\left(e^{i\omega}\right)$ so that

$$\mathscr{F}^d\{\boldsymbol{\ell}\}(\omega) = \left(\frac{1 + e^{-i\omega}}{2} \right)^k Q_{k-1}\left(e^{i\omega}\right) .$$

In particular, we have

$$2 = 2q_k(0) = \mathscr{F}^d\{\mathbf{p}\}(0) = \left| \sqrt{2}\mathscr{F}^d\{\boldsymbol{\ell}\}(0) \right|^2 = 2 \left| \sum_{n=-\infty}^{\infty} \ell_n \right|^2 ,$$

so we can choose the sign of $\boldsymbol{\ell}$ to satisfy

$$\sum_{n=-\infty}^{\infty} \ell_n = 1 .$$

Also

$$0 = 2q_k(1) = \mathscr{F}^d\{\mathbf{p}\}(\pi) = \left| \sqrt{2}\mathscr{F}^d\{\boldsymbol{\ell}\}(\pi) \right|^2 = 2 \left| \sum_{n=-\infty}^{\infty} (-1)^n \ell_n \right|^2 ,$$

so

$$\sum_{n=-\infty}^{\infty} (-1)^n \ell_n = 0 .$$

Thus $\boldsymbol{\ell}$ must be a lowpass filter.

Example 1.8.6 Suppose that we want to find the maxflat filter for $k = 1$. We have $q_1(y) = 1 - y$, so

$$\mathscr{F}^d\{\mathbf{p}\}(\omega) = 2\left(1 - \frac{1 - \cos\omega}{2}\right) = 1 + \cos\omega = 1 + \frac{1}{2}\left[e^{i\omega} + e^{-i\omega}\right]$$

$$= 2\left[\frac{1}{4}e^{-i\omega} + \frac{1}{2} + \frac{1}{4}e^{i\omega}\right] = 2\left|\frac{1 + e^{-i\omega}}{2}\right|^2$$

Thus the product filter has impulse response

$$\mathbf{p}_n = \begin{cases} 1/2, & n = -1 \\ 1, & n = 0 \\ 1/2, & n = 1 \\ 0, \text{ otherwise} \end{cases},$$

and we can take $\boldsymbol{\ell}$ to be determined by its discrete Fourier transform

$$\mathscr{F}^d\{\boldsymbol{\ell}\}(\omega) = \frac{1 + e^{-i\omega}}{2}.$$

Taking the inverse discrete time Fourier transform as in Theorem 1.8.1 gives us

$$\ell_n = \begin{cases} 1/2, & n = 0 \\ 1/2, & n = 1 \\ 0, \text{ otherwise} \end{cases}.$$

Note that any shift of this lowpass filter would also work:

$$\left|\mathscr{F}^d\left\{\mathbf{T}^\sigma\boldsymbol{\ell}\right\}(\omega)\right|^2 = \left|e^{-i\sigma\omega}\mathscr{F}^d\{\boldsymbol{\ell}\}(\omega)\right|^2 = \left|\mathscr{F}^d\{\boldsymbol{\ell}\}(\omega)\right|^2.$$

In order to avoid looking backwards, we normally shift so that the nonzero lowpass filter components have nonzero entries for components 0 and 1; this is the Haar filter. We will choose $\alpha = 1$ in Theorem 1.8.5. Then the impulse response for the highpass filter is $\mathbf{h}$ where

$$\mathbf{h}_n = (-1)^n\overline{\ell_{\alpha-n}} = (-1)^n\ell_{1-n} = \begin{cases} 1/2, & n = 0 \\ -1/2, & n = 1 \\ 0, \text{ otherwise} \end{cases}.$$

Example 1.8.7 Suppose that we want to find the maxflat filter for $k = 2$. We have

$$q_2(y) = (1 - y)^2\left[\binom{1}{0} + \binom{2}{1}y\right] = (1 - y)^2(1 + 2y).$$

It follows that the impulse response for the product filter satisfies

$$\mathscr{F}^d\{\mathbf{p}\}(\omega) = 2\left(1 - \frac{1-\cos\omega}{2}\right)^2\left[1 + 2\frac{1-\cos\omega}{2}\right] = 2\left[\frac{1+\cos\omega}{2}\right]^2[2-\cos\omega] \ .$$

$$= 2\left[\frac{2 + e^{-i\omega} + e^{i\omega}}{4}\right]^2\left[\frac{4 - e^{-i\omega} - e^{i\omega}}{2}\right]$$

$$= \frac{1}{16}\left[e^{-2i\omega} + 4e^{-i\omega} + 6 + 4e^{i\omega} + 2e^{2i\omega}\right]\left[-e^{-i\omega} + 4 - e^{i\omega}\right]$$

$$= \frac{1}{16}\left[-e^{-3i\omega} + 9e^{-i\omega} + 16 + 9e^{i\omega} - e^{3i\omega}\right]$$

Thus the product filter has components

$$\mathbf{p}_n = \begin{cases} -1/16, & n = -3 \\ 9/16, & n = -1 \\ 1, & n = 0 \\ 9/16, & n = 1 \\ -1/16, & n = 3 \\ 0, \text{ otherwise} \end{cases} \ .$$

Now

$$\frac{1+\cos\omega}{2} = \left|\frac{1 + e^{i\omega}}{2}\right|^2 = \left|\frac{1 + e^{-i\omega}}{2}\right|^2$$

is already the square of the modulus of a trigonometric polynomial. In order to factor

$$2 - \cos\omega = \left|a - \frac{e^{i\omega}}{2a}\right|^2 \ ,$$

we expand

$$2-\cos\omega = -\frac{e^{-i\omega}}{2} + 2 - \frac{e^{i\omega}}{2} = \left(a - \frac{e^{i\omega}}{2a}\right)\left(a - \frac{e^{-i\omega}}{2a}\right) = -\frac{e^{-i\omega}}{2} + a^2 + \frac{1}{4a^2} - \frac{e^{i\omega}}{2} \ .$$

Thus, we want to find a scalar a^2 to solve $a^2 + 1/(4a^2) = 2$. This implies that $a^2 = 1 \pm \sqrt{3}/2$. We will choose the positive sign in this expression, and the positive solution for

$$a = \frac{1 + \sqrt{3}}{2} \ .$$

We now have

$$
2 - \cos\omega = \left(a - \frac{e^{-i\omega}}{2a}\right)\left(a - \frac{e^{i\omega}}{2a}\right)
$$

$$
= \left(\frac{1+\sqrt{3}}{2} - \frac{\sqrt{3}-1}{2}e^{-i\omega}\right)\left(\frac{1+\sqrt{3}}{2} - \frac{\sqrt{3}-1}{2}e^{i\omega}\right) .
$$

Thus $\mathscr{F}^d\{\mathbf{p}\}(\omega) = \left|\sqrt{2}\mathscr{F}^d\{\boldsymbol{\ell}\}(\omega)\right|^2$ where

$$
\mathscr{F}^d\{\boldsymbol{\ell}\}(\omega) = \left[\frac{1+e^{-i\omega}}{2}\right]^2\left[a - \frac{e^{-i\omega}}{2a}\right]
$$

$$
= \left[-\frac{1}{8a}e^{-3i\omega} + \left(\frac{a}{4} - \frac{1}{4a}\right)e^{-2i\omega} + \left(\frac{a}{2} - \frac{1}{8a}\right)e^{-i\omega} + \frac{a}{4}\right]
$$

$$
= \frac{1}{8}\left[(1+\sqrt{3}) + (3+\sqrt{3})e^{-i\omega} + (3-\sqrt{3})e^{-2i\omega} + (1-\sqrt{3})e^{-3i\omega}\right] .
$$

Again, any shift of $\boldsymbol{\ell}$ would work as well. Thus the components of $\boldsymbol{\ell}$ are

$$
\ell_n = \begin{cases} (1+\sqrt{3})/8, & n = 0 \\ (3+\sqrt{3})/8, & n = 1 \\ (3-\sqrt{3})/8, & n = 2 \\ (1-\sqrt{3})/8, & n = 3 \\ 0, \text{ otherwise} \end{cases} .
$$

We can check that

$$
(\boldsymbol{\ell}*\boldsymbol{\ell}^*)_n = \sum_{m=0}^{3}\ell_m\ell_{n-m}^* = \sum_{m=0}^{3}\ell_m\ell_{m-n} = \begin{cases} \ell_0\ell_3, n=-3 \\ \ell_0\ell_2 + \ell_1\ell_3, n=-2 \\ \ell_0\ell_1 + \ell_1\ell_2 + \ell_2\ell_3, n=-1 \\ \ell_0\ell_0 + \ell_1\ell_1 + \ell_2\ell_2 + \ell_3\ell_3, & n=0 \\ \ell_1\ell_0 + \ell_2\ell_1 + \ell_3\ell_2, & n=1 \\ \ell_2\ell_0 + \ell_3\ell_1, & n=2 \\ \ell_3\ell_0, & n=3 \end{cases}
$$

$$
= \begin{cases} -1/32, & n=-3 \\ 0, & n=-2 \\ 9/32, & n=-1 \\ 1/2, & n=0 \\ 9/32, & n=1 \\ 0, & n=2 \\ -1/32, & n=3 \\ 0, \text{ otherwise} \end{cases} = \frac{1}{2}\mathbf{p}_n .
$$

We will take the impulse response for the highpass filter to be $\mathbf{h} = \mathbf{RT}^3\boldsymbol{\ell}^*$, which implies that

$$\mathbf{h}_n = (-1)^n\overline{\boldsymbol{\ell}_{3-n}} = \begin{cases} (1 - \sqrt{3})/8, & n = 0 \\ -(3 - \sqrt{3})/8, & n = 1 \\ (3 + \sqrt{3})/8, & n = 2 \\ -(1 + \sqrt{3})/8, & n = 3 \\ 0, \text{ otherwise} \end{cases}$$

Interested readers may view C^{++} code to implement these orthogonal filter banks for arbitrary order in the files DaubechiesFilterBank.H and DaubechiesFilterBank.C. The `DaubechiesFilterBank` class constructor uses the Fejér and Riesz Lemma 1.8.12 to construct the lowpass filter. Furthermore, the C^{++} file testDaubechiesFilterBank.C will use the Daubechies filter bank for arbitrary order k to analyze and then synthesize a signal produced by the exponential function.

Exercise 1.8.9 Find the maxflat product filter $\mathbf{p}$ for $k = 3$. Then find a lowpass filter impulse response $\boldsymbol{\ell}$ so that

$$\mathscr{F}^d\{\mathbf{p}\}(\omega) = \left|\sqrt{2}\mathscr{F}^d\{\boldsymbol{\ell}\}(\omega)\right|^2 .$$

Finally, find the components of $\boldsymbol{\ell}$ and the corresponding highpass filter impulse response $\mathbf{h}$.

1.8.1.9 Maxflat Biorthogonal Filters

We could also use the maxflat product filter to define biorthogonal filters. However, some additional care is needed so that the adjoint lowpass filter will correspond to a continuous scaling function. For more information on this topic, see either Cohen [48, p. 110ff] or Cohen et al. [49].

Recall that for any $k \geq 1$ we defined the maxflat product filter to have impulse response $\mathbf{p}^{(k)}$ where

$$\mathscr{F}^d\{\mathbf{p}^{(k)}\}(\omega) = 2\left[\frac{1 + \cos\omega}{2}\right]^k \sum_{n=0}^{k-1}\binom{k-1+n}{n}\left[\frac{1 - \cos\omega}{2}\right]^n$$

$$= 2\left|\frac{1 + e^{-i\omega}}{2}\right|^{2k} \sum_{n=0}^{k-1}\binom{k-1+n}{n}\left|\frac{1 - e^{-i\omega}}{2}\right|^{2n} .$$

$$= 2\left\{e^{i(j-k)\omega}\left[\frac{1 + e^{-i\omega}}{2}\right]^{2k-j}\sum_{n=0}^{k-1}\binom{k-1+n}{n}\left|\frac{1 - e^{-i\omega}}{2}\right|^{2n}\right\}\left\{\left[\frac{1 + e^{-i\omega}}{2}\right]^j\right\} .$$

$$(1.113)$$

This implies a factorization $\mathbf{p} = 2\boldsymbol{\ell} * \boldsymbol{\lambda}^*$ of the product filter, as required by Eq. (1.106) of the Filter Bank Perfect Synthesis Theorem 1.8.4. Given a positive integer j, the positive integer k could be chosen so that

$$2k - j > \frac{1/2 + (j - 2)\log_2 \sqrt{3}}{1 - \log_2 \sqrt{3}}.$$

This condition is sufficient to make sure that the dual scaling function is square integrable (see Sect. 1.8.2.5).

Equation (1.113) suggests that we might choose the impulse response $\boldsymbol{\ell}^{(j)}$ to satisfy

$$\mathscr{F}^d\{\boldsymbol{\ell}^{(j)}\}(\omega) = \left[\frac{1 + e^{-i\omega}}{2}\right]^j = 2^{-j}\sum_{n=0}^{j}\binom{j}{n}e^{-in\omega}. \tag{1.114}$$

In other words,

$$\ell_n^{(j)} = \begin{cases} 2^{-j}\binom{j}{n}, \ 0 \leq n \leq j \\ \quad\quad 0, \ \text{otherwise} \end{cases}. \tag{1.115}$$

Equation (1.106) from the Filter Bank Perfect Synthesis Theorem 1.8.4 requires that we define the impulse response $\boldsymbol{\lambda}^{(k,j)}$ by

$$\overline{\mathscr{F}^d\left\{\boldsymbol{\lambda}^{(k,j)}\right\}(\omega)} = \mathscr{F}^d\left\{\left[\boldsymbol{\lambda}^{(k,j)}\right]^*\right\}(\omega) = \frac{\mathscr{F}^d\{\mathbf{p}^{(k)}\}(\omega)}{2\mathscr{F}^d\{\boldsymbol{\ell}^{(j)}\}(\omega)}$$

$$= e^{i(j-k)\omega}\left[\frac{1 + e^{i\omega}}{2}\right]^{2k-j}\sum_{n=0}^{k-1}\binom{k-1+n}{n}\left|\frac{1 - e^{-i\omega}}{2}\right|^{2n}.$$

It is easy to see that $\boldsymbol{\lambda}^{(k,j)}$ is the impulse response for a lowpass filter, whenever $2k - j > 0$.

The form of $\mathscr{F}^d\left\{\boldsymbol{\lambda}^{(k,j)}\right\}(\omega)$ suggests that we define an auxiliary impulse response $\boldsymbol{\beta}^{(k)}$ by

$$\mathscr{F}^d\left\{\boldsymbol{\beta}^{(k)}\right\}(\omega) = \sum_{n=0}^{k-1}\binom{k-1+n}{n}\left|\frac{1 - e^{-i\omega}}{2}\right|^{2n}. \tag{1.116}$$

Since Lemma 1.8.6 shows that the discrete Fourier transform converts convolution into multiplication, we find that

$$\lambda^{(k,j)} = \left[\mathbf{T}^{k-j} \boldsymbol{\ell}^{(2k-j)} \right] * \boldsymbol{\beta}^{(k)} \,.$$

Note that if $k = 1$, then $\mathscr{F}^d \left\{ \boldsymbol{\beta}^{(k)} \right\} (\omega) = 1$, so $\boldsymbol{\beta}^{(k)} = \boldsymbol{\delta}$. In order to evaluate the components of $\boldsymbol{\beta}^{(k)}$ for $k > 1$, we will use the following lemma.

Lemma 1.8.13 *Suppose that $0 \leq m \leq n$ are integers Then for $0 \leq k \leq m$ we have*

$$\sum_{i=0}^{k} \binom{m}{i} \binom{n}{k-i} = \binom{m+n}{k} , \tag{1.117a}$$

for $m \leq k \leq n$ we have

$$\sum_{i=0}^{m} \binom{m}{i} \binom{n}{k-i} = \binom{m+n}{k} , \tag{1.117b}$$

and for $n \leq k \leq m+n$ we have

$$\sum_{i=k-n}^{m} \binom{m}{i} \binom{n}{k-i} = \binom{m+n}{k} . \tag{1.117c}$$

Proof If $m \leq n$, then

$$(1 + t)^{m+n} = \sum_{\ell=0}^{m+n} \binom{m+n}{\ell} t^{\ell}$$

$$= (1+t)^m (1+t)^n = \sum_{i=0}^{m} \sum_{j=0}^{n} \binom{m}{i} \binom{n}{j} t^{i+j} = \sum_{i=0}^{m} \sum_{l=i}^{n+i} \binom{m}{i} \binom{n}{k-i} t^k$$

$$= \sum_{k=0}^{m-1} \sum_{i=0}^{k} \binom{m}{i} \binom{n}{k-i} t^k + \sum_{k=m}^{n} \binom{m}{i} \binom{n}{k-i} t^k + \sum_{k=n+1}^{n+m} \binom{m}{i} \binom{n}{k-i} t^k \,.$$

We can equate coefficients of the polynomials to obtain the Eqs. (1.117).

Now we can expand

$$\mathscr{F}^d\left\{\boldsymbol{\beta}^{(k)}\right\}(\omega) = \sum_{n=0}^{k-1}\binom{k-1+n}{n}\left|\frac{1-e^{-i\omega}}{2}\right|^{2n}$$

$$= \sum_{n=0}^{k-1}\binom{k-1+n}{n}\left[\frac{1-e^{-i\omega}}{2}\right]^{n}\left[\frac{1-e^{i\omega}}{2}\right]^{n}$$

and then use binomial expansions to get

$$= \sum_{n=0}^{k-1}\sum_{m=0}^{n}\sum_{\ell=0}^{n}\binom{k-1+n}{n}\binom{n}{m}\binom{n}{\ell}(-1)^{\ell+m}2^{-2n}e^{i\omega(\ell-m)}$$

then we can change summation variables

$$= \sum_{n=0}^{k-1}\sum_{m=0}^{n}\sum_{j=-m}^{n-m}\binom{k-1+n}{n}\binom{n}{m}\binom{n}{j+m}(-1)^{j}2^{-2n}e^{i\omega j}$$

$$= 1 + \sum_{n=1}^{k-1}\sum_{m=0}^{n}\sum_{j=-m}^{n-m}\binom{k-1+n}{n}\binom{n}{m}\binom{n}{j+m}(-1)^{j}2^{-2n}e^{i\omega j}$$

then we can interchange the order of summation to get

$$= 1 + \sum_{n=1}^{k-1}\sum_{j=-n}^{-1}\sum_{m=-j}^{n}\binom{k-1+n}{n}\binom{n}{m}\binom{n}{j+m}(-1)^{j}2^{-2n}e^{i\omega j}$$

$$+ \sum_{n=1}^{k-1}\sum_{m=0}^{n}\binom{k-1+n}{n}\binom{n}{m}\binom{n}{m}2^{-2n}$$

$$+ \sum_{n=1}^{k-1}\sum_{j=1}^{n}\sum_{m=0}^{n-j}\binom{k-1+n}{n}\binom{n}{m}\binom{n}{j+m}(-1)^{j}2^{-2n}e^{i\omega j}$$

$$= 1 + \sum_{j=1-k}^{-1}\sum_{n=-j}^{k-1}\sum_{m=-j}^{n}\binom{k-1+n}{n}\binom{n}{m}\binom{n}{j+m}(-1)^{j}2^{-2n}e^{i\omega j}$$

$$+ \sum_{n=1}^{k-1}\sum_{m=0}^{n}\binom{k-1+n}{n}\binom{n}{m}\binom{n}{m}2^{-2n}$$

$$+ \sum_{j=1}^{k-1}\sum_{n=j}^{k-1}\sum_{m=0}^{n-j}\binom{k-1+n}{n}\binom{n}{m}\binom{n}{j+m}(-1)^{j}2^{-2n}e^{i\omega j}$$

then we can use Lemma 1.8.13

$$= 1 + \sum_{j=1-k}^{-1}\sum_{n=-j}^{k-1}\binom{k-1+n}{n}\binom{2n}{n-j}(-1)^j 2^{-2n}e^{i\omega j} + \sum_{n=1}^{k-1}\binom{k-1+n}{n}\binom{2n}{n}2^{-2n}$$

$$+ \sum_{j=1}^{k-1}\sum_{n=j}^{k-1}\binom{k-1+n}{n}\binom{2n}{n-j}(-1)^j 2^{-2n}e^{i\omega j}$$

$$= \sum_{j=1-k}^{k-1}\sum_{n=|j|}^{k-1}\binom{k-1+n}{n}\binom{2n}{n-|j|}(-1)^j 2^{-2n}e^{i\omega j}$$

It follows that

$$\beta_j^{(k)} = \begin{cases} (-1)^j \sum_{n=|j|}^{k-1}\binom{k-1+n}{n}\binom{2n}{n-|j|}2^{-2n} & ,\, |j| < k \\ 0, \text{ otherwise} \end{cases}.$$

In summary, the filters for the maxflat biorthogonal filter are given by the equations

$$\ell_n^{(k)} = \begin{cases} 2^{-k}\binom{k}{n}, 0 \le n \le k \\ 0, \text{ otherwise} \end{cases},$$

$$\beta_n^{(k)} = \begin{cases} \sum_{m=|n|}^{k-1}\binom{k+m-1}{m}\binom{2m}{m-|n|}, & |n| < k \\ 0, \text{ otherwise} \end{cases},$$

$$\boldsymbol{\lambda}^{(k)} = \boldsymbol{\ell}^{(k)} * \boldsymbol{\beta}^{(k)} \iff \left[\boldsymbol{\lambda}^{(k)}\right]^* = \left[\boldsymbol{\ell}^{(k)}\right]^* * \left[\boldsymbol{\beta}^{(k)}\right]^*,$$

$$\mathbf{h}^{(k)} = \mathbf{RT}^{-\alpha}\left[\boldsymbol{\lambda}^{(k)}\right]^* = \mathbf{RT}^{-\alpha}\left\{\left[\boldsymbol{\ell}^{(k)}\right]^*\left[\boldsymbol{\beta}^{(k)}\right]^*\right\} \text{ and}$$

$$\left[\boldsymbol{\eta}^{(k)}\right]^* = \mathbf{RT}^{\alpha}\boldsymbol{\ell}^{(k)}.$$

Here α is any odd integer. According to Theorem 1.8.4, for some given discrete time signal $\mathbf{x}$ we can compute the lowpass and highpass analysis filters by

$$\mathbf{L}\{\mathbf{x}\} = (\downarrow_2)\left[\boldsymbol{\beta}^{(k)}\right]^* * \left\{\left[\boldsymbol{\ell}^{(k)}\right]^* * \mathbf{x}\right\} \text{ and}$$

$$\mathbf{H}\{\mathbf{x}\} = (\downarrow_2)\mathbf{RT}^{\alpha}\left\{\boldsymbol{\ell}^{(k)} * \mathbf{x}\right\}.$$

We synthesize by computing

$$\mathbf{S}\{\mathbf{x}\} = 2\boldsymbol{\ell}^{(k)} * [(\uparrow_2)\mathbf{L}\{\mathbf{x}\}] + 2\mathbf{RT}^{-\alpha}\left[\boldsymbol{\beta}^{(k)}\right]^* * \left[\boldsymbol{\ell}^{(k)}\right]^* * [(\uparrow_2)\mathbf{H}\{\mathbf{x}\}].$$

Example 1.8.8 If $k = 1 = j$ and α is an odd integer, the theory in Theorem 1.8.4 applied to this maxflat biorthogonal filter tells us to take

$$\ell_n = \begin{cases} 1/2, & n = 0 \text{ or } 1 \\ 0, & \text{otherwise} \end{cases} ,$$

$$\beta = \delta ,$$

$$\lambda^* = (\ell * \beta)^* = \ell^* ,$$

$$\mathbf{h} = \mathbf{RT}^{-\alpha}\lambda^* = \mathbf{RT}^{-\alpha}\ell^* \text{ and}$$

$$\eta^* = \mathbf{RT}^{\alpha}\ell .$$

Here we have dropped the superscripts for convenience. It will be convenient to take $\alpha = 1$, so that the nonzero components of $\mathbf{h}$ have indices zero and one. Then the impulse responses will have components

$$\ell_n = \begin{cases} 1/2, & n = 0 \text{ or } 1 \\ 0, & \text{otherwise} \end{cases} ,$$

$$\lambda^* = \begin{cases} 1/2, & n = -1 \text{ or } 0 \\ 0, & \text{otherwise} \end{cases} ,$$

$$\mathbf{h}_n = \begin{cases} 1/2, & n = 0 \\ -1/2, & n = 1 \\ 0, & \text{otherwise} \end{cases} ,$$

$$\eta^* = \begin{cases} -1/2, & n = -1 \\ 1/2, & n = 0 \\ 0, & \text{otherwise} \end{cases} .$$

This is the Haar filter bank.

Example 1.8.9 If $k = 2 = j$ and α is an odd integer, the maxflat biorthogonal filter takes

$$\ell_n = \begin{cases} 1/4, & n = 0 \\ 1/2, & n = 1 \\ 1/4, & n = 2 \\ 0, & \text{otherwise} \end{cases} ,$$

$$\beta_n = \begin{cases} -1/2, & n = -1 \\ 2, & n = 0 \\ -1/2, & n = 1 \\ 0, & \text{otherwise} \end{cases} ,$$

$$\boldsymbol{\lambda}^* = \boldsymbol{\ell}^* * \boldsymbol{\beta}^* = \begin{cases} \ell_2\beta_1, & n = -3 \\ \ell_1\beta_1 + \ell_2\beta_0, & n = -2 \\ \ell_0\beta_1 + \ell_1\beta_0 + \ell_2\beta_{-1}, & n = -1 \\ \ell_0\beta_0 + \ell_1\beta_{-1}, & n = 0 \\ \ell_0\beta_{-1}, & n = 1 \\ 0, & \text{otherwise} \end{cases} = \begin{cases} -1/8, & n = -3 \\ 1/4, & n = -2 \\ 3/4, & n = -1 \\ 1/4, & n = 0 \\ -1/8, & n = 1 \\ 0, & \text{otherwise} \end{cases},$$

$\mathbf{h} = \mathbf{R}\mathbf{T}^{-\alpha}\boldsymbol{\lambda}^*$ and

$\boldsymbol{\eta}^* = \mathbf{R}\mathbf{T}^{\alpha}\boldsymbol{\ell}$.

Again, we have dropped the superscripts for convenient notation. We will find it helpful to choose $\alpha = 3$. This choice implies that the impulse responses will have components

$$\mathbf{h}_n = (-1)^n\boldsymbol{\lambda}^*_{n-3} = \begin{cases} -1/8, & n = 0 \\ -1/4, & n = 1 \\ 3/4, & n = 2 \\ -1/4, & n = 3 \\ -1/8, & n = 4 \\ 0, & \text{otherwise} \end{cases} \text{ and}$$

$$\boldsymbol{\eta}^*_n = (-1)^n\ell_{n+3} = \begin{cases} -1/4, & n = -3 \\ 1/2, & n = -2 \\ -1/4, & n = -1 \\ 0, & \text{otherwise} \end{cases}.$$

It is straightforward to check that this is a biorthogonal filter bank. However, the dual scaling function defined from the adjoint lowpass filter is not sufficiently smooth to be useful in computation.

Example 1.8.10 If $k = 4, j = 2$ and α is an odd integer, the maxflat biorthogonal filter takes

$$\ell_n = \begin{cases} 1/4, & n = 0 \\ 1/2, & n = 1 \\ 1/4, & n = 2 \\ 0, & \text{otherwise} \end{cases},$$

$$\beta_n = \begin{cases} 3/8, & n = -2 \\ -9/4, & n = -1 \\ 19/4, & n = 0 \\ -9/4, & n = 1 \\ 3/8, & n = 2 \\ 0, & \text{otherwise} \end{cases},$$

$$\lambda^* = \begin{cases} 3/128, & n = -5 \\ -3/64, & n = -4 \\ -1/8, & n = -3 \\ 19/64, & n = -2 \\ 45/64, & n = -1 \\ 19/64, & n = 0 \\ -1/8, & n = 1 \\ -3/64, & n = 2 \\ 3/128, & n = 3 \\ 0, & \text{otherwise} \end{cases},$$

$$\mathbf{h} = \mathbf{R}\mathbf{T}^{-\alpha}\lambda^* \text{ and}$$

$$\eta^* = \mathbf{R}\mathbf{T}^{\alpha}\ell .$$

Again, we have dropped the superscripts for convenient notation. We will find it helpful to choose $\alpha = 3$. This choice implies that the impulse responses will have components

$$\mathbf{h}_n = (-1)^n \lambda^*_{n-3} = \begin{cases} 3/128, & n = 0 \\ 3/64, & n = 1 \\ -1/8, & n = 2 \\ -19/64, & n = 3 \\ 45/64, & n = 4 \\ -19/64, & n = 5 \\ -1/8, & n = 6 \\ 3/64, & n = 7 \\ 3/128, & n = 8 \\ 0, & \text{otherwise} \end{cases} \text{ and}$$

$$\eta^*_n = (-1)^n \ell_{n+3} = \begin{cases} -1/4, & n = -3 \\ 1/2, & n = -2 \\ -1/4, & n = -1 \\ 0, & \text{otherwise} \end{cases}.$$

It is straightforward to check that this is a biorthogonal filter bank. In this case, the dual scaling function defined from the adjoint lowpass filter is continuous.

Interested readers may view C^{++} code to implement these biorthogonal spline filter banks for arbitrary order in the files SplineFilterBank.H and SplineFilter-Bank.C. Furthermore, the C^{++} file testSplineScalingFunction.C will use the spline filter bank for arbitrary product filter order k and arbitrary spline order j to analyze and synthesize a signal produced by the exponential function.

Exercise 1.8.10 Construct the lowpass and highpass filters for the biorthogonal spline filter bank with $k = 1$ and $j = 2$.

1.8.2 Functions on a Continuum

Our next goal is to extend the ideas regarding filter banks for discrete time signals to functions on a continuum. We will see that, under certain conditions, we can use the coefficients from the impulse function of a lowpass filter to define continuous function at binary numbers. Linear combinations of the translations of this function will generate a linear space of functions, and successive dilations of these translations will generate an infinite sequence of linear spaces. Under the right conditions, those linear spaces will have good approximation properties.

1.8.2.1 Continuous Fourier Transform

In Sect. 1.8.1.4, we studied the discrete time Fourier transform, which is designed to operate on discrete time signals. In this section, we will present the continuous Fourier transform, which is designed to operate on functions defined on the real line. We will connect the discrete and continuous Fourier transforms when we discuss aliasing in Theorem 1.8.10, and the Shannon sampling Theorem 1.8.11.

Let us begin by defining our new concept.

Definition 1.8.8 Suppose that f is absolutely integrable in $(-\infty, \infty)$. Then for all real numbers ω, the **Fourier transform** of f is

$$\mathcal{F}\{f\}(\omega) \equiv \int_{-\infty}^{\infty} f(t)e^{-i\omega t} \, dt \,. \tag{1.118}$$

There are a number of important and well-known results regarding this Fourier transform. However, the proofs of these results are more complicated than the corresponding statements for discrete time Fourier series. Our first result, for which a proof can be found in Rudin [153, p. 184], shows that the Fourier transform is a bounded linear operator.

Lemma 1.8.14 *If f is absolutely integrable in $(-\infty, \infty)$, then $\mathcal{F}\{f\}$ is continuous, vanishes at infinity, and*

$$|\mathcal{F}\{f\}(\omega)| \le \int_{-\infty}^{\infty} |f(t)| \, dt \,.$$

for all $\omega \in (-\infty, \infty)$.

The next result establishes the inverse operator for the Fourier transform. A proof of this theorem can be found in Rudin [153, p. 186], Yosida [187, p. 147] or Zygmund [190, vol. II, p. 247ff].

Theorem 1.8.6 (Fourier Inversion) *Suppose that both f and $\mathscr{F}\{f\}$ are absolutely integrable in $(-\infty, \infty)$. Let*

$$\phi(t) = \frac{1}{2\pi} \int_{-\infty}^{\infty} \mathscr{F}\{f\}(\omega) e^{i\omega t} \, d\omega \ .$$

Then ϕ is continuous, vanishes at infinity, and $\phi = f$ almost everywhere.
The next theorem is proved in Rudin [153, p. 189] and in Zygmund [190, vol. II, p. 250].

Theorem 1.8.7 (Parseval Identity) *If f and g are square-integrable in $(-\infty, \infty)$, then*

$$\int_{-\infty}^{\infty} \mathscr{F}\{f\}(\omega) \overline{\mathscr{F}\{g\}(\omega)} \, d\omega = 2\pi \int_{-\infty}^{\infty} f(t) \overline{g(t)} \, dt \ . \tag{1.119}$$

The Parseval identity implies the next result.

Theorem 1.8.8 (Plancherel) *If f is square-integrable in $(-\infty, \infty)$, then $\mathscr{F}\{f\}$ is also square-integrable in $(-\infty, \infty)$, and*

$$\int_{-\infty}^{\infty} |f(t)|^2 \, dt = \frac{1}{2\pi} \int_{-\infty}^{\infty} |\mathscr{F}\{f\}(\omega)|^2 \, d\omega \ .$$

Like the discrete Fourier transform (see Lemma 1.8.6), the continuous Fourier transform converts convolution into multiplication. For a proof of this fact, see Rudin [153, p. 181] or Zygmund [190, vol. II, p. 252].

Lemma 1.8.15 *Suppose that f and g are absolutely integrable on the real line. Then the convolution $f * g$ is absolutely integrable, and*

$$\mathscr{F}\{f * g\}(\omega) = \mathscr{F}\{f\}(\omega) \mathscr{F}\{g\}(\omega) \ .$$

We can also determine the Fourier transform of a derivative.

Lemma 1.8.16 *Suppose that the absolutely integrable function f has an absolutely integrable derivative f'. Then*

$$\mathscr{F}\{f'\}(\omega) = \frac{1}{i\omega} \mathscr{F}\{f\}(\omega) \ .$$

Proof Since f has a derivative, it is continuous. Also, since f is absolutely integrable, it tends to zero as its argument becomes infinite. This implies that we can integrate by parts to find that

$$\mathcal{F}\{f'\}(\omega) = \int_{-\infty}^{\infty} f'(t)e^{-i\omega t}\, dt$$

$$= f(t)e^{-i\omega t}\big|_{-\infty}^{\infty} - \int_{-\infty}^{\infty} f(t)\frac{e^{-i\omega t}}{-i\omega}\, dt = \frac{1}{i\omega}\mathcal{F}\{f\}(\omega)\ .$$

Next, we will examine the Fourier transform of a an operator on functions that is similar to upsampling on discrete time signals; this lemma will lead to a result that is similar to Eq. (1.101).

Lemma 1.8.17 *If α is a nonzero scalar, define the **dilation operator** D_α by*

$$D_\alpha\{f\}(t) = f\left(\frac{t}{\alpha}\right)\ .$$

Then for any absolutely integrable function f we have

$$\mathcal{F}\{D_\alpha\{f\}\}(\omega) = \alpha\mathcal{F}\{f\}(\alpha\omega)\ .$$

Proof We can change variables of integration to get

$$\mathcal{F}\{D_\alpha\{f\}\}(\omega) = \int_{-\infty}^{\infty} f\left(\frac{t}{\alpha}\right)e^{-i\omega t}\, dt = \int_{-\infty}^{\infty} f(y)e^{-i\omega\alpha y}\alpha\, dy = \alpha\mathcal{F}\{f\}(\alpha\omega)\ .$$

The next lemma examines the Fourier transform of the translation operator, leading to a result analogous to (1.99) for the discrete time Fourier transform.

Lemma 1.8.18 *Define the **translation operator** T_τ for any real number τ by $T_\tau\{f\}(t) = f(t - \tau)$. If f is square integrable on $(-\infty, \infty)$, then*

$$\mathcal{F}\{T_\tau\{f\}\}(\omega) = e^{-i\omega\tau}\mathcal{F}\{f\}(\omega)\ . \tag{1.120}$$

*Next, suppose that the sequence $\{a_j\}_{j=-\infty}^{\infty}$ is **square summable**, meaning that*

$$\sum_{j=-\infty}^{\infty} |a_j|^2 < \infty\ .$$

Then

$$\mathcal{F}\left\{\sum_{j=-\infty}^{\infty} a_j T_j\{f\}\right\}(\omega) = \left[\sum_{j=-\infty}^{\infty} a_j e^{-i\omega j}\right]\mathcal{F}\{f\}(\omega) \tag{1.121a}$$

and

$$\int_{-\infty}^{\infty} \left| \sum_{j=-\infty}^{\infty} a_j T_j\{f\}(t) \right|^2 \, dt = \int_0^{2\pi} \left| \sum_{j=-\infty}^{\infty} a_j e^{-i\omega j} \right|^2 \left[\sum_{n=-\infty}^{\infty} |\mathscr{F}\{f\}(\omega + 2\pi n)|^2 \right] \, d\omega .$$

(1.121b)

Proof First we compute

$$\mathscr{F}\{T_\tau\{f\}\}(\omega) \equiv \int_{-\infty}^{\infty} f(t - \tau) e^{-i\omega t} \, dt = \int_{-\infty}^{\infty} f(s) e^{-i\omega[s+\tau]} \, ds = e^{-i\omega\tau} \mathscr{F}\{f\}(\omega) .$$

This proves (1.120), from which (1.121a) follows easily. Then we use this result to show that

$$\int_{-\infty}^{\infty} \left| \sum_{j=-\infty}^{\infty} a_j T_j\{f\}(t) \right|^2 \, dt = \int_{-\infty}^{\infty} \left| \sum_{j=-\infty}^{\infty} a_j f(t - j) \right|^2 \, dt$$

then we return to the beginning and use the Plancherel Theorem 1.8.8 to obtain

$$= \int_{-\infty}^{\infty} \left| \mathscr{F}\left\{ \sum_{j=-\infty}^{\infty} a_j T_j\{f\} \right\}(\omega) \right|^2 \, d\omega = \int_{-\infty}^{\infty} \left| \sum_{j=-\infty}^{\infty} a_j e^{-i\omega j} \right|^2 |\mathscr{F}\{f\}(\omega)|^2 \, d\omega$$

then we decompose the infinite integral into an infinite sum of finite integrals

$$= \sum_{n=-\infty}^{\infty} \int_{2\pi n}^{2\pi(n+1)} \left| \sum_{j=-\infty}^{\infty} a_j e^{-i\omega j} \right|^2 |\mathscr{F}\{f\}(\omega)|^2 \, d\omega$$

then we change variables of integration to get

$$= \sum_{n=-\infty}^{\infty} \int_0^{2\pi} \left| \sum_{j=-\infty}^{\infty} a_j e^{-i\theta j} \right|^2 |\mathscr{F}\{f\}(\theta + 2\pi n)|^2 \, d\theta$$

$$= \int_0^{2\pi} \left| \sum_{j=-\infty}^{\infty} a_j e^{-i\theta j} \right|^2 \sum_{n=-\infty}^{\infty} |\mathscr{F}\{f\}(\theta + 2\pi n)|^2 \, d\theta .$$

A proof of the following important result can be found in Zygmund [190, p. 68].

Theorem 1.8.9 (Poisson Summation Formula) *Suppose that f is absolutely integrable on the real line, that f has bounded variation, and that for all t we have*

$$f(t) = \frac{1}{2} \left[\lim_{\varepsilon \downarrow 0} f(t + \varepsilon) + \lim_{\varepsilon \uparrow 0} f(t + \varepsilon) \right] .$$

Then

$$\lim_{N \to \infty} \sum_{n=-N}^{N} f(t + 2n\pi) = \lim_{N \to \infty} \sum_{n=-N}^{N} \frac{1}{2\pi} \mathscr{F}\{f\}(n) e^{int} .$$

The Poisson Summation Formula has the following useful consequence, which is proved in Chui [42, p. 48].

Corollary 1.8.1 *Suppose that f is a continuous function on the real line, has compact support, is square-integrable and has bounded variation in its support. Then for all real t we have*

$$\sum_{k=-\infty}^{\infty} |\mathscr{F}\{f\}(t + 2k\pi)|^2 = \sum_{k=-\infty}^{\infty} \int_{-\infty}^{\infty} f(y + k)\overline{f(y)} \, dy \, e^{-ikt} . \tag{1.122}$$

Next, we will provide an important connection between the continuous and discrete time Fourier transforms.

Theorem 1.8.10 (Continuous Time Aliasing) *Suppose that f is continuous and absolutely integrable on the real line. Let $\sigma \geq 1$ be an integer, and define the discrete time signal $\mathbf{f}$ by*

$$\mathbf{f}_n = f(n\sigma) .$$

Then for all $\omega \in (-\pi, \pi)$ we have

$$\mathscr{F}^d\{\mathbf{f}\}(\omega) = \frac{1}{\sigma} \sum_{m=-\infty}^{\infty} \mathscr{F}\{f\} \left(\frac{\omega + 2\pi m}{\sigma} \right) .$$

Proof The Fourier inversion Theorem 1.8.6 implies that for all real t

$$f(t) = \frac{1}{2\pi} \int_{-\infty}^{\infty} \mathscr{F}\{f\}(\psi) e^{i\psi t} \, d\psi = \frac{1}{2\pi} \sum_{m=-\infty}^{\infty} \int_{(2m-1)\pi/\sigma}^{(2m+1)\pi/\sigma} \mathscr{F}\{f\}(\psi) e^{i\psi t} \, d\psi$$

$$= \frac{1}{2\pi} \sum_{m=-\infty}^{\infty} \int_{-\pi/\sigma}^{\pi/\sigma} \mathscr{F}\{f\} \left(\frac{2\pi m}{\sigma} + \theta \right) e^{i(2\pi m/\sigma + \theta)t} \, d\theta$$

$$= \frac{1}{2\pi} \int_{-\pi/\sigma}^{\pi/\sigma} e^{i\theta t} \sum_{m=-\infty}^{\infty} \mathscr{F}\{f\} \left(\theta + \frac{2\pi m}{\sigma} \right) e^{i2\pi mt/\sigma} \, d\theta .$$

In particular,

$$
\mathbf{f}_n = f(n\sigma) = \frac{1}{2\pi} \int_{-\pi/\sigma}^{\pi/\sigma} e^{in\sigma\theta} \sum_{m=-\infty}^{\infty} \mathscr{F}\{f\} \left(\theta + \frac{2\pi m}{\sigma} \right) e^{i2\pi mn} \, d\theta
$$

$$
= \frac{1}{2\pi} \int_{-\pi/\sigma}^{\pi/\sigma} e^{in\sigma\theta} \sum_{m=-\infty}^{\infty} \mathscr{F}\{f\} \left(\theta + \frac{2\pi m}{\sigma} \right) \, d\theta
$$

$$
= \frac{1}{2\pi\sigma} \int_{-\pi}^{\pi} e^{in\omega} \sum_{m=-\infty}^{\infty} \mathscr{F}\{f\} \left(\frac{\omega + 2\pi m}{\sigma} \right) \, d\omega \ .
$$

The claimed result follows from the discrete time Fourier inversion Theorem 1.8.1. The aliasing Theorem 1.8.10 allows us to prove the following interesting result.

Theorem 1.8.11 (Shannon Sampling) *Suppose that f is continuous and absolutely integrable on the real line. Let $\sigma \geq 1$ be an integer, and suppose that the Fourier transform of f satisfies*

$$
\mathscr{F}\{f\}(\omega) = 0 \ \text{for all} \ |\omega| \geq \pi/\sigma \ . \tag{1.123}
$$

Then for all real t we have

$$
f(t) = \sum_{n=-\infty}^{\infty} f(n\sigma) \frac{\sin([t - n\sigma]\pi/\sigma)}{[t - n\sigma]\pi/\sigma} \ .
$$

Proof Define the discrete time signal $\mathbf{f}$ by

$$
\mathbf{f}_n = f(n\sigma) \ .
$$

Then the continuous time aliasing Theorem 1.8.10 and assumption (1.123) imply that for all $\omega \in (-\pi/\sigma, \pi/\sigma)$

$$
\mathscr{F}^d\{\mathbf{f}\}(\omega\sigma) = \frac{1}{\sigma} \sum_{m=-\infty}^{\infty} \mathscr{F}\{f\} \left(\omega + \frac{2\pi m}{\sigma} \right) = \frac{1}{\sigma} \mathscr{F}\{f\}(\omega) \ .
$$

Next, the continuous inverse Fourier transform Theorem 1.8.6 implies that for all real t

$$
f(t) = \frac{1}{2\pi} \int_{-\infty}^{\infty} \mathscr{F}\{f\}(\omega) e^{i\omega t} \, d\omega = \frac{1}{2\pi} \int_{-\pi/\sigma}^{\pi/\sigma} \mathscr{F}\{f\}(\omega) e^{i\omega t} \, d\omega
$$

$$
= \frac{\sigma}{2\pi} \int_{-\pi/\sigma}^{\pi/\sigma} \mathscr{F}^d\{\mathbf{f}\}(\omega\sigma) e^{i\omega t} \, d\omega = \frac{\sigma}{2\pi} \int_{-\pi/\sigma}^{\pi/\sigma} \sum_{n=-\infty}^{\infty} \mathbf{f}_n e^{-in\omega\sigma} e^{i\omega t} \, d\omega
$$

$$
= \sum_{n=-\infty}^{\infty} f(n\sigma) \frac{\sigma}{2\pi} \int_{-\pi/\sigma}^{\pi/\sigma} e^{i\omega(t-n\sigma)} \, d\omega = \sum_{n=-\infty}^{\infty} f(n\sigma) \frac{\sigma}{2\pi} \frac{e^{i[t-n\sigma]\pi/\sigma} - e^{-i[t-n\sigma]\pi/\sigma}}{i[t-n\sigma]}
$$

$$
= \sum_{n=-\infty}^{\infty} f(n\sigma) \frac{\sin([t-n\sigma]\pi/\sigma)}{[t-n\sigma]\pi/\sigma} \, .
$$

The Shannon sampling theorem implies that we can recover values of **band-limited** functions by sampling their values at discrete points related to the band width. This motivates the use of band-limited filters for analysis and synthesis of signals. In the next section, we will discuss these ideas in detail.

Exercise 1.8.11 Suppose that $\mathbf{f}$ is a square-summable discrete time signal, and ϕ is a square-integrable function. Compute the continuous Fourier transform of

$$
\Phi(t) = \sum_{n=-\infty}^{\infty} \mathbf{f}_n \phi(t-n)
$$

and express $\mathscr{F}\{\Phi\}$ in terms of $\mathscr{F}^d\{\mathbf{f}\}$ and $\mathscr{F}\{\phi\}$.

Exercise 1.8.12 Suppose that f is a square-integrable function with

$$
\int_{-\infty}^{\infty} |f(t)|^2 \, dt = 1 \, .
$$

Also suppose that f has finite **second moment**

$$
M_2(f) = \int_{-\infty}^{\infty} t^2 \, |f(t)|^2 \, dt \, .
$$

Prove that the second moment of the Fourier transform of f satisfies

$$
M_2(f) M_2 \left(\mathscr{F}\{f\} \right) \geq \frac{1}{16\pi^2} \, .
$$

This is called the **Heisenberg uncertainty principle**.

1.8.2.2 Scaling Functions

We will begin with the following important definition.

Definition 1.8.9 Suppose that ℓ is the impulse response for a lowpass filter, meaning that

$$
\mathscr{F}^d\{\ell\}(0) = \sum_{n=-\infty}^{\infty} \ell_n = 1 \, .
$$

and

$$\mathscr{F}^d\{\boldsymbol{\ell}\}(\pi) = \sum_{n=-\infty}^{\infty} (-1)^n \boldsymbol{\ell}_n = 0 .$$

Then a **scaling function** $\phi(t)$ for $\boldsymbol{\ell}$ is a function that satisfies the **refinement equation** (also called the **dilation equation**)

$$\phi(t) = 2 \sum_{n=-\infty}^{\infty} \boldsymbol{\ell}_n \phi(2t - n) , \qquad (1.124)$$

and the normalizing condition

$$\int_{-\infty}^{\infty} \phi(t) \, dt = 1 .$$

Note that the refinement equation involves two fundamental operators, namely translation and scaling. The effects of these operators on the Fourier transform were examined in Lemmas 1.8.18, and 1.8.17. As a result, we obtain the following lemma.

Lemma 1.8.19 *Suppose that ϕ is the scaling function for a lowpass filter impulse response $\boldsymbol{\ell}$. Then*

$$\mathscr{F}\{\phi\}(\omega) = \mathscr{F}^d\{\boldsymbol{\ell}\} \left(\frac{\omega}{2}\right) \mathscr{F}\{\phi\} \left(\frac{\omega}{2}\right) . \qquad (1.125)$$

Furthermore,

$$\mathscr{F}\{\phi\}(2k\pi) = \delta_{k,0} .$$

Proof The refinement Eq. (1.124) implies that

$$\mathscr{F}\{\phi\}(\omega) = \int_{-\infty}^{\infty} \phi(t) e^{-i\omega t} \, dt = 2 \sum_{n=-\infty}^{\infty} \int_{-\infty}^{\infty} \boldsymbol{\ell}_n \phi(2t - n) e^{-i\omega t} \, dt$$

$$= 2 \sum_{n=-\infty}^{\infty} \boldsymbol{\ell}_n \int_{-\infty}^{\infty} \phi(s) e^{-i\omega(s+n)/2} \frac{ds}{2} = \sum_{n=-\infty}^{\infty} \boldsymbol{\ell}_n e^{-i\omega n/2} \mathscr{F}\{\phi\} \left(\frac{\omega}{2}\right)$$

$$= \mathscr{F}^d\{\boldsymbol{\ell}\} \left(\frac{\omega}{2}\right) \mathscr{F}\{\phi\} \left(\frac{\omega}{2}\right) .$$

Next, we note that the normalizing condition on ϕ implies that

$$\mathscr{F}\{\phi\}(0) = \int_{-\infty}^{\infty} \phi(t) \, dt = 1 .$$

Finally, we will use induction to prove that $\mathscr{F}\{\phi\}(2k\pi) = 0$ for all $k \neq 0$. Since ℓ is a lowpass filter,

$$\mathscr{F}\{\phi\}(\pm 2\pi) = \mathscr{F}^d\{\ell\}(\pm\pi)\mathscr{F}\{\phi\}(\pm\pi) = 0 \times \mathscr{F}\{\phi\}(\pm\pi) = 0 \,.$$

This proves the inductive hypothesis for $|k| = 1$. Assume inductively that $\mathscr{F}\{\phi\}(2j\pi) = 0$ for $0 < |j| < |k|$. Then

$$\mathscr{F}\{\phi\}(2k\pi) = \mathscr{F}^d\{\ell\}(k\pi)\mathscr{F}\{\phi\}(k\pi) = \begin{cases} 0 \times \mathscr{F}\{\phi\}(k\pi), & k \text{ odd} \\ \mathscr{F}^d\{\ell\}(k\pi) \times 0, & k \text{ even} \end{cases} = 0 \,.$$

Lemma 1.8.20 *If the finite impulse response ℓ is such that $\ell_n = 0$ for $n \leq \alpha$ or $n > \alpha + \beta$, then any corresponding scaling function ϕ can be chosen so that $\phi(t) = 0$ for $t \notin [\alpha + 1, \alpha + \beta)$.*

Proof If $t < \alpha + 1$ and $\alpha + 1 \leq n \leq \alpha + \beta$, then

$$2t - n < 2(\alpha + 1) - (\alpha + 1) = \alpha + 1 \,,$$

so we can choose $\phi(t) = 0$ for all $t < \alpha + 1$ in the refinement Eq. (1.124). Similarly, if $t \geq \alpha + \beta$ and $\alpha + 1 \leq n \leq \alpha + \beta$, then

$$2t - n \geq 2(\alpha + \beta) - (\alpha + \beta) = \alpha + \beta \,,$$

so we can choose $\phi(t) = 0$ for all $t \geq \alpha + \beta$ in the refinement equation. Any nonzero coefficients ℓ_n will only multiply zero function values for t outside the claimed interval.

Suppose that $\ell_n = 0$ for $n \leq \alpha$ or $n > \alpha + \beta$, and define

$$\Phi(t) = \phi(t + \alpha + 1) \,.$$

If ϕ has support on $(\alpha + 1, \alpha + \beta]$, then Φ has support on $[0, \beta - 1]$, and

$$\Phi(t) = \phi(t + \alpha + 1) = 2 \sum_{n=\alpha+1}^{\alpha+\beta} \ell_n \phi(2t + 2\alpha + 2 - n)$$

$$= 2 \sum_{n=\alpha+1}^{\alpha+\beta} \ell_n \Phi(2t + \alpha + 1 - n) = 2 \sum_{m=0}^{\beta-1} \ell_m \Phi(2t - m) \,.$$

Since Φ satisfies a refinement equation for ℓ and integrates to one, we can translate the scaling function to have support on $[0, \beta - 1]$.

1.8.2.3 Scaling Function Evaluation

For some scaling functions, we know only the refinement equation and do not have a closed form for evaluating the scaling function directly. In this section, we will show how to use the refinement equation for such scaling functions to evaluate them at binary numbers.

In general, the refinement equation gives us

$$\Phi(t) = 2 \sum_{m=0}^{\beta-1} \ell_{m+\alpha+1} \Phi(2t - m) ,$$

so for integers $k \in [0, \beta - 2]$ we have

$$\Phi(k) = 2 \sum_{m=0}^{\beta-1} \ell_{m+\alpha+1} \Phi(2k - m) = 2 \sum_{m=\max\{0,2k-\beta+2\}}^{\min\{\beta-1,2k\}} \ell_{m+\alpha+1} \Phi(2k - m) .$$

This shows that refinement equation can be used to determine the values of the scaling function at integers within its support. The beginning step in this process uses the following lemma.

Lemma 1.8.21 *Suppose that the finite impulse response* ℓ *satisfies*

$$\mathscr{F}^d\{\ell\}(0) = \sum_{n=-\infty}^{\infty} \ell_n = 1 \text{ and } \mathscr{F}^d\{\ell\}(\pm\pi) = \sum_{n=-\infty}^{\infty} (-1)^n \ell_n = 0 ,$$

i.e., ℓ *is the impulse response for a lowpass filter. Assume that* $\ell_n = 0$ *for* $n \le \alpha$ *or* $n > \alpha + \beta$. *For* $0 \le i \le \beta - 2$, *let*

$$\mathbf{M}_{ij} = \begin{cases} 2\ell_{2i-j+\alpha+1}, & \max\{2i - \beta + 1, 0\} \le j \le \min\{2i, \beta - 2\} \\ 0, & \text{otherwise} \end{cases}$$

be the entries of the $(\beta - 1) \times (\beta - 1)$ *matrix* $\mathbf{M}$. *Then* $\mathbf{M}$ *has at least one eigenvalue equal to one.*

Proof Note that since $\sum_{n=-\infty}^{\infty} (-1)^n \ell_n = 0$, the even and odd entries of ℓ must have the same sum. Thus

$$1 = \sum_{\substack{n=\alpha+1}}^{\alpha+\beta} \ell_n = \sum_{\substack{n=\alpha+1 \\ n \text{ even}}}^{\alpha+\beta} \ell_n + \sum_{\substack{n=\alpha+1 \\ n \text{ odd}}}^{\alpha+\beta} \ell_n = 2 \sum_{\substack{n=\alpha+1 \\ n \text{ even}}}^{\alpha+\beta} \ell_n = 2 \sum_{\substack{n=\alpha+1 \\ n \text{ odd}}}^{\alpha+\beta} \ell_n$$

If j is even, then $m = 2i - j$ is even and the column sum

$$\sum_{i=0}^{\beta-2} \mathbf{M}_{ij} = 2 \sum_{i=j/2}^{j/2+\beta/2+1} 2\ell_{2i-j+\alpha+1} = 2 \sum_{\substack{m=0 \\ m \text{ even}}}^{\beta-2} \ell_{m+\alpha+1} = 1 \,,$$

since this is a sum over all even entries of ℓ when α is odd, and a sum over all odd entries when α is even. On the other hand, if j is odd then $m = 2i - j - 1$ is even and

$$\sum_{i=0}^{\beta-2} \mathbf{M}_{ij} = 2 \sum_{i=(j+1)/2}^{(j-1)/2+\beta/2+1} \ell_{2i-j+\alpha+1} = 2 \sum_{\substack{m=0 \\ m \text{ even}}}^{\beta-2} \ell_{m+\alpha+2} = 1 \,,$$

since this is a sum over all odd entries of ℓ when α is odd, and the sum over all even entries when α is even. Thus all columns of $\mathbf{M}$ sum to one, so $(\mathbf{M})^\top - \mathbf{I}$ is singular. This implies that $\mathbf{M} - \mathbf{I}$ is singular, and $\mathbf{M}$ has at least one eigenvalue equal to one.

Let $\boldsymbol{\Phi}$ be a nonzero $(\beta - 1)$-vector such that $\boldsymbol{\Phi} = \mathbf{M}\boldsymbol{\Phi}$. We assume that $\boldsymbol{\Phi}$ can be normalized so that the sum of its entries is one; otherwise, theorem 1.8.12 shows that the scaling function ϕ will not be useful in approximating arbitrary square-integrable functions. After normalization, we have $\Phi(k) = \boldsymbol{\Phi}_k$ for $0 \le k \le \beta - 2$. This eigenvector computation tells us how to evaluate the scaling function at the integers within its support.

After we determine $\boldsymbol{\Phi}$ at the integers, we can determine it at the half-integers by means of the refinement equation:

$$\Phi\left(k + \frac{1}{2}\right) = 2 \sum_{m=0}^{\beta-1} \ell_{m+\alpha+1} \Phi(2k + 1 - m) \,.$$

For any $j \ge 1$, once we have determined values of $\boldsymbol{\Phi}$ at points $k + 2^{1-j}n$ for k and n integers, then for odd integers n we can compute

$$\Phi\left(k + 2^{-j}n\right) = 2 \sum_{m=0}^{\beta-1} \ell_{m+\alpha+1} \Phi\left(2k + 2^{1-j}n - m\right) \,.$$

This computation uses only previously computed values of $\boldsymbol{\Phi}$.

Example 1.8.11 Suppose that the impulse response ℓ comes from the maxflat product filter with $\beta = 2$; in other words, the impulse response ℓ has only two nonzero entries. In this case, we have

$$\begin{bmatrix} \ell_0 \\ \ell_1 \end{bmatrix} = \begin{bmatrix} 1 \\ 1 \end{bmatrix} \frac{1}{2} \,.$$

This is the lowpass impulse response for the Haar filter. Then the support of Φ is $[0, 1]$, and

$$\Phi(t) = 2\ell_0\Phi(2t) + 2\ell_1\Phi(2t - 1) = \Phi(2t) + \Phi(2t - 1) .$$

Thus $\mathbf{M} = [1]$. We can take $\Phi(0) = 1$ and then find that $\Phi(1/2) = \Phi(0) = 1$, $\Phi(1/4) = \Phi(1/2) = 1$, $\Phi(3/4) = \Phi(1/2) = 1$, and so on. In general, we have $\Phi(t) = 1$ for all binary numbers $t \in [0, 1)$.

Example 1.8.12 The maxflat product filter with $\beta = 4$ has lowpass impulse response

$$\begin{bmatrix} \ell_0 \\ \ell_1 \\ \ell_2 \\ \ell_3 \end{bmatrix} = \begin{bmatrix} 1 + \sqrt{3} \\ 3 + \sqrt{3} \\ 3 - \sqrt{3} \\ 1 - \sqrt{3} \end{bmatrix} \frac{1}{8} ,$$

so

$$\mathbf{M} = \begin{bmatrix} 1 + \sqrt{3} & & \\ 3 - \sqrt{3} & 3 + \sqrt{3} & 1 + \sqrt{3} \\ & 1 - \sqrt{3} & 3 - \sqrt{3} \end{bmatrix} \frac{1}{4} .$$

Note that the eigenvector of $\mathbf{M}$ with eigenvalue one is evident from the equation

$$\mathbf{M}\boldsymbol{\Phi} = \begin{bmatrix} 1 + \sqrt{3} & & \\ 3 - \sqrt{3} & 3 + \sqrt{3} & 1 + \sqrt{3} \\ & 1 - \sqrt{3} & 3 - \sqrt{3} \end{bmatrix} \frac{1}{4} \begin{bmatrix} 0 \\ (1 + \sqrt{3})/2 \\ (1 - \sqrt{3})/2 \end{bmatrix} = \begin{bmatrix} 0 \\ (1 + \sqrt{3})/2 \\ (1 - \sqrt{3})/2 \end{bmatrix}$$

$$= \boldsymbol{\Phi} = \begin{bmatrix} \Phi(0) \\ \Phi(1) \\ \Phi(2) \end{bmatrix} .$$

Afterward, we have

$$\Phi(1/2) = 2\sum_{m=0}^{3} \ell_m\Phi(1 - m) = 2\left[\ell_0\Phi(1) + \ell_1\Phi(0)\right] ,$$

$$\Phi(3/2) = 2\sum_{m=0}^{3} \ell_m\Phi(3 - m) = 2\left[\ell_1\Phi(2) + \ell_2\Phi(1) + \ell_3\Phi(0)\right] \text{ and}$$

$$\Phi(5/2) = 2\sum_{m=0}^{3} \ell_m\Phi(5 - m) = 2\ell_3\Phi(2) .$$

We can continue to find $\Phi(t)$ at finer binary numbers t.

Alternatively, we can use Fourier transforms to determine a scaling function from an impulse response. Equation (1.125) implies that for all ω we have

$$\mathscr{F}\{\phi\}(\omega) = \mathscr{F}^d\{\ell\}\left(2^{-1}\omega\right)\mathscr{F}\{\phi\}\left(2^{-1}\omega\right)$$

$$= \dots = \left\{\prod_{m=1}^{M}\mathscr{F}^d\{\ell\}\left(2^{-m}\omega\right)\right\}\mathscr{F}\{\phi\}\left(2^{-M}\omega\right)$$

$$= \lim_{M\to\infty}\left[\left\{\prod_{m=1}^{M}\mathscr{F}^d\{\ell\}\left(2^{-m}\omega\right)\right\}\mathscr{F}\{\phi\}\left(2^{-M}\omega\right)\right]$$

$$= \left\{\prod_{m=1}^{\infty}\mathscr{F}^d\{\ell\}\left(2^{-m}\omega\right)\right\}\mathscr{F}\{\phi\}(0)$$

$$= \prod_{m=1}^{\infty}\mathscr{F}^d\{\ell\}\left(2^{-m}\omega\right)\ .$$

Example 1.8.13 Consider the maxflat product filter with $\beta = 2$, for which the lowpass filter is $\ell = [\mathbf{T}\boldsymbol{\delta} + \boldsymbol{\delta}]/\sqrt{2}$. Thus

$$\mathscr{F}^d\left\{\frac{\ell}{\sqrt{2}}\right\}(\omega) = \frac{1}{2}\left[e^{-i\omega} + 1\right]\ .$$

We claim that

$$\prod_{m=1}^{M}\frac{1 + e^{-i2^{-m}\omega}}{2} = 2^{-M}\sum_{k=0}^{2^M-1}e^{-ik2^{-M}\omega}\ . = \frac{1 - e^{-i\omega}}{2^M\left[1 - e^{-i2^{-M}\omega}\right]}\ .$$

Since

$$\frac{1 + e^{-i2^{-1}\omega}}{2} = 2^{-1}\sum_{k=0}^{2-1}e^{-ik2^{-1}\omega} = \frac{1 - e^{-i\omega}}{2\left[1 - e^{-i2^{-1}\omega}\right]}\ ,$$

the claim is obviously true for $M = 1$. Using the claim for $M - 1$ as the inductive hypothesis, we get

$$\prod_{m=1}^{M}\frac{1 + e^{-i2^{-m}\omega}}{2} = \left[\prod_{m=1}^{M-1}\frac{1 + e^{-i2^{-m}\omega}}{2}\right]\frac{1 + e^{-i2^{-M}\omega}}{2}$$

$$= \left[2^{-M+1}\sum_{k=0}^{2^{M-1}-1}e^{-ik2^{-M+1}\omega}\right]\frac{1 + e^{-i2^{-M}\omega}}{2}$$

$$= 2^{-M} \sum_{k=0}^{2^{M-1}-1} \left[e^{-ik2^{-M+1}\omega} + e^{-ik2^{-M+1}\omega - i2^{-M}\omega} \right]$$

$$= 2^{-M} \sum_{k=0}^{2^{M-1}-1} \left[e^{-i2k2^{-M}\omega} + e^{-i(2k+1)2^{-M}} \right]$$

$$= 2^{-M} \sum_{j=0}^{2^{M}-1} e^{-ij2^{-M}\omega} = \frac{1 - e^{-i\omega}}{2^M \left[1 - e^{-i2^{-M}\omega} \right]} \cdot$$

The last equation is the formula for the sum of a geometric series. Since

$$2^M \left[1 - e^{-i2^{-M}\omega} \right] = 2^M \left[1 - \sum_{k=0}^{\infty} \frac{(i2^{-M}\omega)^k}{k!} \right] = - \sum_{k=1}^{\infty} 2^{M(1-k)} \frac{(i\omega)^k}{k!} \, ,$$

we see that

$$\lim_{M\to\infty} 2^M \left[1 - e^{-i2^{-M}\omega} \right] = i\omega \, ,$$

and that

$$\mathscr{F}\{\phi\}(\omega) = \lim_{M\to\infty} \prod_{m=1}^{M} \frac{1 + e^{-i2^{-m}\omega}}{2} = \frac{1 - e^{-i\omega}}{i\omega} \cdot$$

It is easy to take the inverse Fourier transform to discover that $\phi(t) = 1$ for $t \in [0, 1)$ and $\phi(t) = 0$ otherwise.

In Fig. 1.12 we show two scaling functions and corresponding wavelets for the Daubechies filter banks. These figures were plotted by executing the C^{++} program testDaubechiesScalingFunction.C. The actual scaling functions are implemented as C^{++} classes in ScalingFunction.H and ScalingFunction.C. Given the impulse response for a lowpass filter, the `ScalingFunction` constructor solves an eigenvector problem to determine the values of the scaling function at the integers. Values of the scaling function at binary numbers are determined by the refinement equation.

Alternatively, interested readers may execute a JavaScript program for computing the **maxflat D4 scaling function.** The eigenvector solution is hardwired in this program, but the algorithm clearly shows how the refinement equation for the scaling function can be used to evaluate it at binary numbers.

In Fig. 1.13 we show two scaling functions with their corresponding wavelets, dual scaling functions and dual wavelets for the spline filter banks. These figures were plotted by executing the C^{++} program testSplineScalingFunction.C.

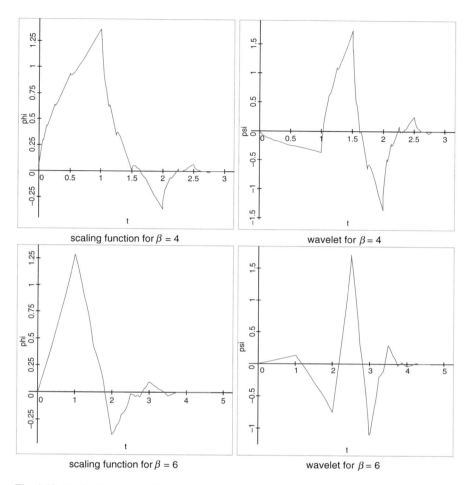

Fig. 1.12 Daubechies scaling functions and wavelets

1.8.2.4 Wavelets

We can use a scaling function and a highpass filter impulse response to define another useful function.

Definition 1.8.10 Suppose that ϕ is the scaling function for a lowpass impulse response ℓ. If $\mathbf{h}$ is the impulse response for a highpass filter, then then the corresponding **wavelet function** ψ is defined by the wavelet refinement equation

$$\psi(t) = 2 \sum_{n=-\infty}^{\infty} \mathbf{h}_n \phi(2t - n) \, . \tag{1.126}$$

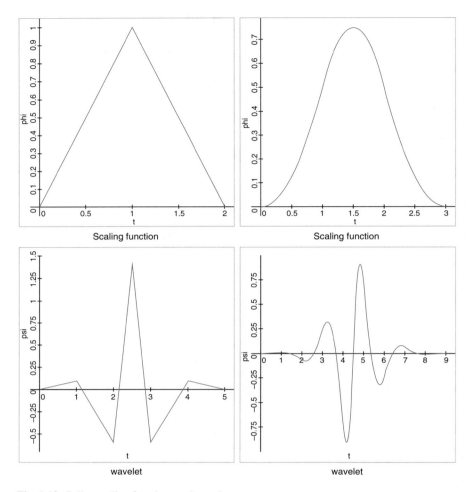

Fig. 1.13 Spline scaling functions and wavelets

Example 1.8.14 For the **Haar wavelet** we chose the lowpass filter to be

$$\ell_n = \begin{cases} 1/2, & n = 0, 1 \\ 0, \text{ otherwise} \end{cases},$$

and the highpass filter to be

$$\mathbf{h}_n = \begin{cases} 1/2, & n = -1 \\ -1/2, & n = 0 \\ 0, \text{ otherwise} \end{cases},$$

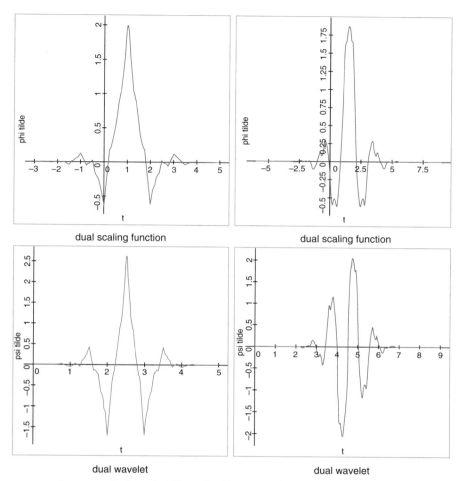

dual scaling function   dual scaling function

dual wavelet   dual wavelet

lowpass order 2, product filter order 4 lowpass order 3, product filter order 6

Fig. 1.13 (continued)

These choices implied that the scaling function is

$$\phi(t) = \begin{cases} 1, \ 0 \le t < 1 \\ 0, \ \text{otherwise} \end{cases} .$$

Thus the corresponding wavelet is

$$\psi(t) = 2\left\{\frac{1}{2}\phi(2t+1) - \frac{1}{2}\phi(2t)\right\} = \begin{cases} 1, \ -1/2 \le t < 0 \\ -1, \ -0 \le t < 1/2 \\ 0, \ \text{otherwise} \end{cases} . \tag{1.127}$$

Note that the integral of $\psi(t)$ over any interval with endpoints equal to an odd multiple of one-half.

The following lemma proves some simple consequences of the definition of a wavelet.

Lemma 1.8.22 *Suppose that* ℓ *is the impulse response for a lowpass filter, and* **h** *is the impulse response for a highpass filter. Let* ϕ *be the scaling function corresponding to* ℓ*, and let* ψ *be the wavelet corresponding to* **h** *and* ϕ*. Then*

$$\mathscr{F}\{\psi\}(\omega) = \mathscr{F}^d\{\mathbf{h}\}\left(\frac{\omega}{2}\right)\mathscr{F}\{\phi\}\left(\frac{\omega}{2}\right) , \tag{1.128}$$

and

$$\int_{-\infty}^{\infty} \psi(t)\, \mathrm{d}t = \mathscr{F}\{\psi\}(0) = 0 .$$

Proof The proof is directly analogous to the proof of Lemma 1.8.19.

Some wavelets were plotted in Figs. 1.12 and 1.13. Interested readers may also view C^{++} code to implement define and evaluate wavelets in Wavelet.H and Wavelet.C. Given the impulse response for a highpass filter and a scaling function, the class constructor uses the refinement equation for the wavelet to determine the values of the wavelet at the integers. The wavelet for Daubechies filter banks of arbitrary order may be plotted by executing testDaubechiesScalingFunction.C. Also testSplineScalingFunction.C will plot the wavelet for the spline filter bank.

1.8.2.5 Dual Scaling Functions

The fundamental obstacle in approximating general functions by translations of scaling functions is the determination of the best coefficients in the approximation. We could return to the ideas in Sect. 1.7 concerning least-squares approximation, but we would generate large linear systems corresponding to ill-conditioned normal equations. Instead, the next definition will give us a useful tool for this computation.

Definition 1.8.11 Suppose that ϕ is square-integrable on the real line. Then the square-integrable function $\widetilde{\phi}$ is **dual** to ϕ if and only if for all integers m and n we have

$$\int_{-\infty}^{\infty} \phi(t-m)\overline{\widetilde{\phi}(t-n)}\, \mathrm{d}t = \delta_{m,n} . \tag{1.129}$$

An interesting result in a very short paper by Lemarié [127] shows that for any scaling function ϕ with compact support, there is a dual scaling function $\widetilde{\phi}$ with compact support.

Given a scaling function ϕ and its dual $\widetilde{\phi}$, the following lemma shows us how to define an oblique projector onto the span of the translations of ϕ.

Lemma 1.8.23 *If ϕ is a scaling function and $\widetilde{\phi}$ is its dual, then the operator P, defined on square-integrable functions f by*

$$\{Pf\}(t) = \sum_{k=-\infty}^{\infty} \int_{-\infty}^{\infty} f(\tau)\overline{\widetilde{\phi}(\tau - k)} \, d\tau \, \phi(t - k) \,,$$

*is an **oblique projector** onto $\operatorname{span}\{T_k\phi\}_{k=-\infty}^{\infty}$, meaning that for all square-integrable functions we have*

$$P(Pf) = Pf \,.$$

Proof We can use the definition of P and the duality Eq. (1.129) to compute

$$\{P(Pf)\}(t) = P\left\{\sum_{m=-\infty}^{\infty} \int_{-\infty}^{\infty} f(\tau)\overline{\widetilde{\phi}(\tau - m)} \, d\tau \, T_m\phi\right\}(t)$$

$$= \sum_{n=-\infty}^{\infty} \int_{-\infty}^{\infty} \left\{\sum_{m=-\infty}^{\infty} \int_{-\infty}^{\infty} f(\tau)\overline{\widetilde{\phi}(\tau - m)} \, d\tau \, \phi(\sigma - m)\overline{\widetilde{\phi}(\sigma - n)} \, d\sigma\right\} \phi(t - n)$$

$$= \sum_{n=-\infty}^{\infty} \sum_{m=-\infty}^{\infty} \int_{-\infty}^{\infty} f(\tau)\overline{\widetilde{\phi}(\tau - m)} \, d\tau \int_{-\infty}^{\infty} \phi(\sigma - m)\overline{\widetilde{\phi}(\sigma - n)} \, d\sigma \, \phi(t - n)$$

$$= \sum_{n=-\infty}^{\infty} \int_{-\infty}^{\infty} f(\tau)\overline{\widetilde{\phi}(\tau - n)} \, d\tau \, \phi(t - n) = \{Pf\}(t) \,.$$

Note that we can generalize the projector to work on different binary scales:

$$\{P_j f\}(t) = \sum_{k=-\infty}^{\infty} \int_{-\infty}^{\infty} f(\tau)\overline{2^{j/2}\widetilde{\phi}\left(2^j\tau - k\right)} \, d\tau \, 2^{j/2}\phi\left(2^j t - k\right) \,. \tag{1.130}$$

This leads to an interesting and important question: if we have an arbitrary square-integrable function f, is there a sufficiently fine scale j so that the span of $\{2^{j/2}\phi\left(2^j t - k\right)\}_{k=-\infty}^{\infty}$ is sufficiently close to f? Cohen [48, p. 60] answers this question by proving the following theorem.

Theorem 1.8.12 *Suppose that ϕ and $\widetilde{\phi}$ are real-valued dual scaling functions with compact support. Then*

$$\lim_{j\to\infty} \int_{-\infty}^{\infty} \left|\{P_j f\}(t) - f(t)\right|^2 \, dt = 0$$

if and only if for almost all real t

$$\left[\int_{-\infty}^{\infty} \overline{\phi(t)}\, dt\right]\left[\sum_{n=-\infty}^{\infty} \phi(t-n)\right] = 1 .$$

Cohen also shows [48, p. 62] that the claim in this theorem is equivalent to

$$\left[\int_{-\infty}^{\infty} \phi(t)\, dt\right]\left[\int_{-\infty}^{\infty} \overline{\phi(t)}\, dt\right] = 1 .$$

Recall that our definition (1.8.9) of a scaling function required that

$$\int_{-\infty}^{\infty} \phi(t)\, dt = 1 .$$

Thus the projection $P_j f$ will approach f as the refinement index j becomes infinite if and only if

$$\int_{-\infty}^{\infty} \widetilde{\phi}(t)\, dt = 1 ,$$

or equivalently

$$\sum_{n=-\infty}^{\infty} \phi(t-n) = 1 .$$

The latter equation says that the translations of ϕ form a **partition of unity**.

In the next lemma, we examine connections between the coefficients in the refinement equations for dual scaling functions. This lemma has been adapted from Cohen [48, pp. 62 and 75].

Lemma 1.8.24 *Suppose that ϕ and $\widetilde{\phi}$ are real-valued functions on the real line, have compact support, and satisfy the refinement equations*

$$\phi(t) = 2 \sum_{n=-\infty}^{\infty} \ell_n \phi(2t-n) \text{ and} \qquad (1.131\text{a})$$

$$\widetilde{\phi}(t) = 2 \sum_{n=-\infty}^{\infty} \lambda_n \widetilde{\phi}(2t-n) . \qquad (1.131\text{b})$$

Also assume that $\widetilde{\phi}$ is the dual of ϕ and

$$\int_{-\infty}^{\infty} \phi(t)\, dt = 1 = \int_{-\infty}^{\infty} \widetilde{\phi}(t)\, dt ,$$

and that for all t we have

$$\sum_{n=-\infty}^{\infty} \phi(t-n) = 1 = \sum_{n=-\infty}^{\infty} \widetilde{\phi}(t-n) .$$

Then

$$\sum_{n=-\infty}^{\infty} \ell_n = 1 = \sum_{n=-\infty}^{\infty} \lambda_n ,$$

$$\sum_{n=-\infty}^{\infty} (-1)^n \ell_n = 0 = \sum_{n=-\infty}^{\infty} (-1)^n \lambda_n \text{ and}$$

$$\frac{1}{2}\delta_{0,k} = \sum_{n=-\infty}^{\infty} \ell_n \overline{\lambda_{n+2k}} = \left[(\downarrow_2) \left(\ell * \lambda^* \right) \right]_{-k} \text{ for all } k .$$

Note that $\sum_{n=-\infty}^{\infty} \ell_n = \mathscr{F}^d\{\ell\}(0)$ and $\sum_{n=-\infty}^{\infty}(-1)^n \ell_n = \mathscr{F}^d\{\ell\}(\pi)$. In other words, the assumptions of this lemma imply that ℓ and λ are impulse responses for lowpass filters.

Proof The refinement equation for ϕ implies that

$$1 = \int_{-\infty}^{\infty} \phi(t) \, dt = \int_{-\infty}^{\infty} 2 \sum_{n=-\infty}^{\infty} \ell_n \phi(2t-n) \, dt = \sum_{n=-\infty}^{\infty} \ell_n \int_{-\infty}^{\infty} \phi(\tau) \, d\tau = \sum_{n=-\infty}^{\infty} \ell_n .$$

A similar computation applies to $\widetilde{\phi}$. This proves the first claim in the lemma.

Next, we note that for all t we have

$$1 = \sum_{m=-\infty}^{\infty} \phi(t-m) = \sum_{m=-\infty}^{\infty} 2 \sum_{n=-\infty}^{\infty} \ell_n \phi(2t-2m-n)$$

$$= 2 \sum_{\substack{m=-\infty \\ n \text{ even}}}^{\infty} \sum_{n=-\infty}^{\infty} \ell_n \phi(2t-2m-n) + 2 \sum_{\substack{m=-\infty \\ n \text{ odd}}}^{\infty} \sum_{n=-\infty}^{\infty} \ell_n \phi(2t-2m-n)$$

$$= 2 \sum_{m=-\infty}^{\infty} \sum_{k=-\infty}^{\infty} \ell_{2k} \phi(2t-2m-2k) + 2 \sum_{m=-\infty}^{\infty} \sum_{k=-\infty}^{\infty} \ell_{2k+1} \phi(2t-2m-2k-1)$$

$$= 2 \sum_{j=-\infty}^{\infty} \sum_{k=-\infty}^{\infty} \ell_{2k} \phi(2t-2j) + 2 \sum_{j=-\infty}^{\infty} \sum_{k=-\infty}^{\infty} \ell_{2k+1} \phi(2t-2j-1) .$$

This equation and the duality condition (1.129) imply that

$$1 = \int_{-\infty}^{\infty} \overline{\phi(2t)} \, dt$$

$$= \int_{-\infty}^{\infty} \left\{ 2 \sum_{j=-\infty}^{\infty} \sum_{k=-\infty}^{\infty} \ell_{2k} \phi(2t - 2j) + 2 \sum_{j=-\infty}^{\infty} \sum_{k=-\infty}^{\infty} \ell_{2k+1} \phi(2t - 2j - 1) \right\}$$

$$\overline{\phi(2t)} \, dt$$

$$= 2 \sum_{k=-\infty}^{\infty} \ell_{2k} \, ,$$

and that

$$1 = \int_{-\infty}^{\infty} \overline{\phi(2t - 1)} \, dt$$

$$= \int_{-\infty}^{\infty} \left\{ 2 \sum_{j=-\infty}^{\infty} \sum_{k=-\infty}^{\infty} \ell_{2k} \phi(2t - 2j) + 2 \sum_{j=-\infty}^{\infty} \sum_{k=-\infty}^{\infty} \ell_{2k+1} \phi(2t - 2j - 1) \right\}$$

$$\overline{\phi(2t - 1)} \, dt$$

$$= 2 \sum_{k=-\infty}^{\infty} \ell_{2k+1} \, .$$

Thus

$$\sum_{n=-\infty}^{\infty} (-1)^n \ell_n = \sum_{\substack{n=-\infty \\ n \text{ even}}}^{\infty} \ell_n - \sum_{\substack{n=-\infty \\ n \text{ odd}}}^{\infty} \ell_n = \frac{1}{2} - \frac{1}{2} = 0 \, .$$

A similar set of computations applies to λ.

Finally, the duality condition (1.129) and the refinement equations for the two dual functions imply that

$$\delta_{0,k} = \int_{-\infty}^{\infty} \phi(t - k) \overline{\widetilde{\phi}(t)} \, dt$$

$$= \int_{-\infty}^{\infty} 2 \left[\sum_{m=-\infty}^{\infty} \ell_m \phi(2t - 2k - m) \right] \overline{2 \left[\sum_{n=-\infty}^{\infty} \lambda_n \widetilde{\phi}(2t - n) \right]} \, dt$$

$$= 4 \sum_{m=-\infty}^{\infty} \sum_{n=-\infty}^{\infty} \ell_m \overline{\lambda_n} \int_{-\infty}^{\infty} \phi(2t - 2k - m) \overline{\widetilde{\phi}(2t - n)} \, dt$$

$$= 2 \sum_{m=-\infty}^{\infty} \sum_{n=-\infty}^{\infty} \ell_m \overline{\lambda_n} \int_{-\infty}^{\infty} \phi(\tau + n - m - 2k) \overline{\widetilde{\phi}(\tau)} \, d\tau = 2 \sum_{m=-\infty}^{\infty} \ell_m \overline{\lambda_{m+2k}} \ .$$

It is interesting to note that Eq. (1.93) shows that the final claim in Lemma 1.8.24 is equivalent to

$$\delta = 2(\uparrow_2)(\downarrow_2)\left[\ell * \lambda^*\right] = \ell * \lambda^* + R\left(\ell * \lambda^*\right) \ .$$

Using this observation, Cohen [48, p. 95] proves the next lemma, which is something like a converse for the previous lemma. The proof uses infinite products of discrete time Fourier transforms of the lowpass filters, as discussed in Sect. 1.8.2.3 on scaling function evaluation.

Lemma 1.8.25 *If the impulse responses ℓ and λ satisfy*

$$\ell * \lambda^* + R(\ell * \lambda^*) = \delta \ ,$$

then the scaling functions ϕ and $\widetilde{\phi}$, determined by the respective refinement Eq. (1.131) involving these two impulse responses, are dual to each other.

Some dual functions for spline scaling functions are plotted in Fig. 1.13. Interested readers may view C^{++} code to plot the dual scaling functions for the spline filter bank of arbitrary order in testSplineScalingFunction.C.

1.8.2.6 Dual Wavelets

Next, we will show that the duality/orthogonality conditions

$$\int_{-\infty}^{\infty} \begin{bmatrix} \phi(t) \\ \psi(t) \end{bmatrix} \overline{\left[\widetilde{\phi}(t-k) \ \widetilde{\psi}(t-k)\right]} \, dt = \begin{bmatrix} \delta_{0,k} & 0 \\ 0 & \delta_{0,k} \end{bmatrix} \text{ for all } k$$

are equivalent to the modulation matrix condition

$$\begin{bmatrix} \ell & R\ell \\ h & Rh \end{bmatrix} * \begin{bmatrix} \lambda^* & \eta^* \\ R\lambda^* & R\eta^* \end{bmatrix} = \begin{bmatrix} \delta & 0 \\ 0 & \delta \end{bmatrix}$$

for the impulse responses in the refinement equations for the scaling functions and their wavelets.

Lemma 1.8.26 *Suppose that the functions ϕ and $\widetilde{\phi}$ satisfy the refinement equations*

$$\phi(t) = 2 \sum_{n=-\infty}^{\infty} \ell_n \phi(2t - n) \text{ and } \widetilde{\phi}(t) = 2 \sum_{n=-\infty}^{\infty} \lambda_n \widetilde{\phi}(2t - n) ,$$

and that the functions ψ and $\widetilde{\psi}$ satisfy the refinement equations

$$\psi(t) = 2 \sum_{n=-\infty}^{\infty} h_n \phi(2t - n) \text{ and } \widetilde{\psi}(t) = 2 \sum_{n=-\infty}^{\infty} \eta_n \widetilde{\phi}(2t - n) .$$

Then ϕ and $\widetilde{\phi}$ are dual if and only if

$$\ell * \lambda^* + \mathbf{R}\left(\ell * \lambda^*\right) = \delta ,$$

and ψ and $\widetilde{\psi}$ are dual if and only if

$$h * \eta^* + \mathbf{R}\left(h * \eta^*\right) = \delta .$$

In addition, if ϕ and $\widetilde{\phi}$ are dual, then

$$\int_{-\infty}^{\infty} \psi(t)\overline{\widetilde{\phi}(t - k)} \, dt = 0 \text{ for all } k \iff h * \lambda^* + \mathbf{R}\left(h * \lambda^*\right) = 0 \text{ and}$$

$$\int_{-\infty}^{\infty} \phi(t)\overline{\widetilde{\psi}(t - k)} \, dt = 0 \text{ for all } k \iff \ell * \eta^* + \mathbf{R}\left(\ell * \eta^*\right) = 0 .$$

Proof First, we note that

$$\int_{-\infty}^{\infty} \phi(t)\overline{\widetilde{\phi}(t - k)} \, dt$$

$$= \int_{-\infty}^{\infty} 2\left[\sum_{m=-\infty}^{\infty} \ell_m \phi(2t - m)\right] \overline{2\left[\sum_{n=-\infty}^{\infty} \lambda_n \widetilde{\phi}(2t - 2k - n)\right]} \, dt$$

$$= 4 \sum_{m=-\infty}^{\infty} \sum_{n=-\infty}^{\infty} \ell_m \overline{\lambda_n} \int_{-\infty}^{\infty} \phi(2t - m)\overline{\widetilde{\phi}(2t - 2k - n)} \, dt$$

$$= 2 \sum_{m=-\infty}^{\infty} \sum_{n=-\infty}^{\infty} \ell_m \overline{\lambda_n} \int_{-\infty}^{\infty} \phi(\tau)\overline{\widetilde{\phi}(\tau + m - 2k - n)} \, d\tau .$$

It is now easy to see that if ϕ and $\widetilde{\phi}$ are dual then

$$\delta_k = 2 \sum_{m=-\infty}^{\infty} \ell_m \overline{\lambda_{m-2k}} = 2 \sum_{m=-\infty}^{\infty} \ell_m \lambda^*_{2k-m} = 2\left[\ell * \lambda^*\right]_{2k} = 2\left[(\downarrow_2)\ell * \lambda^*\right]_k .$$

Since $(\uparrow_2)\delta = \delta$, Eq. (1.93) shows that ϕ and $\widetilde{\phi}$ are dual if and only if

$$\delta = 2(\uparrow_2)(\downarrow_2)\ell * \lambda^* = \ell * \lambda^* + \mathbf{R}\left(\ell * \lambda^*\right) .$$

On the other hand, if $\ell * \lambda^* + \mathbf{R}\left(\ell * \lambda^*\right) = \delta$, then Lemma 1.8.25 proves that ϕ and $\widetilde{\phi}$ are dual. This proves the first claim.

The proof of the second claim is similar. First, we note that

$$\int_{-\infty}^{\infty} \psi(t)\overline{\widetilde{\psi}(t-k)}\, dt$$

$$= \int_{-\infty}^{\infty} 2\left[\sum_{m=-\infty}^{\infty} \mathbf{h}_m\phi(2t-m)\right] 2\overline{\left[\sum_{n=-\infty}^{\infty} \eta_n\widetilde{\phi}(2t-2k-n)\right]} dt$$

$$= 4\sum_{m=-\infty}^{\infty}\sum_{n=-\infty}^{\infty} \mathbf{h}_m\overline{\eta_n}\int_{-\infty}^{\infty} \phi(2t-m)\overline{\widetilde{\phi}(2t-2k-n)}\, dt$$

$$= 2\sum_{m=-\infty}^{\infty}\sum_{n=-\infty}^{\infty} \mathbf{h}_m\overline{\eta_n}\int_{-\infty}^{\infty} \phi(\tau)\overline{\widetilde{\phi}(\tau+m-2k-n)}\, d\tau ,$$

so ψ and $\widetilde{\psi}$ are dual if and only if

$$\delta_k = 2\sum_{m=-\infty}^{\infty} \mathbf{h}_m\overline{\eta_{m-2k}} = 2\sum_{m=-\infty}^{\infty} \mathbf{h}_m\eta^*_{2k-m} = 2\left[\mathbf{h} * \eta^*\right]_{2k} = 2\left[(\downarrow_2)\mathbf{h} * \eta^*\right]_k = \delta_{0,k} .$$

This is equivalent to

$$\delta = 2(\uparrow_2)(\downarrow_2)\mathbf{h} * \eta^* = \mathbf{h} * \eta^* + \mathbf{R}\left(\mathbf{h} * \eta^*\right) .$$

Next, suppose that ϕ and $\widetilde{\phi}$ are dual. Then

$$\int_{-\infty}^{\infty} \psi(t)\overline{\widetilde{\phi}(t-k)}\, dt$$

$$= \int_{-\infty}^{\infty}\left[2\sum_{m=-\infty}^{\infty} \mathbf{h}_m\phi(2t-m)\right] 2\overline{\left[\sum_{n=-\infty}^{\infty} \lambda_n\widetilde{\phi}(2t-2k-n)\right]} dt$$

$$= 4\sum_{m=-\infty}^{\infty}\sum_{n=-\infty}^{\infty} \mathbf{h}_m\overline{\lambda_n}\int_{-\infty}^{\infty} \phi(2t-m)\overline{\widetilde{\phi}(2t-2k-n)}\, dt$$

$$= 2\sum_{m=-\infty}^{\infty}\sum_{n=-\infty}^{\infty} \mathbf{h}_m\overline{\lambda_n}\int_{-\infty}^{\infty} \phi(\tau)\overline{\widetilde{\phi}(\tau+m-2k-n)}\, dt .$$

$$= 2\sum_{m=-\infty}^{\infty} \mathbf{h}_m\overline{\lambda_{m-2k}} = 2\sum_{m=-\infty}^{\infty} \mathbf{h}_m\lambda^*_{2k-m} = 2\left[\mathbf{h} * \lambda^*\right]_{2k} = 2\left[(\downarrow_2)\mathbf{h} * \lambda^*\right]_k$$

It follows that ψ is orthogonal to the span of translations of $\widetilde{\phi}$ if and only if

$$0 = 2(\downarrow_2)\mathbf{h} * \boldsymbol{\lambda}^* .$$

Since $\mathbf{0} = (\uparrow_2)\mathbf{0}$, this is equivalent to

$$0 = 2(\uparrow_2)(\downarrow_2)\mathbf{h} * \boldsymbol{\lambda}^* = \mathbf{h} * \boldsymbol{\lambda}^* + \mathbf{R}\left(\mathbf{h} * \boldsymbol{\lambda}^*\right) .$$

The final claim is proved similarly.

Next, we will use the impulse responses for the dual scaling functions to construct new impulse responses.

Lemma 1.8.27 *Suppose that the impulse responses $\boldsymbol{\ell}$ and $\boldsymbol{\lambda}$ satisfy*

$$\boldsymbol{\ell} * \boldsymbol{\lambda}^* + \mathbf{R}\left(\boldsymbol{\ell} * \boldsymbol{\lambda}^*\right) = \boldsymbol{\delta} .$$

If α is an odd integer, define the impulse responses $\mathbf{h}$ and $\boldsymbol{\eta}$ by

$$\mathbf{h} = \mathbf{RT}^{-\alpha}\boldsymbol{\lambda}^* \text{ and}$$
$$\boldsymbol{\eta}^* = \mathbf{RT}^{\alpha}\boldsymbol{\ell} .$$

Then

$$\begin{bmatrix} \boldsymbol{\ell} & \mathbf{h} \\ \mathbf{R}\boldsymbol{\ell} & \mathbf{R}\mathbf{h} \end{bmatrix} * \begin{bmatrix} \boldsymbol{\lambda}^* & \mathbf{R}\boldsymbol{\lambda}^* \\ \boldsymbol{\eta}^* & \mathbf{R}\boldsymbol{\eta}^* \end{bmatrix} = \begin{bmatrix} \boldsymbol{\delta} & \mathbf{0} \\ \mathbf{0} & \boldsymbol{\delta} \end{bmatrix} ,$$

and

$$\begin{bmatrix} \boldsymbol{\ell} & \mathbf{R}\boldsymbol{\ell} \\ \mathbf{h} & \mathbf{R}\mathbf{h} \end{bmatrix} * \begin{bmatrix} \boldsymbol{\lambda}^* & \boldsymbol{\eta}^* \\ \mathbf{R}\boldsymbol{\lambda}^* & \mathbf{R}\boldsymbol{\eta}^* \end{bmatrix} = \begin{bmatrix} \boldsymbol{\delta} & \mathbf{0} \\ \mathbf{0} & \boldsymbol{\delta} \end{bmatrix} .$$

Proof This argument is similar to that in the proof of the Filter Bank Perfect Synthesis Theorem 1.8.4. We can use identities (1.89) and (1.97) to compute

$$\begin{bmatrix} \boldsymbol{\ell} & \mathbf{h} \\ \mathbf{R}\boldsymbol{\ell} & \mathbf{R}\mathbf{h} \end{bmatrix} * \begin{bmatrix} \boldsymbol{\lambda}^* & \mathbf{R}\boldsymbol{\lambda}^* \\ \boldsymbol{\eta}^* & \mathbf{R}\boldsymbol{\eta}^* \end{bmatrix} = \begin{bmatrix} \boldsymbol{\ell} & \mathbf{RT}^{-\alpha}\boldsymbol{\lambda}^* \\ \mathbf{R}\boldsymbol{\ell} & \mathbf{T}^{-\alpha}\boldsymbol{\lambda}^* \end{bmatrix} * \begin{bmatrix} \boldsymbol{\lambda}^* & \mathbf{R}\boldsymbol{\lambda}^* \\ \mathbf{RT}^{\alpha}\boldsymbol{\ell} & \mathbf{T}^{\alpha}\boldsymbol{\ell} \end{bmatrix}$$

$$= \begin{bmatrix} \boldsymbol{\ell} * \boldsymbol{\lambda}^* + \mathbf{R}\left(\mathbf{T}^{-\alpha}\boldsymbol{\lambda}^* * \mathbf{T}^{\alpha}\boldsymbol{\ell}\right) & \boldsymbol{\ell} * \mathbf{R}\boldsymbol{\lambda}^* - \mathbf{T}^{-\alpha}\mathbf{R}\boldsymbol{\lambda}^* * \mathbf{T}^{\alpha}\boldsymbol{\ell} \\ \mathbf{R}\boldsymbol{\ell} * \boldsymbol{\lambda}^* - \mathbf{T}^{-\alpha}\boldsymbol{\lambda}^* * \mathbf{T}^{\alpha}\mathbf{R}\boldsymbol{\ell} & \mathbf{R}\left(\boldsymbol{\ell} * \boldsymbol{\lambda}^*\right) + \mathbf{T}^{-\alpha}\boldsymbol{\lambda}^* * \mathbf{T}^{\alpha}\boldsymbol{\ell} \end{bmatrix}$$

$$= \begin{bmatrix} \boldsymbol{\ell} * \boldsymbol{\lambda}^* + \mathbf{R}\left(\boldsymbol{\lambda}^* * \boldsymbol{\ell}\right) & \boldsymbol{\ell} * \mathbf{R}\boldsymbol{\lambda}^* - \mathbf{R}\boldsymbol{\lambda}^* * \boldsymbol{\ell} \\ \mathbf{R}\boldsymbol{\ell} * \boldsymbol{\lambda}^* - \boldsymbol{\lambda}^* * \mathbf{R}\boldsymbol{\ell} & \mathbf{R}\left(\boldsymbol{\ell} * \boldsymbol{\lambda}^*\right) + \boldsymbol{\lambda}^* * \boldsymbol{\ell} \end{bmatrix} = \begin{bmatrix} \boldsymbol{\delta} & \mathbf{0} \\ \mathbf{0} & \boldsymbol{\delta} \end{bmatrix} .$$

This proves the first claim.

Next, we compute

$$\begin{bmatrix}\boldsymbol{\ell} & \mathbf{R}\boldsymbol{\ell} \\ \mathbf{h} & \mathbf{R}\mathbf{h}\end{bmatrix} * \begin{bmatrix}\boldsymbol{\lambda}^* & \boldsymbol{\eta}^* \\ \mathbf{R}\boldsymbol{\lambda}^* & \mathbf{R}\boldsymbol{\eta}^*\end{bmatrix} = \begin{bmatrix}\boldsymbol{\ell} & \mathbf{R}\boldsymbol{\ell} \\ \mathbf{R}\mathbf{T}^{-\alpha}\boldsymbol{\lambda}^* & \mathbf{T}^{-\alpha}\boldsymbol{\lambda}^*\end{bmatrix} * \begin{bmatrix}\boldsymbol{\lambda}^* & \mathbf{R}\mathbf{T}^{\alpha}\boldsymbol{\ell} \\ \mathbf{R}\boldsymbol{\lambda}^* & \mathbf{T}^{\alpha}\boldsymbol{\ell}\end{bmatrix}$$

$$= \begin{bmatrix}\boldsymbol{\ell}*\boldsymbol{\lambda}^* + \mathbf{R}\left(\boldsymbol{\ell}*\boldsymbol{\lambda}^*\right) & -\boldsymbol{\ell}*\mathbf{T}^{\alpha}\mathbf{R}\boldsymbol{\ell} + \mathbf{R}\boldsymbol{\ell}*\mathbf{T}^{\alpha}\boldsymbol{\ell} \\ -\mathbf{T}^{-\alpha}\mathbf{R}\boldsymbol{\lambda}^**\boldsymbol{\lambda}^* + \mathbf{T}^{-\alpha}\boldsymbol{\lambda}^**\mathbf{R}\boldsymbol{\lambda}^* & \mathbf{R}\left(\mathbf{T}^{-\alpha}\boldsymbol{\lambda}^**\mathbf{T}^{\alpha}\boldsymbol{\ell}\right) + \mathbf{T}^{-\alpha}\boldsymbol{\lambda}^**\mathbf{T}^{\alpha}\boldsymbol{\ell}\end{bmatrix}$$

$$= \begin{bmatrix}\boldsymbol{\ell}*\boldsymbol{\lambda}^* + \mathbf{R}\left(\boldsymbol{\ell}*\boldsymbol{\lambda}^*\right) & -\boldsymbol{\ell}*\mathbf{T}^{\alpha}\mathbf{R}\boldsymbol{\ell} + \mathbf{T}^{\alpha}\mathbf{R}\boldsymbol{\ell}*\boldsymbol{\ell} \\ -\mathbf{R}\boldsymbol{\lambda}^**\mathbf{T}^{-\alpha}\boldsymbol{\lambda}^* + \mathbf{T}^{-\alpha}\boldsymbol{\lambda}^**\mathbf{R}\boldsymbol{\lambda}^* & \mathbf{R}\left(\boldsymbol{\lambda}^**\boldsymbol{\ell}\right) + \boldsymbol{\lambda}^**\boldsymbol{\ell}\end{bmatrix} = \begin{bmatrix}\boldsymbol{\delta} & \mathbf{0} \\ \mathbf{0} & \boldsymbol{\delta}\end{bmatrix} .$$

Next, we will show how to use the scaling function and wavelet coefficients to reconstruct the scaling function on a coarser scale. The ultimate use of this equation is not to evaluate scaling functions; we discussed their evaluation in Sect. 1.8.2.3. Rather, we will use Eq. (1.132) in Sect. 1.8.2.7 below to develop the cascade algorithm for approximating general functions on coarser scales.

Lemma 1.8.28 *Given the impulse response $\boldsymbol{\ell}$, suppose that ϕ satisfies the refinement equation*

$$\phi(t) = 2 \sum_{n=-\infty}^{\infty} \boldsymbol{\ell}_n \phi(2t-n) .$$

Given impulse responses $\mathbf{h}$, assume that ψ satisfies the refinement equation

$$\psi(t) = 2 \sum_{n=-\infty}^{\infty} \mathbf{h}_n \phi(2t-n) .$$

Next, suppose that the impulse responses $\boldsymbol{\lambda}^*$ and $\boldsymbol{\eta}^*$ satisfy

$$\begin{bmatrix}\boldsymbol{\ell} & \mathbf{h}\end{bmatrix}\begin{bmatrix}\boldsymbol{\lambda}^* & \mathbf{R}\boldsymbol{\lambda}^* \\ \boldsymbol{\eta}^* & \mathbf{R}\boldsymbol{\eta}^*\end{bmatrix} = \begin{bmatrix}\boldsymbol{\delta} & \mathbf{0}\end{bmatrix} .$$

Then for all real t and all integers k we have the **decomposition relation**

$$\phi(2t-k) = \sum_{n=-\infty}^{\infty} \left[\boldsymbol{\lambda}_{2n-k}^*\phi(t-n) + \boldsymbol{\eta}_{2n-k}^*\psi(t-n)\right] . \tag{1.132}$$

Proof First, note that

$$\boldsymbol{\delta} = \boldsymbol{\delta} + \mathbf{0} = \left[\boldsymbol{\ell}*\boldsymbol{\lambda}^* + \mathbf{h}*\boldsymbol{\eta}^*\right] + \left[\boldsymbol{\ell}*\left(\mathbf{R}\boldsymbol{\lambda}^*\right) + \mathbf{h}*\left(\mathbf{R}\boldsymbol{\eta}^*\right)\right]$$

$$= \boldsymbol{\ell}*\left[\boldsymbol{\lambda}^* + \mathbf{R}\boldsymbol{\lambda}^*\right] + \mathbf{h}*\left[\boldsymbol{\eta}^* + \mathbf{R}\boldsymbol{\eta}^*\right]$$

$$= 2\left\{\boldsymbol{\ell}*\left[(\uparrow_2)(\downarrow_2)\boldsymbol{\lambda}^*\right] + \mathbf{h}*\left[(\uparrow_2)(\downarrow_2)\boldsymbol{\eta}^*\right]\right\} ,$$

and

$$\begin{aligned}
\delta = \delta - 0 &= \left[\ell * \lambda^* + h * \eta^* \right] - \left[\ell * (R\lambda^*) + h * (R\eta^*) \right] \\
&= \ell * \left[\lambda^* - R\lambda^* \right] + h * \left[\eta^* - R\eta^* \right] \\
&= \ell * \left[2T^{-1}(\uparrow_2)(\downarrow_2)T\lambda^* \right] + h * \left[2T^{-1}(\uparrow_2)(\downarrow_2)T\eta^* \right] \\
&= 2T^{-1} \left\{ \ell * \left[(\uparrow_2)(\downarrow_2)T\lambda^* \right] + h * \left[(\uparrow_2)(\downarrow_2)T\eta^* \right] \right\} \ .
\end{aligned}$$

It follows that

$$\int_{-\infty}^{\infty} \phi(2t)\mathrm{e}^{-i\omega t}\, dt = \mathscr{F}\left\{ D_{1/2}\phi \right\}(\omega) = \frac{1}{2}\mathscr{F}\{\phi\}\left(\frac{\omega}{2}\right) = \frac{1}{2}\mathscr{F}\{\phi\}\left(\frac{\omega}{2}\right)\mathscr{F}^d\{\delta\}(\omega)$$

$$= \mathscr{F}\{\phi\}\left(\frac{\omega}{2}\right)\left[\mathscr{F}^d\{\ell\}\left(\frac{\omega}{2}\right)\mathscr{F}^d\{(\uparrow_2)(\downarrow_2)\lambda^*\}\left(\frac{\omega}{2}\right) \right.$$

$$\left. + \mathscr{F}^d\{h\}\left(\frac{\omega}{2}\right)\mathscr{F}^d\{(\uparrow_2)(\downarrow_2)\eta^*\}\left(\frac{\omega}{2}\right) \right]$$

$$= \mathscr{F}\{\phi\}(\omega)\mathscr{F}^d\{(\downarrow_2)\lambda^*\}(\omega) + \mathscr{F}\{\psi\}(\omega)\mathscr{F}^d\{(\downarrow_2)\eta^*\}(\omega)$$

$$= \int_{-\infty}^{\infty} \phi(s)\mathrm{e}^{-i\omega s}\, ds \sum_{n=-\infty}^{\infty} \lambda_{2n}{}^*\mathrm{e}^{-i\omega n} + \int_{-\infty}^{\infty} \psi(s)\mathrm{e}^{-i\omega s}\, ds \sum_{n=-\infty}^{\infty} \eta_{2n}{}^*\mathrm{e}^{-i\omega n}$$

$$= \int_{-\infty}^{\infty} \sum_{n=-\infty}^{\infty} \left[\phi(t-n)\lambda_{2n}{}^* + \psi(t-n)\eta_{2n}{}^* \right]\mathrm{e}^{-i\omega t}\, dt$$

and

$$\int_{-\infty}^{\infty} \phi(2t-1)\mathrm{e}^{-i\omega t}\, dt = \mathscr{F}\left\{ D_{1/2}T_1\phi \right\}(\omega) = \frac{1}{2}\mathscr{F}\{T_1\phi\}\left(\frac{\omega}{2}\right)$$

$$= \frac{1}{2}\mathrm{e}^{-i\omega/2}\mathscr{F}\{\phi\}\left(\frac{\omega}{2}\right)$$

$$= \mathscr{F}\{\phi\}\left(\frac{\omega}{2}\right)\left[\mathscr{F}^d\{\ell\}\left(\frac{\omega}{2}\right)\mathscr{F}^d\{(\uparrow_2)(\downarrow_2)T\lambda^*\}\left(\frac{\omega}{2}\right) \right.$$

$$\left. + \mathscr{F}^d\{h\}\left(\frac{\omega}{2}\right)\mathscr{F}^d\{(\uparrow_2)(\downarrow_2)T\eta^*\}\left(\frac{\omega}{2}\right) \right]$$

$$= \mathscr{F}\{\phi\}(\omega)\mathscr{F}^d\{(\downarrow_2)T\lambda^*\}(\omega) + \mathscr{F}\{\psi\}(\omega)\mathscr{F}^d\{(\downarrow_2)T\eta^*\}(\omega)$$

$$= \int_{-\infty}^{\infty} \phi(s)\mathrm{e}^{-i\omega s}\, ds \sum_{n=-\infty}^{\infty} \lambda_{2n-1}{}^*\mathrm{e}^{-i\omega n} + \int_{-\infty}^{\infty} \psi(s)\mathrm{e}^{-i\omega s}\, ds \sum_{n=-\infty}^{\infty} \eta_{2n-1}{}^*\mathrm{e}^{-i\omega n}$$

$$= \int_{-\infty}^{\infty} \sum_{n=-\infty}^{\infty} \left[\phi(t-n)\lambda_{2n-1}{}^* + \psi(t-n)\eta_{2n-1}{}^* \right]\mathrm{e}^{-i\omega t}\, dt \ .$$

By taking inverse Fourier transforms, we can conclude that

$$\phi(2t) = \sum_{n=-\infty}^{\infty} \left[\phi(t-n)\lambda_{2n}{}^* + \psi(t-n)\eta_{2n}{}^*\right] \text{ and}$$

$$\phi(2t-1) = \sum_{n=-\infty}^{\infty} \left[\phi(t-n)\lambda_{2n-1}{}^* + \psi(t-n)\eta_{2n-1}{}^*\right].$$

These two equations imply (1.132).

Recall the oblique projector P_j, defined by (1.130), and define the operator Q_j by

$$Q_j f = P_{j+1}f - P_j f$$

for all square-integrable functions f. Then

$$Q_j Q_j = P_{j+1}P_{j+1} - P_{j+1}P_j - P_j P_{j+1} + P_j P_j = P_{j+1} - P_j - P_j + P_j = P_{j+1} - P_j = Q_j,$$

so Q_j is an oblique projector. Cohen [48, p. 77] shows that, under the first set of assumptions of Lemma 1.8.27, the projector Q_j can be evaluated as

$$\{Q_j f\}(t) = \sum_{k=-\infty}^{\infty} \left[\int_{-\infty}^{\infty} f(\tau)\overline{2^{j/2}\widetilde{\psi}\left(2^j\tau - k\right)}\, d\tau\right] 2^{j/2}\psi\left(2^j t - k\right).$$

A similar proof would draw the same conclusion if we used the alternate definitions of $\mathbf{h}$ and η^* in Lemma 1.8.27.

Some dual wavelet functions for spline wavelet functions are plotted in Fig. 1.13. Interested readers may view C^{++} code to plot the dual wavelet for the spline filter bank of arbitrary order in testSplineScalingFunction.C.

1.8.2.7 Function Decomposition

In this section, we will discuss how to use scaling functions and wavelets to determine a function approximation on some sufficiently fine scale. Suppose that we are given a function $g(t)$, which we would like to approximate on the interval $[a, b]$ on a mesh of width 2^{-N}. In other words, we want to find coefficients $\gamma_n^{(N)}$ so that the approximation

$$g(t) \approx \sum_{n=-\infty}^{\infty} \gamma_n^{(N)}\phi\left(2^N t - n\right) \tag{1.133}$$

is appropriately accurate. For example, if the translations of the scaling functions ϕ span all polynomials of degree at most $k - 1$, then we might like this approximation to be exact for all polynomials of degree at most $k - 1$.

Since we are only interested in approximating g on a bounded interval, we only need to evaluate a finite number of the coefficients $\gamma_n^{(N)}$ for some scale N. If the scaling function has support $[\alpha + 1, \alpha + \beta]$, then

$$2^N t - n \in [\alpha + 1, \alpha + \beta] \text{ if and only if } 2^N t - \alpha - \beta \le n \le 2^N t - \alpha - 1 \ .$$

Since $a \le t \le b$, this implies that n satisfies

$$2^N a - \alpha - \beta \le n \le 2^N b - \alpha - 1 \ .$$

For other values of n, the coefficient $\gamma_n^{(N)}$ will not be needed to evaluate $g(t)$ for any $t \in [a, b]$.

One way to choose the coefficients $\gamma_n^{(N)}$ in the approximation

$$g(t) \approx \sum_{n=-\infty}^{\infty} \gamma_n^{(N)} \phi\left(2^N t - n\right)$$

is to use the projection P_N, which we previously defined in Eq. (1.130):

$$g(t) \approx P_N g(t) = \sum_{n=-\infty}^{\infty} \left[\int_{-\infty}^{\infty} g(\tau) \overline{2^N \widetilde{\phi}\left(2^N \tau - n\right)} \, d\tau \right] \phi\left(2^N t - n\right) \ .$$

In other words, we need to compute the coefficients

$$\gamma_n^{(N)} = \int_{-\infty}^{\infty} g(\tau) \overline{2^N \widetilde{\phi}\left(2^N \tau - n\right)} \, d\tau = \int_{-\infty}^{\infty} g\left(2^{-N}[\sigma + n]\right) \overline{\widetilde{\phi}(\sigma)} \, d\sigma \ .$$

Of course, the integral on the right-hand side of this expression need only be computed over the support of $\widetilde{\phi}$. If the dual scaling function is defined at binary numbers, then we could use any appropriately accurate quadrature rule employing binary numbers to approximate this integral. (Here "appropriately accurate" means that the quadrature rule is exact for all those polynomials that are reproduced exactly by the scaling function.) If the dual scaling function is known only by its refinement equation, then we can use extrapolations of the trapezoidal rule to approximate the integral using evaluations of $\widetilde{\phi}$ and g only at binary numbers. This sort of numerical quadrature technique is discussed in Sect. 2.3.9 below.

After we have approximated $g(t)$ on some scale N as in Eq. (1.133), we may choose to determine similar approximations on nearby scales. We will see that approximations on coarser scales will require that we use extra terms involving the wavelets. We will include such terms in our general discussion.

Given some scale $M \leq N$, suppose that the function $g(t)$ has been approximated by

$$g^{(M)}(t) = \sum_{m=-\infty}^{\infty} \gamma_m^{(M)} \phi\left(2^M t - m\right) + \sum_{K=M}^{N-1} \sum_{m=-\infty}^{\infty} \Gamma_m^{(K)} \psi\left(2^K t - m\right) .$$

Equation (1.132) implies that we can represent $g^{(M)}(t)$ on the next coarser scale as follows:

$$g^{(M)}(t) = \sum_{m=-\infty}^{\infty} \gamma_m^{(M)} \phi\left(2\left[2^{M-1}t\right] - m\right) + \sum_{K=M}^{N-1} \sum_{m=-\infty}^{\infty} \Gamma_m^{(K)} \psi\left(2^K t - m\right)$$

$$= \sum_{m=-\infty}^{\infty} \gamma_m^{(M)} \sum_{n=-\infty}^{\infty} \left[\lambda_{2n-m}^* \phi\left(2^{M-1}t - n\right) + \eta_{2n-m}^* \psi\left(2^{M-1}t - n\right)\right]$$

$$+ \sum_{K=M}^{N-1} \sum_{m=-\infty}^{\infty} \Gamma_m^{(K)} \psi\left(2^K t - m\right)$$

$$= \sum_{n=-\infty}^{\infty} \left[\sum_{m=-\infty}^{\infty} \lambda_{2n-m}^* \gamma_m^{(M)}\right] \phi\left(2^{M-1}t - n\right)$$

$$+ \sum_{n=-\infty}^{\infty} \left[\sum_{m=-\infty}^{\infty} \eta_{2n-m}^* \gamma_m^{(M)}\right] \psi\left(2^{M-1}t - n\right)$$

$$+ \sum_{K=M}^{N-1} \sum_{m=-\infty}^{\infty} \Gamma_m^{(K)} \psi\left(2^K t - m\right)$$

$$\equiv \sum_{n=-\infty}^{\infty} \gamma_n^{(M-1)} \phi\left(2^{M-1}t - n\right) + \sum_{K=M-1}^{N-1} \sum_{m=-\infty}^{\infty} \Gamma_m^{(K)} \psi\left(2^K t - m\right)$$

This computation says that on the coarser scale, $g^{(M)}(t)$ can be represented as a linear combination of translations of dilations of the scaling function and wavelet, with coefficients given by the **cascade algorithm**

$$\gamma^{(M-1)} = (\downarrow_2)\left[\lambda^* * \gamma^{(M)}\right] \text{ and } \Gamma^{(M-1)} = (\downarrow_2)\left[\eta^* * \gamma^{(M)}\right] . \tag{1.134}$$

This cascade algorithm is exactly analogous to the analysis step in the filter bank of Theorem 1.8.4. The cascade algorithm applies the dual lowpass and highpass filters λ^* and η^* to the scaling function coefficients $\gamma^{(M)}$ on the fine scale (this is the input discrete time signal), in order to obtain the scaling function coefficients $\gamma^{(M-1)}$ (the lowpass analysis output) and wavelet coefficients $\Gamma^{(M-1)}$ (the highpass analysis output) on the coarse scale.

On the other hand, suppose that the function $g(t)$ has been approximated on some coarse scale M by

$$g^{(M)}(t) = \sum_{n=-\infty}^{\infty} \gamma_n^{(M)} \phi\left(2^M t - n\right) + \sum_{K=M}^{N-1} \sum_{m=-\infty}^{\infty} \Gamma_m^{(K)} \psi\left(2^K t - m\right) \,.$$

Then the refinement Eqs. (1.124) and (1.126) for ϕ and ψ imply that we can represent $g^{(M)}(t)$ on the next finer scale by

$$g^{(M)}(t) = \sum_{m=-\infty}^{\infty} \gamma_m^{(M)} \phi\left(\frac{1}{2}\left[2^{M+1} t - 2m\right]\right)$$

$$+ \sum_{K=M}^{N-1} \sum_{m=-\infty}^{\infty} \Gamma_m^{(K)} \psi\left(\frac{1}{2}\left[2^{K+1} t - 2m\right]\right)$$

$$= \sum_{m=-\infty}^{\infty} \gamma_m^{(M)} \sum_{n=-\infty}^{\infty} 2\ell_n \phi\left(2^{M+1} t - 2m - n\right)$$

$$+ \sum_{m=-\infty}^{\infty} \Gamma_m^{(M)} \sum_{n=-\infty}^{\infty} 2\mathbf{h}_n \phi\left(2^{M+1} t - 2m - n\right)$$

$$+ \sum_{K=M+1}^{N-1} \sum_{m=-\infty}^{\infty} \Gamma_m^{(K)} \psi\left(2^K t - m\right)$$

$$= \sum_{j=-\infty}^{\infty} \left\{ 2 \sum_{m=-\infty}^{\infty} \left[\ell_{j-2m} \gamma_m^{(M)} + \mathbf{h}_{j-2m} \Gamma_m^{(M)} \right] \right\} \phi\left(2^{M+1} t - j\right)$$

$$+ \sum_{K=M+1}^{N-1} \sum_{m=-\infty}^{\infty} \Gamma_m^{(K)} \psi\left(2^K t - m\right)$$

$$\equiv \sum_{j=-\infty}^{\infty} \gamma_j^{(M+1)} \phi\left(2^{M+1} t - j\right) + \sum_{K=M+1}^{N-1} \sum_{m=-\infty}^{\infty} \Gamma_m^{(K)} \psi\left(2^K t - m\right) \,.$$

Thus, on the next finer scale, $g^{(M)}(t)$ can be represented as a linear combination of translations of dilations of the scaling function alone, with coefficients given by

$$\gamma_j^{(M+1)} = 2 \sum_{m=-\infty}^{\infty} \left[\ell_{j-2m} \gamma_m^{(M)} + \mathbf{h}_{j-2m} \Gamma_m^{(M)} \right] = 2 \sum_{\substack{n=-\infty \\ n \text{ even}}}^{\infty} \left[\ell_{j-n} \gamma_{n/2}^{(N)} + \mathbf{h}_{j-n} \Gamma_{n/2}^{(N)} \right]$$

$$= 2 \left\{ \ell * \left[(\uparrow_2) \gamma^{(M)} \right] + \mathbf{h} * \left[(\uparrow_2) \Gamma^{(M)} \right] \right\}_j \,.$$

In other words, the fine scale coefficients are computed by the **pyramid algorithm**

$$\boldsymbol{\gamma}^{(M+1)} = 2\{\boldsymbol{\ell} * (\uparrow_2)\boldsymbol{\gamma}^{(M)} + \mathbf{h} * (\uparrow_2)\boldsymbol{\Gamma}^{(M)}\} \ . \tag{1.135}$$

The pyramid algorithm is exactly analogous to the filter bank synthesis step in Theorem 1.8.4.

For orthogonal scaling functions, it is common to rewrite the function approximation (1.133) in the alternate form

$$g(t) \approx \sum_{n=-\infty}^{\infty} \gamma_n^{(N)} 2^{N/2} \phi\left(2^N t - n\right) \ . \tag{1.136}$$

This factor introduces some modifications into the computations of the coefficients. First, we would compute

$$\gamma_n^{(N)} = \int_{-\infty}^{\infty} g(\tau)\overline{2^{N/2}\phi\left(2^N \tau - n\right)}\, d\tau \ .$$

The cascade algorithm would be modified to take the form

$$\boldsymbol{\gamma}^{(M-1)} = \sqrt{2}(\downarrow_2)\left[\boldsymbol{\lambda}^* * \boldsymbol{\gamma}^{(N)}\right] \text{ and}$$

$$\boldsymbol{\Gamma}^{(M-1)} = \sqrt{2}(\downarrow_2)\left[\boldsymbol{\eta}^* * \boldsymbol{\gamma}^{(N)}\right] \ ,$$

and the pyramid algorithm would look like

$$\boldsymbol{\gamma}^{(M+1)} = \sqrt{2}\{\boldsymbol{\ell} * (\uparrow_2)\boldsymbol{\gamma}^{(M)} + \mathbf{h} * (\uparrow_2)\boldsymbol{\Gamma}^{(M)}\} \ .$$

In Fig. 1.14 we show approximations to the exponential on the unit interval formed by projecting onto translates of the Daubechies scaling function of order 2. The plot of error versus scale indicates that the approximation is second-order accurate. Similar computations for the Daubechies filter of order 3 are shown in Fig. 1.15, and computations for the spline filter of order 2 with product filter of order 4 are shown in Fig. 1.16.

Interested readers may view C++ code to approximate the exponential function by Daubechies scaling functions of arbitrary order in testDaubechiesFilterBank2.C. For each scale N, this program computes the coefficients $\gamma_n^{(N)}$ by extrapolation of trapezoidal rule integration to compute an integral approximation. (See Sect. 2.3.9 below to understand the numerical integration process used here, and see Sect. 1.8.2.10 below to understand the order of accuracy in approximation by scaling functions.) Afterward, the log base 2 of the maximum pointwise error in the approximation is plotted versus scale. The slope of the curve demonstrates the order of the approximation. Furthermore, the C++ code in testSplineFilterBank2.C will perform similar computations using the spline filter bank.

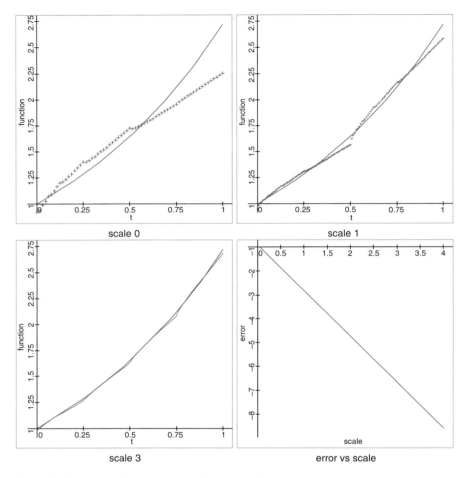

Fig. 1.14 Daubechies filter of order two for exponential

1.8.2.8 Riesz Sequences

In this section, we will examine conditions under which functions can be in the span of the set of translations of some function. Afterward, in Sect. 1.8.2.9, this will lead to the development of a hierarchy of spaces of square-integrable functions with good approximation properties, with the spaces obtained from translations and scalings of a single function.

First, we will need to make some definitions. We begin by specializing some sequences.

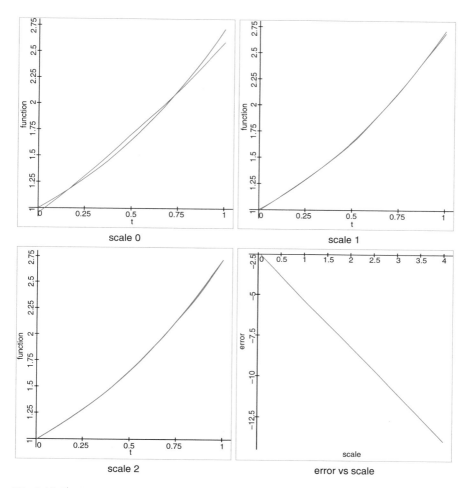

Fig. 1.15 Daubechies filter of order three for exponential

Definition 1.8.12 Suppose that X is a normed linear space. Then the sequence $\{x_n\}_{n=1}^{\infty} \subset X$ is a **Cauchy sequence** if and only if for all $\varepsilon > 0$ there exists an integer $k > 0$ so that for all $n, m > k$ we have $\|x_n - x_m\|_X < \varepsilon$.
Next, we will specialize some linear spaces.

Definition 1.8.13 Suppose that H is a linear space with an inner product. Then H is a **Hilbert space** if and only if for every Cauchy sequence $\{h_n\}_{n=1}^{\infty}$ in H there exists $h \in H$ so that the Cauchy sequence converges to h in the norm generated by the inner product.
For example, it is well-known (see Yosida [187, p. 53]) that the space of all square-integrable functions on $(-\infty, \infty)$ is a Hilbert space. Finally, we will specialize some sequences.

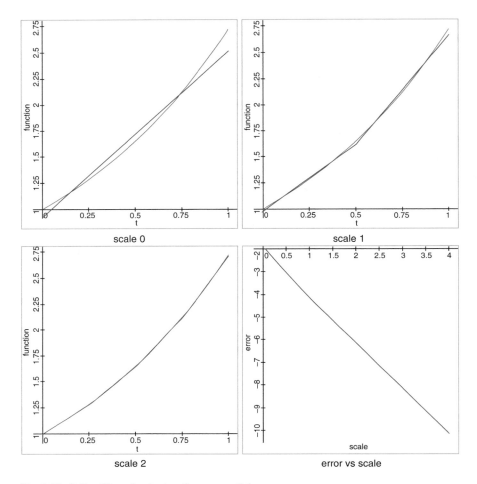

Fig. 1.16 Spline filter of order two for exponential

Definition 1.8.14 Suppose that H is a Hilbert space and J is a countable set. Then $\{x_j\}_{j\in J}$ is a **Riesz sequence** if and only if there exist positive constants $\underline{c} \leq \overline{c}$ such that for all square-summable sequences $\{a_j\}_{j\in J}$ we have

$$\underline{c} \left[\sum_{j\in J} |a_j|^2 \right]^{1/2} \leq \left\| \sum_{j\in J} a_j x_j \right\|_H \leq \overline{c} \left[\sum_{j\in J} |a_j|^2 \right]^{1/2} . \tag{1.137}$$

A Riesz sequence $\{x_j\}_{j\in J}$ is a **Riesz basis** if and only if the span of $\{x_j\}_{j\in J}$ is H.

The following lemma relates the Fourier transform of a function to the condition that its translations form a Riesz sequence.

Lemma 1.8.29 ([184, p. 22])   *Suppose that ϕ is square-integrable, and $0 < \underline{c} \leq \overline{c}$. Then for all square-summable sequences $\{a_j\}_{j=-\infty}^{\infty}$*

$$\underline{c} \left[\sum_{j=-\infty}^{\infty} |a_j|^2 \right]^{1/2} \leq \left[\int_{-\infty}^{\infty} \left| \sum_{j=-\infty}^{\infty} a_j T_j\{\phi\}(t) \right|^2 dt \right]^{1/2} \leq \overline{c} \left[\sum_{j=-\infty}^{\infty} |a_j|^2 \right]^{1/2}$$

(1.138a)

if and only if for almost all real ξ

$$\underline{c}^2 \leq \sum_{n=-\infty}^{\infty} |\mathcal{F}\{\phi\}(\xi + 2\pi n)|^2 \leq \overline{c}^2 .$$

(1.138b)

Proof First, note that since the functions $e^{-i\xi n}$ are mutually orthogonal, we have

$$\int_0^{2\pi} \left| \sum_{n=-\infty}^{\infty} a_n e^{-in\xi} \right|^2 d\xi = \int_0^{2\pi} \left[\sum_{n=-\infty}^{\infty} a_n e^{-in\xi} \right] \overline{\left[\sum_{m=-\infty}^{\infty} a_m e^{-im\xi} \right]} d\xi$$

$$= \sum_{n=-\infty}^{\infty} \sum_{m=-\infty}^{\infty} a_n \overline{a_m} \int_0^{2\pi} e^{i(m-n)\xi} d\xi = 2\pi \sum_{n=-\infty}^{\infty} |a_n|^2 .$$

We will begin the proof by showing that the second condition (1.138b) implies the first condition (1.138a). Note that

$$\frac{\underline{c}^2}{2\pi} \int_0^{2\pi} \left| \sum_{j=-\infty}^{\infty} a_j e^{-ij\xi} \right|^2 d\xi = \underline{c}^2 \sum_{j=-\infty}^{\infty} |a_j|^2 \leq \int_{-\infty}^{\infty} \left| \sum_{j=-\infty}^{\infty} a_j T_j\{\phi\}(t) \right|^2 dt$$

$$\leq \overline{c}^2 \sum_{j=-\infty}^{\infty} |a_j|^2 = \frac{\overline{c}^2}{2\pi} \int_0^{2\pi} \left| \sum_{j=-\infty}^{\infty} a_j e^{-ij\xi} \right|^2 d\xi$$

Also note that Eq. (1.121b) implies that

$$\int_{-\infty}^{\infty} \left| \sum_{j=-\infty}^{\infty} a_j T_j\{\phi\}(t) \right|^2 dt = \int_0^{2\pi} \left| \sum_{j=-\infty}^{\infty} a_j e^{-ij\xi} \right|^2 \sum_{n=-\infty}^{\infty} |\mathcal{F}\{\phi\}(\xi + 2\pi n)|^2 d\xi .$$

If the lower bound in (1.138b) were false, then we could find a scalar $\alpha < \underline{c}^2$ so that the set

$$A_\alpha \equiv \left\{ \xi \in [0, 2\pi] : \sum_{n=-\infty}^{\infty} |\mathcal{F}\{\phi\}(\xi + 2\pi n)|^2 \leq \alpha \right\}$$

has positive measure. Let

$$\chi_{A_\alpha}(\xi) = \begin{cases} 1, \ \xi \in A_\alpha \\ 0, \ \xi \notin A_\alpha \end{cases} ,$$

and let $\{a_n\}_{n=-\infty}^{\infty}$ be the Fourier coefficients for χ_{A_α}. Then for almost all $\xi \in [0, 2\pi]$ we have

$$\chi_{A_\alpha}(\xi) = \sum_{n=-\infty}^{\infty} a_n e^{-in\xi} ,$$

and

$$2\pi \underline{c}^2 \sum_{j=-\infty}^{\infty} |a_j|^2 \le \int_0^{2\pi} \left| \sum_{j=-\infty}^{\infty} a_j e^{-ij\xi} \right|^2 \sum_{n=-\infty}^{\infty} |\mathscr{F}\{\phi\}(\xi + 2\pi n)|^2 \ \mathrm{d}\xi$$

$$= \int_{A_\alpha} \left| \sum_{j=-\infty}^{\infty} a_j e^{-ij\xi} \right|^2 \sum_{n=-\infty}^{\infty} |\mathscr{F}\{\phi\}(\xi + 2\pi n)|^2 \ \mathrm{d}\xi \le \alpha \int_{A_\alpha} \left| \sum_{j=-\infty}^{\infty} a_j e^{-ij\xi} \right|^2 \ \mathrm{d}\xi$$

$$= \alpha \int_0^{2\pi} \left| \sum_{j=-\infty}^{\infty} a_j e^{-ij\xi} \right|^2 \ \mathrm{d}\xi = 2\pi \alpha \sum_{j=-\infty}^{\infty} |a_j|^2 .$$

Since $\alpha < \underline{c}^2$ and

$$2\pi \sum_{j=-\infty}^{\infty} |a_j|^2 = \int_0^{2\pi} \left| \sum_{j=-\infty}^{\infty} a_j e^{ij\xi} \right|^2 \ \mathrm{d}\xi = \int_0^{2\pi} |\chi_{A_\alpha}(\xi)|^2 \ \mathrm{d}\xi = \int_{A_\alpha} \mathrm{d}\xi ,$$

we have a contradiction. We conclude that the lower bound in (1.138b) must be true. A similar approach will prove the upper bound.

Next, suppose that inequality (1.138b) is satisfied. For any $\alpha > 0$, we let

$$A_\alpha \equiv \left\{ \xi \in [0, 2\pi] : \sum_{n=-\infty}^{\infty} |\mathscr{F}\{\phi\}(\xi + 2\pi n)|^2 > \alpha \right\} .$$

If there exists α so that the measure of A_α is positive, let $\{a_n\}_{n=-\infty}^{\infty}$ be such that for almost all $\xi \in [0, 2\pi]$

$$\chi_{A_\alpha}(\xi) = \sum_{n=-\infty}^{\infty} a_n e^{-in\xi} .$$

Since χ_{A_α} is square-integrable, $\{a_n\}_{n=-\infty}^{\infty}$ is square-summable. Note that since $\{e^{-in\xi}\}_{n=-\infty}^{\infty}$ is mutually orthogonal on $[0, 2\pi]$,

$$
\int_{A_\alpha} d\xi = \int_0^{2\pi} \chi_{A_\alpha}(\xi)^2 \, d\xi = \int_0^{2\pi} \left| \sum_{n=-\infty}^{\infty} a_n e^{-in\xi} \right|^2 d\xi = 2\pi \sum_{n=-\infty}^{\infty} |a_n|^2 .
$$

Note that inequality (1.121b) implies that

$$
\frac{1}{2\pi} \int_{-\infty}^{\infty} \left| \sum_{j=-\infty}^{\infty} a_j T_j\{\phi\}(\xi) \right|^2 d\xi
$$

$$
= \frac{1}{2\pi} \int_0^{2\pi} \left| \sum_{j=-\infty}^{\infty} a_j e^{-ij\theta} \right|^2 \sum_{n=-\infty}^{\infty} |\mathcal{F}\{\phi\}(\theta + 2\pi n)|^2 \, d\theta
$$

$$
= \frac{1}{2\pi} \int_{A_\alpha} \sum_{n=-\infty}^{\infty} |\mathcal{F}\{\phi\}(\theta + 2\pi n)|^2 \, d\theta \geq \frac{\alpha}{2\pi} \int_{A_\alpha} d\xi = \alpha \sum_{n=-\infty}^{\infty} |a_n|^2 .
$$

Then (1.138a) implies that $\alpha \leq \bar{c}^2$. In other words, if $\alpha > \bar{c}^2$ then the measure of A_α is zero. This means that for almost all real scalars ξ we have

$$
\sum_{n=-\infty}^{\infty} |\mathcal{F}\{\phi\}(\xi + 2\pi n)|^2 \leq \bar{c}^2 .
$$

To prove the left-hand inequality in (1.138b), we consider

$$
B_\alpha \equiv \left\{ \xi \in [0, 2\pi] : \sum_{n=-\infty}^{\infty} \|\mathcal{F}\{\phi\}(\xi + 2\pi n)\|^2 < \alpha \right\}
$$

in place of A_α in the preceding argument.

1.8.2.9 Multiresolution Analysis

In Sect. 1.8.2.8 we studied Riesz sequences and their connection to linear combinations of translations. In this section, we will examine sequences of linear spaces, with each space formed from translations of some function, and the linear spaces interrelated by dilation. We begin with a fundamental definition, due to Mallat [132].

Definition 1.8.15 A **multiresolution analysis** is a sequence $\{V_j\}_{j=-\infty}^{\infty}$ of closed subspaces of square-integrable functions on the real line so that

1. for all integers j, the "coarser" subspace V_j is contained in the "finer" subspace V_{j+1};

2. $\lim_{j\to\infty} V_j$ is the set of all square-integrable functions; in other words, for every square-integrable function f there is a sequence $\{v_j\}_{j=j_0}^\infty$ so that

$$\lim_{j\to-\infty} \int_{-\infty}^\infty \left|f(t) - v_j(t)\right|^2 \, \mathrm{d}t = 0 \; . \tag{1.139}$$

3. $\bigcap_{j=-\infty}^\infty V_j = \{0\}$;
4. $f \in V_j$ if and only if $D_2\{f\} \in V_{j-1}$, where the dilation operator D_2 is defined by $D_2\{f\}(t) = f(t/2)$; and
5. $f \in V_0$ if and only if for all integers k we have $T_k\{f\} \in V_0$, where the translation operator T_k is defined by $T_k\{f\}(t) = f(t-k)$.

Here is an example of a multiresolution analysis.

Example 1.8.15 The sets V_j of all trigonometric polynomials of degree at most j form a multiresolution analysis, since it is well-known that any square-integrable function can be approximated arbitrarily well by a trigonometric polynomial. Unfortunately, the trigonometric polynomials are not localized well in time. Instead, we will prefer to use scaling functions to form a multiresolution analysis.

Definition 1.8.16 If $\{V_j\}_{j=-\infty}^\infty$ is a multiresolution analysis, then $\phi \in V_0$ is a **scaling function** for this multiresolution analysis if and only if ϕ is absolutely integrable on the real line; the translations of ϕ form a partition of unity, meaning that

$$\sum_{k=-\infty}^\infty \phi(t-k) = 1 \text{ for all } t \; ;$$

the coefficients ℓ in the refinement Eq. (1.124) for ϕ are absolutely summable; and $\{T_k\{\phi\}\}_{k=-\infty}^\infty$ is a Riesz basis for V_0.

In many cases, it is reasonably easy to show that the translations of some scaling function form a Riesz sequence, by applying the ideas in Lemma 1.8.29 to the Fourier transform of ϕ. The condition (1.139) required by the definition of a multiresolution analysis would normally be much more difficult to verify for a specific scaling function. Fortunately, Theorem 1.8.12 deals with that issue, by providing equivalent and simple tests on the scaling function and its dual.

Example 1.8.16 Lemma 1.6.5 showed that the cardinal B-splines, defined in Eq. (1.44), satisfy the refinement equation

$$B_k(t) = 2 \sum_{n=0}^k 2^{-k} \binom{k}{n} B_k(2t-n) \; .$$

Since this refinement equation involves a finite sequence of coefficients, the coefficient sequence is absolutely summable. Also, Lemma 1.6.4 showed that B_k

is absolutely summable and forms a partition of unity. A proof that the cardinal B-spline is a scaling function for a multiresolution analysis can be found in Cohen [48, p. 55].

We will need the following definition, which is a generalization of some ideas we saw in the Fundamental Theorem of Linear Algebra 3.2.3 of Chap. 3 in Volume I.

Definition 1.8.17 If $\mathscr{X}$ and $\mathscr{Y}$ are two subspaces of a linear space $\mathscr{H}$, then $\mathscr{H}$ is the **direct sum** of $\mathscr{X}$ and $\mathscr{Y}$ if and only if $\mathscr{X} \cap \mathscr{Y} = \{0\}$, and for all $h \in \mathscr{H}$ there exist $x \in \mathscr{X}$ and $y \in \mathscr{Y}$ so that $h = x + y$. If $\mathscr{H}$ is the **direct sum** of $\mathscr{X}$ and $\mathscr{Y}$, then we write $\mathscr{H} = \mathscr{X} \oplus \mathscr{Y}$.

Note that if $\mathscr{H} = \mathscr{X} \oplus \mathscr{Y}$ and $h = x + y$, then x and y are unique. To see this fact, note that if $h = x_1 + y_1 = x_2 + y_2$, then $x_1 - x_2 = y_2 - y_1 \in \mathscr{X} \cap \mathscr{Y}$, so $x_1 - x_2 = 0 = y_2 - y_1$.

We will use our new definition of direct sums in the following lemma.

Lemma 1.8.30 *Suppose that ℓ and $\mathbf{h}$ are impulse responses for lowpass and highpass filters in a biorthogonal filter bank. Let the scaling function ϕ have refinement equation using the discrete time signal ℓ, and let ψ be the corresponding wavelet using the discrete time signal $\mathbf{h}$ for its refinement equation. Also suppose that ϕ is the scaling function for a multiresolution analysis $\{V_j\}_{j=-\infty}^{\infty}$. Then $V_{j+1} = V_j \oplus W_j$, where*

$$V_j = span \left\{2^{j/2} \phi \left(2^j t - n\right)\right\}_{n=-\infty}^{\infty} \text{ and } W_j = span \left\{2^{j/2} \psi \left(2^j t - n\right)\right\}_{n=-\infty}^{\infty} .$$

Proof Note that the refinement Eq. (1.126) for the wavelet ψ shows that $D_{2^j} \psi \in V_{j+1}$. Also, the decomposition relation (1.132) for the scaling function shows that $V_{j+1} = V_j + W_j$. All that remains is to show that $V_j \cap W_j = \{0\}$. Equivalently, we can show that if there are coefficients $\{\gamma_n\}_{n=-\infty}^{\infty}$ and $\{\Gamma_n\}_{n=-\infty}^{\infty}$ so that

$$0 = \sum_{n=-\infty}^{\infty} [\gamma_n \phi (\tau - n) + \Gamma_n \psi (\tau - n)] = 0 \text{ for all } \tau = 2^j t$$

then $\gamma_n = 0 = \Gamma_n$ for all n.

We can use the reconstruction algorithm (1.135) to obtain

$$0 = \ell * [(\uparrow_2)\gamma] + \mathbf{h} * [(\uparrow_2)\Gamma] .$$

We can apply a reversal to this equation to get

$$0 = \mathbf{R} \{\ell * [(\uparrow_2)\gamma] + \mathbf{h} * [(\uparrow_2)\Gamma]\} = (\mathbf{R}\ell) * [\mathbf{R}(\uparrow_2)\gamma] + (\mathbf{R}\mathbf{h}) * [\mathbf{R}(\uparrow_2)\Gamma]$$
$$= (\mathbf{R}\ell) * [(\uparrow_2)\gamma] + (\mathbf{R}\mathbf{h}) * [(\uparrow_2)\Gamma]$$

Together, these equations can be written

$$\begin{bmatrix} \boldsymbol{\ell} & \mathbf{h} \\ \mathbf{R\ell} & \mathbf{Rh} \end{bmatrix} * \begin{bmatrix} (\uparrow_2)\boldsymbol{\gamma} \\ (\uparrow_2)\boldsymbol{\Gamma} \end{bmatrix} = \begin{bmatrix} 0 \\ 0 \end{bmatrix} .$$

If we take the discrete Fourier transform of this system of equations, we get

$$\begin{bmatrix} \mathscr{F}^d\{\boldsymbol{\ell}\}(\omega) & \mathscr{F}^d\{\mathbf{h}\}(\omega) \\ \mathscr{F}^d\{\mathbf{R\ell}\}(\omega) & \mathscr{F}^d\{\mathbf{Rh}\}(\omega) \end{bmatrix} \begin{bmatrix} \mathscr{F}^d\{\boldsymbol{\gamma}\}(2\omega) \\ \mathscr{F}^d\{\boldsymbol{\Gamma}\}(2\omega) \end{bmatrix} = \begin{bmatrix} 0 \\ 0 \end{bmatrix} .$$

Since Theorem 1.8.4 shows that the modulation matrix is nonsingular, we conclude that $\boldsymbol{\gamma}_n = 0 = \boldsymbol{\Gamma}_n$ for all n. This proves that $V_0 \cap W_0 = \{0\}$, and completes the proof of the lemma.

1.8.2.10   Error Estimates

Ultimately, we hope to use scaling functions and wavelets to approximate arbitrary functions. This leads to the important question of accuracy in the approximation. The next two lemmas will provide us with a test to determine the order of polynomials that are reproduced exactly by translations of some scaling function. The first of these lemmas can also be found in Cohen [48, p. 89].

Lemma 1.8.31 *Suppose that $\boldsymbol{\ell}$ is the impulse response for a lowpass filter, and suppose that there exists an integer $L > 0$ and a trigonometric polynomial $P(\omega)$ so that*

$$\mathscr{F}^d\{\boldsymbol{\ell}\}(\omega) = 2^{-L}\left[1 + e^{-i\omega}\right]^L P(\omega) .$$

Then the scaling function ϕ with the refinement equation

$$\phi(t) = 2 \sum_{n=-\infty}^{\infty} \ell_n \phi(2t - n)$$

*satisfies the **Strang-Fix conditions** of order $L-1$: for all nonnegative integers $q < L$ and for all nonzero integers n*

$$\frac{d^q \mathscr{F}\{\phi\}}{d\omega^q}(2\pi n) = 0 . \tag{1.140}$$

Proof For any nonzero integer n there exists an integer $k \geq 1$ and an integer m so that the binary representation of n can be written

$$n = (2m + 1)2^{k-1} .$$

It follows that

$$\mathscr{F}^d\{\ell\}\left(2^{-k}[2n\pi]\right) = 2^{-L}\left[1 + e^{-i2^{1-k}n\pi}\right]^L P\left(2^{1-k}n\pi\right)$$

$$= 2^{-L}\left[1 + e^{-i(2m+1)\pi}\right]^L P\left(2^{1-k}n\pi\right) = 2^{-L}\left[1 - 1\right]^L P\left(2^{1-k}n\pi\right) = 0\ .$$

Since

$$\mathscr{F}\{\phi\}(\omega) = \mathscr{F}^d\{\ell\}\left(2^{-1}\omega\right)\mathscr{F}\{\phi\}\left(2^{-1}\omega\right)$$

$$= \ldots = \prod_{\nu=1}^{N}\mathscr{F}^d\{\ell\}\left(2^{-\nu}\omega\right)\mathscr{F}\{\phi\}\left(2^{-N}\omega\right)\ ,$$

it follows that for all nonzero $n = (2m + 1)2^{k-1}$ we can choose $N = k$ to form this product and see that when

$$\omega = 2n\pi + 2^k\varepsilon$$

we have

$$\mathscr{F}\{\phi\}(\omega) = \prod_{\nu=1}^{k}\mathscr{F}^d\{\ell\}\left(2^{-\nu}\omega\right)\mathscr{F}\{\phi\}\left(2^{-k}\omega\right)$$

$$= \mathscr{F}^d\{\ell\}\left(2^{1-k}n\pi + \varepsilon\right)\prod_{\nu=1}^{k-1}\mathscr{F}^d\{\ell\}\left(2^{-\nu}\omega\right)\mathscr{F}\{\phi\}\left(2^{-k}\omega\right)$$

$$= 2^{-L}\left[1 + e^{-i(2m+1)\pi}e^{-i\varepsilon}\right]^L P(\omega)\prod_{\nu=1}^{k-1}\mathscr{F}^d\{\ell\}\left(2^{1-\nu}\omega\right)\mathscr{F}\{\phi\}\left(2^{1-k}\omega\right)$$

$$= 2^{-L}\left[1 - ie^{-i\varepsilon}\right]^L P(\omega)\prod_{\nu=1}^{k-1}\mathscr{F}^d\{\ell\}\left(2^{1-\nu}\omega\right)\mathscr{F}\{\phi\}\left(2^{1-k}\omega\right)$$

$$= 2^{-L}\left[2i\sin(\varepsilon/2)e^{-i\varepsilon/2}\right]^L P(\omega)\prod_{\nu=1}^{k-1}\mathscr{F}^d\{\ell\}\left(2^{1-\nu}\omega\right)\mathscr{F}\{\phi\}\left(2^{1-k}\omega\right)$$

$$= [\sin(\varepsilon/2)]^L e^{-iL[\varepsilon-\pi]/2} P(\omega)\prod_{\nu=1}^{k-1}\mathscr{F}^d\{\ell\}\left(2^{1-\nu}\omega\right)\mathscr{F}\{\phi\}\left(2^{1-k}\omega\right)\ .$$

This proves the claim.

Cohen [48, p. 90] proves the following result using distribution theory.

Lemma 1.8.32 *Suppose that ϕ is absolutely integrable on the real line, has compact support, and*

$$\int_{-\infty}^{\infty} \phi(t)\, dt = 1 .$$

Then ϕ satisfies the Strang-Fix conditions (1.140) *if and only if for all nonnegative integers $q < L$ there is a polynomial r so that for all real t*

$$\sum_{k=-\infty}^{\infty} k^q \phi(t-k) = t^q + r(t) .$$

This lemma says that the Strang-Fix conditions (1.140) are equivalent to polynomial exactness of order $L - 1$. Indeed, we can express $r(t)$ as a linear combination of monomials of order less than q and continue until we obtain

$$t^q = \sum_{k=-\infty}^{\infty} \gamma_q(k)\phi(t-k)$$

for some polynomial γ_q of degree q.

The Bramble-Hilbert lemma [21] (see Sect. 1.6.6 above) may be used to show that whenever a scaling function reproduces all polynomials of degree at most $L-1$ exactly, then there is a constant $C > 0$ so that for all scales $j \geq 0$, all derivative orders $s \in [0, L]$ and all functions g whose derivatives of order at most s are square-integrable, we have

$$\left[\int_{-\infty}^{\infty} \left| g(t) - P_j g(t) \right|^2 dt \right]^{1/2} \leq C 2^{js} \left[\int_{-\infty}^{\infty} \left| \frac{d^s g}{dt^s}(t) \right|^2 dt \right]^{1/2} .$$

Very smooth functions will have their scaling function approximation limited in accuracy by the order $L - 1$ of polynomials reproduced exactly; rough functions will have their **order of accuracy** limited by the number s of square-integrable derivatives that they possess.

Readers may find more detailed discussion of this error estimate in Cohen [48, p. 171].

Finally, we note that some numerical experiments with various scaling functions in Figs. 1.14, 1.15 and 1.16 provide evidence of the validity of these theoretical error estimates.

Exercise 1.8.13 Examine the definition of the discrete Fourier transform of the lowpass filter ℓ for the Daubechies filter in Sect. 1.8.1.8, and compare it to the condition at the beginning of Lemma 1.8.31. Determine the order of the Strang-Fix condition satisfied by the Daubechies filters, and then describe how the approximation error for the Daubechies filter should depend on the scale and the differentiability of the function being approximated.

Chapter 2
Differentiation and Integration

Does anyone believe that the difference between the Lebesgue and Riemann integrals can have physical significance, and that whether say, an airplane would or would not fly could depend on this difference? If such were claimed, I should not care to fly in that plane.

Richard Hammin *[17, p. 16]*

God does not care about our mathematical difficulties. He integrates empirically.

Albert Einstein *[111, p. 279]*

Abstract This chapter develops numerical methods for computing derivatives and integrals. Numerical differentiation of polynomials can be performed by synthetic division, or through special properties of trigonometric polynomials or orthogonal polynomials. For derivatives of more general functions, finite differences lead to difficulties with rounding errors that can be largely overcome by clever post-processing, such as Richardson extrapolation. Integration is a more complicated topic. The Lebesgue integral is related to Monte Carlo methods, and Riemann sums are improved by trapezoidal and midpoint rules. Analysis of the errors leads to the Euler-MacLaurin formula. Various polynomial interpolation techniques lead to specialized numerical integration methods. The chapter ends with discussions of tricks for difficult integrals, adaptive quadrature, and integration in multiple dimensions.

2.1 Overview

Differentiation and integration are two basic transformations in elementary calculus. As a result, they are important to all of those fields that motivated the development of calculus, especially physics and engineering. Since calculus is a prerequisite for most scientific computing courses, the reader will probably begin this chapter with

---

Additional Material: The details of the computer programs referred in the text are available in the Springer website (http://extras.springer.com/2018/978-3-319-69110-7) for authorized users.

J.A. Trangenstein, *Scientific Computing*, Texts in Computational Science and Engineering 20, https://doi.org/10.1007/978-3-319-69110-7_2

many preconceptions regarding how derivatives and integrals should be computed. Although familiarity with the concepts in elementary calculus is helpful to this course, many of the computations in elementary calculus are not very effective in finite precision arithmetic.

Our goals in this chapter are to develop accurate and efficient numerical methods for computing derivatives and integrals. We will begin with numerical differentiation. Our early experience with finite differences in Sect. 2.3.1.2 of Chap. 2 in Volume I showed that these computations are quickly contaminated by rounding errors. In Sect. 2.2.1, we will see that for differentiation of polynomials we can use synthetic division as discussed in Sect. 2.3.1.8 of Chap. 2 in Volume I to avoid the rounding errors in finite differences. We will also discuss differentiation of trigonometric polynomials and orthogonal polynomials. For more general functions, we will discuss some basic ideas regarding finite differences in Sects. 2.2.2 and 2.2.3. Then, we will develop the very important idea of extrapolation in Sects. 2.2.4 and 2.2.5 to overcome catastrophic loss of accuracy in finite differences.

After we discuss numerical differentiation, we will develop methods for numerical integration. Within this topic, we will develop a number of scientific paths. In Sect. 2.3.2, we will discuss Monte Carlo methods, which make use of both random sampling of the integrand and Lebesgue integration. Afterward, in Sects. 2.3.3–2.3.5 we will discuss methods based on integrating spline approximations to the integrand. Error estimates for the trapezoidal and midpoint rules will be discussed in Sect. 2.3.6. This will lead to the very important Euler-MacLaurin formula, and important implications for the integration of periodic functions. These deterministic integration ideas will all make use of the concepts in Riemann integration.

In other applications, the use of equidistant quadrature nodes is important; in such cases, the Newton-Cotes quadrature rules in Sect. 2.3.7 will be useful. Better yet, the Chebyshev nodes can be used to develop Clenshaw-Curtis quadrature in Sect. 2.3.8. We can also develop high-order numerical methods for integration by applying extrapolation. The result will be an approach called Romberg integration in Sect. 2.3.9.

For some applications, especially finite element methods for solving differential equations, the order of accuracy of numerical integration is more important than the absolute error; in these cases we will find that orthogonal polynomials will allow us to develop very efficient Gaussian quadrature and Lobatto quadrature rules beginning with Sect. 2.3.10.

Gaussian quadrature provides guaranteed order of accuracy, but the magnitude of its error is hard to estimate. However, it is possible to extend a Gaussian quadrature rule with an optimal placement of interlacing quadrature nodes to obtain a higher-order quadrature. The combination of the two quadrature rules provides a computable error estimate, as discussed in Sect. 2.3.12.

Integrals of functions involving discontinuities, singularities, unbounded domains or a large number of oscillations present special numerical difficulties. We will present some ideas for numerical integration under these circumstances in Sect. 2.3.13.

These various numerical integration techniques all involve fairly regular placement of function evaluations. However, it would be more efficient to concentrate the function evaluations where needed for guaranteed accuracy. Adaptive quadrature can be added to any of the previous quadrature methods, and provide computable error estimates in Sect. 2.3.14.

We will end this chapter by discussing some methods for computing multi-dimensional integrals in Sect. 2.3.15.

This chapter depends strongly on much of the material in Chap. 1. In particular, we assume that the reader is familiar with the ideas of polynomial interpolation in Sect. 1.2, splines in Sect. 1.6, orthogonal polynomials in Sect. 1.7.3 and trigonometric polynomials in Sect. 1.7.4. The material in this chapter will be crucial to understanding the numerical solution of initial value problems for ordinary differential equations in Chap. 3, and boundary-value problems in Chap. 4.

For more information about numerical integration, we recommend Davis and Rabinowitz [60], Stroud and Secrest [167] and Stroud [166]. A number of numerical analysis books have very good discussions of numerical differentiation and integration, such as Dahlquist and Björck [57], Greenbaum [91], Kincaid and Cheney [119], Press et al. [144], and Ralston and Rabinowitz [149].

For numerical differentiation software, we recommend GSL (GNU Scientific Library) Numerical Differentiation. In MATLAB, the closest commands are diff and gradient, which work on a specified grid of function and argument values.

For numerical integration, we recommend GSL Numerical Integration and quadpack. In MATLAB, the basic integration commands are integral, quadgk and trapz. For Monte Carlo integration, we recommend GSL Monte Carlo Integration.

2.2 Numerical Differentiation

Numerical differentiation is computationally simpler than numerical integration. Nevertheless, we will develop some key ideas in our discussion of numerical differentiation that will assist us in developing numerical integration methods.

Let us recall the definition of a derivative.

Definition 2.2.1 Suppose that f is continuous in the interval (a, b). Then f is **differentiable** at $x \in (a, b)$ if and only if the limit

$$\lim_{\xi \to x} \frac{f(\xi) - f(x)}{\xi - x}$$

exists. If f is differentiable at x, then the value of the limit is called the **derivative** of f at x, and may be written as $f'(x)$, $\frac{df}{dx}(x)$ or $Df(x)$. If f is differentiable at every point of (a, b), then we say that f is **differentiable in** (a, b).

The following more general definition applies to functions of several variables.

Definition 2.2.2 Suppose that $\mathbf{f}$ is a continuous function in some open subset Ω of an n-dimensional space, and has function values in some m-dimensional space. Then $\mathbf{f}$ is **differentiable** at $\mathbf{x} \in \Omega$ if and only if there is a linear mapping $\mathbf{Df}$ from

n-vectors to m-vectors such that the limit

$$\lim_{\xi \to x} \frac{\|\mathbf{f}(\xi) - \mathbf{f}(x) - \mathbf{Df}(\xi - x)\|}{\|\xi - x\|} = 0$$

for any norms on the two spaces. If $\mathbf{f}$ is differentiable at $\mathbf{x}$, then $\mathbf{Df}(\mathbf{x})$ is called the **derivative** of $\mathbf{f}$ at $\mathbf{x}$. Finally, $\mathbf{f}$ is **differentiable in** Ω if and only if it is differentiable at all points of Ω.

Difference quotients are an essential part of the definitions of derivatives, so it is natural to use difference quotients in computations. However, Example 2.3.1.2 of Chap. 2 in Volume I showed that in finite precision arithmetic, difference quotients with small increments can develop large errors. Thus, we should be motivated to find ways to avoid difference quotients when possible.

In many cases, it is possible to examine the computer code for a given function and develop **automatic differentiation** rules to compute the derivative. We will not discuss automatic differentiation further in this text; instead, we refer the reader to Griewank and Walther [92], Neidinger [138] and Rall [148]. There are also automatic differentiation routines available in Fortran and C or C^{++}.

2.2.1 Polynomials

It is possible to compute derivatives of some functions without employing finite differences. We will discuss three cases, namely polynomials, trigonometric polynomials and orthogonal polynomials.

2.2.1.1 Synthetic Division

In Sect. 2.3.1.9 of Chap. 2 in Volume I, we showed how to differentiate a polynomial without forming difference quotients. Let us explain that computation again, in greater detail.

First, we will present Horner's rule for polynomial evaluation in a more formal way.

Lemma 2.2.1 (Horner's Rule) *Given scalars $a_0, \ldots, a_n$, define the polynomial*

$$p_n(x) = \sum_{i=0}^{n} a_{n-i} x^i$$

and for $0 \le k \le n$ let the partial sums be

$$s_{n-k}(x) = \sum_{i=k}^{n} a_{n-i} x^{i-k} \ . \tag{2.1}$$

Then $s_0(x) = a_0$, and for $0 < m \leq n$ we have

$$s_m(x) = xs_{m-1}(x) + a_m \ . \tag{2.2}$$

In particular, $s_n(x) = p_n(x)$.

Proof By taking $k = n$ in (2.1), we can easily see that $s_0(x) = a_0$, and by taking $k = 0$ we can see that $s_n(x) = p_n(x)$.

All that remains is to prove (2.2). If we define $m = n - k$, then the partial sums in Eq. (2.1) are defined for $0 \leq n - k \equiv m \leq n$. By taking $\ell = i - k$ and $m = n - k$ in the sum, we see that

$$s_m(x) = \sum_{\ell=0}^{m} a_{m-\ell} x^\ell \ .$$

If $m > 0$, it follows that

$$s_m(x) = a_m + \sum_{\ell=1}^{m} a_{m-\ell} x^\ell = a_m + x \sum_{\ell=1}^{m} a_{m-\ell} x^{\ell-1} = a_m + x \sum_{k=0}^{m-1} a_{m-1-k} x^k$$

$$= a_m + xs_{m-1}(x) \ .$$

Next, let us find an expression for the coefficients of the derivative of a polynomial.

Lemma 2.2.2 (Synthetic Division)   *Given scalars $a_0, \ldots, a_n$, define the polynomial*

$$p_n(x) = \sum_{i=0}^{n} a_{n-i} x^i \ ,$$

and for $0 \leq j \leq n - 1$ define the polynomials

$$b_{n-j-1}(x) = \sum_{i=j+1}^{n} a_{n-i} x^{i-1-j} \ . \tag{2.3}$$

Then

$$p_n'(x) = \sum_{j=0}^{n-1} b_{n-1-j}(x) x^j \ . \tag{2.4}$$

Proof Note that for $t \neq x$,

$$\frac{p_n(t) - p_n(x)}{t - x} = \sum_{i=0}^{n} a_{n-i} \frac{t^i - x^i}{t - x} = \sum_{i=1}^{n} a_{n-i} \sum_{j=0}^{i-1} t^j x^{i-1-j} = \sum_{j=0}^{n-1} t^j \sum_{i=j+1}^{n} a_{n-i} x^{i-1-j}$$

$$= \sum_{j=0}^{n-1} b_{n-j-1}(x) t^j .$$

In the limit as t approaches x, we obtain (2.4).

Here, it is important to note that if $b_{n-j-1}(x)$ is given by (2.3) and $s_{n-k}(x)$ is given by (2.1), then $b_k(x) = s_k(x)$ for $0 \leq k \leq n - 1$. In other words, the coefficients in the synthetic division formula (2.4) for $p_n'(x)$ are precisely the same polynomials computed as partial sums in Horner's rule.

If we use Horner's rule to evaluate p_n' from the formula (2.4), we will generate partial sums that can be generated by a recurrence similar to (2.2). In order to evaluate $p_n(x)$ and $p_n'(x)$ simultaneously, we can use the following algorithm.

Algorithm 2.2.1 (Repeated Synthetic Division)

$$c_0 = b_0 = a_0$$
$$\text{for } 1 \leq i < n$$
$$b_i = a_i + x * b_{i-1}$$
$$c_i = b_i + x * c_{i-1}$$
$$b_n = a_n + x * b_{n-1} .$$

Then $p_n(x) = b_n$ and $p_n'(x) = c_{n-1}$.

A Fortran program to implement synthetic division can be found in synthetic_division.f.

Exercise 2.2.1 If $q_n(t, x) = p_n(t)$ and

$$q_{n-k-1}(t, x) = \frac{q_{n-k}(t) - q_{n-k}(x)}{t - x} \text{ for } 0 \leq k \leq n ,$$

show that

$$p_n^{(k)}(t) = k! q_{n-k}(t, t) .$$

Exercise 2.2.2 Modify the synthetic division Algorithm 2.2.1 to compute the polynomial value, first derivative and second derivative.

2.2.1.2 Trigonometric Polynomials

The derivative of the trigonometric polynomial

$$p(t) = \sum_{k=0}^{m-1} c_k e^{ikt}$$

is

$$p'(t) = \sum_{k=0}^{m-1} (ikc_k) e^{ikt} .$$

In order to compute p' at all of the points $t_j = 2\pi j/m$ for $0 \leq j < m$, we can take the fast Fourier transform of the values for p on these equidistant points, multiply each Fourier coefficient c_k by ik, and then invert the Fourier transform.

If p is the trigonometric polynomial interpolant for some function f, the derivative of p is not necessarily close to f' unless we impose some restrictions on f. The following result, which is proved in Zygmund [190, p. 40], provides the needed restrictions.

Theorem 2.2.1 (Derivative of Fourier Series) *Suppose that f is a complex-valued 2π-periodic function of a real variable t. Also suppose that f is **absolutely continuous**, meaning that for all $\varepsilon > 0$ there exists $\delta > 0$ so that for all sequences $\{(s_k, t_k)\}_{k=1}^{n}$ of pairwise disjoint intervals satisfying*

$$\sum_{k=1}^{n} |t_k - s_k| < \delta ,$$

we have

$$\sum_{k=1}^{n} |f(t_k) - f(s_k)| < \varepsilon .$$

Then the Fourier series for f' converges to the derivative of the Fourier series for f. It follows from the Dirichlet-Jordan test Theorem 1.7.7 that if f is periodic and absolutely continuous, and f' has bounded variation, then the derivative of the infinite Fourier series converges to f' at all points in $(0, 2\pi)$ where f' is continuous.

Figure 2.1 shows the maximum error in interpolating $e^{-2[x-\pi]^2}$ and its derivative over the interval $[0, 2\pi)$ by Fourier series at various numbers of interpolation points. This figure shows rapid decrease in the errors until they reach a magnitude of about 10^{-8}, followed by no decrease in the error in the derivative and slow decrease in the error in the function. In this case, the function is not quite periodic over this interval; in fact, the derivative is approximately $\pm 3 \times 10^{-8}$ at the endpoints of the interval.

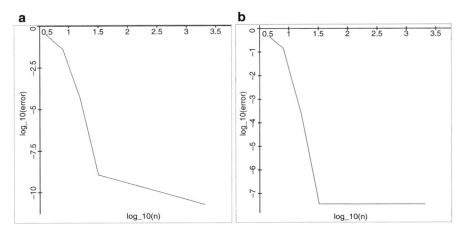

Fig. 2.1 Errors in Fourier series, $f(x) = e^{-2[x-\pi]^2}$, $\log_{10}$ of max error vs $\log_{10}$ of number interpolation points. (**a**) Error in function. (**b**) Error in derivative

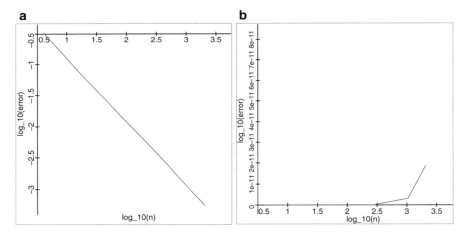

Fig. 2.2 Errors in Fourier series, $f(x) = |x - \pi|$, $\log_{10}$ of max error vs $\log_{10}$ of number interpolation points. (**a**) Error in function. (**b**) Error in derivative

Next, Fig. 2.2 shows the maximum error in interpolating $|x - \pi|$ and its derivative by Fourier series. This figure shows steady decrease in the interpolation error for the function, and a slight *increase* in the interpolation error for the derivative. The errors in the derivative remain near one for all numbers of interpolation points, due to the Gibbs phenomenon. These figures were generated by the C^{++} program GUIFourier.C, In that program, the computation of the fourier series for the derivative is complicated by the fact that the various fast fourier transform software packages store the fourier coefficients in different orders. This makes it a bit tricky to determine the correct factor by which each entry of the Fourier transform return array should be multiplied.

Readers may also execute a JavaScript program to compute a Fourier series derivative. This program plots the derivative of the Fourier series approximation to $f(t) = |t - \pi|$.

Exercise 2.2.3 Examine the errors in the refinement study for the derivative of the Fourier series for $e^{-(x-\pi)^2}$, and explain why the maximum attainable accuracy for the derivative of the trigonometric series is so much poorer than the maximum attainable accuracy for the trigonometric interpolation of the original function.

2.2.1.3 Orthogonal Polynomials

Since our work in Sect. 1.7.3 showed that orthogonal polynomials can be generated by three-term recurrences, it is reasonable to seek three-term recurrences for their derivatives. These recurrences are not necessarily easy to derive. However, the recurrences for the derivatives of orthogonal polynomials will become very important in Sect. 2.3.11 on Lobatto quadrature. For more information on this topic, we recommend that the reader consult either Hildebrand [103], or Isaacson and Keller [112].

First, we note that if the **Legendre polynomials** are generated by the three-term recurrence

Algorithm 2.2.2 (Legendre Polynomials)

$$p_{-1}(x) = 0$$
$$p_0(x) = 1$$
$$\text{for } n = 0, 1, \ldots \ p_{n+1}(x) = \frac{2n+1}{n+1} x p_n(x) - \frac{n}{n+1} p_{n-1}(x) , \qquad (2.5)$$

then their derivatives satisfy the following three-term recurrence

Algorithm 2.2.3 (Legendre Polynomial Derivatives)

$$p_0'(x) = 0$$
$$p_1'(x) = 1$$
$$\text{for } n = 1, 2, \ldots \ p_{n+1}'(x) = \frac{2n+1}{n} x p_n'(x) - \frac{n+1}{n} p_{n-1}'(x) . \qquad (2.6)$$

This recurrence can be proved by differentiating the recurrence (2.5) for the Legendre polynomials to get

$$(n+1)p_{n+1}'(x) = (2n+1)x p_n'(x) - n p_{n-1}'(x) + (2n+1)p_n(x)$$

and subtracting the equation

$$p'_{n+1}(x) = p'_{n-1}(x) + (2n+1)p_n(x) \,,$$

which was proved in Theorem 1.7.5.

The following theorem will be useful for developing Lobatto quadrature in Sect. 2.3.11.

Theorem 2.2.2 (Orthogonality of Legendre Polynomial Derivatives) *The Legendre polynomials, defined by the three-term recurrence (1.67), are orthogonal with respect to the inner product*

$$(f \,,\, g) = \int_{-1}^{1} f(x)g(x)(1-x^2)\,\mathrm{d}x \,.$$

Furthermore,

$$(p'_n, p'_n) = \frac{2n(n+1)}{2n+1} \,. \tag{2.7}$$

Proof Suppose that $n \geq 2$ and q is a polynomial of degree at most $n-2$. Then integration by parts produces

$$\int_{-1}^{1} q(x)p'_n(x)(1-x^2)\,\mathrm{d}x = q(x)p_n(x)(1-x^2)|_{-1}^{1} - \int_{-1}^{1} \frac{\mathrm{d}}{\mathrm{d}x}\left[q(x)(1-x^2)\right]p_n(x)\,\mathrm{d}x = 0 \,,$$

since $\mathrm{d}/\mathrm{d}x[q(x)(1-x^2)]$ is a polynomial of degree at most $n-1$ and the integral of p_n times any polynomial of lower degree is zero. This proves the first claim.

Next, we begin with

$$(2^n n!)^2 \int_{-1}^{1} p'_n(x)^2(1-x^2)\,\mathrm{d}x = -\int_{-1}^{1} (x^2-1)\frac{\mathrm{d}^{n+1}}{\mathrm{d}x^{n+1}}(x^2-1)^n \frac{\mathrm{d}^{n+1}}{\mathrm{d}x^{n+1}}(x^2-1)^n\,\mathrm{d}x$$

then integration by parts produces

$$= -\left[(x^2-1)\frac{\mathrm{d}^n}{\mathrm{d}x^n}(x^2-1)^n \frac{\mathrm{d}^{n+1}}{\mathrm{d}x^{n+1}}(x^2-1)^n\right]_{-1}^{1}$$

$$+ \int_{-1}^{1} \frac{\mathrm{d}^n}{\mathrm{d}x^n}(x^2-1)^n \frac{\mathrm{d}}{\mathrm{d}x}\left\{(x^2-1)\frac{\mathrm{d}^{n+1}}{\mathrm{d}x^{n+1}}(x^2-1)^n\right\}\,\mathrm{d}x$$

$$= \int_{-1}^{1} \frac{\mathrm{d}^n}{\mathrm{d}x^n}(x^2-1)^n \frac{\mathrm{d}}{\mathrm{d}x}\left\{(x^2-1)\frac{\mathrm{d}^{n+1}}{\mathrm{d}x^{n+1}}(x^2-1)^n\right\}\,\mathrm{d}x$$

then another integration by parts gives us

$$= \left[\frac{d^{n-1}}{dx^{n-1}} (x^2 - 1)^n \frac{d}{dx} \left\{ (x^2 - 1) \frac{d^{n+1}}{dx^{n+1}} (x^2 - 1)^n \right\} \right]_{-1}^{1}$$

$$- \int_{-1}^{1} \frac{d^{n-1}}{dx^{n-1}} (x^2 - 1)^n \frac{d^2}{dx^2} \left\{ (x^2 - 1) \frac{d^{n+1}}{dx^{n+1}} (x^2 - 1)^n \right\} \, dx$$

then Eq. (1.76) yields

$$= - \int_{-1}^{1} \frac{d^{n-1}}{dx^{n-1}} (x^2 - 1)^n \frac{d^2}{dx^2} \left\{ (x^2 - 1) \frac{d^{n+1}}{dx^{n+1}} (x^2 - 1)^n \right\} \, dx$$

then repeated integration by parts leads to

$$= \ldots = (-1)^n \int_{-1}^{1} (x^2 - 1)^n \frac{d^{n+1}}{dx^{n+1}} \left\{ (x^2 - 1) \frac{d^{n+1}}{dx^{n+1}} (x^2 - 1)^n \right\} \, dx$$

then pulling the constant derivative term outside the integral gives us

$$= (-1)^n \frac{d^{n+1}}{dx^{n+1}} \left\{ (x^2 - 1) \frac{d^{n+1}}{dx^{n+1}} (x^2 - 1)^n \right\} \int_{-1}^{1} (x^2 - 1)^n \, dx$$

$$= 2 \frac{(2^n n!)^2}{(2n + 1)!} \frac{d^{n+1}}{dx^{n+1}} \left\{ (x^2 - 1) \frac{d^{n+1}}{dx^{n+1}} (x^2 - 1)^n \right\}$$

then the Leibniz rule for differentiation produces

$$= 2 \frac{(2^n n!)^2}{(2n + 1)!} \sum_{k=2}^{2} \binom{n + 1}{k} \frac{d^{2n+2-k}}{dx^{2n+2-k}} (x^2 - 1)^n \frac{d^k}{dx^k} (x^2 - 1)$$

$$= 4 \frac{(2^n n!)^2}{(2n + 1)!} \binom{n + 1}{2} \frac{d^{2n}}{dx^{2n}} (x^2 - 1)^n = 4 \frac{(2^n n!)^2}{(2n + 1)!} \binom{n + 1}{2} (2n)! \; .$$

This result is equivalent to the claim (2.7).

Chebyshev polynomials of the first kind can be defined by

$$t_n(x) = \cos \left(n \cos^{-1}(x) \right) \; ,$$

and can be generated by the three-term recurrence

$$t_0(x) = 1$$

$$t_1(x) = x$$

$$\text{for all } n \geq 1 \; , \; t_{n+1}(x) = 2x t_n(x) - t_{n-1}(x) \; .$$

Chebyshev polynomials of the second kind can be defined by

$$u_n(x) = \frac{\sin\left([n+1]\cos^{-1}(x)\right)}{\sin\left(\cos^{-1}(x)\right)} \, ,$$

and can be generated by the three-term recurrence

$$u_{-1}(x) = 0$$

$$u_0(x) = 1$$

for all $n \geq 0$, $u_{n+1}(x) = 2x u_n(x) - u_{n-1}(x)$,

Since the derivatives of Chebyshev polynomials of the first kind satisfy

$$t'_n(x) = n u_{n-1}(x) \, ,$$

so we can use the recurrence relation for u_n to generate a recurrence for derivatives of t_n:

$$t'_0(x) = 0$$

$$t'_1(x) = 1$$

for all $n \geq 0$, $t'_{n+1}(x) = 2\frac{n+1}{n} x t'_n(x) - \frac{n+1}{n-1} t'_{n-1}(x)$.

As an aside, we remark that Chebyshev polynomials of the first kind have the generating function

$$e^{tx} \cosh\left(t\sqrt{x^2 - 1}\right) = \sum_{n=0}^{\infty} \frac{t_n(x)}{n!} t^n \, ,$$

and Chebyshev polynomials of the second kind have the generating function

$$e^{tx} \left[\cosh\left(t\sqrt{x^2 - 1}\right) + \frac{x}{\sqrt{x^2 - 1}} \sinh\left(t\sqrt{x^2 - 1}\right) \right] = \sum_{n=0}^{\infty} \frac{u_n(x)}{n!} t^n \, .$$

Hermite polynomials can be generated by the three-term recurrence

$$h_{-1}(x) = 0$$

$$h_0(x) = 1$$

for all $n \geq 1$, $h_{n+1}(x) = 2x h_n(x) - 2n h_{n-1}(x)$.

Their derivatives are easily computed from the formula

$$h'_n(x) = 2nh_{n-1}(x) .$$

This formula can be proved by manipulation of the generating function

$$e^{2tx-t^2} = \sum_{n=0}^{\infty} \frac{h_n(x)}{n!} t^n .$$

Laguerre polynomials can be generated by the three-term recurrence

$$\ell_{-1}(x) = 0$$

$$\ell_0(x) = 1$$

for all $n \geq 1$, $\ell_{n+1}(x) = \dfrac{2n+1-x}{n+1}\ell_n(x) - \dfrac{n}{n+1}\ell_{n-1}(x) .$

Their derivatives satisfy

$$\ell'_n(x) = -\ell^{(1)}_{n-1}(x) ,$$

where the generalized Laguerre polynomials $\ell^{(1)}_n$ are generated by

$$\ell^{(1)}_{-1}(x) = 0$$

$$\ell^{(1)}_0(x) = 1$$

for all $n \geq 1$, $\ell^{(1)}_n(x) = \left(2 - \dfrac{x}{n}\right)\ell^{(1)}_{n-1}(x) - \ell^{(1)}_{n-2}(x) .$

Thus the Laguerre polynomial derivatives satisfy the recurrence

$$\ell'_0(x) = 0$$

$$\ell'_1(x) = -1$$

for all $n \geq 1$, $\ell'_{n+1}(x) = \left(2 - \dfrac{x}{n}\right)\ell'_n(x) - \ell'_{n-1}(x) .$

The generating function for the Laguerre polynomials is

$$e^{2tx-t^2} = \sum_{n=0}^{\infty} \ell_n(x)t^n .$$

Exercise 2.2.4 Verify that for $1 \leq n \leq 3$ the Legendre polynomial derivatives, generated by their three-term recurrence, are the derivatives of the Legendre polynomials.

Exercise 2.2.5 The generating function for Legendre polynomials is

$$g(x, t) \equiv (1 + t^2 - 2xt)^{-1/2} .$$

First, show that

$$(x - t)g(x, t) = (1 + t^2 - 2xt)\frac{\partial g}{\partial t}(x, t) .$$

Then, use the series expansion

$$g(x, t) = \sum_{n=0}^{\infty} p_n(x)t^n$$

to replace $g(x, t)$ and $\partial g/\partial t(x, t)$ and show that the coefficients $p_n(x)$ satisfy

$$p_{n+1}(x) = \frac{2n + 1}{n + 1}xp_n(x) - \frac{n}{n + 1}p_{n-1}(x) .$$

Then verify that $p_0(x) = 1$ and $p_1(x) = x$ in the series expansion for $g(x, t)$. Conclude that the coefficients $p_n(x)$ are the Legendre polynomials.

Exercise 2.2.6 If $g(x, t)$ is the generating function for the Legendre polynomials, show that

$$(1 + t^2 - 2xt)\frac{\partial g}{\partial x}(x, t) = tg(x, t) .$$

Then use the series expansion to replace $g(x, t)$ and $\partial g/\partial x(x, t)$ and show that

$$p'_{n+1}(x) + p'_{n-1}(x) = 2xp'_n(x) + p_n(x) .$$

Then differentiate the recurrence relation for the Legendre polynomials, multiply times 2, and add to $(2n + 1)$ times the previous equation, and conclude that

$$p'_{n+1}(x) - p'_{n-1}(x) = (2n + 1)p_n(x) .$$

Exercise 2.2.7 Can you find a three-term recurrence for the integrals $\int_{-1}^{x} p_n(s) \, ds$ of the Legendre polynomials?

2.2.2 One-Sided Differencing

After studying several methods for computing derivatives of various kinds of polynomials, we are ready to turn to more general numerical differentiation methods.
In calculus, the most common technique for computing a derivative is a one-sided
divided difference. In the next lemma, we will analyze the errors associated with
such computations.

Lemma 2.2.3 *Suppose that f is continuously differentiable in the interval $[a, b]$,
and let F and F' be constants so that for all $a \le x \le b$*

$$|f(x)| \le F \text{ and } |f'(x)| \le F' \ .$$

*Assume that f' is Lipschitz continuous with Lipschitz constant F'' in the interval
(a, b):*

$$\left| f'(x) - f'(y) \right| \le F'' \left| x - y \right| \text{ for all } x, y \in (a, b) \ .$$

*Let both x and $x + h$ lie in (a, b). Next, suppose that the values of $f(x)$ and $f(x + h)$
are computed with relative errors ε_f and ε_+, respectively:*

$$fl(f(x)) = f(x)(1 + \varepsilon_f) \text{ and } fl(f(x + h)) = f(x + h)(1 + \varepsilon_+) \ . \tag{2.8}$$

*Assume that the difference quotient $[fl(f(x + h)) - fl(f(x))]/h$ is computed with a
relative error $\varepsilon_\div$. Let ε be an upper bound on all of the relative errors:*

$$\max\{|\varepsilon_f|, |\varepsilon_+|, |\varepsilon_\div|\} \le \varepsilon \ .$$

Then the one-side finite difference approximation

$$f'(x) \approx \frac{f(x + h) - f(x)}{h}$$

has floating point error satisfying

$$\left| fl\left(\frac{fl(f(x + h)) - fl(f(x))}{h} \right) - f'(x) \right| \le \frac{F'' h}{2} + \varepsilon \max_{x < \xi < x+h} |f'(\xi)| + \frac{\varepsilon F}{h} \ .$$

The right-hand side of this expression is minimized when

$$h = \sqrt{\frac{2\varepsilon(1 + \varepsilon)F}{F''}} \ .$$

Proof The errors in floating point arithmetic imply that

$$\left| fl\left(\frac{fl(f(x+h)) - fl(f(x))}{h} \right) - f'(x) \right|$$

$$= \left| \frac{f(x+h) + \varepsilon_+ f(x+h) - f(x) - \varepsilon_f f(x)}{h}(1+\varepsilon_\div) - f'(x) \right|$$

then the fundamental theorem of calculus implies that

$$= \left| \frac{1}{h}\int_0^h f'(x+\xi) - f'(x)\, d\xi + \frac{\varepsilon_\div}{h}\int_0^h f'(x+\xi)\, d\xi \right.$$

$$\left. + \frac{\varepsilon_+(1+\varepsilon_\div)}{h}f(x+h) + \frac{\varepsilon_f(1+\varepsilon_\div)}{h}f(x) \right|$$

then Lipschitz continuity and the bound on the relative errors imply

$$\le \frac{F''}{h}\int_0^h \xi\, d\xi + \varepsilon \max_{y\in(x,x+h)}|f'(y)| + \frac{\varepsilon F}{h} = \frac{F''h}{2} + \varepsilon \max_{y\in(x,x+h)}|f'(y)| + \frac{\varepsilon(1+\varepsilon)F}{h}\ .$$

If we differentiate the right-hand side of this inequality with respect to h and find its zero, we obtain the claimed equation for the value of h that minimizes the right-hand side.

This lemma shows that the error in computing a derivative by a difference quotient involves both a **truncation error** that can be bounded by $F''h/2$, and rounding errors that are bounded by terms proportional to ε. The rounding errors in computing the function values are divided by the finite difference increment h, and these rounding errors eventually dominates all other errors in the computation as the increment becomes infinitesimal. The **maximum attainable accuracy** in computing a derivative by a one-sided finite difference occurs at the optimal increment, and is on the order of the square root of machine precision. In general, on a machine with floating point computations accurate to 16 digits we can expect at most 8 accurate digits in one-sided finite differences.

Figure 2.3 shows the logarithm of the error in one-sided differences versus the logarithm of the increment h for the derivative of $\log x$ at $x = 3$. Note that the smallest error of roughly 10^{-8} is achieved at roughly $h = 10^{-8}$. This figure was generated by the C^{++} program GUIFiniteDifference.C.

Exercise 2.2.8 Describe how to compute a second derivative by a one-sided difference of one-sided differences. Then analyze the rounding errors in the computation and determine the size of the optimal increment h.

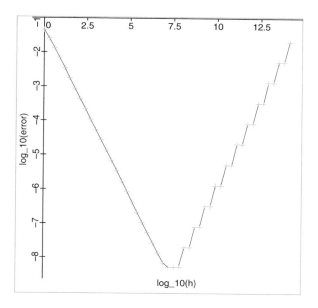

Fig. 2.3 Errors in one-sided differences, $f(x) = \log(x)$, $x = 3$

2.2.3   Centered Differencing

It is reasonable to hope that a more accurate approximation to the derivative might improve the **maximum attainable accuracy**. The next lemma describes how this goal can be achieved with centered differences.

Lemma 2.2.4 *Suppose that f has three continuous derivatives in the interval* $[a, b]$, *and let both* $x-h$ *and* $x+h$ *lie in this interval. Assume that there are constants* F, F' *and* F''' *so that for all* $x \in (a, b)$

$$|f(x)| \leq F, \; |f'(x)| \leq F' \text{ and } |f'''(x)| \leq F''' .$$

Suppose that the values of $f(x+h)$ *and* $f(x-h)$ *are computed with relative errors* ε_+ *and* ε_-, *respectively. Assume that the difference quotient* $[f(x + h) - f(x - h)]/[2h]$ *is computed with a relative error* $\varepsilon_\div$. *Let* ε *be an upper bound on all of the relative errors:*

$$\max\{|\varepsilon_f|, |\varepsilon_+|, |\varepsilon_\div|\} \leq \varepsilon .$$

Then

$$\left| fl\left(\frac{fl(f(x + h)) - fl(f(x - h))}{2h} \right) - f'(x) \right|$$
$$\leq \frac{h^2}{6} F''' + \varepsilon F' + \frac{\varepsilon(1 + \varepsilon)F}{2h} .$$

The right-hand side of this expression is minimized when

$$h = \sqrt[3]{\frac{3}{2} \frac{\varepsilon(1+\varepsilon)F}{F'''}} \; .$$

Proof First, we note that the fundamental theorem of calculus implies that

$$f(x+h) - f(x-h) - 2hf'(x) = \int_{x-h}^{x+h} f'(\xi) - f'(x) \; d\xi$$

$$= \int_{x-h}^{x+h} \int_{x}^{\xi} f''(\eta) \; d\eta \; d\xi$$

then a change of order of integration gives us

$$= -\int_{x-h}^{x} \int_{x-h}^{\eta} f''(\eta) \; d\xi \; d\eta + \int_{x}^{x+h} \int_{\eta}^{x+h} f''(\eta) \; d\xi \; d\eta$$

$$= \int_{x}^{x+h} (h + x - \eta)f''(\eta) \; d\eta - \int_{x-h}^{x} (h + \eta - x)f''(\eta) \; d\eta$$

then integration by parts leads to

$$= \frac{1}{2}(h + x - \eta)^2 |_{x}^{x+h} - \int_{x}^{x+h} \frac{1}{2}(h + x - \eta)^2 f'''(\eta) \; d\eta$$

$$+ \frac{1}{2}(h + \eta - x)^2 |_{x-h}^{x} + \int_{x-h}^{x} \frac{1}{2}(h + \eta - x)^2 f'''(\eta) \; d\eta$$

$$= -\int_{x}^{x+h} \frac{1}{2}(h + x - \eta)^2 f'''(\eta) \; d\eta + \int_{x-h}^{x} \frac{1}{2}(h + \eta - x)^2 f'''(\eta) \; d\eta \; .$$

Then the errors in floating point arithmetic imply

$$\left| fl\left(\frac{fl(f(x+h)) - fl(f(x-h))}{2h} \right) - f'(x) \right|$$

$$= \left| \frac{f(x+h)(1 + \varepsilon_+) - f(x-h)(1 + \varepsilon_-)}{2h} (1 + \varepsilon_\div) - f'(x) \right|$$

$$\leq \left| -\frac{1}{2h} \int_{x}^{x+h} \frac{1}{2}(h + x - \eta)^2 f'''(\eta) \; d\eta + \frac{1}{2h} \int_{x-h}^{x} \frac{1}{2}(h + \eta - x)^2 f'''(\eta) \; d\eta \right|$$

$$+ \left| \frac{\varepsilon_\div}{2h} \int_{x-h}^{x+h} f'(\xi) \; d\xi + \frac{\varepsilon_+(1 + \varepsilon_\div)}{2h} f(x+h) + \frac{\varepsilon_-(1 + \varepsilon_\div)}{2h} f(x-h) \right|$$

$$\leq \frac{h^2}{6} F''' + \varepsilon F' + \frac{\varepsilon(1+\varepsilon)}{2h} F \; .$$

Fig. 2.4 Errors in centered differences, $f(x) = \log(x)$, $x = 3$

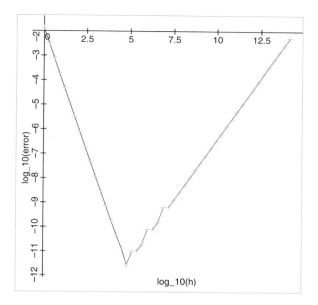

We can differentiate the right-hand side of this inequality with respect to h (ignoring the bounds on the maximum of the third derivative) and find its zero, to obtain the claimed expression for the value of h that minimizes the right-hand side.

This lemma shows that for centered differences, the optimal increment h is on the order of $\varepsilon^{1/3}$, and the error in centered differences with the optimal h is on the order of $\varepsilon^{2/3}$. In general, on a machine with 16 digits of relative accuracy we can expect at most 11 or 12 accurate digits in centered differences.

Figure 2.4 shows the logarithm of the error in centered differences versus the logarithm of the increment h for the derivative of $\log x$ at $x = 3$. Note that the smallest error of roughly 10^{-12} is achieved at roughly $h = 10^{-6}$. This figure was generated by the C^{++} program GUIFiniteDifference.C.

2.2.4 Richardson Extrapolation

At this point in our study of finite difference approximations to a derivative, we have seen that one-sided differences have first-order truncation error and a **maximum attainable accuracy** of approximately the $1/2$ power of machine precision. We have also shown that centered differences have second-order truncation error, and a maximum attainable accuracy of approximately the $2/3$ power of machine precision. It is reasonable to expect that difference approximations with even higher-order truncation errors might have even better maximum attainable accuracy. This motivates us to search for methods to generate higher-order difference approximations. One technique for achieving this goal is provided by the next theorem.

Theorem 2.2.3 (Richardson Extrapolation)   *Suppose that $\mathscr{L}\{f\}$ is some linear operator on functions. Also suppose that for some $n \geq 1$ we have an approximation $\widetilde{\mathscr{L}}_n\{f; h\}$ such that for all functions f with k continuous derivatives*

$$\widetilde{\mathscr{L}}_n\{f; h\} - \mathscr{L}\{f\} = \sum_{j=n+m,\ell}^{k-1} c_j^{(n)}(f)h^{j-m} + O\left(h^{k-m}\right) . \tag{2.9}$$

Here, the subscript $j = n+m, \ell$ on the summation indicates that the sum begins with $j = n + m$, and then increments by ℓ until the upper bound is reached or exceeded. Then if $n + \ell < k - m$ and $\alpha \neq 1$, the **Richardson extrapolant**

$$\widetilde{\mathscr{L}}_{n+\ell}\{f; h\} \equiv \begin{cases} \widetilde{\mathscr{L}}_n\{f; h\} + \frac{1}{\alpha^n - 1}\left[\widetilde{\mathscr{L}}_n\{f; h\} - \widetilde{\mathscr{L}}_n\{f; \alpha h\}\right], \alpha > 1 \\ \widetilde{\mathscr{L}}_n\{f; \alpha h\} + \frac{1}{1/\alpha^n - 1}\left[\widetilde{\mathscr{L}}_n\{f; \alpha h\} - \widetilde{\mathscr{L}}_n\{f; h\}\right], \alpha < 1 \end{cases} \tag{2.10}$$

satisfies

$$\widetilde{\mathscr{L}}_{n+\ell}\{f; h\} - \mathscr{L}\{f\} = \sum_{j=n+\ell+m,\ell}^{k-1} c_j^{(n)}(f)\frac{\alpha^n - \alpha^{k-m}}{\alpha^n - 1}h^{j-m} + O(h^{k-m}) .$$

Furthermore,

$$\widetilde{\mathscr{L}}_n\{f; h\} - \mathscr{L}\{f\} = \frac{\widetilde{\mathscr{L}}_n\{f; \alpha h\} - \widetilde{\mathscr{L}}_n\{f; h\}}{\alpha^n - 1} + O\left(h^{n+\ell}\right) . \tag{2.11}$$

In this chapter, we will be interested primarily in two possible linear operators, namely $\mathscr{L}\{f\} = f'$ and $\mathscr{L}\{f\} = \int f$.

Proof We can use the definition (2.10) of $\widetilde{\mathscr{L}}_{n+\ell}\{f; h\}$ to see that

$$\widetilde{\mathscr{L}}_{n+\ell}\{f; h\} - \mathscr{L}\{f\} = \frac{\alpha^n \left[\widetilde{\mathscr{L}}_n\{f; h\} - \mathscr{L}\{f\}\right] - \left[\widetilde{\mathscr{L}}_n\{f; \alpha h\} - \mathscr{L}\{f\}\right]}{\alpha^n - 1}$$

then we can use the expansion (2.9) for $\widetilde{\mathscr{L}}_{n+\ell}\{f; h\} - \mathscr{L}\{f\}$ to get

$$= \sum_{j=n+m,\ell}^{k-1} c_j^{(n)}\frac{\alpha^n - \alpha^{j-m}}{\alpha^n - 1}h^{j-m} + O\left(h^{k-m}\right)$$

and finally, since the leading term is zero, we can simplify the sum to obtain

$$= \sum_{j=n+\ell+m,\ell}^{k-1} c_j^{(n)}\frac{\alpha^n - \alpha^{j-m}}{\alpha^n - 1}h^{j-m} + O\left(h^{k-m}\right) . \tag{2.12}$$

Thus the error in $\widetilde{\mathscr{L}}_{n+\ell}\{f;h\}$ has the same form as the error in $\widetilde{\mathscr{L}}_n\{f;h\}$, with coefficients

$$c_j^{(n+\ell)} = c_j^{(n)}\frac{\alpha^n - \alpha^{j-m}}{\alpha^n - 1}.$$

Next, we note that

$$\widetilde{\mathscr{L}}_n\{f;h\} - \mathscr{L}\{f\} = \left[\widetilde{\mathscr{L}}_n\{f;h\} - \widetilde{\mathscr{L}}_{n+\ell}\{f;h\}\right] + \left[\widetilde{\mathscr{L}}_{n+\ell}\{f;h\} - \mathscr{L}\{f\}\right]$$

then the definition of $\widetilde{\mathscr{L}}_{n+\ell}\{f;h\}$, and the leading term in Eq. (2.12) yield

$$= \frac{\widetilde{\mathscr{L}}_n\{f;\alpha h\} - \widetilde{\mathscr{L}}_n\{f;h\}}{\alpha^n - 1} + O\left(h^{n+\ell}\right).$$

Example 2.2.1 For extrapolation of one-sided differences, we take

$$\widetilde{\mathscr{L}}_1\{f;h\} = \frac{f(x+h) - f(x)}{h} \quad \text{and} \quad \mathscr{L}\{f\} = f'(x).$$

If f has k derivatives, then the Taylor expansion (1.26) gives us

$$f(x+h) = \sum_{j=0}^{k-1}\frac{f^{(j)}(x)}{j!}h^j + \frac{h^k}{(k-1)!}\int_0^1 s^{k-1}f^{(k)}(x+h[1-s])\,\mathrm{d}s,$$

so

$$\widetilde{\mathscr{L}}_1\{f;h\} - \mathscr{L}\{f\} = \frac{\left[\sum_{j=0}^{k-1}f^{(j)}(x)h^j/j!\right] - f(x)}{h} - f'(x)$$

$$+ \frac{h^{k-1}}{(k-1)!}\int_0^1 s^{k-1}f^{(k)}(x+h[1-s])\,\mathrm{d}s$$

$$= \sum_{j=2}^{k-1}\frac{f^{(j)}(x)}{j!}h^{j-1} + \frac{h^{k-1}}{(k-1)!}\int_0^1 s^{k-1}f^{(k)}(x+h[1-s])\,\mathrm{d}s.$$

In this case, we see that we can apply the Richardson extrapolation theorem with $n = 1, m = 1$ and $\ell = 1$. Suppose that we take $\alpha = 1/2$. Then our first extrapolant is

$$\widetilde{\mathscr{L}}_2\{f;h\} = \widetilde{\mathscr{L}}_1\{f;h/2\} + \frac{1/2}{1-1/2}\left[\widetilde{\mathscr{L}}_1\{f;h/2\} - \widetilde{\mathscr{L}}_1\{f;h\}\right]$$

$$= \widetilde{\mathscr{L}}_1\{f;h/2\} + \left[\widetilde{\mathscr{L}}_1\{f;h/2\} - \widetilde{\mathscr{L}}_1\{f;h\}\right],$$

and our next extrapolant is

$$\widetilde{\mathscr{L}}_3\{f;h\} = \widetilde{\mathscr{L}}_2\{f;h/2\} + \frac{(1/2)^2}{1-(1/2)^2}\left[\widetilde{\mathscr{L}}_2\{f;h/2\} - \widetilde{\mathscr{L}}_2\{f;h\}\right]$$

$$= \widetilde{\mathscr{L}}_1\{f;h/2\} + \frac{1}{3}\left[\widetilde{\mathscr{L}}_2\{f;h/2\} - \widetilde{\mathscr{L}}_2\{f;h\}\right] .$$

We can continue in this way, building extrapolants of increasing order, until f runs out of derivatives to support the extrapolation, or we have found a sufficiently accurate extrapolant.

Example 2.2.2 For extrapolation of centered differences, we take

$$\widetilde{\mathscr{L}}_1\{f;h\} = \frac{f(x+h)-f(x-h)}{2h} \text{ and } \mathscr{L}\{f\} = f'(x) .$$

If f has k derivatives, then a Taylor expansion (1.26) gives us

$$f(x+h) = \sum_{j=0}^{k-1}\frac{f^{(j)}(x)}{j!}h^j + \frac{h^k}{(k-1)!}\int_0^1 s^{k-1}f^{(k)}(x+h[1-s])\,ds , \text{ and}$$

$$f(x-h) = \sum_{j=0}^{k-1}\frac{f^{(j)}(x)}{j!}(-h)^j + \frac{(-h)^k}{(k-1)!}\int_0^1 s^{k-1}f^{(k)}(x-h[1-s])\,ds ,$$

so

$$\widetilde{\mathscr{L}}_1\{f;h\} - \mathscr{L}\{f\} = \frac{1}{2h}\left[\sum_{j=0}^{k-1}\frac{f^{(j)}(x)}{j!}h^j - \sum_{j=0}^{k-1}\frac{f^{(j)}(x)}{j!}(-h)^j\right] - f'(x)$$

$$+ \frac{1}{2(k-1)!}h^{k-1}\int_0^1 s^{k-1}f^{(k)}(x+h[1-s])\,ds$$

$$- \frac{1}{2(k-1)!}(-h)^{k-1}\int_0^1 s^{k-1}f^{(k)}(x-h[1-s])\,ds$$

$$= \sum_{j=1}^{k-1}\frac{f^{(j)}(x)}{j!}\frac{h^j-(-h)^j}{2h} - f'(x) + \frac{1}{2(k-1)!}h^{k-1}\int_0^1 s^{k-1}f^{(k)}(x+h[1-s])\,ds$$

$$- \frac{1}{2(k-1)!}(-h)^{k-1}\int_0^1 s^{k-1}f^{(k)}(x-h[1-s])\,ds$$

$$= \sum_{j=3,2}^{k-1}\frac{f^{(j)}(x)}{j!}h^{j-1} + \frac{1}{2(k-1)!}h^{k-1}\int_0^1 s^{k-1}f^{(k)}(x+h[1-s])\,ds$$

$$- \frac{1}{2(k-1)!}(-h)^{k-1}\int_0^1 s^{k-1}f^{(k)}(x-h[1-s])\,ds .$$

In this case, we see that we can apply the Richardson extrapolation theorem with $n = 2$, $m = 1$ and $\ell = 2$. Suppose that we take $\alpha = 1/2$. Then our first extrapolant is

$$\widetilde{\mathscr{L}}_2\{f; h\} = \widetilde{\mathscr{L}}_1\{f; h/2\} + \frac{(1/2)^2}{1 - (1/2)^2} \left[\widetilde{\mathscr{L}}_1\{f; h/2\} - \widetilde{\mathscr{L}}_1\{f; h\}\right]$$

$$= \widetilde{\mathscr{L}}_1\{f; h/2\} + \frac{1}{3}\left[\widetilde{\mathscr{L}}_1\{f; h/2\} - \widetilde{\mathscr{L}}_1\{f; h\}\right],$$

and our next extrapolant is

$$\widetilde{\mathscr{L}}_3\{f; h\} = \widetilde{\mathscr{L}}_2\{f; h/2\} + \frac{(1/2)^4}{1 - (1/2)^4} \left[\widetilde{\mathscr{L}}_2\{f; h/2\} - \widetilde{\mathscr{L}}_2\{f; h\}\right]$$

$$= \widetilde{\mathscr{L}}_1\{f; h/2\} + \frac{1}{15}\left[\widetilde{\mathscr{L}}_2\{f; h/2\} - \widetilde{\mathscr{L}}_2\{f; h\}\right].$$

Our development of Richardson extrapolation for centered differences leads to the following

Algorithm 2.2.4 (Richardson Extrapolation)

$$\text{for } j = 2, 4, 6, \ldots$$
$$\delta_o = (f(x + h) - f(x - h))/(2h)$$
$$\beta = \alpha^2$$
$$\text{for } 1 \leq i < j$$
$$e = \delta_o - \widetilde{\mathscr{L}}_i\{f, h/\alpha\}$$
$$\widetilde{\mathscr{L}}_i\{f, h\} = \delta_o$$
$$\delta_n = \delta_o + e * \beta/(1 - \beta)$$
$$\text{if } |e| \leq \text{ tolerance} * \delta_o \text{ then stop with answer } = \delta_n$$
$$\delta_o = \delta_n$$
$$\beta = \beta * \alpha^\ell$$
$$h = h * \alpha$$

Rounding errors become less of a problem as the order of the extrapolant increases. This is because of the form of the extrapolant (2.10). As the order $n + \ell$ increases, cancellation errors in forming $\widetilde{\mathscr{L}}_n\{f, \alpha h\} - \widetilde{\mathscr{L}}_n\{f, h\}$ are multiplied by the numbers $1/(\alpha^n - 1)$ (if $\alpha > 1$) or $\alpha^n/(1 - \alpha^n)$ (if $\alpha < 1$). In the former case, we can guarantee that the multiplicative factor is at most one by requiring $\alpha^n \geq 2$ for $n = 2, 4, \ldots$. In the latter case, we would require $\alpha^n \leq 1/2$ for $n = 2, 4, \ldots$.

In practice, it is convenient to take $\alpha = 1/2$ for computing derivatives. If we take α to be very small, then h decreases rapidly and substantial rounding errors can occur in computing the centered differences needed for high-order extrapolants.

Within the GNU Scientific Library, the C routines gsl_deriv_central, gsl_deriv_forward and gsl_deriv_backward perform a couple of steps of Richardson extrapolation to compute derivatives. In the Harwell library, the Fortran routine td01 will use Richardson extrapolation to compute a derivative. Readers may also view a C^{++} program to compute derivatives of $log(x)$ in GUIFiniteDifference.C. Finally, readers may also execute a JavaScript routine for Richardson extrapolation with One-Sided Differences or with Centered Differences.

Exercise 2.2.9 Use centered differences and Richardson extrapolation to compute the derivatives of

1. $\ln x$ at $x = 3$
2. $\tan x$ at $x = \arcsin(0.8)$
3. $\sin(x^2 + x/3)$ at $x = 0$.

Exercise 2.2.10 Use centered differences and Richardson extrapolation to compute the second derivatives of the functions in the previous exercise.

Exercise 2.2.11 Estimate the truncation error in the following approximations

1.

$$f'''(x) \approx \frac{1}{h^3} \left[f(x + 3h) - 3f(x + 2h) + 3f(x + h) - f(x) \right]$$

2.

$$f'''(x) \approx \frac{1}{2h^3} \left[f(x + 2h) - 2f(x + h) + 2f(x - h) - f(x - 2h) \right]$$

2.2.5 Wynn's Epsilon Algorithm

Another extrapolation algorithm for convergent sequences is due to Wynn [185]. Given a positive integer N and a sequence $\{s_n\}_{n=0}^{\infty}$ of scalars (possibly complex), this algorithm takes the form

Algorithm 2.2.5 (Wynn's Epsilon)

$$\varepsilon_0^{(0)} = s_0$$

for $k = 1, 2, \ldots$

$$\varepsilon_{-1}^{(k+1)} = 0$$

$$\varepsilon_0^{(k)} = s_k$$

for $j = 1, \ldots, k$

$$\varepsilon_j^{(k-j)} = \varepsilon_{j-2}^{(k-j+1)} + \frac{1}{\varepsilon_{j-1}^{(k-j+1)} - \varepsilon_{j-1}^{(k-j)}}$$

This algorithm is currently used in QUADPACK for numerical integration.

We would like to relate Wynn's epsilon algorithm to Padé approximation. Let us define the sequence $\{\gamma_k\}_{k=0}^{\infty}$ by

$$\gamma_k = \begin{cases} s_0, & k = 0 \\ s_k - s_{k-1}, & k \geq 1 \end{cases}$$

and the function

$$f(z) = \sum_{k=0}^{\infty} \gamma_k z^k = \lim_{n \to \infty} \sum_{k=0}^{n} \gamma_k z^k . \tag{2.13}$$

Then

$$f(1) = \lim_{n \to \infty} \sum_{k=0}^{n} \gamma_k = \lim_{n \to \infty} s_n .$$

The next theorem will prove a very interesting connection between the entries $\varepsilon_{2i}^{(\ell)}$ and Padé approximations to $f(1)$.

Theorem 2.2.4 *Given a sequence $\{s_n\}_{n=0}^{\infty}$, define the scalars $\varepsilon_j^{(k)}$ for $j \geq -1$ and $k \geq 0$ by Wynn's epsilon Algorithm 2.2.5. Then for $i \geq 0$ and $\ell \geq 0$, $\varepsilon_{2i}^{(\ell)}$ is the $(i + \ell)/i$ Padé approximation to $f(1)$, where $f(z)$ is defined by (2.13).*

Proof Following Baker and Graves-Morris [12, p. 84f], we will prove this theorem by induction on i. First, consider the case with $i = 0$. Wynn's epsilon Algorithm 2.2.5 shows that

$$\varepsilon_0^{(\ell)} = s_\ell .$$

Next, we recall our proof of the Padé approximation Theorem 1.4.1 to see that the $\ell/0$ Padé approximation to $f(1)$ is $a_{\ell/0}(1)/b_{\ell/0}(1)$ where

$$a_{\ell/0}(z) = \det\left[\sum_{k=0}^{\ell} \gamma_k z^k\right] = s_0 + \sum_{k=1}^{\ell} (s_k - s_{k-1}) z^k \implies a_{\ell/0}(1) = s_\ell \text{ and}$$

$$b_{\ell/0}(z) = [1] \implies b_{\ell/0}(1) = 1 .$$

Thus the claim is true for $i = 0$.

Next, consider the case with $i = 1$. Wynn's epsilon Algorithm 2.2.5 shows that

$$\varepsilon_1^{(\ell)} = \frac{1}{\varepsilon_0^{(\ell+1)} - \varepsilon_0^{(\ell)}} = \frac{1}{s_{\ell+1} - s_\ell}$$

and that

$$
\begin{aligned}
\varepsilon_2^{(\ell)} &= \varepsilon_0^{(\ell+1)} + \frac{1}{\varepsilon_1^{(\ell+1)} - \varepsilon_1^{(\ell)}} = s_{\ell+1} + \frac{1}{\frac{1}{s_{\ell+2}-s_{\ell+1}} - \frac{1}{s_{\ell+1}-s_\ell}} \\
&= s_{\ell+1} - \frac{(s_{\ell+2} - s_{\ell+1})(s_{\ell+1} - s_\ell)}{(s_{\ell+2} - s_{\ell+1}) - (s_{\ell+1} - s_\ell)} \\
&= s_\ell + (s_{\ell+1} - s_\ell) - \frac{(s_{\ell+2} - s_{\ell+1})(s_{\ell+1} - s_\ell)}{(s_{\ell+2} - s_{\ell+1}) - (s_{\ell+1} - s_\ell)} \\
&= s_\ell + \frac{s_{\ell+1} - s_\ell}{(s_{\ell+2} - s_{\ell+1}) - (s_{\ell+1} - s_\ell)} \left[(s_{\ell+2} - s_{\ell+1}) - (s_{\ell+1} - s_\ell) - (s_{\ell+2} - s_{\ell+1}) \right] \\
&= s_\ell - \frac{(s_{\ell+1} - s_\ell)^2}{(s_{\ell+2} - s_{\ell+1}) - (s_{\ell+1} - s_\ell)} \ .
\end{aligned}
$$

This is more commonly known as the **Aitken $\triangle^2$ extrapolation** of the sequence $\{s_n\}_{n=0}^\infty$. For $\ell \geq 0$, the $(\ell + 1)/1$ Padé approximations to $f(z)$ use the polynomials $a_{(\ell+1)/1}(z)$ and $b_{(\ell+1)/1}(z)$. Recall from Eqs. (1.34) and (1.32) that

$$
a_{(\ell+1)/1}(z) = \det \begin{bmatrix} \gamma_{\ell+1} & \gamma_{\ell+2} \\ \sum_{k=0}^\ell \gamma_k z^{k+1} & \sum_{k=0}^{\ell+1} \gamma_k z^k \end{bmatrix} ,
$$

which implies that

$$
\begin{aligned}
a_{(\ell+1)/1}(1) &= \det \begin{bmatrix} s_{\ell+1} - s_\ell & s_{\ell+2} - s_{\ell+1} \\ s_\ell & s_{\ell+1} \end{bmatrix} = s_{\ell+1}(s_{\ell+1} - s_\ell) - s_\ell(s_{\ell+2} - s_{\ell+1}) \\
&= s_{\ell+1}^2 - s_\ell s_{\ell+2} = (s_{\ell+1} - s_\ell)^2 + 2s_\ell s_{\ell+1} - s_\ell^2 - s_\ell s_{\ell+2} \\
&= (s_{\ell+1} - s_\ell)^2 - s_\ell([s_{\ell+2} - s_\ell + 1] - [s_{\ell+1} - s_\ell]) \ .
\end{aligned}
$$

Also recall from Eqs. (1.33) and (1.31) that

$$
b_{(\ell+1)/1}(z) = \det \begin{bmatrix} \gamma_{\ell+1} & \gamma_{\ell+2} \\ z & 1 \end{bmatrix} ,
$$

which implies that

$$
b_{(\ell+1)/1}(1) = \gamma_{\ell+1} - \gamma_{\ell+2} = -([s_{\ell+2} - s_{\ell+1}] - [s_{\ell+1} - s_\ell]) \ .
$$

Thus the $(\ell + 1)/1$ Padé approximation to $f(1)$ is

$$\frac{a_{(\ell+1)/1}(1)}{b_{(\ell+1)/1}(1)} = \frac{s_\ell([s_{\ell+2} - s_{\ell+1}] - [s_{\ell+1} - s_\ell]) - (s_{\ell+1} - s_\ell)^2}{[s_{\ell+2} - s_{\ell+1}] - [s_{\ell+1} - s_\ell]}$$

$$= s_\ell - \frac{(s_{\ell+1} - s_\ell)^2}{[s_{\ell+2} - s_{\ell+1}] - [s_{\ell+1} - s_\ell]} = \varepsilon_2^\ell .$$

This verifies the claimed result for $i = 1$.

Inductively, let us assume that for $0 \leq j < i$ and $k \geq 1$ the Wynn extrapolant $\varepsilon_{2j}^{(k)}$ is the $(k+j)/j$ Padé approximation to $f(1)$. Wynn's epsilon algorithm implies that

$$\varepsilon_{2i+1}^{(\ell-1)} - \varepsilon_{2i-1}^{(\ell)} = \frac{1}{\varepsilon_{2i}^{(\ell)} - \varepsilon_{2i}^{(\ell-1)}} ,$$

$$\varepsilon_{2i+1}^{(\ell)} - \varepsilon_{2i-1}^{(\ell+1)} = \frac{1}{\varepsilon_{2i}^{(\ell+1)} - \varepsilon_{2i}^{(\ell)}} ,$$

$$\varepsilon_{2i}^{(\ell)} - \varepsilon_{2i-2}^{(\ell+1)} = \frac{1}{\varepsilon_{2i-1}^{(\ell+1)} - \varepsilon_{2i-1}^{(\ell)}} \quad \text{and}$$

$$\varepsilon_{2i+2}^{(\ell-1)} - \varepsilon_{2i}^{(\ell)} = \frac{1}{\varepsilon_{2i+1}^{(\ell)} - \varepsilon_{2i+1}^{(\ell-1)}} .$$

We can use the first two of these equations to get

$$-\frac{1}{\varepsilon_{2i}^{(\ell)} - \varepsilon_{2i}^{(\ell-1)}} + \frac{1}{\varepsilon_{2i}^{(\ell+1)} - \varepsilon_{2i}^{(\ell)}} = -\left[\varepsilon_{2i+1}^{(\ell-1)} - \varepsilon_{2i-1}^{(\ell)}\right] + \left[\varepsilon_{2i+1}^{(\ell)} - \varepsilon_{2i-1}^{(\ell+1)}\right]$$

$$= -\left[\varepsilon_{2i-1}^{(\ell+1)} - \varepsilon_{2i-1}^{(\ell)}\right] + \left[\varepsilon_{2i+1}^{(\ell)} - \varepsilon_{2i+1}^{(\ell-1)}\right]$$

then we use the reciprocals of the last two equations from Wynn's epsilon algorithm to obtain

$$= -\frac{1}{\varepsilon_{2i}^{(\ell)} - \varepsilon_{2i-2}^{(\ell+1)}} + \frac{1}{\varepsilon_{2i+2}^{(\ell-1)} - \varepsilon_{2i}^{(\ell)}} .$$

The inductive hypothesis allows us to rewrite this equation in the form

$$\frac{1}{\dfrac{a_{(\ell+i-1)/i}(1)}{b_{(\ell+i-1)/i}(1)} - \dfrac{a_{(\ell+i)/i}(1)}{b_{(\ell+i)/i}(1)}} + \frac{1}{\dfrac{a_{(\ell+i+1)/i}(1)}{b_{(\ell+i+1)/i}(1)} - \dfrac{a_{(\ell+i)/i}(1)}{b_{(\ell+i)/i}(1)}}$$

$$= \frac{1}{\dfrac{a_{\ell/(i-1)}(1)}{b_{\ell/(i-1)}(1)} - \dfrac{a_{(\ell+i i)/i}(1)}{b_{(\ell+i i)/i}(1)}} + \frac{1}{\varepsilon_{2i+2}^{(\ell-1)} - \dfrac{a_{(\ell+i)/i}(1)}{a_{(\ell+i)/i}(1)}} .$$

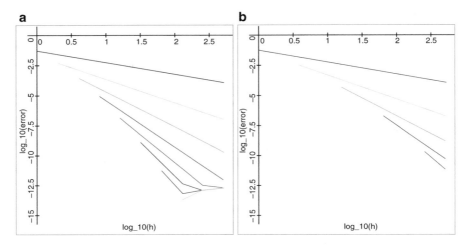

Fig. 2.5 Extrapolation errors in one-sided finite differences for $f'(3)$ with $f(x) = \log(x)$. (**a**) Richardson extrapolation. (**b**) Wynn epsilon algorithm

Wynn's identity (1.35) now shows that $\varepsilon_{2i+2}^{(\ell-1)}$ is the $\ell/(i+1)$ Padé approximation to $f(1)$.

Richardson extrapolation requires some prior knowledge of the asymptotic behavior of the sequence $\{s_n\}_{n=0}^{\infty}$ in order to perform the extrapolation properly. On the other hand, Wynn's epsilon algorithm does not require any information about the behavior of the original sequence. Richardson extrapolation is less sensitive to rounding errors in the computation of the extrapolants than Wynn's epsilon algorithm. The latter should be stopped as soon as two entries $\varepsilon_{2i}^{(\ell)}$ and $\varepsilon_{2i}^{(\ell+1)}$ agree, in order to avoid rounding errors dominating the computation of $\varepsilon_{2i+1}^{(\ell)}$.

Figure 2.5 shows the errors in various columns of the Richardson extrapolation table and the Wynn epsilon algorithm for one-sided differences to compute $f'(3)$ with $f(x) = \log(x)$. The graph on the right in this figure shows that the errors in column $2j$ decrease at a rate roughly proportional to h^{j+1}. These figures were generated by the C^{++} program GUIExtrapolate.C.

Wynn's epsilon algorithm has been implemented in Fortran as QUADPACK routine dqelg. Readers may also execute a JavaScript routine for Wynn's epsilon algorithm to compute a derivative with Centered Differences.

2.3 Numerical Integration

Numerical integration requires a more elaborate discussion than numerical differentiation. This is due to the fact that the mathematical concept of integration is more complicated than differentiation.

We will present the two fundamental ideas for integration in Sect. 2.3.1, namely Riemann integrals and Lebesgue integrals. We will use Lebesgue integrals to develop Monte Carlo integration in Sect. 2.3.2. Afterward, we will use Riemann integral ideas to present Riemann sums in Sect. 2.3.3. Section 2.3.4 will show that the midpoint rule is a specific case of Riemann sums that corresponds to computing the exact integral of a piecewise constant interpolation to the integrand. On the other hand, the trapezoidal rule will be developed in Sect. 2.3.5 as the exact integral of a piecewise linear interpolation to the integrand. The Euler-MacLaurin formula in Sect. 2.3.6 will show, surprisingly, that the midpoint rule is asymptotically more accurate than the trapezoidal rule. We will also discuss Newton-Cotes quadrature (which involves integrating the polynomial interpolant to the integrand at equidistant lattice points) in Sect. 2.3.7, and Clenshaw-Curtis quadrature (which involves integrating the polynomial interpolant to the integrand at the Chebyshev points) in Sect. 2.3.8. We will also notice that both the midpoint and trapezoidal rules have an asymptotic error expansion involving only even powers of the mesh width h, so extrapolation will be shown in Sect. 2.3.9 to be especially efficient.

After these developments, we will approach numerical integration from a different viewpoint. Instead of varying the order of the method to achieve a given accuracy (as in extrapolation), we will maximize the order of the method for a given number of quadrature nodes. This will be accomplished by Gaussian quadrature in Sect. 2.3.10. A related idea, Lobatto quadrature in Sect. 2.3.11, requires two of the quadrature nodes to be located at the ends of the integration interval and chooses the remaining nodes to maximize the order of the method. For both Gaussian and Lobatto quadrature, we will see that the nodes are always contained within the integration region, and the weights given to the function values at these nodes are always positive. We will also develop Gauss-Kronrod quadrature rules, which extend given Gaussian quadrature rules to higher order, thereby providing an error estimate for the Gaussian quadrature rule (or both).

We will end our treatment of numerical integration with three additional topics. First, we will discuss integrals with singularities, unbounded integration regions and oscillatory integrands in Sect. 2.3.13. After that, we will attempt to minimize the number of function evaluations via adaptive quadrature in Sect. 2.3.14. We will end the section with some ideas on multidimensional integration in Sect. 2.3.15.

2.3.1 Fundamental Concepts

There are several ways to define an integral, each leading to different computational strategy. We will begin with the definition that is most commonly found in elementary calculus.

Definition 2.3.1 Let (a, b) be an interval of real numbers. Then

$$\mathscr{P} = \{x_i\}_{i=0}^n$$

is a **partition** of (a, b) if and only if

$$a = x_0 < x_1 < \ldots < x_n = b .$$

If $\mathscr{P}$ is a partition of (a, b), then the **partition width** is

$$h = \max_{1 \le i \le n} [x_i - x_{i-1}] .$$

Suppose that f is continuous almost everywhere in (a, b), meaning that the set of points where it is not continuous has zero measure. Then f is **Riemann integrable** in (a, b) if and only if there is a number I so that for any tolerance $\varepsilon > 0$ there exists a maximum partition width $\delta > 0$ so that for all partitions of $\mathscr{P} = \{x_i\}_{i=0}^n$ of (a, b) with width $h < \delta$ and for all choices of the evaluation points $\xi_i \in [x_{i-1}, x_i]$ for $1 \le i \le n$ we have

$$\left| \sum_{i=1}^n f(\xi_i)(x_i - x_{i-1}) - I \right| < \varepsilon .$$

If f is Riemann integrable in (a, b), then I is called the **Riemann integral** of f, and we write

$$I = \int_a^b f(x) \, dx .$$

Riemann integrals in multiple dimensions are constructed as follows. Given a bounded set Ω in d dimensions, find d-vectors $\mathbf{a}$ and $\mathbf{b}$ so that the box $(\mathbf{a}_1, \mathbf{b}_1) \times \ldots (\mathbf{a}_d, \mathbf{b}_d)$ that contains Ω. Partition each of the intervals $(\mathbf{a}_i, \mathbf{b}_i)$ that define the box, choose points in each of the boxes defined by the partition, and form the sum of the function value at those point times the multi-dimensional volume of the partition box, over all boxes that intersect Ω. If there is some number I that is arbitrarily close to a sum of this form, for all sufficiently fine partitions and any choice of the points in the partition boxes, then I is the integral of f over Ω.

For more information regarding the Riemann integral, see for example Apostol [6], Buck [25, Chapter 3] or Shilov and Gurevich [159, Chapter 1]. In particular, it is known that f has a Riemann integral over (a, b) whenever f is bounded and continuous almost everywhere in (a, b) (i.e., except on a set of measure zero).

An alternative definition of the integral is due to Lebesgue. Its notions are more technical than those for the Riemann integral. However, the Lebesgue integral will be important to our work, because many of its ideas have been adopted by probability theory. First, we will define the class of functions for which we can define a Lebesgue integral.

Definition 2.3.2 Let Ω be a set of real d-vectors. A collection of subsets $\mathscr{M}$ of Ω is a σ-**algebra** if and only if

1. $\Omega \in \mathscr{M}$,
2. for all subsets A of Ω, if $A \in \mathscr{M}$ then the **complement of** A **relative to** Ω

$$A^c \equiv \{\mathbf{x} \in \Omega : \mathbf{x} \notin A\}$$

   satisfies $A^c \in \mathscr{M}$, and
3. if $A_1, A_2, \ldots$ is a countable collection of members of $\mathscr{M}$, then

$$\bigcup_{i=1}^{\infty} A_i \in \mathscr{M} .$$

The **Borel** σ-**algebra** is the smallest σ-algebra that contains all open subsets of Ω. If Ω has a Borel σ-algebra, then Ω is said to be a **measurable space**, and the members of $\mathscr{M}$ are the **measurable sets**.

If f is a real-valued function on Ω, and for every open subset V of the real line the set

$$f^{-1}(V) \equiv \{\mathbf{x} \in \Omega : f(\mathbf{x}) \in V\}$$

is measurable, then f is said to be a **measurable function**. If $\mathscr{M}$ is a Borel σ-algebra on Ω and μ is a non-negative function on $\mathscr{M}$, then μ is a **positive measure** if and only if

1. there exists $A \in \mathscr{M}$ so that $\mu(A) < \infty$, and
2. for all countable collections $\{A_i\}_{i=1}^{\infty}$ of disjoint measurable sets we have

$$\mu\left(\bigcup_{i=1}^{\infty} A_i\right) = \sum_{i=1}^{n} \mu(A_i) .$$

Next, we will define a class of functions that are easy to integrate.

Definition 2.3.3 Let Ω be a measurable set in d dimensions, and let s map Ω into the real line. Then s is a **simple function** if and only if there is a finite set $\{\alpha_1, \ldots, \alpha_n\}$ such that for all $\mathbf{x} \in \Omega$, $s(\mathbf{x}) \in \{\alpha_1, \ldots, \alpha_n\}$.

Note that every real-valued function f on Ω can be written as $f = f^+ + f^-$ where

$$f^+(\mathbf{x}) = \max\{f(\mathbf{x}), 0\} \text{ and } f^-(\mathbf{x}) = \min\{f(\mathbf{x}), 0\} .$$

Also note that for every non-negative measurable function f, we can construct a strictly increasing sequence $\{s_n\}$ of simple functions that converge to f at every point of Ω. Specifically, for each n let

$$B_n = f^{-1}([n, \infty])$$

and for each $1 \le i \le n2^n$ let

$$A_{n,i} = f^{-1}\left(\left[\frac{i-1}{2^n}, \frac{i}{2^n}\right)\right).$$

Define the **characteristic function**

$$\chi_A(\mathbf{x}) = \begin{cases} 1, \mathbf{x} \in A \\ 0, \mathbf{x} \notin A \end{cases},$$

and let

$$s_n(\mathbf{x}) = \sum_{i=1}^{n2^n} \frac{i-1}{2^n} \chi_{A_{n,i}}(\mathbf{x}) + n\chi_{B_n}(\mathbf{x}).$$

Then $0 \le s_1 \le \ldots \le s_n \le \ldots \le f$ and $s_n(\mathbf{x}) \to f(x)$ for all $\mathbf{x} \in \Omega$. We are now ready to define the Lebesgue integral.

Definition 2.3.4 Let Ω be a measurable set of real d-vectors with σ-algebra $\mathcal{M}$ and positive measure μ. Suppose that s is a simple function taking values $\alpha_1, \ldots, \alpha_n$ on Ω. Then for all measurable subsets $A \in \mathcal{M}$ the **Lebesgue integral** of s over A is

$$\int_A s \, \mathrm{d}\mu = \sum_{i=1}^{n} \alpha_i \mu \left(A \cap \{\mathbf{x} : s(\mathbf{x}) = \alpha_i\}\right).$$

Here, the contribution to the sum is zero whenever $\alpha_i = 0$, even if the measure of the corresponding set is infinite. Next, if f is a nonnegative measurable function on Ω and A is a measurable set, then

$$\int_A f \, \mathrm{d}\mu = \sup_{\substack{0 \le s \le f \\ s \text{ simple}}} \int_A s \, \mathrm{d}\mu.$$

If $f = f^+ + f^-$ where $f^+ \ge 0$ and $f^- \le 0$, then for all measurable sets A

$$\int_A f \, \mathrm{d}\mu = \int_A f^+ \, \mathrm{d}\mu - \int_A (-f^-) \, \mathrm{d}\mu.$$

Readers should notice that any Riemann integral will partition the *domain* of f, while a Lebesgue integral will partition the *range* of f. For more details on the Lebesgue integral, see Folland [81], Royden [152], Rudin [154] or Shilov and Gurevich [159, Chap. 2].

There are several strategies for approximating $\int_a^b f(x)\,dx$. We could randomly sample values of f in $[a, b]$, and multiply their average by the measure of the domain of integration. With this approach, we get more **accuracy** by increasing the sample size. This approach corresponds roughly to the Lebesgue integral. Alternatively, we could approximate f by a function ϕ that is easy to integrate, and compute $\int_a^b \phi(x)dx$. With this approach, we get more accuracy by choosing functions ϕ that approximate f well. Or, we could break the interval $[a, b]$ into elements and use the previous approach on each piece. Here, we get more accuracy by using more mesh elements. This third approach corresponds roughly to the Riemann integral.

2.3.2 Monte Carlo

The **Monte Carlo method** is very effective for computing integrals in high dimensional spaces. It uses the laws of probability to provide assurance that the computed results are reasonably accurate. To explain the Monte Carlo method, we will begin by introducing some basic ideas from probability.

Definition 2.3.5 Suppose that Ω has Borel σ-algebra $\mathcal{M}$ and positive measure μ with $\mu(\Omega) < \infty$. If P is a nonnegative function on $\mathcal{M}$, then P is a **probability function** if and only if $P(\Omega) = 1$ and for all pairwise disjoint countable collections $\{A_i\}_{i=1}^\infty$ of measurable sets we have

$$P\left(\bigcup_{i=1}^\infty A_i\right) = \sum_{i=1}^\infty P(A_i) .$$

For example, if Ω has Borel σ-algebra $\mathcal{M}$ and $P(A) = \mu(A)/\mu(\Omega)$ is a probability function on $\mathcal{M}$.

A **random variable** is a real-valued measurable function defined on a measurable set Ω. The **expected value** of a random variable is

$$E_\Omega(f) \equiv \frac{\int_\Omega f\,d\mu}{\mu(\Omega)} \equiv \int_\Omega f\,dP ,$$

and the **variance** of f is

$$Var_\Omega(f) \equiv \frac{\int_\Omega [f - E_\Omega(f)]^2\,d\mu}{\mu(\Omega)} = \int_\Omega [f - E_\Omega(f)]^2\,dP .$$

The set $\{f_i\}_{i=1}^n$ of random variables is **independent** if and only if for all measurable subsets $R_1, \ldots, R_n$ of the real line we have

$$P\left(\bigcap_{i=1}^n f_i^{-1}(R_i)\right) = \prod_{i=1}^n P\left(f_i^{-1}(R_i)\right) .$$

The set $\{f_i\}_{i=1}^n$ of random variables is **identically distributed** if and only there is a function f, called the **probability measure function**, so that for all $1 \le i \le n$ and all measurable subsets R of the real line we have

$$P\left(f_i^{-1}(R)\right) = P\left(f^{-1}(R)\right) .$$

The following result is well-known in statistics; a proof can be found in Casella and Berger [32, p. 208].

Theorem 2.3.1 (Sample Variance Is Unbiased Estimate of Variance) *Suppose that $f_1, \ldots, f_n$ are independent identically distributed random variables on some set Ω, with probability measure function f. Then*

$$E_\Omega \left(\frac{1}{n-1} \sum_{i=1}^n \left[f_i - \frac{1}{n} \sum_{j=1}^n f_j \right]^2 \right) = Var_\Omega(f) .$$

We will also make use of the following important result, which is proved in Casella and Berger [32, p. 214] and in Feller [77, p. 246ff, vol. I].

Theorem 2.3.2 (Weak Law of Large Numbers) *Suppose that Ω is a set in d dimensions with Borel σ-algebra $\mathcal{M}$ and corresponding probability function P. Let $\{f_i\}_{i=1}^\infty$ be a sequence of independent identically distributed random variables with probability measure function f on Ω. Then for all $\varepsilon > 0$*

$$\lim_{n \to \infty} P\left(\left\{ \mathbf{x} : \left| \frac{1}{n} \sum_{i=1}^n f_i(\mathbf{x}) - E_\Omega(f) \right| > \varepsilon \right\} \right) = 0 .$$

Let us apply these theoretical results to numerical integration. Suppose that f is a real-valued function defined on some measurable set Ω of d-vectors. Suppose that we can randomly select points $\mathbf{x}_i \in \Omega$, in such a way that for any measure $m \le \mu(\Omega)$ there is a probability p so that for any measurable subset $A \subset \Omega$ with $\mu(A) = m$ the probability that we select point from A is p. In this case, we say that the points $\mathbf{x}_i$ are drawn from a **uniform probability distribution**. Let

$$V = \mu(\Omega)$$

be the measure of Ω (i.e., its d-dimensional volume), and

$$Q_n = \frac{V}{n} \sum_{i=1}^n f(\mathbf{x}_i) .$$

Then the weak law of large numbers (Theorem 2.3.2) says that

$$Q_n \to \int_\Omega f \, d\mu \, ,$$

and Theorem 2.3.1 says that the variance of Q_n is

$$Var(Q_n) = \frac{V^2}{n^2} \sum_{i=1}^{n} Var(f) = \frac{V^2}{n} Var(f) \, .$$

Theorem 2.3.1 shows that

$$Var(f) \approx \sigma_n \, ,$$

where

$$\sigma_n^2 \equiv \frac{1}{n-1} \sum_{i=1}^{n} \left(f(\mathbf{x}_i) - \frac{1}{n} \sum_{j=1}^{n} f(\mathbf{x}_j) \right)^2 \, .$$

In other words, the mean-square error in Q_n as an approximation to the integral of f is inversely proportional to the square root of the sample size (minus one), *independent of the number of spatial dimensions d.*

Monte Carlo integration is generally very easy to program. GNU Scientific Library users should examine its Monte Carlo Integration offerings. MATLAB does not offer its own Monte Carlo integration procedure, but does provide a link to a contributed routine in its file exchange. Alternatively, readers may execute a JavaScript Monte Carlo program. The program plots the log of the error in the computed integral $\int_0^1 x^2 \, dx$ versus the log of the number of function evaluations. For comparison purposes, the program also plots the logarithm of the estimated error.

For more information regarding Monte Carlo methods, we recommend that readers read Caflisch [30], or Press et al. [144].

Exercise 2.3.1 Suppose that we want to compute the area of some bounded region Ω in two dimensions, and that we have a function $\chi_\Omega(\mathbf{x})$ that is one if and only if $\mathbf{x} \in \Omega$. If Ω is contained in a rectangle R, describe how to use the area of R and the function χ_Ω to construct a Monte Carlo method for computing the area of Ω. Use your method to compute the area of the unit circle.

Exercise 2.3.2 Suppose that the random variable X is drawn from a uniform distribution on $[0, 1)$. (This is the case with most compiler-supplied functions that generate random numbers.) In other words, for any $y \in [0, 1)$ the probability that $X \le y$ is y. If f is a strictly increasing function taking values in $[0, 1)$, what is the probability that $f(X)$ is less than or equal to y?

Exercise 2.3.3 Suppose that we want to compute the integral of a function $f(x, y)$ over the unit circle, by using the Monte Carlo method. We would like to generate a random variable that is uniformly distributed over the unit circle. We could take $X = R \cos \Theta$ and $Y = R \sin \Theta$, where Θ is a random variable drawn from a uniform distribution over $[0, 2\pi)$. How should we define the random variable R, if we are given only a function that generates a random number that is uniformly distributed over $[0, 1)$? Use your random variables R and Θ to program a Monte Carlo method to compute the volume of the unit sphere by approximating

$$\int_0^{2\pi} \int_0^1 \sqrt{1 - r^2} \, r \, dr \, d\theta \ .$$

2.3.3 Riemann Sums

Next, we turn to the simplest deterministic method for computing an integral, namely Riemann sums. This numerical integration technique is familiar to beginning calculus readers.

Suppose that we are given a function f defined on some interval (a, b), and we want to compute $\int_a^b f(x) \, dx$. The following lemma describes the numerical method and provides an upper bound for its error.

Lemma 2.3.1 *Suppose that f is continuous and $f\prime$ is bounded and continuous almost everywhere on the interval (a, b). Choose a partition*

$$a = x_0 < x_1 < \ldots < x_{n-1} < x_n = b \ ,$$

and define its width to be

$$h = \max_{0 \le i < n} (x_{i+1} - x_i) \ .$$

Also compute the **Riemann sum**

$$R_f(h) = \sum_{i=0}^{n-1} f(x_i)(x_{i+1} - x_i) \ .$$

Then

$$\left| \int_a^b f(x) \, dx - R_f(h) \right| \le \frac{b - a}{2} \left[\max_{0 \le i < n} (x_{i+1} - x_i) \right] \left[\max_{a \le x \le b} \left| f'(x) \right| \right] \ .$$

Proof For each $0 \leq i < n$,

$$\int_{x_i}^{x_{i+1}} f(x) - f(x_i)dx = \int_{x_i}^{x_{i+1}} \int_{x_i}^{x} f'(s) \, ds \, dx$$

$$= \int_{x_i}^{x_{i+1}} \int_{s}^{x_{i+1}} f'(s) \, dx \, ds = \int_{x_i}^{x_{i+1}} (x_{i+1} - s)f'(s) \, ds \, .$$

It follows that

$$\left| \int_{x_i}^{x_{i+1}} f(x) - f(x_i) \, dx \right| = \left| \int_{x_i}^{x_{i+1}} (x_{i+1} - s)f'(s) ds \right|$$

$$\leq \max_{x_i \leq s \leq x_{i+1}} |f'(s)| \int_{x_i}^{x_{i+1}} (x_{i+1} - s) ds = \max_{x_i \leq s \leq x_{i+1}} |f'(s)| \frac{(x_{i+1} - x_i)^2}{2} \, .$$

We can sum this result over the partition to get an expression for the error in the Riemann sum:

$$\left| \int_a^b f(x) dx - R_f(h) \right| \leq \sum_{i=0}^{N-1} \left| \int_{x_i}^{x_{i+1}} f(x) - f(x_i) dx \right|$$

$$\leq \sum_{i=0}^{N-1} \max_{x_i \leq s \leq x_{i+1}} |f'(s)| \frac{(x_{i+1} - x_i)^2}{2} \leq \max_{a \leq s \leq b} |f'(s)| \sum_{i=0}^{N-1} \frac{(x_{i+1} - x_i)^2}{2}$$

$$\leq \frac{b-a}{2} \left[\max_{0 \leq i < N} (x_{i+1} - x_i) \right] \left[\max_{a \leq s \leq b} |f'(s)| \right] \, .$$

Figure 2.6 shows the base 10 logarithm of the error in Riemann sums for approximating $\int_0^1 e^x \, dx$, versus the base 10 logarithm of the number of partition intervals. This figure was generated by the C^{++} program GUIQuadrature.C.

To experiment with Riemann sums for computing $\int_0^1 e^x \, dx$, readers may also execute a JavaScript quadrature program. This program plots the log of the error various Riemann sums versus the log of the partition width.

Exercise 2.3.4 Suppose that we use a uniform partition with n elements to compute a Riemann sum. If we approximate the integral of a continuously differentiable function, how does the error in the Riemann sum behave as n increases? How does this compare with the Monte Carlo method?

Exercise 2.3.5 Program Riemann sums on a uniform partition to compute the integral of $\sqrt{x}$ over $(0, 1)$. How does the error behave as the number of partition points increases?

Fig. 2.6 Errors in Riemann sums for $\int_0^1 e^x \, dx$

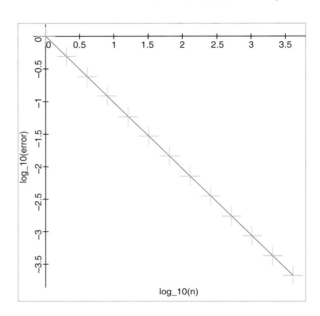

2.3.4 Midpoint Rule

Our error estimate for Riemann sums in Lemma 2.3.1 would be much the same if we had used the function value at the right-hand end of each partition interval, instead of the left. However, our next lemma shows that we can obtain a more accurate approximation to the integral by evaluating f at the *midpoint* of each partition interval.

Lemma 2.3.2 *Suppose that f is continuously differentiable on the interval (a, b), and that f'' is bounded and continuous almost everywhere on (a, b). Choose a partition*

$$a = x_0 < x_1 < \ldots < x_{n-1} < x_n = b ,$$

and define its width to be

$$h = \max_{0 \leq i < n} (x_{i+1} - x_i) .$$

Also compute the **midpoint rule quadrature**

$$M_f(h) = \sum_{i=0}^{n-1} f\left(\frac{x_{i+1} + x_i}{2}\right)(x_{i+1} - x_i) .$$

Then

$$\left| \int_a^b f(x) \, dx - M_f(h) \right| \leq \frac{b-a}{24} \left[\max_{0 \leq i < n} (x_{i+1} - x_i) \right]^2 \max_{x \in (a,b)} |f''(x)| \ .$$

Proof Let us write

$$x_{i+1/2} = \frac{1}{2}(x_i + x_{i+1}) \ .$$

Then for all $0 \leq i < n$

$$\int_{x_i}^{x_{i+1}} f(x) - f(x_{i+1/2}) \, dx = - \int_{x_i}^{x_{i+1/2}} \int_x^{x_{i+1/2}} f'(s) \, ds \, dx + \int_{x_{i+1/2}}^{x_{i+1}} \int_{x_{i+1/2}}^x f'(s) \, ds \, dx$$

$$= - \int_{x_i}^{x_{i+1/2}} \int_{x_i}^s f'(s) \, dx \, ds + \int_{x_{i+1/2}}^{x_{i+1}} \int_s^{x_{i+1}} f'(s) \, dx \, ds$$

$$= - \int_{x_i}^{x_{i+1/2}} (s - x_i) f'(s) \, ds + \int_{x_{i+1/2}}^{x_{i+1}} (x_{i+1} - s) f'(s) \, ds$$

$$= - \frac{1}{2}(s - x_i)^2 f'(s)|_{x_i}^{x_{i+1/2}} + \int_{x_i}^{x_{i+1/2}} \frac{1}{2}(s - x_i)^2 f''(s) \, ds$$

$$- \frac{1}{2}(x_{i+1} - s)^2 f'(s)|_{x_{i+1/2}}^{x_{i+1}} + \int_{x_{i+1/2}}^{x_{i+1}} \frac{1}{2}(x_{i+1} - s)^2 f''(s) \, ds$$

$$= \int_{x_i}^{x_{i+1}} \chi_i(s) f''(s) \, ds \ ,$$

where

$$\chi_i(s) \equiv \begin{cases} (s - x_i)^2/2, & s < x_{i+1/2} \\ (x_{i+1} - s)^2/2, & s > x_{i+1/2} \end{cases} \ .$$

Since

$$\int_{x_i}^{x_{i+1}} \chi_i(s) \, ds = \frac{(x_{i+1} - x_i)^3}{24} \ ,$$

we see that the error in the midpoint rule satisfies

$$\left| \int_a^b f(x) dx - M_f(h) \right| \leq \sum_{i=0}^{N-1} \left| \int_{x_i}^{x_{i+1}} f(x) - f(x_{i+1/2}) \, dx \right|$$

$$\leq \sum_{i=0}^{N-1} \max_{x_i < s < x_{i+1}} |f''(s)| \int_{x_i}^{x_{i+1}} \chi_i(s) \, ds \leq \max_{s \in (a,b)} |f''(s)| \sum_{i=0}^{N-1} \frac{|x_{i+1} - x_i|^3}{24}$$

$$\leq \frac{b-a}{24} \left[\max_{0 \leq i < n} (x_{i+1} - x_i)^2 \right] \max_{s \in (a,b)} |f''(s)| \ .$$

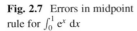

Fig. 2.7 Errors in midpoint rule for $\int_0^1 e^x\, dx$

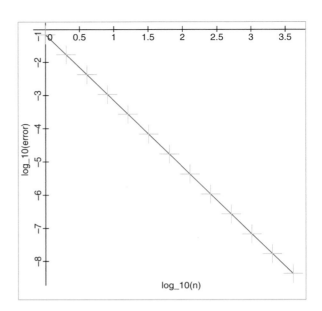

Let us compare the numerical integration techniques we have studied so far. If we use n function evaluations to compute an approximate integral, then the error in the Monte Carlo method is roughly proportional to $n^{-1/2}$, the error in a Riemann sum is proportional to n^{-1} and the error in the midpoint rule is proportional to n^{-2}. This statement assumes, of course, that the integrand f is sufficiently smooth for the error estimates to apply. Under such circumstances, we expect the midpoint rule to converge much more quickly than the other two methods for integrating functions of a single variable.

Figure 2.7 shows the base 10 logarithm of the error in the midpoint rule for approximating $\int_0^1 e^x\, dx$, versus the base 10 logarithm of the number of partition intervals. This figure was generated by the C^{++} program GUIQuadrature.C.

In order to experiment with the midpoint rule for computing $\int_0^1 e^x\, dx$, readers may also execute a JavaScript quadrature program.

Exercise 2.3.6 Use the error estimate in Lemma 2.3.2 to determine how many partition intervals are needed to guarantee that the midpoint rule approximation to $\int_1^e \ln(x)\, dx$ has an error of at most 10^{-8}. Then write a program to compute this midpoint rule approximation, and verify that the results are at least as accurate as predicted.

Exercise 2.3.7 Describe how to use the midpoint rule to compute $\int_0^1 \int_0^1 e^{x+y}\, dx\, dy$. How would the error decrease as a function of the number of evaluations of the integrand?

2.3.5 Trapezoidal Rule

Let us step back and take a different look at the numerical quadrature methods we have developed. With either Riemann sums or the midpoint rule, we interpolate the integrand f with a piecewise constant function, then integrate the piecewise constant function exactly. Based on our study of splines in Sect. 1.6, it seems reasonable to expect that if we interpolate the integrand f with a piecewise *linear* function, then we should obtain greater accuracy. The next lemma shows that this expectation is false.

Lemma 2.3.3 *Suppose that f is continuously differentiable on the interval (a, b), and that f'' is bounded and continuous almost everywhere on (a, b). Choose a partition*

$$a = x_0 < x_1 < \ldots < x_{n-1} < x_n = b ,$$

and define its width to be

$$h = \max_{0 \le i < n} (x_{i+1} - x_i) .$$

Also compute the **trapezoidal rule quadrature**

$$T_f(h) = \sum_{i=0}^{n-1} \frac{1}{2} [f(x_{i+1}) + f(x_i)] (x_{i+1} - x_i)$$

$$= \frac{1}{2} \left\{ f(x_0)(x_1 - x_0) + \sum_{i=1}^{n-1} f(x_i)(x_{i+1} - x_{i-1}) + f(x_n)(x_n - x_{n-1}) \right\} .$$

Then

$$\left| \int_a^b f(x) \, dx - T_f(h) \right| \le \frac{b-a}{12} \left[\max_{0 \le i < n} (x_{i+1} - x_i) \right]^2 \max_{x \in (a,b)} |f''(x)| .$$

Proof Let us write

$$x_{i+1/2} = \frac{1}{2}(x_i + x_{i+1}) .$$

For each $0 \leq i < n$,

$$\int_{x_i}^{x_{i+1}} f(x) - \frac{1}{2}[f(x_i) + f(x_{i+1})] \, dx$$

$$= \frac{1}{2}\left[\int_{x_i}^{x_{i+1}} f(x) - f(x_i) \, ds + \int_{x_i}^{x_{i+1}} f(x) - f(x_{i+1}) \, dx\right]$$

$$= \frac{1}{2}\left[\int_{x_i}^{x_{i+1}} \int_{x_i}^{x} f'(s) \, ds \, dx - \int_{x_i}^{x_{i+1}} \int_{x}^{x_{i+1}} f'(s) \, ds \, dx\right]$$

$$= \frac{1}{2}\left[\int_{x_i}^{x_{i+1}} \int_{s}^{x_{i+1}} f'(s) \, dx \, ds - \int_{x_i}^{x_{i+1}} \int_{x_i}^{s} f'(s) \, dx \, ds\right]$$

$$= \frac{1}{2}\left[\int_{x_i}^{x_{i+1}} (x_{i+1} - s)f'(s) \, ds - \int_{x_i}^{x_{i+1}} (s - x_i)f'(s) \, ds\right] = \int_{x_i}^{x_{i+1}} (x_{i+\frac{1}{2}} - s)f'(s) \, ds$$

$$= \frac{1}{2}(x_{i+1} - s)(s - x_i)f'(s)\Big|_{x_i}^{x_{i+1}} - \int_{x_i}^{x_{i+1}} \frac{1}{2}(x_{i+1} - s)(s - x_i)f''(s) \, ds$$

$$= \int_{x_i}^{x_{i+1}} \chi_i(s)f''(s) \, ds \, ,$$

where

$$\chi_i(s) \equiv -\frac{1}{2}(x_{i+1} - s)(s - x_i) \, .$$

Note that

$$\int_{x_i}^{x_{i+1}} \chi_i(s) ds = \frac{(x_{i+1} - x_i)^3}{12} \, ,$$

so we estimate the error in the trapezoidal rule to be

$$\left|\int_a^b f(x) dx - T_f(h)\right| = \left|\sum_{i=0}^{N-1} \int_{x_i}^{x_{i+1}} \chi_i(s)f''(s) ds\right|$$

$$\leq \sum_{i=0}^{N-1} \max_{s \in (x_i, x_{i+1})} |f''(s)| \int_{x_i}^{x_{i+1}} \chi_i(s) ds = \sum_{i=0}^{N-1} \frac{(x_{i+1} - x_i)^3}{12} \max_{s \in (x_i, x_{i+1})} |f''(s)|$$

$$\leq \frac{(x_{i+1} - x_i)^2}{12} \max_{a < s < b} |f(s)| \sum_{i=0}^{N-1} (x_{i+1} - x_i) = \frac{(x_{i+1} - x_i)^2}{12} (b - a) \max_{s \in (a,b)} |f''(s)| \, .$$

Note that this error estimate says that the midpoint rule is asymptotically twice as accurate as the trapezoidal rule. This is possible because the error in the piecewise constant interpolant to the integrand tends to oscillate in sign within each partition

Fig. 2.8 Errors in trapezoidal
rule for $\int_0^1 e^x \, dx$

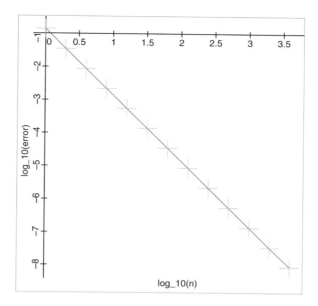

interval. For the trapezoidal rule, the error in the piecewise linear interpolant to the integrand tends to have the same sign across each partition intervals, and even across many partition intervals. Since the error in the integral approximation is the average of the errors in polynomial interpolation on the partition intervals, the error in the midpoint rule is almost always smaller than for the error with the trapezoidal rule.

Figure 2.8 shows the base 10 logarithm of the error in the trapezoidal rule for approximating $\int_0^1 e^x \, dx$, versus the base 10 logarithm of the number of partition intervals. This figure was generated by the C^{++} program GUIQuadrature.C.

MATLAB users can use the command trapz to compute trapezoidal rule approximations to integrals. To experiment with the trapezoidal rule for computing $\int_0^1 e^x \, dx$, readers may also execute a JavaScript quadrature program.

Here is an interesting connection between the midpoint rule and the trapezoidal rule.

Lemma 2.3.4 *Suppose that f is bounded and continuous almost everywhere on (a, b). Choose a partition*

$$a = x_0 < x_1 < \ldots < x_{n-1} < x_n = b \,,$$

and define its width to be

$$h = \max_{0 \le i < n} (x_{i+1} - x_i) \,.$$

Also define the midpoint rule quadrature

$$M_f(h) = \sum_{i=0}^{n-1} f\left(\frac{x_{i+1} + x_i}{2}\right)(x_{i+1} - x_i)$$

and the trapezoidal rule quadrature

$$T_f(h) = \sum_{i=0}^{n-1} \frac{1}{2}[f(x_{i+1}) + f(x_i)](x_{i+1} - x_i) \ .$$

Then the trapezoidal rule quadrature $T_f(h/2)$ on the fine partition

$$a = x_0 < x_{1/2} = \frac{x_1 + x_0}{2} < x_1 < \ldots < x_{n-1} < x_{n-1/2} = \frac{x_n + x_{n-1}}{2} < x_n = b$$

satisfies

$$T_f(h/2) = \frac{1}{2}\{T_f(h) + M_f(h)\} \ .$$

Proof Let us write

$$x_{i+1/2} = \frac{1}{2}(x_i + x_{i+1})$$

for the midpoint of the *i*th element. Then

$$T_f(h/2) = \sum_{i=0}^{n-1} \left\{\frac{x_{i+1/2} - x_i}{2}[f(x_{i+1/2}) + f(x_i)] + \frac{x_{i+1} - x_{i+1/2}}{2}[f(x_{i+1}) + f(x_{i+1/2})]\right\}$$

$$= \frac{1}{2}\sum_{i=0}^{n-1} \left\{\frac{x_{i+1} - x_i}{2}[f(x_{i+1}) + f(x_i)] + f(x_{i+1/2})(x_{i+1} - x_i)\right\}$$

$$= \frac{1}{2}\{T_f(h) + M_f(h)\} \ . \tag{2.14}$$

In other words, the trapezoidal rule $T_f(h/2)$ with bisected partition intervals is the average of the original trapezoidal rule $T_f(h)$ and the original midpoint rule $M_f(h)$.

Exercise 2.3.8 Write a program to compute both the midpoint rule and the trapezoidal rule approximations to $I \equiv \int_0^1 \sinh(x) \ dx$ using equidistant partitions with $n = 2^k$ partition intervals. Plot $M_f(h) - I$ and $T_f(h) - I$ versus h. Comment on the magnitude and the signs of these two errors as a function of n.

2.3.6 Euler-MacLaurin Formula

In Lemma 2.3.3, our error estimate for the trapezoidal rule assumed that the integrand has a bounded second derivative that is continuous almost everywhere. In our next result, we will develop an expansion for the error in the trapezoidal rule applied on a uniform partition to a function with an arbitrarily large number of continuous derivatives. This error expansion will tell us how to apply Richardson extrapolation to trapezoidal rule quadratures. This theorem will have some other very important consequences.

Theorem 2.3.3 (Euler-MacLaurin) *Suppose that f has infinitely many continuous derivatives on the interval $[a, b]$. Given an integer n, let the partition width be $h = (b - a)/n$, and define the partition points to be $x_i = a + ih$ for $0 \le i \le n$. Then the error in the trapezoidal rule approximation*

$$T_f(h) = \frac{h}{2} \sum_{i=0}^{n-1} [f(x_{i+1}) + f(x_i)]$$

satisfies

$$\int_a^b f(x)dx - T_f(h) = \sum_{\substack{j \ge 2 \\ j \text{ even}}} c_j h^j \left[f^{(j-1)}(b) - f^{(j-1)}(a) \right] , \qquad (2.15)$$

where the constants c_j are independent of f, and are the coefficients in the Taylor expansion for

$$1 + \frac{1 + e^h}{1 - e^h} \frac{h}{2} = 1 - \frac{h}{2} \coth \frac{h}{2} = \sum_{\substack{j \ge 2 \\ j \text{ even}}} c_j h^j = -\frac{h^2}{12} + \frac{h^4}{720} - \frac{h^6}{30240} + \frac{h^8}{1209600} - \cdots .$$

Proof Suppose that the function ϕ is infinitely many times continuously differentiable on the interval $[-h/2, h/2]$. Then the error in the trapezoidal rule for ϕ on this element is

$$e(h) \equiv \int_{-h/2}^{h/2} \phi(\xi) \, d\xi - \frac{h}{2} \left[\phi\left(-\frac{h}{2}\right) + \phi\left(\frac{h}{2}\right) \right] .$$

It is easy to see that e is an odd function of h:

$$e(-h) = \int_{h/2}^{-h/2} \phi(\xi) \, d\xi + \frac{h}{2} \left[\phi\left(\frac{h}{2}\right) + \phi\left(-\frac{h}{2}\right) \right] = -e(h) .$$

Lemma 2.3.3 also showed that $e(h)/h \to 0$ as $h \to 0$. Thus if $e(h)$ has a Taylor expansion, it must involve only odd positive powers of h. If ϕ has the Taylor series

$$\phi(\xi) = \sum_{k=0}^{\infty} \frac{\phi^{(k)}(0)}{k!} \xi^k ,$$

then

$$\frac{h}{2} \left[\phi\left(-\frac{h}{2}\right) + \phi\left(\frac{h}{2}\right) \right] = h \sum_{k=0}^{\infty} \frac{\phi^{(k)}(0)}{k!} \frac{1}{2} \left[\left(-\frac{h}{2}\right)^k + \left(\frac{h}{2}\right)^k \right]$$

$$= h \sum_{\substack{k \geq 0 \\ k \text{ even}}} \frac{\phi^{(k)}(0)}{k!} \left(\frac{h}{2}\right)^k ,$$

and

$$\int_{-h/2}^{h/2} \phi(\xi) \, d\xi = \sum_{k=0}^{\infty} \frac{\phi^{(k)}(0)}{k!} \int_{-h/2}^{h/2} \xi^k \, d\xi$$

$$= \sum_{k=0}^{\infty} \frac{\phi^{(k)}(0)}{k!} \frac{1}{k+1} \left[\left(\frac{h}{2}\right)^{k+1} - \left(-\frac{h}{2}\right)^{k+1} \right]$$

$$= h \sum_{\substack{k \geq 0 \\ k \text{ even}}} \frac{\phi^{(k)}(0)}{(k+1)!} \left(\frac{h}{2}\right)^k .$$

It follows that

$$e(h) \equiv \int_{-h/2}^{h/2} \phi(\xi) \, d\xi - \frac{h}{2} \left[\phi\left(-\frac{h}{2}\right) + \phi\left(\frac{h}{2}\right) \right]$$

$$= h \sum_{\substack{k \geq 0 \\ k \text{ even}}} \frac{\phi^{(k)}(0)}{k!} \left(\frac{h}{2}\right)^k \left[\frac{1}{k+1} - 1 \right]$$

$$= -h \sum_{\substack{k \geq 2 \\ k \text{ even}}} \frac{\phi^{(k)}(0)}{(k+1)!} k \left(\frac{h}{2}\right)^k$$

For each k, we can write

$$\frac{1}{h} \left[\phi^{(k)}\left(\frac{h}{2}\right) - \phi^{(k)}\left(-\frac{h}{2}\right) \right] = \sum_{j=0}^{\infty} \frac{\phi^{(j+k)}(0)}{j!} \frac{(h/2)^j - (-h/2)^j}{h}$$

$$= \sum_{\substack{j \geq 1 \\ j \text{ odd}}} \frac{\phi^{(j+k)}(0)}{j!} \left(\frac{h}{2}\right)^{j-1} = \sum_{\substack{j \geq 0 \\ j \text{ even}}} \frac{\phi^{(j+k+1)}(0)}{(j+1)!} \left(\frac{h}{2}\right)^j .$$

In other words,

$$\phi^{(k+1)}(0) = \frac{\phi^{(k)}(h/2) - \phi^{(k)}(-h/2)}{h} - \sum_{\substack{j \geq 2 \\ j \text{ even}}} \frac{\phi^{(j+k+1)}(0)}{(j+1)!} \left(\frac{h}{2}\right)^j .$$

Since this expression involves only even powers of h, we can repeatedly substitute back into the expansion for $e(h)$ and obtain

$$e(h) = \sum_{\substack{j \geq 2 \\ j \text{ even}}} c_j h^j \left[\phi^{(j-1)}\left(\frac{h}{2}\right) - \phi^{(j-1)}\left(-\frac{h}{2}\right)\right] .$$

For each $0 \leq i < n$ we can take $\phi(\xi) = f(a + ih/2 + \xi)$, and apply our previous work to the error in the trapezoidal rule for the original integrand f:

$$\sum_{i=0}^{n-1} \left[\int_{a+ih}^{a+ih+h} f(x)\, dx - \frac{h}{2}\{f(a+ih) + f(a+ih+h)\}\right]$$

$$= \sum_{i=0}^{n-1} \left\{\int_{-h/2}^{h/2} f(a+ih+h/2+\xi)\, d\xi - \frac{h}{2}[f(a+ih) + f(a+ih+h)]\right\}$$

$$= \sum_{\substack{j \geq 2 \\ j \text{ even}}} c_j h^j \left\{\sum_{i=0}^{n-1} \left[f^{(j-1)}(a+ih+h) - f^{(j-1)}(a+ih)\right]\right\}$$

$$= \sum_{\substack{j \geq 2 \\ j \text{ even}}} c_j h^j \left[f^{(j-1)}(b) - f^{(j-1)}(a)\right] .$$

Now, consider the case when $f(x) = e^x$. In this case, $f^{(j)}(x) = e^x$ for all $j \geq 0$, so the Euler-MacLaurin formula gives us

$$[e^b - e^a] \sum_{\substack{j \geq 2 \\ j \text{ even}}} c_j h^j = \sum_{\substack{j \geq 2 \\ j \text{ even}}} c_j h^j \left[\mathbf{D}^{j-1} e^x(b) - \mathbf{D}^{j-1} e^x(a)\right]$$

$$= \int_a^b e^x\, dx - \frac{h}{2} \sum_{i=0}^{n-1} \left[e^{a+ih} + e^{a+ih+h}\right] = e^b - e^a - \frac{h}{2}\left[e^a + e^{a+h}\right] \sum_{i=0}^{n-1} e^{ih}$$

$$= e^b - e^a - \frac{h}{2}\left[e^a + e^{a+h}\right]\frac{1 - e^{nh}}{1 - e^h} = e^b - e^a - \frac{h}{2}(1 + e^h)\frac{e^a - e^b}{1 - e^h}$$

$$= [e^b - e^a]\left[1 + \frac{1 + e^h}{1 - e^h}\frac{h}{2}\right] .$$

Of course, if f does not have an infinite number of derivatives, then the expansion is only valid for as many terms as f has continuous derivatives.

Next, we note that it is easy to find a similar expansion for the error in the midpoint rule.

Corollary 2.3.1 *Suppose that f has infinitely many continuous derivatives on the interval $[a, b]$. Given an integer n, let the element width be $h = (b-a)/n$, and define the partition points to be $x_i = a + ih$ for $0 \leq i \leq n$. Then the error in the midpoint rule approximation*

$$M_f(h) = h \sum_{i=0}^{n-1} f\left(\frac{x_{i+1} + x_i}{2}\right)$$

satisfies

$$\int_a^b f(x)\, dx - M_f(h) = -\sum_{\substack{j \geq 2 \\ j\ even}} c_j h^j \left[1 - 2^{1-j}\right]\left[f^{(j-1)}(b) - f^{(j-1)}(a)\right],$$

where the constants c_j are the same as in the Euler-MacLaurin formula (2.15). Alternatively, we can write

$$\int_a^b f(x)\, dx - M_f(h) = -\sum_{\substack{j \geq 2 \\ j\ even}} d_j h^j \left[f^{(j-1)}(b) - f^{(j-1)}(a)\right], \tag{2.16}$$

where the constants d_j are the coefficients in the Taylor expansion for

$$1 - \frac{h/2}{\sinh(h/2)} = \sum_{\substack{j \geq 2 \\ j\ even}} d_j h^j = \frac{1}{24}h^2 - \frac{7}{5760}h^4 + \frac{31}{967680}h^6 - \frac{127}{154828800}h^8 + \cdots.$$

Proof We can use the Euler-MacLaurin formula (2.15) and the relationship (2.14) between the trapezoidal and midpoint rules to write

$$\int_a^b f(x)dx - M_f(h) = 2\left[\int_a^b f(x)dx - T_f(h/2)\right] - \left[\int_a^b f(x)dx - T_f(h)\right].$$

$$= 2\sum_{\substack{j \geq 2 \\ j\ even}} c_j (h/2)^j \left[f^{(j-1)}(b) - f^{(j-1)}(a)\right] - \sum_{\substack{j \geq 2 \\ j\ even}} c_j h^j \left[f^{(j-1)}(b) - f^{(j-1)}(a)\right]$$

$$= -\sum_{\substack{j \geq 2 \\ j\ even}} c_j h^j \left[1 - 2^{1-j}\right]\left[f^{(j-1)}(b) - f^{(j-1)}(a)\right]$$

$$\equiv \sum_{\substack{j \geq 2 \\ j\ even}} d_j h^j \left[f^{(j-1)}(b) - f^{(j-1)}(a)\right].$$

We can use $f(x) = e^x$ to evaluate the constants d_j:

$$\left[e^b - e^a\right] \sum_{\substack{j \geq 2 \\ j \text{ even}}} d_j h^j = \sum_{\substack{j \geq 2 \\ j \text{ even}}} d_j h^j \left[\mathbf{D}^{j-1} e^x(b) - \mathbf{D}^{j-1} e^x(a)\right]$$

$$= \int_a^b e^x \, dx - h \sum_{i=0}^{n-1} e^{a+ih+h/2} = e^b - e^a - h e^{a+h/2} \sum_{i=0}^{n-1} e^{ih}$$

$$= e^b - e^a - h e^{a+h/2} \frac{1 - e^{nh}}{1 - e^h} = e^b - e^a - h e^{h/2} \frac{e^a - e^b}{1 - e^h}$$

$$= [e^b - e^a] \left[1 + \frac{h e^{h/2}}{1 - e^h}\right] = [e^b - e^a] \left[1 - \frac{h/2}{\sinh(h/2)}\right] .$$

Note that if we double n (i.e., halve h), then for sufficiently large n the error in either the midpoint rule $M_f(h)$ or the trapezoidal rule $T_f(h)$ should be divided by 4. Also note that for small values of h, the midpoint rule is twice as accurate as the trapezoidal rule. As $h \to 0$, the error in the midpoint rule will tend to have the opposite sign of the error in the trapezoidal rule. Thus either the trapezoidal rule converges from above and the midpoint rule from below, or *vice versa*. If follows that for h sufficiently small, the midpoint rule $M_f(h)$ and trapezoidal rule $T_f(h)$ are close to the true integral when they are close to each other.

Also note that if $f'(b) = f'(a)$, $f^{(3)}(b) = f^{(3)}(a)$ and so on, then both the trapezoidal rule and the midpoint rule converge faster than any power of h. Thus either of these two rules is especially accurate for periodic functions. In particular, the discrete Fourier coefficients defined in Sect. 1.7.4.1 by

$$c_k = \frac{1}{2\pi m} \sum_{j=0}^{m-1} f\left(\frac{2\pi j}{m}\right) e^{-i2\pi kj/m}$$

are trapezoidal rule approximations to the continuous Fourier coefficients

$$c(k) = \int_0^{2\pi} f(t) e^{-ikt} \, dt .$$

For periodic functions f, the discrete Fourier coefficients are extremely accurate approximations to the continuous Fourier coefficients.

To experiment with the quadrature rules for computing $\int_0^{2\pi} e^{1/\pi^2 - 1/[x(2\pi - x)]} \, dx$, readers may execute the JavaScript quadrature program. This integrand in this program is periodic on $[0, 2\pi]$, so the trapezoidal and midpoint rule approximations to the integral converge faster than any power of the mesh width h.

Exercise 2.3.9 Consider trapezoidal and midpoint rule approximations to compute $\int_0^\pi \sin^2(x) \, dx$. What are the errors for partitions with 1, 2 and 4 intervals?

Exercise 2.3.10 Program the trapezoidal and midpoint rules to integrate $f(x) = \sqrt{x}$ over $(0, 1)$. Plot the logarithms of the errors in both rules versus the logarithm of the number of mesh elements. What is the apparent order of convergence of these two rules?

2.3.7 Newton-Cotes Quadrature

Given a positive integer m, define the equidistant nodes

$$\xi_j = j/m \text{ for } 0 \le j \le m .$$

Newton-Cotes quadrature rules (see Abramowitz and Stegun [1, p. 886]) on the reference interval $(0, 1)$ take the form

$$\int_0^1 f(\xi) \, d\xi \approx \int_0^1 \sum_{j=0}^m f(\xi_j) \prod_{\substack{i=0 \\ i \ne j}}^m \frac{\xi - \xi_i}{\xi_j - \xi_i} \, d\xi = \sum_{j=0}^m f(\xi_j) \int_0^1 \prod_{\substack{i=0 \\ i \ne j}}^m \frac{\xi - \xi_i}{\xi_j - \xi_i} \, d\xi \equiv \sum_{j=0}^m f\left(\xi_j\right) \alpha_j .$$

In other words, the Newton-Cotes rule approximates the integral of f by the exact integral of the Lagrange polynomial interpolant to f. The quadrature weights α_j are such that $\alpha_j = \alpha_{m-j}$, and the rule is exact for all polynomials of degree at most m. The latter claim is possible because Lagrange polynomial interpolation with $m + 1$ points is exact for all polynomials of degree at most m.

The quadrature weights for the first nine Newton-Cotes quadrature rules can be found in Table 2.1. For $m = 8$ and $m > 9$ the Newton-Cotes quadrature rules involve negative weights, so we should not use Newton-Cotes quadrature rules with $m \ge 8$. Newton-Cotes quadratures are obviously exact for polynomials of degree $m + 1$, and fortuitously exact for polynomials of degree $m + 2$ if m is even [8, p. 226]. Thus Newton-Cotes quadrature rules with $m = 2k$ achieve the same order with

Table 2.1 Newton-Cotes quadrature rules

m	α_0	α_1	α_2	α_3	α_4	Rule
1	1 / 2					Trapezoidal
2	1 / 6	2 / 3				Simpson's
3	1 / 8	3 / 8				
4	7 / 90	16 / 45	2 / 15			
5	19 / 288	25 / 96	25 / 144			
6	41 / 840	9 / 35	9 / 280	34 / 105		
7	751 / 17280	3577 / 17280	49 / 640	2989 / 17280		
8	989 / 28350	2944 / 14175	-464 / 14175	5248 / 14175	454 / 2835	
9	2857 / 89600	15741 / 89600	27 / 2240	1209 / 5600	2889 / 44800	

fewer function evaluations than the $2k + 1$ rule. In general, it is not efficient to use Newton-Cotes quadrature rules with m odd, except for the trapezoidal rule ($m = 1$).

The files `Quadrature.H` and `Quadrature.C`, available from Cambridge University Press, implement various C^{++} classes for quadrature rules, including Newton-Cotes quadrature. The weights and nodes for a Newton-Cotes quadrature rule with $m \leq 8$ can are determined in the `NewtonCotesQuadrature` constructor.

2.3.8 Clenshaw-Curtis Quadrature

Let us discuss another useful quadrature rule due to Clenshaw and Curtis [46]. Further discussion of this quadrature rule has been provided by Imhof [109] and Trefethen [175]. The basic idea is to compute

$$\int_{-1}^{1} f(x) \, dx = \int_{0}^{\pi} f(\cos \theta) \sin \theta \, d\theta$$

via a truncation of the **Fourier cosine series**

$$f(\cos \theta) = \frac{a_0}{2} + \sum_{k=1}^{\infty} a_k \cos(k\theta) \ .$$

Integration of the Fourier cosine series gives us

$$\int_{0}^{\pi} f(\cos \theta) \sin \theta \, d\theta = a_0 - 2 \sum_{k=1}^{\infty} \frac{a_{2k}}{4k^2 - 1} \ ,$$

so only even-indexed Fourier coefficients are needed. We approximate the Fourier coefficient

$$a_{2k} = \frac{2}{\pi} \int_{0}^{\pi} f(\cos \theta) \cos(2k\theta) \, d\theta$$

by the trapezoidal rule evaluation of f at $N + 1$ **Chebyshev points**:

$$a_{2k} \approx \widetilde{a}_{2k} \equiv \frac{2}{N} \left[\frac{f(-1)}{2} + \sum_{n=1}^{N-1} f\left(\cos \frac{n\pi}{N} \right) \cos \left(\frac{2kn\pi}{N} \right) + \frac{f(1)}{2} \right] \ .$$

The quadrature rule uses these approximate Fourier coefficients and truncates the Fourier cosine series as follows:

$$\int_0^\pi f(\cos\theta)\sin\theta\ d\theta \approx \widetilde{a}_0 - 2 \sum_{k=1}^{\lfloor N/2-1\rfloor} \frac{\widetilde{a}_{2k}}{4k^2-1} - \begin{cases} 0, & N=1 \\ \dfrac{a_{2\lfloor N/2\rfloor}}{(2\lfloor N/2\rfloor)^2-1}, & N\ge 2 \text{ even} \\ 2\dfrac{a_{2\lfloor N/2\rfloor}}{(2\lfloor N/2\rfloor)^2-1}, & N\ge 3 \text{ odd} \end{cases}.$$

The case $N = 1$ is the trapezoidal rule, and $N = 2$ produces Simpson's rule. These quadrature rules have order $2\lfloor N/2\rfloor + 2$. If desired, the quadrature rules can be rewritten as sums of function values times quadrature weights.

A Fortran implementation of Clenshaw-Curtis quadrature is available from netlib as TOMS 424. An implementation in C^{++} can be found in files `Quadrature.H` and `Quadrature.C` from Cambridge University Press.

2.3.9 Romberg Integration

A second look at the Euler-MacLaurin expansion (2.15) will enable us to find a strategy for constructing higher-order approximations to integrals. This formula shows that if we could know the odd-order derivatives of f at the endpoints a and b, then we could generate a highly accurate integration formula, beginning with either the trapezoidal rule or the midpoint rule. This approach would also require knowledge of the values of the coefficients in the Euler-MacLaurin expansion. A better approach would use the *existence of the error expansion* to apply Richardson extrapolation (see Sect. 2.2.4), without knowing either the expansion coefficients or the odd-order derivatives of f.

Let us apply Richardson extrapolation analytically to the trapezoidal rule on a uniform partition. From Eq. (2.10) and the refinement Eq. (2.14) for the trapezoidal rule, we see that the Richardson extrapolant is

$$T_f^{(4)}(h/2) = T_f(h/2) + \frac{1}{2^2-1}\left[T_f(h/2) - T_f(h)\right] = \frac{4}{3}T_f(h/2) - \frac{1}{3}T_f(h)$$

$$= \frac{2}{3}\left[T_f(h) + M_f(h)\right] - \frac{1}{3}T_f(h) = \frac{1}{3}T_f(h) + \frac{2}{3}M_f(h)\ .$$

This is commonly called **Simpson's rule**. Because the Euler-MacLaurin formula (2.15) shows that both the trapezoidal rule and the midpoint rule have error expansions involving only even powers of the partition width h, we conclude that Simpson's rule has an error expansion involving only even powers of h, and the lowest-order term in this expansion is on the order of h^4.

We can continue this extrapolation process to get even higher-order approximations. No additional evaluations of f are required to compute any of the extrapolants of the trapezoidal rule. If f is expensive to evaluate, this means that the extrapolants are relatively inexpensive.

If some of the coefficients in the expansion for the trapezoidal rule are zero, then some of the extrapolants will be more accurate than expected. After reaching this point in the extrapolation process, there will be no increase in accuracy until Richardson extrapolation uses a coefficient that is appropriate for the current **order of accuracy**.

A rounding error analysis of the extrapolation shows that if the rounding errors in each trapezoidal rule computation are bounded by ε for all partition widths h under consideration, then the rounding errors in the extrapolants are bounded by 2ε. Thus extrapolation is numerically stable.

Example 2.3.1 Suppose that we want to compute $\int_0^1 e^x dx$. Of course, the exact value is $e - 1 \approx 1.718281828$.

We have the extrapolations of the trapezoidal rule:

$$T_f^{(4)}(h) = T_f(h) + \left(T_f(h) - T_f(2h)\right)/3 \,,$$

$$T_f^{(6)}(h) = T_f^{(4)}(h) + \left(T_f^{(4)}(h) - T_f^{(4)}(2h)\right)/15 \text{ and}$$

$$T_f^{(8)}(h) = T_f^{(6)}(h) + \left(T_f^{(6)}(h) - T_f^{(6)}(2h)\right)/63 \,.$$

Similarly, we have the following extrapolations of the midpoint rule:

$$M_f^{(4)}(h) = M_f(h) + \left(M_f(h) - M_f(2h)\right)/3 \,,$$

$$M_f^{(6)}(h) = M_f^{(4)}(h) + \left(M_f^{(4)}(h) - M_f^{(4)}(2h)\right)/15 \text{ and}$$

$$M_f^{(8)}(h) = M_f^{(6)}(h) + \left(M_f^{(6)}(h) - M_f^{(6)}(2h)\right)/63 \,.$$

We can also use the formula

$$T_f(h/2) = \frac{1}{2}(T_f(h) + M_f(h))$$

to compute the new trapezoidal rule approximation from the two previous trapezoidal and midpoint rule approximations.

It may be useful to make tables of the errors in the approximations to $I = \int_a^b f(x)dx$. For the trapezoidal rule, the errors are

h	$I - T_f(h)$	$I - T_f^{(4)}(h)$	$I - T_f^{(6)}(h)$	$I - T_f^{(8)}(h)$
2^0	-1.41×10^{-1}			
2^{-1}	-3.56×10^{-2}	-5.79×10^{-4}		
2^{-2}	-8.94×10^{-3}	-3.70×10^{-5}	-8.59×10^{-7}	
2^{-3}	-2.24×10^{-3}	-2.33×10^{-6}	-1.38×10^{-8}	-3.35×10^{-10}
2^{-4}	-5.59×10^{-4}	-1.46×10^{-7}	-2.16×10^{-10}	-1.34×10^{-12}

For the midpoint rule, the errors are

h	$I - M_f(h)$	$I - M_f^{(4)}(h)$	$I - M_f^{(6)}(h)$	$I - M_f^{(8)}(h)$
2^0	6.96×10^{-2}			
2^{-1}	1.78×10^{-2}	5.05×10^{-4}		
2^{-2}	4.47×10^{-3}	3.24×10^{-5}	8.32×10^{-7}	
2^{-3}	1.12×10^{-3}	2.04×10^{-6}	1.33×10^{-8}	3.33×10^{-10}
2^{-4}	2.80×10^{-4}	1.27×10^{-7}	2.10×10^{-10}	1.33×10^{-12}

Note from the tables that

$$I - M_f(h) \approx -\frac{1}{2}(I - T_f(h)) \,,$$

$$I - M_f(h) \approx \frac{1}{4}(I - M_f(2h)) \,, \ I - T_f(h) \approx \frac{1}{4}(I - T_f(2h)) \,,$$

$$I - M_f^{(4)}(h) \approx \frac{1}{16}(I - M_f^{(4)}(2h)) \,, \ I - T_f^{(4)}(h) \approx \frac{1}{16}(I - T_f^{(4)}(2h)) \,,$$

and so on. Also note that the errors in the midpoint rule have opposite sign of the errors in the trapezoidal rule.

Figure 2.9 shows the errors in the diagonal entries of the Romberg extrapolation table for both the midpoint rule and the trapezoidal rule for this example. Note that

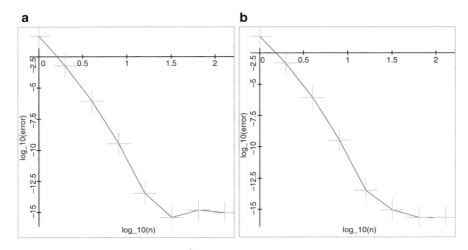

Fig. 2.9 Romberg integration for $\int_0^1 e^x \, dx$: log of error vs. minus log of partition width. (**a**) Trapezoidal rule. (**b**) Midpoint rule

the error decreases rapidly until it reaches the order of machine precision. This figure was generated by the C^{++} program GUIQuadrature.C. This program allows the user to choose one of three basic integration techniques (Riemann sums, midpoint rule or trapezoidal rule) for one of two integrands (e^x or $\sqrt{x}$).

To experiment with Richardson extrapolation of either Riemann sums, the midpoint rule or the trapezoidal rule for computing $\int_0^1 e^x \, dx$, readers may execute a JavaScript Romberg integration program

We also remark that quadpack uses extrapolation to handle oscillatory integrands, integrands with singularities, and infinite integration regions. However, QUAD-PACK uses Wynn's epsilon algorithm rather than Richardson extrapolation.

Exercise 2.3.11 Note that

$$I \equiv \int_0^1 \frac{4}{1+x^2} dx = \pi \; .$$

1. Use the midpoint and trapezoidal rules to approximate I for various step sizes h. Plot the log of the error versus $\log h$ for each. Describe the effect of rounding errors as $h \to 0$.
2. Implement Romberg integration to compute I. Plot the log of the error in the diagonal terms in the extrapolation table versus $\log h$.

Exercise 2.3.12 Repeat the previous exercise for

$$I \equiv \int_0^1 \sqrt{x} \log(x) dx = -\frac{4}{9} \; .$$

2.3.10 Gaussian Quadrature

The extrapolation process in Romberg integration showed us how to achieve arbitrarily high order accuracy in integrating smooth functions. This process allows us to manipulate quadratures from low-order rules into high-order results. For example, with Romberg integration applied to the midpoint rule we can achieve order $h^{2(n+1)}$ accuracy with 2^n nodes.

However, it is possible to find quadrature rules that achieve a given **order of accuracy** with far fewer quadrature nodes than this example of Romberg integration. If function evaluations are expensive, it may be essential to achieve a given order of accuracy with the fewest possible function evaluations. This approach is useful in cases where the control of the errors occurs somewhere else, such as in the use of a numerical scheme of some given order for solving a differential equation, especially by finite element methods (see Sect. 4.6).

Here is a mathematical description of our new goal. Given some nonnegative integer m, we would like to find a set $\{x_0, \ldots, x_m\}$ of nodes and corresponding set

$\{\omega_0, \ldots, \omega_m\}$ of quadrature weights so that the approximation

$$\int_a^b f(x)w(x)dx \approx \sum_{j=0}^m f(x_j)\omega_j$$

has the highest possible order for arbitrary smooth functions f. The solution of this problem is contained in the following theorem.

Theorem 2.3.4 (Gaussian Quadrature) *Suppose that the function w is positive on the open interval (a, b), and that the set of polynomials $\{\phi_j\}_{j=0}^\infty$ is orthogonal with respect to the inner product*

$$(f, g) = \int_a^b f(x)g(x)w(x)dx .$$

Then for any integer $m \geq 0$, the zeros $x_0, \ldots, x_m$ of $\phi_{m+1}(x)$ are real, distinct and lie in the open interval (a, b).

Next, let the Lagrange interpolation polynomial at these zeros be

$$\lambda_{j,m}(x) = \prod_{\substack{i=0 \\ i \neq j}}^m (x - x_i) / \prod_{\substack{i=0 \\ i \neq j}}^m (x_j - x_i) ,$$

and define the scalars ω_j by

$$\omega_j = \int_a^b \lambda_{j,m}(x)w(x)dx . \tag{2.17}$$

Then the equation

$$\int_a^b p(x)w(x)\, dx = \sum_{j=0}^m p(x_j)\omega_j$$

is satisfied for all polynomials p of degree at most $2m + 1$.

Proof We will prove the first claim by contradiction. Since ϕ_{m+1} is nonzero and has degree $m + 1$, it cannot have more than $m + 1$ zeros in (a, b), and thus it cannot have more than $m + 1$ sign changes in this open interval. Suppose that ϕ_{m+1} has n sign changes on (a, b), where $0 \leq n \leq m$. If $n = 0$, then $\phi_{m+1}(x)$ has constant sign; without loss of generality we will assume that $\phi_{m+1} > 0$. Then the orthogonality of ϕ_{m+1} to any polynomial of degree at most m leads to

$$0 = (\phi_{m+1}, 1) = \int_a^b \phi_{m+1}(x)w(x)dx .$$

However, the integrand on the right is positive, so its integral must also be positive, giving us a contradiction. We are left with the possibility that there are between 1 and m sign changes. Let $t_1, \ldots, t_n$ be the points where ϕ_{m+1} changes sign, and let

$$p(x) = \prod_{i=1}^{n} (x - t_i) \, .$$

Then p is a polynomial of degree $n \leq m$, so ϕ_{m+1} is orthogonal to $p(x)$. Also, $\phi_{m+1}(x)p(x)$ has constant sign, which without loss of generality we may assume is positive. It follows that

$$0 = (\phi_{m+1}, p) = \int_a^b \phi_{m+1}(x)p(x)w(x)dx \, .$$

Again, the integrand on the right is positive, so its integral is positive, giving us another contradiction. We conclude that ϕ_{m+1} must have $m + 1$ sign changes in (a, b), which in turn implies that it has $m + 1$ distinct zeros in (a, b).

To prove the second conclusion, let p be a polynomial of degree $2m + 1$. Define the polynomials q and r of at most degree m by

$$p(x) = q(x)\phi_{m+1}(x) + r(x) \, .$$

In other words, q is the quotient in dividing p by ϕ_{m+1}, and r is the remainder. Since ϕ_{m+1} is orthogonal to q,

$$\int_a^b p(x)w(x)dx = \int_a^b q(x)\phi_{m+1}(x)w(x)dx + \int_a^b r(x)w(x)dx$$

$$= \int_a^b r(x)w(x)dx \, .$$

Since r is a polynomial of degree at most m, Lagrange polynomial interpolation (see Sect. 1.2.3) at the zeros $x_0, \ldots, x_m$ of ϕ_{m+1} shows us that

$$r(x) = \sum_{j=0}^{m} \varrho_j \lambda_{j,m}(x) \, ,$$

where for $0 \leq j \leq m$

$$\varrho_j = r(x_j) = q(x_j)\phi_{m+1}(x_j) + r(x_j) = p(x_j) \, .$$

Then

$$\int_a^b p(x)w(x)dx = \int_a^b r(x)w(x)dx = \sum_{j=0}^{m} \varrho_j \int_a^b \lambda_{j,m}(x)w(x)dx = \sum_{j=0}^{m} p(x_j)\omega_j \, .$$

Next, we will show that the weights ω_j in Gaussian quadrature are all positive. This fact is important to the numerical stability of Gaussian quadrature.

Lemma 2.3.5 *Suppose that the function w is positive on the closed interval $[a, b]$. Let the sequence of polynomials $\{p_j\}_{j=0}^{\infty}$ be orthogonal with respect to the inner product*

$$(f, g) = \int_a^b f(x)g(x)w(x)dx ,$$

and assume that these polynomials satisfy the three-term recurrence

$$p_{-1}(x) = 0 , \; p_0(x) = c_0 , \; p_k(x) = (c_k x - a_k)p_{k-1}(x) - b_k p_{k-2}(x) \text{ for } k \geq 1 .$$

Let $x_0, \ldots, x_m$ be the zeros of p_{m+1}, and let the Lagrange interpolation polynomial at these points be

$$\lambda_{j,m}(x) = \prod_{\substack{0 \leq i \leq m \\ i \neq j}} (x - x_i) / \prod_{\substack{0 \leq i \leq m \\ i \neq j}} (x_j - x_i) .$$

Then for $0 \leq j \leq m$ the Gaussian quadrature weights

$$\omega_j = \int_a^b \lambda_{j,m}(x)w(x)dx$$

are all positive. Furthermore,

$$\omega_j = c_{m+1} \frac{\int_a^b p_m(x)^2 w(x) \, dx}{p_m(x_j)p'_{m+1}(x_j)} . \tag{2.18}$$

Proof For each $0 \leq k \leq m$, $\lambda_{k,m}(x)$ is a polynomial of degree m, so $\lambda_{k,m}(x)^2$ is a polynomial of degree $2m$. Since Theorem 2.3.4 showed that Gaussian quadrature using the nodes $x_0, \ldots, x_m$ and corresponding weights $\omega_0, \ldots, \omega_m$ is exact for polynomials of degree at most $2m + 1$, we must have

$$0 < \int_a^b \lambda_{k,m}(x)^2 w(x)dx = \sum_{j=0}^m \lambda_{k,m}(x_j)^2 \omega_j = \omega_k .$$

Next, let us prove (2.18). It is easy to prove by induction that for $k \geq 0$ the leading coefficient of $p_k(x)$ is

$$\gamma_k = \prod_{i=0}^k c_i .$$

We note that for $0 \le j \le m$ and $x \ne x_j$ we have

$$\gamma_{m+1} \prod_{\substack{0 \le i \le m \\ i \ne j}} (x - x_i) = \frac{p_{m+1}(x)}{(x - x_j)} = \frac{p_{m+1}(x) - p_{m+1}(x_j)}{x - x_j} . \tag{2.19}$$

We can take the limit as $x \to x_j$ to get

$$\gamma_{m+1} \prod_{\substack{0 \le i \le m \\ i \ne j}} (x_j - x_i) = p'_{m+1}(x_j) . \tag{2.20}$$

Thus

$$\omega_j = \int_a^b \lambda_{j,m}(x) w(x) dx = \int_a^b \frac{\prod_{\substack{0 \le i \le m \\ i \ne j}} (x - x_i)}{\prod_{\substack{0 \le i \le m \\ i \ne j}} (x_j - x_i)} w(x) dx$$

$$= \frac{1}{p'_{m+1}(x_j)} \int_a^b \frac{p_{m+1}(x)}{x - x_j} w(x) \, dx . \tag{2.21}$$

Next, we note that for all $k \ge 0$ we have

$$\frac{1}{x - x_j} = \frac{1 - (x/x_j)^k}{x - x_j} + \left(\frac{x}{x_j}\right)^k \frac{1}{x - x_j} = -\frac{1}{x_j} \sum_{i=0}^{k-1} \left(\frac{x}{x_j}\right)^i + \left(\frac{x}{x_j}\right)^k \frac{1}{x - x_j} .$$

Since the geometric series on the right is a polynomial of degree $k - 1$ and therefore orthogonal to p_{m+1} for $0 \le k \le m + 1$, we get

$$x_j^k \int_a^b \frac{p_{m+1}(x)}{x - x_j} w(x) \, dx$$

$$= -\sum_{i=0}^{k-1} x_j^{k-i-1} \int_a^b x^i p_{m+1}(x) w(x) \, dx + \int_a^b \frac{x^k p_{m+1}(x)}{x - x_j} w(x) \, dx$$

$$= \int_a^b \frac{x^k p_{m+1}(x)}{x - x_j} w(x) \, dx .$$

We can take linear combinations of this equation to see that for any polynomial q of degree at most $m + 1$ we have

$$q(x_j) \int_a^b \frac{p_{m+1}(x)}{x - x_j} w(x) \, dx = \int_a^b \frac{q(x) p_{m+1}(x)}{x - x_j} w(x) \, dx . \tag{2.22}$$

Since $p_{m+1}(x_j) = 0$, it follows that

$$\frac{p_{m+1}(x)}{x - x_j} = \frac{p_{m+1}(x) - p_{m+1}(x_j)}{x - x_j} = \gamma_{m+1} \prod_{\substack{0 \le i \le m \\ i \ne j}} (x - x_i) = \frac{\gamma_{m+1}}{\gamma_m} p_m(x) + r(x)$$

(2.23)

where r is a polynomial of degree at most $m - 1$. We recall Eq. (2.19)

$$\omega_j = \frac{1}{p'_{m+1}(x_j)} \int_a^b \frac{p_{m+1}(x)}{x - x_j} w(x) \, dx$$

then Eq. (2.22) with $q = p$ produces

$$= \frac{1}{p_m(x_j) p'_{m+1}(x_j)} \int_a^b \frac{p_m(x) p_{m+1}(x)}{x - x_j} w(x) \, dx$$

then Eq. (2.23) gives us

$$= \frac{1}{p_m(x_j) p'_{m+1}(x_j)} \int_a^b p_m(x) \left[\frac{\gamma_{m+1}}{\gamma_m} p_m(x) + r(x) \right] w(x) \, dx$$

$$= \frac{c_{m+1}}{p_m(x_j) p'_{m+1}(x_j)} \int_a^b p_m(x)^2 w(x) \, dx \; .$$

This proves (2.18).

Next, we will find an expression for the error in Gaussian quadrature.

Theorem 2.3.5 (Gaussian Quadrature Error) *Suppose that the function w is positive on the closed interval $[a, b]$, and that the set of polynomials $\{\phi_j\}_{j=0}^{\infty}$ is orthogonal with respect to the inner product*

$$(f, g) = \int_a^b f(x) g(x) w(x) dx \; .$$

Assume that the orthogonal polynomials ϕ_j are normalized to have coefficient equal to one for the highest-order term. If f is $2m + 2$ times continuously differentiable in $[a, b]$, then there exists $\xi \in (a, b)$ so that the error in Gaussian quadrature satisfies

$$\int_a^b f(x) w(x) dx - \sum_{j=0}^m f(x_j) \omega_j = \frac{f^{(2m+2)}(\xi)}{(2m + 2)!} \int_a^b \phi_{m+1}(x)^2 w(x) dx \; .$$

Proof Let p be the polynomial of degree at most $2m + 1$ that is determined by Hermite interpolation (see Sect. 1.2.4) to f at the zeros $x_0, \ldots, x_m$ of ϕ_{m+1}. In other words, for all $0 \leq j \leq m$,

$$p(x_j) = f(x_j) \text{ and } p'(x_j) = f'(x_j) \ .$$

Since Theorem 2.3.4 showed that Gaussian quadrature is exact for p, we have

$$\int_a^b p(x)w(x)dx = \sum_{j=0}^m p(x_j)\omega_j = \sum_{j=0}^m f(x_j)\omega_j \ .$$

By the formula (1.2) for the error in polynomial interpolation, for any $a < x < b$ there exists $\xi_x \in (a, b)$ so that

$$f(x) - p(x) = \frac{1}{(2m+2)!}f^{(2m+2)}(\xi_x)\prod_{j=0}^m (x - x_j)^2 = \frac{1}{(2m+2)!}f^{(2m+2)}(\xi_x)\phi_{m+1}(x)^2 \ .$$

Since ξ_x depends continuously on x, the mean-value theorem for integrals implies that

$$\int_a^b [f(x) - p(x)]w(x)dx = \int_a^b \frac{1}{(2m+2)!}f^{(2m+2)}(\xi_x)\phi_{m+1}(x)^2 w(x)dx$$

$$= \frac{1}{(2m+2)!}f^{(2m+2)}(\xi)\int_a^b \phi_{m+1}(x)^2 w(x)dx \ .$$

Let us pause to consider the implications of Gaussian quadrature. Theorem 2.3.5 shows that we can achieve order $2n + 2$ with just $n + 1$ Gaussian quadrature nodes. To achieve the same order with Romberg integration, we would use 2^n nodes.

In order to make Gaussian quadrature practical, we need to find efficient ways to compute the nodes and weights. The next lemma, due to Golub and Welsch [88], will give us a simple way to use linear algebra to find the nodes, since they are zeros of orthogonal polynomials.

Theorem 2.3.6 (Golub-Welsch)   *Suppose that the sequence of polynomials $\{\phi_k\}$ is generated by the three-term recurrence*

$$\phi_0(x) = 1$$

$$\phi_1(x) = x - \alpha_1$$

$$\text{for } 2 \leq k \ , \ \phi_k(x) = (x - \alpha_k)\phi_{k-1}(x) - \beta_k\phi_{k-2}(x) \ .$$

Then the zeros of ϕ_n are the eigenvalues of the $n \times n$ tridiagonal matrix

$$\mathbf{T} = \begin{bmatrix} \alpha_1 & 1 & & & \\ \beta_2 & \alpha_2 & & & \\ & \ddots & \ddots & \ddots & \\ & & \ddots & \alpha_{n-1} & 1 \\ & & & \beta_n & \alpha_n \end{bmatrix} .$$

If we also have $\beta_k > 0$ for all $k \geq 2$, then the zeros of ϕ_n are also eigenvalues of the corresponding symmetric tridiagonal matrix

$$\mathbf{S} = \begin{bmatrix} \alpha_1 & \sqrt{\beta_2} & & & \\ \sqrt{\beta_2} & \alpha_2 & \ddots & & \\ & \ddots & \ddots & \ddots & \\ & & \ddots & \alpha_{n-1} & \sqrt{\beta_n} \\ & & & \sqrt{\beta_n} & \alpha_n \end{bmatrix} ,$$

*which is similar to $\mathbf{T}$, and is called the **Jacobi matrix**. These eigenvalues are the nodes for the corresponding Gaussian quadrature rule. For $1 \leq j \leq n$, if $\mathbf{q}_j$ is the unit eigenvector of $\mathbf{S}$ corresponding to eigenvalue x_j, then the Gaussian quadrature weight for node x_j is*

$$\omega_j = \left(\mathbf{e}_0 \cdot \mathbf{q}_j \right)^2 \int_a^b w(x) \, dx .$$

Formula (2.3.6) is generally preferable to Eq. (2.18).

Proof If the n-vector $\mathbf{f}$ has entries $f_0, \ldots, f_{n-1}$, we can easily see that

$$\mathbf{f}\lambda - \mathbf{T}\mathbf{f} = \begin{bmatrix} (\lambda - \alpha_0)f_0 - f_1 \\ -\beta_2 f_0 + (\lambda - \alpha_1)f_1 - f_2 \\ \vdots \\ -\beta_n f_{n-2} + (\lambda - \alpha_n)f_{n-1} \end{bmatrix} .$$

If this vector is zero, then we must have

$$f_1 = (\lambda - \alpha_0)f_0$$
$$\text{for } 2 \leq k < n \, , \; f_k = (\lambda - \alpha_k)f_{k-1} - \beta_k f_{k-2}$$
$$f_n \equiv (\lambda - \alpha_n)f_{n-1} - \beta_n f_{n-2} = 0 \, .$$

If we take $f_0 = 1$, then we see that these equations are the same as the three-term recurrence for the polynomials ϕ_k, and with the last equation requiring λ to be a zero of ϕ_n.

Next, we note that $\mathbf{S} = \mathbf{D}^{-1}\mathbf{T}\mathbf{D}$, where the $n \times n$ diagonal matrix $\mathbf{D}$ has k-th diagonal entry

$$\mathbf{D}_{kk} = \sqrt{\prod_{j=2}^{k} \beta_j}$$

for $1 \leq k \leq n$. (Here the empty product for $k = 1$ is assumed to be one.) Also, note that the diagonal entries of $\mathbf{D}$ are real because Eq. (1.65) showed that the coefficients β_k are all positive. This equation also shows that

$$\mathbf{D}_{kk}^2 = \beta_2 \dots \beta_k = \frac{\int_a^b \phi_1(x)^2 w(x)\,dx}{\int_a^b \phi_0(x)^2 w(x)\,dx} \dots \frac{\int_a^b \phi_{k-1}(x)^2 w(x)\,dx}{\int_a^b \phi_{k-2}(x)^2 w(x)\,dx} = \frac{\int_a^b \phi_{k-1}(x)^2 w(x)\,dx}{\int_a^b \phi_0(x)^2 w(x)\,dx}.$$

We have shown that

$$\mathbf{T}\mathbf{f}_j = \mathbf{f}_j x_j$$

where the eigenvector $\mathbf{f}_j$ has entries

$$\mathbf{f}_j = \begin{bmatrix} \phi_0(x_j) \\ \vdots \\ \phi_{n-1}(x_j) \end{bmatrix}$$

and eigenvalue x_j. It follows that

$$\mathbf{p}_j \equiv \mathbf{D}^{-1}\mathbf{f}_j \frac{1}{\sqrt{\int_a^b \phi_0(x)^2 w(x)\,dx}} = \begin{bmatrix} \phi_0(x_j)/\sqrt{\int_a^b \phi_0(x)^2 w(x)\,dx} \\ \vdots \\ \phi_{n-1}(x_j)/\sqrt{\int_a^b \phi_{n-1}(x)^2 w(x)\,dx} \end{bmatrix}$$

is an eigenvector of $\mathbf{S} = \mathbf{D}^{-1}\mathbf{T}\mathbf{D}$ with eigenvalue x_j.

Now we recall the Christoffel-Darboux identity (1.70) in the form

$$\sum_{k=0}^{n-1} \frac{\phi_k(x)\phi_k(y)}{\int_a^b \phi_k(z)^2 w(z)\,dz} = \frac{\phi_n(x)\phi_{n-1}(y) - \phi_{n-1}(x)\phi_n(y)}{(x-y)\int_a^b \phi_{n-1}(z)^2 w(z)\,dz}.$$

We can take the limit as $y \to x$ to get

$$
\begin{aligned}
\sum_{k=0}^{n-1} \frac{\phi_k(x)^2}{\int_a^b \phi_k(z)^2 w(z) \, dz} &= \frac{1}{\int_a^b \phi_{n-1}(z)^2 w(z) \, dz} [\phi_n \phi_{n-1}]'(x) \\
&= \frac{\phi_n'(x)\phi_n(x) + \phi_n(x)\phi_{n-1}'(x)}{\int_a^b \phi_{n-1}(z)^2 w(z) \, dz} .
\end{aligned}
$$

Next, we can let $x = x_j$ to obtain

$$
\|\mathbf{p}_j\|_2^2 = \sum_{k=0}^{n-1} \frac{\phi_k(x_j)^2}{\int_a^b \phi_k(z)^2 w(z) \, dz} = \frac{\phi_n'(x_j)\phi_{n-1}(x_j)}{\int_a^b \phi_{n-1}(z)^2 w(z) \, dz} .
$$

then we use Eq. (2.18) for the weight in Gaussian quadrature to get

$$
= \frac{1}{\omega_j} .
$$

Let

$$
\mathbf{q}_j = \mathbf{p}_j / \|\mathbf{p}_j\|_2
$$

be the unit eigenvector of $\mathbf{S}$ for eigenvalue x_j. Then the square of the first entry of this vector is

$$
(\mathbf{e}_0 \cdot \mathbf{q}_j)^2 = \frac{(\mathbf{e}_0 \cdot \mathbf{p}_j)^2}{\|\mathbf{p}_j\|_2^2} = \frac{\phi_0(x_j)^2 / \int_a^b \phi_0(x)^2 w(x) \, dx}{1/\omega_j} = \frac{\omega_j}{\int_a^b w(x) \, dx} .
$$

We conclude that

$$
\omega_j = (\mathbf{e}_0 \cdot \mathbf{q}_j)^2 \int_a^b w(x) \, dx .
$$

As a result of this lemma, we see that we can easily compute all of the zeros of a degree n orthogonal polynomial by computing all of the eigenvalues of a symmetric tridiagonal matrix. For more information about computing the eigenvalues of a symmetric tridiagonal matrix, see Sects. 1.3.7, 1.3.8 or 1.3.9 in Chap. 1 of Volume II.

Alternatively, we can find the quadrature weights for low-order Gaussian quadrature schemes by examining the requirements that the quadrature rules be exact for polynomials of appropriate degrees. This approach is illustrated by the following example.

Example 2.3.2 Suppose that we want to compute $\int_{-1}^{1} f(x)dx$. The orthogonal polynomials for this interval and weight function are the Legendre polynomials

$$p_k(x) = \frac{2k-1}{k}xp_{k-1}(x) - \frac{k-1}{k}p_{k-2}(x) \text{ where } \|p_k\|^2 = \frac{2}{2k+1}.$$

We can use the three-term recurrence to generate the polynomials, and easily find their zeros for degree at most three:

$$p_0(x) \equiv 1,$$

$$p_1(x) = x \Longrightarrow x_0^{(1)} = 0,$$

$$p_2(x) = \frac{3}{2}x^2 - \frac{1}{2} \Longrightarrow x_0^{(2)} = -\sqrt{\frac{1}{3}}, \ x_1^{(2)} = \sqrt{\frac{1}{3}},$$

$$p_3(x) = \frac{5}{2}x^3 - \frac{3}{2}x \Longrightarrow x_0^{(3)} = -\sqrt{\frac{3}{5}}, \ x_1^{(3)} = 0, \ x_2^{(3)} = \sqrt{\frac{3}{5}}.$$

Now, let us determine the quadrature weights for several cases of Gaussian quadrature.

$m = 0$: We want $\int_{-1}^{1} p(x)dx = p(0)\omega_0$ to be exact for all polynomials of degree at most one. The choice $p(x) \equiv 1$ shows that we must have $\omega_0 = 2$.

$m = 1$: We want

$$\int_{-1}^{1} p(x)dx = p(-\sqrt{1/3})\omega_0 + p\left(\sqrt{1/3}\right)\omega_1$$

to be exact for all polynomials of degree at most three. If we choose $p(x) = x$, we see that $\omega_0 = \omega_1$; if we choose $p(x) = 1$ we see that $\omega_0 = 1 = \omega_1$.

$m = 2$: We want

$$\int_{-1}^{1} p(x)dx = p\left(-\sqrt{\frac{3}{5}}\right)\omega_0 + p(0)\omega_1 + p\left(\sqrt{\frac{3}{5}}\right)\omega_2$$

to be exact for all polynomials p of degree at most 5. If we choose $p(x) = x$, we see that $\omega_0 = \omega_2$; if we choose $p(x) = x^2$ we see that $\omega_0 = 5/9 = \omega_2$. If we choose $p(x) = 1$, we see that $\omega_1 = 8/9$;

In summary, our first three Gaussian quadrature formulas are

$$\int_{-1}^{1} f(x)dx \approx \begin{cases} 2f(0), \text{ exact for degree 1} \\ f(-\sqrt{1/3}) + f(\sqrt{1/3}), \text{ exact for degree 3} \\ \left[5f(-\sqrt{3/5}) + 8f(0) + 5f(\sqrt{3/5})\right]/9, \text{ exact for degree 5} \end{cases}.$$

Example 2.3.3 Suppose that we want to approximate $\int_{-h}^{h} f(y)dy$. We first perform a change of variables $y = xh$. Then we could approximate

$$\int_{-h}^{h} f(y)dy = h \int_{-1}^{1} f(hx)dx \approx h \left[f\left(-h/\sqrt{3}\right) + f\left(h/\sqrt{3}\right) \right] .$$

Example 2.3.4 Suppose that we want to compute $\int_{a}^{b} f(z)dz$. Let

$$c = \frac{b+a}{2} \text{ and } h = \frac{b-a}{2} .$$

We can perform the change of variables

$$\xi = -1 + \frac{x-a}{h} \iff x = c + h\xi .$$

Then we could approximate

$$\int_{a}^{b} f(x) \, dx = \int_{-1}^{1} f(c + h\xi) \, h \, d\xi$$

$$\approx h \left\{ \frac{5}{9}f\left(c - h\sqrt{\frac{3}{5}}\right) + \frac{8}{9}f(c) \frac{5}{9}f\left(c + h\sqrt{\frac{3}{5}}\right) \right\} .$$

The files `Polynomial.H` and `Polynomial.C` from Cambridge University Press implement various C^{++} classes for polynomials, including Legendre polynomials. In particular, the procedures `LegendrePolynomial::values` and `LegendrePolynomial::derivatives` will compute all of the Legendre polynomial values and derivatives, respectively, up to some given order at a specified point. The files `Quadrature.H` and `Quadrature.C`, also from Cambridge University Press implement various C^{++} classes for quadrature rules, including Gaussian quadrature. The weights and nodes for a Gaussian quadrature rule of arbitrary order can are determined in the `GaussianQuadrature` constructor. This procedure creates the appropriate symmetric tridiagonal matrix for the Legendre polynomials, and calls LAPACK routine `dsterf` to find the eigenvalues, which are the zeros of the appropriate Legendre polynomial.

Exercise 2.3.13 The gamma function is defined by

$$\Gamma(x) \equiv \int_{0}^{\infty} t^{x-1}e^{-t}dt , \ x > 0 .$$

Use the Laguerre polynomials to determine Gaussian quadrature rule approximations to $\Gamma(x)$ that are exact for $\mathscr{P}^{2m+1}$, $0 \leq m \leq 2$. (Note: do not use Legendre polynomials to develop the quadrature rules.)

Exercise 2.3.14 Gaussian quadrature rules can be determined by applying symmetry conditions to the solution of polynomial equations. For example, suppose we want to find a quadrature rule that is exact for cubics on the interval $[-1, 1]$. Then for $0 \leq j \leq 3$ we require

$$\frac{1 - (-1)^{j+1}}{j + 1} = \int_{-1}^{1} x^j \, \mathrm{d}x = \xi_0^j \omega_0 + \xi_1^j \omega_1 .$$

In the beginning, this looks possible because there are four values of j and four unknowns ξ_0, ξ_1, ω_0 and ω_1. However, the interval $[-1, 1]$ is symmetric about the origin, and the quadrature rule should be symmetric as well:

$$\xi_0 = -\xi_1 \text{ and } \omega_0 = \omega_1 .$$

This leaves us with only two free parameters. However, when j is odd, the integral is zero and the symmetry conditions on the quadrature rule produce a zero value. We are left with only two constraints, for $j = 0$ and $j = 2$, to determine values for $\xi_0 = -\xi_1$ and $\omega_0 = \omega_1$:

$$\frac{2}{1} = (-\xi)^0 \omega + \xi^0 \omega = 2\omega \text{ and}$$

$$\frac{2}{3} = (-\xi)^2 \omega + \xi^2 \omega = 2\xi^2 \omega .$$

The first equation implies that $\omega_0 = \omega_1 = 1$ and the second equation implies that $-\xi_0 = \xi_1 = \sqrt{1/3}$. Use this approach to find the Gaussian quadrature rule that is exact for polynomials of degree 5 on $[-1, 1]$.

2.3.11   Lobatto Quadrature

There are important quadratures in which we want to use function values at the endpoints of the interval of integration. These function values may have already been evaluated for a quadrature in neighboring intervals, or may already known (possibly as initial or boundary conditions for a differential equation, as in Sect. 3.5.2). At any rate, the optimal placement of additional quadrature nodes gives rise to a slightly different quadrature rule, known as **Gauss-Lobatto quadrature** or **Radau quadrature**. The following theorem describes the process.

Theorem 2.3.7 (Lobatto Quadrature)   *Let* $\{p_j\}_{j=0}^{\infty}$ *be the sequence of Legendre polynomials, which are defined by the three-term recurrence (1.67) and are orthogonal in the inner product*

$$(f , g) = \int_{-1}^{1} f(x)g(x) \, \mathrm{d}x .$$

For any integer $m \geq 2$ the zeros $x_1, \ldots, x_{m-1}$ of $p'_m(x)$ are real, distinct and lie in the open interval $(-1, 1)$. Let $x_0 = -1$, $x_m = 1$ and the Lagrange interpolation polynomial at the nodes $x_0, \ldots, x_m$ be

$$\lambda_{j,m}(x) = \prod_{\substack{0 \leq i \leq m \\ i \neq j}} (x - x_i) / \prod_{\substack{0 \leq i \leq m \\ i \neq j}} (x_j - x_i) .$$

Define the scalars ω_j for $0 \leq j \leq m$ by

$$\omega_j = \int_{-1}^{1} \lambda_{j,m}(x) \, dx .$$

Then the equation

$$\int_{-1}^{1} p(x) \, dx = \sum_{j=0}^{m} p(x_j) \omega_j$$

is satisfied for all polynomials p of degree at most $2m - 1$.

Proof Recall that the proof of Theorem 2.3.4 showed that the zeros of p_m are real, distinct and lie in $(-1, 1)$. The zeros of p'_m must lie between the zeros of p_m, and therefore also lie in the interior of $(-1, 1)$.

Lagrange interpolation (see Sect. 1.2.3) shows that any polynomial p of degree at most m can be written

$$p(x) = \sum_{j=0}^{m} p(x_j) \lambda_{j,m}(x) .$$

The definition of the coefficients ω_j now shows that

$$\int_{-1}^{1} p(x) \, dx = \sum_{j=0}^{m} p(x_j) \int_{-1}^{1} \lambda_{j,m}(x) \, dx = \sum_{j=0}^{m} p(x_j) \omega_j$$

is exact for any polynomial p of degree at most m.

Note that since $m \geq 1$, p_m is orthogonal to a constant, and since the choice of the ω_j makes the quadrature rule exact for polynomials of degree at most m,

$$0 = \int_{-1}^{1} p_m(x) \, dx = -[(1 - x)p_m(x)]_{-1}^{1} + \int_{-1}^{1} (1 - x)p'_m(x) \, dx$$

$$= 2p_m(-1) + \int_{-1}^{1} (1 - x)p'_m(x) \, dx = 2p_m(-1) + \sum_{j=0}^{m} (1 - x_j)p'_m(x_j)\omega_j$$

$$= 2p_m(-1) + 2p'_m(-1)\omega_0 .$$

Also,

$$p_m(1) - p_m(-1) = \int_{-1}^{1} p'_m(x) \, dx = \sum_{j=0}^{m} p'_m(x_j)\omega_j = p'_m(-1)\omega_0 + p'_m(1)\omega_m .$$

Thus

$$0 = p_m(-1) + p'_m(-1)\omega_0 = p_m(1) - p'_m(1)\omega_m .$$

If p is a polynomial of degree at most $2m - 1$, then we can write

$$p = qp'_m + r$$

where q and r are polynomials of degree at most m. Then

$$\int_{-1}^{1} p(x) \, dx = \int_{-1}^{1} q(x)p'_m(x) \, dx + \int_{-1}^{1} r(x) \, dx$$

$$= q(1)p_m(1) - q(-1)p_m(-1) - \int_{-1}^{1} q'(x)p_m(x) \, dx + \sum_{j=1}^{m} r(x_j)\omega_j$$

$$= q(1)p_m(1) - q(-1)p_m(-1) + \sum_{j=0}^{m} \left[q(x_j)p'(x_j) + r(x_j) \right] \omega_j$$

$$- q(-1)p'_m(-1)\omega_0 - q(1)p'_m(1)\omega_m$$

$$= \sum_{j=0}^{m} p(x_j)\omega_j + q(1) \left[p_m(1) - p'_m(1)\omega_m \right] - q(-1) \left[p_m(-1) + p'_m(-1)\omega_0 \right]$$

$$= \sum_{j=0}^{m} p(x_j)\omega_j .$$

Note that Gauss-Lobatto quadrature is exact for polynomials of degree two less than the corresponding Gaussian quadrature rule with the same number of nodes.

Next, we will show that the weights in Lobatto quadrature are all positive.

Lemma 2.3.6 *Let $\{p_j\}_{j=0}^{\infty}$ be the sequence of Legendre polynomials, which are defined by the three-term recurrence (1.67) and are orthogonal in the inner product*

$$(f , g) = \int_{-1}^{1} f(x)g(x) \, dx .$$

Let $x_0 = -1$, $x_m = 1$ and let $x_1, \ldots, x_{m-1}$ be the zeros of ϕ'_m. Define the Lagrange interpolation polynomial at these nodes to be

$$\lambda_{j,m}(x) = \prod_{\substack{0 \leq i \leq m \\ i \neq j}} (x - x_i) / \prod_{\substack{0 \leq i \leq m \\ i \neq j}} (x_j - x_i) \,.$$

Then the Lobatto quadrature weights

$$\omega_j = \int_{-1}^{1} \lambda_{j,m}(x) dx$$

are all positive. In fact, we have

$$\omega_0 = \omega_m = \frac{2}{m(m+1)} \quad and \tag{2.24}$$

$$\omega_k = \frac{2m}{(1 - x_k^2) p''_m(x_k) p'_{m-1}(x_k)} = \frac{2}{m(m+1) p_m(x_k)^2} \quad for \ 1 \leq k < m \,. \tag{2.25}$$

Proof Let us define

$$\ell_0(x) = \left[\prod_{i=1}^{m-1} \frac{x - x_i}{-1 - x_i} \right]^2 \frac{1 - x}{2} \,,$$

$$\ell_k(x) = \frac{x + 1}{x_k + 1} \left[\prod_{\substack{i=1 \\ i \neq k}}^{m-1} \frac{x - x_i}{x_k - x_i} \right]^2 \frac{1 - x}{1 - x_k} \quad for \ 1 \leq k < m \,, \ and$$

$$\ell_m(x) = \frac{x + 1}{2} \left[\prod_{i=1}^{m-1} \frac{x - x_i}{1 - x_i} \right]^2 \,.$$

Since Theorem 2.3.7 showed that the Lobatto nodes are contained in the open interval $(-1, 1)$, it follows that for each $0 \leq k \leq m$ the polynomial $\ell_k(x)$ is nonnegative and has degree at most $2m - 1$. Also, for each $0 \leq k \leq m$ the definition of ℓ_k shows that for $0 \leq j \leq m$ we have $\ell_k(x_j) = \delta_{jk}$. Since Theorem 2.3.7 showed that Lobatto quadrature using the nodes $x_0, \ldots, x_m$ and corresponding weights $\omega_0, \ldots, \omega_m$ is exact for polynomials of degree at most $2m - 1$, for all $0 \leq k \leq m$ we have

$$0 < \int_{-1}^{1} \ell_k(x)^2 \, dx = \sum_{j=0}^{m} \ell_k(x_j)^2 \omega_j = \omega_k \,.$$

For $m \geq 1$, let

$$p_m'(x) = \gamma_m \prod_{j=1}^{m-1} (x - x_j)$$

be the derivative of the m-th Legendre polynomial. Note that for $1 \leq k < m$

$$\gamma_m \prod_{\substack{1 \leq j < m \\ j \neq k}} (x - x_j) = \frac{p_m'(x)}{x - x_k} = \frac{p_m'(x) - p_m'(x_k)}{x - x_k} ,$$

so we can take the limit as $x \to x_k$ to get

$$\gamma_m \prod_{\substack{1 \leq j < m \\ j \neq k}} (x_k - x_j) = p_m''(x_k) .$$

Let $\lambda_{j,m}(x)$ be the Lagrange interpolation polynomial at node j for $0 \leq j \leq m$. For $m \geq 1$ we see that

$$\omega_0 = \int_{-1}^{1} \lambda_{0,m}(x) \, dx = \frac{1}{2p_m'(-1)} \int_{-1}^{1} (1 - x)p_m'(x) \, dx$$

then integration by parts produces

$$= \frac{1}{2p_m'(-1)} \left\{ (1 - x)p_m(x)|_{-1}^{1} + \int_{-1}^{1} p_m(x) \, dx \right\} = -\frac{p_m(-1)}{p_m'(-1)}$$

then Theorem 1.7.5 implies that

$$= -\frac{(-1)^m}{(-1)^m m(m + 1)/2} = \frac{2}{m(m + 1)} .$$

The computation for ω_m is similar.

If $1 \leq k < m$ then the Lobatto quadrature weight is

$$\omega_k = \int_{-1}^{1} \lambda_{k,m}(x) \, dx = \frac{1}{(1 - x_k^2)p_m''(x_k)} \int_{-1}^{1} \frac{(1 - x^2)p_m'(x)}{x - x_k} \, dx . \qquad (2.26)$$

Note that for $0 \leq \ell$ we can write

$$\frac{1}{x - x_k} = -\frac{1}{x_k} \left\{ \frac{1 - (x/x_k)^\ell}{1 - x/x_k} + \frac{(x/x_k)^\ell}{1 - x/x_k} \right\}$$

$$= -\frac{1}{x_k} \sum_{j=0}^{\ell-1} \left(\frac{x}{x_k} \right)^j - \frac{1}{x_k} \frac{(x/x_k)^\ell}{1 - x/x_k} .$$

Thus if $\ell < m - 1$ we have

$$\int_{-1}^{1} \frac{(1-x^2)p_m'(x)}{x-x_k}\, dx$$

$$= -\sum_{j=0}^{\ell-1} \frac{1}{x_k^{j+1}} \int_{-1}^{1} (1-x^2)x^j p_m'(x)\, dx + \frac{1}{x_k^{\ell}} \int_{-1}^{1} \frac{(1-x^2)x^{\ell}p_m'(x)}{x-x_k}\, dx$$

so integration by parts leads to

$$= -\sum_{j=0}^{\ell-1} \frac{1}{x_k^{j+1}} \left[(1-x^2)x^j p_m(x)\right]_{-1}^{1}$$

$$+ \sum_{j=0}^{\ell-1} \frac{1}{x_k^{j+1}} \int_{-1}^{1} \frac{d}{dx}\left[(1-x^2)x^j\right]p_m(x)\, dx$$

$$+ \frac{1}{x_k^{\ell}} \int_{-1}^{1} \frac{(1-x^2)x^{\ell}p_m'(x)}{x-x_k}\, dx$$

and since $d/dx[(1-x^2)x^j]$ is a polynomial of degree at most $\ell + 1 < m$

$$= \frac{1}{x_k^{\ell}} \int_{-1}^{1} \frac{(1-x^2)x^{\ell}p_m'(x)}{x-x_k}\, dx \ .$$

We can rewrite this equation as

$$x_k^{\ell} \int_{-1}^{1} \frac{(1-x^2)p_m'(x)}{x-x_k}\, dx = \int_{-1}^{1} \frac{(1-x^2)x^{\ell}p_m'(x)}{x-x_k}\, dx$$

for all $\ell < m - 1$. Then we can take linear combinations of this equation to see that

$$q(x_k) \int_{-1}^{1} \frac{(1-x^2)p_m'(x)}{x-x_k}\, dx = \int_{-1}^{1} \frac{(1-x^2)q(x)p_m'(x)}{x-x_k}\, dx \qquad (2.27)$$

for all polynomials of degree at most $m - 2$.

If γ_m is the leading coefficient of $p_m(x)$, then the three-term recurrence (1.67) for the Legendre polynomials shows that

$$\frac{\gamma_m}{\gamma_{m-1}} = \frac{2m-1}{m} \ .$$

Since $p'_m(x)/(x - x_k)$ is a polynomial of degree $m - 2$ with leading coefficient $m\gamma_m$, we can write

$$\frac{p'_m(x)}{x - x_k} = \frac{m\gamma_m}{(m-1)\gamma_{m-1}} p'_{m-1}(x) + \text{ lower order terms }.$$

We can take $q = p'_{m-1}$ in Eq. (2.27) to get

$$\int_{-1}^{1} \frac{p'_m(x)(1 - x^2)}{x - x_k} \, dx = \frac{1}{p'_{m-1}(x_k)} \int_{-1}^{1} \frac{p'_{m-1}(x)p'_m(x)(1 - x^2)}{x - x_k} \, dx$$

$$= \frac{1}{p'_{m-1}(x_k)} \frac{m\gamma_m}{(m-1)\gamma_{m-1}} \int_{-1}^{1} p'_{m-1}(x)^2 (1 - x^2) \, dx$$

then Eq. (2.7) produces

$$= \frac{1}{p'_{m-1}(x_k)} \frac{m\gamma_m}{(m-1)\gamma_{m-1}} \frac{2(m-1)m}{2m-1} = \frac{2m}{p'_{m-1}(x_k)} .$$

Combining this result with (2.26) proves the first equation in (2.25).

The second equation in (2.25) follows from the first, once we apply the Legendre differential Eq. (1.75) to get

$$(1 - x_k^2)p''_m(x_k) = 2x_k p'_m(x_k) - m(m + 1)p_m(x_k) = -m(m + 1)p_m(x_k)$$

and Eq. (1.73e) in the form

$$p'_{m-1}(x_k) = x_k p'_m(x_k) - mp_m(x_k) = -mp_m(x_k) .$$

Next, we will provide an error estimate for Lobatto quadrature.

Theorem 2.3.8 (Lobatto Quadrature Error) *Suppose that the set of polynomials* $\{\phi_j\}_{j=0}^{\infty}$ *is orthogonal with respect to the inner product*

$$(f, g) = \int_{-1}^{1} f(x)g(x)dx .$$

Assume that the orthogonal polynomials are normalized so that for $j \geq 1$, ϕ'_j has leading coefficient equal to one. If f is $2m$ times continuously differentiable in $[-1, 1]$, then there exists $\xi \in (-1, 1)$ so that the error in Gaussian quadrature satisfies

$$\int_{-1}^{1} f(x)dx - \sum_{j=0}^{m} f(x_j)\omega_j = -\frac{f^{(2m)}(\xi)}{(2m)!} \int_{-1}^{1} \phi'_m(x)^2 \left(1 - x^2\right) \, dx .$$

Proof Let the polynomial p of degree at most $2m-1$ be determined by interpolating f at $x_0, \ldots, x_m$ and f' at $x_1, \ldots, x_{m-1}$. Since Theorem 2.3.7 showed that Lobatto quadrature is exact for p,

$$\int_{-1}^1 p(x)dx = \sum_{j=0}^m p(x_j)\omega_j = \sum_{j=0}^m f(x_j)\omega_j .$$

By the formula (1.2) for the error in polynomial interpolation, for any $-1 < x < 1$ there exists $\xi_x \in (-1, 1)$ so that

$$f(x) - p(x) = \frac{1}{(2m)!} f^{(2m)}(\xi_x)(x+1) \left[\prod_{j=1}^m (x - x_j)^2 \right] (x-1)$$

$$= -\frac{1}{(2m)!} f^{(2m)}(\xi_x)\phi_m'(x)^2(x+1)(1-x) .$$

By the mean-value theorem for integrals,

$$\int_{-1}^1 [f(x) - p(x)] \, dx = -\int_{-1}^1 \frac{1}{(2m)!} f^{(2m)}(\xi_x)\phi_m'(x)^2(x+1)(1-x) \, dx$$

$$= -\frac{1}{(2m)!} f^{(2m)}(\xi) \int_{-1}^1 \phi_m'(x)^2(x+1)(1-x) \, dx .$$

The files `Quadrature.H` and `Quadrature.C` from Cambridge University Press implement a C^{++} class for Lobatto quadrature. The weights and nodes for a Lobatto quadrature rule of arbitrary order can are determined in the class constructor. This procedure creates the appropriate symmetric tridiagonal matrix for the Legendre polynomial derivatives, and calls LAPACK routine `dsterf` to find the eigenvalues, which are the zeros of the appropriate Legendre polynomial derivative.

Exercise 2.3.15 Show that the Lobatto quadrature rule using 2 nodes is the trapezoidal rule.

Exercise 2.3.16 Show that the Lobatto quadrature rule using 3 nodes is Simpson's rule.

Exercise 2.3.17 Show that the Lobatto quadrature rule using 4 nodes is

$$\int_{-1}^1 f(x) \, dx \approx \frac{1}{6}[f(-x) + f(1)] + \frac{5}{6}\left[f\left(-\sqrt{1/5}\right) + f\left(\sqrt{1/5}\right) \right] .$$

2.3.12 Gauss-Kronrod Quadrature

Let us review some of the integration methods we have discussed so far. We have seen that Romberg integration beginning with the midpoint rule uses 2^k nodes to achieve order $2k + 2$. None of the nodes are shared by successive midpoint rule quadratures. Another nice feature of the Romberg extrapolation table is that we can estimate the error in an extrapolant $M_f^{(2k)}(h)$ by

$$I - M_f^{(2k)}(h) \approx M_f^{(2k+2)}(h) - M_f^{(2k)}(h) = \frac{M_f^{(2k)}(h) - M_f^{(2k)}(2h)}{2^{2k} - 1} .$$

On the other hand, Clenshaw-Curtis quadrature with $N + 1$ nodes is exact for polynomials of degree at most $2\lfloor N/2 \rfloor + 1$. Some nodes may be shared by quadratures with $N + 1$ and $2N + 1$ nodes. Clenshaw-Curtis quadrature achieves high order with relatively fewer nodes than Romberg integration, but does not have an easily computable error estimate. Finally, we have seen that Gaussian quadrature with m nodes has order $2m$. Almost no nodes are shared by Gaussian quadrature rules of different orders, and there are no easily computable error estimates for Gaussian quadrature.

In order to provide error estimates for Gaussian quadrature rules, Kronrod [120] developed the following idea. Given a Gaussian quadrature rule using m nodes, he showed how to choose an optimal set of $m + 1$ additional nodes in order that the new quadrature rule is exact for polynomials of degree at most $3m + 1$ (if m is even) or $3m + 2$ (if m is odd). Subsequently, Patterson [140] showed how to extend the idea to Lobatto quadrature. Subsequently, Kronrod's ideas were related to Stieltjes polynomials, whose zeros were shown to be usefully located by Szegö [168]. These ideas and their connection to numerical quadrature have been presented in a survey paper by Monegato [135], whose presentation we will summarize next.

We begin with the following existence theorem.

Theorem 2.3.9 (Stieltjes-Szegö) *If p_n is a Legendre polynomial of degree $n \geq 0$, then there is a polynomial s_{n+1} of degree $n + 1$ so that for all $0 \leq k \leq n$*

$$\int_{-1}^{1} p_n(x) s_{n+1}(x) x^k \, dx = 0 . \tag{2.28}$$

In addition, the zeros of s_{n+1} are all contained in $(-1, 1)$, and the zeros of p_n interlace the zeros of s_{n+1}. Finally, if $n + 1$ is even then s_{n+1} is an even function, and if $n + 1$ is odd then s_{n+1} is an odd function.

Proof See Monegato [135].

Next, let us see how this theorem could be useful in quadrature.

Lemma 2.3.7 *Given an integer $n \geq 0$, let p_n be the Legendre polynomial of degree n and s_{n+1} be the Stieltjes polynomial of degree $n + 1$. Suppose that $\{x_j^{GK}\}_{j=-n}^{n}$ is* *the strictly increasing set of zeros of either p_n or s_{n+1}, and suppose that the scalars ω_j^{GK} for $-n \leq j \leq n$ are such that*

$$\int_{-1}^{1} r(x) \, dx = \sum_{j=-n}^{n} r\left(x_j^{GK}\right) \omega_j^{GK}$$

for all polynomials r of degree at most 2n. Then

$$x_{-j}^{GK} = x_j^{GK} \text{ and } \omega_{-j}^{GK} = \omega_j^{GK}$$

for $0 \leq j \leq n$, and

$$\int_{-1}^{1} p(x) \, dx = \sum_{j=-n}^{n} p\left(x_j^{GK}\right) \omega_j^{GK}$$

for all polynomials p of degree at most $3n + 1$ if n is even, and $3n + 2$ if n is odd.

Proof Since both Legendre and Stieltjes polynomials are even if their order is even, and odd if their order is odd, we must have

$$x_{-j}^{GK} = -x_j^{GK}$$

for $0 \leq j \leq n$. Let

$$\lambda_{j,2n}(x) = \prod_{\substack{i=-n \\ i \neq j}}^{n} \frac{x - x_i^{GK}}{x_j^{GK} - x_i^{GK}}$$

be the Lagrange interpolation polynomial at x_j^{GK}. Then for all x we have

$$\lambda_{-j,2n}(-x) = \prod_{\substack{k=-n \\ k \neq -j}}^{n} \frac{-x - x_k^{GK}}{x_{-j}^{GK} - x_k^{GK}} = \prod_{\substack{i=n \\ i \neq j}}^{n} \frac{-x - x_{-i}^{GK}}{x_{-j}^{GK} - x_{-i}^{GK}} = \prod_{\substack{i=n \\ i \neq j}}^{n} \frac{-x + x_i^{GK}}{-x_j^{GK} + x_i^{GK}}$$

$$= \prod_{\substack{i=n \\ i \neq j}}^{n} \frac{x - x_i^{GK}}{x_j^{GK} - x_i^{GK}} = \lambda_{j,2n}(x) \ .$$

As a result, for $0 \leq j \leq n$ we have

$$\omega_{-j}^{GK} = \sum_{k=-n}^{n} \lambda_{-k,2n}(x)\omega_k^{GK} = \int_{-1}^{1} \lambda_{-k,2n}(x) \, dx = \int_{-1}^{1} \lambda_{k,2n}(-x) \, dx$$

$$= \int_{-1}^{1} \lambda_{k,2n}(x) \, dx = \omega_j^{GK} \ .$$

Let p be a polynomial of degree at most $3n + 1$. Then we can find a quotient polynomial q of degree at most n and a remainder polynomial r of degree at most $2n$ so that

$$p = p_n s_{n+1} q + r .$$

Since $p_n s_{n+1}$ is orthogonal to all polynomials of degree at most n, we see that

$$\int_{-1}^{1} p(x) \, dx = \int_{-1}^{1} p_n(x) s_{n+1}(x) q(x) \, dx + \int_{-1}^{1} r(x) \, dx .$$

Since r has degree at most $2n$, it lies in a space of degree at most $2n + 1$, and it should be possible to choose the $2n + 1$ weights ω_j^{GK} for $0 \le j \le 2n$ so that

$$\int_{-1}^{1} r(x) \, dx = \sum_{j=0}^{2n} r\left(x_j^{GK}\right) \omega_j^{GK} ,$$

where $\left\{x_j^{GK}\right\}_{j=0}^{2n}$ is the set of all zeros of $p_n s_{n+1}$. We conclude that

$$\sum_{j=0}^{2n} p\left(x_j^{GK}\right) \omega_j^{GK} = \sum_{j=0}^{2n} \left[p_n\left(x_j^{GK}\right) s_{n+1}\left(x_j^{GK}\right) + r\left(x_j^{GK}\right) \right] \omega_j^{GK}$$

$$= \sum_{j=0}^{2n} r\left(x_j^{GK}\right) \omega_j^{GK} = \int_{-1}^{1} r(x) \, dx = \int_{-1}^{1} p_n(x) s_{n+1}(x) q(x) + r(x) \, dx$$

$$= \int_{-1}^{1} p(x) \, dx .$$

If n is even, we are done. If n is odd and p is a polynomial of degree $3n + 2$, then we can write

$$p(x) = c x^{3n+2} + \ell(x)$$

where ℓ is a polynomial of degree at most $3n + 1$. Then

$$\int_{-1}^{1} p(x) \, dx = \int_{-1}^{1} c x^{3n+2} \, dx + \int_{-1}^{1} \ell(x) \, dx = \int_{-1}^{1} \ell(x) \, dx = \sum_{j=-n}^{n} \ell\left(x_j^{GK}\right) \omega_j^{GK}$$

$$= c \sum_{j=0}^{n} \left[\left(-x_j^{GK}\right)^{3n+2} \right) + \left(x_j^{GK}\right)^{3n+2} \right] \omega_j^{GK} + \sum_{j=-n}^{n} \ell\left(x_j^{GK}\right) \omega_j^{GK}$$

$$= \sum_{j=-n}^{n} p\left(x_j^{GK}\right) \omega_j^{GK} .$$

In order to compute the nodes and weights for **Gauss-Kronrod quadrature** rules, we will present an approach due to Calvetti et al. [31]. Their ideas depend on the following observation regarding Jacobi matrices.

Theorem 2.3.10 *Given a set $\{x_j\}_{j=1}^n$ of real quadrature nodes and a set $\{\omega_j\}_{j=0}^n$ of positive quadrature weights such that*

$$\sum_{j=1}^n \omega_j = 1 \, ,$$

there is an $n \times n$ real symmetric tridiagonal Jacobi matrix $\mathbf{T}$ and a real orthogonal matrix $\mathbf{W}$ so that

$$\mathbf{TW} = \mathbf{W} \, diag(x_1, \ldots x_n) \text{ and } \mathbf{W}^\top \mathbf{e}_1 = \begin{bmatrix} \sqrt{\omega_1} \\ \vdots \\ \sqrt{\omega_n} \end{bmatrix} .$$

Conversely, if $\mathbf{T}$ is a real symmetric tridiagonal matrix, then there is a quadrature rule with real quadrature nodes given by the eigenvalues of $\mathbf{T}$ and a set $\{\omega_j\}_{j=0}^n$ of positive quadrature weights given by the squares of the entries of the first row of the orthogonal matrix of eigenvectors for $\mathbf{T}$; furthermore, if the sub-diagonal entries of $\mathbf{T}$ are all nonzero, then the quadrature nodes are distinct.

Proof Let the unit vector $\mathbf{w}_1$ have entries given by the square roots of the quadrature weights ω_j. We can apply the Lanczos process of Sect. 1.3.12 of Chap. 1 in Volume II to find a tridiagonal matrix $\mathbf{T}$ and an orthogonal matrix $\mathbf{W}$ so that

$$diag(x_1, \ldots x_n)\mathbf{W}^\top = \mathbf{W}^\top \mathbf{T} \text{ and } \mathbf{W}^\top \mathbf{e}_1 = \begin{bmatrix} \sqrt{\omega_1} \\ \vdots \\ \sqrt{\omega_n} \end{bmatrix} .$$

The converse follows from the Golub-Welsch Theorem 2.3.6.
This theorem has the following easy consequence, due to Laurie [126, p. 1136].

Corollary 2.3.2 *If the Gauss-Kronrod quadrature rule with nodes $\{x_j^{GK}\}_{j=-n}^n$ has all positive weights $\{\omega_j^{GK}\}_{j=-n}^n$ that sum to one, then there is a symmetric tridiagonal Jacobi matrix $\mathbf{T}_{2n+1}^{GK}$ whose eigenvalues are the Gauss-Kronrod nodes and for which there is an orthogonal matrix of eigenvectors with first row entries given by the square roots of the Gauss-Kronrod weights.*

Some information about the signs of the Gauss-Kronrod weights is known *a priori*. Monegato [134, p. 813] proved a more general theorem which showed that the Gauss-Kronrod weights corresponding to the zeros of the Stieltjes polynomial s_{n+1} are positive if and only if the zeros of the Stieltjes polynomial s_{n+1} interlace the zeros of the Legendre polynomial p_n. Since the Stieltjes-Szegö Theorem 2.3.9

showed that these zeros in fact interlace, we are uncertain only about the signs of the weights for the Gaussian nodes in a Gauss-Kronrod quadrature.

Also, some information about the entries of the symmetric tridiagonal Jacobi matrix $\mathbf{T}_{2n+1}^{GK}$ is known *a priori* from Laurie [125, p. 742]. To understand this new information, it will be useful to develop the symmetric tridiagonal matrix whose eigenvectors are the *monic* Legendre polynomials. Using the definition (1.67) of the Legendre polynomials that satisfy $p_n(1) = 1$ for all n, together with Lemma 1.7.2 to relate general orthogonal polynomials to monic polynomials, and then Lemma 2.3.6 to find a similar symmetric tridiagonal matrix, we see that the $n \times n$ symmetric tridiagonal Jacobi matrix for the monic Legendre polynomials is

$$\mathbf{T}_n = \begin{bmatrix} 0 & \frac{1}{\sqrt{1\cdot 3}} & & & & \\ \frac{1}{\sqrt{1\cdot 3}} & 0 & \frac{2}{\sqrt{3\cdot 5}} & & & \\ & \frac{2}{\sqrt{3\cdot 5}} & 0 & \ddots & & \\ & & \ddots & \ddots & \frac{n-1}{\sqrt{(2n-3)\cdot(2n-1)}} & \\ & & & \frac{n-1}{\sqrt{(2n-3)\cdot(2n-1)}} & 0 \end{bmatrix}. \tag{2.29}$$

Next, we will revisit the Sturm sequence polynomials of Sect. 1.3.4 of Chap. 1 in Volume II, and connect them to the quadrature rule associated with the symmetric tridiagonal matrix.

Lemma 2.3.8 *Suppose that tridiagonal matrix*

$$\mathbf{T} = \begin{bmatrix} \alpha_0 & \beta_1 & & \\ \beta_1 & \alpha_2 & \ddots & \\ & \ddots & \ddots & \beta_n \\ & & \beta_n & \alpha_n \end{bmatrix}$$

has the spectral decomposition (see Theorem 1.3.1 of Chap. 1 in Volume II)

$$\mathbf{TQ} = \mathbf{Q}\Lambda \text{ where } \Lambda = \text{diag}(\lambda_0, \dots, \lambda_n) \text{ and } \mathbf{Q}^\top \mathbf{Q} = \mathbf{I}.$$

Define the quadrature weights

$$\omega_j = \left(\mathbf{e}_0{}^\top \mathbf{Q}\mathbf{e}_j\right)^2 \text{ for } 0 \le j \le n$$

and the discrete inner product

$$(f, g)_\mathbf{T} = \sum_{k=0}^{n} f(\lambda_k)g(\lambda_k)\omega_j.$$

Then the Sturm sequence polynomials $\pi_i(x)$, *defined by the three-term recurrence*

$$\pi_{-1}(x) = 0 \;,\;\; \pi_0(x) = 1 \text{ and } \pi_{i+1}(x) = (x - \alpha_k)\pi_i(x) - \beta_i^2 \pi_{i-1}(x) \text{ for } 0 \le i \le n \tag{2.30}$$

satisfy

$$\pi_{n+1}(\lambda_k) = 0 \text{ for } 0 \le k \le n \;, \tag{2.31a}$$

$$(\pi_i \;,\; \pi_j)_{\mathbf{T}} = 0 \text{ for } 0 \le i \ne j \le n \;, \tag{2.31b}$$

$$(\pi_j \;,\; \pi_j)_{\mathbf{T}} = \prod_{\ell=1}^{j} \beta_\ell^2 \text{ for } 0 \le j \le n \;, \tag{2.31c}$$

$$\alpha_j = \frac{(x\pi_j \;,\; \pi_j)_{\mathbf{T}}}{(\pi_j \;,\; \pi_j)_{\mathbf{T}}} \text{ for } 0 \le j \le n \text{ and} \tag{2.31d}$$

$$\beta_j^2 = \frac{(\pi_j \;,\; \pi_j)_{\mathbf{T}}}{(\pi_{j-1} \;,\; \pi_{j-1})_{\mathbf{T}}} \text{ for } 1 \le j \le n \;. \tag{2.31e}$$

Proof Define the polynomials $q_i(x)$ by the three-term recurrence

$$q_{-1}(x) = 0 \;,\;\; q_0(x) = 1 \text{ and } q_{i+1}(x) = [(x - \alpha_i)q_i(x) - \beta_i q_{i-1}(x)]/\beta_{i+1} \text{ for } 0 \le i \le n \;.$$

Then the equation

$$\mathbf{T}\mathbf{Q}\mathbf{e}_j = \mathbf{Q}\mathbf{\Lambda}\mathbf{e}_j$$

implies that

$$\mathbf{Q}\mathbf{e}_j = \begin{bmatrix} q_0(\lambda_j) \\ \vdots \\ q_n(\lambda_j) \end{bmatrix} \mathbf{e}_0^{\top}\mathbf{Q}\mathbf{e}_j \;,$$

and

$$q_{n+1}(\lambda_j) = 0 \;.$$

We will prove by induction that

$$q_k(x) = \pi_k(x)/\prod_{i=1}^{k} \beta_i \tag{2.32}$$

for $0 \le k \le n + 1$. Note that

$$q_0(x) = \pi_0(x) \text{ and } q_1(x) = \pi_1(x)/\beta_1 \;.$$

Assume that (2.32) is true for all $0 \leq j \leq k$. Then

$$q_{k+1}(x) = [(x - \alpha_k)q_k(x) - \beta_k q_{k-1}(x)] / \beta_{k+1}$$

$$= \left[(x - \alpha_k) \frac{\pi_k(x)}{\prod_{i=1}^{k} \beta_i} - \frac{\beta_k \pi_{k-1}(x)}{\prod_{i=1}^{k-1} \beta_i} \right] / \beta_{k+1}$$

$$= \left[(x - \alpha_k)\pi_k(x) - \beta_k^2 \pi_{k-1}(x) \right] / \prod_{i=1}^{k+1} \beta_i = \pi_{k+1}(x) / \prod_{i=1}^{k+1} \beta_i .$$

This completes the inductive proof of (2.32). Since $q_{n+1}(\lambda_j) = 0$ for $0 \leq j \leq n$, Eq. (2.32) implies the claim (2.31a).

Since the rows of $\mathbf{Q}$ are orthonormal,

$$\delta_{ij} = (\mathbf{Q}^\top \mathbf{e}_i)^\top \mathbf{Q}^\top \mathbf{e}_j = \sum_{k=0}^{n} q_i(\lambda_k)q_j(\lambda_k)\omega_k = \frac{(\pi_i , \pi_j)_\mathbf{T}}{\left(\prod_{m=1}^{i} \beta_m \right) \left(\prod_{\ell=1}^{j} \beta_\ell \right)}$$

The case $i \neq j$ proves (2.31b), and the case $i = j$ proves (2.31c).

For $0 \leq j \leq n$ we have

$$\alpha_j = \mathbf{e}_j^\top \mathbf{T} \mathbf{e}_j = (\mathbf{Q}^\top \mathbf{e}_j)^\top \mathbf{\Lambda} \mathbf{Q}^\top \mathbf{e}_j = \sum_{k=0}^{n} q_j(\lambda_k)\lambda_k q_j(\lambda_k)\omega_k = (xq_j , q_j)_\mathbf{T}$$

$$= \frac{(x\pi_j , \pi_j)_\mathbf{T}}{\prod_{\ell=1}^{j} \beta_\ell^2} = \frac{(x\pi_j , \pi_j)_\mathbf{T}}{(\pi_j , \pi_j)_\mathbf{T}} .$$

This proves (2.31d). For $1 \leq j \leq n$ we also have

$$\beta_j = \mathbf{e}_j^\top \mathbf{T} \mathbf{e}_{j-1} = (\mathbf{Q}^\top \mathbf{e}_j)^\top \mathbf{\Lambda} \mathbf{Q}^\top \mathbf{e}_{j-1} = \sum_{k=0}^{n} q_j(\lambda_k)\lambda_k q_{j-1}(\lambda_k)\omega_k = (xq_j , q_{j-1})_\mathbf{T}$$

$$= \frac{(\pi_j , x\pi_{j-1})_\mathbf{T}}{\left(\prod_{\ell=1}^{j-1} \beta_\ell \right) \left(\prod_{m=1}^{j} \beta_m \right)} = \frac{(\pi_j , \pi_j + \alpha_{j-1}\pi_{j-1} + \beta_{j-1}^2 \pi_{j-2})_\mathbf{T}}{\beta_j(\pi_{j-1} , \pi_{j-1})_\mathbf{T}}$$

then the orthogonality (2.31b) of the polynomials π_j and π_i for $i < j$ implies that

$$= \frac{(\pi_j , \pi_j)_\mathbf{T}}{\beta_j(\pi_{j-1} , \pi_{j-1})_\mathbf{T}} .$$

This equation is equivalent to (2.31e).

Next, we will prove the following theorem, which has been adapted from a discussion in Laurie [125, p. 742].

Theorem 2.3.11 *Suppose that the Gauss-Kronrod quadrature rule involving $2n+1$ nodes has all positive weights, so there is a symmetric tridiagonal Jacobi matrix*

$$
\mathbf{T}^{GK}_{2n+1} = \begin{bmatrix} \alpha^{GK}_0 & \beta^{GK}_1 & & \\ \beta^{GK}_1 & \alpha^{GK}_1 & \ddots & \\ & \ddots & \ddots & \beta^{GK}_{2n} \\ & & \beta^{GK}_{2n} & \alpha^{GK}_{2n} \end{bmatrix}
$$

whose eigenvalues are the quadrature nodes and the quadrature weights are the squares of the first entries of the eigenvectors. Let the monic Legendre polynomials be generated by the three-term recurrence

$$
\phi_{-1}(x) = 0 , \ \phi_0(x) = 1 \ and \ \phi_{j+1}(x) = (x - \alpha_j)\phi_j(x) - \beta_j^2\phi_{j-1}(x) \ for \ 0 \le j \le 2n .
$$

Then if n is even, then

$$
\alpha_j^{GK} = \alpha_j \ for \ 0 \le j \le \frac{3n}{2} \ and \ \beta_j^{GK} = \beta_j \ for \ 1 \le j \le \frac{3n}{2} .
$$

If n is odd, then

$$
\alpha_j^{GK} = \alpha_j \ for \ 0 \le j \le \frac{3n+1}{2} \ and \ \beta_j^{GK} = \beta_j \ for \ 1 \le j \le \frac{3n+1}{2} .
$$

Proof Define the continuous inner product

$$
(f , g) = \int_{-1}^{1} f(x)g(x) \ dx .
$$

Lemma 1.7.1 shows that

$$
\alpha_j = \frac{(x\pi_j , \pi_j)}{(\pi_j , \pi_j)} \ for \ 0 \le j \le 2n \ and
$$

$$
\beta_j^2 = \frac{(\pi_j , \pi_j)}{(\pi_{j-1} , \pi_{j-1})} \ for \ 1 \le j \le 2n .
$$

Thus the k-th entry of the sequence

$$
\{\alpha_0, \beta_1, \alpha_1, \dots, \beta_{2n}, \alpha_{2n}\}
$$

is the first to involve a polynomial of degree k.

If $n > 0$ is even, Lemma 2.3.7 shows that Gauss-Kronrod quadrature is exact for polynomials of degree at most $3n + 1$. Note that if the Sturm sequence polynomials

π_j^{GK} for $\mathbf{T}_{2n+1}^{GK}$ are defined by (2.30), then

$$\phi_0(x) = \pi_0^{GK}(x) .$$

Equation (2.31d) shows that

$$\alpha_0^{GK} = \frac{(\pi_0^{GK} , x\pi_0^{GK})_{\mathbf{T}_{2n+1}^{GK}}}{(\pi_0^{GK} , \pi_0^{GK})_{\mathbf{T}_{2n+1}^{GK}}} = \frac{(\phi_0 , x\phi_0)_{\mathbf{T}_{2n+1}^{GK}}}{(\phi_0 , \phi_0)_{\mathbf{T}_{2n+1}^{GK}}} = \frac{(\phi_j , x\phi_j)}{(\phi_j , \phi_j)} = \alpha_0 ,$$

so $\phi_1 = \pi_1^{GK}$. Then Eq. (2.31e) shows that

$$\beta_1^{GK} = \frac{(\pi_1^{GK} , \pi_1^{GK})_{\mathbf{T}_{2n+1}^{GK}}}{(\pi_0^{GK} , \pi_0^{GK})_{\mathbf{T}_{2n+1}^{GK}}} = \frac{(\pi_1^{GK} , \pi_1^{GK})}{(\pi_0^{GK} , \pi_0^{GK})} = \frac{(\phi_1 , \phi_1)}{(\phi_0 , \phi_0)} = \beta_1 .$$

We can continue in this way to show that

$$\alpha_j^{GK} = \alpha_j \text{ for } 0 \leq j \leq (3n)/2 ,$$
$$\beta_j^{GK} = \beta_j \text{ for } 1 \leq j \leq (3n)/2 \text{ and}$$
$$\pi_j^{GK} = \phi_j \text{ for } 0 \leq j \leq (3n)/2 .$$

A similar argument applies when n is odd.

Example 2.3.5 If we begin with a one-point Gaussian quadrature rule and extend it to a three-point Gauss-Kronrod rule, then Theorem 2.3.11 shows that

$$\mathbf{T}_3^{GK} = \mathbf{T}_3 .$$

Thus the three-point Gauss-Kronrod quadrature rule is the 3-point Gaussian quadrature rule. The nodes and weights for this rule were discovered in Example 2.3.2.

Next, we will prove a result due to Laurie [126, p. 1137], which gives us more information about the Jacobi matrix for Gauss-Kronrod rules.

Theorem 2.3.12 *Suppose that the Gauss-Kronrod quadrature rule involving $2n+1$ nodes has all positive weights, so we can partition its Jacobi matrix in the form*

$$\mathbf{T}_{2n+1}^{GK} = \begin{bmatrix} \mathbf{S}_n^{GK} & \mathbf{e}_n\beta_n^{GK} & \\ \beta_n^{GK}\mathbf{e}_n^\top & \alpha_n^{GK} & \beta_{n+1}^{GK}\mathbf{e}_1^\top \\ & \mathbf{e}_1\beta_{n+1}^{GK} & \mathbf{U}_n^{GK} \end{bmatrix} .$$

Then the eigenvalues of $\mathbf{S}_n^{GK}$ are the same as the eigenvalues of $\mathbf{U}_n^{GK}$.

Proof Let us partition

$$\mathbf{S}_n^{GK} = \begin{bmatrix} \mathbf{S}_{n-1}^{GK} & \mathbf{e}_{n-1}\beta_{n-1}^{GK} \\ \beta_{n-1}^{GK}\mathbf{e}_{n-1}^\top & \alpha_{n-1}^{GK} \end{bmatrix} \text{ and } \mathbf{U}_n^{GK} = \begin{bmatrix} \alpha_{n+1}^{GK} & \beta_{n+2}^{GK}\mathbf{e}_1^\top \\ \mathbf{e}_1\beta_{n+2}^{GK} & \mathbf{U}_{n-1}^{GK} \end{bmatrix} .$$

We can apply expansion by minors (3.8) in Chap. 3 of Volume I to row $n+1$ of $\mathbf{T}_{2n+1}^{GK}$ to get

$$\det\left(\mathbf{T}_{2n+1}^{GK} - \mathbf{I}\lambda\right) = -\beta_n^{GK}\det \begin{bmatrix} \mathbf{S}_{n-1}^{GK} - \mathbf{I}\lambda & & & \\ \beta_{n-1}^{GK}\mathbf{e}_{n-1}^\top & \beta_n^{GK} & & \\ & \beta_{n+1}^{GK} & \alpha_{n+1}^{GK} - \lambda & \beta_{n+2}^{GK}\mathbf{e}_1^\top \\ & & \mathbf{e}_1\beta_{n+2}^{GK} & \mathbf{U}_{n-1}^{GK} - \mathbf{I}\lambda \end{bmatrix}$$

$$+ \left(\alpha_n^{GK} - \lambda\right)\det \begin{bmatrix} \mathbf{S}_{n-1}^{GK} - \mathbf{I}\lambda & \mathbf{e}_{n-1}\beta_{n-1}^{GK} & & \\ \beta_{n-1}^{GK}\mathbf{e}_{n-1}^\top & \alpha_{n-1}^{GK} - \lambda & & \\ & & \alpha_{n+1}^{GK} - \lambda & \beta_{n+2}^{GK}\mathbf{e}_1^\top \\ & & \mathbf{e}_1\beta_{n+2}^{GK} & \mathbf{U}_{n-1}^{GK} - \mathbf{I}\lambda \end{bmatrix}$$

$$- \beta_{n+1}^{GK}\det \begin{bmatrix} \mathbf{S}_{n-1}^{GK} - \mathbf{I}\lambda & \mathbf{e}_{n-1}\beta_{n-1}^{GK} & & \\ \beta_{n-1}^{GK}\mathbf{e}_{n-1}^\top & \alpha_{n-1}^{GK} - \lambda & \beta_n^{GK} & \\ & & \beta_{n+1}^{GK} & \beta_{n+2}^{GK}\mathbf{e}_1^\top \\ & & & \mathbf{U}_{n-1}^{GK} - \mathbf{I}\lambda \end{bmatrix}$$

$$= -\left(\beta_n^{GK}\right)^2 \det\left(\mathbf{S}_{n-1}^{GK} - \mathbf{I}\lambda\right)\det\left(\mathbf{U}_n^{GK} - \mathbf{I}\lambda\right)$$

$$+ \left(\alpha_n^{GK} - \lambda\right)\det\left(\mathbf{S}_n^{GK} - \mathbf{I}\lambda\right)\det\left(\mathbf{U}_n^{GK} - \mathbf{I}\lambda\right)$$

$$- \left(\beta_{n+1}^{GK}\right)^2 \det\left(\mathbf{S}_n^{GK} - \mathbf{I}\lambda\right)\det\left(\mathbf{U}_{n-1}^{GK} - \mathbf{I}\lambda\right)$$

Since the Gauss-Kronrod quadrature rule shares n nodes with the Gaussian quadrature rule having n nodes, the characteristic polynomial $\det\left(\mathbf{T}_{2n+1}^{GK} - \mathbf{I}\lambda\right)$ must be divisible by the characteristic polynomial $\det\left(\mathbf{T}_n - \mathbf{I}\lambda\right)$. Furthermore, since Theorem 2.3.11 shows that

$$\mathbf{S}_n^{GK} = \mathbf{T}_n ,$$

we see that

$$\frac{\det\left(\mathbf{T}_{2n+1}^{GK} - \mathbf{I}\lambda\right)}{\det\left(\mathbf{T}_n - \mathbf{I}\lambda\right)} = -\left(\beta_n^{GK}\right)^2 \det\left(\mathbf{T}_{n-1} - \mathbf{I}\lambda\right)\frac{\det\left(\mathbf{U}_n^{GK} - \mathbf{I}\lambda\right)}{\det\left(\mathbf{T}_n - \mathbf{I}\lambda\right)}$$

$$+ \left(\alpha_n^{GK} - \lambda\right)\det\left(\mathbf{U}_n^{GK} - \mathbf{I}\lambda\right) - \left(\beta_{n+1}^{GK}\right)^2 \det\left(\mathbf{U}_{n-1}^{GK} - \mathbf{I}\lambda\right) .$$

Since Lemma 1.3.9 of Chap. 1 in Volume II showed that the eigenvalues of $\mathbf{T}_{n-1}$ are distinct from the eigenvalues of $\mathbf{T}_n$, we conclude that the characteristic polynomial

$\det\left(\mathbf{U}_n^{GK} - \mathbf{I}\lambda\right)$ must be divisible by the characteristic polynomial $\det\left(\mathbf{T}_n - \mathbf{I}\lambda\right)$. Since both are polynomials of degree n, they must have the same zeros.

Example 2.3.6 If we begin with a two-point Gaussian quadrature rule and extend it to a five-point Gauss-Kronrod rule, then Theorem 2.3.11 shows that

$$\mathbf{T}_5^{GK} = \begin{bmatrix} \mathbf{T}_4 & \beta_4^{GK} \\ \beta_4^{GK} & \alpha_4^{GK} \end{bmatrix},$$

where $\mathbf{T}_4$ is the Jacobi matrix for the four-point Gaussian quadrature rule. Recall from Eq. (2.29) that the diagonal entries of $\mathbf{T}_4$ are all zero. Next, Theorem 2.3.12 shows that the eigenvalues of the leading and trailing 2×2 submatrices are the same. Lemma 1.2.8 of Chap. 1 in Volume II shows that

$$0 = \text{tr}\,(\mathbf{T}_2) = \text{tr}\left(\mathbf{U}_2^{GK}\right) = \alpha_3^{GK} + \alpha_4^{GK} = \alpha_3 + \alpha_4^{GK} = \alpha_4^{GK}$$

and that

$$-\frac{1}{3} = -\beta_1^2 = \det\left(\mathbf{T}_2\right) = \det\left(\mathbf{U}_2^{GK}\right) = \det\begin{bmatrix} \alpha_3^{GK} & \beta_4^{GK} \\ \beta_4^{GK} & \alpha_4^{GK} \end{bmatrix}$$

$$= \det\begin{bmatrix} 0 & \beta_4^{GK} \\ \beta_4^{GK} & 0 \end{bmatrix} = -\left(\beta_4^{GK}\right)^2.$$

Here $\text{tr}(\mathbf{A})$ represents the **trace** of the matrix $\mathbf{A}$. Thus the Jacobi matrix for the Gauss-Kronrod quadrature rule with five nodes is

$$\mathbf{T}_5^{GK} = \begin{bmatrix} 0 & 1/\sqrt{1 \cdot 3} & & & \\ 1/\sqrt{1 \cdot 3} & 0 & 2/\sqrt{3 \cdot 5} & & \\ & 2/\sqrt{3 \cdot 5} & 0 & 3/\sqrt{5 \cdot 7} & \\ & & 3/\sqrt{5 \cdot 7} & 0 & 1/\sqrt{1 \cdot 3} \\ & & & 1/\sqrt{1 \cdot 3} & 0 \end{bmatrix}.$$

Using the expansion by minors in the proof of Theorem 2.3.12 we can easily see that the Gauss-Kronrod quadrature nodes are $x_0 = 0$, $x_1 = -x_{-1} = 1/\sqrt{3}$ and $x_2 = -x_{-2} = \sqrt{6/7}$. Then we can find that the weights are $\omega_0 = 28/45$, $\omega_1 = \omega_{-1} = 27/55$ and $\omega_2 = \omega_{-2} = 98/495$.

Example 2.3.7 Next, suppose that we begin with a three-point Gaussian quadrature rule and extend it to a seven-point Gauss-Kronrod rule. Theorem 2.3.11 shows that

$$\mathbf{T}_7^{GK} = \begin{bmatrix} \mathbf{T}_6 & \beta_6^{GK} \\ \beta_6^{GK} & \alpha_6^{GK} \end{bmatrix},$$

where $\mathbf{T}_6$ is the Jacobi matrix for the four-point Gaussian quadrature rule. Recall from Eq. (2.29) that the diagonal entries of $\mathbf{T}_6$ are all zero. Next, Theorem 2.3.12 shows that the eigenvalues of the leading and trailing 3×3 submatrices are the same. Lemma 1.2.8 of Chap. 1 in Volume II shows that

$$0 = \operatorname{tr}(\mathbf{T}_3) = \operatorname{tr}\left(\mathbf{U}_3^{GK}\right) = \alpha_4^{GK} + \alpha_5^{GK} + \alpha_6^{GK} = \alpha_4 + \alpha_5 + \alpha_6^{GK} = \alpha_6^{GK} .$$

Since

$$\det\left(\begin{bmatrix} 0 & \gamma_2 & \\ \gamma_2 & 0 & \gamma_3 \\ & \gamma_3 & 0 \end{bmatrix} - \mathbf{I}\lambda\right) = -\lambda^3 + \lambda\left(\gamma_2^2 + \gamma_3^2\right) ,$$

we must also have

$$\frac{1}{3} + \frac{4}{15} = \beta_2^2 + \beta_3^3 = \left(\beta_6^{GK}\right)^2 + \left(\beta_7^{GK}\right)^2 = \beta_6^2 + \left(\beta_7^{GK}\right)^2 = \frac{25}{99} + \left(\beta_7^{GK}\right)^2$$

from which we conclude that

$$\left(\beta_7^{GK}\right)^2 = \frac{1}{3} + \frac{4}{15} - \frac{25}{99} = \frac{172}{495} .$$

and that Thus the Jacobi matrix for the Gauss-Kronrod quadrature rule with seven nodes is

$$\mathbf{T}_7^{GK} = \begin{bmatrix} 0 & 1/\sqrt{1 \cdot 3} & & & & & \\ 1/\sqrt{1 \cdot 3} & 0 & 2/\sqrt{3 \cdot 5} & & & & \\ & 2/\sqrt{3 \cdot 5} & 0 & 3/\sqrt{5 \cdot 7} & & & \\ & & 3/\sqrt{5 \cdot 7} & 0 & 4/\sqrt{7 \cdot 9} & & \\ & & & 4/\sqrt{7 \cdot 9} & 0 & 5/\sqrt{9 \cdot 11} & \\ & & & & 5/\sqrt{9 \cdot 11} & 0 & \sqrt{172}/\sqrt{495} \\ & & & & & \sqrt{172}/\sqrt{495} & 0 \end{bmatrix} .$$

The crux of the Calvetti et al. development at this point depends on the following analogue of the Gaussian quadrature theorem for discrete inner products.

Theorem 2.3.13 (Gaussian Quadrature with Discrete Inner Product) *Let $[a, b]$ be a closed interval. Given a positive integer N, assume that the scalars $w_1, \ldots, w_N$ are all positive, and that the set of scalars $\{\xi_k\}_{k=1}^N$ are all distinct and contained in (a, b). Suppose that the set of polynomials $\{\phi_j\}_{j=0}^{N-1}$ is orthogonal with respect to the inner product*

$$(f, g) = \sum_{k=1}^N f(\xi_k) g(\xi_k) w_k .$$

Then for any integer $1 \leq m \leq N - 1$, the zeros $x_1, \ldots, x_m$ of $\phi_m(x)$ are real, distinct and lie in the open interval (a, b). Let the Lagrange interpolation polynomial at these zeros be

$$\lambda_{j,m}(x) = \prod_{\substack{1 \leq i \leq m \\ i \neq j}} (x - x_i) / \prod_{\substack{1 \leq i \leq m \\ i \neq j}} (x_j - x_i) ,$$

and define the scalars ω_j for $1 \leq j \leq m$ by

$$\omega_j = \sum_{k=1}^{N} \lambda_{j,m}(\xi_k) w_k . \tag{2.33}$$

Then the equation

$$\sum_{k=1}^{N} p(\xi_k) w_k = \sum_{j=1}^{m} p(x_j) \omega_j$$

is satisfied for all polynomials p of degree at most $2m - 1$. Furthermore, the scalars ω_j are all nonnegative.

Proof We will prove the first claim by contradiction. Since ϕ_m is nonzero and has degree m, it cannot have more than m zeros in (a, b), and thus it cannot have more than m sign changes in this open interval. Suppose that ϕ_m has n sign changes on (a, b), where $0 \leq n < m$. If $n = 0$, then $\phi_m(x)$ has constant sign; without loss of generality we will assume that $\phi_m > 0$. Sine $m \geq 1$, the polynomial ϕ_m must be orthogonal to a constant polynomial, so

$$0 = (\phi_m, 1) = \sum_{k=1}^{N} \phi_m(\xi_k) w_k .$$

The terms in the sum on the right are all positive, so the sum must also be positive, giving us a contradiction. We are left with the possibility that there are between 1 and $m - 1$ sign changes in (a, b). Let $t_1, \ldots, t_n$ be the points where ϕ_m changes sign, and let

$$p(x) = \prod_{i=1}^{n} (x - t_i) .$$

Then p is a polynomial of degree $n < m$, so ϕ_m is orthogonal to $p(x)$. Also, $\phi_m(x)p(x)$ has constant sign in (a, b), which without loss of generality we may assume is nonnegative. It follows that

$$0 = (\phi_m, p) = \sum_{k=1}^{N} \phi_m(\xi_k) p(\xi_k) w_k \; .$$

If $m < N$, the sum on the right involves all nonnegative terms and at least one positive term, so its value is positive, giving us another contradiction. We conclude that for $m \leq N - 1$ the polynomial ϕ_m must have m sign changes in (a, b), which in turn implies that it has m distinct zeros in (a, b).

To prove the second conclusion, let p be a polynomial of degree $2m - 1$. Define the polynomials q and r of at most degree $m - 1$ by

$$p(x) = q(x)\phi_m(x) + r(x) \; .$$

In other words, q is the quotient in dividing p by ϕ_m, and r is the remainder. Since ϕ_m is orthogonal to q,

$$\sum_{k=1}^{N} p(\xi_k) w_k = \sum_{k=1}^{N} q(\xi_k)\phi_m(\xi_k) w_k + \sum_{k=1}^{N} r(\xi_k) w_k$$

$$= \sum_{k=1}^{N} r(\xi_k) w_k \; .$$

Since r is a polynomial of degree at most $m - 1$, Lagrange polynomial interpolation (see Sect. 1.2.3) at the zeros $x_1, \ldots, x_m$ of ϕ_m shows us that

$$r(x) = \sum_{j=1}^{m} \varrho_j \lambda_{j,m}(x) \; ,$$

where for $1 \leq j \leq m$

$$\varrho_j = r(x_j) = q(x_j)\phi_m(x_j) + r(x_j) = p(x_j) \; .$$

Then

$$\sum_{k=1}^{N} p(\xi_k) w_k = \sum_{k=1}^{N} r(\xi_k) w_k = \sum_{j=1}^{m} \varrho_j \sum_{k=1}^{N} \lambda_{j,m}(\xi_k) w_k = \sum_{j=1}^{m} p(x_j)\omega_j \; .$$

To prove that ω_j is nonnegative for $1 \leq j \leq m$, note that $\lambda_{j,m}^2$ is a polynomial of degree $2m - 2$, so

$$\omega_j = \sum_{i=1}^{m} \lambda_{j,m}(\xi_i)^2 \omega_i = \sum_{k=1}^{N} \lambda_{j,m}(x_k)^2 w_k \; .$$

The latter is a sum of nonnegative terms, and since $m < N$ it must involve at least one nonzero term.

The process we are about to discuss is nearly the same for even and odd values of n. Since the Gauss-Kronrod using $2n + 1$ nodes with n odd is exact for polynomials of degree $3n + 2$, but only for degree $3n + 1$ when n is even, we will restrict our discussion to the cases when $n \geq 3$ is odd. Calvetti et al. [31]. present the computations for n even.

Let us partition the $\left(\frac{3n+1}{2}\right) \times \left(\frac{3n+1}{2}\right)$ Jacobi matrix for the monic Legendre polynomials in the form

$$
\mathbf{T}_{(3n+3)/2} = \begin{bmatrix} \mathbf{T}_n & \mathbf{e}_n \beta_n & \\ \beta_n \mathbf{e}_n^\top & \alpha_n & \beta_{n+1} \mathbf{e}_1^\top \\ & \mathbf{e}_1 \beta_{n+1} & \mathbf{U}_{(n+1)/2} \end{bmatrix} .
$$

Then Theorem 2.3.11 shows that we can partition the Jacobi matrix for the Gauss-Kronrod quadrature rule in the form

$$
\mathbf{T}_{2n+1}^{GK} = \begin{bmatrix} \mathbf{T}_n & \mathbf{e}_n \beta_n & \\ \beta_n \mathbf{e}_n^\top & \alpha_n & \beta_{n+1} \mathbf{e}_1^\top \\ & \mathbf{e}_1 \beta_{n+1} & \mathbf{U}_n^{GK} \end{bmatrix} .
$$

Theorem 2.3.11 shows that we can partition

$$
\mathbf{U}_n^{GK} \equiv \begin{bmatrix} \mathbf{U}_{(n+1)/2} & \mathbf{e}_{(n+1)/2} \beta_{(3n+3)/2}^{GK} \\ \beta_{(3n+3)/2}^{GK} \mathbf{e}_{(n+1)/2}^\top & \mathbf{W}_{(n-1)/2}^{GK} \end{bmatrix} ,
$$

and Theorem 2.3.12 shows that the eigenvalues of this matrix are the same as the eigenvalues of $\mathbf{T}_n$. Thus the scalars $\beta_{(3n+3)/2}^{GK}$ and tridiagonal matrix $\mathbf{W}_{(n-1)/2}^{GK}$ are unknown; furthermore, they will not need to be computed.

Since $\mathbf{T}_n$ is the leading $n \times n$ submatrix of the Jacobi matrix for the monic Legendre polynomials, it has a knowable spectral decomposition

$$
\mathbf{T}_n \mathbf{Q}_n = \mathbf{Q}_n \Lambda_n \text{ where } \mathbf{Q}_n^\top \mathbf{Q}_n = \mathbf{I} \text{ and } \Lambda_n \text{ is diagonal .}
$$

We will require the computation of Λ_n, as well as the *last* row of $\mathbf{Q}_n$. Since $\mathbf{U}_{(n+1)/2}$ is the trailing $\left(\frac{n+1}{2}\right) \times \left(\frac{n+1}{2}\right)$ submatrix of the Jacobi matrix $\mathbf{T}_{(3n+3)/2}$, it has a knowable spectral decomposition

$$
\mathbf{U}_{\frac{n+1}{2}} \mathbf{P}_{\frac{n+1}{2}} = \mathbf{P}_{\frac{n+1}{2}} \mathbf{D}_{\frac{n+1}{2}} \text{ where } \mathbf{P}_{\frac{n+1}{2}}^\top \mathbf{P}_{\frac{n+1}{2}} = \mathbf{I} \text{ and } \mathbf{D}_{\frac{n+1}{2}} \text{ is diagonal .}
$$

We will require the computation of the eigenvalues $\mathbf{D}_{(n+1)/2}$, as well as the *first* row of the eigenvectors $\mathbf{P}_{(n+1)/2}$.

At this point, the spectral decomposition

$$\mathbf{U}_n^{GK}\mathbf{P}_n^{GK} = \mathbf{P}_n^{GK}\boldsymbol{\Lambda}_n \text{ where } \left(\mathbf{P}_n^{GK}\right)^{\top}\mathbf{P}_n^{GK} = \mathbf{I}$$

has known eigenvalues $\boldsymbol{\Lambda}_n$ but unknown eigenvectors $\mathbf{P}_n^{GK}$, as well as some unknown entries in $\mathbf{U}_n^{GK}$. However, Theorem 2.3.6 shows that the diagonal entries $\lambda_{j,n}$ of $\boldsymbol{\Lambda}_n$ are the nodes for the quadrature rule associated with the Jacobi matrix $\mathbf{U}_n^{GK}$, and the squares of the entries of the first row of $\mathbf{P}_n^{GK}$ must be the weights:

$$\omega_{j,n}^* = \left(\mathbf{e}_1^{\top}\mathbf{P}_n^{GK}\mathbf{e}_j\right)^2 \text{ for } 1 \le j \le n .$$

Also, Theorem 2.3.13 shows that the discrete inner product Gaussian quadrature rule using $(n+1)/2 < n$ nodes exactly integrates all polynomials of degree at most n in the discrete quadrature rule for $\mathbf{U}_n^{GK}$. Let $\{\delta_k\}_{k=1}^{(n+1)/2}$ and $\{\pi_j\}_{j=1}^{(n+1)/2}$ be the nodes and weights for the Gaussian quadrature rule associated with the Jacobi matrix $\mathbf{U}_{(n+1)/2}$; in other words,

$$\mathbf{D}_{(n+1/2)} = \text{diag}(\delta_1, \ldots, \delta_{(n+1)/2}) \text{ and } \pi_j = (\mathbf{e}_1^{\top}\mathbf{P}_{(n+1)/2}\mathbf{e}_j)^2 \text{ for } 1 \le j \le (n+1)/2 .$$

For $1 \le j \le n$, we can define the Lagrange interpolation polynomials

$$\ell_{j,n}(x) = \prod_{\substack{1 \le i \le n \\ i \ne j}} \frac{x - \lambda_{i,n}}{\lambda_{j,n} - \lambda_{i,n}} .$$

Each of these is a polynomial of degree $n-1$, so discrete Gaussian quadrature using $(n+1)/2$ quadrature points will sum this polynomial exactly in the quadrature rule determined by the Jacobi matrix $\mathbf{U}_n^{GK}$.. Consequently, for $1 \le j \le n$ we must have

$$\omega_{j,n}^* \equiv \sum_{k=1}^{n} \ell_{j,n}(\lambda_{k,n})\omega_{k,n}^* = \sum_{k=1}^{(n+1)/2} \ell_{j,n}(\delta_k)\pi_k .$$

The second sum is computable from the Gaussian quadrature rule associated with the Jacobi matrix $\mathbf{U}_{(n+1)/2}$. If we find that these new weights $\omega_{j,n}^*$ are all positive, we can take the square roots of these weights to get the first row of $\mathbf{P}_n^{GK}$.

The Jacobi matrix $\mathbf{T}_{2n+1}^{GK}$ for the Gauss-Kronrod quadrature rule using $2n+1$ nodes is similar to

$$
\begin{bmatrix} \mathbf{Q}_n{}^\top & & \\ & (\mathbf{P}_n^{GK})^\top & \\ & & 1 \end{bmatrix}
\begin{bmatrix} \mathbf{I} & & \\ & \mathbf{I} & \\ & & 1 \end{bmatrix}
\begin{bmatrix} \mathbf{T}_n & \mathbf{e}_n\beta_n & \\ \beta_n\mathbf{e}_n{}^\top & \alpha_n & \beta_{n+1}\mathbf{e}_1{}^\top \\ & \mathbf{e}_1\beta_{n+1} & \mathbf{U}_n^{GK} \end{bmatrix}
\begin{bmatrix} \mathbf{I} & & \\ & 1 & \\ & & \mathbf{I} \end{bmatrix}
\begin{bmatrix} \mathbf{Q}_n & & \\ & \mathbf{P}_n^{GK} & \\ & & 1 \end{bmatrix}
$$

$$
= \begin{bmatrix} \boldsymbol{\Lambda}_n & & \mathbf{Q}_n{}^\top\mathbf{e}_n\beta_n \\ & \boldsymbol{\Lambda}_n & (\mathbf{P}_n^{GK})^\top\mathbf{e}_1\beta_{n+1} \\ \beta_n\mathbf{e}_n{}^\top\mathbf{Q}_n & \beta_{n+1}\mathbf{e}_1{}^\top\mathbf{P}_n^{GK} & \alpha_n \end{bmatrix} .
$$

All of the entries of the matrix on the right-hand side are known, even though $\mathbf{T}_{2n+1}^{GK}$ has some unknown entries. Using the ideas in Sect. 6.9 of Chap. 6 in Volume I, we can find elementary rotations $\mathbf{G}_{1,n+1}, \ldots, \mathbf{G}_{n,2n}$ so that

$$
\mathbf{G}_{n,2n}{}^\top \ldots \mathbf{G}_{1,n+1}{}^\top = \begin{bmatrix} \boldsymbol{\Lambda}_n & & \mathbf{Q}_n{}^\top\mathbf{e}_n\beta_n \\ & \boldsymbol{\Lambda}_n & (\mathbf{P}_n^{GK})^\top\mathbf{e}_1\beta_{n+1} \\ \beta_n\mathbf{e}_n{}^\top\mathbf{Q}_n & \beta_{n+1}\mathbf{e}_1{}^\top\mathbf{P}_n^{GK} & \alpha_n \end{bmatrix} \mathbf{G}_{1,n+1} \ldots \mathbf{G}_{n,2n}
$$

$$
= \begin{bmatrix} \boldsymbol{\Lambda}_n & & \\ & \boldsymbol{\Lambda}_n & \mathbf{c} \\ & \mathbf{c}^\top & \alpha_n \end{bmatrix} .
$$

The entries γ_j of $\mathbf{c}$ can also be computed by the equation

$$
\gamma_j^2 \equiv \left(\mathbf{e}_j{}^\top\mathbf{c}\right)^2 = \left(\mathbf{e}_n{}^\top\mathbf{Q}_n\mathbf{e}_j\beta_n\right)^2 + \left(\mathbf{e}_1{}^\top\mathbf{P}_n^{GK}\mathbf{e}_j\beta_{n+1}\right)^2 .
$$

Because $\mathbf{e}_1{}^\top\mathbf{P}_n^{GK}\mathbf{e}_j\beta_{n+1}$ is a positive vector, $\mathbf{c}$ is a positive n-vector.

We claim that the eigenvalues of the trailing $(n+1) \times (n+1)$ diagonal submatrix are distinct from the eigenvalues of $\boldsymbol{\Lambda}_n$. Indeed, if

$$
\begin{bmatrix} \boldsymbol{\Lambda}_n & \mathbf{c} \\ \mathbf{c}^\top & \alpha_n \end{bmatrix} \begin{bmatrix} \mathbf{x} \\ \xi \end{bmatrix} = \begin{bmatrix} \mathbf{x} \\ \xi \end{bmatrix} \lambda ,
$$

and λ is equal to the jth diagonal entry of $\boldsymbol{\Lambda}_n$, then

$$
(\boldsymbol{\Lambda}_n - \mathbf{I}\lambda)\mathbf{x} = -\mathbf{c}\xi
$$

implies that $\mathbf{e}_j{}^\top\mathbf{c}\xi = 0$. Since $\mathbf{c} > \mathbf{0}$, we conclude that $\xi = 0$. Since $\boldsymbol{\Lambda}_n$ is the diagonal matrix of zeros of the Legendre polynomial p_b, Theorem 2.3.4 shows that its diagonal entries are distinct, and we must have

$$
(\boldsymbol{\Lambda}_n - \mathbf{I}\lambda)\mathbf{x} = \mathbf{0} .
$$

This implies that $\mathbf{e}_i^{\top}\mathbf{x} = 0$ for all $i \neq j$. But we also have

$$0 = (\alpha_n - \lambda)\xi = -\mathbf{c}^{\top}\mathbf{x} = -\mathbf{c}^{\top}\mathbf{e}_j\mathbf{e}_j^{\top}\mathbf{x},$$

from which we conclude that $\mathbf{x} = \mathbf{0}$.

If $\boldsymbol{\Lambda}_n - \mathbf{I}\lambda$ is nonsingular, we can factor

$$\begin{bmatrix} \boldsymbol{\Lambda}_n & \mathbf{c} \\ \mathbf{c}^{\top} & \alpha_n \end{bmatrix} - \mathbf{I}\lambda = \begin{bmatrix} \mathbf{I} & \\ \mathbf{c}^{\top}(\boldsymbol{\Lambda}_n - \mathbf{I}\lambda)^{-1} & 1 \end{bmatrix} \begin{bmatrix} \boldsymbol{\Lambda}_n - \mathbf{I}\lambda & \mathbf{c} \\ & \alpha_n - \lambda - \mathbf{c}^{\top}(\boldsymbol{\Lambda}_n - \mathbf{I}\lambda)^{-1}\mathbf{c} \end{bmatrix}.$$

Since the eigenvalues of the given matrix are distinct from the eigenvalues of $\boldsymbol{\Lambda}_n$, the final diagonal entry of the right-hand factor must be zero when λ is an eigenvalue of the given $(n+1) \times (n+1)$ symmetric tridiagonal matrix. This equation can be written in the form

$$\lambda - \alpha_n + \sum_{j=1}^{n} \frac{\gamma_j^2}{\lambda_{j,n} - \lambda} = 0. \tag{2.34}$$

This is very similar to the secular Eq. (1.33) in Chap. 1 of Volume II, and can be solved very quickly because its poles are known. It is also easy to see that Eq. (2.34) has $n+1$ real roots that interlace the zeros $\lambda_{j,n}$ of the Legendre polynomial of degree n.

In the process of solving Eq. (2.34), we can compute the vector $(\boldsymbol{\Lambda}_n - \mathbf{I}\lambda)^{-1}\mathbf{c}$ in order to form an eigenvector of the matrix that generated the secular Eq. (2.34). The corresponding eigenvector equation would take the form

$$\begin{bmatrix} \boldsymbol{\Lambda}_n & \mathbf{c} \\ \mathbf{c}^{\top} & \alpha_n \end{bmatrix} \begin{bmatrix} -(\boldsymbol{\Lambda}_n - \mathbf{I}\lambda)^{-1}\mathbf{c} \\ 1 \end{bmatrix} = \begin{bmatrix} -(\boldsymbol{\Lambda}_n - \mathbf{I}\lambda)^{-1}\mathbf{c} \\ 1 \end{bmatrix}\lambda.$$

We can normalize these eigenvectors to find an orthogonal matrix of eigenvectors:

$$\begin{bmatrix} \boldsymbol{\Lambda}_n & \mathbf{c} \\ \mathbf{c}^{\top} & \alpha_n \end{bmatrix} \begin{bmatrix} \mathbf{W} \\ \mathbf{w}^{\top} \end{bmatrix} = \begin{bmatrix} \mathbf{W} \\ \mathbf{w}^{\top} \end{bmatrix} \begin{bmatrix} \boldsymbol{\Lambda}_n^{GK} & \\ & \lambda_{n+1}^{GK} \end{bmatrix}.$$

Afterward, the similarity transformation for $\mathbf{T}_{2n+1}^{GK}$ can be used to find the first row of the matrix of eigenvectors for $\mathbf{T}_{2n+1}^{GK}$. Since

$$\begin{bmatrix} \mathbf{T}_n & \mathbf{e}_n\beta_n & \\ \beta_n\mathbf{e}_n^{\top} & \alpha_n & \beta_{n+1}\mathbf{e}_1^{\top} \\ & \mathbf{e}_1\beta_{n+1} & \mathbf{U}_n^{GK} \end{bmatrix} \begin{bmatrix} \mathbf{I} & & \\ & 1 & \\ & & \mathbf{I} \end{bmatrix} \begin{bmatrix} \mathbf{Q}_n & & \\ & \mathbf{P}_n^{GK} & \\ & & 1 \end{bmatrix} \mathbf{G}_{1,n+1} \cdots \mathbf{G}_{n,2n} \begin{bmatrix} \mathbf{I} & \\ & \mathbf{W} \\ & \mathbf{w}^{\top} \end{bmatrix}$$

$$= \begin{bmatrix} \mathbf{I} & & \\ & 1 & \\ & & \mathbf{I} \end{bmatrix} \begin{bmatrix} \mathbf{Q}_n & & \\ & \mathbf{P}_n^{GK} & \\ & & 1 \end{bmatrix} \mathbf{G}_{1,n+1} \cdots \mathbf{G}_{n,2n} \begin{bmatrix} \mathbf{I} & \\ & \mathbf{W} \\ & \mathbf{w}^{\top} \end{bmatrix} \begin{bmatrix} \boldsymbol{\Lambda}_n & \\ & \boldsymbol{\Lambda}_{n+1}^{GK} \end{bmatrix},$$

we can form the unit vector

$$\begin{bmatrix} \mathbf{p}^\top \ \mathbf{q}^\top \end{bmatrix} = \begin{bmatrix} \mathbf{e}_1{}^\top \mathbf{Q}_n, \ \mathbf{0}^\top \end{bmatrix} \mathbf{G}_{1,n+1} \ldots \mathbf{G}_{n,2n} \,,$$

and see that the first row of the matrix of eigenvectors for $\mathbf{T}_{2n+1}^{GK}$ is

$$\begin{bmatrix} \mathbf{e}_1{}^\top, \ \mathbf{0}^\top, \ \mathbf{0}^\top \end{bmatrix} \begin{bmatrix} \mathbf{I} & & \\ & 1 & \\ & & \mathbf{I} \end{bmatrix} \begin{bmatrix} \mathbf{Q}_n & \\ & \mathbf{P}_n^{GK} & \\ & & 1 \end{bmatrix} \mathbf{G}_{1,n+1} \ldots \mathbf{G}_{n,2n} \begin{bmatrix} \mathbf{I} & \\ & \mathbf{W} \\ & \mathbf{w}^\top \end{bmatrix} = \begin{bmatrix} \mathbf{p}^\top, \ \mathbf{q}^\top \mathbf{W} \end{bmatrix} \,.$$

In summary, we have the following algorithm for finding a Gauss-Kronrod quadrature rule using $2n + 1$ nodes with $n \geq 3$ odd:

Algorithm 2.3.1 (Determine Gauss-Kronrod Quadrature Rule)

1. Find the eigenvalues $\boldsymbol{\Lambda}_n = \operatorname{diag}(\lambda_{1,n}, \ldots, \lambda_{n,n})$ of the Jacobi matrix $\mathbf{T}_n$ for the monic Legendre polynomials in Eq. (2.29), and compute the last row $\mathbf{e}_n{}^\top \mathbf{Q}_n$ of its matrix of eigenvectors.
2. If the Jacobi matrix for the monic Legendre polynomials of degree $(3n + 3)/2$ is partitioned

$$\mathbf{T}_{(3n+3)/2} = \begin{bmatrix} \mathbf{T}_n & \mathbf{e}_n \beta_n & \\ \beta_n \mathbf{e}_n{}^\top & \alpha_n & \beta_{n+1} \mathbf{e}_1{}^\top \\ & \mathbf{e}_1 \beta_{n+1} & \mathbf{U}_{(n+1)/2} \end{bmatrix} \,,$$

find the eigenvalues $\mathbf{D}_{(n+1)/2} = \operatorname{diag}(\delta_1, \ldots, \delta_{(n+1)/2})$ and the first row $\mathbf{e}_1{}^\top \mathbf{P}_{(n+1)/2}$ of the matrix of eigenvectors for $\mathbf{U}_{(n+1)/2}$.
3. For $1 \leq k \leq (n + 1)/2$, compute the weights

$$\pi_j = \left(\mathbf{e}_1{}^\top \mathbf{P}_{(n+1)/2} \mathbf{e}_j \right)^2 \,,$$

then for $1 \leq j \leq n$ use the Lagrange interpolation polynomials

$$\ell_{j,n}(x) = \prod_{\substack{1 \leq i \leq n \\ i \neq j}} \frac{x - \lambda_{i,n}}{\lambda_{j,n} - \lambda_{i,n}}$$

to compute the weights

$$\omega_j^* = \sum_{k=1}^{(n+1)/2} \ell_{j,n}(\delta_k) \pi_j \,.$$

If these weights are all positive, then compute

$$\mathbf{e}_1{}^\top \mathbf{P}_n^{GK} = \left[\sqrt{\omega_1^*}, \ldots, \sqrt{\omega_n^*} \right] .$$

4. Compute the n-vector $\mathbf{c}$ for which the squares of the entries are

$$\gamma_j^2 = \left(\mathbf{e}_n{}^\top \mathbf{Q}_n \mathbf{e} \right) j \beta_n \right)^2 + \left(\mathbf{e}_1{}^\top \mathbf{P}_n^{GK} \mathbf{e} \right) j \beta_{n+1} \right)^2 .$$

Then for $1 \leq j \leq n+1$ find the distinct real zeros λ_j^{GK} of the secular equation

$$\lambda - \alpha_n + \sum_{j=1}^{n} \frac{\gamma_j^2}{\lambda_{j,n} - \lambda} = 0 .$$

For each secular equation solution λ_j^{GK}, find the corresponding eigenvector

$$\begin{bmatrix} -\left(\boldsymbol{\Lambda}_n - \mathbf{I}\lambda_j^{GK} \right)^{-1} \mathbf{c} \\ 1 \end{bmatrix}$$

and normalize it to find the jth column of the orthogonal matrix

$$\begin{bmatrix} \mathbf{W} \\ \mathbf{w}^\top \end{bmatrix} .$$

5. Use the elementary rotations $\mathbf{G}_{1,n+1}, \ldots, \mathbf{G}_{n,2n}$ defined by

$$\mathbf{G}_{n,2n}{}^\top \ldots \mathbf{G}_{1,n+1}{}^\top \begin{bmatrix} \mathbf{Q}_n{}^\top \mathbf{e}_n \beta_n \\ \mathbf{P}_n^{GK}{}^\top \mathbf{e}_1 \beta_{n+1} \end{bmatrix} = \begin{bmatrix} \mathbf{0} \\ \mathbf{c} \end{bmatrix}$$

to compute the transposed unit vector

$$\left[\mathbf{p}^\top, \mathbf{q}^\top \right] = \left[\mathbf{e}_1{}^\top \mathbf{Q}_n, \mathbf{0}^\top \right] \mathbf{G}_{1,n+1} \ldots \mathbf{G}_{n,2n} .$$

6. Compute the first row

$$\mathbf{e}_1{}^\top \mathbf{Q}_{2n+1}^{GK} = \left[\mathbf{p}^\top, \mathbf{q}^\top \mathbf{W} \right]$$

of the orthogonal matrix of eigenvectors of $\mathbf{T}_{2n+1}^{GK}$. Then compute the weights for the Gauss-Kronrod quadrature rule with $2n + 1$ nodes as

$$\omega_j^{GK} = \left(\mathbf{p}^\top \mathbf{e}_j \right)^2 \text{ for } 1 \leq j \leq n \text{ and } \omega_{j+n}^{GK} = \left(\mathbf{q}^\top \mathbf{W} \mathbf{e}_j \right)^2 \text{ for } 1 \leq j \leq n+1 .$$

The first n weights correspond to the original Gaussian quadrature nodes, and the latter correspond to the zeros of the Stieltjes polynomial s_{n+1}.

Gauss-Kronrod quadrature rules are available from netlib as QUADPACK dqng, or from the GNU Scientific Library (GSL) as QNG non-adaptive Gauss-Kronrod integration. In particular, QUADPACK routine dqk15 implements both a seven-point Gaussian quadrature and a 15-point Gauss-Kronrod quadrature (i.e., $n = 7$). QUADPACK Routines dqk21, dqk31, dqk41, dqk51 and dqk61 implement both Gaussian quadrature and Gauss-Kronrod quadrature for $n = 10, 15, 20, 25$ and 30, respectively.

MATLAB users may consider the command quadgk to perform Gauss-Kronrod quadrature.

2.3.13 Difficult Integrals

So far in this chapter, we have limited our discussion to the integration of continuous functions on a bounded interval. We have been avoiding integrands that are discontinuous, because standard numerical quadratures may converge slowly to the correct integrals. Even worse, if an integrand has a singularity then its numerical quadrature may generate floating point exceptions, or the convergence of the quadrature may again be slow. Unbounded integration regions may also cause difficulties, as do highly oscillatory integrands.

2.3.13.1 Integrands with Singularities

If f or one of its derivatives is singular in $[a, b]$, then the numerical methods we have developed may not work well. In the following examples, we will illustrate some techniques that may help.

Example 2.3.8 Suppose that we want to compute

$$I = \int_0^1 \frac{e^x}{\sqrt{x}} \, dx \ .$$

If we substitute $x = t^2$, then we get

$$I = 2 \int_0^1 e^{t^2} \, dt \ .$$

This integral has no singularity.

Example 2.3.9 Since the first integral in the previous example has an integrable singularity, we can integrate by parts to get:

$$I = 2\sqrt{x}e^x\big|_0^1 - 2\int_0^1 \sqrt{x}e^x \, dx = 2e - 2\left\{\frac{2}{3}x^{3/2}e^x\Big|_0^1 - \frac{2}{3}\int_0^1 x^{3/2}e^x \, dx\right\}$$

$$= \frac{2}{3}e + \frac{4}{3}\int_0^1 x^{3/2}e^x \, dx \; .$$

The last integral involves a continuously differentiable function. Additional integrations by parts could push the singularity into higher derivatives.

Example 2.3.10 In order to compute the same integral, we could subtract some number of leading order terms at the singularity:

$$\int_0^1 \frac{1}{\sqrt{x}}e^x \, dx = \int_0^1 \frac{1}{\sqrt{x}}\left(1 + x + \frac{x^2}{2}\right) \, dx + \int_0^1 \frac{1}{\sqrt{x}}(e^x - 1 - x - \frac{x^2}{2}) \, dx \; .$$

The first integral on the right can be computed analytically, and the second integral on the right involves a twice continuously differentiable function.

The QUADPACK Fortran routine dqagp or the GSL C routine gsl_integration_qagp will approximate the integral $\int_a^b f(x) \, dx$ where f has discontinuities or singularities at a specified set of points in (a, b). On the other hand, the QUADPACK Fortran routine dqags or the GSL C routine gsl_integration_qags will approximate an integral $\int_a^b f(x) \, dx$ in which f has a singularity at either a or b, while the QUADPACK Fortran routine dqaws or the GSL C routine gsl_integration_qaws will approximate an integral
$\int_a^b f(x)w(x) \, dx$ in which w has a singularity at either a or b. The QUADPACK Fortran routine dqawc or the GSL C routine gsl_integration_qawc will compute the **Cauchy principal value**

$$PV \int_a^b \frac{f(x)}{x - c} \, dx = \lim_{\varepsilon \downarrow 0}\left[\int_a^{c-\varepsilon} \frac{f(x)}{x - c} \, dx + \int_{c+\varepsilon}^b \frac{f(x)}{x - c} \, dx\right] \; .$$

2.3.13.2 Indefinite Integrals

If f and its derivatives are small outside some interval $[-R, R]$, then the Euler-MacLaurin formula shows that either the trapezoidal or midpoint rule will give good results in approximating $\int_{-R}^R f(x) \, dx$. However, it can be tricky to estimate the truncation error

$$\int_{-\infty}^{\infty} f(x)\, dx - \int_{-R}^{R} f(x)\, dx = \int_{-\infty}^{-R} f(x)\, dx + \int_{R}^{\infty} f(x)\, dx \ .$$

We will provide two examples of approaches to indefinite integrals.

Example 2.3.11   Suppose that we want to compute

$$\int_{-\infty}^{\infty} e^{-x^2}\, dx \ .$$

Note that for $|x| \geq 4$, we have $e^{-x^2} \leq 0.5 \times 10^{-6}$. We could approximate $\int_{-4}^{4} e^{-x^2}\, dx$ by a quadrature rule; for this integral, either the trapezoidal rule or midpoint rule (and their extrapolants) will be very accurate. Next, let us estimate the truncation error in replacing the indefinite integral by this definite integral:

$$\left| \int_{-\infty}^{\infty} e^{-x^2}\, dx - \int_{-4}^{4} e^{-x^2}\, dx \right| = 2 \int_{4}^{\infty} e^{-x^2}\, dx = 2 \int_{16}^{\infty} \frac{e^{-t}}{2\sqrt{t}}\, dt$$

$$< \frac{1}{4} \int_{16}^{\infty} e^{-t}\, dt = \frac{e^{-16}}{4} < 10^{-7} \ .$$

Example 2.3.12   Sometimes a substitution can help. If $x = \tan\theta$ then

$$\int_{0}^{\infty} (1 + x^2)^{-4/3}\, dx = \int_{0}^{\pi/2} (\cos\theta)^{2/3}\, d\theta \ .$$

The integral on the right involves an integrand with infinitely many continuous derivatives.

The QUADPACK Fortran routine dqagi or the GSL C routines gsl_integration_qagi, gsl_integration_qagiu and gsl_integration_qagil compute an indefinite integral by the substitution $x = (1 - t)/t$, leading to the computations

$$\int_{0}^{\infty} f(x)\, dx = \int_{0}^{1} f\left(\frac{1-t}{t}\right) \frac{dt}{t^2}$$

and

$$\int_{-\infty}^{\infty} f(x)\, dx = \int_{0}^{1} f\left(\frac{1-t}{t}\right) + f\left(-\frac{1-t}{t}\right) \frac{dt}{t^2} \ .$$

2.3.13.3 Oscillatory Integrals

Integrals involving products with sines or cosines arise in Fourier series, particularly during the solution of partial differential equations such as the wave equation, heat equation or Laplace's equation. Numerical evaluation of such integrals could produce relatively large rounding errors, if many function evaluations with nearly equal values and opposite signs are summed.

The QUADPACK Fortran routine dqawo or the GSL C routine gsl_integration_qawo uses an approach described by Plessens [143] to compute $\int_a^b f(x) \cos(\omega x) \, dx$ or $\int_a^b f(x) \sin(\omega x) \, dx$. After performing a change of variables to integration over $(-1, 1)$, these routines use an approximation of the form

$$\int_{-1}^1 f(x) w(x) \, dx \approx \sum_{n=0}^N c_n M_n$$

where f is approximated by Chebyshev polynomials

$$f(x) \approx \sum_{n=0}^N c_n t_n(x)$$

and the related **Chebyshev moments** of the weight function are

$$M_n = \int_{-1}^1 t_n(x) w(x) \, dx = \int_0^\pi w(\cos \theta) \sin \theta \cos(n\theta) \, d\theta \ .$$

The Chebyshev moments are computed recursively in QUADPACK routine dqmomo.

Exercise 2.3.18 Discuss how to evaluate the following integrals. Describe how to preprocess the integrals analytically, and how you would compute the transformed integrals numerically. If your approach has a truncation error involving a small unbounded integral, show how to bound the truncation error.

1. $\int_0^1 x^{-1/4} \sin x \, dx$,
2. $\int_0^1 x^{-2/5} e^{2x} \, dx$,
3. $\int_1^2 (x-1)^{-1/5} \ln x \, dx$,
4. $\int_0^1 x^{-1/3} \cos(2x) \, dx$,
5. $\int_0^1 (1-x)^{-1/2} e^{-x} \, dx$,
6. $\int_0^{\frac{1}{2}} (2x-1)^{-1/3} \, dx$,
7. $\int_{-1}^0 (3x+1)^{-1/3} \, dx$,
8. $\int_0^2 (x-1)^{2/3} x e^x \, dx$,
9. $\int_1^\infty (x^2+9)^{-1} \, dx$,
10. $\int_1^\infty (1+x^4)^{-1} \, dx$,

11. $\int_1^\infty x^{-3} \cos(x) \, dx$,
12. $\int_1^\infty x^{-4} \sin(x) \, dx$,
13. $\int_0^\infty \frac{x^3}{e^x - 1} \, dx$,
14. $\int_{-1}^1 \int_{-1}^1 \frac{1}{\sqrt{(x-\xi)^2 + (y-\eta)^2}} \, d\xi \, d\eta$ for $2 \le x, y \le 10$.

2.3.14 Adaptive Quadrature

It can be difficult to integrate functions with very localized behavior, such as a local smooth spike. Even though the integrands may be smooth, with methods such as Romberg integration we will need a large number of nodes to capture the region of rapid change. The problem is that the extrapolation in Romberg integration involves repeated refinement of a uniform mesh. This means that computational effort will be wasted on function evaluations outside the localized behavior.

In this section, we will develop a way to concentrate the integration work where it is needed. Given a user-specified tolerance ε, our goal is to design an **adaptive quadrature** rule $Q(f; a, b)$ so that

$$\left| \int_a^b f(x) \, dx - Q(f; a, b) \right| \le \varepsilon \int_a^b |f(x)| \, dx .$$

At its finest level of operation, our quadrature rule should have some identifiable order k; in other words, for any $[z - h/2, z + h/2] \subset [a, b]$,

$$\int_{z-h/2}^{z+h/2} f(x) dx - Q(f; z - h/2, z + h/2) = O(h^k) .$$

It is also helpful if the user gives us an estimate of the minimal number of nodes to use, or if the user tells us how to look for the localized behavior. We want to be sure that our adaptive algorithm samples enough to see the abrupt behavior before we stop.

Our adaptive quadrature algorithm will use **recursion**. Given some subinterval $[c, c + h]$ of $[a, b]$, we will bisect this interval and apply the adaptive algorithm to both sub-intervals. Ideally, this process will continue until the integral on each interval satisfies

$$\left| \int_c^{c+h} f(x) \, dx - Q(f; c, c + h) \right| \le \varepsilon \frac{h}{b - a} \int_a^b |f(x)| \, dx . \tag{2.35}$$

If this process succeeds, then the recursive algorithm will have generated a mesh $a = x_0 < \ldots < x_n = b$ so that

$$\left| \int_a^b f(x)\,dx - \sum_{j=1}^m Q(f; x_{j-1}, x_j) \right| \leq \sum_{j=1}^m \left| \int_{x_{j-1}}^{x_j} f(x)\,dx - Q(f; x_{j-1}, x_j) \right|$$

$$\leq \sum_{j=1}^m \left[\varepsilon \frac{x_j - x_{j-1}}{b-a} \int_a^b |f(x)|\,dx \right] = \varepsilon \int_a^b |f(x)|\,dx .$$

On an interval $[c, c + h]$, the recursion will approximate $\int_c^{c+h} f(x)dx$ by two quadratures. First, we will use our quadrature rule $Q(f; c, c + h)$ to approximate $\int_c^{c+h} f(x)\,dx$. Afterward, we will use the quadrature rule on the two sub-intervals $[c, c+h/2]$ and $[c+h/2, c+h]$ to approximate $\int_c^{c+h} f(x)\,dx$ by $Q(f; c, c+h/2) + Q(f; c+h/2, c+h)$. Assuming that our quadrature rule has order k, we should have

$$\int_c^{c+h} f(x)dx - Q(f; c, c + h) \approx ah^k \text{ and}$$

$$\int_c^{c+h} f(x)dx - Q(f; c, c + h/2) - Q(f; c + h/2, c + h) \approx a(h/2)^k .$$

We can subtract the first approximation from the second to get

$$[Q(f; c, c + h/2) + Q(f; c + h/2, c + h)] - Q(f; c, c + h) \approx ah^k(1 - 1/2^k) .$$

This allows us to eliminate the unknown error constant a and estimate the error in the finer quadrature by

$$\int_c^{c+h} f(x)\,dx - [Q(f; c, c + h/2) + Q(f; c + h/2, c + h)]$$

$$\approx \frac{1}{2^k - 1} \{ [Q(f; c, c + h/2) + Q(f; c + h/2, c + h)] - Q(f; c, c + h) \} .$$

If the right-hand side of this approximation is less than the bound in our convergence criterion (2.35), in other words, if

$$|[Q(f; c, c + h/2) + Q(f; c + h/2, c + h)] - Q(f; c, c + h)|$$

$$\leq \varepsilon \left(2^k - 1 \right) \frac{h}{b-a} \int_a^b |f(x)|\,dx ,$$

then we stop the recursion on the current interval $[c, c + h]$.

The integral $\int_a^b |f(x)| dx$ can be estimated by some quadrature process that is maintained during the course of adaptive integration. In other words, while we are computing quadratures for the integral of f, we can easily compute a second quadrature for $|f|$.

QUADPACK performs adaptive quadrature using Gauss-Kronrod quadrature rules in Fortran routine dqag. The GNU Scientific Library (GSL) provides the same algorithm in the C routine gsl_integration_qag. Most of the other QUADPACK and GSL integration routines use adaptive quadrature as well.

The MATLAB command integral also uses adaptive quadrature.

A C++ program to perform adaptive quadrature can be found in GUIAdaptive-Quadrature.C. Alternatively, readers may execute a JavaScript program for adaptive quadrature.

Exercise 2.3.19 Use adaptive quadrature to compute the following integrals such that

$$\left| \int_a^b f(x)dx - Q(f; a, b) \right| \leq \varepsilon \int_a^b |f(x)| dx$$

where $\varepsilon = 10^{-10}$. Tell how many function evaluations were required for each integral.

1. $\int_1^3 e^{2x} \sin(3x) dx$
2. $\int_1^3 e^{3x} \sin(2x) dx$
3. $\int_0^5 \{2x \cos(2x) - (x-2)^2\} dx$
4. $\int_0^5 \{4x \cos(2x) - (x-2)^2\} dx$
5. $\int_{0.1}^2 \sin(1/x) dx$
6. $\int_{0.1}^2 \cos(1/x) dx$
7. $\int_0^1 \sqrt{x} dx$
8. $\int_0^1 \sqrt{1-x} dx$
9. $\int_0^1 (1-x)^{1/4} dx$

2.3.15 Multiple Dimensions

Integration in multiple dimensions is often far more difficult than in one dimension. An integration domain with curved boundaries may need to be approximated by a **tessellation**, involving simpler sub-domains of particular shapes. In two dimensions, the sub-domains are typically triangles and quadrilaterals. In three dimensions, it is common to use tetrahedra, wedges and octahedra. If the boundary of the original domain is curved, then tessellations formed from a union of these sub-domains would provide a low-order approximation to the original domain. To deal with curved boundaries, it is common to define coordinate mappings from certain canonical sub-domains to portions of the original domain, and then integrate on the mapped sub-domain via a change of variables. Such an approach is common in the **finite element method**, which is used to solve boundary-value problems in ordinary and partial differential equations. For more information about finite element methods, we recommend books by Braess [20], Brenner and Scott [23], Chen [37], Ciarlet [43], Hughes [106], Johnson [116], Strang and Fix [163], Szabó and Babuška [169], Trangenstein [174], Wait and Mitchell [180] or Zienkiewicz [189].

2.3.15.1 Tensor Product Quadrature

In order to compute integrals on a unit square or a unit cube, we can use products of one-dimensional rules. Given a 1D rule

$$\int_0^1 f(\xi) \, d\xi \approx \sum_{q=0}^{Q} f(\xi_q) w_q \, ,$$

we can approximate a 2D integral over the unit square by

$$\int_0^1 \int_0^1 f(\xi_0, \xi_1) \, d\xi_0 \, d\xi_1 \approx \sum_{q_1=0}^{Q} \sum_{q_0=0}^{Q} f(\xi_{q_0}, \xi_{q_1}) w_{q_0} w_{q_1}$$

or approximate a 3D integral over the unit cube by

$$\int_0^1 \int_0^1 \int_0^1 f(\xi_0, \xi_1, \xi_2) \, d\xi_0 \, d\xi_1 \, d\xi_2 \approx \sum_{q_2=0}^{Q} \sum_{q_1=0}^{Q} \sum_{q_0=0}^{Q} f(\xi_{q_0}, \xi_{q_1}, \xi_{q_2}) w_{q_0} w_{q_1} w_{q_2} \, .$$

If the weights in the 1D quadrature rule are nonnegative, so are the weights in the product rule. Furthermore, if the 1D quadrature rule is exact for polynomials of degree k, then the product rule is exact for polynomials of degree k in several variables.

The MATLAB commands integral2 and integral3 will integrate a function over a rectangular region in two or three dimensions.

2.3.15.2 Integrals Over Simplices

By using **barycentric coordinates** (described in Definition 1.3.4), the interior of a triangle can be written

$$\mathcal{R} = \{\boldsymbol{\xi} : \mathbf{b}(\boldsymbol{\xi}) > 0\} \, .$$

Any function $g(\xi)$ defined on the reference triangle $\mathcal{R}$ can be written in terms of another function of barycentric coordinates:

$$g(\xi) = \widetilde{g}(\mathbf{b}(\xi)) .$$

This representation is not unique.

Integrals of barycentric monomials are easy to compute exactly. The next lemma, which is stated without proof in Hughes [106, p. 172], validates this claim. The lemma uses multi-indices, which were first described in Definition 1.3.1.

Lemma 2.3.9 *Suppose that* $d \geq 1$, *and that* α *is a multi-index with* $d + 1$ *components. Then*

$$\int_{\mathbf{b}(\xi)>0} \mathbf{b}(\xi)^{\alpha} \, d\xi = \frac{\alpha!}{(d + |\alpha|)!} . \tag{2.36}$$

Proof It is well-known (see, for example, Feller [77, v. II p. 47]) that

$$\int_0^1 x^m (1-x)^n \, dx = \frac{\Gamma(m+1)\Gamma(n+1)}{\Gamma(m+n+2)} ,$$

where the Γ function is defined by $\Gamma(n) \equiv \int_0^\infty x^{n-1} e^{-x} \, dx$. Further, for all integers $n \geq 0$ we have $\Gamma(n+1) = n!$. Finally, if m and n are integers, then the binomial expansion Theorem 1.6.1 of Chap. 1 in Volume II implies that

$$\frac{m!n!}{(m+n+1)!} = \int_0^1 x^m (1-x)^n \, dx = \sum_{k=0}^n \binom{n}{k} \int_0^1 x^m (-x)^k \, dx$$

$$= \sum_{k=0}^n \binom{n}{k} \frac{(-1)^k}{m+k+1} . \tag{2.37}$$

Thus the claimed result is easy to prove in one dimension:

$$\int_{\mathbf{b}(\xi)>0} \mathbf{b}(\xi)^{\alpha} \, d\xi = \int_0^1 \xi_1^{\alpha_1} (1-\xi_1)^{\alpha_2} \, d\xi_1 = \frac{\alpha_1! \, \alpha_2!}{(\alpha_1 + \alpha_2 + 1)!} = \frac{\alpha!}{(1 + |\alpha|)!} .$$

Inductively, let us assume that the claim holds for all d-vectors ξ. Define the $(d+1)$-vector $\overline{\xi}$, the multi-index $\overline{\alpha}$ with $d+2$ components, and the barycentric coordinates $\overline{\beta}\left(\overline{\xi}\right)$ by

$$\overline{\xi} = \begin{bmatrix} \xi \\ \xi_d \end{bmatrix} , \quad \overline{\alpha} = \begin{bmatrix} \alpha \\ \alpha_d \\ \alpha_{d+1} \end{bmatrix} \quad \text{and} \quad \overline{\beta}\left(\overline{\xi}\right) = \begin{bmatrix} \xi \\ \xi_d \\ 1 - |\xi| - \xi_d \end{bmatrix} .$$

First we expand

$$\int_{\overline{\mathbf{b}}(\overline{\xi})>0} \overline{\mathbf{b}}\left(\overline{\xi}\right)^{\overline{\alpha}} d\xi = \int_{\beta(\xi)>0} \int_0^{1-|\xi|} \xi^\alpha \xi_d^{\alpha_d} (1-|\xi|-\xi_d)^{\alpha_{d+1}} \, d\xi_d \, d\xi$$

then we use the multinomial expansion (1.17) to get

$$= \int_{\beta(\xi)>0} \int_0^{1-|\xi|} \xi^\alpha \sum_{k=0}^{\alpha_{d+1}} \binom{\alpha_{d+1}}{k} (-1)^k (1-|\xi|)^{\alpha_{d+1}-k} \xi_d^{\alpha_d+k} \, d\xi_d \, d\xi$$

$$= \sum_{k=0}^{\alpha_{d+1}} \binom{\alpha_{d+1}}{k} \frac{(-1)^k}{\alpha_d+k+1} \int_{\beta(\xi)>0} \xi^\alpha (1-|\xi|-\xi_d)^{\alpha_d+\alpha_{d+1}+1} \, d\xi$$

then we use the inductive hypothesis to obtain

$$= \sum_{k=0}^{\alpha_{d+1}} \binom{\alpha_{d+1}}{k} \frac{(-1)^k}{\alpha_d+k+1} \frac{\alpha!\,(\alpha_d+\alpha_{d+1}+1)!}{(d-1+|\alpha|+\alpha_d+\alpha_{d+1}+1)!}$$

and finally we use Eq. (2.37)

$$= \frac{\alpha_d!\alpha_{d+1}!}{(\alpha_d+\alpha_{d+1}+1)!} \frac{\alpha!\,(\alpha_d+\alpha_{d+1}+1)!}{(d+|\alpha|+\alpha_d+\alpha_{d+1})!} = \frac{\overline{\alpha}!}{(d+|\overline{\alpha}|)!} \ .$$

This lemma gives us analytical results for integrals of polynomials against which we can check our quadrature rules for order of accuracy.

2.3.15.3 Quadratures on Triangles

We will define the **reference triangle** to have vertices $(0,0)$, $(1,0)$ and $(0,1)$. One approach to numerical quadrature on the reference triangle is to approximate a given function by its Lagrange polynomial interpolant at equally spaced lattice points in the reference triangle, and then integrate the Lagrange interpolation polynomials exactly using Eq. (2.36). Here are a few of these quadrature rules:

$$\int_{\mathbf{b}(\xi)>0} f(\xi) \, d\xi \approx \frac{1}{6}[f(1,0)+f(0,1)+f(0,0)] \text{ (exact for degree 1) ,}$$

$$\int_{\mathbf{b}(\xi)>0} f(\xi) \, d\xi \approx \frac{1}{6}\left[f\left(\frac{1}{2},\frac{1}{2}\right)+f\left(0,\frac{1}{2}\right)+f\left(\frac{1}{2},0\right)\right] \text{ (exact for degree 2) ,}$$

$$\int_{\mathbf{b}(\xi)>0} f(\xi) \, d\xi \approx \frac{1}{24}\left[f(1,0)+f(0,1)+f(0,0)+9f\left(\frac{1}{3},\frac{1}{3}\right)\right] \text{ (exact for degree 2)}$$

The first rule involves the same number of quadrature points as the second rule, but the second rule is exact for polynomials of higher degree. As a result, there is no good reason to use the first rule. Similarly, the third rule is superseded by the second rule.

Next, let us discuss generalizations of Gaussian quadrature to triangles. Cowper [52] derived **symmetric** Gaussian quadrature rules for triangles. Symmetric quadrature rules are such that if $\mathbf{b}(\xi) \in \mathbb{R}^{d+1}$ are the barycentric coordinates for a quadrature point with weight w, then all $(d + 1)!$ permutations of the entries of $\mathbf{b}(\xi)$ are also quadrature points in the same quadrature rule with the same weight.

The simplest symmetric rule involves one quadrature point, and has the form

$$\int_{\mathbf{b}(\xi)>0} f(\xi) \, d\xi \approx f(\beta, \beta)w \, .$$

Symmetry requires that all entries of β are equal, so $\beta = 1/3$. In order for the quadrature rule to be exact for constants, we must have $w = 1/2$. The resulting symmetric quadrature rule is

$$\int_{\mathbf{b}(\xi)>0} f(\xi) \, d\xi \approx \frac{1}{2} f\left(\frac{1}{3}, \frac{1}{3}\right) \, .$$

This quadrature rule is exact for polynomials of degree at most one.

The next symmetric quadrature rule involves three quadrature points, and has the form

$$\int_{\mathbf{b}(\xi)>0} f(\xi) \, d\xi \approx w \left[f(\beta_1, \beta_2) + f(\beta_2, \beta_1) + f(\beta_2, \beta_2) \right] \, ,$$

where $\beta_1 + 2\beta_2 = 1$. There is one solution that is exact for polynomials of degree 2, namely

$$\int_{\mathbf{b}(\xi)>0} f(\xi) \, d\xi \approx \frac{1}{6} \left[f\left(0, \frac{1}{2}\right) + f\left(\frac{1}{2}, 0\right) + f\left(\frac{1}{2}, \frac{1}{2}\right) \right] \, .$$

The next symmetric quadrature rule involves four quadrature points, and has the form

$$\int_{\mathbf{b}(\xi)>0} f(\xi) \, d\xi \approx w_0 f\left(\frac{1}{3}, \frac{1}{3}\right) + w_1 \left[f(\beta_1, \beta_2) + f(\beta_2, \beta_1) + f(\beta_2, \beta_2) \right] \, ,$$

where $\beta_1 + 2\beta_2 = 1$. There is one solution that is exact for polynomials of degree 3, namely

$$\int_{\mathbf{b}(\xi)>0} f(\xi) \, d\xi \approx -\frac{9}{16} f\left(\frac{1}{3}, \frac{1}{3}\right) + \frac{25}{48} \left[f\left(\frac{3}{5}, \frac{1}{5}\right) + f\left(\frac{1}{5}, \frac{3}{5}\right) + f\left(\frac{1}{5}, \frac{1}{5}\right) \right] \, .$$

There are several difficulties with this approach for generating symmetric Gaussian quadratures on triangles. First, there is no general formula for arbitrarily high-order quadrature points and weights. Next, there is no guarantee that the quadrature weights are positive. Finally, there is no guarantee that the quadrature points have nonnegative barycentric coordinates.

Dunavant [70] determined symmetric Gaussian quadrature rules of order up to 20 for triangles. Some of these rules involve either negative quadrature weights, or quadrature points outside the reference triangle. Most are efficient, meaning that they use fewer points for the same order than competing symmetric Gaussian quadrature rules. Wandzura and Xiao [181] used symmetry properties and group theory to develop symmetric Gaussian quadratures exact for polynomials of degree $5k$, and computed the quadrature rules for $1 \leq k \leq 6$. All of these rules involve positive quadrature weights and quadrature points inside the reference triangle. Each of their rules are more efficient than the comparable Dunavant rule.

It is also possible to formulate generalizations of Lobatto quadrature rules for triangles. The simplest of these rules is

$$\int_{\mathbf{b}(\xi)>0} f(\xi) \, d\xi \approx \frac{1}{6} \left[f(1,0) + f(0,1) + f(0,0) \right] ,$$

which is exact for polynomials of degree 1. The next rule is

$$\int_{\mathbf{b}(\xi)>0} f(\xi) \, d\xi \approx \frac{1}{24} \left[f(1,0) + f(0,1) + f(0,0) \right] + \frac{3}{8} f \left(\frac{1}{3}, \frac{1}{3} \right) ,$$

which is exact for polynomials of degree 2. Taylor et al. [170] developed Fekete points for symmetric Gaussian quadratures on triangles. The Fekete points are known to coincide with the Lobatto points on the sides of the triangles. These authors computed the quadrature points and weights for quadrature rules exact for polynomials of degree $3k$ with $1 \leq k \leq 6$.

In order to generate arbitrarily high-order quadrature rules on triangles, so that the quadrature weights are positive and the quadrature points are inside the triangle, it was formerly popular to map tensor product quadrature rules from squares to triangles. If (ξ_0, ξ_1) is an arbitrary point inside the reference square, it was typically mapped to the point

$$\begin{bmatrix} \mathbf{x}_0 \\ \mathbf{x}_1 \end{bmatrix} (\xi) = \begin{bmatrix} \xi_0(1 - \xi_1/2) \\ \xi_1(1 - \xi_0/2) \end{bmatrix} .$$

In this way, the vertex $(1, 1)$ of the reference square is mapped to the midpoint of the long side of the reference triangle. The **Jacobian** 563 of this coordinate transformation is

$$\mathbf{J} = \begin{bmatrix} 1 - \xi_1/2 & -\xi_0/2 \\ -\xi_1/2 & 1 - \xi_0/2 \end{bmatrix} ,$$

and the determinant of the Jacobian is

$$\det \mathbf{J} = 1 - \frac{\xi_0 + \xi_1}{2} .$$

Note that this determinant is linear, and vanishes at the vertex $(1, 1)$ of the reference square. The quadrature rule takes the form

$$\int_{\mathbf{x}_0>0, \mathbf{x}_1>0, \mathbf{x}_0+\mathbf{x}_1<1} f(\mathbf{x}) \, d\mathbf{x} = \int_0^1 \int_0^1 f(\mathbf{x}(\xi)) \det \mathbf{J}(\xi) \, d\xi_0 \, d\xi_1$$

$$\approx \sum_{q_1=0}^{Q} \sum_{q_0=0}^{Q} f\left(\xi_{q_0}\left[1 - \frac{\xi_{q_1}}{2}\right], \xi_{q_1}\left[1 - \frac{\xi_{q_0}}{2}\right]\right)\left[1 - \frac{\xi_{q_0} + \xi_{q_1}}{2}\right] w_{q_0} w_{q_1} .$$

Note that if $f(\mathbf{x}) = \mathbf{x}^\alpha$ is a monomial, then

$$f\left(\xi_{q_0}\left[1 - \frac{\xi_{q_1}}{2}\right], \xi_{q_1}\left[1 - \frac{\xi_{q_0}}{2}\right]\right) = \left[\xi_0^{\alpha_0}(1 - \xi_0/2)^{\alpha_1}\right]\left[\xi_1^{\alpha_1}(1 - \xi_1/2)^{\alpha_0}\right]$$

is a product of polynomials in the individual reference square coordinates, and each of the polynomials in the product has degree at most $|\alpha|$. Since the determinant of the Jacobian is linear, the tensor product Gaussian quadrature rule applied to a polynomial of degree k is being applied to a linear function times a linear combination of products of polynomials of degree k in the individual coordinates. It follows that the product Gaussian quadrature rule on triangles has order one less than the corresponding Gaussian quadrature rule on a square. This process, of course, generates an asymmetric quadrature rule, with one coordinate being treated differently from the other two. For this reason, we do not recommend this approach.

A survey of triangle quadrature rules of order up to 12 can be found in [131]. Several C^{++} classes to implement various quadrature rules on triangles can be found in files `Quadrature.H` and `Quadrature.C` available from Cambridge University Press. In particular, the class `TriangleGaussianQuadrature` selects the most efficient quadrature rule from the various options that have been discussed.

2.3.15.4 Quadratures on Tetrahedra

Using barycentric coordinates, the interior of a tetrahedron can be written

$$\mathscr{R} = \{\xi : \mathbf{b}(\xi) > 0\} .$$

Any function $g(\xi)$ defined on the reference tetrahedron can be written in terms of the barycentric coordinates: $g(\xi) = \widetilde{g}(\mathbf{b}(\xi))$. This representation is not unique.

Newton-Cotes rules for tetrahedra use the equally spaced lattice points for the quadrature points, with the quadrature weights chosen so that the quadrature rule is exact for all polynomials of degree at most n with n as large as possible. Some of these results can be obtained by integrating the Lagrange interpolant. Here are three of these quadrature rules:

$$\int_{\mathbf{b}(\xi)>0} f(\xi) \, d\xi \approx \frac{1}{24}[f(1,0,0) + f(0,1,0) + f(0,0,1) + f(0,0,0)]$$

(exact for degree 1),

$$\int_{\mathbf{b}(\xi)>0} f(\xi) \, d\xi \approx \frac{1}{240}[f(1,0,0) + f(0,1,0) + f(0,0,1) + f(0,0,0)]$$

$$+ \frac{3}{80}\left[f\left(\frac{1}{3},\frac{1}{3},\frac{1}{3}\right) + f\left(0,\frac{1}{3},\frac{1}{3}\right) + f\left(\frac{1}{3},0,\frac{1}{3}\right) + f\left(\frac{1}{3},\frac{1}{3},0\right)\right]$$

(exact for degree 2) and

$$\int_{\mathbf{b}(\xi)>0} f(\xi) \, d\xi \approx \frac{1}{360}[f(1,0,0) + f(0,1,0) + f(0,1,0) + f(0,0,0)]$$

$$+ \frac{1}{90}\left[f\left(\frac{1}{2},0,0\right) + f\left(0,\frac{1}{2},0\right) + f\left(0,0,\frac{1}{2}\right) + f\left(0,\frac{1}{2},\frac{1}{2}\right)\right.$$

$$\left. + f\left(\frac{1}{2},0,\frac{1}{2}\right) + f\left(\frac{1}{2},\frac{1}{2},0\right)\right] + \frac{8}{15}f\left(\frac{1}{4},\frac{1}{4},\frac{1}{4}\right) \quad \text{(exact for degree 3)}.$$

Many of the Newton-Cotes rules for tetrahedra involve negative weights, and should not be used.

Yu [188] described some symmetric Gaussian quadrature rules for tetrahedra. Here are two such rules:

$$\int_{\mathbf{b}(\xi)>0} f(\xi) \, d\xi \approx \frac{1}{6}f\left(\frac{1}{4},\frac{1}{4},\frac{1}{4}\right) \quad \text{and}$$

$$\int_{\mathbf{b}(\xi)>0} f(\xi) \, d\xi \approx \frac{1}{24}[f(\beta_1,\beta_2,\beta_2) + f(\beta_2,\beta_1,\beta_2) + f(\beta_2,\beta_2,\beta_1) + f(\beta_2,\beta_2,\beta_2)]$$

$$\text{where } \beta_1 = \frac{1 + 3\sqrt{1/5}}{4} \text{ and } \beta_2 = \frac{1 - \sqrt{1/5}}{4},$$

These rules are exact for polynomials of degree 1 and 2, respectively. The symmetric nine-point rules

$$\int_{\mathbf{b}(\xi)>0} f(\xi) \, d\xi \approx w_1\left[f\left(\beta_1,\beta_2,\beta_2\right) + f\left(\beta_2,\beta_1,\beta_2\right) + f\left(\beta_2,\beta_2,\beta_1\right) + f\left(\beta_2,\beta_2,\beta_2\right)\right]$$

$$+ w_2\left[f\left(\gamma_1,\gamma_2,\gamma_2\right) + f\left(\gamma_2,\gamma_1,\gamma_2\right) + f\left(\gamma_2,\gamma_2,\gamma_1\right) + f\left(\gamma_2,\gamma_2,\gamma_2\right)\right]$$

that are exact for as many polynomials as possible either have a negative weight or a quadrature point outside the tetrahedron. The symmetric 12-point rules of the form

$$\int_{b(\xi)>0} f(\xi)\,d\xi \approx \frac{1}{72}[f(\beta_1,\beta_2,\beta_3)+f(\beta_1,\beta_3,\beta_2)+f(\beta_1,\beta_3,\beta_3)+f(\beta_2,\beta_1,\beta_3)$$

$$+f(\beta_3,\beta_1,\beta_2)+f(\beta_3,\beta_1,\beta_3)+f(\beta_2,\beta_3,\beta_1)+f(\beta_3,\beta_2,\beta_1)$$

$$+f(\beta_3,\beta_3,\beta_1)+f(\beta_2,\beta_3,\beta_3)+f(\beta_3,\beta_2,\beta_3)+f(\beta_3,\beta_3,\beta_3)]$$

are all exact for polynomials of degree at most 3, and are therefore superseded by the eight-point Newton-Cotes rules. The symmetric 12-point rules of the form

$$\int_{b(\xi)>0} f(\xi)\,d\xi \approx w_1[f(\beta_1,\beta_2,\beta_2)+f(\beta_2,\beta_1,\beta_2)+f(\beta_2,\beta_2,\beta_1)+f(\beta_2,\beta_2,\beta_2)]$$

$$+ w_2[f(\gamma_1,\gamma_2,\gamma_2)+f(\gamma_2,\gamma_1,\gamma_2)+f(\gamma_2,\gamma_2,\gamma_1)+f(\gamma_2,\gamma_2,\gamma_2)]$$

$$+ w_3[f(\delta_1,\delta_2,\delta_2)+f(\delta_2,\delta_1,\delta_2)+f(\delta_2,\delta_2,\delta_1)+f(\delta_2,\delta_2,\delta_2)]$$

that are exact for polynomials of degree 4 either have negative weights or quadrature points outside the reference tetrahedron. Yu [188] recommended symmetric Gaussian quadrature rules with either 4, 5, 16, 17 or 29 points. Other symmetric quadrature rules for tetrahedra can be found in papers by Felippa [76], Keast [117], and Liu and Vinokur [129].

In order to generate arbitrarily high-order quadrature rules on tetrahedra, so that the quadrature weights are positive and the quadrature points are inside the triangle, it was popular to map tensor product quadrature rules from cubes to tetrahedra. If (ξ_0,ξ_1,ξ_2) is an arbitrary point inside the reference cube, it was mapped to the point

$$\begin{bmatrix} x_0 \\ x_1 \\ x_2 \end{bmatrix}(\xi) = \begin{bmatrix} \xi_0(1-\xi_1/2-\xi_2/2+\xi_1\xi_2/3) \\ \xi_1(1-\xi_2/2-\xi_0/2+\xi_2\xi_0/3) \\ \xi_2(1-\xi_0/2-\xi_1/2+\xi_0\xi_1/3) \end{bmatrix}.$$

Thus the vertex $(1,1,1)$ in the reference cube was mapped to the point $(1/3,1/3,1/3)$ in the center of the face opposite the origin in the reference tetrahedron. Evaluation of the Jacobian of this coordinate transformation shows that it is a polynomial of degree at most 4. The quadrature rule takes the form

$$\int_{x_0>0,x_1>0,x_2>0,x_0+x_1+x_2>0} f(\mathbf{x})\,d\mathbf{x} = \int_0^1\int_0^1\int_0^1 f(\mathbf{x}(\xi))\det\frac{\partial\mathbf{x}}{\partial\xi}\,d\xi_0\,d\xi_1\,d\xi_2$$

$$\approx \sum_{q_2=0}^Q \sum_{q_1=0}^Q \sum_{q_2=0}^Q f\left(\xi_{q_0}\left[1-\frac{\xi_{q_1}}{2}-\frac{\xi_{q_2}}{2}+\frac{\xi_{q_1}\xi_{q_2}}{3}\right],\xi_{q_1}\left[1-\frac{\xi_{q_2}}{2}-\frac{\xi_{q_0}}{2}+\frac{\xi_{q_2}\xi_{q_0}}{3}\right]\right.$$

$$\left.,\xi_{q_2}\left[1-\frac{\xi_{q_0}}{2}-\frac{\xi_{q_1}}{2}+\frac{\xi_{q_0}\xi_{q_1}}{3}\right]\right)\det\left[\frac{\partial\mathbf{x}}{\partial\xi}(\xi_{q_0},\xi_{q_1},\xi_{q_2})\right]w_{q_0}w_{q_1}w_{q_2}.$$

In practice, this quadrature rule has order two less than the corresponding Gaussian quadrature rule on the cube. In particular, the single-point rule is not exact for constants, and should not be used. Also note that this process generates an asymmetric quadrature rule, with one coordinate being treated differently from the other three.

Several C^{++} classes to implement various quadrature rules on tetrahedra can be found in files Quadrature.H and Quadrature.C available from Cambridge University Press. In particular, the class TetrahedronGaussianQuadrature selects the most efficient quadrature rule from the various options that have been discussed.

Chapter 3
Initial Value Problems

> *Differential equations first appeared in the late seventeenth century in the work of Isaac Newton, Gottfried Wilhelm Leibniz, and the Bernoulli brothers, Jakob and Johann. They occurred as a natural consequence of the efforts of these great scientists to apply the new ideas of the calculus to certain problems in mechanics, such as the motion of celestial bodies... For over 300 years, differential equations have served as an essential tool for describing and analyzing problems in many scientific disciplines. Their importance has motivated generations of mathematicians and other scientists to develop methods of studying properties of their solutions... Moreover, they have played a central role in the development of mathematics itself since questions about differential equations have spawned new areas of mathematics and advances in analysis, topology, algebra and geometry have often offered new perspectives for differential equations.*
>
> Walter Kelley and Allan Peterson *[118, p. ix]*

Abstract This chapter is devoted to initial values problems for ordinary differential equations. It discusses theory for existence, uniqueness and continuous dependence on the data of the problem. Special techniques for linear ordinary differential equations with constant coefficients are discussed in terms of matrix exponentials and their approximations. Next, linear multistep methods are introduced and analyzed, leading to a presentation of important families of linear multistep methods and their stability. These methods are implemented through predictor-corrector methods, and techniques for automatically selecting stepsize and order are discussed. Afterwards, deferred correction and Runge-Kutta methods are examined. The chapter ends with the selection of numerical methods for stiff problems, and a discussion of nonlinear stability.

Additional Material: The details of the computer programs referred in the text are available in the Springer website (http://extras.springer.com/2018/978-3-319-69110-7) for authorized users.

3.1 Overview

A number of interesting scientific problems relate the rate of change in some quantities to their current values. Examples include such different physical situations as compound interest in economics, population dynamics in biology, and motion under the influence of gravity and drag in physics. Consequently, the numerical solution of initial value problems for ordinary differential equation is an important topic in scientific computing. This computational task combines ideas from numerical differentiation (after all, there are derivatives in differential equations), numerical integration (that is how we invert derivatives), and the solution of both linear and nonlinear equations. We will also use our knowledge of eigenvalues, interpolation and approximation to solve ordinary differential equations. In other words, the numerical solution of ordinary differential equations will give us motivation to review and apply nearly all of our previous topics in scientific computing. Looking beyond this topic, readers will find that techniques for solving initial value problems are useful in solving boundary value problems for ordinary differential equations, and many partial differential equations.

There are many good texts available to readers on this topic. We recommend books by Brenan et al. [22], Deuflhard and Bornemann [67], Gear [86], Hairer et al. [94] and [93], Henrici [100, 101] and Lambert [124]. Some classical texts on the theory of ordinary differential equation are by Birkhoff and Rota [16], Coddington and Levinson [47], Hartman [98], Hurewicz [108] and Ince [110]. We also recommend a recent book by Kelley and Peterson [118]. For deeper discussion of nonlinear ordinary differential equations, see Drazin [69] or Teschl [171].

There are also a number of software packages available for solving initial value problems. One of the earliest publicly available programs, particularly for stiff ordinary differential equations, was the **DIFSUB** package published by Gear in his book [86]. This software package set a high standard, because it allowed the user to specify a problem and a desired accuracy, from which it chose the order of the method and the size of the time step to achieve the desired accuracy with the minimum computational work. Currently, the Fortran packages daskr, rksuite, vode, vodpk and symbolic are all available from Netlib. The GNU Scientific Library (GSL), contains C programs to implement the Bulirsch-Stoer extrapolation routine `gsl_odeiv2_step_bsimp`, two linear multistep routines, as well as eight Runge-Kutta routines. MATLAB provides two Runge-Kutta commands and six linear multistep commands for solving ordinary differential equations.

Our goals in this chapter are to introduce the reader to the basic theory of initial value problems for ordinary differential equations, to use that theory to understand the convergence and stability of numerical methods for initial value problems, and to develop efficient algorithm for solving these initial value problems. Readers should learn how to choose an appropriate algorithm for a particular problem, and how to apply that algorithm reliably.

Recall that in Chap. 1 of Volume I, we examined some simple ordinary differential equations and developed some basic numerical methods to solve these problems. We presented some elementary theory to aid our understanding of the mathematical problem, and some simple techniques of numerical analysis to explain the performance of the methods. That analysis helped us to understand how to apply the methods to individual ordinary differential equations.

We will begin this chapter by examining the theory of initial value problems for ordinary differential equations in Sect. 3.2. After that, we will use the theory to develop numerical methods. Some of these methods will be highly specialized, such as the linear algebraic techniques in Sect. 3.3 for differential equations with constant coefficients. Other numerical techniques will be very general. In particular, we will study linear multistep methods in Sect. 3.4. We will see that there are hierarchies of linear multistep methods that allow us to vary both method order and step size in order to minimize the estimated work in solving a particular initial value problem. This automatic selection of order and step size results in powerful software packages, that are capable of delivering very accurate results for a wide range of problems with very high efficiency. In some cases it is more important to choose the solver for an initial value problem so that it delivers a result of a desired order of accuracy with guaranteed stability. Both deferred correction (discussed in Sect. 3.5) and Runge-Kutta methods (see Sect. 3.6) are useful for this purpose. Stiffness is a common obstacle to the successful application of numerical methods for solving ordinary differential equations. Because it is important that readers understand this issue, and how to work around it, we will examine this topic in Sect. 3.7. We will end the chapter with a discussion of nonlinear stability in Sect. 3.8.

3.2   Theory

Let t_0 be a scalar and $\mathbf{y}_0$ be a vector. The **ordinary differential equation**

$$\mathbf{y}'(t) = \mathbf{f}(\mathbf{y}(t), t) \text{ for } t \geq t_0 \tag{3.1a}$$

becomes an **initial value problem** when we also require that

$$\mathbf{y}(t_0) = \mathbf{y}_0 . \tag{3.1b}$$

Here $\mathbf{f}$ is some function that returns vectors. Equation (3.1) constitute a mathematical model that we would like analyze in this section. Following the steps in scientific computing, as described in Sect. 1.3 of Chap. 1 in Volume I, we would like to discuss the circumstances under which this problem has a solution and the solution is unique. We would also like to develop some sense in which the solution of the initial value problem depends continuously on its data. At the end of this discussion, in Sect. 3.2.4 we will re-examine all of these issues for the special case of linear ordinary differential equations.

Recall that we provided radioactive decay as an example of an initial values problems in Sect. 1.2 of Chap. 1 in Volume I. Our first example in this chapter shows us how to convert a high-order differential equation into a system of ordinary differential equations.

Example 3.2.1 Let $p > 1$ be an integer and $y_0, \ldots, y_{p-1}$ be scalars. Suppose that we want to solve a p-th order ordinary differential equation of the form

$$D^p y(t) = f\left(y(t), Dy(t), \ldots, D^{p-1} y(t), t\right)$$

with initial values

$$y(t_0) = y_0 \ , \ Dy(t_0) = y_1 \ , \ \ldots \ , \ D^{p-1} y(t_0) = y_{p-1} \ .$$

Let us define the p-vectors

$$\mathbf{y}(t) = \begin{bmatrix} y(t) \\ Dy(t) \\ \vdots \\ D^{p-1} y(t) \end{bmatrix} , \ \mathbf{y}_0 = \begin{bmatrix} y_0 \\ y_1 \\ \vdots \\ y_{p-1} \end{bmatrix} \text{ and } \mathbf{f}\left(\mathbf{y}(t) \ , \ t\right)$$

$$= \begin{bmatrix} \mathbf{e}_1{}^\top \mathbf{y}(t) \\ \vdots \\ \mathbf{e}_{p-1}{}^\top \mathbf{y}(t) \\ f\left(\mathbf{e}_0{}^\top \mathbf{y}(t) \ , \ \ldots \ , \ \mathbf{e}_{p-1}{}^\top \mathbf{y}(t) \ , \ t\right) \end{bmatrix} .$$

These definitions allow us to write the original pth order ordinary differential equation can be written in the forms (3.1a) and (3.1b).

3.2.1   Existence

Very few conditions are needed to guarantee the existence of a solution to an initial value problem for an ordinary differential equation. For example, the following theorem is proved in Coddington and Levinson [47, p. 6].

Theorem 3.2.1 (Cauchy-Peano Existence)   *Let* $\mathbf{y}$ *be a vector,* t *be a scalar, and assume that* $\mathbf{f}(\mathbf{y}, t)$ *produces vectors for its values. Suppose that there is a scalar* t_0, *a vector* $\mathbf{y}_0$, *and positive scalars* η *and* τ *so that* $\mathbf{f}$ *is continuous on the set of points*

$$R = \{(\mathbf{y}, t) : \|\mathbf{y} - \mathbf{y}_0\| \leq \eta \text{ and } |t - t_0| \leq \tau\} \ .$$

Then there is a scalar $\tau_0 \leq \tau$ *and a continuously differentiable function* $\mathbf{y}(t)$ *so that*

$$\mathbf{y}'(t) = \mathbf{f}(\mathbf{y}(t), t) \text{ for all } |t - t_0| \leq \tau_0 \text{ and}$$

$$\mathbf{y}(t_0) = \mathbf{y}_0 \ .$$

3.2.2 Uniqueness

Unfortunately, this Cauchy-Peano existence theorem is inadequate for our computing needs. One major problem with this theorem is that its assumptions are too
weak to guarantee that the solution is unique. This could be a problem because, if
the solution is not unique, then we cannot expect any given numerical method to
find the solution we want.

We have already presented some examples of initial value problems with multiple
solutions in Examples 1.3.2 and 1.3.3 in Chap. 1 of Volume I. The following
additional example can be found in Kelley and Peterson [118, p. 3].

Example 3.2.2 Given an initial height h and an acceleration due to gravity g,
consider the initial-value problem

$$y'(t) = -\sqrt{2g\,|h - y(t)|}\,,\ y(0) = h$$

for the position $y(t)$ of an object as a function of time t. In order to apply the Cauchy-
Peano existence theorem, we could take $f(y, t) = -\sqrt{2g|1 - y|}$, $y_0 = h$ and $t_0 = 0$.
Both η and τ can be chosen to be arbitrarily large for the proposed region of
continuity for f. By inspection, we note that both $y(t) \equiv h$ and $y(t) = h - gt^2/2$ are
solutions.

In order to overcome the problem with non-uniqueness of the solution in
Theorem 3.2.1, we will need to impose an additional assumption on the problem.
We offer following result, also proved in Coddington and Levinson [47, p. 10].

Theorem 3.2.2 (Existence for Initial-Value Problems) *Suppose that* $\mathbf{y}$ *is a vector,*
t *is a scalar, and* $\|\cdot\|$ *is a norm on vectors. Assume that there is a scalar* t_0, *a vector*
$\mathbf{y}_0$, *and positive scalars* η *and* τ *so that the function* $\mathbf{f})\mathbf{y}, t)$ *mapping pairs of vectors
and scalars to vectors is continuous on the set of points*

$$R = \{(\mathbf{y}, t) : \|\mathbf{y} - \mathbf{y}_0\| \le \eta \text{ and } |t - t_0| \le \tau\}\ .$$

Also suppose that $\mathbf{f}$ *is* **Lipschitz continuous** *in* $\mathbf{y}$ *on* R, *meaning that there is a
positive scalar* Φ *so that for all* $|t - t_0| \le \tau$, *all* $\|\mathbf{y}_1 - \mathbf{y}_0\| \le \eta$ *and* $\|\mathbf{y}_2 - \mathbf{y}_0\| \le \eta$
we have

$$\|\mathbf{f}(\mathbf{y}_1, t) - \mathbf{f}(\mathbf{y}_2, t)\| \le \Phi \|\mathbf{y}_1 - \mathbf{y}_2\|\ .$$

Let

$$\phi = \max\{\|\mathbf{f}(\mathbf{y}, t)\| : (\mathbf{y}, t) \in R\}$$

and

$$\tau_0 = \min\{\eta/\phi, \tau\}\ .$$

Then there is a continuously differentiable function $\mathbf{y}(t)$ *so that* $\mathbf{y}(t)$ *is the unique solution to the initial value problem*

$$\mathbf{y}'(t) = \mathbf{f}(\mathbf{y}(t), t) \text{ for all } |t - t_0| \leq \tau_0 \text{ and}$$

$$\mathbf{y}(t_0) = \mathbf{y}_0 .$$

3.2.3   Perturbations

Our next goal is to determine how perturbations in the data for an initial value problem affect the solution. To estimate such perturbations, we will use one of the following two lemmas.

Lemma 3.2.1 (Gronwall's Inequality) *Suppose that* λ *is a positive real number, and* $\delta(t)$ *is absolutely integrable for* $0 \leq t \leq T$. *If the function* $\theta(t)$ *satisfies the differential inequality*

$$\theta'(t) \leq \lambda\theta(t) + \delta(t) ,$$

then for all $0 \leq t < T$

$$\theta(t) \leq e^{\lambda t}\theta(0) + \int_0^t e^{\lambda(t-s)}\delta(s) \, ds . \tag{3.2}$$

Proof If we multiply the given inequality for θ by $e^{-\lambda t}$ and rearrange terms, we get

$$\frac{d}{dt}\left[e^{-\lambda t}\theta(t)\right] \leq e^{-\lambda t}\delta(t) .$$

Since δ is absolutely integrable on $[0, T]$ and $e^{-\lambda t}$ is bounded on this interval, $e^{-\lambda t}\delta(t)$ is integrable on this interval. By integrating both sides with respect to t, we obtain

$$e^{-\lambda t}\theta(t) - \theta(0) \leq \int_0^t e^{-\lambda s}\delta(s) \, ds .$$

We can solve this inequality for $\theta(t)$ to obtain the claimed result.

Here is another form of Gronwall's inequality, in a form that will be useful for studying the deferred correction algorithm in Sect. 3.5.

Lemma 3.2.2 ([118, p. 383]) *Suppose that* θ *and* β *are nonnegative continuous functions on* $[t_0, t_0 + \tau]$, *and assume that there is a nonnegative constant* C *so that for all* $t \in [t_0, t_0 + \tau]$

$$\theta(t) \leq C + \int_{t_0}^t \theta(s)\beta(s) \, ds . \tag{3.3}$$

Then for all t ∈ [t_0, t_0 + τ]

$$\theta(t) \le C e^{\int_{t_0}^t \beta(s)\, ds}\ .$$

Proof Denote the right-hand side of inequality (3.3) by

$$\zeta(t) = C + \int_{t_0}^t \theta(s)\beta(s)\, ds\ .$$

Then assumption (3.3) implies that $\theta(t) \le \zeta(t)$ for all $t \in [t_0, t_0 + τ]$ and

$$\zeta'(t) = \theta(t)\beta(t) \le \zeta(t)\beta(t)\ .$$

This is equivalent to

$$\frac{\mathrm{d}\log\zeta}{\mathrm{d}t} \le \beta\ ,$$

so we can take integrals of both sides of this inequality to get

$$\log\zeta(t) - \log\zeta(t_0) \le \int_{t_0}^t \beta(s)\, ds\ .$$

Since $\zeta(t_0) = C$, we can solve for $\zeta(t)$ to get

$$\zeta(t) \le C e^{\int_{t_0}^t \beta(s)\, ds}\ .$$

Since assumption (3.3) implies that $\theta(t) \le \zeta(t)$, the claimed result follows.

Gronwall's inequality 3.2.1 can be used to estimate how the solution of an initial value problem depends on its data.

Theorem 3.2.3 (Continuous Dependence for Initial-Value Problems) *Suppose that **f** and **y** are vectors, and t is a scalar. Assume that there is a scalar t_0, a vector y_0, and positive scalars η and $τ$ so that **f** is continuous on the set of points*

$$R = \{(\mathbf{y}, t) : \|\mathbf{y} - \mathbf{y}_0\| \le \eta\ and\ |t - t_0| \le τ\}\ .$$

*Also suppose that **f** is Lipschitz continuous in **y** on R, with Lipschitz constant Φ. Let ϵ be a vector, choose $τ_0 \le τ$, and let the function $\delta(t)$ be bounded for $t \in [t_0, t_0 + τ]$. Further suppose that **y**(t) solves the initial value problem*

$$\mathbf{y}'(t) = \mathbf{f}(\mathbf{y}(t), t)\ ,\ \ \mathbf{y}(t_0) = \mathbf{y}_0\ ,$$

and $\widetilde{\mathbf{y}}(t)$ *solves the perturbed initial value problem*

$$\widetilde{\mathbf{y}}'(t) = \mathbf{f}(\widetilde{\mathbf{y}}(t), t) + \boldsymbol{\delta}(t) , \ \widetilde{\mathbf{y}}(0) = \mathbf{y}_0 + \boldsymbol{\epsilon}$$

for $|t - t_0| \leq \tau_0$. *If* $\mathbf{y}(t) \neq \widetilde{\mathbf{y}}(t)$ *for all* $t \in [t_0, \tau_0]$, *then*

$$\max_{t_0 \leq t \leq \tau_0} \|\widetilde{\mathbf{y}}(t) - \mathbf{y}(t)\| \leq e^{\Phi(t-t_0)} \|\boldsymbol{\epsilon}\| + \frac{e^{\Phi(t-t_0)} - 1}{\Phi} \max_{t_0 \leq t \leq \tau_0} \|\boldsymbol{\delta}(t)\| .$$

Proof First, we note that

$$\frac{\mathrm{d}\|\widetilde{\mathbf{y}} - \mathbf{y}\|}{\mathrm{d}t}(t) = \frac{\widetilde{\mathbf{y}}(t) - \mathbf{y}(t)}{\|\widetilde{\mathbf{y}}(t) - \mathbf{y}(t)\|} \cdot \frac{\mathrm{d}\widetilde{\mathbf{y}} - \mathbf{y}}{\mathrm{d}t}(t)$$

$$= \frac{\widetilde{\mathbf{y}}(t) - \mathbf{y}(t)}{\|\widetilde{\mathbf{y}}(t) - \mathbf{y}(t)\|} \cdot [\mathbf{f}(\widetilde{\mathbf{y}}(t), t) - \mathbf{f}(\mathbf{y}(t), t) + \boldsymbol{\delta}(t)]$$

$$\leq \|\mathbf{f}(\widetilde{\mathbf{y}}(t), t) - \mathbf{f}(\mathbf{y}(t), t) + \boldsymbol{\delta}(t)\| \leq \Phi\|\widetilde{\mathbf{y}}(t) - \mathbf{y}(t)\| + \|\boldsymbol{\delta}(t)\| .$$

Then Gronwall's inequality (3.2) implies that for $t \in [t_0, \tau_0]$

$$\|\widetilde{\mathbf{y}}(t) - \mathbf{y}(t)\| \leq e^{\Phi(t-t_0)} \|\boldsymbol{\epsilon}\| + \int_{t_0}^{t} e^{\Phi(t-s)} \|\boldsymbol{\delta}(s)\| \, \mathrm{d}s$$

$$\leq e^{\Phi(t-t_0)} \|\boldsymbol{\epsilon}\| + \frac{e^{\Phi(t-t_0)} - 1}{\Phi} \max_{t_0 \leq t \leq \tau_0} \|\boldsymbol{\delta}(t)\| .$$

Exercise 3.2.1 Show that the function f in Example 3.2.2 is not Lipschitz continuous in y.

Exercise 3.2.2 Try to modify Theorem 3.2.3 to bound the error in the solution for $t < t_0$. Do you get an upper bound for the error, or a lower bound?

3.2.4   Linear Equations

Next, we would like to discuss the **linear ordinary differential equation**

$$\mathbf{y}'(t) = \mathbf{A}(t)\mathbf{y}(t) + \mathbf{b}(t) . \tag{3.4}$$

We can relate this equation to the general initial value problem (3.1a) by taking $\mathbf{f}(\mathbf{y}, t) = \mathbf{A}(t)\mathbf{y} + \mathbf{b}(t)$. The following theorem provides an approach for solving this problem.

Theorem 3.2.4 (Fundamental Solution for Linear ODE)   *Given scalars* $t_0 < t_1$, *assume that* $\mathbf{A}(t)$ *is a matrix for each* $t \in [t_0, t_1]$, *and that* $\mathbf{A}(t)$ *is continuous in* *this interval. Then there is a unique **fundamental matrix** $\mathbf{F}(t)$ mapping scalars to matrices such that*

$$\mathbf{F}'(t) = \mathbf{A}(t)\mathbf{F}(t) \text{ for } t \in [t_0, t_1] \text{ and } \mathbf{F}(t_0) = \mathbf{I} . \tag{3.5}$$

*The determinant and trace of the fundamental matrix satisfy the **Liouville formula***

$$\det \mathbf{F}(t) = \det \mathbf{F}(\tau) e^{\int_\tau^t \mathrm{tr}\mathbf{A}(\sigma)\, d\sigma} \text{ for all } t_0 \le \tau < t \le t_1 .$$

Furthermore, $\mathbf{F}(t)$ is nonsingular for all $t \in [t_0, t_1]$.

Proof The existence and uniqueness of $\mathbf{F}(t)$ is proved in Coddington and Levinson [47, p. 20]. The determinant of $\mathbf{F}$ satisfies the ordinary differential equation

$$D\left(\det \mathbf{F}\right)(t) = [\, \mathrm{tr}\mathbf{A}(t)]\, [\det \mathbf{F}(t)] .$$

This is proved in Coddington and Levinson [47, p. 28] as a consequence of the ordinary differential equation $\mathbf{F}' = \mathbf{A}\mathbf{F}$ and the Laplace expansion (3.7) in Chap. 3 of Volume I. Since $\mathbf{A}$ is continuous on $[t_0, t_1]$, it is bounded on this interval, and the Liouville formula (3.2.4) follows by solving the scalar linear ordinary differential equation for $\det \mathbf{F}$. Recall that $\mathbf{F}(t_0) = \mathbf{I}$, so $\det \mathbf{F}(t_0) = 1$. The Liouville formula (3.2.4) now shows that

$$\det \mathbf{F}(t) = e^{\int_{t_0}^t \mathrm{tr}\mathbf{A}(\sigma)\, d\sigma} > 0$$

for all $t \in (t_0, t_1]$, so $\mathbf{F}(t)$ is nonsingular.

Next, let us see how to use the fundamental matrix to solve a linear initial value problem.

Theorem 3.2.5 (General Solution of Linear IVP) *Given scalars $t_0 < t_1$, assume that $\mathbf{A}(t)$ is a square matrix for each $t \in [t_0, t_1]$, and that $\mathbf{A}(t)$ is continuous in this interval. Also suppose that the function $\mathbf{b}(t)$ maps scalars to vectors and is continuous for all $t \in [t_0, t_1]$. Let $\mathbf{F}(t)$ solve the initial value problem*

$$\mathbf{F}'(t) = \mathbf{A}(t)\mathbf{F}(t) \text{ for } t \in [t_0, t_1] \text{ and } \mathbf{F} = \mathbf{I} .$$

for the fundamental matrix. Then the solution of the linear initial value problem

$$\mathbf{y}'(t) = \mathbf{A}(t)\mathbf{y}(t) + \mathbf{b}(t) \text{ for } t \in (t_0, t_1) \text{ with } \mathbf{y}(t_0) = \mathbf{y}_0$$

is

$$\mathbf{y}(t) = \mathbf{F}(t)\mathbf{y}_0 + \mathbf{F}(t) \int_{t_0}^t \mathbf{F}(s)^{-1}\mathbf{b}(s)\, ds .$$

Proof Theorem 3.2.4 shows that $\Phi(t)$ is nonsingular for all $t \in [t_0, t_1]$. We can differentiate the equation $\mathbf{F}(t)^{-1}\mathbf{F}(t) = \mathbf{I}$ with respect to t to get

$$\mathbf{0} = D\mathbf{F}^{-1}(t)\mathbf{F}(t) + \mathbf{F}(t)^{-1}\mathbf{F}'(t) ,$$

which we can solve to get

$$D\mathbf{F}^{-1}(t) = -\mathbf{F}(t)^{-1}\mathbf{F}'(t)\mathbf{F}(t)^{-1} .$$

Next, we can multiply (3.4) by $\mathbf{F}(t)^{-1}$ to get

$$
\begin{aligned}
\mathbf{F}(t)^{-1}\mathbf{b}(t) &= \mathbf{F}(t)^{-1}\mathbf{y}'(t) - \mathbf{F}(t)^{-1}\mathbf{A}(t)\mathbf{y}(t) \\
&= \mathbf{F}(t)^{-1}\mathbf{y}'(t) - \mathbf{F}(t)^{-1}\mathbf{A}(t)\mathbf{F}(t)\mathbf{F}(t)^{-1}\mathbf{y}'(t) \\
&= \mathbf{F}(t)^{-1}\mathbf{y}'(t) - \mathbf{F}(t)^{-1}\mathbf{F}'(t)\mathbf{F}(t)^{-1}\mathbf{y}'(t) = \mathbf{F}(t)^{-1}\mathbf{y}'(t) + D\mathbf{F}^{-1}(t)\mathbf{y}'(t) \\
&= D\left[\mathbf{F}^{-1}\mathbf{y}\right](t) .
\end{aligned}
$$

We can integrate this ordinary differential equation to obtain

$$\int_{t_0}^{t} \mathbf{F}(s)^{-1}\mathbf{b}(s) \, \mathrm{d}s = \mathbf{F}(t)^{-1}\mathbf{y}(t) - \mathbf{F}(t_0)\mathbf{y}(t_0) = \mathbf{F}(t)^{-1}\mathbf{y}(t) - \mathbf{y}_0 .$$

Finally, we can solve for $\mathbf{y}(t)$ to get the claimed result.

Exercise 3.2.3 If $\mathbf{y}_1(t)$ and $\mathbf{y}_2(t)$ are two solutions of the linear ordinary differential equation

$$\mathbf{y}'(t) = \mathbf{A}(t)\mathbf{y}(t) + \mathbf{b}(t) ,$$

show that $\mathbf{y}_1(t) - \mathbf{y}_2(t)$ satisfies the homogeneous linear ordinary differential equation

$$\mathbf{y}'(t) = \mathbf{A}(t)\mathbf{y}(t) .$$

Conversely, if $\mathbf{y}_1(t)$ is a solution of the linear ordinary differential equation

$$\mathbf{y}'(t) = \mathbf{A}(t)\mathbf{y}(t) + \mathbf{b}(t) ,$$

show that any other solution of this equation is of the form $\mathbf{y}_1(t) + \mathbf{z}(t)$ where $\mathbf{z}(t)$ solves the homogeneous linear ordinary differential equation

$$\mathbf{z}'(t) = \mathbf{A}(t)\mathbf{z}(t) .$$

Exercise 3.2.4 Suppose that the function $\mathbf{A}(t)$ maps scalars to matrices, and this function is periodic with period ω. In other words, for all t

$$\mathbf{A}(t + \omega) = \mathbf{A}(t) .$$

Let $\mathbf{F}(t)$ be the fundamental matrix for $\mathbf{A}(t)$. Show that there is a function $\mathbf{P}(t)$ mapping scalars to matrices with

$$\mathbf{P}(t + \omega) = \mathbf{P}(t) ,$$

and there is a matrix $\mathbf{R}$ so that

$$\mathbf{F}(t) = \mathbf{P}(t) \exp{(t\mathbf{R})} .$$

3.3 Linear Equations with Constant Coefficients

In the previous section, we presented some important theory about the well-posedness of initial value problems for ordinary differential equations. In this section, we will discuss some analytical results for a simple class of initial value problems, namely linear ordinary differential equations involving constant coefficients. Our theory will involve the matrix exponential. We will see that approximations to the matrix exponential correspond to various interesting numerical methods for solving initial value problems.

3.3.1 Matrix Exponentials

Suppose $\mathbf{A}$ is a matrix, $\mathbf{b}$ and $\mathbf{y}_0$ are vectors, and we want to solve the initial value problem

$$\mathbf{y}'(t) = \mathbf{A}\mathbf{y}(t) + \mathbf{b} , \ \mathbf{y}(t_0) = \mathbf{y}_0 \tag{3.6}$$

for $\mathbf{y}(t)$. The fundamental matrix for this problem is $\mathbf{F}(t)$ where

$$\mathbf{F}'(t) = \mathbf{A}\mathbf{F}(t) \text{ and } \mathbf{F}(t_0) = \mathbf{I} .$$

Theorem 1.7.1 of Chap. 1 in Volume II shows that

$$\mathbf{F}(t) = \exp{(\mathbf{A}[t - t_0])} ,$$

and Theorem 3.2.5 shows us that the solution of the initial value problem is

$$\mathbf{y}(t) = \exp\left(\mathbf{A}[t - t_0]\right)\mathbf{y}_0 + \int_{t_0}^{t} \exp\left(\mathbf{A}[t - s]\right)\mathbf{b}\,\mathrm{d}s$$

$$= \exp\left(\mathbf{A}[t - t_0]\right)\mathbf{y}_0 + \int_{0}^{t-t_0} \exp(\mathbf{A}\tau)\,\mathrm{d}\tau\,\mathbf{b}\ . \tag{3.7}$$

If $\mathbf{A}$ is diagonalizable, then it is relatively easy to compute its exponential. Suppose that there is a nonsingular matrix $\mathbf{X}$ and a diagonal matrix Λ so that

$$\mathbf{AX} = \mathbf{X}\Lambda\ .$$

Then

$$\exp(\mathbf{A}t) = \mathbf{X}\exp(\Lambda t)\mathbf{X}^{-1}$$

and $\exp(\Lambda t)$ is the diagonal matrix of exponentials of the diagonal entries of Λt. Let us define

$$\mathbf{z}(t) \equiv \mathbf{X}^{-1}\mathbf{y}(t)\ ,\ \ \mathbf{z}_0 \equiv \mathbf{X}^{-1}\mathbf{y}_0\ \text{and}\ \mathbf{c} \equiv \mathbf{X}^{-1}\mathbf{b}\ .$$

Then we can multiply the ordinary differential Eq. (3.6) by $\mathbf{X}^{-1}$ to get

$$\mathbf{z}'(t) = \mathbf{X}^{-1}\left[\mathbf{Ay}(t) + \mathbf{b}\right] = \Lambda\mathbf{X}^{-1}\mathbf{y}(t) + \mathbf{X}^{-1}\mathbf{b} = \Lambda\mathbf{z}(t) + \mathbf{c}\ .$$

Of course, the initial value is

$$\mathbf{z}(t_0) = \mathbf{z}_0\ .$$

These equations show that the components of $\mathbf{z}(t)$ satisfy separate initial value problems. We can solve for each component of $\mathbf{z}(t)$ individually, then re-assemble to get

$$\mathbf{y}(t) = \mathbf{X}\mathbf{z}(t)\ .$$

Example 3.3.1 **Harmonic Oscillator** Suppose that we want to solve the second-order initial-value problem

$$\eta''(t) = -\omega^2\eta\ ,\ \ \eta(0) = \xi, \eta'(0) = \xi'\ .$$

This can be written in matrix-vector form as

$$\frac{\mathrm{d}}{\mathrm{d}t}\begin{bmatrix} \eta \\ \eta' \end{bmatrix} = \begin{bmatrix} 0 & 1 \\ -\omega^2 & 0 \end{bmatrix}\begin{bmatrix} \eta \\ \eta' \end{bmatrix}\ ,\ \ \begin{bmatrix} \eta(0) \\ \eta'(0) \end{bmatrix} = \begin{bmatrix} \xi \\ \xi' \end{bmatrix}\ .$$

We begin by finding the eigenvalues and eigenvectors of $\mathbf{A}$:

$$\mathbf{AX} = \mathbf{A} \begin{bmatrix} 1 & 1 \\ i\omega & -i\omega \end{bmatrix} = \begin{bmatrix} 1 & 1 \\ i\omega & -i\omega \end{bmatrix} \begin{bmatrix} i\omega & 0 \\ 0 & -i\omega \end{bmatrix} = \mathbf{X}\Lambda .$$

Next we solve for $\mathbf{z}_0$:

$$\mathbf{Xz}_0 = \begin{bmatrix} \xi \\ \xi' \end{bmatrix} \Longrightarrow \mathbf{z}_0 = \begin{bmatrix} \xi - i\xi'/\omega \\ \xi + i\xi'/\omega \end{bmatrix} \frac{1}{2} .$$

Given a value for t, we compute $\mathbf{z}(t)$ by

$$\mathbf{z}(t) = \begin{bmatrix} e^{i\omega t}\zeta \\ e^{-i\omega t}\zeta' \end{bmatrix} .$$

Finally, we form $\mathbf{y}(t) = \mathbf{Xz}(t)$:

$$\begin{bmatrix} \eta(t) \\ \eta'(t) \end{bmatrix} \equiv \mathbf{y}(t) = \begin{bmatrix} 1 & 1 \\ i\omega & -i\omega \end{bmatrix} \begin{bmatrix} e^{i\omega t} & 0 \\ 0 & e^{-i\omega t} \end{bmatrix} \begin{bmatrix} \zeta - i\zeta'/\omega \\ \zeta + i\zeta'/\omega \end{bmatrix} \frac{1}{2}$$

$$= \begin{bmatrix} e^{i\omega t} & e^{-i\omega t} \\ i\omega e^{i\omega t} & -i\omega e^{-i\omega t} \end{bmatrix} \begin{bmatrix} \zeta - i\zeta'/\omega \\ \zeta + i\zeta'/\omega \end{bmatrix} \frac{1}{2} = \begin{bmatrix} \zeta \cos \omega t + \frac{\zeta'}{\omega} \sin \omega t \\ -\zeta\omega \sin \omega t + \zeta' \cos \omega t \end{bmatrix}$$

It is more difficult to solve constant coefficient initial value problems in which the coefficient matrix is not diagonalizable. Most analytical treatments of ordinary differential equations, such as Coddington and Levinson [47, p. 63ff] or Kelley and Peterson [118, p. 61ff], depend on the Jordan Decomposition Theorem 1.4.6 of Chap. 1 in Volume II. However, the Jordan canonical form is not useful for scientific computing, because it is known to be unstable with respect to numerical perturbations (see Demmel [64] or Wilkinson [183, p. 77ff]). In the presence of rounding errors, scientific computation of the eigenvalues and eigenvectors of a general square matrix $\mathbf{A}$ will almost surely find a diagonal matrix of eigenvalues and a nonsingular matrix of eigenvectors. The problem is that the matrix of computed eigenvectors for a nondiagonalizable matrix may be nearly singular, and solutions of linear equations involving this matrix may be inaccurate.

One alternative that does not use the Jordan decomposition is due to Putzer [146]. Given an $m \times m$ matrix $\mathbf{A}$, let its eigenvalues be $\lambda_1, \ldots, \lambda_m$. Putzer's idea is described by the following

Algorithm 3.3.1 (Putzer's)

$$\text{solve } \mathbf{P}'_1(t) = \lambda_1 \mathbf{P}_1(t) , \; \mathbf{P}_1(0) = \mathbf{I}$$

$$\text{for } k = 2 \ldots m$$

$$\text{solve } \mathbf{P}'_k(t) = \mathbf{P}_{k-1}(t) + \lambda_k \mathbf{P}_k(t) , \; \mathbf{P}_k(0) = \mathbf{0}$$

$$\mathbf{F}_m(t) = \mathbf{P}_m(t)$$

for $k = m \ldots 1$

$$\mathbf{F}_k(t) = \mathbf{P}_k(t) + (\mathbf{A} - \mathbf{I}\lambda_k)\mathbf{F}_{k+1}(t)$$

Note that Putzer's algorithm requires storage of each of the matrices $\mathbf{P}_1(t)$ through $\mathbf{P}_m(t)$; however, the arrays $\mathbf{F}_k(t)$ can all overwrite the last matrix $\mathbf{P}_m(t)$. The algorithm does not specify how the initial value problems should be solved, but the original intention was that these problems would be solved analytically. The following theorem explains the connection between this algorithm and the matrix exponential.

Theorem 3.3.1 (Putzer's)   *Let the $m \times m$ matrix $\mathbf{A}$ have eigenvalues $\lambda_1, \ldots, \lambda_m$ (not necessarily distinct). Compute the matrices $\mathbf{P}_k$ for $1 \le k \le m$ and the matrices $\mathbf{F}_k(t)$ for $m \ge k \ge 1$ by Putzer's Algorithm 3.3.1. Then*

$$\mathbf{F}_1(t) = \exp(\mathbf{A}t) .$$

Proof It is equivalent to show that $\mathbf{F}_1(t)$ is the fundamental matrix for the initial value problem

$$\mathbf{F}'(t) = \mathbf{A}\mathbf{F}(t) , \ \mathbf{F}(t) = \mathbf{I} .$$

We will prove by induction that $\mathbf{F}_k(0) = \mathbf{0}$ for $2 \le k \le m$. Note that

$$\mathbf{F}_m(0) = \mathbf{P}_m(0) = \mathbf{0} .$$

For $1 < k < m$, assume inductively that $\mathbf{P}_{+1}(0) = \mathbf{0}$. Then

$$\mathbf{F}_k(0) = \mathbf{P}_k(0) + (\mathbf{A} - \mathbf{I}\lambda_k)\mathbf{F}_{k+1}(0) = \mathbf{0} + (\mathbf{A} - \mathbf{I}\lambda_k)\mathbf{0} = \mathbf{0} .$$

This completes the induction. We are left with

$$\mathbf{F}_1(0) = \mathbf{P}_1(0) + (\mathbf{A} - \mathbf{I}\lambda_1)\mathbf{F}_2(0) = \mathbf{I} + (\mathbf{A} - \mathbf{I}\lambda_1)\mathbf{0} = \mathbf{I} .$$

Thus $\mathbf{F}_1(t)$ satisfies the initial value condition for a fundamental matrix.
   Next, we will prove by induction that

$$\mathbf{F}_1'(t) - \mathbf{A}\mathbf{F}_1(t) = \prod_{j=1}^{k}(\mathbf{A} - \mathbf{I}\lambda_j)\left[-\mathbf{P}_k(t) + \mathbf{F}_{k+1}'(t) - \mathbf{A}\mathbf{F}_{k+1}(t)\right] \text{ for } 1 \le k < m .$$

Note that

$$
\begin{aligned}
\mathbf{F}_1'(t) - \mathbf{A}\mathbf{F}_1(t) &= \left[\mathbf{P}_1'(t) + (\mathbf{A} - \mathbf{I}\lambda_1)\mathbf{F}_2'(t)\right] - \mathbf{A}\left[\mathbf{P}_1(t) + (\mathbf{A} - \mathbf{I}\lambda_1)\mathbf{F}_2(t)\right] \\
&= \lambda_1\mathbf{P}_1(t) + (\mathbf{A} - \mathbf{I}\lambda_1)\mathbf{F}_2'(t) - \mathbf{A}\left[\mathbf{P}_1(t) + (\mathbf{A} - \mathbf{I}\lambda_1)\mathbf{F}_2(t)\right] \\
&= (\mathbf{A} - \mathbf{I}\lambda_1)\left[-\mathbf{P}_1(t) + \mathbf{F}_2'(t) - \mathbf{A}\mathbf{F}_2(t)\right] .
\end{aligned}
$$

This verifies the claim for $k = 1$. Assume inductively that the claim is true for $k - 1$ with $1 \leq k < m$. Then

$$
\mathbf{F}_1'(t) - \mathbf{A}\mathbf{F}_1(t) = \prod_{j=1}^{k-1}(\mathbf{A} - \mathbf{I}\lambda_j)\left[-\mathbf{P}_{k-1}(t) + \mathbf{F}_k'(t) - \mathbf{A}\mathbf{F}_k(t)\right]
$$

$$
= \prod_{j=1}^{k-1}(\mathbf{A} - \mathbf{I}\lambda_j)\left[-\mathbf{P}_{k-1}(t) + \mathbf{P}_k'(t) + (\mathbf{A} - \mathbf{I}\lambda_k)\mathbf{F}_{k+1}'(t)\right.
$$

$$
\left. -\mathbf{A}\mathbf{P}_k(t) - \mathbf{A}(\mathbf{A} - \mathbf{I}\lambda_k)\mathbf{F}_{k+1}(t)\right]
$$

$$
= \prod_{j=1}^{k-1}(\mathbf{A} - \mathbf{I}\lambda_j)\left[+\lambda_k\mathbf{P}_k(t) + (\mathbf{A} - \mathbf{I}\lambda_k)\mathbf{F}_{k+1}'(t) - \mathbf{A}\mathbf{P}_k(t) - \mathbf{A}(\mathbf{A} - \mathbf{I}\lambda_k)\mathbf{F}_{k+1}(t)\right]
$$

$$
= \prod_{j=1}^{k}(\mathbf{A} - \mathbf{I}\lambda_j)\left[-\mathbf{P}_k(t) + \mathbf{F}_{k+1}'(t) - \mathbf{A}\mathbf{P}_{k+1}(t)\right] .
$$

This completes the inductive proof. We conclude that

$$
\mathbf{F}_1'(t) - \mathbf{A}\mathbf{F}_1(t) = \prod_{j=1}^{m-1}(\mathbf{A} - \mathbf{I}\lambda_j)\left[-\mathbf{P}_{m-1}(t) + \mathbf{F}_m'(t) - \mathbf{A}\mathbf{F}_m(t)\right]
$$

$$
= \prod_{j=1}^{m-1}(\mathbf{A} - \mathbf{I}\lambda_j)\left[-\mathbf{P}_{m-1}(t) + \mathbf{P}_m'(t) - \mathbf{A}\mathbf{P}_m(t)\right]
$$

$$
= \prod_{j=1}^{m-1}(\mathbf{A} - \mathbf{I}\lambda_j)\left[\lambda_m\mathbf{P}_m(t) - \mathbf{A}\mathbf{P}_m(t)\right] = \prod_{j=1}^{m}(\mathbf{A} - \mathbf{I}\lambda_j)\mathbf{P}_m(t) .
$$

The Cayley-Hamilton Theorem 1.4.4 of Chap. 1 in Volume II proves that the right-hand side of this equation is zero.

Another alternative to the Jordan canonical form is to use the Schur Decomposition Theorem 1.4.2 of Chap. 1 in Volume II. Suppose that we want to solve

$$
\mathbf{y}'(t) = \mathbf{A}\mathbf{y} + \mathbf{b} , \quad \mathbf{y}(t_0) = \mathbf{y}_0 .
$$

The Schur Decomposition Theorem 1.4.2 of Chap. 1 in Volume II allows us to write

$$\mathbf{AU} = \mathbf{UR} ,$$

where $\mathbf{R}$ is right-triangular and $\mathbf{U}$ is unitary. Let $\mathbf{R}$ have components ϱ_{ij}, and let us define

$$\mathbf{U}^H \mathbf{b} \equiv \mathbf{c} = \begin{bmatrix} \gamma_1 \\ \vdots \\ \gamma_m \end{bmatrix} \text{ and } \mathbf{U}^H \mathbf{y}_0 \equiv \mathbf{z}_0 = \begin{bmatrix} \zeta_{1,0} \\ \vdots \\ \zeta_{m,0} \end{bmatrix} .$$

Then $\mathbf{z}(t) = \mathbf{U}^H \mathbf{y}(t)$ satisfies

$$\mathbf{z}'(t) = \mathbf{Rz}(t) + \mathbf{c} , \ \mathbf{z}(t_0) = \mathbf{z}_0 .$$

We can back-solve this system of linear ordinary differential equations to find the individual components of $\mathbf{z}(t)$. We begin by solving

$$\zeta_m'(t) = \varrho_{nn} \zeta_m(t) + \gamma_m , \ \zeta_m(0) = \eta_{m,0}$$

to obtain

$$\zeta_m(t) = \begin{cases} e^{\varrho_{nn}[t-t_0]} \zeta_{m,0} + \gamma_m (e^{\varrho_{nn}[t-t_0]} - 1)/\varrho_{nn}, \ \varrho_{nn} \neq 0 \\ \zeta_{m,0} + \gamma_m[t - t_0], \ \varrho_{nn} = 0 \end{cases} .$$

Assume inductively that we have computed $\zeta_m(t), \ldots, \zeta_{i+1}(t)$. Then the solution of

$$\zeta_i'(t) = \varrho_{ii} \zeta_i(t) + \gamma_i + \sum_{j=i+1}^{m} \varrho_{ij} \zeta_j(t) , \ \zeta_i(0) = \zeta_{i,0}$$

is

$$\zeta_i(t) = \begin{cases} e^{\varrho_{ii}[t-t_0]} \zeta_{i,0} + \gamma_i(e^{\varrho_{ii}[t-t_0]} - 1)/\varrho_{ii} + \sum_{j=i+1}^{m} \varrho_{ij} \int_{t_0}^{t} e^{\varrho_{ii}(t-s)} \zeta_j(s) \, ds, \ \varrho_{ii} \neq 0 \\ \zeta_i + \mathbf{c}_i[t - t_0] + \sum_{j=i+1}^{m} \mathbf{R}_{ij} \int_{0}^{t} \zeta_j(s) \, ds, \ \varrho_{ii} = 0 \end{cases} .$$

We can continue in this way until all of the components of $\mathbf{z}(t)$ have been determined. Note that each component of $\mathbf{z}(t)$ is a linear combination of exponentials and polynomials in t, so their integrals can be computed analytically and numerically.

Example 3.3.2 Suppose we want to solve $\mathbf{y}'(t) = \mathbf{Jy}(t)$ with $\mathbf{y}(0) = \mathbf{y}_0$, where

$$\mathbf{J} = \begin{bmatrix} \lambda & 1 \\ 0 & \lambda \end{bmatrix} .$$

Since $\mathbf{A}$ is already right-triangular, we can take $\mathbf{U} = \mathbf{I}$ in the Schur decomposition. Then

$$\zeta_2'(t) = \lambda \zeta_2(t) \ , \ \zeta_2(0) = b f e_2^\top \mathbf{y}_0 \equiv \zeta_{2,0} \ ,$$

so

$$\zeta_2(t) = \begin{cases} e^{\lambda t} \eta_{2,0}, \ \lambda \neq 0 \\ \eta_{2,0}, \ \lambda = 0 \end{cases} .$$

This implies that

$$\zeta_1'(t) = \lambda \zeta_1(t) + \zeta_2(t) \ , \ \zeta_1(0) = \zeta_{1,0} \equiv e_1^\top \mathbf{y}_0 \ ,$$

so

$$\zeta_1(t) = \begin{cases} e^{\lambda t} \zeta_{1,0} + \int_0^t e^{\lambda(t-s)} \zeta_2(s) \, ds, \ \lambda \neq 0 \\ \zeta_{1,0} + \int_0^t \zeta_2(s) \, ds, \ \lambda = 0 \end{cases} = \begin{cases} e^{\lambda t} \zeta_{1,0} + t e^{\lambda t} \zeta_{2,0}, \ \lambda \neq 0 \\ \zeta_{1,0} + \zeta_{2,0} t, \ \lambda = 0 \end{cases} .$$

In summary, matrix exponentials are useful for scientific computation of solutions to initial value problems primarily when the matrix $\mathbf{A}$ is diagonalizable, and the matrix of eigenvectors $\mathbf{X}$ is well-conditioned.

3.3.2 Linear Stability

The analytical solution (3.7) to the initial value problem with constant coefficients is useful for understanding how perturbations in the solution depend on perturbations in the data. Given two initial values $\mathbf{y}_0$ and $\widetilde{\mathbf{y}}_0$, as well as two inhomogeneities $\mathbf{b}$ and $\widetilde{\mathbf{b}}$, the initial value problems

$$\mathbf{y}'(t) = \mathbf{A}\mathbf{y}(t) + \mathbf{b} \ , \ \mathbf{y}(t_0) = \mathbf{y}_0 \text{ and}$$
$$\widetilde{\mathbf{y}}'(t) = \mathbf{A}\widetilde{\mathbf{y}}(t) + \widetilde{\mathbf{b}} \ , \ \widetilde{\mathbf{y}}(t_0) = \widetilde{\mathbf{y}}_0$$

have solutions whose difference is

$$\widetilde{\mathbf{y}}(t) - \mathbf{y}(t) = \exp\left(\mathbf{A}[t - t_0]\right)\left[\widetilde{\mathbf{y}}_0 - \mathbf{y}_0\right] + \int_0^{t-t_0} \exp\left(\mathbf{A}s\right) \, ds \left[\widetilde{\mathbf{b}} - \mathbf{b}\right] \ .$$

In order to determine how the perturbation in the solution behaves for large time, we need to understand how the matrix exponential behaves for large time. To assist this discussion, we provide the following definition.

Definition 3.3.1 Let $\mathbf{A}$ be a square matrix and let $\mathbf{y}(t)$ satisfy the homogeneous linear ordinary differential equation

$$\mathbf{y}'(t) = \mathbf{A}\mathbf{y}(t) \text{ for } t > 0 . \tag{3.8}$$

Then the ordinary differential Eq. (3.8) is **stable** if and only if for any $\varepsilon > 0$ there exists $\delta > 0$ so that for all initial values $\mathbf{y}(0)$ satisfying

$$\|\mathbf{y}(0)\| \leq \delta$$

we have

$$\|\mathbf{y}(t)\| \leq \varepsilon \text{ for all } t \geq 0 .$$

The ordinary differential Eq. (3.8) is **unstable** if and only if it is not stable. Finally, the ordinary differential Eq. (3.8) is **globally asymptotically stable** if and only if it is stable and for all initial values $\mathbf{y}(0)$ we have

$$\lim_{t \to \infty} \mathbf{y}(t) = \mathbf{0} .$$

In order to understand the stability of a linear homogeneous ordinary differential equation with constant coefficients, we will study the behavior of matrix exponentials for large time. For purposes of this analysis, we will use the Jordan canonical form. We will present two lemmas considering separate situations regarding the eigenvalues of the coefficient matrix in the differential equation.

Lemma 3.3.1 *If $\mathbf{A}$ is a square matrix and all of the eigenvalues of $\mathbf{A}$ have negative real part, then for any consistent matrix norm $\|\cdot\|$ there are positive scalars σ, $C_{(3.9)}$ and $C_{(3.9)}$ so that for all $t \geq 0$*

$$\|\exp(\mathbf{A}t)\| \leq C_{(3.9)}e^{-\sigma t} \text{ and} \tag{3.9}$$

$$\left\| \int_0^t \exp(\mathbf{A}\tau) \, d\tau \right\| \leq C_{(3.10)} . \tag{3.10}$$

Proof Let

$$\sigma = -\min_i \Re e(\lambda_i)/2$$

where λ_i is any eigenvalue of $\mathbf{A}$. If $\mathbf{J}$ is a Jordan block of $\mathbf{A}$ corresponding to an eigenvalue λ, then

$$\exp(\mathbf{J}t) = \sum_{j=0}^{m-1} \mathbf{N}^j \frac{t^j}{j!} e^{\lambda t} ,$$

where $\mathbf{N}$ is the nilpotent matrix in Definition 1.4.5 of Chap. 1 in Volume II. As a result,

$$\|\exp(\mathbf{J}t)\|\, e^{\sigma t} = \left\| \sum_{j=0}^{m-1} \mathbf{N}^j \frac{t^j}{j!} e^{(\lambda+\sigma)t} \right\| \leq \left| e^{(\lambda+\sigma)t} \right| \sum_{j=0}^{m-1} \frac{t^j}{j!} \|\mathbf{N}\|^j \equiv \phi(t) \,.$$

Note that $\phi(t)$ is continuous for all t. Further, since $\lambda + \sigma$ has negative real part, we see that $\phi(t) \to 0$ as $t \to \infty$. Thus there is a positive constant C_ϕ so that for all $t \geq 0$ and all Jordan blocks in the canonical decomposition of $\mathbf{A}$ we have

$$\phi(t) \leq C_\phi \,.$$

If $\mathbf{AX} = \mathbf{XJ}$ where $\mathbf{J}$ is a Jordan canonical form for $\mathbf{A}$, then

$$\|\exp(\mathbf{A}t)\| = \left\| \mathbf{X} \exp(\mathbf{J}t)\, \mathbf{X}^{-1} \right\| \leq \|\mathbf{X}\|\, \|\mathbf{X}^{-1}\|\, \|\exp(\mathbf{J}t)\|$$

$$\leq \|\mathbf{X}\|\, \|\mathbf{X}^{-1}\|\, C_\phi e^{-\sigma t} \equiv C_{(3.9)} e^{-\sigma t} \,.$$

This proves the first claim.

Similarly, if $\mathbf{J}$ is a Jordan block with eigenvalue λ having negative real part, then

$$\left\| \int_0^t \exp(\mathbf{J}\tau)\, d\tau \right\| = \left\| \sum_{j=0}^{m-1} \mathbf{N}^j \frac{1}{j!} \int_0^t \tau^j e^{\lambda\tau}\, d\tau \right\|$$

$$= \left\| \sum_{j=0}^{m-1} \mathbf{N}^j \frac{1}{j!} \left[e^{\lambda\tau} \sum_{i=0}^{j} (-1)^i \frac{j!\,\tau^{j-i}}{(j-i)!\lambda^{i+1}} \right]_0^t \right\|$$

$$\leq \sum_{j=0}^{m-1} \|\mathbf{N}^j\| \left| (-1)^{j+1} \frac{1 - e^{\lambda t}}{\lambda^{j+1}} + e^{\lambda t} \sum_{i=0}^{j-1} (-1)^i \frac{t^{j-i}}{(j-i)!\lambda^{i+1}} \right|$$

$$\leq \sum_{j=0}^{m-1} \|\mathbf{N}^j\| \left\{ \frac{|1 - e^{\lambda t}|}{|\lambda|^{j+1}} + |e^{\lambda t}| \sum_{i=0}^{j-1} \frac{t^{j-i}}{(j-i)!|\lambda|^{i+1}} \right\} \equiv \psi(t) \,.$$

Again, $\psi(t)$ is continuous for all t, and

$$\psi(t) \to \sum_{j=0}^{m-1} \|\mathbf{N}^j\| \frac{1}{|\lambda|^{j+1}} \quad \text{as } t \to \infty \,.$$

Thus there is a positive constant C_ψ so that for all $t \geq 0$ and all Jordan blocks in the canonical decomposition of $\mathbf{A}$ we have

$$\psi(t) \leq C_\psi \,.$$

Then

$$\left\| \int_0^t \exp(\mathbf{A}\tau) \, d\tau \right\| = \left\| \mathbf{X} \int_0^t \exp(\mathbf{J}\tau) \, d\tau \mathbf{X}^{-1} \right\| \le \|\mathbf{X}\| \, \|\mathbf{X}^{-1}\| \left\| \int_0^t \exp(\mathbf{J}\tau) \, d\tau \right\|$$

$$\le \|\mathbf{X}\| \, \|\mathbf{X}^{-1}\| \, C_\psi \equiv C_{(3.10)} \, .$$

Lemma 3.3.2 *Suppose that* $\mathbf{A}$ *is a square matrix, all of the eigenvalues of* $\mathbf{A}$ *have nonpositive real part, and the largest Jordan block for an eigenvalue of* $\mathbf{A}$ *with zero real part is* $m \times m$. *Then for any consistent matrix norm* $\|\cdot\|$ *there is a positive polynomial (meaning that all of the coefficients in its monomial expansion are positive)* $p_{(3.11)}(t)$ *of degree at most* $m - 1$ *and a positive polynomial* $p_{(3.12)}(t)$ *of degree at most* m *so that for all* $t \ge 0$

$$\|\exp(\mathbf{A}t)\| \le p_{(3.11)}(t) \ and \tag{3.11}$$

$$\left\| \int_0^t \exp(\mathbf{A}\tau) \, d\tau \right\| \le p_{(3.12)}(t) \, . \tag{3.12}$$

Proof If $\mathbf{J}$ is a $m \times m$ Jordan block of $\mathbf{A}$ corresponding to eigenvalue λ, then

$$\exp(\mathbf{J}t) = \sum_{j=0}^{m-1} \mathbf{N}^j \frac{t^j}{j!} e^{\lambda t} \, ,$$

where $\mathbf{N}$ is the nilpotent matrix in Definition 1.4.5 of Chap. 1 in Volume II. If λ has zero real part, then $|e^{\lambda t}| = 1$ for all t and

$$\|\exp(\mathbf{J}t)\| = \left\| \sum_{j=0}^{m-1} \mathbf{N}^j \frac{t^j}{j!} e^{\lambda t} \right\| \le \sum_{j=0}^{m-1} \frac{t^j}{j!} \|\mathbf{N}\|^j \equiv \phi(t) \, .$$

Note that $\phi(t)$ is a positive polynomial of degree $m - 1$. Combining these ideas with the proof of Lemma 3.3.1, we see that there is a positive polynomial $\phi(t)$ of degree $m - 1$ so that for all $t \ge 0$ and all Jordan blocks in the canonical decomposition of $\mathbf{A}$ we have

$$\|\exp(\mathbf{J}t)\| \le \phi(t) \, .$$

If $\mathbf{AX} = \mathbf{XJ}$ where $\mathbf{J}$ is a Jordan canonical form for $\mathbf{A}$, then

$$\|\exp(\mathbf{A}t)\| = \left\| \mathbf{X} \exp(\mathbf{J}t) \, \mathbf{X}^{-1} \right\| \le \|\mathbf{X}\| \, \|\mathbf{X}^{-1}\| \, \|\exp(\mathbf{J}t)\|$$

$$\le \|\mathbf{X}\| \, \|\mathbf{X}^{-1}\| \, \phi(t) \equiv p_{(3.11)}(t) \, .$$

This proves the first claim.

Similarly, if $\mathbf{J}$ is a $m \times m$ Jordan block with eigenvalue λ having zero real part, then

$$\left\| \int_0^t \exp(\mathbf{J}\tau)\, d\tau \right\| = \left\| \sum_{j=0}^{m-1} \mathbf{N}^j \frac{1}{j!} \int_0^t \tau^j e^{\lambda\tau}\, d\tau \right\|$$

$$= \begin{cases} \left\| \sum_{j=0}^{m-1} \mathbf{N}^j \frac{1}{j!} \left[e^{\lambda\tau} \sum_{i=0}^j (-1)^i \frac{j!\tau^{j-i}}{(j-i)!\lambda^{i+1}} \right]_0^t \right\|, & \lambda \neq 0 \\ \left\| \sum_{j=0}^{m-1} \mathbf{N}^j \frac{t^{j+1}}{(j+1)!} \right\|, & \lambda = 0 \end{cases}$$

$$\leq \begin{cases} \sum_{j=0}^{m-1} \left\| \mathbf{N}^j \right\| \left| (-1)^{j+1} \frac{1-e^{\lambda t}}{\lambda^{j+1}} + e^{\lambda t} \sum_{i=0}^{j-1} (-1)^i \frac{t^{j-i}}{(j-i)!\lambda^{i+1}} \right|, & \lambda \neq 0 \\ \sum_{j=0}^{m-1} \left\| \mathbf{N}^j \right\| \frac{t^{j+1}}{(j+1)!}, & \lambda = 0 \end{cases}$$

$$\leq \begin{cases} \sum_{j=0}^{m-1} \left\| \mathbf{N}^j \right\| \left\{ \frac{2}{|\lambda|^{j+1}} + \sum_{i=0}^{j-1} \frac{t^{j-i}}{(j-i)!|\lambda|^{i+1}} \right\}, & \lambda \neq 0 \\ \sum_{j=0}^{m-1} \left\| \mathbf{N}^j \right\| \frac{t^{j+1}}{(j+1)!}, & \lambda = 0 \end{cases} \equiv \psi(t) .$$

Again, $\psi(t)$ is a positive polynomial of degree m. Thus there is a positive polynomial $\psi(t)$ of degree at most m so that for all $t \geq 0$ and all Jordan blocks in the canonical decomposition of $\mathbf{A}$ we have

$$\left\| \int_0^t \exp(\mathbf{J}\tau)\, d\tau \right\| \leq \psi(t) .$$

Then

$$\left\| \int_0^t \exp(\mathbf{A}\tau)\, d\tau \right\| = \left\| \mathbf{X} \int_0^t \exp(\mathbf{J}\tau)\, d\tau \mathbf{X}^{-1} \right\| \leq \|\mathbf{X}\| \, \|\mathbf{X}^{-1}\| \left\| \int_0^t \exp(\mathbf{J}\tau)\, d\tau \right\|$$

$$\leq \|\mathbf{X}\| \, \|\mathbf{X}^{-1}\| \, \psi(t) \equiv p_{(3.12)}(t) .$$

Lemmas 3.3.1 and 3.3.2 have the following easy consequence.

Corollary 3.3.1 *Let* $\mathbf{A}$ *be a square matrix. Then the homogeneous linear ordinary differential equation*

$$\mathbf{y}'(t) = \mathbf{A}\mathbf{y}(t)$$

is stable if and only if all eigenvalues of $\mathbf{A}$ *have nonpositive real part and all eigenvalues of* $\mathbf{A}$ *with zero real part correspond to* 1×1 *Jordan blocks. Also, this homogeneous linear ordinary differential equation is globally asymptotically stable if and only if all eigenvalues of* $\mathbf{A}$ *have negative real part. Finally, if any eigenvalue of* $\mathbf{A}$ *has positive real part or any eigenvalue of* $\mathbf{A}$ *with zero real part corresponds to a Jordan block that is larger than* 1×1, *then this homogeneous linear ordinary differential equation is unstable.*

Proof The first two claims are easy consequences of the previous two lemmas. For the third claim, there are two cases to consider. First, if $\mathbf{x}$ is an eigenvector of $\mathbf{A}$ with eigenvalue λ having positive real part, then the solution of

$$\mathbf{y}'(t) = \mathbf{A}\mathbf{y}(t) \text{ for } t > 0 , \ \mathbf{y}(0) = \mathbf{x}$$

is

$$\mathbf{y}(t) = \exp(\mathbf{A}t)\mathbf{x} = \mathbf{x}e^{\lambda t} .$$

This solution is unbounded as $t \to \infty$. Second, suppose that $\mathbf{A}$ has a zero eigenvalue with Jordan block that is larger than 1×1. Then there are nonzero vectors $\mathbf{x}_1$ and $\mathbf{x}_2$ so that

$$\mathbf{A}\mathbf{x}_1 = \mathbf{0} \text{ and } \mathbf{A}\mathbf{x}_2 = \mathbf{x}_1 .$$

It follows that

$$\exp(\mathbf{A}t)\mathbf{x}_2 = \sum_{k=0}^{\infty} \mathbf{A}^k \mathbf{x}_2 \frac{t^k}{k!} = \mathbf{x}_2 + \mathbf{A}\mathbf{x}_2 t .$$

This vector is unbounded as $t \to \infty$.

3.3.3   *Approximate Matrix Exponentials*

Because matrix exponential are difficult to compute, it is reasonable to search for approximations to matrix exponentials. One common approach is to choose a polynomial (or rational polynomial) $p(t)$ such that $p(0) = 1$, and such that for some given timestep h we have that $p(t)$ approximates e^t well near the eigenvalues of $\mathbf{A}h$. Then we approximate $\exp(\mathbf{A}kh)$ by $p(\mathbf{A}h)^k$.

Example 3.3.3 If we approximate the exponential by the first two terms in its Taylor series, we get $e^t \approx 1 + t$. This suggests the matrix exponential approximation

$$\exp(\mathbf{A}kh) \approx (\mathbf{I} + \mathbf{A}h)^k .$$

We would normally implement this approximation as a linear recurrence. If $\mathbf{y}_k \approx \mathbf{y}(kh)$ then the corresponding numerical method is

$$\mathbf{y}_{k+1} = (\mathbf{I} + \mathbf{A}h)\mathbf{y}_k = \mathbf{y}_k + h\mathbf{A}\mathbf{y}_k .$$

This is **Euler's method**.

Example 3.3.4 If we approximate e^t by the first three terms in the Taylor series, we obtain $e^t \approx 1 + t + t^2/2$. Again, if $\mathbf{y}_k \approx \mathbf{y}(kh)$ then we approximate the solution of the initial value problem by

$$\mathbf{y}_{k+1} = \left(\mathbf{I} + \mathbf{A}h + \mathbf{A}^2 \frac{h^2}{2} \right) \mathbf{y}_k .$$

This can be computed in two stages:

$$\mathbf{y}_{k+1/2} = \mathbf{y}_k + \mathbf{A}\mathbf{y}_k \frac{h}{2}$$

$$\mathbf{y}_{k+1} = \mathbf{y}_k + \mathbf{A}\mathbf{y}_{k+1/2}h .$$

This method is commonly called the **midpoint method**.

Example 3.3.5 Let us consider the Padé approximation

$$e^t \approx \frac{1 + t/2}{1 - t/2} .$$

When applied to an initial value problem, this approximation leads to the recurrence

$$\left(\mathbf{I} - \mathbf{A}\frac{h}{2} \right) \mathbf{y}_{k+1} = \left(\mathbf{I} + \mathbf{A}\frac{h}{2} \right) \mathbf{y}_k .$$

Usually, this is written in the implicit form

$$\mathbf{y}_{k+1} = \mathbf{y}_k + \mathbf{A}(\mathbf{y}_{k+1} + \mathbf{y}_k)\frac{h}{2} , \qquad (3.13)$$

which is called the **trapezoidal method**.

In general, the order of accuracy of the approximation $e^t \approx p(t)$ at $t = 0$ tells us the order of the numerical method for the ordinary differential equation. Further, the location of the zeros and poles of $p(t)$ tells us about the stability of the numerical method.

For an interesting discussion about the use of rational approximations to exponentials in the course of solving differential equations, we recommend reading Thomée [172, p. 96ff]. Rather than spending additional time to study matrix exponentials, we will examine methods for solving initial value problems directly.

Exercise 3.3.1 The backward Euler method for $\mathbf{y}'(t) = \mathbf{A}\mathbf{y}(t)$ is

$$\mathbf{y}_k = \mathbf{y}_{k-1} + \mathbf{A}\mathbf{y}_k h .$$

What is the matrix exponential corresponding to this method?

Exercise 3.3.2 Find a Padé approximation to the exponential of the form

$$e^t \approx \frac{1}{1 + at + bt^2} \; .$$

Use this Padé approximation to develop a two-stage numerical method for solving $\mathbf{y}'(t) = \mathbf{A}\mathbf{y}(t)$.

3.4   Linear Multistep Methods

In this section, we will develop a general class of methods for solving initial value problems for ordinary differential equations. We will begin by describing the idea in general form, and investigating the conditions under which the corresponding methods are consistent and convergent. Consistency is usually easy to determine, but convergence is more difficult. The difficulty lies in choosing timesteps so that the method is numerically stable, in some appropriate sense. If the ordinary differential equation is linear, then stability is relatively easy to discuss. If the ordinary differential equation is nonlinear, the discussion is more complicated.

We will eventually learn that it is generally more efficient to use high-order methods for smooth problems in which the solution is varying slowly, and low-order methods for rough problems, or intervals in time where the solution is varying rapidly. With that aim in mind, our goal in this section is to develop a hierarchy of numerical methods, such that it is easy to vary the order of the method and the step size. We will also want to apply different hierarchies of linear multistep methods to stiff and non-stiff problems.

3.4.1   Introduction

Suppose that we want to solve the initial value problem

$$\mathbf{y}'(t) = \mathbf{f}(\mathbf{y}(t), t) \text{ for } t > t_0 \; , \; \mathbf{y}(t_0) = \mathbf{y}_0 \; .$$

We could formally integrate this equation to obtain

$$\int_t^{t+h} y'(s) \, \mathrm{d}s = \int_t^{t+h} \mathbf{f}(\mathbf{y}(s), s) ds \; .$$

In order to approximate the integrals, we could determine polynomial interpolants to $\mathbf{y}$ and $\mathbf{f}$ and integrate these polynomials. The result would be a formula of the form

$$\sum_{j=0}^{k} \mathbf{y}_{n+1-j} \alpha_{k-j} = h \sum_{j=0}^{k} \mathbf{f}\left(\mathbf{y}_{n+1-j}, t_n + [1-j]h\right) \beta_{k-j} . \tag{3.14}$$

Numerical methods for initial value problems that can be written in this form are called **linear multistep methods**.

Since we are using Eq. (3.14) to determine $\mathbf{y}_{n+1}$, we will normalize the equation by requiring that

$$\alpha_k = 1 .$$

Because there is no point in disguising the number of terms being used in the linear multistep method, we also require that

$$|\alpha_0| + |\beta_0| \neq 0 .$$

The linear multistep method is **explicit** if $\beta_k = 0$; otherwise it is **implicit**. Note that linear multistep methods, by design, can be implicit in at most a single vector $\mathbf{y}_{n+1}$; this is very different from Runge-Kutta methods, which are described in Sect. 3.6.

Linear multistep methods will be organized around various interpolation strategies. Such a strategy might be Newton interpolation of the forcing function in the integral form of the initial value problem *over the current timestep* for the Adams-Bashforth and Adams-Moulton methods in Sects. 3.4.7.1 and 3.4.7.2. An alternative might be Newton interpolation of the forcing function in the integral form of the initial value problem *over the current and previous timestep* by Nyström methods in Sect. 3.4.7.3. Yet another strategy is to differentiate the Newton interpolation polynomial to the solution vector over the current and previous timesteps, as in backward differentiation formulas, which are described in Sect. 3.4.7.4.

Example 3.4.1 Euler's method

$$\mathbf{y}_{n+1} = \mathbf{y}_n + h\mathbf{f}(\mathbf{y}_n, t_n)$$

is an explicit linear multistep method with $k = 1$, $\alpha_0 = -1$, $\beta_0 = 1$ and $\beta_1 = 0$. Here we understand that $\mathbf{y}_n \approx \mathbf{y}(t_n)$ and $\mathbf{y}_{n+1} \approx \mathbf{y}(t_n + h)$.

Example 3.4.2 The backward Euler method

$$\mathbf{y}_{n+1} = \mathbf{y}_n + h\mathbf{f}(\mathbf{y}_{n+1}, t_n + h)$$

is an implicit linear multistep method with $k = 1$, $\alpha_0 = -1$, $\beta_0 = 0$ and $\beta_1 = 1$.

Example 3.4.3 The trapezoidal rule

$$\mathbf{y}_{n+1} = \mathbf{y}_n + \frac{h}{2}\left[\mathbf{f}(\mathbf{y}_n \, , \, t_n) + \mathbf{f}(\mathbf{y}_{n+1} \, , \, t_n + h)\right]$$

is an implicit linear multistep method with $k = 1$, $\alpha_0 = -1$, $\beta_0 = 1/2$ and $\beta_1 = 1/2$.

Example 3.4.4 The modified midpoint method

$$\mathbf{y}_{n+1} = \mathbf{y}_{n-1} + 2h\mathbf{f}(\mathbf{y}_n \, , \, t_n)$$

is an explicit linear multistep method with $k = 2$, $\alpha_0 = -1$, $\alpha_1 = 0$, $\beta_0 = 0$, $\beta_1 = 2$ and $\beta_2 = 0$. Here we understand that $\mathbf{y}_{n+1} \approx \mathbf{y}(t_n + h)$, $\mathbf{y}_n \approx \mathbf{y}(t_n)$ and $\mathbf{y}_{n-1} \approx \mathbf{y}(t_n - h)$.

3.4.2 Consistency and Convergence

We would like to determine conditions under which linear multistep methods provide accurate approximations to solutions of initial value problems. We will begin with the following definition.

Definition 3.4.1 If the linear multistep method

$$\sum_{j=0}^{k} \mathbf{y}_{n+1-j}\alpha_{k-j} = h\sum_{j=0}^{k} \mathbf{f}_{n+1-j}\beta_{k-j}$$

approximates the solution of the initial value problem

$$\mathbf{y}'(t) = \mathbf{f}(\mathbf{y}, t) \text{ for } t > t_0 \text{ with } \mathbf{y}(t_0) = \mathbf{y}_0 \, ,$$

then the **local truncation error** is

$$r(t, h) \equiv \sum_{j=0}^{k} \mathbf{y}(t - jh)\alpha_{k-j} - h\sum_{j=0}^{k} \mathbf{f}(\mathbf{y}(t - jh), t - jh)\beta_{k-j} \, .$$

The linear multistep method is **consistent** if and only if its local truncation error satisfies

$$\lim_{h \to 0} \frac{r(t, h)}{h} = 0 \, .$$

The next lemma provides the conditions under which a linear multistep method is consistent.

Lemma 3.4.1 (Sufficient Condition for Consistency) *Suppose that*

$$\mathbf{y}'(t) = \mathbf{f}(\mathbf{y}(t), t) \text{ for } t > t_0 \text{ with } \mathbf{y}(t_0) = \mathbf{y}_0$$

has a continuously differentiable solution $\mathbf{y}(t)$, *and that* $\mathbf{f}$ *is continuous in both* $\mathbf{y}$ *and* t. *If the linear multistep method*

$$\sum_{j=0}^{k} \mathbf{y}_{n+1-j} \alpha_{k-j} = h \sum_{j=0}^{k} \mathbf{f}\left(\mathbf{y}_{n+1-j}, \ t_n + [1-j]h\right) \beta_{k-j}$$

satisfies

$$\sum_{j=0}^{k} \alpha_j = 0 = \sum_{j=0}^{k} \left\{ j\alpha_j - \beta_j \right\} \ ,$$

then the method is consistent.

Proof By the mean-value theorem,

$$\mathbf{y}(t + jh) - \mathbf{y}(t) = jh\mathbf{y}'(\tau_j) \ ,$$

where τ_j lies between t and $t + jh$. Thus the local truncation error satisfies

$$\frac{r(t + kh, h)}{h} = \frac{1}{h} \sum_{j=0}^{k} \mathbf{y}\left(t + [k-j]h\right) \alpha_{k-j} - \sum_{j=0}^{k} \mathbf{f}\left(\mathbf{y}(t + [k-j]h), t + [k-j]h\right) \beta_{k-j}$$

$$= \frac{1}{h} \sum_{j=0}^{k} \left\{\mathbf{y}(t) + \mathbf{y}'(\tau_j)[k-j]h\right\} \alpha_{k-j} - \sum_{j=0}^{k} \mathbf{f}\left(\mathbf{y}(t + [k-j]h), t + [k-j]h\right) \beta_{k-j}$$

$$= \mathbf{y}(t)\frac{1}{h} \sum_{j=0}^{k} \alpha_{k-j} + \sum_{j=0}^{k} \left\{\mathbf{y}'(\tau_j)[k-j]\alpha_{k-j} - \mathbf{f}(\mathbf{y}(t + [k-j]h), t + [k-j]h)\beta_{k-j}\right\}$$

$$= \sum_{j=0}^{k} \left\{\mathbf{y}'(\tau_j)[k-j]\alpha_{k-j} - \mathbf{f}(\mathbf{y}(t + [k-j]h), t + [k-j]h)\beta_j\right\} \ .$$

As $h \to 0$, we have $\mathbf{y}'(\tau_j) \to \mathbf{y}'(t) = \mathbf{f}(\mathbf{y}(t), t)$ and $\mathbf{f}(\mathbf{y}(t + [k-j]h), t + [k-j]h) \to \mathbf{f}(\mathbf{y}(t), t)$. Thus as $h \to 0$ we find that

$$\frac{r(t + kh, h)}{h} \to \mathbf{f}(\mathbf{y}(t), t) \sum_{j=0}^{k} \left\{[k-j]\alpha_{k-j} - \beta_{k-j}\right\} = 0 \ .$$

Next, we will define convergence for a linear multistep method.

Definition 3.4.2 Let $\mathbf{y}$ be a vector, and let t, η and τ be scalars. Assume that there is a vector $\mathbf{y}_0$ and a scalar t_0 so that the function $\mathbf{f}(\mathbf{y}, t)$ is Lipschitz continuous on the region

$$R = \{(\mathbf{y}, t) : \|\mathbf{y} - \mathbf{y}_0\| \le \eta \text{ and } |t - t_0| \le \tau\} .$$

Suppose that the initial value problem

$$\mathbf{y}'(t) = \mathbf{f}(\mathbf{y}(t), t) \text{ for } t > t_0 \text{ and } \mathbf{y}(t_0) = \mathbf{y}_0$$

has a solution that exists for $t_0 \le t < t_1$. Then a linear multistep method

$$\sum_{j=0}^{k} \mathbf{y}_{n+1-j}\alpha_{k-j} = h \sum_{j=0}^{k} \mathbf{f}(\mathbf{y}_{n+1-j}, t_0 + [1-j]h)\beta_{k-j}$$

is **convergent** if and only if for all $t \in (t_0, t_1)$ and for all $\varepsilon > 0$ there is an integer $N > 0$ so that for all $n \ge N$ the linear multistep method using stepsize $h = (t - t_0)/n$ and the exact initial condition satisfies

$$\|\mathbf{y}_n - \mathbf{y}(t_0 + nh)\| \le \varepsilon .$$

The following lemma serves as a converse for Lemma 3.4.1

Lemma 3.4.2 (Necessary Condition for Consistency)   *If the linear multistep method*

$$\sum_{j=0}^{k} \mathbf{y}_{n+1-j}\alpha_{k-j} = h \sum_{j=0}^{k} \mathbf{f}\left(\mathbf{y}_{n+1-j}, t_n + [1-j]h\right) \beta_{k-j}$$

is convergent, then

$$\sum_{j=0}^{k} \alpha_j = 0 .$$

If, in addition, the linear multistep method converges to a solution of $\mathbf{y}'(t) = \mathbf{f}(\mathbf{y}(t), t)$ *where* $\mathbf{f}$ *is continuous in both arguments, then*

$$\sum_{j=0}^{k} \{j\alpha_j - \beta_j\} = 0 .$$

Proof By definition of the linear multistep method,

$$\sum_{j=0}^{k} \mathbf{y}_{n+1-j}\alpha_{k-j} = h\sum_{j=0}^{k} \mathbf{f}\left(\mathbf{y}_{n+1-j}\,,\;t_n + [1-j]h\right)\beta_{k-j} \to 0 \text{ as } h \to 0\,.$$

Suppose that the linear multistep scheme converges. Then $\mathbf{y}_{n+1-j} \to \mathbf{y}(t)$ as $h \to 0$ with $t = nh$ held fixed and $0 \le j \le k$, and

$$\sum_{j=0}^{k} \mathbf{y}_{n+1-j}\alpha_{k-j} \to \mathbf{y}(t)\sum_{j=0}^{k}\alpha_{k-j} \text{ as } h \to 0\,.$$

Thus convergence requires that $\sum_{j=0}^{k}\alpha_j = 0$.

Next, suppose that the scheme converges to $\mathbf{y}(t)$ where $\mathbf{y}'(t) = \mathbf{f}(\mathbf{y}(t), t)$. Then

$$\frac{1}{h}\sum_{j=0}^{k}\mathbf{y}_{n+1-j}\alpha_{k-j} = \frac{1}{h}\sum_{j=0}^{k}[\mathbf{y}_{n+1-j} - \mathbf{y}_{n+1-k}]\alpha_{k-j}$$

$$= \sum_{j=0}^{k}\frac{\mathbf{y}_{n+1-j} - \mathbf{y}_{n+1-k}}{(k-j)h}(k-j)\alpha_{k-j} \to \mathbf{y}'(t)\sum_{j=0}^{k}(k-j)\alpha_{k-j}\,.$$

By the definition of the linear multistep method,

$$\frac{1}{h}\sum_{j=0}^{k}\mathbf{y}_{n+1-j}\alpha_{k-j} = \sum_{j=0}^{k}\mathbf{f}\left(\mathbf{y}_{n+1-j}\,,\;t_n + [1-j]h\right)\beta_{k-j} \to \mathbf{y}'(t)\sum_{j=0}^{k}\beta_{k-j}\,.$$

We can combine these two limits to obtain the second claim in the lemma.

3.4.3 Characteristic Polynomials

If we apply the linear multistep method

$$0 = \sum_{j=0}^{k}\mathbf{y}_{n+1-j}\alpha_{k-j} - h\sum_{j=0}^{k}\mathbf{f}\left(\mathbf{y}_{n+1-j}\,,\;t_n + [1-j]h\right)\beta_{k-j}$$

to the ordinary differential equation

$$y' = \lambda y\,,$$

we get

$$0 = \sum_{j=0}^{k} y_{n+1-j} \left[\alpha_{k-j} - h\lambda\beta_{k-j} \right] = y_{n+1}[\alpha_k - h\lambda\beta_k] + \sum_{j=1}^{k} y_{n+1-j} \left[\alpha_{k-j} - h\lambda\beta_{k-j} \right] .$$

Using the assumption that $\alpha_k = 1$, we can solve for y_{n+1} to get

$$y_{n+1} = -\sum_{j=1}^{k} \gamma_{k-j}(h\lambda) y_{n+1-j} ,$$

where the functions γ_j are defined by

$$\gamma_j(h\lambda) \equiv \frac{\alpha_j - h\lambda\beta_j}{1 - h\lambda\beta_j} .$$

Thus the numerical solution y_{n+k} is determined by a **linear recurrence**. In matrix-vector form, this recurrence can be written

$$\mathbf{z}_{n+1} \equiv \begin{bmatrix} y_{n+1} \\ \vdots \\ y_{n+2-k} \end{bmatrix} = \begin{bmatrix} -\gamma_{k-1} & \cdots & -\gamma_1 & -\gamma_0 \\ 1 & \cdots & 0 & 0 \\ \vdots & \ddots & \vdots & \vdots \\ 0 & \cdots & 1 & 0 \end{bmatrix} \begin{bmatrix} y_n \\ \vdots \\ y_{n+1-k} \end{bmatrix} \equiv \mathbf{C}\mathbf{z}_n . \qquad (3.15)$$

The matrix $\mathbf{C}$ in this recurrence is a **companion matrix**. We have studied companion matrices previously in Sect. 1.4.1.5 of Chap. 1 in Volume II. The eigenvectors $\mathbf{z}$ and eigenvalues ζ of the companion matrix satisfy

$$\mathbf{C}\mathbf{z} \equiv \begin{bmatrix} -\gamma_{k-1} & \cdots & -\gamma_1 & -\gamma_0 \\ 1 & \cdots & 0 & 0 \\ \vdots & \ddots & \vdots & \vdots \\ 0 & \cdots & 1 & 0 \end{bmatrix} \begin{bmatrix} \mathbf{z}_{k-1} \\ \mathbf{z}_{k-2} \\ \vdots \\ \mathbf{z}_0 \end{bmatrix} = \mathbf{z}\zeta ,$$

which is equivalent to the equations

$$-\sum_{j=0}^{k-1} \gamma_j \mathbf{z}_j = \mathbf{z}_{k-1}\zeta ,$$

$$\mathbf{z}_j = \mathbf{z}_{j-1}\zeta \text{ for } 1 \leq j \leq k - 1 .$$

These equations imply that

$$\mathbf{z}_j = \mathbf{z}_0\zeta^j \text{ for } 0 \leq j \leq k - 1$$

and

$$0 = \zeta^k + \sum_{j=0}^{k-1} \gamma_j \zeta^j = \zeta^k + \sum_{j=0}^{k-1} \frac{\alpha_j - h\lambda\beta_j}{1 - h\lambda\beta_k}\zeta^j = \frac{1}{1 - h\lambda\beta_k}\sum_{j=0}^{k}\frac{\alpha_j - h\lambda\beta_j}{\zeta}^j .$$

Let us define the **first characteristic polynomial** $a(\zeta)$ by

$$a(\zeta) \equiv \sum_{j=0}^{k} \alpha_j \zeta^j , \qquad (3.16)$$

and the **second characteristic polynomial** by

$$b(\zeta) \equiv \sum_{j=0}^{k} \beta_j \zeta^j . \qquad (3.17)$$

Then the eigenvalues ζ of $\mathbf{C}$ satisfy

$$0 = a(\zeta) - h\lambda b(\zeta) ,$$

and the corresponding eigenvector has the form

$$\mathbf{z} = \begin{bmatrix} \zeta^{k-1} \\ \vdots \\ \zeta \\ 1 \end{bmatrix} .$$

Recall that Lemma 1.4.6 of Chap. 1 in Volume II proved that for each distinct eigenvalue the matrix $\mathbf{C}$ has a single eigenvector.

From the consistency Lemmas 3.4.1 and 3.4.2, and from the definitions (3.16) and (3.17) of the characteristic polynomials, we know the following three facts.

Lemma 3.4.3 *Let a and b be the first and second characteristic polynomials for a linear multistep method, as defined in Eqs. (3.16) and (3.17). If both $a(1) = 0$ and $a'(1) = b(1)$, then the corresponding linear multistep method is consistent. If the corresponding linear multistep method is convergent, then $a(1) = 0$. Finally, if the corresponding linear multistep method converges to a solution of the related initial value problem, then $a'(1) = b(1)$.*

Example 3.4.5 Euler's method

$$\mathbf{y}_{n+1} - \mathbf{y}_n = h\mathbf{f}(\mathbf{y}_n , t_n)$$

has first characteristic polynomial $a(\zeta) = \zeta - 1$ and second characteristic polynomial $b(\zeta) = 1$. Since $a(1) = 0$ and $a'(1) - b(1) = 0$, this method is consistent.

Example 3.4.6 The backward Euler method

$$\mathbf{y}_{n+1} - \mathbf{y}_n = h\mathbf{f}(\mathbf{y}_{n+1} , t_{n+1})$$

has characteristic polynomials $a(\zeta) = \zeta - 1$ and $b(\zeta) = \zeta$. Since $a(1) = 0$ and $a'(1) - b(1) = 0$, this method is consistent.

Example 3.4.7 The trapezoidal rule

$$\mathbf{y}_{n+1} - \mathbf{y}_n = \frac{h}{2} \left[\mathbf{f}(\mathbf{y}_{n+1} , t_{n+1}) + \mathbf{f}(\mathbf{y}_n , t_n) \right]$$

has characteristic polynomials $a(\zeta) = \zeta - 1$ and $b(\zeta) = \frac{1}{2}(\zeta + 1)$. Since $a(1) = 0$ and $a'(1) - b(1) = 0$, this method is consistent.

Example 3.4.8 The modified midpoint method

$$\mathbf{y}_{n+2} - \mathbf{y}_n = 2h\mathbf{f}(\mathbf{y}_{n+1} , t_{n+1})$$

has characteristic polynomials $a(\zeta) = \zeta^2 - 1$ and $b(\zeta) = 2\zeta$. Since $a(1) = 0$ and $a'(1) - b(1) = 0$, this method is consistent.

Exercise 3.4.1 Find the characteristic polynomials for **Simpson's rule**

$$\mathbf{y}_{n+1} - \mathbf{y}_{n-1} = \frac{h}{3} \left[\mathbf{f}(\mathbf{y}_{n+1} , t_n + h) + 4\mathbf{f}(\mathbf{y}_n , t_n) + \mathbf{f}(\mathbf{y}_{n-1} , t_n - h) \right] ,$$

then determine if this method is consistent.

3.4.4   Zero Stability

There are many different notions of stability regarding numerical methods for initial value problems. Our first stability notion basically requires that a sufficiently small perturbation in the initial data leads to bounded growth of the perturbation at later time.

Definition 3.4.3 A numerical method for solving $\mathbf{y}' = \mathbf{f}(\mathbf{y}, t)$ with initial value $\mathbf{y}(t_0) = \mathbf{y}_0$ is **zero-stable** for $t_0 < t < t_1$ if and only if there is a perturbation bound $\varepsilon > 0$, a growth factor $\mathbf{C}$ and an upper bound $\overline{h}$ on the step size such that for all $0 < h < \overline{h}$ and for all step numbers $0 < n \le (t_1 - t_0)/h$ if the perturbed initial condition $\widetilde{\mathbf{y}}_0$ satisfies

$$\| \widetilde{\mathbf{y}}_0 - \mathbf{y}_0 \| \le \varepsilon$$

then the perturbed subsequent numerical solutions satisfy

$$\|\widetilde{\mathbf{y}}_n - \mathbf{y}_n\| \leq C\varepsilon .$$

All of the linear multistep methods to be discussed below are zero-stable. Simple examples of linear multistep methods that are not zero-stable are usually contrived, and not based on any reasonable interpolation to the original ordinary differential equation or its integral form. For example, see an example provided by Lambert [123, p. 34]. On the other hand, there are some higher-order linear multistep methods that are not zero-stable, for example in the Stórmer and Cowell families [123, p. 255].

The next theorem completely characterizes the linear multistep methods that are zero-stable.

Theorem 3.4.1 (Necessary and Sufficient Condition for Zero Stability)   *A linear multistep method is zero-stable if and only if its first characteristic polynomial a satisfies the **root condition**: each zero ζ of a has modulus $|\zeta| \leq 1$, and if $|\zeta| = 1$ then $a'(\zeta) \neq 0$.*

Proof A proof can be found in Henrici [101, p. 17], Isaacson and Keller [112], or in a slightly more restrictive form as Theorem 3.4.7 below.
The following convergence theorem is due to Dahlquist [55]. References to proofs of generalizations may be found in Lambert [124, p. 36].

Theorem 3.4.2 (Necessary and Sufficient Conditions for Convergence)   *A linear multistep method is convergent if and only if it is both consistent and zero-stable.*

Example 3.4.9 The trapezoidal rule

$$\mathbf{y}_{n+1} - \mathbf{y}_n = \frac{h}{2} [\mathbf{f}(\mathbf{y}_n , t_n) + \mathbf{f}(\mathbf{y}_{n+1} , t_n + h)]$$

has characteristic polynomials $a(\zeta) = \zeta - 1$ and $b(\zeta) = (\zeta + 1)/2$. Since $a(1) = 0$ and $a'(1) = b(1)$, the trapezoidal method is consistent. Since the only zero of a is $\zeta = 1$, the trapezoidal method is zero-stable. Theorem 3.4.2 guarantees that the trapezoidal method is convergent.

Example 3.4.10 The modified midpoint rule

$$\mathbf{y}_{n+1} - \mathbf{y}_{n-1} = 2hf(\mathbf{y}_n , t_n)$$

has first characteristic polynomials $a(\zeta) = \zeta^2 - 1$ and second characteristic polynomial $b(\zeta) = 2\zeta$. Since $a(1) = 0$ and $a'(1) = b(1)$, the modified midpoint rule is consistent. Since the zeros of a are $\zeta = \pm 1$, the trapezoidal method is zero-stable. Theorem 3.4.2 also guarantees that the trapezoidal method is convergent.

3.4.5   Order

Next, we would like to discuss the **order of accuracy** of linear multistep methods.

Definition 3.4.4 The linear multistep method

$$\sum_{j=0}^{k} \mathbf{y}_{n+1-j}\alpha_{k-j} = h \sum_{j=0}^{k} \mathbf{f}_{n+1-j}\beta_{k-j}$$

has **linear difference operator**

$$\mathscr{L}_h\{\mathbf{z}\}(t) \equiv \sum_{j=0}^{k} \left[\mathbf{z}(t-jh)\alpha_{k-j} - \mathbf{z}'(t-jh)h\beta_{k-j} \right] \ . \tag{3.18}$$

This linear multistep method has **order** r if and only if both

$$\mathscr{L}_h\{\mathbf{z}\}(t) = O(h^{r+1}) \text{ as } h \to 0$$

for all functions $\mathbf{z}(t)$ that have $r+1$ continuous derivatives at t, and

$$\lim_{h \to 0} \left[h^{-r-1} \mathscr{L}_h\{\mathrm{e}^t\}(t) \right] \neq \mathbf{0} \ .$$

A simple calculation will quantify the order of a linear multistep method.

Lemma 3.4.4 *The linear multistep method*

$$\sum_{j=0}^{k} \mathbf{y}_{n+1-j}\alpha_{k-j} = h \sum_{j=0}^{k} \mathbf{f}\left(\mathbf{y}_{n+1-j}, \ t_n + [1-j]h\right)\beta_{k-j}$$

has order at least r if and only if

$$\sum_{j=0}^{k} \alpha_j = 0 \ ,$$

and

$$\sum_{j=0}^{k} \left[\frac{j^q}{q!}\alpha_j - \frac{j^{q-1}}{(q-1)!}\beta_j \right] = 0$$

for all $1 \leq q \leq r$.

Proof Let $\mathbf{z}(t)$ be $r + 1$ times continuously differentiable at t. Then

$$\mathbf{z}(t + jh) = \sum_{q=0}^{r} \frac{(jh)^q}{q!} D^q \mathbf{z}(t) + O(h^{r+1})$$

and

$$\mathbf{z}'(t + jh) = \sum_{q=1}^{r} \frac{(jh)^{q-1}}{(q-1)!} D^q \mathbf{z}(t) + O(h^{r+1}) .$$

It follows that

$$\mathscr{L}_h\{\mathbf{z}\}(t + kh) = \sum_{j=0}^{k} \left[\mathbf{z}(t + [k - j]h)\alpha_{k-j} - \mathbf{z}'(t + [k - j]h)h\beta_{k-j} \right]$$

$$= \sum_{j=0}^{k} \left[\alpha_{k-j} \sum_{q=0}^{r} \frac{([k - j]h)^q}{q!} D^q \mathbf{z}(t) - h\beta_{k-j} \sum_{q=1}^{r} \frac{([k - j]h)^{q-1}}{(q-1)!} D^q \mathbf{z}(t) \right] + O(h^{r+1})$$

$$= \mathbf{z}(t) \sum_{j=0}^{k} \alpha_j + \sum_{q=1}^{r} h^q D^q \mathbf{z}(t) \sum_{j=0}^{k} \left[\alpha_j \frac{j^q}{q!} - \beta_j \frac{j^{q-1}}{(q-1)!} \right] + O(h^{r+1}) .$$

The linear multistep method has order at least r if and only if the coefficient of $D^q \mathbf{z}(t)$ vanishes for $0 \le q \le r$. This statement is equivalent to the claimed result.

The following theorem, also due to Dahlquist [55], shows that a zero-stable linear multistep method can achieve high order only by employing a sufficient number of previous steps.

Theorem 3.4.3 (First Dahlquist Barrier) *A zero-stable linear multistep method involving $k + 1$ steps has order at most $k + 1$ when k is odd, and order $k + 2$ when k is even.*

Exercise 3.4.2 Show that Euler's method and the backward Euler method both have order 1.

Exercise 3.4.3 Show that the trapezoidal method

$$\mathbf{y}_{n+1} = \mathbf{y}_n + \frac{h}{2} [\mathbf{f}(\mathbf{y}_n , t_n) + \mathbf{f}(\mathbf{y}_{n+1} , t_n + h)]$$

has order 2.

Exercise 3.4.4 Show that Simpson's rule

$$\mathbf{y}_{n+1} - \mathbf{y}_{n-1} = \frac{h}{3} [\mathbf{f}(\mathbf{y}_{n+1} , t_n + h) + 4\mathbf{f}(\mathbf{y}_n , t_n) + \mathbf{f}(\mathbf{y}_{n-1} , t_n - h)] ,$$

has order 4.

3.4.6   Other Stability Notions

We defined our zero stability in Sect. 3.4.4. This stability notion requires that perturbations in the method lead to bounded growth. However growing perturbations
may not appear to be all that stable, especially if the growth bound is large.

At the very beginning of this book, in Sect. 1.3.5.2 of Chap. 1 in Volume I, we
defined the region of absolute stability. For convenience, we will present the idea
again.

Definition 3.4.5 A numerical method for the linear ordinary differential equation
$y'(t) = \lambda y(t)$ is **absolutely stable** if and only if a perturbation in the numerical
solution does not increase from one step to the next.

Absolute stability is probably closer to our intuitive idea of stability. However,
we should be careful to relate absolute stability of a numerical method to the
true solution of the original problem. The differential equation $y'(t) = \lambda y(t)$ has
analytical solution

$$y(t) = e^{\lambda t} y(0) \ .$$

Thus the analytical solution displays absolute stability, meaning that $|y(t)| \le |y(0)|$,
if and only if

$$\left| e^{\lambda t} \right| \le 1 \ .$$

In other words, the absolute stability region for the original problem consists of
all problems for which the real part of λ is nonpositive. The extent to which a
numerical method fails to mimic the same region for its absolute stability indicates
the inappropriateness of the method. For example, a numerical method that is
absolutely stable for large regions of $\Re e(h\lambda) > 0$ should not be used to solve
problems with growing analytical solutions. Similarly, a numerical method that
is absolutely stable for small regions of $\Re e(h\lambda) < 0$ should not be used to solve
problems with decaying analytical solutions.

The following lemma describes the precise conditions under which a linear
multistep method is absolutely stable.

Lemma 3.4.5 *Given the linear multistep method*

$$\sum_{j=0}^{k} \mathbf{y}_{n+1-j} \alpha_{k-j} = h \sum_{j=0}^{k} \mathbf{f}\left(\mathbf{y}_{n+1-j} \ , \ t_n + [1-j]h\right) \beta_{k-j} \ ,$$

denote its first and second characteristic polynomials by

$$a(\zeta) \equiv \sum_{j=0}^{k} \alpha_j \zeta^j \ and \ b(\zeta) \equiv \sum_{j=0}^{k} \beta_j \zeta^j \ ,$$

respectively. Then this linear multistep method is absolutely stable if and only if for all roots ζ of $a(\zeta) - h\lambda b(\zeta) = 0$ we have both

1. $|\zeta| \leq 1$, and
2. $|\zeta| = 1$ implies that ζ is a root of multiplicity one.

Proof If we apply a linear multistep method to $y' = \lambda y$, we get

$$0 = \sum_{j=0}^{k} y_{n+1-j}\alpha_{k-j} - \lambda h \sum_{j=0}^{k} y_{n+1-j}\beta_{k-j} .$$

This linear recurrence can be written in matrix-vector form using the companion matrix $\mathbf{C}$ in Eq. (3.15). Thus the linear multistep method is absolutely stable if and only if $\|\mathbf{C}^j\| \leq 1$ for all j. Since Lemma 1.2.15 of Chap. 1 in Volume II shows that any consistent norm of a matrix provides an upper bound on the modulus of its eigenvalues, this requires all of the eigenvalues of $\mathbf{C}$ to lie inside the unit circle. Since $\mathbf{C}$ is a companion matrix, Lemma 1.4.6 of Chap. 1 in Volume II and Example 1.4.8 of Chap. 1 in Volume II show that a repeated eigenvalue leads to polynomial growth in $\mathbf{C}^j$ if the eigenvalue has modulus 1.

Example 3.4.11 Euler's method has characteristic polynomials

$$a(\zeta) = \zeta - 1 \text{ and } b(\zeta) = 1 .$$

Thus absolute stability of Euler's method depends on the location of the zeros of

$$a(\zeta) - h\lambda b(\zeta) = \zeta - 1 - h\lambda .$$

If $\zeta = \varrho e^{i\theta}$ is a zero of this polynomial and $\varrho \leq 1$, then $h\lambda = -1 + \varrho e^{i\theta}$. The set of all such points $h\lambda$ is therefore a unit circle with center -1. This is the absolute stability region for Euler's method, and it is displayed in graph (a) of Fig. 3.4. Euler's method is not absolutely stable for any problem with $\Re e\lambda > 0$, which is appropriate. On the other hand, Euler's method is absolutely stable for only a small region where $\Re e\lambda < 0$, which indicates that it may have limited usefulness for problems with decaying solutions.

Example 3.4.12 The backward Euler method has characteristic polynomials

$$a(\zeta) = \zeta - 1 \text{ and } b(\zeta) = \zeta ,$$

so absolute stability of the backward Euler method is determined by the location of the zeros of $a(\zeta) - h\lambda b(\zeta) = \zeta - 1 - h\lambda\zeta$. If $\zeta = \varrho e^{i\theta}$ is a zero of this polynomial and $\varrho \leq 1$, then $h\lambda = 1 - e^{-i\theta}/\varrho$ lies in the exterior of the unit circle with center 1. This is the absolute stability region for the backward Euler method, and it is displayed in graph (a) of Fig. 3.5. The backward Euler method is absolutely stable for all problems with $\Re e\lambda < 0$, which is appropriate. On the other hand, the backward

Euler method is also absolutely stable for a large region where $\Re e\lambda > 0$, which indicates that it may have limited usefulness for problems with growing solutions.

Example 3.4.13 The trapezoidal rule has characteristic polynomials

$$a(\zeta) = \zeta - 1 \text{ and } b(\zeta) = (\zeta + 1)/2 \,,$$

so absolute stability of the trapezoidal rule can be determined by the location of the zeros of $a(\zeta) - h\lambda b(\zeta) = \zeta - 1 - \frac{h\lambda}{2}(1 + \zeta)$. If $\zeta = \varrho e^{i\theta}$ is a zero of this polynomial and $\varrho \le 1$, then

$$h\lambda = 2\frac{\varrho e^{i\theta} - 1}{\varrho e^{i\theta} + 1}$$

When $\varrho = 0$ then we have $h\lambda = -2$, so this point must fall in the region of absolute stability. The boundary of the region of absolute stability satisfies

$$h\lambda = 2\frac{e^{i\theta/2} - e^{-i\theta/2}}{e^{i\theta/2} + e^{-i\theta/2}} = 2\frac{2i\sin\theta}{2\cos\theta} = 2i\tan\theta \,.$$

This is the imaginary axis, so the region of absolute stability for the trapezoidal rule is the left half-plane. This is the same as the absolute stability region of the original problem.

Example 3.4.14 The characteristic polynomials for the modified midpoint rule are

$$a(\zeta) = \zeta^2 - 1 \text{ and } b(\zeta) = 2\zeta \,.$$

If $\zeta = \varrho e^{i\theta}$ is a zero of $a(\zeta) - h\lambda b(\zeta) = \zeta^2 - 1 - 2h\lambda\zeta$, then

$$h\lambda = \frac{1}{2}[\varrho e^{i\theta} - e^{-i\theta}/\varrho]$$

On the boundary of the region of absolute stability we have $\varrho = 1$ and

$$h\lambda = \frac{1}{2}[e^{i\theta} - e^{-i\theta}] = i\sin\theta$$

This represents an interval of length 2 on the imaginary axis. Thus the modified midpoint method may not produce useful results for initial value problems with decaying solutions.

In order to match decaying numerical solutions with decaying analytical solutions, we will offer a new stability notion.

Definition 3.4.6 A linear multistep method is **A-stable** if and only if the negative half-plane is contained in its region of absolute stability.

Here are some useful results regarding A-stability that were proved by Dahlquist [56].

Theorem 3.4.4 (Second Dahlquist Barrier) *An explicit linear multistep method cannot be A-stable. Also, the order of an A-stable linear multistep method cannot exceed 2. Finally, the second-order A-stable linear multistep method with smallest error constant is the trapezoidal rule.*

Theorem 3.4.4 indicates that A-stability imposes to great a burden on linear multistep methods. Instead of requiring that the entire negative half-plane be in the stability region, we might require that some region surrounding the negative real axis lie in the stability region. In that way, initial value problems with decay rates having large negative real parts and moderate imaginary parts may still be treated well.

Definition 3.4.7 A linear multistep method is **A(0) stable** if and only if some angle $0 < \alpha < \pi/2$, the complex numbers $\zeta = re^{i\theta}$ with $|\theta - \pi| < \pi/2 - \alpha$ are contained in the absolute stability region.

The following theorem due, to Widlund [182], provides some restrictions on linear multistep methods for A(0) stability.

Theorem 3.4.5 (Restrictions for A(0) Stability) *An explicit linear multistep method cannot be A(0)-stable. Also, there is no A(0) stable linear k-step method with order greater than k.*

We will consider one more relaxation of the definition of stability, in which we require only that all real negative decay rates produce stable results.

Definition 3.4.8 A linear multistep method is A_0 **stable** if and only if the negative real axis is contained in the absolute stability region.

The following result, due to Cryer [54], provides some restrictions on linear multistep methods for this new stability notion.

Theorem 3.4.6 (Restrictions for A_0 Stability) *An explicit linear multistep method cannot be A_0-stable. However, there exist A_0 stable linear multistep methods of arbitrary order.*

Exercise 3.4.5 Consider the differential equation $\mathbf{y}'(t) = \lambda y(t)$ with $y(0)$ given and λ complex. Write down the analytical solution to this initial value problem. How does the solution behave if the real part of λ is positive, negative, or zero? For which values of λ do we want a numerical method for this problem to be absolutely stable, A-stable, $A(0)$-stable or A_0 stable?

Exercise 3.4.6 Consider Euler's method for $\mathbf{y}'(t) = \lambda y(t)$ with $y(0)$ given and λ complex. Under what conditions on the timestep h is this method zero-stable? Under what conditions is it absolutely stable, A-stable, $A(0)$-stable or A_0 stable?

Exercise 3.4.7 Repeat the previous exercise for the backward Euler method.

Exercise 3.4.8 Consider a linear multistep method (3.14) for $\mathbf{y}'(t) = \lambda y(t)$ where λ is real. Let us define a linear multistep method for this problem to be

monotonicity-preserving if and only if whenever the members of the set $\{y_{n+j}\}_{j=0}^{k-1}$ are monotone (either increasing or decreasing), then $\{y_{n+j}\}_{j=0}^{k}$ is also monotone. Under what conditions on the timestep h is Euler's method monotonicity-preserving? How about the backward Euler method?

Exercise 3.4.9 Suppose that we want to solve $y'(t) = -48y(t)$ with $y(0) = 1$, for $0 \le t \le 1$. Program Euler's method for this problem. Take 3, 6, 12, 24, 48 and 96 timesteps with this method. Which of these timesteps are absolutely stable? Which are monotonicity-preserving?

Exercise 3.4.10 Suppose that we want to solve $y'(t) = 8y(t)$ with $y(0) = 1$, for $0 \le t \le 1$. Program the backward Euler method for this problem. Take 3, 6, 12, 24, 48 and 96 timesteps with this method. Which of these timesteps are absolutely stable? Which are monotonicity-preserving?

Exercise 3.4.11 Suppose that we want to solve $y'(t) = iy(t)$ with $y(0) = 1$, for $0 \le t \le 10$. Program Euler's method for this problem, using complex variables. Take 3, 6, 12, 24, 48 and 96 timesteps with this method. Which of these timesteps are absolutely stable? What happens to the growth of the numerical solution as the timestep is refined?

Exercise 3.4.12 Repeat the previous exercise, using the backward Euler method.

3.4.7 Families

We will identify three broad families of linear multistep methods. Methods in the Adams family all have first characteristic polynomial of the form

$$a(\zeta) = \zeta^k - \zeta^{k-1} .$$

Note that this polynomial has an isolated zero at $\zeta = 1$, and a multiple zero at $\zeta = 0$, so Lemma 3.4.1 proves that Adams family methods are all zero-stable. Explicit Adams methods (those with the degree of the second characteristic polynomial b less than k) are called **Adams-Bashforth methods**. Implicit Adams methods are called **Adams-Moulton methods**.

  Nyström methods all have first characteristic polynomial of the form

$$a(\zeta) = \zeta^k - \zeta^{k-2} .$$

Note that this polynomial has isolated roots at $\zeta = \pm 1$, and all other roots at $\zeta = 0$ when $k \ge 3$, so Lemma 3.4.1 proves that Nyström family methods are also zero-stable. Nyström methods are explicit: the degree of the second characteristic polynomial b is assumed to be less than k.

Backward differentiation formulas have second characteristic polynomial of the form

$$b(\zeta) = \beta_k \zeta^k .$$

Backward differential formulas are always implicit. Later, we will see that backward differentiation formulas are zero-stable only for $k \leq 6$.

We will now develop each of these families of linear multistep methods in greater detail.

3.4.7.1   Adams-Bashforth

Adams-Bashforth methods approximate the ordinary differential equation in the form

$$\mathbf{y}(t_n + h) - \mathbf{y}(t_n) = \int_{t_n}^{t_n+h} \mathbf{f}(\mathbf{y}(t), t) \, dt$$

by integrating the Newton interpolation polynomial to $\mathbf{f}$ at the previous times $t_n, \ldots, t_{n-k+1}$. If the timesteps are equally-spaced, meaning that

$$t_{n+j} = t_n + jh ,$$

then the implementation of the method can make use of the following lemma.

Lemma 3.4.6 *Define the sequence $\{\gamma_j\}_{j=0}^{\infty}$ by the recurrence*

$$\gamma_0 = 1 , \text{ and } \gamma_j = 1 - \sum_{i=0}^{j-1} \frac{\gamma_i}{j-i+1} \text{ for } j > 0 . \tag{3.19}$$

Suppose that we are given a positive integer k, a time t_n, a positive timestep h and a sequence of vectors $\{\mathbf{f}_{n-j}\}_{j=0}^{k-1}$. Define the differences $\nabla^j \mathbf{f}_n$ by the recursion

$$\nabla^0 \mathbf{f}_n \equiv \mathbf{f}_n , \text{ and } \nabla^j \mathbf{f}_n \equiv \nabla^{j-1} \mathbf{f}_n - \nabla^{j-1} \mathbf{f}_{n-1} \text{ for } 0 < j < k , \tag{3.20}$$

and define the scalars

$$t_{n-j} = t_n - jh \text{ for } 0 \leq j < k .$$

Then the differences $\nabla^j \mathbf{f}_n$ are related to the function values $\{\mathbf{f}_{n-j}\}_{j=0}^{k-1}$ and to their divided differences $\mathbf{f}[t_n, \ldots, t_{n-j}]$ (defined in Eq. (1.6)) for $0 \le j < k$ by

$$\nabla^j \mathbf{f}_n = \mathbf{f}[t_n, \ldots, t_{n-j}] h^j j! \tag{3.21}$$

$$= \sum_{i=0}^{j} \binom{j}{i} (-1)^i \mathbf{f}_{n-i} . \tag{3.22}$$

Then the polynomial interpolant to the sequence $\{\mathbf{f}_{n-j}\}_{j=0}^{k-1}$ and the corresponding times $\{t_{n-j}\}_{j=0}^{k-1}$ is

$$\mathbf{p}_n(t) = \sum_{j=0}^{k-1} \mathbf{f}[t_n, \ldots, t_{n-j}] \prod_{i=0}^{j-1} (t - t_{n-i}) = \sum_{j=0}^{k-1} \frac{\nabla^j \mathbf{f}_n}{j!} \prod_{i=0}^{j-1} \frac{t - t_{n-i}}{h} , \tag{3.23}$$

and its integral over $(t_n, t_n + h)$ is

$$\int_{t_n}^{t_n+h} \mathbf{p}_n(t) \, dt = h \sum_{j=0}^{k-1} \gamma_j \nabla^j \mathbf{f}_n . \tag{3.24}$$

Proof First, we will prove (3.21) for all $0 \le j < k$. It is easy to see that

$$\nabla^0 \mathbf{f}_n = \mathbf{f}_n = \mathbf{f}[t_n] h^0 0! ,$$

and

$$\nabla^1 \mathbf{f}_n = \mathbf{f}_n - \mathbf{f}_{n-1} = \mathbf{f}[t_n, t_{n-1}](t_n - t_{n-1}) = \mathbf{f}[t_n, t_{n-1}] h .$$

This proves (3.22) for $j = 0$ and 1. Inductively, let us assume that this equation holds for $j \ge 1$, and prove it holds for $j + 1$. The inductive hypothesis implies that

$$\nabla^{j+1} \mathbf{f}_n = \nabla^j \mathbf{f}_n - \nabla^j \mathbf{f}_{n-1}$$

$$= \mathbf{f}[t_n, \ldots, t_{n-j}] h^j j! - \mathbf{f}[t_{n-1}, \ldots, t_{n-j-1}] h^j j!$$

then the definition (1.6) of the divided difference gives us

$$= \mathbf{f}[t_n, \ldots, t_{n-j-1}](t_n - t_{n-j-1}) h^j j!$$

and the assumption that the times are equally spaced leads to

$$= \mathbf{f}[t_n, \ldots, t_{n-j-1}](j+1) h h^j j! = \mathbf{f}[t_n, \ldots, t_{n-j-1}] h^{j+1} (j+1)!$$

This completes the inductive proof of the claim (3.22).

Next, let us prove (3.22). Since $\nabla^0 \mathbf{f}_n = \mathbf{f}_n$, the claim is true for $j = 0$. Assume that the claim is true for $j - 1 \geq 0$. Then the definition (3.20) of the differences implies that

$$\nabla^j \mathbf{f}_n = \nabla^{j-1} \mathbf{f}_n - \nabla^{j-1} \mathbf{f}_{n-1} = \sum_{\ell=0}^{j-1} \binom{j-1}{\ell}(-1)^\ell \mathbf{f}_{n-\ell} - \sum_{i=0}^{j-1} \binom{j-1}{i}(-1)^i \mathbf{f}_{n-1-i}$$

$$= \sum_{\ell=0}^{j-1} \binom{j-1}{\ell}(-1)^\ell \mathbf{f}_{n-\ell} - \sum_{M=1}^{j} \binom{j-1}{M-1}(-1)^{M-1} \mathbf{f}_{n-M}$$

$$= \binom{j-1}{0} \mathbf{f}_n + \sum_{i=1}^{j-1} \left[\binom{j-1}{i} + \binom{j-1}{i-1} \right](-1)^i \mathbf{f}_{n-i} + \binom{j-1}{j-1}(-1)^j \mathbf{f}_{n-j}$$

$$= \mathbf{f}_n + \sum_{i=1}^{j-1} \binom{j}{i}(-1)^i \mathbf{f}_{n-i} + (-1)^j \mathbf{f}_{n-j}$$

$$= \sum_{i=0}^{j} \binom{j}{i}(-1)^i \mathbf{f}_{n-i} .$$

This completes the inductive proof of (3.22).

Our discussion in Sect. 1.2.2 shows that $\mathbf{p}_n$ defined by (3.23) is the Newton interpolation polynomial for the given data. Its integral is

$$\int_{t_n}^{t_n+h} \sum_{j=0}^{k-1} \mathbf{f}[t_n, \ldots, t_{n-j}] \prod_{i=0}^{j-1} (t - t_{n-i}) \, dt = \int_{t_n}^{t_n+h} \sum_{j=0}^{k-1} \frac{\nabla^j \mathbf{f}_n}{j!} \prod_{i=0}^{j-1} \frac{t - t_{n-i}}{h} \, dt$$

$$= \sum_{j=0}^{k-1} \frac{\nabla^j \mathbf{f}_n}{j!} h \int_0^1 \prod_{i=0}^{j-1} (s + i) \, ds = h \sum_{j=0}^{k-1} \nabla^j \mathbf{f}_n (-1)^j \int_0^1 \binom{-s}{j} \, ds .$$

This suggests that we define

$$\gamma_j = (-1)^j \int_0^1 \binom{-s}{j} \, ds .$$

In order to find a recurrence for the values of γ_j, we will examine the generating function

$$G(x) \equiv \sum_{j=0}^{\infty} \gamma_j x^j = \sum_{j=0}^{\infty} \int_0^1 (-x)^j \binom{-s}{j} ds = \int_0^1 \left[\sum_{j=0}^{\infty} (-x)^j \binom{-s}{j} \right] ds$$

$$= \int_0^1 (1-x)^{-s} ds = -\left. \frac{(1-x)^{-s}}{\ln(1-x)} \right|_{s=0}^1 = -\frac{x}{(1-x)\ln(1-x)} .$$

It follows that

$$\sum_{j=0}^{\infty} x^j = \frac{1}{1-x} = G(x)\left[-\frac{\ln(1-x)}{x}\right] = \left[\sum_{j=0}^{\infty} \gamma_j x^j\right]\left[\sum_{i=0}^{\infty} \frac{x^i}{i+1}\right]$$

$$= \sum_{\ell=0}^{\infty} x^\ell \sum_{j=0}^{\ell} \frac{\gamma_j}{\ell-j+1} .$$

Equating coefficients in the series gives us

$$1 = \sum_{j=0}^{\ell} \frac{\gamma_j}{\ell-j+1} = \gamma_\ell + \sum_{j=0}^{\ell-1} \frac{\gamma_j}{\ell-j+1} .$$

This gives us $\gamma_0 = 1$ and the specified recursion for the other γ_j.

The first six coefficients for the Adams-Bashforth scheme with uniform timesteps are

$$\gamma_0 = 1 , \gamma_1 = 1 - \frac{\gamma_0}{2} = \frac{1}{2} , \gamma_2 = 1 - \frac{\gamma_0}{3} - \frac{\gamma_1}{2} = \frac{5}{12}, \gamma_3 = \frac{3}{8}, \gamma_4 = \frac{251}{720} \text{ and } \gamma_5 = \frac{95}{288}.$$

The first-three Adams-Bashforth schemes are

$$\mathbf{y}_{n+1} - \mathbf{y}_n = h\mathbf{f}_n \text{ (order 1 : Euler's method)}$$

$$\mathbf{y}_{n+1} - \mathbf{y}_n = h\left[\mathbf{f}_n + \frac{1}{2}\nabla^1\mathbf{f}_n\right] \text{ (order 2)}$$

$$\mathbf{y}_{n+1} - \mathbf{y}_n = h\left[\mathbf{f}_n + \frac{1}{2}\nabla^1\mathbf{f}_n + \frac{5}{12}\nabla^2\mathbf{f}_n\right] \text{ (order 3) .}$$

The difference form of the Adams-Bashforth method allows us to vary the order of the scheme just by changing the number of terms carried in the sum. Also note that if $\{\mathbf{y}_n\}$ is generated by a k-step Adams-Bashforth method and $\{\tilde{\mathbf{y}}_n\}$ is generated by a $k + 1$-step method, then

$$\tilde{\mathbf{y}}_{n+1} - \mathbf{y}_{n+1} = h\gamma_k\nabla^k\mathbf{f}_n = h^{k+1}\gamma_k\frac{d^{k+1}\mathbf{y}}{dt^{k+1}}(t_n) + O(h^{k+2}) .$$

This shows that the k-step method has order k and error constant γ_k.

Example 3.4.15 Suppose that we have stored $\mathbf{f}_{n-1}$, $\nabla^1\mathbf{f}_{n-1}$, $\nabla^2\mathbf{f}_{n-1}$ and $\mathbf{y}_n$. In order to advance the third-order Adams-Bashforth scheme to t_{n+1}, we compute

$$
\begin{aligned}
\mathbf{f}_n &= \mathbf{f}(\mathbf{y}_n, t_n) \\
\nabla^1\mathbf{f}_n &= \mathbf{f}_n - \mathbf{f}_{n-1} \ (\mathbf{f}_{n-1} \text{ no longer needed}) \\
\nabla^2\mathbf{f}_n &= \nabla^1\mathbf{f}_n - \nabla^1\mathbf{f}_{n-1} \ (\nabla^1\mathbf{f}_{n-1} \text{ no longer needed}) \\
\nabla^3\mathbf{f}_n &= \nabla^2\mathbf{f}_n - \nabla^2\mathbf{f}_{n-1} \ (\nabla^2\mathbf{f}_{n-1} \text{ no longer needed})
\end{aligned}
$$

Then we compute

$$
\mathbf{y}_{n+1} = \mathbf{y}_n + h\left[\mathbf{f}_n + \frac{1}{2}\nabla^1\mathbf{f}_n + \frac{5}{12}\nabla^2\mathbf{f}_n + \frac{3}{8}\nabla^3\mathbf{f}_n\right] .
$$

After this calculation is performed, $\mathbf{y}_n$ is no longer needed.

An alternative form of the Adams-Bashforth method, due to Nordsieck [139], will allow us to vary the stepsize. The approach will differ from Newton interpolation as discussed in Sect. 1.2.2. We will operate as if previous timesteps had been equally spaced, but discover how to adjust prior information from timesteps that were not equally spaced. We will see how this is possible after we prove the following lemma.

Lemma 3.4.7 *Suppose that we are given a positive integer k, a time t_n, a timestep h and a sequence of vectors $\{\mathbf{f}_{n-j}\}_{j=0}^{k-1}$. Define the differences $\nabla^j\mathbf{f}_n$ for $0 \leq j < k$ by the recursion (3.20), and use Eq. (3.21) to define the vector polynomial $\mathbf{p}_n(t)$ of degree $k-1$ that interpolates the sequence $\{\mathbf{f}_{n-j}\}_{j=0}^{k-1}$ at the corresponding equally-spaced times $t_{n-j} = t_n - jh$ by*

$$
\mathbf{p}_n(t) = \sum_{j=0}^{k-1} \mathbf{f}\left[t_n, \ldots, t_{n-j}\right] \prod_{i=0}^{j-1}(t - t_{n-i}) = \sum_{j=0}^{k-1} \frac{\nabla^j\mathbf{f}_n}{j!} \prod_{i=0}^{j-1} \frac{t - t_{n-i}}{h} .
$$

Define the scalars $\alpha_{i,j}$ for $1 \leq j \leq k$ and $0 \leq i \leq j$ by

$$
\alpha_{0,1} = 1
$$

for $2 \leq j \leq k$

$$
\alpha_{0,j} = (j-1)\alpha_{0,j-1}
$$

for $0 \leq i < j$

$$
\alpha_{i,j} = (j-1)\alpha_{i,j-1} + i\alpha_{i-1,j-1}
$$

$$
\alpha_{j,j} = j\alpha_{j-1,j-1} .
$$

Then for all $1 \leq j \leq k$ and all $0 \leq i \leq j$ we have

$$\alpha_{i,j} = \frac{\mathrm{d}^i}{\mathrm{d}s^i} \left[\prod_{\ell=0}^{j-1} (s+\ell) \right]_{s=0} , \qquad (3.25)$$

and for all $0 \leq i < k$

$$h^i D^i \mathbf{p}_n(t_n) = \sum_{j=i}^{k-1} \frac{\alpha_{i,j}}{j!} \nabla^j \mathbf{f}_n . \qquad (3.26)$$

Proof Define the polynomial $\phi_j(s)$ of degree j by

$$\phi_j(s) = \prod_{\ell=0}^{j-1} (s+\ell) .$$

Then

$$\phi_1(0) = 1 = \alpha_{0,1} .$$

This proves (3.25) for $j = 0$. Assume inductively that (3.25) is true for some $j-1 \geq 1$ and all $0 \leq i \leq j-1$. We will prove that (3.25) is true for j, and for all $0 \leq i \leq j$. When $i = 0$ we see that

$$\phi_j(0) = \left[\prod_{\ell=0}^{j-1} (s+\ell) \right]_{s=0} = \left[(s+j-1)\phi_{j-1}(s) \right]_{s=0} = (j-1)\phi_{j-1}(0)$$

$$= (j-1)\alpha_{0,j-1} .$$

Next, recall Leibniz's formula

$$D^i(uv) = \sum_{j=0}^{i} \binom{i}{j} D^{i-j} u D^j v .$$

For $1 \leq i < j$, Leibniz's formula yields

$$\frac{\mathrm{d}^i \phi_j}{\mathrm{d}s^i}(0) = \frac{\mathrm{d}^i}{\mathrm{d}s^i} \left[(s+j-1)\phi_{j-1}(s) \right]_{s=0}$$

$$= \left[(s+j-1) \frac{\mathrm{d}^i \phi_{j-1}}{\mathrm{d}s^i} \right]_{s=0} + i \left[\frac{\mathrm{d}(s+j-1)}{\mathrm{d}s} \frac{\mathrm{d}^{i-1} \phi_{j-1}}{\mathrm{d}s^{i-1}} \right]_{s=0}$$

$$= (j-1) \frac{\mathrm{d}^i \phi_{j-1}}{\mathrm{d}s^i}(0) + i \frac{\mathrm{d}^{i-1} \phi_{j-1}}{\mathrm{d}s^{i-1}}(0) = (j-1)\alpha_{i,j-1} + i\alpha_{i-1,j-1} .$$

Finally, Leibniz's formula produces

$$\frac{\mathrm{d}^j \phi_j}{\mathrm{d}s^j}(0) = \left[(s+j-1)\frac{\mathrm{d}^j \phi_{j-1}}{\mathrm{d}s^i}\right]_{s=0} + j\left[\frac{\mathrm{d}(s+j-1)}{\mathrm{d}s}\frac{\mathrm{d}^{j-1}\phi_{j-1}}{\mathrm{d}s^{j-1}}\right]_{s=0}$$

$$= j\frac{\mathrm{d}^{j-1}\phi_{j-1}}{\mathrm{d}s^{j-1}}(0) = j\alpha_{j-1,j-1}\ .$$

These prove that (3.25) is true for j and all $0 \le i \le j$.

To prove (3.26), we compute

$$h^i \frac{\mathrm{d}^i \mathbf{p}}{\mathrm{d}t^i}(t_n) = h^i \sum_{j=0}^{k-1} \frac{\nabla^j \mathbf{f}_n}{j!}\frac{\mathrm{d}^i}{\mathrm{d}t^i}\left[\prod_{\ell=0}^{j-1}\frac{t-t_{n-\ell}}{h}\right]_{t=t_n}$$

$$= \sum_{j=i}^{k-1}\frac{\nabla^j \mathbf{f}_n}{j!}\frac{\mathrm{d}^i}{\mathrm{d}s^i}\left[\prod_{\ell=0}^{j-1}(s+\ell)\right]_{s=0} = \sum_{j=i}^{k-1}\alpha_{i,j}\frac{\nabla^j \mathbf{f}_n}{j!}$$

In the algorithm to solve the initial-value problem, we can work with the differences $\nabla^i \mathbf{f}_n$ to determine the solution at the new time via (3.24) by

$$\mathbf{y}_{n+1} = \mathbf{y}_n + \int_{t_n}^{t_n+h}\mathbf{p}_n(t)\ \mathrm{d}t = \mathbf{y}_n + h\sum_{j=0}^{k-1}\gamma_j\nabla^j f_n\ ,$$

or to update these values after stepping to a new time via (3.20) by

$$\nabla^j \mathbf{f}_{n+1} = \nabla^{j-1}\mathbf{f}_{n+1} - \nabla^{j-1}\mathbf{f}_n\ .$$

On the other hand, if we want to scale the timestep by some factor σ, we can compute the scaled derivatives $h^i \mathrm{d}^i \mathbf{p}_n / \mathrm{d}t^i$ of the current interpolating polynomial from the differences $\nabla^j \mathbf{f}_n / j!$ via (3.26), then scale the derivatives by computing

$$(\sigma h)^i \frac{\mathrm{d}^i \mathbf{p}_n}{\mathrm{d}t^i}(t_n) = \sigma^i\left[h^i \frac{\mathrm{d}^i \mathbf{p}_n}{\mathrm{d}t^i}(t_n)\right]\ ,$$

and transform back to create a fictitious array of differences. The array $\alpha_{i,j}$ used in the transformation is upper triangular, and independent of the timestep h or the timestep index n.

An alternative form of the Adams-Bashforth method is due to Gear [85].

Lemma 3.4.8 *Suppose that we are given a positive integer k, a time t_n, a timestep h and a sequence of vectors $\{\mathbf{f}_{n-j}\}_{j=0}^{k-1}$. Define the vector polynomial $\mathbf{q}_n$ of degree k by*

$$\mathbf{q}_n(t) = \mathbf{y}_n + \int_{t_n}^{t} \sum_{\ell=0}^{k-1} \mathbf{f}[t_n, \dots, t_{n-\ell}] \prod_{i=0}^{\ell-1} (\tau - t_{n-i}) \, d\tau \; . \tag{3.27}$$

Then

$$\mathbf{y}_n = \mathbf{q}_n(t_n) \text{ and}$$

$$h\mathbf{f}_{n-i} = \sum_{j=1}^{k} \left[j(-i)^{j-1} \right] \left[\frac{h^j}{j!} D^j \mathbf{q}_n(t_n) \right] \text{ for } 0 \le i < k \; .$$

Furthermore,

$$\mathbf{q}_n(t_n + h) = \sum_{j=0}^{k} \frac{h^j}{j!} D^j \mathbf{q}_n(t_n) \; .$$

Proof It is trivial from the definition (3.27) of $\mathbf{q}_n$ that $\mathbf{y}_n = \mathbf{q}_n(t_n)$. Since $\mathbf{q}_n'$ interpolates $\mathbf{f}$ at t_{n-i} for $0 \le i < k$, a Taylor expansion gives us

$$h\mathbf{f}_{n-i} = h\mathbf{q}_n'(t_{n-i}) = h \sum_{j=0}^{k-1} D^{j+1} \mathbf{q}_n(t_n) \frac{(t_{n-i} - t_n)^j}{j!}$$

$$= \sum_{j=0}^{k-1} \frac{(-i)^j h^{j+1}}{j!} D^{j+1} \mathbf{q}_n(t_n) = \sum_{j=1}^{k} \left[j(-i)^{j-1} \right] \left[\frac{h^j}{j!} D^j \mathbf{q}_n(t_n) \right] \; .$$

This proves the second claim. Finally, the Adams-Bashforth method $\mathbf{y}_{n+1} \equiv \mathbf{q}_n(t_n + h)$ and a Taylor expansion gives us

$$\mathbf{y}_{n+1} = \mathbf{q}_n(t_n + h) = \sum_{j=0}^{k} \frac{h^j}{j!} D^j \mathbf{q}_n(t_n) \; .$$

In Gear's approach to the Adams-Bashforth method, we store the values $(h^j/j!)(d^j\mathbf{q}_n/dt^j)(t_n)$ for $0 \le j \le k$ at each timestep. These are easy to scale, if we want to replace the timestep h by σh. It is also easy to recover the sequence $\{\mathbf{f}_{n-i}\}_{\ell=0}^{k-1}$ from the derivatives of $\mathbf{q}_n$, as we will see in the next lemma.

Lemma 3.4.9 *Given a time t_n, a timestep h, a vector $\mathbf{y}_n$ and a sequence $\{\mathbf{f}_{n-\ell}\}_{\ell=0}^{k-1}$ of vectors each the same size as $\mathbf{y}_n$, define the vector*

$$
\boldsymbol{\eta} = \begin{bmatrix} \mathbf{y}_n \\ h\mathbf{f}_n \\ \vdots \\ h\mathbf{f}_{n+1-k} \end{bmatrix}.
$$

Using the polynomial $\mathbf{q}_n$ of degree k that was given by (3.27), define

$$
\boldsymbol{\zeta} = \begin{bmatrix} \mathbf{q}_n(t_n) \\ hD\mathbf{q}_n(t_n) \\ \vdots \\ \frac{h^{k-1}}{(k-1)!}D^{k-1}\mathbf{q}_n(t_n) \end{bmatrix}.
$$

For $0 \le i, j \le k$ let the (i,j) block of the matrix $\boldsymbol{\Omega}$ be

$$
\boldsymbol{\Omega}_{ij} = \begin{cases} \mathbf{I}j(1-i)^{j-1}, & 1 \le i,j \le k \\ \mathbf{I}\delta_{ij}, & \text{otherwise} \end{cases},
$$

and let the (i,j) block of the matrix $\mathbf{Q}$ be

$$
\mathbf{Q}_{ij} = \begin{cases} \mathbf{I}\sum_{\ell=j-1}^{k-1} \frac{\alpha_{i-1,\ell}}{i!\ell!}\binom{\ell}{j-1}(-1)^j, & 1 \le i,j \le k \\ \mathbf{I}\delta_{ij}, & \text{otherwise} \end{cases},
$$

where the scalars α_{ij} were defined in Eq. (3.25). Then

$$
\boldsymbol{\eta} = \boldsymbol{\Omega}\boldsymbol{\zeta} \text{ and } \boldsymbol{\Omega}\mathbf{Q} = \mathbf{I}.
$$

Proof Note that the first block of $\boldsymbol{\eta}$ is

$$
\boldsymbol{\eta}_0 = \mathbf{y}_n = \frac{h^0}{0!}D^0\mathbf{q}_n(t_n) = \boldsymbol{\zeta}_0,
$$

and for $1 \le i \le k$ the remaining blocks of $\boldsymbol{\eta}$ are

$$
\boldsymbol{\eta}_i = h\mathbf{f}_{n+1-i} = h\mathbf{q}_n'(t_{n+1-i}) = h\sum_{\ell=0}^{k-1}D^{\ell+1}\mathbf{q}_n(t_n)\frac{(t_n - t_{n+1-i})^\ell}{\ell!}
$$

$$
= \sum_{\ell=0}^{k-1}\frac{(1-i)^\ell h^{\ell+1}}{\ell!}D^{\ell+1}\mathbf{q}_n(t_n) = \sum_{j=1}^{k}j(1-i)^{j-1}\boldsymbol{\zeta}_j.
$$

The coefficients in these equations are the block entries of $\boldsymbol{\Omega}$. This proves that $\eta = \boldsymbol{\Omega}\zeta$.

Let us verify directly that $\boldsymbol{\Omega}\,\mathbf{Q} = \mathbf{I}$. Note that for $0 \leq j \leq k$

$$[\boldsymbol{\Omega}\,\mathbf{Q}]_{0j} = \sum_{m=0}^{k} \boldsymbol{\Omega}_{0m}\mathbf{Q}_{mj} = \mathbf{Q}_{0j} = \delta_{0j}\mathbf{I}\,,$$

and for $0 \leq i \leq k$ we have

$$[\boldsymbol{\Omega}\,\mathbf{Q}]_{i0} = \sum_{m=0}^{k} \boldsymbol{\Omega}_{im}\mathbf{Q}_{m0} = \boldsymbol{\Omega}_{i0} = \delta_{i0}\mathbf{I}\,.$$

Next, for $1 \leq i,j \leq k$ we have

$$[\boldsymbol{\Omega}\,\mathbf{Q}]_{ij} = \sum_{m=0}^{k} \boldsymbol{\Omega}_{im}\mathbf{Q}_{mj} = \sum_{m=1}^{k} \boldsymbol{\Omega}_{im}\mathbf{Q}_{mj}$$

$$= \mathbf{I}\sum_{m=1}^{k} m(1-i)^{m-1} \sum_{\ell=j-1}^{k-1} \frac{\alpha_{m-1,\ell}}{m!\ell!} \binom{\ell}{j-1}(-1)^{j-1}$$

then the definition (3.25) of $\alpha_{m-1,\ell}$ implies

$$= \mathbf{I}\frac{(-1)^{j-1}}{(j-1)!} \sum_{m=1}^{k} \frac{(1-i)^{m-1}}{(m-1)!} \sum_{\ell=j-1}^{k-1} \frac{1}{(\ell+1-j)!} \frac{d^{m-1}}{ds^{m-1}} \left[\prod_{\alpha=0}^{\ell-1}(s+\alpha)\right]_{s=0}$$

$$= \mathbf{I}\frac{(-1)^{j-1}}{(j-1)!} \sum_{\ell=j-1}^{k-1} \frac{1}{(\ell+1-j)!} \sum_{m=1}^{k} \frac{(1-i)^{m-1}}{(m-1)!} \frac{d^{m-1}}{ds^{m-1}} \left[\prod_{\alpha=0}^{\ell-1}(s+\alpha)\right]_{s=0}$$

$$= \mathbf{I}\frac{(-1)^{j-1}}{(j-1)!} \sum_{\ell=j-1}^{k-1} \frac{1}{(\ell+1-j)!} \sum_{M=0}^{k-1} \frac{(1-i)^{M}}{M!} \frac{d^{M}}{ds^{M}} \left[\prod_{\alpha=0}^{\ell-1}(s+\alpha)\right]_{s=0}$$

then a Taylor expansion leads to

$$= \mathbf{I}\frac{(-1)^{j-1}}{(j-1)!} \sum_{\ell=j-1}^{k-1} \frac{1}{(\ell+1-j)!} \left[\prod_{\alpha=0}^{\ell-1}(s+\alpha)\right]_{s=1-i}$$

$$= \mathbf{I}\frac{(-1)^{j-1}}{(j-1)!} \sum_{\ell=j-1}^{k-1} \frac{1}{(\ell+1-j)!} \prod_{\alpha=0}^{\ell-1}(\alpha+1-i)$$

$$= \mathbf{I}\frac{(-1)^{j-1}}{(j-1)!} \sum_{\ell=j-1}^{k-1} \frac{1}{(\ell+1-j)!}(i-1)\cdot\ldots\cdot(i-\ell)\,.$$

Note that if $j - 1 > i$, then for all $\ell \in [j-1, k-1]$ we have

$$i - \ell \leq i + 1 - j < 0 ,$$

from which it follows that $[\boldsymbol{\varOmega}\mathbf{Q}]_{ij} = \mathbf{0}$. On the other hand, if $j < i$ then

$$[\boldsymbol{\varOmega}\mathbf{Q}]_{ij} = \mathbf{I}\frac{(-1)^{j-1}}{(j-1)!} \sum_{\ell=j-1}^{i-1} \frac{1}{(\ell+1-j)!} \frac{(i-1)!}{(i-\ell-1)!}$$

$$= \mathbf{I}\frac{(-1)^{j-1}}{(j-1)!} \frac{(i-1)!}{(i-j)!} \sum_{\ell=j-1}^{i-1} \binom{i-j}{\ell+1-j}(-1)^\ell$$

and then a binomial expansion gives us

$$= \mathbf{I}\frac{1}{(j-1)!} \frac{(i-1)!}{(i-j)!} \sum_{m=0}^{i-j} \binom{i-j}{m}(-1)^m = \mathbf{I}\frac{1}{(j-1)!} \frac{(i-1)!}{(i-j)!}(1-1)^{i-j} = \mathbf{0} .$$

Finally, if $i = j$ then

$$[\boldsymbol{\varOmega}\mathbf{Q}]_{jj} = \mathbf{I}\frac{(-1)^{j-1}}{(j-1)!} \sum_{\ell=j-1}^{j-1} \frac{(-1)^\ell}{(\ell+1-j)!}(j-1)! = \mathbf{I}\frac{(-1)^{j-1}}{(j-1)!}\frac{(-1)^{j-1}}{0!}(j-1)! = \mathbf{I} .$$

Note that the block array $\mathbf{A}$ with block entries

$$\mathbf{A}_{ij} = \begin{cases} \mathbf{I}\alpha_{i-1,j-1}/[i!(j-1)!], & 1 \leq i \leq j \leq k \\ \mathbf{I}\delta_{ij}, & \text{otherwise} \end{cases}$$

is block right-triangular, and the block array $\mathbf{P}$ with block entries

$$\mathbf{P}_{ij} = \begin{cases} \mathbf{I}\binom{i-1}{j-1}(-1)^{j-1}, & 1 \leq j \leq i \leq k \\ \mathbf{I}\delta_{ij}, & \text{otherwise} \end{cases}$$

is block left-triangular. Furthermore, we have the R-L factorization

$$\mathbf{Q} = \mathbf{AP} .$$

Subroutine `dvindy` of the Brown, Hindmarsh and Byrne program vode computes the vector $\boldsymbol{\zeta}$ from the vector $\boldsymbol{\eta}$ as described in Lemma 3.4.9. The coefficients γ_j in Lemma 3.4.6 are computed in subroutine `dvset` of this same program.

It is common to test numerical methods for initial value problems on the problem $y'(t) = \lambda y(t)$ with $y(0) = 1$. For $\lambda > 0$, Adams-Bashforth methods produce approximations without numerical oscillations, and converge nicely as the timestep

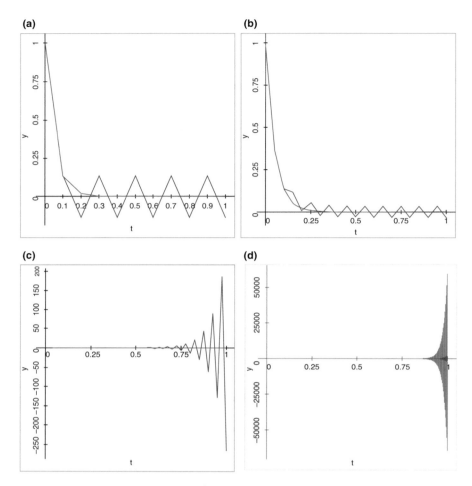

Fig. 3.1 Adams-Bashforth method for $y' = -20y, y(0) = 1$; insufficiently small timesteps lead to oscillation and instability, and higher-order methods require much smaller timesteps for stability. (**a**) Order 1, 10 steps. (**b**) Order 2, 20 steps. (**c**) Order 4, 20 steps. (**d**) Order 8, 320 steps

is decreased or the order is increased. However, Fig. 3.1 shows that for $\lambda < 0$ the Adams-Bashforth schemes can produce oscillations and instability unless the timestep is chosen sufficiently small. High-order Adams-Bashforth schemes must choose timesteps that are much smaller than timesteps for low-order Adams-Bashforth schemes. Figure 3.2 shows mesh refinement studies for Adams-Bashforth methods of various orders. Note that the fourth and eighth order methods exhibit growth in the solution until the timestep becomes sufficiently small, corresponding to stable timesteps.

Figures 3.1 and 3.2 were generated by the C^{++} program GUILinearMultistep.C and the Fortran program integrate.f, which implement a simple linear

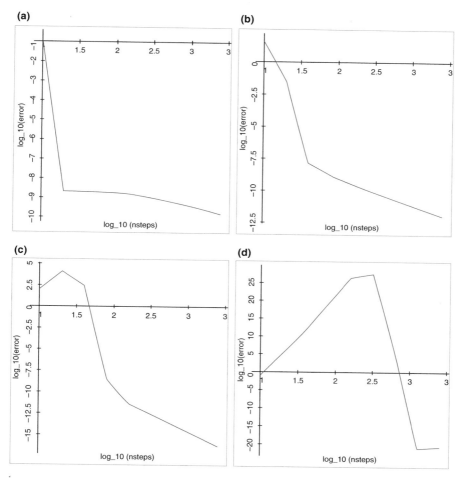

Fig. 3.2 Adams-Bashforth mesh refinement study for $y' = -20\mathbf{y}, \mathbf{y}(0) = 1$; high-order methods require much smaller timesteps to avoid instability. (**a**) Order 1. (**b**) Order 2. (**c**) Order 4. (**d**) Order 8

multistep algorithm. The latter Fortran file contains routines `adams_bashforth_startup` and `adams_bashforth`, which perform the initialization and integration stages of an Adams-Bashforth method.

To solve $y'(t) = \lambda y(t), y(0) = 1$ for real λ, readers may execute the JavaScript program for **real linear multistep methods.** Alternatively, to solve $y'(t) = \lambda y(t), y(0) = 1$ for complex λ, readers can execute the JavaScript program for **complex linear multistep methods.**

3.4.7.2   Adams-Moulton

Adams-Moulton methods approximate an ordinary differential equation in the form

$$\mathbf{y}(t_{n+1}) - \mathbf{y}(t_n) = \int_{t_n}^{t_{n+1}} \mathbf{f}(\mathbf{y}(t), t) dt$$

by integrating the Newton interpolation polynomial to $\mathbf{f}$ at the times $\{t_{n+1}, t_n, \dots,$
t_{n-k+1}. Because t_{n+1} is one of the interpolation points and $\mathbf{f}(\mathbf{y}(t_{n+1}), t_{n+1})$ depends
on the unknown solution value $\mathbf{y}(t_{n+1})$, this method is implicit. If the timesteps
are equally-spaced, then the implementation of the method can make use of the
following lemma.

Lemma 3.4.10 *Define the sequence $\{\gamma_j\}_{j=0}^{\infty}$ by the recursion*

$$\gamma_0 = 1 , \text{ and } \gamma_j = -\sum_{i=0}^{j-1} \frac{\gamma_i}{j-i+1} \text{ for } j > 0 . \tag{3.28}$$

*Suppose that we are given a nonnegative integer k, a time t_n, a timestep h and
a sequence of vectors $\{\mathbf{f}_{n+1-j}\}_{j=0}^{k}$. Define the differences $\nabla^j \mathbf{f}_{n+1}$ recursively by
Eq. (3.20). Define the scalars*

$$t_{n+1-j} = t_n - (j-1)h$$

*for $0 \leq j \leq k$. Then the polynomial interpolant to the sequence $\{\mathbf{f}_{n+1-j}\}_{j=0}^{k}$ at the
corresponding times $\{t_{n+1-j}\}_{j=0}^{k}$ is*

$$\mathbf{p}_{n+1}(t) = \sum_{j=0}^{k} \mathbf{f}\left[t_{n+1}, \dots, t_{n+1-j}\right] \prod_{i=0}^{j-1} (t - t_{n+1-i}) , \tag{3.29}$$

and

$$\int_{t_n}^{t_n+h} \mathbf{p}_{n+1}(t) \, dt = h \sum_{j=0}^{k-1} \gamma_j \nabla^j \mathbf{f}_{n+1} . \tag{3.30}$$

Proof Newton interpolation of the values $\{\mathbf{f}_{n+1-j}\}_{j=0}^{k}$ at the points $\{t_{n+1-j}\}_{j=0}^{k}$ gives
us the polynomial

$$\mathbf{p}_{n+1}(t) = \sum_{j=0}^{k} \mathbf{f}[t_{n+1}, \dots, t_{n+1-j}] \prod_{i=0}^{j-1} (t - t_{n+1-i})$$

then we use Eq. (3.21) to write

$$= \sum_{j=0}^{k} \frac{\nabla^j \mathbf{f}_{n+1}}{h^j j!} \prod_{i=0}^{j-1} (t - t_{n+1-i})$$

then we substitute $t = t_{n+1} + sh$ to get

$$= \sum_{j=0}^{k} \frac{\nabla^j \mathbf{f}_{n+1}}{h^j j!} h^j (-1)^j \binom{-s}{j} .$$

By taking the required integral of the interpolating polynomial, we obtain

$$\int_{t_n}^{t_{n+1}} \mathbf{p}_{n+1}(t) \, dt = h \sum_{j=0}^{k} \nabla^j \mathbf{f}_{n+1} (-1)^j \int_{-1}^{0} \binom{-s}{j} \, ds .$$

We would like to find a recurrence for evaluating

$$\gamma_j = (-1)^j \int_{-1}^{0} \binom{-s}{j} \, ds .$$

Using the binomial expansion (2.1) in Chap. 2 of Volume I, we see that the generating function for the coefficients is

$$G(x) \equiv \sum_{j=0}^{\infty} \gamma_j x^j = \sum_{j=0}^{\infty} \int_{-1}^{0} (-x)^j \binom{-s}{j} \, ds = \int_{-1}^{0} \left[\sum_{j=0}^{\infty} (-x)^j \binom{-s}{j} \right] \, ds$$

$$= \int_{-1}^{0} (1-x)^{-s} \, ds = -\frac{(1-x)^{-s}}{\ln(1-x)} \bigg|_{s=-1}^{0} = -\frac{x}{\ln(1-x)} .$$

It follows that

$$1 = G(x) \left[-\frac{\ln(1-x)}{x} \right] = \left[\sum_{j=0}^{\infty} \gamma_j x^j \right] \left[\sum_{i=0}^{\infty} \frac{x^i}{i+1} \right] = \sum_{\ell=0}^{\infty} x^\ell \sum_{j=0}^{\ell} \frac{\gamma_j}{\ell - j + 1} .$$

Equating coefficients in the series gives us $\gamma_0 = 1$ and for $\ell > 0$

$$0 = \sum_{j=0}^{\ell} \frac{\gamma_j}{\ell - j + 1} = \gamma_\ell + \sum_{j=0}^{\ell-1} \frac{\gamma_j}{\ell - j + 1} .$$

The claimed result follows immediately.

The first four values for the Adams-Moulton coefficients are

$$\gamma_0 = 1 , \gamma_1 = -\frac{\gamma_0}{2} = -\frac{1}{2} , \gamma_2 = -\frac{\gamma_0}{3} - \frac{\gamma_1}{2} = -\frac{1}{12}, \gamma_3 = -\frac{\gamma_0}{4} - \frac{\gamma_1}{3} - \frac{\gamma_2}{2} = -\frac{1}{24}.$$

In particular, the first three Adams-Moulton methods are

$$\mathbf{y}_{n+1} - \mathbf{y}_n = h\mathbf{f}_{n+1} \text{ (order 1 : backward Euler)}$$

$$\mathbf{y}_{n+1} - \mathbf{y}_n = h \left[\mathbf{f}_{n+1} - \frac{1}{2}\nabla^1\mathbf{f}_{n+1} \right]$$

$$= h \left[\frac{1}{2}\mathbf{f}_{n+1} + \frac{1}{2}\mathbf{f}_n \right] \text{ (order 2 : trapezoidal rule)}$$

$$\mathbf{y}_{n+1} - \mathbf{y}_n = h \left[\mathbf{f}_{n+1} - \frac{1}{2}\nabla^1\mathbf{f}_{n+1} - \frac{1}{12}\nabla^2\mathbf{f}_{n+1} \right]$$

$$= h \left[\frac{5}{12}\mathbf{f}_{n+1} + \frac{2}{3}\mathbf{f}_n - \frac{1}{12}\mathbf{f}_{n-1} \right] \text{ (order 3) .}$$

Again, we emphasize that the Adams-Moulton methods are implicit. Thus, we typically require a numerical method for solving nonlinear equations in order to compute $\mathbf{y}_{n+1}$. We could use the techniques in Chaps. 5 of Volume I or 3 of Volume II for this purpose. We will see another approach in Sect. 3.4.10 below.

The difference form of the Adams-Moulton method allows us to vary the order of the scheme by changing the number of terms carried in the sum. Also note that if $\{y_n\}$ is generated by a k-step Adams-Moulton method and $\{\tilde{y}_n\}$ is generated by a $k + 1$-step method, then

$$\tilde{y}_{n+1} - y_{n+1} = h\gamma_k\nabla^k f_n = h^{k+1}\gamma_k\frac{d^{k+1}y}{dt^{k+1}}(t_n) + O(h^{k+2}) .$$

This shows that the k-step method has order k and error constant γ_k.

As with the Adams-Bashforth methods, we can use Lemma 3.4.9 to transform the sequence of values for $\mathbf{f}$ to a sequence of derivatives of an interpolating polynomial. This transformation makes it easier to scale the values for a change in timestep size.

Example 3.4.16 Suppose that we have stored $\mathbf{f}_n$, $\nabla^1\mathbf{f}_n$, $\nabla^2\mathbf{f}_n$ and $\mathbf{y}_n$. Given an initial guess for $\mathbf{y}_{n+1}$, we advance the third-order Adams-Moulton scheme to t_{n+1} by computing

$$\mathbf{f}_{n+1} = \mathbf{f}(\mathbf{y}_{n+1}, t_{n+1})$$
$$\nabla^1\mathbf{f}_{n+1} = \mathbf{f}_{n+1} - \mathbf{f}_n \text{ (}\mathbf{f}_n \text{ needed until converged)}$$
$$\nabla^2\mathbf{f}_{n+1} = \nabla^1\mathbf{f}_{n+1} - \nabla^1\mathbf{f}_n \text{ (}\nabla^1\mathbf{f}_n \text{ needed until converged)}$$
$$\nabla^3\mathbf{f}_{n+1} = \nabla^2\mathbf{f}_{n+1} - \nabla^2\mathbf{f}_n \text{ (}\nabla^2\mathbf{f}_n \text{ needed until converged)}$$

Then we compute

$$\mathbf{y}_{n+1} = \mathbf{y}_n + h\left[\mathbf{f}_{n+1} - \frac{1}{2}\nabla^1\mathbf{f}_{n+1} - \frac{1}{12}\nabla^2\mathbf{f}_{n+1} - \frac{1}{24}\nabla^3\mathbf{f}_{n+1}\right]$$

through some convergent iteration for solving the nonlinear equation, in which $\mathbf{f}_{n+1}$ depends on $\mathbf{y}_{n+1}$. After this calculation is converged, $\mathbf{y}_n$ and the differences $\nabla^j\mathbf{f}_n$ are no longer needed.

Some numerical results with Adams-Moulton methods for $y' = 8y$ are shown in Fig. 3.3. These methods can fail to produce monotonically increasing numerical solutions unless the timestep is chosen sufficiently small. The lack of monotonicity

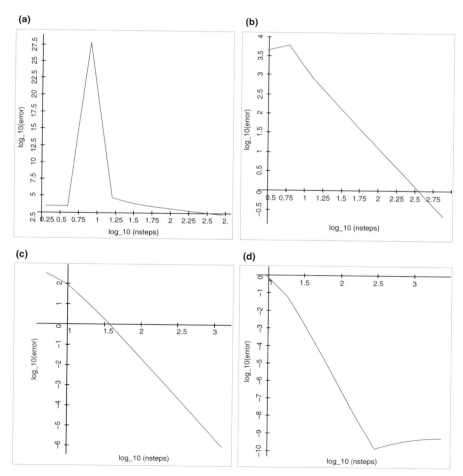

Fig. 3.3 Adams-Moulton mesh refinement study for $y' = 8y, y(0) = 1$. (**a**) Order 1. (**b**) Order 2. (**c**) Order 4. (**d**) Order 8

shows up indirectly in the mesh refinement study, particularly for order 1. In this case, the error increases as the mesh is refined, then decreases appropriately once monotonicity is established.

Figure 3.3 was generated by the C^{++} program GUILinearMultistep.C and the Fortran program integrate.f. These implement a simple form of the Adams-Moulton method in routines `adams_moulton_startup` and `adams_moulton`. There are several old publicly available Fortran programs for integrating initial value problems via Adams methods, such as Netlib routine ode, ODEPACK routine lsode, TOMS routine difsub (#407),and SLATEC routines deabm and sdriv1. The more recent Brown, Hindmarsh and Byrne program vode implements Adams-Moulton methods as a user option. GNU Scientific Library users may be interested in the C routine gsl_odeiv2_step_msadams. MATLAB users could consider the command ode113.

To solve $y'(t) = \lambda y(t), y(0) = 1$ for real λ, readers may execute the JavaScript program for **real linear multistep methods.** Alternatively, to solve $y'(t) = \lambda y(t), y(0) = 1$ for complex λ, readers can execute the JavaScript program for **complex linear multistep methods.**

3.4.7.3   Nyström

Nyström methods approximate

$$\mathbf{y}(t_n + h) - \mathbf{y}(t_n - h) = \int_{t_n-h}^{t_n+h} \mathbf{f}(\mathbf{y}(t), t)dt$$

by integrating an interpolating polynomial for $\mathbf{f}$. Because the first characteristic polynomial for Nyström methods has roots ± 1, at least one of the roots of the characteristic polynomial $a(\zeta) - h\lambda b(\zeta)$ is almost always outside the unit circle and the scheme is almost never absolutely stable.

Explicit Nyström methods take the form

$$\mathbf{y}_{n+1} = \mathbf{y}_{n-1} + h \sum_{j=0}^{k-1} \gamma_j \nabla^j \mathbf{f}_n$$

with

$$\gamma_j = (-1)^j \int_{-1}^{1} \binom{-s}{j} ds .$$

These coefficients have the generating function

$$\sum_{j=0}^{\infty} \gamma_j x^j = G(x) = -\frac{x(2-x)}{(1-x)\log(1-x)}$$

and satisfy the recurrence relation

$$\gamma_0 = 2 \, , \; \gamma_j = 1 - \sum_{m=0}^{j-1} \frac{\gamma_m}{j+1-m} \; \text{ for } j \geq 2 \, .$$

Implicit Nyström methods are also called **Milne-Simpson methods**, and take the form

$$\mathbf{y}_{n+1} = \mathbf{y}_{n-1} + h \sum_{j=0}^{k} \gamma_j \nabla^j \mathbf{f}_{n+1}$$

with

$$\gamma_j = (-1)^j \int_{-1}^{1} \binom{-s+1}{j} ds \, .$$

These coefficients have the generating function

$$\sum_{j=0}^{\infty} \gamma_j x^j = G(x) = -\frac{x(2-x)}{\log(1-x)}$$

and satisfy the recurrence

$$\gamma_0 = 2 \, , \; \gamma_1 = -2 \, , \; \gamma_j = -\sum_{m=0}^{j-1} \frac{\gamma_m}{j+1-m} \; \text{ for } j \geq 2 \, .$$

The only Nyström method that is used much in practice is the **explicit midpoint method**

$$\mathbf{y}_{n+1} = \mathbf{y}_{n-1} + 2h \mathbf{f} \left(\mathbf{y}_n \, , \, t_n \right) \, .$$

The advantage of the explicit midpoint method is that with equally spaced timesteps its error has been proved by Gragg [90] to involve only even powers of h:

$$\mathbf{y}_n = \mathbf{y}(t_n) + \sum_{j=2,2}^{\infty} c_j h^j \, .$$

Thus this Nyström method mimics the behavior of the midpoint method for integration. As a result, it is reasonable to construct Richardson extrapolants of this method to produce methods of high order.

The difficulty with the explicit midpoint method is that the characteristic polynomial, $\varrho(\zeta) - h\lambda\sigma(\zeta) = \zeta^2 - 1 - h\lambda\zeta$, has zeros

$$\zeta = \pm\sqrt{1 + (h\lambda/2)^2} + h\lambda/2 \ .$$

Unless $\lambda = 0$, at least one of the zeros of the characteristic polynomial will lie outside the unit circle, and the method will be (weakly) unstable.

The weak instability can be reduced somewhat by **damping**. This resulting scheme for solving $\mathbf{y}'(t) = \mathbf{f}(\mathbf{y}(t), t)$ for $t > t_0$ with $\mathbf{y}(t_0) = \mathbf{y}_0$ is

Algorithm 3.4.1 (Modified Midpoint Method with Damping)

$$\widetilde{\mathbf{y}}_0 = \mathbf{y}_0$$

$$\widetilde{\mathbf{y}}_1 = \mathbf{y}_0 + \mathbf{f}(\widetilde{\mathbf{y}}_0, t_0)\, h$$

$$\text{for } 1 \le k \le 2n$$

$$\widetilde{\mathbf{y}}_{k+1} = \widetilde{\mathbf{y}}_{k-1} + \mathbf{f}(\widetilde{\mathbf{y}}_k, t_k)\, 2h$$

$$\mathbf{y}_{2n} = (\widetilde{\mathbf{y}}_{2n-1} + 2\widetilde{\mathbf{y}}_{2n} + \widetilde{\mathbf{y}}_{2n+1})\frac{1}{4} \ .$$

Numerical results $\mathbf{y}_{2n}$ corresponding to the same time but increasingly smaller timesteps can be computed and entered into a Richardson extrapolation. Afterward, the integration can begin again with the extrapolated solution as the initial value.

Bader and Deuflhard [11] have suggested the following semi-implicit modification of the midpoint method to solve $\mathbf{y}'(t) = \mathbf{f}(\mathbf{y}(t), t)$ for $t > t_0$ with $\mathbf{y}(t_0) = \mathbf{y}_0$:

Algorithm 3.4.2 (Bader-Deuflhard Semi-Implicit Midpoint Method)

$$\mathbf{J}_0 = \frac{\partial\mathbf{f}}{\partial\mathbf{y}}(\mathbf{y}_0, t_0)$$

$$\widetilde{\mathbf{y}}_0 = \mathbf{y}_0$$

$$\text{solve } (\mathbf{I} - \mathbf{J}_0 h)\widetilde{\mathbf{y}}_1 = \widetilde{\mathbf{y}}_0 + [\mathbf{f}(\widetilde{\mathbf{y}}_0, t_0) - \mathbf{J}_0\mathbf{y}_0]\, h \text{ for } \widetilde{\mathbf{y}}_1$$

$$\text{for } 1 \le k \le 2n$$

$$\text{solve } (\mathbf{I} - \mathbf{J}_0 h)\widetilde{\mathbf{y}}_{k+1} = (\mathbf{I} + \mathbf{J}_0 h)\widetilde{\mathbf{y}}_{k-1} + [\mathbf{f}(\widetilde{\mathbf{y}}_k, t_k) - \mathbf{J}_0\mathbf{y}_k]\, h \text{ for } \widetilde{\mathbf{y}}_{k+1}$$

$$\mathbf{y}_{2n} = (\widetilde{\mathbf{y}}_{2n+1} + \mathbf{y}_{2n-1})\frac{1}{2} \ .$$

As justification for this algorithm, note that

$$\frac{\widetilde{\mathbf{y}}_1 - \widetilde{\mathbf{y}}_0}{h} = \mathbf{f}(\widetilde{\mathbf{y}}_0, t_0) + \mathbf{J}_0[\widetilde{\mathbf{y}}_1 - \widetilde{\mathbf{y}}_0]$$

is consistent with the original differential equation. For subsequent steps, we have

$$\frac{\widetilde{\mathbf{y}}_{k+1} - \widetilde{\mathbf{y}}_{k-1}}{2h} = \frac{[\mathbf{f}\,(\widetilde{\mathbf{y}}_k\,,\,t_k) + \mathbf{J}_0\,(\mathbf{y}_{k+1} - \mathbf{y}_k)] + [\mathbf{f}\,(\widetilde{\mathbf{y}}_k\,,\,t_k) + \mathbf{J}_0\,(\mathbf{y}_{k-1} - \mathbf{y}_k)]}{2}\,,$$

which is also close to the original differential equation if $\mathbf{f}$ does not vary rapidly with respect to $\mathbf{y}$.

For more details regarding extrapolation methods for the midpoint method, see Deuflhard [66], Deuflhard and Bornemann [67, p. 168ff] or Hairer and Wanner [94, p. 224ff].

There are several publicly available programs for implementing extrapolation of the modified midpoint method to solve initial value problems. Readers may consider the Hairer and Wanner [94] Fortran program odex. In C, Gerard Jungman has written bsimp, which is an implementation within the GNU Scientific Library of the Bulirsch and Stoer version of the Bader and Deuflhard extrapolation method. This code is also available in the GNU Scientific Library as routine gsl_odeiv2_step_bsimp.

A simple implementation of the explicit midpoint method can be found in routine explicit_midpoint in file integrate.f. This routine is called from the main program in GUILinearMultistep.C. Alternatively, readers may execute JavaScript programs for real linear multistep methods or for complex linear multistep methods.

Exercise 3.4.13 The generalized Milne-Simpson method [124, p. 47] is the Nyström method

$$\mathbf{y}_{n+1} = \mathbf{y}_{n-1} + \frac{h}{3}\,[\mathbf{f}_{n+1} + 4\mathbf{f}_n + \mathbf{f}_{n-1}]\ .$$

Show that this method is fourth-order, and determine its absolute stability region.

3.4.7.4 Backward Differentiation

Backward differentiation formulas approximate the differential equation

$$\mathbf{y}'(t_{n+1}) = \mathbf{f}(\mathbf{y}(t_{n+1}), t_{n+1})$$

by differentiating the Newton interpolation polynomial to $\mathbf{y}$ at the times $t_{n+1}, t_n \ldots, t_{n-k+1}$. The following lemma describes how to compute this polynomial derivative.

Lemma 3.4.11 *Let k be a positive integer, t_n be a time, h be a timestep and $\{\mathbf{y}_{n+1-j}\}_{j=0}^k$ be a sequence of vectors of a fixed size. Then the polynomial $\mathbf{p}_{n+1}$ of degree k determined by the interpolation constraints*

$$\mathbf{p}_{n+1}(t_n + (1-j)h) = \mathbf{y}_{n+1-j}\,,\ 0 \le j \le k$$

satisfies

$$\mathbf{p}'_{n+1}(t_{n+1}) = \frac{1}{h} \sum_{j=1}^{k} \frac{1}{j} \nabla^j \mathbf{y}_{n+1} \ .$$

Proof From the Newton interpolation formula in Lemma 1.2.3, we have that

$$\mathbf{p}_{n+1}(t) = \sum_{j=0}^{k} \mathbf{y}[t_{n+1}, \dots, t_{n+1-j}] \prod_{i=0}^{j-1} (t - t_{n+1-i})$$

then we can use Eq. (3.21) to get

$$= \sum_{j=0}^{k} \frac{\nabla^j \mathbf{y}_{n+1}}{j!} \prod_{i=0}^{j-1} \frac{t - t_{n+1-i}}{h}$$

then we substitute $t = t_{n+1} + sh$ to obtain

$$= \sum_{j=0}^{k} \nabla^j \mathbf{y}_{n+1} \frac{1}{j!} \prod_{i=0}^{j-1} (s + i) \ .$$

We can differentiate this expression to get

$$\mathbf{p}'_{n+1}(t_{n+1}) = \frac{1}{h} \frac{d}{ds} \left[\sum_{j=0}^{k} \nabla^j \mathbf{y}_{n+1} \frac{1}{j!} \prod_{i=0}^{j-1} (s + i) \right]_{s=0}$$

$$= \frac{1}{h} \sum_{j=0}^{k} \nabla^j \mathbf{y}_{n+1} \frac{1}{j!} \sum_{i=0}^{j-1} \prod_{\substack{0 \le \ell < j \\ \ell \ne i}} (0 + \ell) = \frac{1}{h} \sum_{j=0}^{k} \nabla^j \mathbf{y}_{n+1} \frac{1}{j!} \prod_{\ell=1}^{j-1} \ell = \frac{1}{h} \sum_{j=0}^{k} \frac{1}{j} \nabla^j \mathbf{y}_{n+1} \ .$$

Thus the backward difference schemes have the form

$$\mathbf{f}(\mathbf{y}_{n+1}, t_{n+1}) h = \sum_{j=1}^{k} \nabla^j \frac{\mathbf{y}_{n+1}}{j} = \sum_{j=1}^{k} \frac{1}{j} \sum_{\ell=0}^{j} \mathbf{y}_{n+1-\ell} \binom{j}{\ell} (-1)^\ell$$

$$= \mathbf{y}_{n+1} \sum_{j=1}^{k} \frac{1}{j} + \sum_{\ell=1}^{k} \mathbf{y}_{n+1-\ell} \sum_{j=\ell}^{k} \frac{(-1)^\ell}{j} \binom{j}{\ell}$$

This implies that the first characteristic polynomial is

$$a(\zeta) = \sum_{m=0}^{k-1} \frac{\sum_{j=k-m}^{k} \binom{j}{k-m}(-1)^{k-m}/j}{\sum_{j=1}^{k} 1/j} \zeta^m + \zeta^k$$

$$= \frac{\sum_{j=1}^{k} \sum_{m=k-j}^{k-1} \binom{j}{k-m}(-1)^{k-m}\zeta^m/j + \zeta^k}{\sum_{j=1}^{k} 1/j} = \frac{\sum_{j=1}^{k} \sum_{\ell=0}^{j}(-1)^{\ell}\binom{j}{\ell}\zeta^{k-\ell}/j}{\sum_{j=1}^{k} 1/j}$$

$$= \zeta^k \frac{\sum_{j=1}^{k}(1-1/\zeta)^j/j}{\sum_{j=1}^{k} 1/j} = \frac{\sum_{j=1}^{k}\zeta^{k-j}(\zeta-1)^j/j}{\sum_{j=1}^{k} 1/j} ,$$

and the second characteristic polynomial is obviously

$$b(\zeta) = \frac{\zeta^k}{\sum_{j=1}^{k} 1/j} .$$

Note that

$$a(1) = \frac{1^k \sum_{j=1}^{k} \frac{1}{j} 0^j}{\sum_{j=1}^{k} 1/j} = 0 .$$

Since

$$a'(\zeta) \sum_{j=1}^{k} 1/j = \sum_{j=1}^{k} \left[(k-j)\zeta^{k-j-1}(\zeta-1)^j + j\zeta^{k-j}(\zeta-1)^{j-1} \right]/j$$

$$= \sum_{j=1}^{k} \zeta^{k-j-1}(\zeta-1)^{j-1} [k\zeta + j - k]/j$$

we also have

$$a'(1) \sum_{j=1}^{k} 1/j = 1 = b(1) \sum_{j=1}^{k} 1/j .$$

Lemma 3.4.3 now implies that backward differentiation schemes are consistent. However, Theorem 3.4.1 and an examination of the zeros of the first characteristic polynomial shows that backward differentiation schemes are zero-stable only for $k \leq 6$. A proof of this fact can be found in Hairer et al. [94, p. 329].

The following lemma will describe the order and local truncation error for backward differentiation formulas.

Lemma 3.4.12 *If* **y** *is* $k + 2$ *times continuously differentiable, then the local truncation error in the backward differentiation formula applied to* **y** *is*

$$\mathbf{y}(t_n + h) \sum_{j=1}^{k} \frac{1}{j} + \sum_{\ell=1}^{k} \mathbf{y}(t_n + [1 - \ell]h) \sum_{j=\ell}^{k} \frac{1}{j} \binom{j}{\ell} (-1)^\ell - h\mathbf{y}'(t_{n+1})$$

$$= D^{k+1}\mathbf{y}(t_{n+1}) \frac{h^{k+1}}{k + 1} + O\left(h^{k+2}\right) \ .$$

Proof Let $\mathbf{q}_{n+1}(t)$ be the polynomial of degree $k + 1$ that interpolates **y** at $t_{n+1}, \dots, t_{n-k}$, and define the vectors

$$\mathbf{y}_{n-j} = \mathbf{y}(t_n - jh) \text{ for } -1 \le j \le k + 1 \ .$$

Then Theorem 1.3.2 implies that

$$\mathbf{q}'_{n+1}(t_{n+1}) - \mathbf{y}'(t_{n+1}) = O(h^{k+1}) \ ,$$

and Lemma 3.4.11 implies that

$$\mathbf{q}'_{n+1}(t_{n+1}) = \sum_{j=1}^{k+1} \frac{\nabla^j \mathbf{y}_{n+1}}{jh} \ .$$

If $\mathbf{p}_{n+1}$ is the polynomial of degree k that interpolates **y** at $t_{n+1}, \dots, t_{n-k+1}$, it follows that

$$h\mathbf{p}'_{n+1}(t_{n+1}) - h\mathbf{y}'(t_{n+1}) = \sum_{j=1}^{k} \frac{1}{j} \nabla^j \mathbf{y}_{n+1} - h\mathbf{y}'(t_{n+1})$$

$$= -\frac{1}{k + 1} \nabla^{k+1} \mathbf{y}_{n+1} + \left[\sum_{j=1}^{k+1} \frac{1}{j} \nabla^j \mathbf{y}_{n+1} - h\mathbf{y}'(t_{n+1}) \right]$$

$$= -\frac{1}{k + 1} \nabla^{k+1} \mathbf{y}_{n+1} + \left[\mathbf{q}'_{n+1}(t_{n+1}) - h\mathbf{y}'(t_{n+1}) \right]$$

then Eq. (3.21) gives us

$$= -\frac{1}{k + 1} \mathbf{f}[t_{n+1}, \dots, t_{n-k}] h^{k+1}(k + 1)! + O(h^{k+2})$$

and finally Eq. (1.9) produces

$$= -\frac{h^{k+1}}{k + 1} D^{k+1} \mathbf{y}(t_{n+1}) + O(h^{k+2}) \ .$$

Example 3.4.17 Suppose that we have stored $\mathbf{y}_n$, $\nabla^1\mathbf{y}_n$ and $\nabla^2\mathbf{y}_n$. In order to advance the third-order backward differentiation formula to t_{n+1}, we guess a value for $\mathbf{y}_{n+1}$ and compute

$$\mathbf{f}_{n+1} = \mathbf{f}(\mathbf{y}_{n+1}, t_{n+1})$$
$$\nabla^1\mathbf{y}_{n+1} = \mathbf{y}_{n+1} - \mathbf{y}_n \ (\mathbf{y}_n \text{ needed until iteration converged})$$
$$\nabla^2\mathbf{y}_{n+1} = \nabla^1\mathbf{y}_{n+1} - \nabla^1\mathbf{y}_n \ (\nabla^1\mathbf{y}_n \text{ needed until iteration converged})$$
$$\nabla^3\mathbf{y}_{n+1} = \nabla^2\mathbf{y}_{n+1} - \nabla^2\mathbf{y}_n \ (\nabla^2\mathbf{y}_n \text{ needed until iteration converged})$$

Then we solve the nonlinear equation

$$\nabla^1\mathbf{y}_{n+1} + \frac{1}{2}\nabla^2\mathbf{y}_{n+1} + \frac{1}{3}\nabla^3\mathbf{y}_{n+1} = h\mathbf{f}_{n+1}$$

for $\mathbf{y}_{n+1}$. After the iteration converges, $\mathbf{y}_n$, $\nabla^1\mathbf{y}_n$ and $\nabla^2\mathbf{y}_n$ are no longer needed.

There are several programs available to the public for integrating initial value problems via backward differentiation formulas, including the Byrne and Hindmarsh routine epsode, the Brown, Byrne and Hindmarsh routine svsode, the Shampine and Gordon routine ode and the Shampine and Gordon routine sode. MATLAB users should become familiar with the commands ode15s and ode15i. In addition, a useful source of test problems for initial value problems is Hull et al. [107].

A simple implementation of backward differentiation methods can be found in routines bdf_startup and bdf in file integrate.f. This is called from the main program in file GUILinearMultistep.C. Alternatively, readers may execute a JavaScript program for real linear multistep methods or for complex linear multistep methods.

Exercise 3.4.14 Show that the backward differentiation formula for $k = 1$ is the backward Euler method.

3.4.8 Absolute Stability

As we saw in Sect. 3.4.6, absolute stability is a very useful for schemes that may be applied to problems with decaying solutions. The absolute stability regions for the Adams-Bashforth methods are easy to determine, as the following lemma shows.

Lemma 3.4.13 *The Adams-Bashforth scheme*

$$\mathbf{y}_{n+1} - \mathbf{y}_n = h \sum_{j=0}^{k-1} \gamma_j \nabla^j \mathbf{f}_n$$

is absolutely stable for $\mathbf{y}' = \lambda\mathbf{y}$ *if and only if the zeros of the polynomial equation*

$$\zeta - 1 = h\lambda \sum_{j=0}^{k-1} \gamma_j (1 - 1/\zeta)^j \tag{3.31}$$

all satisfy $|\zeta| \leq 1$. *If*

$$h\lambda = \frac{e^{i\theta} - 1}{\sum_{j=0}^{k-1} \gamma_j (1 - e^{-i\theta})^j} \quad \text{where } 0 \leq \theta < 2\pi,$$

then $h\lambda$ *corresponds to a point where the polynomial in Eq.* (3.31) *has a root of modulus one.*

Proof In Sect. 3.4.3 we showed that linear multistep methods for $\mathbf{y}' = \lambda\mathbf{y}$ generate recurrence relations involving a companion matrix with coefficients γ_j, and that the dominant eigenvector of this recurrence is given by successive powers of the eigenvalue ζ. This proves that

$$\zeta^{n+1} - \zeta^n = h\lambda \sum_{j=0}^{k-1} \gamma_j \nabla^j \zeta^n \tag{3.32}$$

Next, we use Eq. (3.22) and the binomial expansion (2.1) in Chap. 2 of Volume I to write

$$\zeta^{n+1} - \zeta^n = h\lambda \sum_{j=0}^{k-1} \gamma_j \sum_{i=0}^{j} \zeta^{n-i}(-1)^i \binom{j}{i} = h\lambda \sum_{j=0}^{k-1} \gamma_j \zeta^n \sum_{i=0}^{j} (-1/\zeta)^i \binom{j}{i}$$

$$= h\lambda\zeta^n \sum_{j=0}^{k-1} \gamma_j (1 - 1/\zeta)^j.$$

By solving this equation for $h\lambda$ and using Euler's identity, we obtain the formula for the boundary of the absolute stability region.

Absolute stability regions for Adams-Bashforth schemes of orders 1 through 4 are shown in Fig. 3.4. Notice that the region of absolute stability shrinks as the order of the Adams-Bashforth method increases. The curves in Fig. 3.4 were created by executing the C^{++} program GUIAbsoluteStability.C.

Absolute stability regions for the Adams-Moulton methods are also easy to determine.

Lemma 3.4.14 *The Adams-Moulton scheme*

$$y_{n+1} - y_n = h \sum_{j=0}^{k-1} \gamma_j \nabla^j f_{n+1}$$

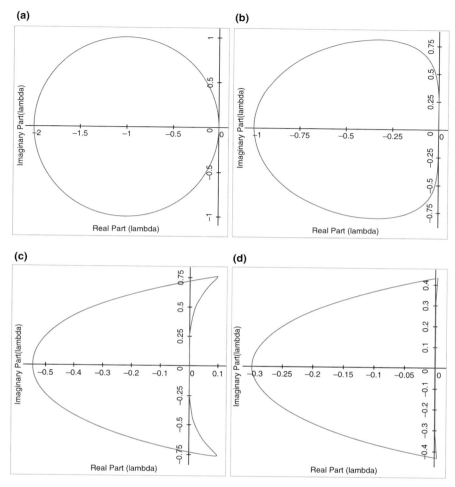

Fig. 3.4 Adams-Bashforth absolute stability regions. (**a**) Order 1. (**b**) Order 2. (**c**) Order 3. (**d**) Order 4

is absolutely stable for $y' = \lambda y$ if and only if all of the zeros of the polynomial equation

$$1 - 1/\zeta = h\lambda \sum_{j=0}^{k-1} \gamma_j (1 - 1/\zeta)^j \tag{3.33}$$

satisfy $|\zeta| \le 1$. If

$$h\lambda = \frac{1 - e^{-i\theta}}{\sum_{j=0}^{k-1} \gamma_j (1 - e^{-i\theta})^j} \quad where \; 0 \le \theta < 2\pi \; ,$$

then $h\lambda$ corresponds to a point where the polynomial in Eq. (3.33) has a root of modulus one.

Proof In Sect. 3.4.3 we showed that linear multistep methods for $y' = \lambda y$ generate recurrence relations involving a companion matrix with coefficients γ_j, and that the dominant eigenvector of this recurrence is given by successive powers of the eigenvalue ζ. This proves that

$$\zeta^{n+1} - \zeta^n = h\lambda \sum_{j=0}^{k-1} \gamma_j \nabla^j \zeta^{n+1}$$

Next, we use Eq. (3.22) and the binomial expansion (2.1) in Chap. 2 of Volume I to write

$$\zeta^{n+1} - \zeta^n = h\lambda \sum_{j=0}^{k-1} \gamma_j \sum_{i=0}^{j} \zeta^{n+1-i}(-1)^i \binom{j}{i} = h\lambda \sum_{j=0}^{k-1} \gamma_j \zeta^{n+1} \sum_{i=0}^{j} (-1/\zeta)^i \binom{j}{i}$$

$$= h\lambda \zeta^{n+1} \sum_{j=0}^{k-1} \gamma_j (1 - 1/\zeta)^j .$$

By solving this equation for $h\lambda$ and using Euler's identity, we obtain the formula for the boundary of the absolute stability region.

Absolute stability regions for Adams-Moulton methods of orders 1, 3, 4 and 8 are shown in Fig. 3.5. The absolute stability region for order 1 is the exterior of the circle, while the absolute stability regions in the other graphs lie inside the curves. Note that the absolute stability regions shrink as the order of the Adams-Moulton method increases. Also note that the absolute stability region of the Adams-Moulton method of order k is larger than the absolute stability region of the Adams-Bashforth method of the same order. The curves in Fig. 3.5 were created by executing the C^{++} program GUIAbsoluteStability.C.

Finally, the absolute stability regions for the backward differentiation formulas can be described by the following lemma.

Lemma 3.4.15 *The backward differentiation formula*

$$\sum_{j=1}^{k} \frac{1}{j} \nabla^j \mathbf{y}_{n+1} = h\mathbf{f}(\mathbf{y}_{n+1} , t_n + h)$$

is absolutely stable for $y' = \lambda y$ if and only if the zeros of the polynomial equation

$$\sum_{j=1}^{k} \frac{1}{j} (1 - 1/\zeta)^j = h\lambda$$

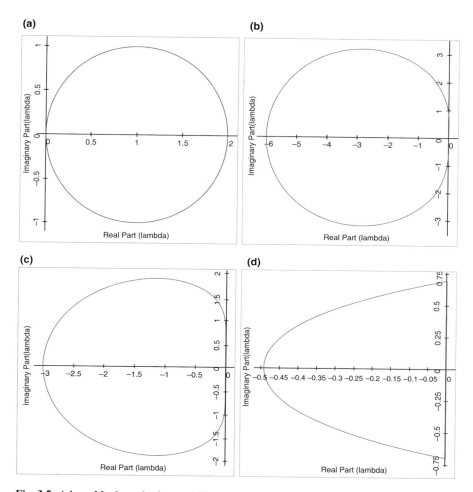

Fig. 3.5 Adams-Moulton absolute stability regions. (**a**) Order 1. (**b**) Order 3. (**c**) Order 4. (**d**) Order 8

satisfy $|\zeta| \leq 1$. *If*

$$h\lambda = \sum_{j=1}^{k} \frac{1}{j}(1 - e^{-i\theta})^j \text{ where } 0 \leq \theta < 2\pi \,,$$

then $h\lambda$ *corresponds to a point where the polynomial in Eq.* (3.4.15) *has a root of modulus one.*

Proof If the numerical solution $\mathbf{y}_n$ is generated by the backward differentiation formula with a fixed timestep h, then $\mathbf{y}_n$ satisfies a linear recurrence; if the initial data is chosen to produce the dominant growth of this linear recurrence, then $\mathbf{y}_n = \zeta^n$ for some complex number ζ. Then

$$\nabla^j \mathbf{y}_n = \sum_{i=0}^{j} \binom{j}{i} (-1)^i \mathbf{y}_{n-i} = \sum_{i=0}^{j} \binom{j}{i} (-1)^i \zeta^{n-i} = \zeta^{n-j} (\zeta - 1)^j$$

The scheme now implies that

$$\zeta^{n+1} \sum_{j=1}^{k} \frac{1}{j} (1 - 1/\zeta)^j = \sum_{j=1}^{k} \frac{1}{j} \nabla^j \mathbf{y}_{n+1} = h\mathbf{f}_{n+1} = h\lambda \mathbf{y}_{n+1} = h\lambda \zeta^{n+1}$$

Solving for $h\lambda$ produces

$$h\lambda = \sum_{j=1}^{k} \frac{1}{j} (1 - 1/\zeta)^j .$$

Absolute stability regions for orders 1 through 4 are shown in Fig. 3.6. In each case, the absolute stability region lies outside the region bounded by the curve. Note that the absolute stability regions shrink as the order of the backward differentiation formula increases. Also note that for all orders of the BDF method, the absolute stability region includes the entire negative real axis. Finally, we remark that the absolute stability region for the BDF method of order k is much larger than that of the Adams-Moulton method of the same order. The curves in Figs. 3.4, 3.5 and 3.6 were created by executing the C^{++} program GUIAbsoluteStability.C.

Readers may also plot the absolute stability region for backward differentiation formulas of various orders by executing the JavaScript **absolute stability program.** This program uses a simpler technique to generate the absolute stability region, and avoids computing the eigenvalues of companion matrices. Instead, the program generates a blue curve of points where the companion matrix has an eigenvalue of modulus 1, a green curve of points where the companion matrix has an eigenvalue of modulus 0.9, and a red curve of points where this matrix has an eigenvalue of modulus 1.1.

It is also crucial to understand the conditions under which backward differentiation formulas are zero-stable. Recall that Theorem 3.4.2 showed that a linear multistep is convergent if and only if it is consistent and zero-stable, and Theorem 3.4.1 showed that a linear multistep method is zero-stable if and only if its first characteristic polynomial has zeros within the closed unit circle, and zeros on the boundary of the unit circle have multiplicity one. The first characteristic polynomial for the backward differentiation formula of degree k is

$$\zeta^k \sum_{j=0}^{k} \frac{1}{j} (1 - 1/\zeta)^j .$$

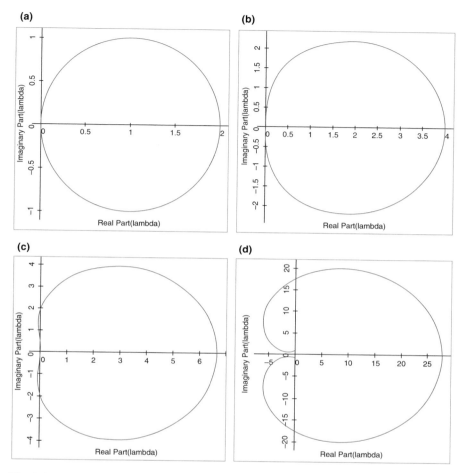

Fig. 3.6 Backward differentiation formulas absolute stability regions. (**a**) Order 1. (**b**) Order 2. (**c**) Order 3. (**d**) Order 6

Cryer [53] has shown that the roots of this polynomial satisfy the conditions for zero stability if and only if $1 \leq k \leq 6$.

3.4.9 *Error Bounds*

In Sect. 3.4.2, we defined the local truncation error of a linear multistep method, and developed necessary and sufficient conditions for consistency. In lemma 3.4.2, we also found a necessary condition for convergence. These conditions were reinterpreted in terms of characteristic polynomials in Sect. 3.4.3.

In this section, we would like to analyze linear multistep methods to determine conditions under which the numerical solutions exist and converge to the true solution of the original initial value problem. Single-step methods, such as Euler's method, are easy to analyze for convergence. In fact, we proved the uniform convergence of Euler's method in Sect. 1.3.5 of Chap. 1 in Volume I. Analysis of linear multistep methods is harder, however. We must warn, however, that the error bounds we will obtain in this section are far too large for practical use in computing.

Let us begin by discussing existing of the numerical solution. Given an explicit linear multistep method

$$\sum_{j=0}^{k} \mathbf{y}_{n+1-j}\alpha_{k-j} = h \sum_{j=1}^{k} \mathbf{f}\left(\mathbf{y}_{n+1-j}, \; t_n + [1-j]h\right)\beta_{k-j} \,,$$

we can use the assumption that $\alpha_k = 1$ to solve for $\mathbf{y}_{n+1}$:

$$\mathbf{y}_{n+1} = -\sum_{j=1}^{k} \mathbf{y}_{n+1-j}\alpha_{k-j} + h \sum_{j=1}^{k} \mathbf{f}\left(\mathbf{y}_{n+1-j}, \; t_n + [1-j]h\right)\beta_{k-j} \,.$$

This formula shows that as long as the terms on the right can be evaluated, the next solution value $\mathbf{y}_{n+1}$ will exist. The situation is more complicated for implicit linear multistep methods, however.

Lemma 3.4.16 *Suppose that the function* $\mathbf{f}$ *mapping pairs of vectors and scalars to vectors is Lipschitz continuous with Lipschitz constant* Φ. *Given a positive integer* k, *an initial time* t_0 *and a positive timestep* h, *let*

$$a(\zeta) = \sum_{j=0}^{k} \alpha_j \zeta^j \; and \; b(\zeta) = \sum_{j=0}^{k} \beta_j \zeta^j$$

be first and second characteristic polynomials for the linear multistep method

$$\sum_{j=0}^{k} \mathbf{y}_{n+1-j}\alpha_{k-j} - h \sum_{j=0}^{k} \mathbf{f}(\mathbf{y}_{n+1-j}, t_0 + [n+1-j]h)\beta_{k-j} = \mathbf{0} \,.$$

Assume that $\alpha_k = 1$ *and* $\beta_k \neq 0$. *If the timestep satisfies*

$$h < \frac{1}{|\beta_k|\Phi} \,,$$

then there exists a unique vector $\mathbf{y}_{n+1}$ *satisfying the implicit equation for the linear multistep method.*

Proof Define

$$\boldsymbol{\phi}(\mathbf{y}) = \mathbf{f}(\mathbf{y}, \ t_0 + [n+1]h)h\beta_k + \sum_{j=1}^{k} \mathbf{f}(\mathbf{y}_{n+1-j}, t_0 + [n+1-j]h)h\beta_{k-j} - \sum_{j=1}^{k} \mathbf{y}_{n+1-j}\alpha_{k-j} .$$

Then $\mathbf{y}_{n+1}$ satisfies the fixed-point equation

$$\mathbf{y}_{n+1} = \boldsymbol{\phi}(\mathbf{y}_{n+1}) .$$

Note that

$$\|\boldsymbol{\phi}(\widetilde{\mathbf{y}}) - \boldsymbol{\phi}(\mathbf{y})\| = h|\beta_k| \, \|\mathbf{f}(\widetilde{\mathbf{y}}, t_0 + [n + 1]h) - \mathbf{f}(\mathbf{y}, t_0 + [n + 1]h)\|$$
$$\leq h|\beta_k|\Phi \, \|\widetilde{\mathbf{y}} - \mathbf{y}\| .$$

Since $h|\beta_k|\Phi < 1$, we see that $\boldsymbol{\phi}$ is contractive, so the contractive mapping Theorem 3.2.2 of Chap. 3 in Volume II implies that the fixed point equation for $\mathbf{y}_{n+1}$ has a unique solution.

Note that Lemma 3.4.16 places an upper bound on the timestep size for implicit multistep methods. It is possible that an implicit linear multistep method has a solution for the numerical solution at the new time when this timestep restriction is violated, but such a solution may not be unique and could therefore be suspicious.

In the remainder of this section, we will follow the book by Henrici [101, p. 11ff] to study the convergence of linear multistep methods. We will begin with the following utility lemma.

Lemma 3.4.17 *Given a positive integer k, let*

$$a(\zeta) = \sum_{j=0}^{k} \alpha_j \zeta^j$$

be a polynomial of degree k; since this implies that $\alpha_k \neq 0$, we can normalize a so that $\alpha_k = 1$. We also assume that a has at least one nonzero root. Assume that the positive scalar ϱ is such that

$$a(\zeta) = 0 \Longrightarrow |\zeta| \leq \varrho$$

and that

$$a(\zeta) = 0 \ and \ |\zeta| = \varrho \Longrightarrow a'(\zeta) \neq 0 .$$

Given a scalar

$$h \in (0, 1/\phi_0) ,$$

and given sequences $\{\boldsymbol{\psi}_m\}_{0 \leq m}$ *and* $\{\mathbf{d}_j\}_{j=0}^{k-1}$ *of vectors, assume that the sequence* $\{\mathbf{d}_j\}_{j=k}^{\infty}$ *satisfies*

$$\sum_{j=0}^{k} \mathbf{d}_{m-j} \alpha_{k-j} = h \sum_{j=0}^{k} \boldsymbol{\phi}_{m,j}(\mathbf{d}_{m-j}) + \boldsymbol{\psi}_m . \tag{3.34}$$

Suppose that there is a sequence of positive scalars $\{\phi_j\}_{j=0}^{k}$ *so that the sequence* $\{\boldsymbol{\phi}_{m,j}\}_{\substack{0 \leq m \\ 0 \leq j \leq k}}$ *of functions mapping vectors to vectors satisfies*

$$\|\boldsymbol{\phi}_{m,j}(\mathbf{d})\| \leq \phi_j \|\mathbf{d}\| \text{ for } 0 \leq m \text{ and } 0 \leq j \leq k . \tag{3.35}$$

Define the scalars

$$\phi = \sum_{j=0}^{k} \phi_j \varrho^{k-j} , \ \psi_n = \sum_{m=k}^{n} \|\boldsymbol{\psi}_m\| \varrho^{k-m} \text{ and } \alpha = \sum_{j=0}^{k-1} |\alpha_j| \varrho^j . \tag{3.36}$$

Then there exists $\Gamma > 0$ *so that for all* $n \geq k$ *we have*

$$\|\mathbf{d}_n\| \leq \frac{\Gamma \varrho^n}{1 - \phi_0 h} \left[\psi_n + (\alpha + \phi h) \sum_{j=0}^{k-1} \|\mathbf{d}_j\| \varrho^{-j} \right] \left[1 + \frac{\Gamma \phi h}{1 - \phi_0 h} \right]^{n-k} . \tag{3.37}$$

Proof Our proof follows Henrici [101, p. 14f]. Define the sequence $\{\gamma_j\}_{j=0}^{\infty}$ by

$$1 = \left[\sum_{j=0}^{k} \varrho^{k-j} \alpha_{k-j} \zeta^j \right] \left[\sum_{\ell=0}^{\infty} \gamma_\ell \zeta^\ell \right] = \sum_{j=0}^{k} \sum_{\ell=0}^{\infty} \varrho^{k-j} \alpha_{k-j} \gamma_\ell \zeta^{j+\ell}$$

$$= \sum_{j=0}^{k} \sum_{m=j}^{\infty} \varrho^{k-j} \alpha_{k-j} \gamma_{m-j} \zeta^m$$

$$= \sum_{m=0}^{k} \left[\sum_{j=0}^{m} \varrho^{k-j} \alpha_{k-j} \gamma_{m-j} \right] \zeta^m + \sum_{m=k+1}^{\infty} \left[\sum_{j=0}^{k} \varrho^{k-j} \alpha_{k-j} \gamma_{m-j} \right] \zeta^m .$$

Equating coefficients of powers of ζ on both sides gives us

$$\varrho^k \gamma_0 = 1 \tag{3.38}$$

and

$$\sum_{i=\max\{0,k-m\}}^{k} \varrho^i \alpha_i \gamma_{m+i-k} = 0 \text{ for } 1 \leq m \tag{3.39}$$

The fundamental theorem of Algebra 1.2.1 of Chap. 1 in Volume II shows that there is an integer r, complex roots ζ_i for $1 \leq i \leq r$ and multiplicities m_i for $1 \leq i \leq r$ so that

$$a(\zeta) = \prod_{i=1}^{r} (\zeta - \zeta_i)^{m_i} .$$

Recall that we have assumed that $|\zeta_i| \leq \varrho$ for $1 \leq i \leq r$, and that $|\zeta_i| = \varrho$ implies that $m_i = 1$. Let us consider the function

$$\widehat{a}(z) = z^k a\left(\frac{\varrho}{z}\right) = z^k \prod_{i=1}^{r} \left(\frac{\varrho}{z} - \zeta_i\right)^{m_i} .$$

If $\zeta = 0$ is a root of $a(\zeta)$, without loss of generality we will assume that it is ordered first. Since a is assumed to have a nonzero root, we also have $r \geq 2$. Then

$$\widehat{a}(z) = z^k \left(\frac{\varrho}{z}\right)^{m_1} \prod_{i=2}^{r} \left(\frac{\varrho}{z} - \zeta_i\right)^{m_i} = \varrho^{m_1} z^{k-m_1} \prod_{i=2}^{r} \left(\frac{\varrho}{z} - \zeta_i\right)^{m_i}$$

$$= \varrho^{m_1} \prod_{i=2}^{r} (\varrho - z\zeta_i)^{m_i}$$

is a polynomial of degree $k - m_1$ with zeros

$$z_i = \frac{\varrho}{\zeta_i} \text{ for } 2 \leq i \leq r .$$

Otherwise zero is not a root of $a(\zeta)$ and

$$\widehat{a}(z) = z^k \prod_{i=1}^{r} \left(\frac{\varrho}{z} - \zeta_i\right)^{m_i} = \prod_{i=1}^{r} (\varrho - z\zeta_i)^{m_i}$$

is a polynomial of degree k with zeros

$$z_i = \frac{\varrho}{\zeta_i} \text{ for} 1 \leq i \leq r .$$

In either case, $\widehat{a}$ is a nonzero polynomial, the zeros of $\widehat{a}(z)$ satisfy $|z_i| \geq 1$ for $1 \leq i \leq r$, and $|z_i| = 1$ implies that $\widehat{a}'(z_i) \neq 0$.

We can use a partial fractions expansion to find constants $A_{i,j}$ for $1 \leq i \leq r$ and $1 \leq j \leq m_i$ so that

$$\sum_{\ell=0}^{\infty} \gamma_\ell z^\ell \equiv \frac{1}{\widehat{a}(z)} = \sum_{i=1}^{r} \sum_{j=1}^{m_i} \frac{A_{i,j}}{(z - z_i)^j} .$$

For those terms with $j = 1$, we let

$$\tau_{i,1,n} = -\frac{1}{z_i^{n+1}} \text{ and } T_{i,1} = 1 ,$$

then we observe that

$$\frac{1}{z - z_i} = -\frac{1}{z_i} \frac{1}{1 - z/z_i} = \sum_{n=0}^{\infty} \tau_{i,1,n} z^n$$

Since $|z_i| \geq 1$, we also see that

$$|\tau_{i,1,n}| \leq T_{i,1} .$$

On the other hand, if $j > 1$ then ζ_i is a root of multiplicity greater than one, so $|\zeta_i| < \varrho$ and therefore $|z_i| > 1$. Since the series

$$\frac{1}{(z - z_i)^j} = \sum_{n=0}^{\infty} \tau_{i,j,n} z^n$$

converges absolutely for $|z| < |z_i|$, we can choose $z = 1$ and see that

$$\sum_{n=0}^{\infty} |\tau_{i,j,n}| < \infty .$$

This in turn implies that the terms in this infinite series tend to zero as $n \to \infty$. Thus for each $1 \leq i \leq r$ with $m_i > 1$, and for each $1 < j \leq m_i$ there exists $T_{i,j} < \infty$ so that for all $n \geq 0$ we have

$$|\tau_{i,j,n}| \leq T_{i,j} .$$

Collecting our results in this paragraph, we see that for all $n \geq 0$ we have

$$|\gamma_n| = \left| \sum_{i=1}^{r} \sum_{j=1}^{m_i} A_{i,j} \tau_{i,j,n} \right| \leq \sum_{i=1}^{r} \sum_{j=1}^{m_i} |A_{i,j}| T_{i,j} \equiv \Gamma . \tag{3.40}$$

If we define the vectors $\boldsymbol{\delta}_j$ by

$$\boldsymbol{\delta}_j = \mathbf{d}_j \varrho^{-j} \text{ for } 0 \le j ,$$

then the recurrence (3.34) implies that

$$\sum_{j=0}^{k} \boldsymbol{\delta}_{m-j} \varrho^{m-j} \alpha_{k-j} = h \sum_{j=0}^{k} \boldsymbol{\phi}_{m,j} \left(\boldsymbol{\delta}_{m-j} \varrho^{m-j} \right) + \boldsymbol{\psi}_m .$$

We can scale these equations and sum to get

$$\sum_{\ell=0}^{n-k} \left[\sum_{j=0}^{k} \boldsymbol{\delta}_{k+\ell-j} \varrho^{k+\ell-j} \alpha_{k-j} \right] \varrho^{-\ell} \gamma_{n-k-\ell}$$

$$= \sum_{\ell=0}^{n-k} \left[h \sum_{j=0}^{k} \boldsymbol{\phi}_{k+\ell,j} \left(\boldsymbol{\delta}_{k+\ell-j} \varrho^{k+\ell-j} \right) + \boldsymbol{\psi}_{k+\ell} \right] \varrho^{-\ell} \gamma_{n-k-\ell} . \qquad (3.41)$$

To bound terms given by this equation, we will consider two cases. Our first case assumes that $k \le n < 2k$. If so, then the left-hand side in (3.41) is

$$\sum_{\ell=0}^{n-k} \sum_{j=0}^{k} \boldsymbol{\delta}_{k+\ell-j} \varrho^{k-j} \alpha_{k-j} \gamma_{n-k-\ell}$$

then we change summation variables to get

$$= \sum_{\ell=0}^{n-k} \sum_{i=\ell}^{k+\ell} \boldsymbol{\delta}_i \varrho^{i-\ell} \alpha_{i-\ell} \gamma_{n-k-\ell}$$

then we interchange the sums to obtain

$$= \sum_{i=0}^{n-k} \boldsymbol{\delta}_i \sum_{\ell=0}^{i} \varrho^{i-\ell} \alpha_{i-\ell} \gamma_{n-k-\ell} + \sum_{i=n-k+1}^{k-1} \boldsymbol{\delta}_i \sum_{\ell=0}^{n-k} \varrho^{i-\ell} \alpha_{i-\ell} \gamma_{n-k-\ell}$$

$$+ \sum_{i=k}^{n} \boldsymbol{\delta}_i \sum_{\ell=i-k}^{n-k} \varrho^{i-\ell} \alpha_{i-\ell} \gamma_{n-k-\ell}$$

then we change summation variables to get

$$
= \sum_{i=0}^{n-k} \delta_i \sum_{I=0}^{i} \varrho^I \alpha_I \gamma_{n+I-k-i} + \sum_{i=n-k+1}^{k-1} \delta_i \sum_{I=i+k-n}^{i} \varrho^I \alpha_I \gamma_{n+I-k-i}
$$

$$
+ \sum_{i=k}^{n} \delta_i \sum_{I=i+k-n}^{k} \varrho^I \alpha_I \gamma_{n+I-k-i}
$$

and finally we use Eqs. (3.38) and (3.39) to arrive at

$$
= \sum_{i=0}^{n-k} \delta_i \sum_{I=0}^{i} \varrho^I \alpha_I \gamma_{n+I-k-i} + \sum_{i=n-k+1}^{k-1} \delta_i \sum_{I=i+k-n}^{i} \varrho^I \alpha_I \gamma_{n+I-k-i} + \delta_n . \qquad (3.42)
$$

On the other hand, the right-hand side of (3.41) is

$$
h \sum_{\ell=0}^{n-k} \sum_{j=0}^{k} \phi_{k+\ell,j} \left(\delta_{k+\ell-j} \varrho^{k+\ell-j} \right) \varrho^{-\ell} \gamma_{n-k-\ell} + \sum_{\ell=0}^{n-k} \psi_{k+\ell} \varrho^{-\ell} \gamma_{n-k-\ell}
$$

then we change summation variables to get

$$
= h \sum_{\ell=0}^{n-k} \sum_{i=\ell}^{k+\ell} \phi_{k+\ell,k+\ell-i} \left(\delta_i \varrho^i \right) \varrho^{-\ell} \gamma_{n-k-\ell} + \sum_{\ell=0}^{n-k} \psi_{k+\ell} \varrho^{-\ell} \gamma_{n-k-\ell}
$$

then we interchange the sums to obtain

$$
= h \sum_{i=0}^{n-k} \sum_{\ell=0}^{i} \phi_{k+\ell,k+\ell-i} \left(\delta_i \varrho^i \right) \varrho^{-\ell} \gamma_{n-k-\ell}
$$

$$
+ h \sum_{i=n-k+1}^{k-1} \sum_{\ell=0}^{n-k} \phi_{k+\ell,k+\ell-i} \left(\delta_i \varrho^i \right) \varrho^{-\ell} \gamma_{n-k-\ell}
$$

$$
+ h \sum_{i=k}^{n} \sum_{\ell=i-k}^{n-k} \phi_{k+\ell,k+\ell-i} \left(\delta_i \varrho^i \right) \varrho^{-\ell} \gamma_{n-k-\ell}
$$

$$
+ \sum_{\ell=0}^{n-k} \psi_{k+\ell} \varrho^{-\ell} \gamma_{n-k-\ell}
$$

then we change summation variables to get

$$= h \sum_{i=0}^{n-k} \sum_{l=0}^{i} \boldsymbol{\phi}_{k+i-l,k-l} \left(\boldsymbol{\delta}_i \varrho^i \right) \varrho^{l-i} \gamma_{n+l-k-i}$$

$$+ h \sum_{i=n-k+1}^{k-1} \sum_{l=i}^{i+k-n} \boldsymbol{\phi}_{k+i-l,k-l} \left(\boldsymbol{\delta}_i \varrho^i \right) \varrho^{l-i} \gamma_{n+l-k-i}$$

$$+ h \sum_{i=k}^{n} \sum_{l=i+k-n}^{k} \boldsymbol{\phi}_{k+i-l,k-l} \left(\boldsymbol{\delta}_i \varrho^i \right) \varrho^{l-i} \gamma_{n+l-k-i}$$

$$+ \sum_{m=k}^{n} \boldsymbol{\psi}_m \varrho^{k-m} \gamma_{n-m} \tag{3.43}$$

We can substitute (3.42) and (3.43) into (3.41) and rearrange the terms to get

$$\boldsymbol{\delta}_n = h \sum_{i=0}^{n-k} \sum_{l=0}^{i} \boldsymbol{\phi}_{k+i-l,k-l} \left(\boldsymbol{\delta}_i \varrho^i \right) \varrho^{l-i} \gamma_{n+l-k-i}$$

$$+ h \sum_{i=n-k+1}^{k-1} \sum_{l=i}^{i+k-n} \boldsymbol{\phi}_{k+i-l,k-l} \left(\boldsymbol{\delta}_i \varrho^i \right) \varrho^{l-i} \gamma_{n+l-k-i}$$

$$+ h \sum_{i=k}^{n} \sum_{l=i+k-n}^{k} \boldsymbol{\phi}_{k+i-l,k-l} \left(\boldsymbol{\delta}_i \varrho^i \right) \varrho^{l-i} \gamma_{n+l-k-i}$$

$$+ \sum_{m=k}^{n} \boldsymbol{\psi}_m \varrho^{k-m} \gamma_{n-m} - \sum_{i=0}^{n-k} \boldsymbol{\delta}_i \sum_{l=0}^{i} \varrho^l \alpha_l \gamma_{n+l-k-i}$$

$$- \sum_{i=n-k+1}^{k-1} \boldsymbol{\delta}_i \sum_{l=i+k-n}^{i} \varrho^l \alpha_l \gamma_{n+l-k-i} .$$

Next, we can take norms of both sides to get

$$\| \boldsymbol{\delta}_n \| \leq h \left\| \boldsymbol{\phi}_{n,0} \left(\boldsymbol{\delta}_n \varrho^n \right) \varrho^{k-n} \gamma_0 \right\| + h \sum_{i=0}^{n-1} \sum_{l=k+i-n}^{k} \left\| \boldsymbol{\phi}_{k+i-l,k-l} \left(\boldsymbol{\delta}_i \varrho^i \right) \varrho^{l-i} \gamma_{n+l-k-i} \right\|$$

$$+ h \sum_{i=n-k+1}^{k-1} \sum_{l=i}^{i+k-n} \left\| \boldsymbol{\phi}_{k+i-l,k-l} \left(\boldsymbol{\delta}_i \varrho^i \right) \varrho^{l-i} \gamma_{n+l-k-i} \right\|$$

$$+ h \sum_{i=k}^{n} \sum_{I=i+k-n}^{k} \left\| \boldsymbol{\phi}_{k+i-I,k-I} \left(\boldsymbol{\delta}_i \varrho^i \right) \varrho^{I-i} \gamma_{n+I-k-i} \right\| + \sum_{m=k}^{n} \left\| \boldsymbol{\psi}_m \varrho^{k-m} \gamma_{n-m} \right\|$$

$$+ \sum_{i=0}^{n-k} \| \boldsymbol{\delta}_i \| \sum_{I=0}^{i} \varrho^I |\alpha_I| \; |\gamma_{n+I-k-i}| + \sum_{i=n-k+1}^{k-1} \| \boldsymbol{\delta}_i \| \sum_{I=i+k-n}^{i} \varrho^I |\alpha_I| \; |\gamma_{n+I-k-i}|$$

then we use inequality (3.35) to obtain

$$\le h\phi_0 \, \| \boldsymbol{\delta}_n \| \, \varrho^k \, |\gamma_0| + h \sum_{i=0}^{n-1} \sum_{I=k+i-n}^{k} \phi_{k-I} \, \| \boldsymbol{\delta}_i \| \, \varrho^I \, |\gamma_{n+I-k-i}|$$

$$+ h \sum_{i=n-k+1}^{k-1} \sum_{I=i}^{i+k-n} \phi_{k-I} \, \| \boldsymbol{\delta}_i \| \, \varrho^I \, |\gamma_{n+I-k-i}| + h \sum_{i=k}^{n} \sum_{I=i+k-n}^{k} \phi_{k-I} \, \| \boldsymbol{\delta}_i \| \, \varrho^I \, |\gamma_{n+I-k-i}|$$

$$+ \sum_{m=k}^{n} \| \boldsymbol{\psi}_m \| \, \varrho^{k-m} \, |\gamma_{n-m}| + \sum_{i=0}^{n-k} \| \boldsymbol{\delta}_i \| \sum_{I=0}^{i} \varrho^I \, |\alpha_I| \; |\gamma_{n+I-k-i}|$$

$$+ \sum_{i=n-k+1}^{k-1} \| \boldsymbol{\delta}_i \| \sum_{I=i+k-n}^{i} \varrho^I \, |\alpha_I| \; |\gamma_{n+I-k-i}|$$

then we use Eq. (3.38) and inequality (3.40), plus a change of summation variables to get

$$\le \| \boldsymbol{\delta}_n \| \, h\phi_0 + \Gamma h \sum_{i=0}^{n-1} \| \boldsymbol{\delta}_i \| \sum_{m=0}^{n-1} \phi_m \varrho^{k-m} + \Gamma h \sum_{i=n-k+1}^{k-1} \| \boldsymbol{\delta}_i \| \sum_{m=n-i}^{k-i} \phi_m \varrho^{k-m}$$

$$+ \Gamma h \sum_{i=k}^{n} \| \boldsymbol{\delta}_i \| \sum_{m=k-i}^{k} \phi_m \varrho^{k-m} + \Gamma \sum_{m=k}^{n} \| \boldsymbol{\psi}_m \| \, \varrho^{k-m} + \Gamma \sum_{i=0}^{n-k} \| \boldsymbol{\delta}_i \| \sum_{I=0}^{i} \varrho^I |\alpha_I|$$

$$+ \Gamma \sum_{i=n-k+1}^{k-1} \| \boldsymbol{\delta}_i \| \sum_{I=i+k-n}^{i} \varrho^I |\alpha_I|$$

and finally we use the definitions (3.36) to write

$$\le \| \boldsymbol{\delta}_n \| \, h\phi_0 + \Gamma h\phi \sum_{i=k}^{n-1} \| \boldsymbol{\delta}_i \| + \Gamma \psi_n + \Gamma \, (h\phi + \alpha) \sum_{i=0}^{k-1} \| \boldsymbol{\delta}_i \| \; . \tag{3.44}$$

Next, we will consider the case when $n \geq 2k$. The left-hand side in (3.41) is

$$\sum_{\ell=0}^{n-k}\sum_{j=0}^{k} \delta_{k+\ell-j}\varrho^{k-j}\alpha_{k-j}\gamma_{n-k-\ell}$$

then we change summation variables to get

$$= \sum_{\ell=0}^{n-k}\sum_{i=\ell}^{k+\ell} \delta_{i}\varrho^{i-\ell}\alpha_{i-\ell}\gamma_{n-k-\ell}$$

then we interchange the sums to obtain

$$= \sum_{i=0}^{k-1}\delta_{i}\sum_{\ell=0}^{i} \varrho^{i-\ell}\alpha_{i-\ell}\gamma_{n-k-\ell} + \sum_{i=k}^{n-k}\delta_{i}\sum_{\ell=i-k}^{i} \varrho^{i-\ell}\alpha_{i-\ell}\gamma_{n-k-\ell}$$
$$+ \sum_{i=n-k+1}^{n}\delta_{i}\sum_{\ell=i-k}^{n-k} \varrho^{i-\ell}\alpha_{i-\ell}\gamma_{n-k-\ell}$$

then we change summation variables to get

$$= \sum_{i=0}^{k-1}\delta_{i}\sum_{I=0}^{i} \varrho^{I}\alpha_{I}\gamma_{n+I-k-i} + \sum_{i=k}^{n-k}\delta_{i}\sum_{I=0}^{k} \varrho^{I}\alpha_{I}\gamma_{n+I-k-i}$$
$$+ \sum_{i=n-k+1}^{n}\delta_{i}\sum_{I=i+k-n}^{k} \varrho^{I}\alpha_{I}\gamma_{n+I-k-i}$$

and finally we use Eqs. (3.38) and (3.39) to arrive at

$$= \sum_{i=0}^{k-1}\delta_{i}\sum_{I=0}^{i} \varrho^{I}\alpha_{I}\gamma_{n+I-k-i} + \delta_{n} . \tag{3.45}$$

On the other hand, the right-hand side in (3.41) is

$$\sum_{\ell=0}^{n-k}h\sum_{j=0}^{k} \phi_{k+\ell,j}\left(\delta_{k+\ell-j}\varrho^{k+ell-j}\right)\varrho^{-\ell}\gamma_{n-k-\ell} + \sum_{\ell=0}^{n-k} \psi_{k+\ell}\varrho^{-\ell}\gamma_{n-k-\ell}$$

then we change summation variables to get

$$= h\sum_{\ell=0}^{n-k}\sum_{i=\ell}^{k+\ell} \phi_{k+\ell,k+\ell-i}\left(\delta_{i}\varrho^{i}\right)\varrho^{-\ell}\gamma_{n-k-\ell} + \sum_{\ell=0}^{n-k} \psi_{k+\ell}\varrho^{-\ell}\gamma_{n-k-\ell}$$

then we interchange the sums to obtain

$$= h \sum_{i=0}^{k-1} \sum_{\ell=0}^{i} \boldsymbol{\phi}_{k+\ell,k+\ell-i} \left(\boldsymbol{\delta}_i \varrho^i\right) \varrho^{-\ell} \gamma_{n-k-\ell} + h \sum_{i=k}^{n-k} \sum_{\ell=i-k}^{i} \boldsymbol{\phi}_{k+\ell,k+\ell-i} \left(\boldsymbol{\delta}_i \varrho^i\right) \varrho^{-\ell} \gamma_{n-k-\ell}$$

$$+ h \sum_{i=n-k+1}^{n} \sum_{\ell=i-k}^{n-k} \boldsymbol{\phi}_{k+\ell,k+\ell-i} \left(\boldsymbol{\delta}_i \varrho^i\right) \varrho^{-\ell} \gamma_{n-k-\ell} + \sum_{\ell=0}^{n-k} \boldsymbol{\psi}_{k+\ell} \varrho^{-\ell} \gamma_{n-k-\ell}$$

then we change summation variables to get

$$= h \sum_{i=0}^{k-1} \sum_{I=0}^{i} \boldsymbol{\phi}_{k+i-I,k-I} \left(\boldsymbol{\delta}_i \varrho^i\right) \varrho^{I-i} \gamma_{n+I-k-i}$$

$$+ h \sum_{i=k}^{n-k} \sum_{I=0}^{k} \boldsymbol{\phi}_{k+i-I,k-I} \left(\boldsymbol{\delta}_i \varrho^i\right) \varrho^{I-i} \gamma_{n+I-k-i}$$

$$+ h \sum_{i=n-k+1}^{n} \sum_{I=i+k-n}^{k} \boldsymbol{\phi}_{k+i-I,k-I} \left(\boldsymbol{\delta}_i \varrho^i\right) \varrho^{I-i} \gamma_{n+I-k-i} + \sum_{m=k}^{n} \boldsymbol{\psi}_m \varrho^{k-m} \gamma_{n-m} . \qquad (3.46)$$

We can substitute (3.45) and (3.46) into (3.41) and rearrange the terms to get

$$\boldsymbol{\delta}_n = h \sum_{i=0}^{k-1} \sum_{I=0}^{i} \boldsymbol{\phi}_{k+i-I,k-I} \left(\boldsymbol{\delta}_i \varrho^i\right) \varrho^{I-i} \gamma_{n+I-k-i}$$

$$+ h \sum_{i=k}^{n-k} \sum_{I=0}^{k} \boldsymbol{\phi}_{k+i-I,k-I} \left(\boldsymbol{\delta}_i \varrho^i\right) \varrho^{I-i} \gamma_{n+I-k-i}$$

$$+ h \sum_{i=n-k+1}^{n} \sum_{I=i+k-n}^{k} \boldsymbol{\phi}_{k+i-I,k-I} \left(\boldsymbol{\delta}_i \varrho^i\right) \varrho^{I-i} \gamma_{n+I-k-i} + \sum_{m=k}^{n} \boldsymbol{\psi}_m \varrho^{k-m} \gamma_{n-m}$$

$$- \sum_{i=0}^{k-1} \boldsymbol{\delta}_i \sum_{I=0}^{i} \varrho^I \alpha_I \gamma_{n+I-k-i} .$$

Next, we can take norms of both sides to get

$$\|\boldsymbol{\delta}_n\| \leq h \left\| \boldsymbol{\phi}_{n,0} \left(\boldsymbol{\delta}_n \varrho^n\right) \varrho^{k-n} \gamma_0 \right\|$$

$$+ h \sum_{i=n-k+1}^{n-1} \sum_{I=i+k-n}^{k} \left\| \boldsymbol{\phi}_{k+i-I,k-I} \left(\boldsymbol{\delta}_i \varrho^i\right) \varrho^{I-i} \gamma_{n+I-k-i} \right\|$$

$$+ h \sum_{i=k}^{n-k} \sum_{I=0}^{k} \left\| \phi_{k+i-I,k-I} \left(\delta_i \varrho^i \right) \varrho^{I-i} \gamma_{n+I-k-i} \right\|$$

$$+ h \sum_{i=0}^{k-1} \sum_{I=0}^{i} \left\| \phi_{k+i-I,k-I} \left(\delta_i \varrho^i \right) \varrho^{I-i} \gamma_{n+I-k-i} \right\|$$

$$+ \sum_{m=k}^{n} \left\| \psi_m \varrho^{k-m} \gamma_{n-m} \right\| + \sum_{i=0}^{k-1} \left\| \delta_i \right\| \sum_{I=0}^{i} \left| \varrho^I \alpha_I \gamma_{n+I-k-i} \right|$$

then we use inequality (3.35) to obtain

$$\leq h \phi_0 \left\| \delta_n \right\| \varrho^k \gamma_0 + h \sum_{i=n-k+1}^{n-1} \sum_{I=i+k-n}^{k} \phi_{k-I} \left\| \delta_i \right\| \varrho^I \left| \gamma_{n+I-k-i} \right|$$

$$+ h \sum_{i=k}^{n-k} \sum_{I=0}^{k} \phi_{k-I} \left\| \delta_i \right\| \varrho^I \left| \gamma_{n+I-k-i} \right| + h \sum_{i=0}^{k-1} \sum_{I=0}^{i} \phi_{k-I} \left\| \delta_i \right\| \varrho^I \left| \gamma_{n+I-k-i} \right|$$

$$+ \sum_{m=k}^{n} \left\| \psi_m \right\| \varrho^{k-m} \left| \gamma_{n-m} \right| + \sum_{i=0}^{k-1} \left\| \delta_i \right\| \sum_{I=0}^{i} \varrho^I \left| \alpha_I \right| \left| \gamma_{n+I-k-i} \right|$$

then we use Eq. (3.38) and inequality (3.40), plus a change of summation variables to get

$$\leq \left\| \delta_n \right\| \phi_0 h + \Gamma h \sum_{i=n-k+1}^{n-1} \left\| \delta_i \right\| \sum_{m=0}^{n-i} \phi_m \varrho^{k-m} + \Gamma h \sum_{i=k}^{n-k} \left\| \delta_i \right\| \sum_{m=0}^{k} \phi_m \varrho^{k-m}$$

$$+ \Gamma h \sum_{i=0}^{k-1} \left\| \delta_i \right\| \sum_{m=k-i}^{k} \phi_m \varrho^{k-m} + \Gamma \sum_{m=k}^{n} \left\| \psi_m \right\| \varrho^{k-m} + \Gamma \sum_{i=0}^{k-1} \left\| \delta_i \right\| \sum_{I=0}^{i} \varrho^I \left| \alpha_I \right|$$

and finally we use the definitions (3.36) to write

$$\leq \left\| \delta_n \right\| \phi_0 h + \Gamma h \phi \sum_{i=k}^{n-1} \left\| \delta_i \right\| + \Gamma \psi_n + \Gamma (h\phi + \alpha) \sum_{i=0}^{k-1} \left\| \delta_i \right\| . \tag{3.47}$$

This is the same bound as in inequality (3.44). Since we have assumed that

$$h \phi_0 < 1 ,$$

we conclude that for $n \geq k$ we have

$$\left\| \delta_n \right\| \leq \frac{\Gamma}{1 - \phi_0 h} \left\{ \phi h \sum_{i=k}^{n-1} \left\| \delta_i \right\| + (h\phi + \alpha) \sum_{i=0}^{k-1} \left\| \delta_i \right\| + \psi_n \right\} . \tag{3.48}$$

We will prove inductively that

$$\|\boldsymbol{\delta}_n\| \le \frac{\Gamma}{1-\phi_0 h}\left\{\psi_n + (\alpha + h\phi)\sum_{i=0}^{k-1}\|\boldsymbol{\delta}_i\|\right\}\left(1 + \frac{\Gamma\phi h}{1-\phi_0 h}\right)^{n-k}.$$

This is identical to (3.48) for $n = k$. We will assume that the inductive hypothesis is true for $n - 1$. Then the bound (3.48) gives us

$$\|\boldsymbol{\delta}_n\| \le \frac{\Gamma}{1-\phi_0 h}\left\{\phi h\sum_{i=k}^{n-1}\|\boldsymbol{\delta}_i\| + (\alpha + h\phi)\sum_{i=0}^{k-1}\|\boldsymbol{\delta}_i\| + \psi_n\right\}$$

then the inductive hypothesis produces

$$\le \frac{\Gamma\phi h}{1-\phi_0 h}\sum_{i=k}^{n-1}\frac{\Gamma}{1-\phi_0 h}\left[\psi_i + (\alpha + \phi h)\sum_{i=0}^{k-1}\|\boldsymbol{\delta}_i\|\right]\left(1 + \frac{\Gamma\phi h}{1-\phi_0 h}\right)^{i-k}$$

$$+ \frac{\Gamma}{1-\phi_0 h}(\alpha + \phi h)\sum_{i=0}^{k-1}\|\boldsymbol{\delta}_i\| + \frac{\Gamma}{1-\phi_0 h}\psi_n$$

$$\le \left[\frac{\Gamma\phi h}{1-\phi_0 h}\frac{\Gamma}{1-\phi_0 h}\sum_{m=0}^{n-k-1}\left(1 + \frac{\Gamma\phi h}{1-\phi_0 h}\right)^m + \frac{\Gamma}{1-\phi_0 h}\right]\psi_n$$

$$+ \left[\frac{\Gamma\phi h}{1-\phi_0 h}(\alpha + \phi h)\frac{\Gamma}{1-\phi_0 h}\sum_{m=0}^{n-k-1}\left(1 + \frac{\Gamma\phi h}{1-\phi_0 h}\right)^m + \frac{\Gamma}{1-\phi_0 h}(\alpha + \phi h)\right]$$

$$\sum_{i=0}^{k-1}\|\boldsymbol{\delta}_i\|$$

then we sum the geometric series to get

$$= \frac{\Gamma}{1-\phi_0 h}\left[\frac{\Gamma\phi h}{1-\phi_0 h}\frac{\left(1 + \frac{\Gamma\phi h}{1-\phi_0 h}\right)^{n-k} - 1}{\frac{\Gamma\phi h}{1-\phi_0 h}} + 1\right]\psi_n$$

$$+ \frac{\Gamma}{1-\phi_0 h}(\alpha + \phi h)\left[\frac{\Gamma\phi h}{1-\phi_0 h}\frac{\left(1 + \frac{\Gamma\phi h}{1-\phi_0 h}\right)^{n-k} - 1}{\frac{\Gamma\phi h}{1-\phi_0 h}} + 1\right]\sum_{i=0}^{k-1}\|\boldsymbol{\delta}_i\|$$

$$= \frac{\Gamma}{1-\phi_0 h}\left[\psi_n + (\alpha + \phi h)\sum_{i=0}^{k-1}\|\boldsymbol{\delta}_i\|\right]\left(1 + \frac{\Gamma\phi h}{1-\phi_0 h}\right)^{n-k}.$$

This completes the inductive proof of an inequality that is equivalent to (3.37).

We can use Lemma 3.4.17 to prove the following theorem, which was previously stated in Sect. 3.4.4.

Theorem 3.4.7 (Necessary and Sufficient Condition for Zero Stability) *If a linear multistep method is zero-stable then its first characteristic polynomial a satisfies the root condition: each zero ζ of the first characteristic polynomial a has modulus $|\zeta| \leq 1$, and if $|\zeta| = 1$ then $a'(\zeta) \neq 0$. Conversely, if the root condition is satisfied and $a(1) = 0$, then the linear multistep method is zero-stable for each initial value problem $y'(t) = f(y(t) , t)$ where f is Lipschitz continuous in y and $f(0, t)$ is bounded for $t \in [t_0, t_0 + \tau]$.*

Proof We will begin by proving that zero stability implies the root condition. It is equivalent to prove the contrapositive, that whenever the root condition is not satisfied then the method is not zero-stable. Whenever the root condition is violated, the first characteristic polynomial either has a zero ζ with $|\zeta| > 1$, or it has a zero ζ with $|\zeta| = 1$ and $a'(\zeta) = 0$.

Consider the initial value problem $y'(t) = 0$ with $y(0) = 1$. The obvious solution is $y(t) = 1$. A linear multistep method for this initial value problem will take the form

$$y_{n+1} = -\sum_{j=1}^{k} y_{n+1-j}\alpha_{k-j}$$

If the first characteristic polynomial

$$a(\zeta) = \zeta^k + \sum_{j=0}^{k-1} \alpha_j \zeta^j$$

has a zero ζ with $|\zeta| > 1$, then the discussion in Sect. 1.4.1.6 of Chap. 1 in Volume II shows that $y_j = \zeta^j$ satisfies the equation for the linear multistep method. Since $|y_j| \to \infty$ as $j \to \infty$, the linear multistep method cannot be zero-stable. If a has a root ζ with $|\zeta| = 1$ and multiplicity greater than one, then the discussion in Sect. 1.4.1.6 of Chap. 1 in Volume II shows that $y_j = j\zeta^j$ also satisfies the equation for the linear multistep method. Since $|y_j| = j \to \infty$ as $j \to \infty$, the linear multistep method also cannot be zero-stable in this situation.

Next, we will prove that the method is zero-stable when the root condition is satisfied, $a(1) = 0$ and f is Lipschitz continuous. We can rewrite the linear multistep method in the form

$$\sum_{j=0}^{k} y_{n+1-j}\alpha_{k-j}$$

$$= h\sum_{j=0}^{k} \left[f\left(y_{n+1-j} , t_0 + [n + 1 - j]h\right) - f\left(0 , t_0 + [n + 1 - j]h\right) \right] \beta_{k-j}$$

$$+ h \sum_{j=0}^{k} \mathbf{f}\left(\mathbf{0}\,,\, t_0 + [n+1-j]h\right) \beta_{k-j}$$

$$\equiv h \sum_{j=0}^{k} \boldsymbol{\phi}_{n+1,j}(\mathbf{y}_{n+1-j}) + \boldsymbol{\psi}_{n+1}$$

where

$$\boldsymbol{\phi}_{n+1,j}(\mathbf{y}) \equiv \left[\mathbf{f}\left(\mathbf{y}\,,\, t_0 + [n+1-j]h\right) - \mathbf{f}\left(\mathbf{0}\,,\, t_0 + [n+1-j]h\right)\right] \beta_{k-j}$$

and

$$\boldsymbol{\psi}_{n+1} \equiv h \sum_{j=0}^{k} \mathbf{f}\left(\mathbf{0}\,,\, t_0 + [n+1-j]h\right)\,.$$

If $\mathbf{f}$ is Lipschitz continuous with Lipschitz constant Φ, then

$$\left\|\boldsymbol{\phi}_{n+1,j}(\mathbf{y})\right\| = \left\|\mathbf{f}\left(\mathbf{y}\,,\, t_0 + [n+1-j]h\right) - \mathbf{f}\left(\mathbf{0}\,,\, t_0 + [n+1-j]h\right)\right\| \, |\beta_{k-j}|$$

$$\leq \Phi |\beta_{k-j}| \, \|\mathbf{y}\| \equiv \phi_j \, \|\mathbf{y}\|\,.$$

Since the maximum modulus of a root of the first characteristic polynomial a is $\varrho = 1$, it follows that

$$\sum_{j=0}^{k} \phi_j \varrho^{k-j} = \Phi \sum_{j=0}^{k} |\beta_j| \equiv \phi\,,$$

$$\sum_{j=0}^{k-1} |\alpha_j| \, \varrho^{j} = \sum_{j=0}^{k-1} |\alpha_j| \equiv \alpha$$

and

$$\sum_{m=k}^{n} \|\boldsymbol{\psi}_m\| \varrho^{k-m} = \sum_{m=k}^{n} \|\boldsymbol{\psi}_m\|$$

$$\leq h \sum_{j=0}^{k} \left\|\mathbf{f}\left(\mathbf{0}\,,\, t_0 + [n+1-j]h\right)\right\| \leq h(n-k+1)(k+1) \max_{t \in [t_0, t_0 + \tau]} \|\mathbf{f}(\mathbf{0}, t)\|$$

$$\leq \tau(k+1) \max_{t \in [t_0, t_0 + \tau]} \|\mathbf{f}(\mathbf{0}, t)\| \equiv \psi\,.$$

If we choose the timestep h so that

$$\phi_0 h < 1/2 \,,$$

then Lemma 3.4.17 proves that there exists $\Gamma > 0$ so that for all $n \geq k$

$$\|\mathbf{y}_n\| \leq \frac{\Gamma \varrho^n}{1 - \phi_0 h} \left[\psi + (\alpha + \phi h) \sum_{j=0}^{k-1} \|\mathbf{y}_j\| \varrho^{-j} \right] \left[1 + \frac{\Gamma \phi h}{1 - \phi_0 h} \right]^{n-k}$$

$$\leq 2\Gamma \left[\psi + (\alpha + \phi h) \sum_{j=0}^{k-1} \|\mathbf{y}_j\| \right] e^{2\Gamma \phi (n-k)h} \leq 2\Gamma \left[\psi + (\alpha + \phi h) \sum_{j=0}^{k-1} \|\mathbf{y}_j\| \right] e^{2\Gamma \phi \tau} \,.$$

Since this bound is independent of n, we have proved zero stability.

Next, we would like to characterize convergent linear multistep methods. We begin with the following theorem.

Theorem 3.4.8 (Convergence Implies Consistency and Zero Stability) *Suppose that the function* $\mathbf{f}$ *mapping pairs of vectors and scalars to vectors is Lipschitz continuous. Given a positive integer k, an initial time t_0 and a positive timestep h, let*

$$a(\zeta) = \sum_{j=0}^{k} \alpha_j \zeta^j \text{ and } b(\zeta) = \sum_{j=0}^{k} \beta_j \zeta^j$$

be first and second characteristic polynomials for the linear multistep method

$$\sum_{j=0}^{k} \mathbf{y}_{n+1-j} \alpha_{k-j} = h \sum_{j=0}^{k} \mathbf{f}\left(\mathbf{y}_{n+1-j} \,, \ t_0 + [n + 1 - j]h\right) \,.$$

Assume that $\alpha_k = 1$ and $|\alpha_0| + |\beta_0| \neq 0$. If the solution of the linear multistep method converges to a solution of the initial value problem

$$\mathbf{y}'(t) = \mathbf{f}(\mathbf{y}(t), t) \text{ for } t > t_0 \text{ with } \mathbf{y}(t_0) = \mathbf{y}_0 \,,$$

then $a(1) = 0$, $a'(1) = b(1)$ (i.e., the method is consistent) and a satisfies the root condition (i.e., the method is zero-stable).

Proof Lemma 3.4.3 proves that $a(1) = 0$ and $a'(1) = b(1)$. It remains to show that convergence to the solution of any initial value problem with a unique solution implies the root condition. Consider the initial value problem $y'(t) = 0$ for $t > 0$ with $y(0) = 0$. The obvious solution is $y(t) = 0$ for all $t \geq 0$. The corresponding linear multistep method is

$$\sum_{j=0}^{k} y_{n+1-j}\alpha_{k-j} = 0 \ .$$

If ζ is a zero of the first characteristic polynomial a, then the initial values

$$y_j = \zeta^j \sqrt{h} \text{ for } 0 \leq j < k$$

converge to the true initial condition as $h \to 0$. Our discussion in Sect. 1.4.1.6 of Chap. 1 in Volume II shows that the corresponding numerical solution is

$$y_j = \zeta^j h \text{ for } k \leq j \leq n \ .$$

Given a time $t > 0$ and an integer $n > 0$, let $h = t/n$. Then for fixed t we must have

$$y_n = h\zeta^{t/h} \to y(t) = 0 \text{ as } n \to \infty \ .$$

This condition implies that $|\zeta| \leq 1$.

Next, if ζ is a zero of a with multiplicity greater than one, then $a'(\zeta) = 0$. We claim that the numerical solution with initial values

$$y_j = j\zeta^j h \text{ for } 0 \leq j < k$$

is

$$y_j = j\zeta^j h \text{ for } 0 \leq j \leq n \ .$$

This is because

$$\sum_{j=0}^{k} y_{m+1-j}\alpha_{k-j} = \sum_{j=0}^{k}(m+1-j)\zeta^{m+1-j}\alpha_{k-j}h$$

$$= \sum_{j=0}^{k} [(m+1-k)+(k-j)]\,\zeta^{m+1-k}\zeta^{k-j}\alpha_{k-j}h$$

$$= (m+1-k)\zeta^{m+1-k}h\sum_{j=0}^{k}\zeta^{k-j}\alpha_{k-j} + \zeta^{m+2-k}h\sum_{j=0}^{k}(k-j)\zeta^{k-j-1}\alpha_{k-j}$$

$$= (m+1-k)\zeta^{m+1-k}ha(\zeta) + \zeta^{m+2-k}ha'(\zeta) = 0$$

for all $k \leq m < n$. However, for fixed $t = nh$ we must have

$$y_n = t\zeta^{t/h} \to y(t) = 0 \text{ as } h \to 0 .$$

This implies that $|\zeta| < 1$.

The next theorem is a converse of the previous theorem.

Theorem 3.4.9 (Consistency and Zero Stability Imply Convergence) *Suppose that the function **f** mapping pairs of vectors and scalars to vectors is Lipschitz continuous with Lipschitz constant Φ. Given a positive integer k, an initial time t_0 and a positive timestep h, let*

$$a(\zeta) = \sum_{j=0}^{k} \alpha_j \zeta^j \text{ and } b(\zeta) = \sum_{j=0}^{k} \beta_j \zeta^j$$

be first and second characteristic polynomials for the linear multistep method

$$\sum_{j=0}^{k} \mathbf{y}_{n+1-j}\alpha_{k-j} = h \sum_{j=0}^{k} \mathbf{f}\left(\mathbf{y}_{n+1-j} , t_0 + [n + 1 - j]h\right) .$$

Assume that $\alpha_k = 1$, $|\alpha_0| + |\beta_0| \neq 0$, $a(1) = 0$, $a'(1) = b(1)$ and a satisfies the root condition. Assume that the solution of the initial value problem

$$\mathbf{y}'(t) = \mathbf{f}(\mathbf{y}(t), t) \text{ for } t > t_0 \text{ with } \mathbf{y}(t_0) = \mathbf{y}_0 ,$$

has uniformly continuous first derivative for $t \in [t_0, t_0 + \tau]$. Suppose that $a(1) = 0$, $a'(1) = b(1)$ and a satisfies the root condition. Then there is a constant Γ (depending only on a) so that for all $t \in (t_0, t_0 + \tau]$ and all $h = (t - t_0)/n < 1/(2\Phi)$ we have

$$\|\mathbf{y}_n - \mathbf{y}(t)\| \leq 2\Gamma \left[(t - t_0) \sum_{j=0}^{k-1} \{(k - j)|\alpha_j| + |\beta_j|\} \max_{\substack{t_0 \leq s, t \leq t_0 + \tau \\ |t-s| \leq kh}} \|\mathbf{y}'(t) - \mathbf{y}'(s)\| \right.$$
$$\left. + \left(\sum_{j=0}^{k} |\alpha_j| + (k + 1)\Phi h\right) \sum_{j=0}^{k-1} \|\mathbf{y}_j - \mathbf{y}(t_0 + jh)\| \right] e^{2\Gamma(k+1)\Phi(t-t_0)} .$$

If in addition the linear multistep method has order p, then for each initial value problem for which the solution $\mathbf{y}(t)$ has $p + 1$ continuous derivatives there is a *constant C so that for all $t \in (t_0, t_0 + \tau]$ and all $h = (t - t_0)/n < 1/(2\Phi)$ we have*

$$\|\mathbf{y}_n - \mathbf{y}(t)\| \leq 2\Gamma \left[C(t - t_0)h^p + \left(\sum_{j=0}^{k} |\alpha_j| + (k+1)\Phi h \right) \sum_{j=0}^{k-1} \|\mathbf{y}_j - \mathbf{y}(t_0 + jh)\| \right]$$

$$e^{2\Gamma(k+1)\Phi(t-t_0)} .$$

Proof Choose $t \in (t_0, t_0 + \tau]$ and $n > 2\Phi(t - t_0)$. Let $h = (t - t_0)/n$ and

$$\mathbf{d}_m = \mathbf{y}_m - \mathbf{y}(t_0 + mh) \text{ for } 0 \leq m \leq n .$$

Then for $k \leq m \leq n$ we have

$$\sum_{j=0}^{k} \mathbf{d}_{m-j}\alpha_{k-j}$$

$$- h \sum_{j=0}^{k} \left\{ \mathbf{f}(\mathbf{y}_{m-j} , \ t_0 + [m-j]h) - \mathbf{f}(\mathbf{y}(t_0 + [m-j]h) , \ t_0 + [m-j]h) \right\} \beta_{k-j}$$

$$= \left\{ \sum_{j=0}^{k} \mathbf{y}_{m-j}\alpha_{k-j} - h \sum_{j=0}^{k} \mathbf{f}(\mathbf{y}_{m-j} , \ t_0 + [m-j]h)\beta_{k-j} \right\}$$

$$- \left\{ \sum_{j=0}^{k} \mathbf{y}(t_0 + [m-j]h)\alpha_{k-j} - h \sum_{j=0}^{k} \mathbf{f}(\mathbf{y}(t_0 + [m-j]h) , \ t_0 + [m-j]h)\beta_{k-j} \right\}$$

$$= - \sum_{j=0}^{k} \mathbf{y}(t_0 + [m-j]h)\alpha_{k-j} + h \sum_{j=0}^{k} \mathbf{f}(\mathbf{y}(t_0 + [m-j]h) , \ t_0 + [m-j]h)\beta_{k-j} \equiv \boldsymbol{\psi}_m .$$

Note that $-\boldsymbol{\psi}_m = \mathbf{r}(t_0 + mh, h)$ is the local truncation error:

$$- \boldsymbol{\psi}_m = \sum_{j=0}^{k} \mathbf{y}(t_0 + [m+1-j]h)\alpha_{k-j} - h \sum_{j=0}^{k} \mathbf{y}'(t_0 + [m+1-j]h)\beta_{k-j}$$

$$= \mathbf{y}(t_0 + mh) \sum_{j=0}^{k} \alpha_{k-j} - \mathbf{y}'(t_0 + mh) \sum_{j=0}^{k} (j\alpha_{k-j} + \beta_{k-j})$$

$$+ \sum_{j=1}^{k} \left\{ \mathbf{y}(t_0 + [m-j]h) - \mathbf{y}(t_0 + mh) - \mathbf{y}'(t_0 + mh) \right\} \alpha_{k-j}$$

$$-h \sum_{j=0}^{k} \left\{ \mathbf{y}'(t_0 + [m-j]h) - \mathbf{y}'(t_0 + mh) \right\} \beta_{k-j}$$

$$= \mathbf{y}(t_0 + mh)a(1) - \mathbf{y}'(t_0 + mh)h \left[ka(1) - a'(1) + b(1) \right]$$

$$+ \sum_{j=1}^{k} \int_{-jh}^{0} \mathbf{y}'(t_0 + mh)i - \mathbf{y}'(t_0 + mh + \sigma) \, d\sigma \alpha_{k-j}$$

$$+ h \sum_{j=1}^{k} \left\{ \mathbf{y}'(t_0 + mh) - \mathbf{y}'(t_0 + [m-j]h) \right\} \beta_{k-j} ,$$

so

$$\|\boldsymbol{\psi}_m\| \leq h \left\{ \sum_{j=1}^{k} (j|\alpha_{k-j}| + |\beta_{k-j}|) \right\} \max_{\substack{t_0 \leq s,t \leq t_0+\tau \\ |t-s| \leq kh}} \|\mathbf{y}'(t) - \mathbf{y}'(s)\| .$$

We can also define

$$\boldsymbol{\phi}_{m,j}(\mathbf{d}) = \mathbf{f}(\mathbf{y}(t_0 + [m-j]h) + \mathbf{d} , \ t_0 + [m-j]h) - \mathbf{f}(\mathbf{y}(t_0 + [m-j]h) , \ t_0 + [m-j]h)$$

and use the Lipschitz continuity of $\mathbf{f}$ to bound

$$\|\boldsymbol{\phi}_{m,j}(\mathbf{d})\| \leq \Phi \|\mathbf{d}\|$$

for all $k \leq m \leq n$, all $0 \leq j \leq k$ and all vectors $\mathbf{d}$. Lemma 3.4.17 implies that

$$\|\mathbf{y}_n - \mathbf{y}(t)\| \leq \frac{\Gamma}{1 - \Phi h} \left\{ (n-k)h \sum_{j=0}^{k-1} \left[(k-j)|\alpha_j| + |\beta_j| \right] \max_{\substack{t_0 \leq s,t \leq t_0+\tau \\ |t-s| \leq kh}} \|\mathbf{y}'(t) - \mathbf{y}'(s)\| \right.$$

$$\left. + \left(\sum_{j=0}^{k} |\alpha_j| + (k+1)\Phi h \right) \sum_{j=0}^{k-1} \|\mathbf{y}_j - \mathbf{y}(t_0 + jh)\| \right\} \left[1 + \frac{\Gamma \Phi h}{1 - \Phi h} \right]^{n-k}$$

$$\leq 2\Gamma \left\{ (t - t_0) \sum_{j=0}^{k-1} \left[(k-j)|\alpha_j| + |\beta_j| \right] \max_{\substack{t_0 \leq s,t \leq t_0+\tau \\ |t-s| \leq kh}} \|\mathbf{y}'(t) - \mathbf{y}'(s)\| \right.$$

$$\left. + \left(\sum_{j=0}^{k} |\alpha_j| + (k+1)\Phi h \right) \sum_{j=0}^{k-1} \|\mathbf{y}_j - \mathbf{y}(t_0 + jh)\| \right\} e^{2\Gamma \Phi t} .$$

The error estimate can be improved if we consider the order of the linear multistep method. If the method has order p, then for each initial value problem with solution having $p + 1$ continuous derivatives on $[t_0, t_0 + \tau]$ there is a constant C so that

$$\|\boldsymbol{\psi}_m\| \leq Ch^{p+1} \ .$$

Lemma 3.4.17 can now be used to produce the final result claimed in the theorem.

Theorem 3.4.8 shows that consistent and zero-stable linear multistep methods converge to the true solution of the initial value problem, provided that the first derivative of the true solution is uniformly continuous and the initial values for the linear multistep method converge to the initial value for the specified problem. Furthermore, if the linear multistep method has order p, then the error will be on the order of h^p provided that the initial values for the linear multistep method are correspondingly accurate. The reader may recall that the discussions of the various families of linear multistep methods in Sect. 3.4.7 related the local truncation error to various polynomial interpolations, and that the error in polynomial interpolation was bounded in Theorem 1.3.2. Finally, we note that the bound on the error in linear multistep methods allows for exponential growth of the error in time, which may be especially pessimistic for initial value problems with decaying solutions.

Exercise 3.4.15 Suppose that $\{\mathbf{y}_n\}_{n=k}^{\infty}$ is the true numerical solution determined by a linear multistep method for some initial value problem, and $\{\widetilde{\mathbf{y}}_n\}_{n=k}^{\infty}$ is the numerical solution computed subject to floating point errors. If the linear multistep method is explicit, show how to modify the proof of Theorem 3.4.8 to bound the difference $\|\widetilde{\mathbf{y}}_n - \mathbf{y}_n\|$. Readers may examine an analysis of rounding errors in implicit linear multistep methods by Henrici [100, p. 262ff].

3.4.10   Predictor-Corrector Methods

So far, we have developed several families of linear multistep methods. We have analyzed their consistency, stability and convergence. We have also determined conditions that guarantee the existence of solutions to implicit linear multistep methods. However, there are several practical issues that must be resolved before we are ready to develop useful software programs to implement linear multistep methods.

For example, we need reasonable criteria for choosing the number k of previous steps that are used to determine the linear multistep method. We will discuss this topic in Sect. 3.4.12. We also need criteria for choosing the timestep h. Although the error bound in the convergence Theorem 3.4.8 could provide some guidance, we will develop a more practical technique for choosing the timestep in Sect. 3.4.11. We will also need to provide vectors $\mathbf{y}_j$ for $0 \leq j < k$ to initialize the linear multistep method. These will be developed as a consequence of our discussion about choosing k and h.

But first, we would like to discuss the numerical solution of a system of nonlinear equations for an implicit linear multistep method. For initial value problems with rapid decay, it is common to use Adams-Moulton methods or backward differentiation formulas. This is because these implicit methods have much larger regions of absolute stability than competing explicit methods, such as Adams-Bashforth methods.

Suppose that we want to use an implicit linear multistep method of the form

$$
\mathbf{y}_{n+1} + \sum_{j=0}^{k-1} \mathbf{y}_{n+1-j}\alpha_{k-j} = h\mathbf{f}(\mathbf{y}_{n+1}, t_{n+1})\beta_k + h\sum_{j=0}^{k-1}\mathbf{f}_{n+1-j}\beta_{k-j} .
$$

Because this equation is implicit, we need to find a way to solve the nonlinear equation for $\mathbf{y}_{n+1}$. If $\mathbf{f}$ is Lipschitz continuous with Lipschitz constant Φ, and if the timestep h satisfies

$$
h|\beta_k|\Phi < 1 ,
$$

sufficiently small, then Lemma 3.4.16 showed that there is a unique solution of the **fixed point equation**

$$
\mathbf{y}_{n+1} = \phi\,(\mathbf{y}_{n+1}) \equiv h\beta_k\mathbf{f}(\mathbf{y}_{n+1}, t_{n+1}) + h\sum_{j=0}^{k-1}\mathbf{f}_{n+1-j}\beta_{k-j} - \sum_{j=0}^{k-1}\mathbf{y}_{n+1-j}\alpha_{k-j} ,
$$

and that the solution can be computed by the **fixed-point iteration**

$$
\text{choose } \mathbf{y}_{n+1}^{(0)}
$$
$$
\text{for } (0 \le \nu) , \; \mathbf{y}_{n+k}^{(\nu+1)} = \phi\left(\mathbf{y}_{n+k}^{(\nu)}\right) .
$$

Note that the restriction on the timestep h is sufficient to guarantee existence of the next solution $\mathbf{y}_{n+1}$. This condition is separate from absolute stability restrictions discussed in Sect. 3.4.8.

In order to get an initial guess for $\mathbf{y}_{n+1}$ in an Adams-Moulton method, we can use an Adams-Bashforth method employing k or fewer steps. Such an approach is called a **predictor-corrector method**. For a backward differentiation formula, we could use the extrapolated value of the previous interpolation polynomial p_n, as described in Lemma 3.4.11.

Example 3.4.18 Suppose that we want to use a two-step Adams Moulton scheme. We could use the two-step Adams-Bashforth predictor

$$
\mathbf{y}_{n+2}^{(0)} = \mathbf{y}_{n+1} + h\left[\mathbf{f}_{n+1} + \frac{1}{2}\nabla^1\mathbf{f}_{n+1}\right]
$$

followed by repeated application of the two-step Adams-Moulton corrector

$$\mathbf{y}_{n+2}^{(v+1)} = \mathbf{y}_{n+1} + h\left[\mathbf{f}(\mathbf{y}_{n+2}^{(v)}, t_{n+2}) - \frac{1}{2}\nabla^1\mathbf{f}_{n+2}\right], 0 \le v .$$

Both methods have the same order, but the Adams-Moulton method has a larger region of absolute stability.

In order to obtain a numerical method with the stability properties of the implicit method, we might want to perform the fixed point iteration until we achieve convergence, In such a case, we could estimate the error between the current iterate and the fixed point by

$$\left\|\mathbf{y}_{n+1}^{(v)} - \mathbf{y}_{n+1}^{(\infty)}\right\| \le \sum_{j=0}^{\infty}\left\|\mathbf{y}_{n+1}^{(v+j+1)} - \mathbf{y}_{n+1}^{(v+j)}\right\| \le \sum_{j=0}^{\infty}(h|\beta_k|L)^j\left\|\mathbf{y}_{n+1}^{(v+1)} - \mathbf{y}_{n+1}^{(v)}\right\|$$

$$= \frac{\left\|\mathbf{y}_{n+1}^{(v+1)} - \mathbf{y}_{n+1}^{(v)}\right\|}{1 - h|\beta_k|L} .$$

Given a tolerance ε on the error $\left\|\mathbf{y}_{n+k}^{(v)} - \mathbf{y}_{n+k}^{(\infty)}\right\|$, this would suggest that we require

$$\left\|\mathbf{y}_{n+k}^{(v+1)} - \mathbf{y}_{n+k}^{(v)}\right\| \le \varepsilon(1 - h|\beta_k|L) .$$

However, this is neither cost-effective nor necessary. Each iterate $\mathbf{y}_{n+k}^{(v)}$ requires another evaluation of $\mathbf{f}$, adding to the cost of the iteration. Further, it is possible to show that at some point the extra iterations do not improve the order of the scheme, and help the stability of the scheme only marginally.

The more common approach is to decide in advance how many fixed-point iterations will be performed. If

P   corresponds to predicting $\mathbf{y}_{n+k}$,
E   corresponds to evaluating $\mathbf{f}_{n+k}$, and
C   corresponds to correcting $\mathbf{y}_{n+k}$,

then the predictor-corrector iterations usually have the form $P(EC)^\mu E$ or $P(EC)^\mu$. In other words, the algorithm looks like

$$\mathbf{y}_{n+k}^{(0)} = -\sum_{j=0}^{k-1}\mathbf{y}_{n+j}\alpha_j^* + h\sum_{j=0}^{k-1}\mathbf{f}_{n+j}\beta_j^*$$
$$\text{for } (0 \le v < \mu - 1)$$
$$\mathbf{f}_{n+k}^{(v)} = \mathbf{f}(\mathbf{y}_{n+k}^{(v)}, t_{n+k})$$
$$\mathbf{y}_{n+k}^{(v+1)} = -\sum_{j=0}^{k-1}\mathbf{y}_{n+j}\alpha_j h\mathbf{f}_{n+k}^{(v)}\beta_k + h\sum_{j=0}^{k-1}\mathbf{f}_{n+j}\beta_j$$
$$\mathbf{f}_{n+k} = \mathbf{f}(\mathbf{y}_{n+k}^{(\mu)}, t_{n+k})$$

or

$$\mathbf{y}_{n+k}^{(0)} = -\sum_{j=0}^{k-1} \mathbf{y}_{n+j}\alpha_j^* + h\sum_{j=0}^{k-1} \mathbf{f}_{n+j}\beta_j^*$$

for $(0 \le v < \mu - 1)$

$$\mathbf{f}_{n+k}^{(v)} = \mathbf{f}(\mathbf{y}_{n+k}^{(v)}, t_{n+k})$$

$$\mathbf{y}_{n+k}^{(v+1)} = -\sum_{j=0}^{k-1} \mathbf{y}_{n+j}\alpha_j + h\mathbf{f}_{n+k}^{(v)}\beta_k + h\sum_{j=0}^{k-1} \mathbf{f}_{n+j}\beta_j .$$

Suppose that the order of the predictor is no greater than the order of the corrector. Analysis of predictor-corrector methods shows that

- Each application of the corrector improves the order of the scheme, until the order of the corrector is reached; see Lambert [124, p. 105ff].
- As the number of corrector iterations $\mu \to \infty$, the characteristic polynomials ϱ and σ of the predictor-corrector scheme tend to those of the corrector; see Lambert [124, p. 117ff].
- For any number of corrector iterations μ, the predictor-corrector scheme $P(EC)^\mu$ has smaller region of absolute stability than $P(EC)^\mu E$; see Lambert [124, p. 122ff].

On the other hand, for very stiff problems (see Sect. 3.7), the predictor step can require far too small a timestep in order to give a reasonable and stable initial guess for the corrector. In such cases, it is best to apply nonlinear solvers to the implicit method directly.

It is interesting to note that if the predictor and corrector have the same order, then we can use **Milne's method** to estimate the local truncation error.

Lemma 3.4.18 *Suppose that* $\mathbf{y}(t) \in C^{p+2}$ *is approximated by predictor and corrector schemes of the same order:*

$$\mathbf{y}(t_{n+k}) - \mathbf{y}_{n+k}^{(0)} = C_{p+1}^* h^{p+1}\mathbf{y}^{(p+1)}(t_{n+k}) + O(h^{p+2}) ,$$

$$\mathbf{y}(t_{n+k}) - \mathbf{y}_{n+k}^{(\mu)} = C_{p+1} h^{p+1}\mathbf{y}^{(p+1)}(t_{n+k}) + O(h^{p+2}) .$$

Then

$$\mathbf{y}(t_{n+k}) - \mathbf{y}_{n+k}^{(\mu)} = (\mathbf{y}_{n+k}^{(\mu)} - \mathbf{y}_{n+k}^{(0)})\frac{C_{p+1}}{C_{p+1}^* - C_{p+1}} + O(h^{p+2}) . \tag{3.49}$$

Proof Subtracting the local error expansion for the corrector from the local error expansion for the predictor gives us

$$\mathbf{y}_{n+k}^{(\mu)} - \mathbf{y}_{n+k}^{(0)} = (C_{p+1}^* - C_{p+1})h^{p+1}\mathbf{y}^{(p+1)}(t_{n+k}) + O(h^{p+2}) .$$

Rearranging terms leads to

$$h^{p+1} \mathbf{y}^{(p+1)}(t_{n+k}) = \frac{\mathbf{y}_{n+k}^{(\mu)} - \mathbf{y}_{n+k}^{(0)}}{C_{p+1}^* - C_{p+1}} + O(h^{p+2}) \ .$$

The result follows by replacing the appropriate expression in the error expansion for the corrector.

Finally, note that with Milne's method we can compute a high-order extrapolant

$$\tilde{\mathbf{y}}_{n+k} = \mathbf{y}_{n+k}^{(\mu)} + (\mathbf{y}_{n+k}^{(\mu)} - \mathbf{y}_{n+k}^{(0)}) \frac{C_{p+1}}{C_{p+1}^* - C_{p+1}} = \mathbf{y}(t_{n+1}) + O(h^{p+1}) \ .$$

Of course, Milne's method and the extrapolation require that we know the error constants for the predictor and corrector. For a k-step Adams-Bashforth method, the error constant is γ_k, where γ_k is computed by the recursion (3.19). For a k-step Adams-Moulton method, the error constant is γ_k, where γ_k is computed by the recursion (3.28).

3.4.11 Choosing the Step Size

We can use Milne's method to judge the acceptability of the stepsize h for a predictor-corrector method in which the predictor and corrector have the same order. Suppose that we are given some tolerance ε on the global error in the solution. We estimate that the global error is approximately the local error times the number of timesteps remaining (although this estimate is often either too optimistic or too pessimistic). With predictor and corrector of the same order, stepsize h is acceptable at step n if

$$\left\| \mathbf{y}_{n+1}^{(\mu)} - \mathbf{y}_{n+1}^{(0)} \right\| \frac{C_{p+1}}{C_{p+1}^* - C_{p+1}} \frac{t_{final} - t_{n+1}}{h} \le \varepsilon \ .$$

Unfortunately, this estimate does not tell us how to pick the stepsize; rather, it allows us to verify the suitability of the timestep h that was used to compute $\mathbf{y}_{n+1}^{(\mu)}$. A better way to choose the timestep is described in the next lemma.

Lemma 3.4.19 *Suppose that the solution $\mathbf{y}(t)$ of an initial value problem has $p + 2$ continuous derivatives for $t \in [t_0, t_0 + \tau]$. Assume that $\mathbf{y}(t_n)$ is approximated by a predictor-corrector scheme using a predictor $\mathbf{y}_n^{(0)}$ and corrector $\mathbf{y}_n^{(\mu)}$ of the same order p, with error constants C_{p+1}^* and C_{p+1}, respectively. Suppose that we most recently took a step of size $h = t_n - t_{n-1}$, and wish to take N steps of size γh to reach time*

$$t_{final} = t_n + N\gamma h \ .$$

Suppose that the remaining steps will each produce a local error of the same size, and that the numerical method will magnify these local errors by a factor that is small compared to N. Then given a tolerance ε, the final error will satisfy

$$\left\| \mathbf{y}\left(t_{final}\right) - \mathbf{y}_{n+N} \right\| \leq \varepsilon ,$$

provided that

$$0 < \gamma^p \leq \frac{h}{t_{final} - t_n} \frac{\varepsilon}{\left\| \mathbf{y}_n^{(\mu)} - \mathbf{y}_n^{(0)} \right\|} \frac{\left| C_{p+1}^* - C_{p+1} \right|}{\left| C_{p+1} \right|}$$

Proof Suppose that we make an error of size at most δ at each remaining step, and each error is magnified by a factor $1 + \varrho$. Then

$$\left\| \mathbf{y}(t_n) - \mathbf{y}_n \right\| \leq \delta \text{ and } \left\| \mathbf{y}(t_{n+j}) - \mathbf{y}_{n+j} \right\| \leq \delta + (1+\varrho) \left\| \mathbf{y}(t_{n+j-1}) - \mathbf{y}_{n+j-1} \right\| \text{ for } 1 \leq j < N .$$

If $N\varrho \ll 1$, it follows that

$$\left\| \mathbf{y}(t_n + Nh) - \mathbf{y}_{n+N} \right\| \leq \delta \sum_{j=0}^{N-1} (1+\varrho)^j = \delta \frac{(1+\varrho)^N - 1}{\varrho} = N\delta \left[1 + O(N\varrho) \right] .$$

Under these circumstances, it is sufficient to require that if each remaining step uses timestep γh then each step has error at most ε/N:

$$\frac{\varepsilon}{N} \geq (h\gamma)^{p+1} \left| C_{p+1} \right| \left\| D^{p+1}\mathbf{y}(t_{n+j}) \right\| + O(h^{p+2}) \geq \left\| \mathbf{y}(t_{n+j}) - \mathbf{y}_{n+j}^{(\mu)} \right\|$$

for $1 \leq j \leq N$. Thus we require

$$\varepsilon \frac{h\gamma}{t_{final} - t_n} \geq \gamma^{p+1} h^{p+1} \left| C_{p+1} \right| \left\| D^{p+1}\mathbf{y}(t_n) \right\| .$$

Using Milne's estimate (3.49) we can rewrite this inequality as a bound on gamma:

$$\gamma^p \leq \varepsilon \frac{h}{t_{final} - t_n} \frac{1}{h^{p+1} \left| C_{p+1} \right| \left\| D^{p+1}\mathbf{y}(t_n) \right\|}$$

$$\approx \varepsilon \frac{h}{t_{final} - t_n} \frac{\left| C_{p+1}^* - C_{p+1} \right|}{\left| C_{p+1} \right|} \frac{1}{\left\| \mathbf{y}_n^{(\mu)} - \mathbf{y}_n^{(0)} \right\|} .$$

It is more common for numerical methods to request a bound ε on the *local truncation error*, and to choose a stepsize $h\gamma$ so that the scheme will produce a local truncation error of at most ε. This is because the local truncation error estimates are reasonably reliable; the assumption in the previous lemma that the global error is the local error times the number of remaining timesteps is not as reliable. In this approach, we choose the stepsize ratio γ so that

$$\varepsilon = \left| C_{p+1} \right| (h\gamma)^{p+1} \left\| D^{p+1} \mathbf{y}(t_n) \right\| \approx \gamma^{p+1} \left\| \mathbf{y}_n^{(\mu)} - \mathbf{y}_n^{(0)} \right\| \frac{C_{p+1}}{C_{p+1}^* - C_{p+1}} .$$

Solving for γ gives us

$$\gamma = \left[\frac{\varepsilon}{\left\| \mathbf{y}_n^{(\mu)} - \mathbf{y}_n^{(0)} \right\|} \frac{C_{p+1}^* - C_{p+1}}{C_{p+1}} \right]^{1/(p+1)} .$$

In practice, this factor is multiplied by a safety factor slightly less than one, so that there is a good chance that the next step will be acceptable.

Once a suitable stepsize has been chosen, we need to determine the representation for the scheme with the new timestep. We can use Eq. (3.26) or the ideas in Lemma 3.4.9 to transform differences of function values, or function values themselves, to coefficients in a Taylor expansion. It is easy to adjust the stepsize for the Taylor expansion coefficients. After the timestep adjustment, we can transform back to function values or differences corresponding to the new timestep size. Note that these transformations are independent of the differential equation being solved, and independent of the timestep. For more details, we recommend that the reader read Byrne and Hindmarsh [29, pp. 82–84].

3.4.12 Choosing the Multistep Number

Suppose that the time remaining to the end of the integration is $\tau = t_{\text{final}} - t_n$, and we are performing μ corrector iterations with a scheme of order p. Then we can estimate the desirable timestep $(\gamma h)_p$ as in Sect. 3.4.11. We can also estimate the total work to complete the integration to be

$$\text{remaining work} = \frac{t_{\text{final}} - t_n}{(\gamma h)_p} (\mu + 1) , \tag{3.50}$$

where the units of work is the number of evaluations of the function $\mathbf{f}$. In order to minimize the estimated work, we can vary the order p of the scheme to make this work estimate as small as possible. Typically, p is moved up or down by at most one with each timestep.

3.4.13 Choosing the Initial Values

One of the understated obstacles to using a k-step linear multistep method is the selection of the initial vectors $\mathbf{y}_j$ for $0 < j < k$. The error estimate in Theorem 3.4.8 suggests that in order for the error in all solution values to converge at some order p, both the local truncation error *and* the error in the initial values must converge at order p. But there does not seem to be any way within the linear multistep method framework to achieve order $p > 2$ error in the second step initial value $\mathbf{y}_1$. For example, in order to obtain the predicted convergence rates with linear multistep methods by the programs GUILinearMultistep.C and integrate.f, we initialized the method with the exact solution at the discrete times. Readers with some experience in numerical methods for initial value problems may suggest the use of Runge-Kutta methods, which are discussed in Sect. 3.6. Instead, we will note that existing software packages employ a different approach.

There are several publicly available programs for solving initial value problems via linear multistep methods, such as the Fortran program vode by Brown, Hindmarsh and Byrne, the C program cvode by Cohen and Hindmarsh, the Fortran program odepack by Hindmarsh, the GNU Scientific Library routines gsl_odeiv2_step_msadams and gsl_odeiv2_step_msbdf, and the MATLAB commands ode113, ode15s and ode15i. These programs use Adams-Bashforth/Adams-Moulton predictor/corrector methods of order at most 12 for non-stiff problems, and backward differentiation formula of order at most 5 for stiff problems. The methods are initialized by taking steps of increasing order with timesteps chosen to produce a final error within a given tolerance, and not of a given order. In other words, the variable-order variable-timestep linear multistep method is initialized by selecting a sufficiently small initial timestep and the initial number of steps to be $k = 1$.

Figure 3.7 shows some numerical results with the cvode program of Cohen and Hindmarsh applied to the initial value problem

$$\begin{bmatrix} y_1'(t) \\ y_2'(t) \\ y_3'(t) \end{bmatrix} = \begin{bmatrix} -y_1(t) \\ y_1(t) - y_2(t)^2 \\ y_2(t)^2 \end{bmatrix} \text{ for } t > 0, \quad \begin{bmatrix} y_1(0) \\ y_2(0) \\ y_3(0) \end{bmatrix} = \begin{bmatrix} 1 \\ 0 \\ 0 \end{bmatrix}.$$

This is test problem B3 (a nonlinear chemical reaction) in Hull et al. [107, p. 618]. The separate graphs of $\log_{10}(h)$ versus t in this figure correspond to different relative error tolerances, with absolute error tolerances set to machine precision. Within each graph, the width of the curve is chosen to be proportional to the order of the Adams family integration scheme used by cvode. The graphs show that for all error tolerances CVODE begins with very small timesteps. The graphs also show that large relative error tolerances lead to low-order computations, and small relative error tolerances lead to high-order computations except in regions of rapid change in the solution. This figure was generated by the C^{++} program GUICVode.C.

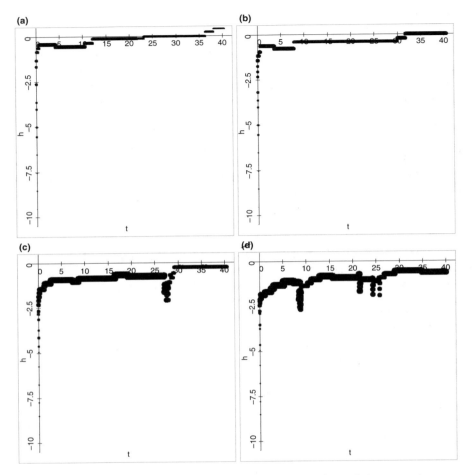

Fig. 3.7 Nonlinear chemical reaction solved by CVODE with various relative error tolerances: $log_{10}(h)$ versus t. (a) 10^{-1}. (b) 10^{-4}. (c) 10^{-9}. (d) 10^{-16}

Exercise 3.4.16 Suppose that you want to integrate the equation

$$mL\theta'' + cL\theta' + mg \sin \theta = \gamma \cos(\omega t)$$

for a driven pendulum. Here L is the length of the pendulum, m is the mass, g is the acceleration due to gravity, c is the drag coefficient and θ is the angle between the pendulum and vertical. We assume that $c^2 < 4gL$.

1. If θ is small, then we can approximate $\sin \theta \approx \theta$. Show that if $\gamma = 0$, then the general solution of the pendulum equation is

$$\theta(t) \approx e^{-ct/(2L)} \{A \sin(\omega t) + B \cos(\omega t)\}$$

where

$$\omega = \sqrt{\frac{g}{L} - \left(\frac{c}{2L}\right)^2}$$

is the fundamental frequency of the undriven pendulum.

2. Choose a numerical method and describe how to apply it to the pendulum problem.
3. If $c = 0$, $g/L = 6/11$ and $\gamma = 0$, perform a refinement study with your numerical method to verify that it converges at the expected rate.
4. If ω is equal to the fundamental frequency for this problem and $\gamma = 1$, perform a refinement study to determine the order of convergence of your method.

Exercise 3.4.17 Consider a resister-inductor-capacitor circuit, modeled by the differential equation

$$Lq'' + Rq' + \frac{1}{C}q = \mathscr{E}(t) , \quad q(0) = q_0 , \quad q'(0) = I_0 .$$

Here L is the inductance, R is the resistance, C is the capacitance, $\mathscr{E}(t)$ is the applied voltage, q_0 is the initial charge and I_0 is the initial current.

1. Write this second-order equation as a first-order system.
2. Find the eigenfunctions of the first-order system.
3. Show that if $\mathscr{E}(t) = 0$ and $4 < R^2C/L$, then $q(t)$ decays; also show that the problem is stiff if $4 \ll R^2C/L$.
4. Show that the solution oscillates if $4 > R^2C/L$ and $\mathscr{E}(t) = 0$.
5. Find the solution of this problem if $4 = R^2C/L$ and $\mathscr{E}(t) = 0$.
6. Suppose that $4 < R^2C/L$ and $\mathscr{E}(t) = Ae^{i\omega t}$. Show that the steady-state solution is

$$q(t) = \frac{Ae^{i\omega t}}{(1/C - \omega^2 L) + i\omega R}$$

Show that the maximum amplitude of the steady-state solution occurs when the capacitance is chosen so that $\omega^2 = 1/LC$.
7. Obtain an initial value problem solver and write a computer program to use it to model this electrical circuit.
8. Suppose that $R = 1$ Ohm, $L = 1$ Henry and $C = 10^4$ Farads, and $q_0 = 1$, $I_0 = 0$ amps. What kind of scheme should we use to solve this problem? Program second-order Adams-Bashforth and second-order Adams-Moulton for this problem (with $\mathscr{E}(t) = 0$). Describe how these methods perform for various timestep sizes. Also describe how they perform in comparison to your solver from the web. (Plots of the results would be nice, for some representative timesteps sizes of the Adams family schemes).

9. Suppose that $R = 1$ Ohm, $L = 1$ Henry and $C = 4$ Farads. Repeat the numerical experiments to compare Adams-Bashforth, Adams-Moulton and the other solver.

10. Suppose that $R = 1$ Ohm, $L = 1$ Henry and $C = 1$ Farad. Program the second-order BDF for this problem and compare it to your solver from the web.

Exercise 3.4.18 Consider the orbit equations

$$x''(t) = -\frac{x(t)}{[x(t)^2 + y(t)^2]^{3/2}} \text{ for } t > 0 \text{ with } x(0) = 1 - \varepsilon \text{ and } x'(0) = 0$$

$$y''(t) = -\frac{y(t)}{[x(t)^2 + y(t)^2]^{3/2}} \text{ for } t > 0 \text{ with } y(0) = 0 \text{ and } y'(0) = \sqrt{\frac{1 + \varepsilon}{1 - \varepsilon}} .$$

Here ε is the eccentricity of the orbit.

1. Write these equations as an initial value problem involving a system of first-order ordinary differential equations.
2. If $u(t)$ is defined implicitly by

$$u(t) - \varepsilon \sin u(t) = t ,$$

show that

$$x(t) = \cos u(t) - \varepsilon \text{ and } y(t) = \sqrt{1 - \varepsilon^2} \sin u(t) .$$

3. Using $\varepsilon = 0.1$, use an initial value problem solver to solve the orbit equations. For desired relative errors of 10^{ℓ} with $1 \leq \ell \leq 10$, plot elapsed computer time and the actual relative errors in $x(t)$ and $y(t)$ at $t = \pi$. How close are the actual relative errors to the desired relative error?

3.5 Deferred Correction

In Sect. 3.4.7.3, we mentioned that extrapolation can be used with the modified midpoint method to construct an arbitrarily high-order method for solving initial value problems. Our goal in this section is to examine another approach for constructing high-order methods. The new approach will be similar to the **iterative improvement** algorithm for solving a linear system $\mathbf{Ax} = \mathbf{b}$, as described in Sect. 3.9.2 of Chap. 3 in Volume I. Given an approximation $\widetilde{\mathbf{A}} \approx \mathbf{A}$, compute an initial guess $\widetilde{\mathbf{x}}^{(0)}$ by solving

$$\widetilde{\mathbf{A}}\widetilde{\mathbf{x}}^{(0)} = \mathbf{b} .$$

Then the error in the approximate solution is

$$\mathbf{x} - \widetilde{\mathbf{x}}^{(0)} = \left(\mathbf{A}^{-1} - \widetilde{\mathbf{A}}^{-1}\right)\mathbf{b} = \left(\mathbf{I} - \widetilde{\mathbf{A}}^{-1}\mathbf{A}\right)\mathbf{x} .$$

Afterward, we will compute

$$\mathbf{r}^{(k+1)} = (\mathbf{A} + \delta\mathbf{A})\,\widetilde{\mathbf{x}}^{(k)} ,$$

solve

$$\widetilde{\mathbf{A}}\widetilde{\Delta\mathbf{x}}^{(k)} = \mathbf{r}^{(k+1)}$$

and update

$$\widetilde{\mathbf{x}}^{(k+1)} = \widetilde{\mathbf{x}}^{(k)} - \widetilde{\Delta\mathbf{x}}^{(k)} .$$

At this point, the error in the approximate solution is

$$\mathbf{x} - \widetilde{\mathbf{x}}^{(k+1)} = \mathbf{x} - \widetilde{\mathbf{x}}^{(k)} + \widetilde{\Delta\mathbf{x}}^{(k)} = \mathbf{x} - \widetilde{\mathbf{x}}^{(k)} + \widetilde{\mathbf{A}}^{-1}\left[(\mathbf{A} + \delta\mathbf{A})\,\widetilde{\mathbf{x}}^{(k)} - \mathbf{b}\right]$$
$$= \mathbf{x} - \widetilde{\mathbf{x}}^{(k)} + \widetilde{\mathbf{A}}^{-1}\,(\mathbf{A} + \delta\mathbf{A})\,\widetilde{\mathbf{x}}^{(k)} - \widetilde{\mathbf{A}}^{-1}\mathbf{A}\mathbf{x} = \left(\mathbf{I} - \widetilde{\mathbf{A}}^{-1}\mathbf{A}\right)\left(\mathbf{x} - \widetilde{\mathbf{x}}^{(k)}\right) + \widetilde{\mathbf{A}}^{-1}\delta\mathbf{A}\,\widetilde{\mathbf{x}}^{(k)} .$$

Normally, we will have

$$\left\|\widetilde{\mathbf{A}}^{-1}\delta\mathbf{A}\right\| \ll \left\|\mathbf{I} - \widetilde{\mathbf{A}}^{-1}\mathbf{A}\right\| .$$

If so, each iteration will reduce the error by the order of $\left\|\mathbf{I} - \widetilde{\mathbf{A}}^{-1}\mathbf{A}\right\|$ until the error reaches the order of $\left\|\widetilde{\mathbf{A}}^{-1}\delta\mathbf{A}\right\|$.

For initial value problems, the corresponding approach is called the **Picard iteration**. Suppose that we want to solve

$$\mathbf{y}'(t) = \mathbf{f}(\mathbf{y}(t), t) \text{ for } t > 0 \text{ with } \mathbf{y}(0) = \mathbf{y}_0 ,$$

This problem is equivalent to the equation

$$\mathbf{y}(t) = \mathbf{y}_0 + \int_0^t \mathbf{f}(\mathbf{y}(\tau), \tau)\,\mathrm{d}\tau .$$

Given an approximate solution

$$\mathbf{y}^{[0]}(t) \approx \mathbf{y}(t) ,$$

we compute a sequence $\{\mathbf{y}^{[k]}(t)\}$ of approximate solutions by the iteration

$$\text{for } k \geq 0 , \quad \mathbf{y}^{[k+1]}(t) = \mathbf{y}_0 + \int_0^t \mathbf{f}(\mathbf{y}^{[k]}(\tau), \tau) \, d\tau .$$

The reader can find a straightforward proof of convergence for the Picard iteration in Coddington and Levinson [47, p. 12].

Example 3.5.1 Suppose that $\mathbf{f}(\mathbf{y}, t) = \lambda \mathbf{y}$; then the dependence of $\mathbf{f}$ on $\mathbf{y}$ is easy to evaluate. Let us take $\mathbf{y}^{[0]}(t) = \mathbf{y}_0$. Then

$$\mathbf{y}^{[1]}(t) = \mathbf{y}_0 + \int_0^t \lambda \mathbf{y}^{[0]}(\tau) \, d\tau = \mathbf{y}_0 \left[1 + \lambda t\right]$$

$$\mathbf{y}^{[2]}(t) = \mathbf{y}_0 + \int_0^t \lambda \mathbf{y}^{[1]}(\tau) \, d\tau = \mathbf{y}_0 \left[1 + \lambda t + \lambda^2 t^2/2\right]$$

$$\vdots$$

$$\mathbf{y}^{[k]}(t) = \mathbf{y}_0 \sum_{j=0}^{k} \frac{(\lambda t)^j}{j!}$$

Of course, the solution of the initial value problem $\mathbf{y}'(t) = \lambda \mathbf{y}(t), \mathbf{y}(0) = \mathbf{y}_0$ is $\mathbf{y}(t) = \mathbf{y}_0 e^{\lambda t}$. Thus for this initial value problem, the Picard iteration with initial approximation given by the initial value produces the Taylor expansion for the solution of the initial value problem.

In the form we have presented, the Picard iteration is more effective as an analytical tool than as a computational procedure. To make the Picard iteration useful for computation, we will need to find ways to approximate the numerical solution at discrete times, to approximate the integral, and to modify the approximation so that errors in the approximations are corrected. We will present two different approaches to these approximations.

3.5.1   Classical

Suppose that we have approximated the solution to an initial value problem, using m equally-spaced timesteps on some interval $[0, T]$ with a k'th order numerical method. Then the numerical solution satisfies

$$\text{for } 0 \leq i \leq m , \quad \mathbf{y}_i = \mathbf{y}(t_i) + O(h^k) .$$

Let

$$L\left(\{y_i\}_{i=0}^m \, , \, \{t_i\}_{i=0}^m \, , \, t\right) = \sum_{i=0}^m y_i \prod_{j\neq i} \frac{t - t_j}{t_i - t_j}$$

be the Lagrange form of the interpolating polynomial to the discrete solution. Then the error

$$\mathbf{d}(t) \equiv \mathbf{y}(t) - L\left(\{y_i\}_{i=0}^m \, , \, \{t_i\}_{i=0}^m \, , \, t\right)$$

satisfies the initial value problem

$$\mathbf{d}'(t) = \mathbf{f}(\mathbf{y}, t) - \frac{\mathrm{d}}{\mathrm{d}t} L\left(\{y_i\}_{i=0}^m \, , \, \{t_i\}_{i=0}^m \, , \, t\right)$$

$$= \mathbf{f}\left(\mathbf{d}(t) + L\left(\{y_i\}_{i=0}^m \, , \, \{t_i\}_{i=0}^m \, , \, t\right)\right) - \frac{\mathrm{d}}{\mathrm{d}t} L\left(\{y_i\}_{i=0}^m \, , \, \{t_i\}_{i=0}^m \, , \, t\right)$$

$$\mathbf{d}(0) = 0 \, .$$

We could approximate the solution of this differential equation by using the same numerical method that was used to generate the values y_i, obtaining

$$\mathbf{d}_i = \mathbf{d}(t_i) + O(h^k) \, .$$

Then if $m \geq 2k$ we have

$$\text{for } 0 \leq i \leq m \, , \quad \mathbf{y}_i + \mathbf{d}_i = \mathbf{y}(t_i) + O(h^{2k})$$

After j such steps, the error will have order $\min\{(j+1)k, m\}$.

However, this form of the deferred correction is seldom used in practice. The method suffers from numerical issues related to the Runge phenomenon (see Sect. 1.2.5), partially due to the use of polynomial interpolation on a uniform grid, but to a much greater extent due to differentiation of the Lagrange interpolation polynomial.

3.5.2   Spectral

An alternative approach due to Dutt et al. [71] uses a different formulation of the deferred correction iteration and a different mesh of timesteps. Let $\{t_i\}_{i=0}^m$ be the Lobatto quadrature points (see Sect. 2.3.11) in the interval $[0, T]$. Specifically,

$t_1, \ldots, t_{m-1}$ are the zeros of the first derivative of the Legendre polynomial of order m, mapped from $[-1, 1]$ to $[0, T]$. Let

$$\mathbf{y}_i = \mathbf{y}(t_i) + O(h^k) \, , \ 0 < i \le m$$

be generated by some k'th order method; typically we will choose $k = 1$ or 2. If $k = 1$, we will use either the forward or backward Euler method, since both work easily with variable-size timesteps and require little startup information. Let

$$L\left(\{\mathbf{y}_i\}_{i=0}^m \, , \ \{t_i\}_{i=0}^m \, , \ t\right) = \sum_{j=0}^m \mathbf{y}_j \prod_{i \ne j} \frac{t - t_i}{t_j - t_i}$$

be the Lagrange form of the interpolating polynomial to the discrete solution. The error

$$\mathbf{d}(t) \equiv \mathbf{y}(t) - L\left(\{\mathbf{y}_i\}_{i=0}^m \, , \ \{t_i\}_{i=0}^m \, , \ t\right)$$

satisfies the following Picard form of a differential equation:

$$\mathbf{d}(t) = \int_0^t \mathbf{f}\left(\mathbf{d}(\tau) + L\left(\{\mathbf{y}_i\}_{i=0}^m, \{t_i\}_{i=0}^m, \tau\right) \, , \ \tau\right) \, \mathrm{d}\tau - L\left(\{\mathbf{y}_i\}_{i=0}^m \, , \ \{t_i\}_{i=0}^m \, , \ t\right)$$

$$+ L\left(\{\mathbf{y}_i\}_{i=0}^m \, , \ \{t_i\}_{i=0}^m \, , \ 0\right)$$

$$= \int_0^t \mathbf{f}\left(\mathbf{d}(\tau) + L\left(\{\mathbf{y}_i\}_{i=0}^m, \{t_i\}_{i=0}^m, \tau\right) \, , \ \tau\right) - \mathbf{f}\left(L\left(\{\mathbf{y}_i\}_{i=0}^m, \{t_i\}_{i=0}^m, \tau\right) \, , \ \tau\right) \, \mathrm{d}\tau$$

$$+ \left[\int_0^t \mathbf{f}\left(L\left(\{\mathbf{y}_i\}_{i=0}^m, \{t_i\}_{i=0}^m, \tau\right) \, , \ \tau\right) \, \mathrm{d}\tau - L\left(\{\mathbf{y}_i\}_{i=0}^m \, , \ \{t_i\}_{i=0}^m \, , \ t\right) + \mathbf{y}_0\right] \, .$$

At times t_i in the timestep mesh, we can approximate

$$\int_0^{t_i} \mathbf{f}\left(L\left(\{\mathbf{y}_i\}_{i=0}^m, \{t_i\}_{i=0}^m, \tau\right) \, , \ \tau\right) \, \mathrm{d}\tau \approx \int_0^{t_i} L\left(\{\mathbf{y}_i\}_{i=0}^m, \{t_i\}_{i=0}^m, \tau\right) \, \mathrm{d}\tau \, .$$

Then the term in the square brackets for the evaluation for $\mathbf{d}(t)$ can be approximated by

$$\int_0^{t_i} \mathbf{f}\left(L\left(\{\mathbf{y}_i\}_{i=0}^m, \{t_i\}_{i=0}^m, \tau\right) \, , \ \tau\right) \, \mathrm{d}\tau - L\left(\{\mathbf{y}_i\}_{i=0}^m, \{t_i\}_{i=0}^m, t_i\right) + \mathbf{y}_0$$

$$\approx \varepsilon_i \equiv \int_0^{t_i} L\left(\{\mathbf{y}_i\}_{i=0}^m, \{t_i\}_{i=0}^m, \tau\right) \, \mathrm{d}\tau - \mathbf{y}_i + \mathbf{y}_0 \, .$$

Let us summarize the approach. If we use Euler's method, then the spectral deferred correction method takes the form

Algorithm 3.5.1 (Spectral Deferred Correction Using Forward Euler)

> for $0 \le i < m$
>
> $\quad \mathbf{y}_{i+1} = \mathbf{y}_i + (t_{i+1} - t_i)\mathbf{f}(\mathbf{y}_i, t_i)$
>
> for $1 \le j < m$
>
> $\quad \varepsilon_0 = 0$
>
> $\quad \mathbf{d}_0 = 0$
>
> $\quad$ for $0 \le i < m$
>
> $$\varepsilon_{i+1} = \int_0^{t_i} L\left(\{\mathbf{y}_i\}_{i=0}^m, \{t_i\}_{i=0}^m, \tau\right) d\tau - \mathbf{y}_{i+1} + \mathbf{y}_0$$
>
> $$\mathbf{d}_{i+1} = \mathbf{d}_i + (t_{i+1} - t_i)\left[\mathbf{f}(\mathbf{d}_i + \mathbf{y}_i, t_i) - \mathbf{f}(\mathbf{y}_i, t_i)\right] + [\varepsilon_{i+1} - \varepsilon_i]$$
>
> $\quad$ for $0 < i \le m$
>
> $\quad\quad \mathbf{y}_i = \mathbf{y}_i + \mathbf{d}_i$.

Each step of deferred correction improves the order of the method by one until the limiting order m of the Lagrange interpolation is reached.

Use of the backward Euler method in spectral deferred correction is similar:

Algorithm 3.5.2 (Spectral Deferred Correction Using Backward Euler)

for $0 \le i < m$

$\quad$ solve $\mathbf{y}_{i+1} = \mathbf{y}_i + (t_{i+1} - t_i)\mathbf{f}(\mathbf{y}_{i+1}, t_{i+1})$ for $\mathbf{y}_{i+1}$

for $1 \le j < m$

$\quad \varepsilon_0 = 0$

$\quad \mathbf{d}_0 = 0$

$\quad$ for $0 \le i < m$

$$\varepsilon_{i+1} = \int_0^{t_i} L\left(\{\mathbf{y}_i\}_{i=0}^m, \{t_i\}_{i=0}^m, \tau\right) d\tau - \mathbf{y}_{i+1} - \mathbf{y}_0$$

$\quad$ solve $\mathbf{d}_{i+1} = \mathbf{d}_i + (t_{i+1} - t_i)[\mathbf{f}(\mathbf{d}_{i+1} + \mathbf{y}_{i+1}, t_{i+1}) - \mathbf{f}(\mathbf{y}_{i+1}, t_{i+1})] + [\varepsilon_{i+1} - \varepsilon_i]$

$\quad\quad$ for $\mathbf{d}_{i+1}$

for $0 < i \le m$, $\mathbf{y}_i = \mathbf{y}_i + \mathbf{d}_i$

Note that the Lobatto quadrature points are initially determined in the interval $[-1, 1]$, from which they should be mapped to some given time interval for the initial value problem. If $i - 1 = s_0 < \ldots < s_m = 1$ are the Lobatto quadrature points in $[-1, 1]$, then the Lobatto quadrature points in $[t_0, t_m]$ are

$$t_i = \frac{t_0 + t_m}{2} + \frac{t_m - t_0}{2} s_i \,,$$

so the integrals of the Lagrange interpolating polynomials are

$$\int_{t_0}^{t_k} \prod_{i \neq j} \frac{t - t_i}{t_j - t_i} \, dt = \frac{t_m - t_0}{2} \int_{s_0}^{s_k} \prod_{i \neq j} \frac{s - s_i}{s_j - s_i} \, ds \,.$$

The integrals with respect to s should be computed once and stored.

Example 3.5.2 Suppose that we want to apply the spectral deferred correction algorithm of order 2 to solve $\mathbf{y}'(t) = \mathbf{y}(t)$ for $t \in (0, h)$. The Lobatto nodes are $t_0 = 0$, $t_1 = h/2$ and $t_2 = h$. The forward Euler scheme produces

$$\mathbf{y}_0 = 1$$

$$\mathbf{y}_1 = \mathbf{y}_0 + (h/2)\mathbf{y}_0 = 1 + h/2$$

$$\mathbf{y}_2 = \mathbf{y}_1 + (h/2)\mathbf{y}_1 = (1 + h/2) + (h/2)(1 + h/2) = 1 + h + h^2/4 \,.$$

Then deferred correction produces

$$\varepsilon_1 = \frac{h}{2}\left[\mathbf{f}_0 \frac{5}{12} + \mathbf{f}_1 \frac{2}{3} + \mathbf{f}_2\left(-\frac{1}{12}\right)\right] - \mathbf{y}_1 + \mathbf{y}_0$$

$$= \frac{h}{2}\left[\frac{5}{12} + (1 + \frac{h}{2})\frac{2}{3} + (1 + h + \frac{h^2}{4})(-\frac{1}{12})\right] - (1 + \frac{h}{2}) + 1 = \frac{h^2}{8}\left[1 - \frac{h}{12}\right]$$

$$\mathbf{d}_1 = \frac{h}{2}[\mathbf{f}(\mathbf{y}_0 + 0) - \mathbf{f}_0] + \varepsilon_1 = \varepsilon_1 = \frac{h^2}{8}\left[1 - \frac{h}{12}\right]$$

$$\varepsilon_2 = \frac{h}{2}\left[\mathbf{f}_0 \frac{1}{3} + \mathbf{f}_1 \frac{4}{3} + \mathbf{f}_2 \frac{1}{3}\right] - \mathbf{y}_2 + \mathbf{y}_0$$

$$= \frac{h}{2}\left[\frac{1}{3} + (1 + \frac{h}{2})\frac{4}{3} + (1 + h + \frac{h^2}{4})\frac{1}{3}\right] - \left[1 + h + h^2/4\right] + 1$$

$$= \frac{h^2}{4}\left[1 + \frac{h}{6}\right]$$

$$\mathbf{d}_2 = \mathbf{d}_1 + \frac{h}{2} \left[\mathbf{f}(\mathbf{y}_1 + \mathbf{d}_1) - \mathbf{f}_1 \right] + [\varepsilon_2 - \varepsilon_1]$$

$$= \frac{h^2}{8} \left[1 - \frac{h}{12} \right] + \frac{h}{2} \left[1 + \frac{h}{2} + \frac{h^2}{8}(1 - \frac{h}{12}) - 1 - \frac{h}{2} \right]$$

$$+ \frac{h^2}{4} \left[1 + \frac{h}{6} \right] - \frac{h^2}{8} \left[1 - \frac{h}{12} \right]$$

$$= \frac{h^2}{4} \left[1 + \frac{5h}{12} - \frac{h^2}{24} \right] .$$

We use these to correct

$$\mathbf{y}_1 \leftarrow \mathbf{y}_1 + \mathbf{d}_1 = 1 + \frac{h}{2} + \frac{h^2}{8} \left[1 - \frac{h}{12} \right]$$

$$\mathbf{y}_2 \leftarrow \mathbf{y}_2 + \mathbf{d}_2 = 1 + h + \frac{h^2}{4} + \frac{h^2}{4} \left[1 + \frac{5h}{12} - \frac{h^2}{24} \right] = 1 + h + \frac{h^2}{2} + \frac{5h^3}{48} - \frac{h^4}{96} .$$

These two values are second-order approximations to $\mathbf{y}(t) = e^t$ at $t = h/2$ and $t = h$, respectively.

A proof of the convergence of spectral deferred correction can be found in Hansen and Strain [97]. Their proof shows that p corrections with a one-step solver of order r produces a spectral deferred correction algorithm of order $\min\{m, r(p + 1)\}$, where m is the order of the Lagrange interpolation. Thus, there is an advantage to using a one-step solver of order greater than one, in order to accelerate the convergence of the overall method.

A program to compute integrals of Lagrange interpolation polynomials and to apply spectral deferred correction is available in deferred_correction.f. The subroutines in this file are called from GUIDeferredCorrection.C. The main program constructs a `LobattoQuadrature` to compute the Lobatto nodes and weights for the rule that will be used. Afterward, it calls `lagrange_polynomial_integrals` to compute the integrals of the Lagrange interpolation polynomials.

Dutt et al. [71] also showed that spectral deferred correction possesses superior stability properties in comparison to linear multistep methods. Absolute stability regions for the spectral deferred correction method can be drawn by executing the code in GUIDCAbsoluteStability.C. Some absolute stability regions for spectral deferred correction using the forward Euler method are shown in Fig. 3.8, and some absolute stability regions for spectral deferred correction using the backward Euler method are shown in Fig. 3.9.

Exercise 3.5.1 Develop a spectral deferred correction algorithm using the trapezoidal method (3.13). What order of accuracy should we expect from each pass of deferred correction?

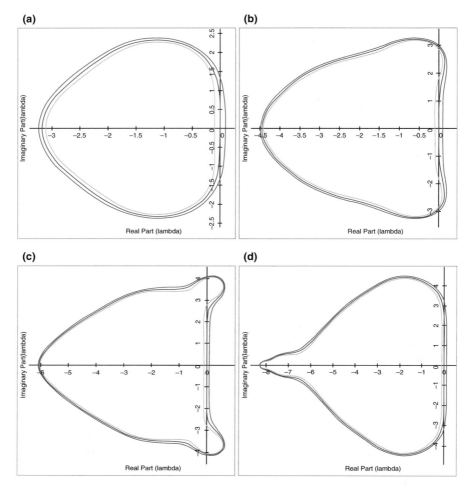

Fig. 3.8 Absolute stability regions for spectral deferred correction using forward Euler method; absolute stability boundary in blue, boundary of region with 10% growth in red, boundary of region with 10% decay in green. (**a**) Order 2. (**b**) Order 3. (**c**) Order 4. (**d**) Order 5

3.6   Runge-Kutta Methods

Previously, we have developed one large class of methods for solving initial value problems, namely the linear multistep methods in Sect. 3.4. Linear multistep methods have several advantages and disadvantages. One disadvantage is that they require a startup procedure to generate values of either **f** or **y** at previous steps. A significant advantage is that high-order linear multistep methods require very few new evaluations of **f**. For example, explicit multistep methods require the computation of only one new **f** value, and predictor-corrector schemes require

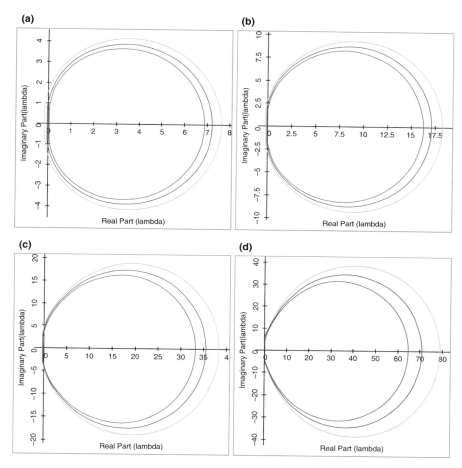

Fig. 3.9 Absolute stability regions for spectral deferred correction using backward Euler method; absolute stability boundary in blue, boundary of region with 10% growth in red, boundary of region with 10% decay in green. (**a**) Order 2. (**b**) Order 3. (**c**) Order 4. (**d**) Order 5

a predetermined number μ of function values. Another useful feature of linear multistep methods is that their local error estimates have a simple structure. A disadvantage is that stepsize changes with linear multistep methods are somewhat complicated; see Lemma 3.4.9 and the surrounding discussion to recall how this process operates.

In this section, we will develop an entirely new class of methods for initial value problems, namely **Runge-Kutta methods**. The advantages and disadvantages of Runge-Kutta methods typically complement those of the linear multistep methods. Runge-Kutta methods require only one previous value of **y**, so they are easy to start. However, Runge-Kutta methods request several new values of **f** to build higher-order approximations. The Runge-Kutta approximations are nonlinear, which will complicate discussions of their stability properties. Finally, it is not always easy to estimate the local truncation error of Runge-Kutta schemes, but it is easy to change the stepsize.

3.6.1   General Principles

To simplify the discussion of Runge-Kutta methods, we will consider initial-value problems in a special form.

Definition 3.6.1 Let $\mathbf{y}_0$ be a vector, t_0 be a scalar, and let $\mathbf{f}$ map vectors to vectors. Then the initial value problem

$$\mathbf{y}'(t) = \mathbf{f}\left(\mathbf{y}(t)\right) \text{ for } t > t_0 \text{ with } \mathbf{y}(t_0) = \mathbf{y}_0$$

is said to be in **autonomous form**.

It is easy to rewrite a general initial value problem in autonomous form. If

$$\mathbf{y}'(t) = \mathbf{f}(\mathbf{y}(t), t) \text{ for } t > t_0 \text{ with } \mathbf{y}(t_0) = \mathbf{y}_0 ,$$

then we can define

$$\overline{\mathbf{y}}(t) = \begin{bmatrix} \mathbf{y}(t) \\ t \end{bmatrix} , \ \overline{\mathbf{y}}_0 = \begin{bmatrix} \mathbf{y}_0 \\ t_0 \end{bmatrix} \text{ and } \overline{\mathbf{f}}\left(\overline{\mathbf{y}}(t)\right) = \begin{bmatrix} \mathbf{f}(\mathbf{y}(t), t) \\ 1 \end{bmatrix}$$

and then observe that

$$\overline{\mathbf{y}}'(t) = \overline{\mathbf{f}}\left(\overline{\mathbf{y}}(t)\right) \text{ for } t > t_0 \text{ with } \overline{\mathbf{y}}(t_0) = \overline{\mathbf{y}}_0 .$$

Next, we will define our new class of methods.

Definition 3.6.2 Let s be a positive integer, h be a timestep, $\mathbf{y}$ be a vector and let $\mathbf{f}$ map vectors to vectors. Then an s-stage Runge-Kutta method takes the form

$$\mathbf{y}_{n+1} = \mathbf{y}_n + h \sum_{i=1}^{s} \beta_i \mathbf{f}_i \text{ where} \tag{3.51a}$$

$$\mathbf{z}_i = \mathbf{y}_n + h \sum_{j=1}^{s} \alpha_{ij} \mathbf{f}_j \text{ for } 1 \leq i \leq s \text{ and} \tag{3.51b}$$

$$\mathbf{f}_i = \mathbf{f}(\mathbf{z}_i) \text{ for } 1 \leq i \leq s . \tag{3.51c}$$

If $\alpha_{ij} = 0$ for $j \geq i$, then the method is **explicit**. If $\alpha_{ij} = 0$ for $j > i$, then the method is **semi-implicit**. If the $s \times s$ matrix $\mathbf{A}$ has components α_{ij} and the s-vector $\mathbf{b}$ has components β_i, then we define the vector $\mathbf{c}$ by

$$\mathbf{c} = \mathbf{A}\mathbf{e} .$$

The corresponding **Butcher array** is

$$\begin{array}{c|c} \mathbf{c} & \mathbf{A} \\ \hline & \mathbf{b}^\mathsf{T} \end{array} .$$

We will also define the vector $\mathbf{c}$ by where $\mathbf{e}$ is the vector of ones.

Lambert [124, p. 157ff] shows that Runge-Kutta schemes must satisfy certain conditions in order to achieve a specified **order of accuracy**.

Lemma 3.6.1 *Suppose that s is a positive integer, that $\mathbf{A}$, $\mathbf{b}$ and $\mathbf{c}$ are the coefficients in the Butcher array for an s-stage Runge-Kutta scheme. Let $\mathbf{A}$ have components α_{ij} and $\mathbf{b}$ have components β_i. Then the components of the s-vector $\mathbf{c}$ are given by*

$$\gamma_i = \sum_{j=1}^{s} \alpha_{ij} .$$

The Runge-Kutta scheme is has order at least one if and only if

$$\sum_{i=1}^{s} \beta_i = 1 . \tag{3.52}$$

The Runge-Kutta scheme has order at least two if and only if conditions (3.52) and

$$\sum_{i=1}^{s} \beta_i \gamma_i = \frac{1}{2} . \tag{3.53}$$

are satisfied. The Runge-Kutta scheme has order at least three if and only if conditions (3.52), (3.53) and

$$\sum_{i=1}^{s} \beta_i \gamma_i^2 = \frac{1}{3} \text{ and } \sum_{j=1}^{s} \sum_{i=1}^{s} \beta_i \alpha_{i,j} \gamma_j = \frac{1}{6} \tag{3.54}$$

are satisfied. For order 4, four additional conditions

$$\sum_{i=1}^{s} \beta_i \gamma_i^3 = \frac{1}{4} , \tag{3.55a}$$

$$\sum_{j=1}^{s} \sum_{i=1}^{s} \beta_i \alpha_{i,j} \gamma_i \gamma_j = \frac{1}{8} , \tag{3.55b}$$

$$\sum_{j=1}^{s}\sum_{i=1}^{s}\beta_i\alpha_{i,j}\gamma_j^2 = \frac{1}{12} \ and \qquad\qquad (3.55c)$$

$$\sum_{k=1}^{s}\sum_{j=1}^{s}\sum_{i=1}^{s}\beta_i\alpha_{i,j}\alpha_{j,k}\gamma_k = \frac{1}{24} \qquad\qquad (3.55d)$$

must be satisfied.

These conditions apply to both explicit and implicit Runge-Kutta methods. It is common to require that all of the coefficients in the Butcher array are nonnegative; such schemes generally have better stability properties.

Exercise 3.6.1 Prove that the s-stage Runge-Kutta scheme in (3.51) has order at least one if and only if Eq. (3.52) is satisfied.

Exercise 3.6.2 Prove that the s-stage Runge-Kutta scheme in (3.51) has order at least two if and only if Eqs. (3.52) and (3.53) are satisfied.

3.6.2   Explicit Methods

Here are some examples of Runge-Kutta schemes. The only explicit first-order one-stage Runge-Kutta scheme is the **forward Euler** method:

$$\mathbf{y}_{n+1} = \mathbf{y}_n + h\mathbf{f}(\mathbf{y}_n) \ .$$

This has Butcher array

$$\begin{array}{c|c} 0 & 0 \\ \hline & 1 \end{array}$$

The **modified Euler method** has Butcher array

$$\begin{array}{c|cc} 0 & & \\ 1/2 & 1/2 & \\ \hline & 0 & 1 \end{array}$$

The scheme can be written as a half-step of the forward Euler method followed by a full step of the midpoint rule:

$$\mathbf{y}_{n+\frac{1}{2}} = \mathbf{y}_n + \frac{h}{2}\mathbf{f}(\mathbf{y}_n) \ ,$$

$$\mathbf{y}_{n+1} = \mathbf{y}_n + h\mathbf{f}(\mathbf{y}_{n+\frac{1}{2}}) \ .$$

In order to understand absolute stability of this method, we can apply the method to $\mathbf{y}'(t) = \lambda \mathbf{y}(t)$ and find that the numerical solution satisfies

$$\mathbf{y}_{n+1} = \mathbf{y}_n \left[1 + h\lambda + \frac{1}{2}(h\lambda)^2 \right] .$$

For a constant-coefficient differential equation, the modified Euler method corresponds to approximating the matrix exponential by the first three terms in the Taylor series for the exponential. If we write

$$h\lambda = -1 + re^{i\theta} ,$$

then we can evaluate the **amplification factor** as

$$1 + h\lambda(1 + h\lambda/2) = 1 + \frac{1}{2} \left[re^{i\theta} - 1 \right] \left[re^{i\theta} + 1 \right] = \frac{1}{2} \left[r^2 e^{2i\theta} + 1 \right] .$$

For absolute stability, we want

$$1 \geq \left| \frac{1}{2} \left[r^2 e^{2i\theta} + 1 \right] \right| ,$$

which is equivalent to

$$r^4 + 2r^2 \cos 2\theta - 3 \leq 0 .$$

This quadratic always has one negative root, so we are lead to the following absolute stability condition:

$$r \leq \sqrt{ \sqrt{3 + \cos^2 2\theta} - \cos 2\theta } .$$

Readers can view this absolute stability region in graph (b) of Fig. 3.10. The graphs in this figure have been generated by GUIRKAbsoluteStability.C.

The **improved Euler method** has Butcher array

$$\begin{array}{c|cc} 0 & & \\ 1 & 1 & \\ \hline & 1/2 & 1/2 \end{array}$$

This corresponds taking $\beta_2 = 1/2$, which implies $\beta_1 = 1/2$ and $\alpha_{21} = 1$. The scheme can be written as a predictor-corrector scheme in which we use a forward

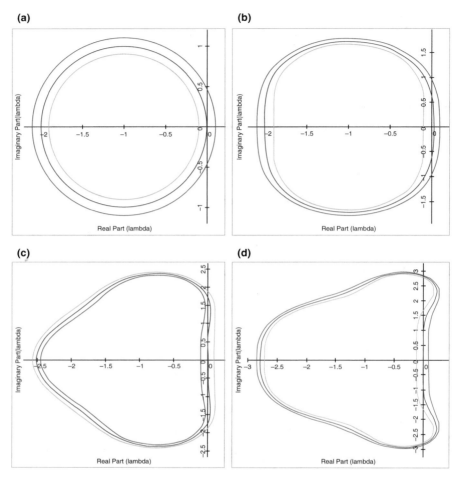

Fig. 3.10 Absolute stability regions for explicit Runge-Kutta methods. (**a**) Order 1. (**b**) Order 2. (**c**) Order 3. (**d**) Order 4

Euler predictor followed by a trapezoidal rule corrector:

$$\tilde{\mathbf{y}}_{n+1} = \mathbf{y}_n + h\mathbf{f}(\mathbf{y}_n) \ ,$$

$$\mathbf{y}_{n+1} = \mathbf{y}_n + \frac{h}{2}\left[\mathbf{f}(\mathbf{y}_n) + \mathbf{f}(\tilde{\mathbf{y}}_{n+1})\right] \ .$$

This method has the same absolute stability region as the modified Euler method.

Heun's third-order method has Butcher array

$$
\begin{array}{c|ccc}
0 & & & \\
1/3 & 1/3 & & \\
2/3 & 0 & 2/3 & \\
\hline
& 1/4 & 0 & 3/4
\end{array}
$$

It can be implemented in the form

$$\mathbf{f}_1 = \mathbf{f}(\mathbf{y}_n)$$

$$\mathbf{f}_2 = \mathbf{f}\left(\mathbf{y}_n + \frac{h}{3}\mathbf{f}_1\right)$$

$$\mathbf{f}_3 = \mathbf{f}\left(\mathbf{y}_n + \frac{2h}{3}\mathbf{f}_2\right)$$

$$\mathbf{y}_{n+1} = \mathbf{y}_n + \frac{h}{4}[\mathbf{f}_1 + 3\mathbf{f}_3] \ .$$

In order to understand absolute stability of this method, we can apply the method to $\mathbf{y}'(t) = \lambda \mathbf{y}(t)$ and find that the numerical solution satisfies

$$\mathbf{y}_{n+1} = \mathbf{y}_n\left[1 + h\lambda + \frac{1}{2}(h\lambda)^2 + \frac{1}{6}(h\lambda)^3\right] \ .$$

For a constant-coefficient differential equation, the modified Euler method corresponds to approximating a matrix exponential by the first four terms in the Taylor series for the exponential. Readers can view this absolute stability region in graph (c) of Fig. 3.10.

  Kutta's third-order method has Butcher array

$$
\begin{array}{c|ccc}
0 & & & \\
1/2 & 1/2 & & \\
1 & -1 & 2 & \\
\hline
& 1/6 & 2/3 & 1/6
\end{array}
$$

It can be implemented in the form

$$\mathbf{f}_1 = \mathbf{f}(\mathbf{y}_n)$$

$$\mathbf{f}_2 = \mathbf{f}\left(\mathbf{y}_n + \frac{h}{2}\mathbf{f}_1\right)$$

$$\mathbf{f}_3 = \mathbf{f}(\mathbf{y}_n - h\mathbf{f}_1 + 2h\mathbf{f}_2)$$

$$\mathbf{y}_{n+1} = \mathbf{y}_n + \frac{h}{6}[\mathbf{f}_1 + 4\mathbf{f}_2 + \mathbf{f}_3] \ .$$

This method has the same absolutely stability region as Heun's third-order method.
The **classical fourth-order Runge-Kutta method** has Butcher array

$$
\begin{array}{c|cccc}
0 \\
1/2 & 1/2 \\
1/2 & 0 & 1/2 \\
1 & 0 & 0 & 1 \\
\hline
& 1/6 & 1/3 & 1/3 & 1/6
\end{array}
$$

It can be implemented in the form

$$\mathbf{f}_1 = \mathbf{f}(\mathbf{y}_n)$$

$$\mathbf{f}_2 = \mathbf{f}\left(\mathbf{y}_n + \frac{h}{2}\mathbf{f}_1\right)$$

$$\mathbf{f}_3 = \mathbf{f}\left(\mathbf{y}_n + \frac{h}{2}\mathbf{f}_2\right)$$

$$\mathbf{f}_4 = \mathbf{f}(\mathbf{y}_n + h\mathbf{f}_3)$$

$$\mathbf{y}_{n+1} = \mathbf{y}_n + \frac{h}{6}[\mathbf{f}_1 + 2\mathbf{f}_2 + 2\mathbf{f}_3 + \mathbf{f}_4] \ .$$

GNU Scientific Library users can access this method via the C routine
gsl_odeiv2_step_rk4. In order to understand absolute stability of this method,
we can apply the method to $\mathbf{y}'(t) = \lambda\mathbf{y}(t)$ and find that the numerical solution
satisfies

$$\mathbf{y}_{n+1} = \mathbf{y}_n\left[1 + h\lambda + \frac{1}{2}(h\lambda)^2 + \frac{1}{6}(h\lambda)^3 + \frac{1}{24}(h\lambda)^4\right] \ .$$

For a constant-coefficient differential equation, the modified Euler method corresponds to approximating a matrix exponential by the first five terms in the Taylor
series for the exponential. Readers can view this absolute stability region in graph
(d) of Fig. 3.10.

The following theorem provides upper bounds on the orders of explicit Runge-
Kutta schemes.

Theorem 3.6.1 (Butcher Barrier) *An explicit Runge-Kutta methods involving s
stages cannot have order greater than s. If $s \geq 5$, then an explicit Runge-Kutta
method involving s stages has order at most $s - 1$. If $s \geq 7$, then an explicit Runge-
Kutta method involving s stages has order at most $s - 2$. Finally, if $s \geq 8$, then an
explicit Runge-Kutta method involving s stages has order at most $s - 3$.*

Proof The first claim is proved in Butcher [27, p. 414]; another proof is also
available in Lambert [124, p. 177]. The second claim is also proved in Butcher
[27, p. 415]; this claim is also proved in Hairer et al. [94, p. 185f]. A proof of the

third claim can be found in Butcher [27, p. 417], and the final claim is proved by Butcher in [28].

Exercise 3.6.3   What is the order of the modified Euler method?

Exercise 3.6.4   What is the order of the improved Euler method?

Exercise 3.6.5   Show that Heun's third-order method has order three.

Exercise 3.6.6   Show that the classical fourth-order Runge-Kutta scheme has order four.

3.6.3   Implicit Methods

The **backward Euler method** is the first-order one-stage implicit Runge-Kutta method

$$\mathbf{f}_1 = \mathbf{f}(\mathbf{y}_n + h\mathbf{f}_1)$$

$$\mathbf{y}_{n+1} = \mathbf{y}_n + h\mathbf{f}_1 \ .$$

This has Butcher array

$$\begin{array}{c|c} 1 & 1 \\ \hline & 1 \end{array}$$

It is more common to implement this scheme in the form

$$\mathbf{y}_{n+1} = \mathbf{y}_n + h\mathbf{f}(\mathbf{y}_{n+1}) \ .$$

GNU Scientific Library users can access this method via the C routine gsl_odeiv2_step_rk1imp. As we saw in Example 3.4.12, the absolute stability region of the backward Euler method is the exterior of the unit circle with center 1.

Butcher [26], and Ceschino and Kuntzman [36, p. 106], have shown that for all $s \geq 1$ there are implicit Runge-Kutta methods involving s stages with order $2s$. We will show how to construct such methods, after we present some examples.

The **implicit midpoint rule** is the one-stage fully implicit Runge-Kutta scheme

$$\mathbf{f}_1 = \mathbf{f}\left(\mathbf{y}_n + \frac{h}{2}\mathbf{f}_1\right)$$

$$\mathbf{y}_{n+1} = \mathbf{y}_n + h\mathbf{f}_1$$

has order 2. This claim is easy to verify by means of Lemma 3.6.1. The scheme also
has Butcher array

$$\frac{\begin{array}{c|c} 1/2 & 1/2 \end{array}}{\ \ 1}$$

This scheme can be rewritten as

$$\mathbf{y}_{n+1} = \mathbf{y}_n + h\mathbf{f}\left(\mathbf{y}_n + \frac{1}{2}\left[\mathbf{y}_{n+1} - \mathbf{y}_n\right]\right) = \mathbf{y}_n + h\mathbf{f}\left(\frac{\mathbf{y}_n + \mathbf{y}_{n+1}}{2}\right).$$

GNU Scientific Library users can access this method via the C routine
gsl_odeiv2_step_rk2imp. If we apply the implicit midpoint rule to the initial value
problem $\mathbf{y}'(t) = \lambda\mathbf{y}(t)$, we find that

$$y_{n+1} = y_n \frac{1 + h\lambda/2}{1 - h\lambda/2}.$$

Note that the **linear fractional transformation**

$$\phi(z) = \frac{1 + z}{1 - z}$$

maps the left half-plane to the interior of the unit circle. Consequently, $|\phi(h\lambda/2)| \leq$
1 if and only if $\mathfrak{Re}(h\lambda/2) \leq 0$. In other words, the region of absolute stability for
the implicit midpoint method is the left half-plane.

The **Hammer-Hollingsworth method** [96] is the two-stage fourth-order fully
implicit Runge-Kutta scheme

$$\mathbf{f}_1 = \mathbf{f}\left(\mathbf{y}_n + h\left[\frac{1}{4}\mathbf{f}_1 + \left\{\frac{1}{4} - \frac{\sqrt{3}}{6}\right\}\mathbf{f}_2\right]\right)$$

$$\mathbf{f}_2 = \mathbf{f}\left(\mathbf{y}_n + h\left[\left\{\frac{1}{4} + \frac{\sqrt{3}}{6}\right\}\mathbf{f}_1 + \frac{1}{4}\mathbf{f}_2\right]\right)$$

$$\mathbf{y}_{n+1} = \mathbf{y}_n + \frac{h}{2}\left[\mathbf{f}_1 + \mathbf{f}_2\right].$$

This scheme has Butcher array

$$\frac{\begin{array}{c|cc} 1/2 - \sqrt{3}/6 & 1/4 & 1/4 - \sqrt{3}/6 \\ 1/2 + \sqrt{3}/6 & 1/4 + \sqrt{3}/6 & 1/4 \end{array}}{\phantom{1/2 + \sqrt{3}/6}\ \ \ 1/2 \qquad\qquad 1/2}$$

GNU Scientific Library users can access this method via the C routine gsl_odeiv2_step_rk4imp. In order to understand absolute stability of this method, we will apply the method to $y'(t) = \lambda y(t)$ and find that the numerical solution satisfies

$$y_{n+1} = y_n \frac{1 + (h\lambda)/2 + (h\lambda)^2/12}{1 - (h\lambda)/2 + (h\lambda)^2/12} .$$

Thus the amplification factor for this method is $\varrho(h\lambda)$ where

$$\varrho(z) = \frac{1 + z/2 + z^2/12}{1 - z/2 + z^2/12} .$$

Note that the linear fractional transformation

$$\phi(\zeta) = \frac{1 + \zeta}{1 - \zeta}$$

maps the unit circle into the right half-plane. We can compute

$$\phi(\varrho(z)) = -\frac{2}{z} - \frac{z}{6} .$$

If $z = x + iy$ then

$$\Re\left(\phi(\varrho(z))\right) = -x\left(\frac{2}{x^2 + y^2} + \frac{1}{6}\right)$$

is nonnegative if and only if $0 > x = \Re(z)$. We conclude that the region of absolute stability for this two-stage fourth-order Runge-Kutta method is the left half-plane.

More generally, we can derive implicit Runge-Kutta methods as special cases of **collocation methods**. Suppose that we want to solve an autonomous initial value problem

$$y'(t) = f(y(t)) \text{ for } t > t_0 \text{ with } y(t_0) = y_0 .$$

Given a positive integer s, a positive timestep h and scalars

$$0 \leq \tau_0 < \tau_1 < \ldots < \tau_s \leq 1 ,$$

the corresponding collocation method finds a vector-valued polynomial $\widetilde{y}(t)$ of degree at most s so that

$$\widetilde{y}(t_0) = y_0 \text{ and} \tag{3.56}$$

$$\widetilde{y}'(t_0 + \tau_i h) = f(\widetilde{y}(t_0 + \tau_i h)) \text{ for } 1 \leq i \leq s . \tag{3.57}$$

Since the collocation polynomial derivative $\widetilde{\mathbf{y}}'$ is a polynomial of degree at most $s-1$ that satisfies s interpolation conditions, we can write it in terms of its Lagrange interpolation polynomial. Let

$$\lambda_j(\tau) = \prod_{\substack{1 \leq \ell \leq s \\ \ell \neq j}} \frac{\tau - \tau_\ell}{\tau_j - \tau_\ell}$$

be the Lagrange interpolation basis polynomial, as described in Sect. 1.2.3. Then the interpolation conditions imply that

$$\widetilde{\mathbf{y}}'(t) = \sum_{j=1}^{s} \widetilde{\mathbf{y}}'(t_0 + \tau_j h) \lambda_j \left(\frac{t - t_0}{h} \right) = \sum_{j=1}^{s} \mathbf{f}\left(\widetilde{\mathbf{y}}(t_0 + \tau_j h) \right) \lambda_j \left(\frac{t - t_0}{h} \right) .$$

We can integrate to obtain

$$\widetilde{\mathbf{y}}(t_0 + \tau_i h) = \mathbf{y}_0 + \int_{t_0}^{t_0 + \tau_i h} \widetilde{\mathbf{y}}'(t) \, dt$$

$$= \mathbf{y}_0 + \int_{t_0}^{t_0 + \tau_i h} \sum_{j=1}^{s} \mathbf{f}\left(\widetilde{\mathbf{y}}(t_0 + \tau_j h) \right) \lambda_j \left(\frac{t - t_0}{h} \right) dt$$

$$= \mathbf{y}_0 + h \sum_{j=1}^{s} \mathbf{f}\left(\widetilde{\mathbf{y}}(t_0 + \tau_j h) \right) \int_0^{\tau_i} \lambda_j(\tau) \, d\tau .$$

This equation suggests that we define the scalars

$$\alpha_{ij} = \int_0^{\tau_i} \lambda_j(\tau) \, d\tau \text{ for } 1 \leq i, j \leq s \text{ and}$$

$$\beta_j = \int_0^1 \lambda_j(\tau) \, d\tau \text{ for } 1 \leq j \leq s$$

and the vectors

$$\mathbf{f}_j = \mathbf{f}\left(\widetilde{\mathbf{y}}(t_0 + \tau_j h) \right) \text{ for } 1 \leq j \leq s .$$

Then for $1 \leq i \leq s$ we have

$$\mathbf{f}_i = \mathbf{f}\left(\widetilde{\mathbf{y}}(t_0 + \tau_i h) \right) = \mathbf{f}\left(\mathbf{y}_0 + h \sum_{j=1}^{s} \alpha_{ij} \mathbf{f}\left(\widetilde{\mathbf{y}}(t_0 + \tau_j h) \right) \right) = \mathbf{f}\left(\mathbf{y}_0 + h \sum_{j=1}^{s} \alpha_{ij} \mathbf{f}_j \right)$$

and

$$\widetilde{\mathbf{y}}(t_0 + h) = \mathbf{y}_0 + h \sum_{j=1}^{s} \beta_j \mathbf{f} \left(\widetilde{\mathbf{y}}(t_0 + \tau_j h) \right) = \mathbf{y}_0 + h \sum_{j=1}^{s} \beta_j \mathbf{f}_j .$$

This shows that collocation is equivalent to an s-stage implicit Runge-Kutta scheme. The next lemma provides an error estimate for collocation.

Lemma 3.6.2 *Let t_0 be a time and $\mathbf{y}_0$ be a vector. Assume that the function $\mathbf{f}$ mapping vectors to vectors is Lipschitz continuous with Lipschitz constant Φ. Suppose that the solution $\mathbf{y}$ of the initial value problem*

$$\mathbf{y}'(t) = \mathbf{f}(\mathbf{y}(t)) \text{ for } t > t_0 \text{ with } \mathbf{y}(t_0) = \mathbf{y}_0$$

has a bounded derivative of order $s + 1$ on $[t_0, t_0 + h]$. Let $0 \leq \tau_1 < \ldots < \tau_s \leq 1$ be distinct scalars, and define the Lagrange interpolation basis polynomials

$$\lambda_j(\tau) = \prod_{\substack{1 \leq \ell \leq s \\ \ell \neq j}} \frac{\tau - \tau_\ell}{\tau_j - \tau_\ell}$$

for $1 \leq j \leq s$. Assume that $\widetilde{\mathbf{y}}$ is the collocation polynomial defined by the conditions

$$\widetilde{\mathbf{y}}(t_0) = \mathbf{y}_0 \text{ and}$$

$$\widetilde{\mathbf{y}}'(t_0 + \tau_i h) = \mathbf{f} \left(\widetilde{\mathbf{y}}(t_0 + \tau_i h) \right) \text{ for } 1 \leq i \leq s .$$

If the timestep h satisfies

$$2h\Phi \sum_{j=1}^{s} \int_0^1 \left| \lambda_j(\tau) \right| \, d\tau < 1 ,$$

then the error in the collocation method satisfies

$$\max_{0 \leq \tau \leq 1} \| \mathbf{y}(t_0 + \tau h) - \widetilde{\mathbf{y}}(t_0 + \tau h) \|_\infty \leq \frac{2}{s!} h^{s+1} \left\{ \max_{t \in [t_0, t_0 + h]} \| D^{s+1} \mathbf{y}(t) \|_\infty \right\}$$

$$\sum_{j=1}^{s} \int_0^\tau \left| \lambda_j(\sigma) \right| \, d\sigma .$$

Proof The Lagrange polynomial interpolant to $\mathbf{y}'$ is

$$\widetilde{\mathbf{f}}(t) = \sum_{j=1}^{s} \mathbf{f} \left(\mathbf{y}(t_0 + \tau_j h) \right) \lambda_j \left(\frac{t - t_0}{h} \right) ,$$

and Lemma 1.2.2 shows that the error in polynomial interpolation satisfies

$$\left\|\mathbf{y}'(t_0 + \tau h) - \widetilde{\mathbf{f}}(t_0 + \tau h)\right\|_\infty \leq \frac{h^s}{s!} \max_{t \in [t_0, t_0 + h]} \left\|D^{s+1}\mathbf{y}(t)\right\|_\infty .$$

The initial value problem and collocation conditions imply that for all $\tau \in [0, 1]$

$$\mathbf{y}(t_0 + \tau h) - \widetilde{\mathbf{y}}(t_0 + \tau h) = h \int_0^\tau \left[\mathbf{f}\left(\mathbf{y}(t_0 + \sigma h)\right) - \sum_{j=1}^s \mathbf{f}\left(\widetilde{\mathbf{y}}(t_0 + \tau_j h)\right) \lambda_j(\sigma) \right] d\sigma$$

then we can add and subtract the interpolant to $\mathbf{y}'$ to get

$$= h \int_0^\tau \left[\widetilde{\mathbf{f}}\left(\mathbf{y}(t_0 + \sigma h)\right) - \sum_{j=1}^s \mathbf{f}\left(\widetilde{\mathbf{y}}(t_0 + \tau_j h)\right) \lambda_j(\sigma) \right] d\sigma$$

$$+ h \int_0^\tau \left[\mathbf{f}\left(\mathbf{y}(t_0 + \sigma h)\right) - \widetilde{\mathbf{f}}\left(\mathbf{y}(t_0 + \sigma h)\right) \right] d\sigma$$

$$= h \sum_{j=1}^s \left[\mathbf{f}\left(\mathbf{y}(t_0 + \tau_j h)\right) - \mathbf{f}\left(\widetilde{\mathbf{y}}(t_0 + \tau_j h)\right) \right] \int_0^\tau \lambda_j(\sigma) \, d\sigma$$

$$+ h \int_0^\tau \left[\mathbf{y}'(t_0 + \sigma h) - \widetilde{\mathbf{f}}\left(\mathbf{y}(t_0 + \sigma h)\right) \right] d\sigma .$$

We can take norms of both sides to obtain

$$\max_{\tau \in [0,1]} \left\|\mathbf{y}(t_0 + \tau h) - \widetilde{\mathbf{y}}(t_0 + \tau h)\right\|_\infty$$

$$\leq h \sum_{j=1}^s \left\| \mathbf{f}\left(\mathbf{y}(t_0 + \tau_j h)\right) - \mathbf{f}\left(\widetilde{\mathbf{y}}(t_0 + \tau_j h)\right) \right\| \int_0^\tau \left|\lambda_j(\sigma)\right| \, d\sigma$$

$$+ h \int_0^\tau \left\|\mathbf{y}'(t_0 + \sigma h) - \widetilde{\mathbf{f}}\left(\mathbf{y}(t_0 + \sigma h)\right)\right\|_\infty d\sigma$$

then we use Lipschitz continuity and the interpolation error estimate to bound

$$\leq \Phi h \sum_{j=1}^s \left\|\mathbf{y}(t_0 + \tau_j h) - \widetilde{\mathbf{y}}(t_0 + \tau_j h)\right\| \int_0^\tau \left|\lambda_j(\sigma)\right| \, d\sigma$$

$$+ \frac{h^{s+1}}{s!} \left\{ \max_{t \in [t_0, t_0 + h]} \left\|D^{s+1}\mathbf{y}(t)\right\|_\infty \right\} \sum_{j=1}^s \int_0^\tau \left|\lambda_j(\sigma)\right| \, d\sigma$$

The assumed bound on h allows us to solve this inequality and obtain the claimed result.

Of course, we still need to determine the interpolation points τ_j for $1 \leq j \leq s$. To do so, we will relate the order conditions for collocation to order conditions for numerical quadrature. If collocation has order p for the given initial value problem, then it must be exact for all initial value problems with solutions $y(t + \tau h) = \tau^k$ and right-hand sides $f(y(t + \tau h)) = k\tau^{k-1}$ for $0 \leq k \leq p$. It is easy to see that the corresponding right-hand side is

$$t(y(t + \tau h)) = y'(t) = \frac{d\tau^k}{d\tau}\frac{d\tau}{dt} = \frac{k}{h}\tau_{k-1}$$

for $1 \leq k \leq p$. The collocation interpolation conditions imply that for $1 \leq k \leq p$ we have

$$y(t_0 + h) = 1^k = h\sum_{j=1}^{s}\beta_j f(y(t + \tau_j h)) = k\sum_{j=1}^{s}\beta_j \tau_j^{k-1} .$$

In other words, the collocation method has order p if and only if

$$\int_0^1 \tau^{k-1}\, d\tau = \sum_{j=1}^{s}\beta_j\tau_j^{k-1}$$

for all $1 \leq k \leq p$. Gaussian quadrature satisfies these conditions for the highest possible order $p = 2s$, and Theorem 2.3.4 shows that the Gaussian quadrature weights are given by the integrals β_j of the Lagrange interpolation polynomials. This discussion shows that the optimal choice for the collocation points τ_j are the zeros of the Legendre polynomial of degree s on $[0, 1]$.

For an alternative discussion of the connection between collocation and implicit Runge-Kutta schemes, see Hairer et al. [94, p. 206ff].

Exercise 3.6.7 What quadrature rule corresponds to the collocation method associated with the backward Euler method?

Exercise 3.6.8 What Gaussian quadrature rule corresponds to the collocation method associated with the implicit midpoint method?

Exercise 3.6.9 Show that the Hammer-Hollingsworth method is a collocation method, and find the associated Gaussian quadrature rule.

Exercise 3.6.10 The **Kuntzmann-Butcher method** is the three-stage implicit Runge-Kutta scheme corresponding to collocation at the zeros of the third-order Legendre polynomial on $[0, 1]$.

1. By finding the nodes and weights for Gaussian quadrature using three quadrature points, show that the Kuntzmann-Butcher method is

$$\mathbf{f}_1 = \mathbf{f}\left(\mathbf{y}_n + h\left[\frac{5}{36}\mathbf{f}_1 + \left\{\frac{2}{9} - \frac{\sqrt{15}}{15}\right\}\mathbf{f}_2 + \left\{\frac{5}{36} - \frac{\sqrt{15}}{30}\right\}\mathbf{f}_3\right]\right)$$

$$\mathbf{f}_2 = \mathbf{f}\left(\mathbf{y}_n + h\left[\left\{\frac{5}{36} + \frac{\sqrt{15}}{24}\right\}\mathbf{f}_1 + \frac{2}{9}\mathbf{f}_2 + \left\{\frac{5}{36} - \frac{\sqrt{15}}{24}\right\}\mathbf{f}_3\right]\right)$$

$$\mathbf{f}_3 = \mathbf{f}\left(\mathbf{y}_n + h\left[\left\{\frac{5}{36} + \frac{\sqrt{15}}{30}\right\}\mathbf{f}_1 + \left\{\frac{2}{9} - \frac{\sqrt{15}}{25}\right\}\mathbf{f}_2 + \frac{5}{36}\mathbf{f}_3\right]\right)$$

$$\mathbf{y}_{n+1} = \mathbf{y}_n + \frac{h}{18}\left[5\mathbf{f}_1 + 8\mathbf{f}_2 + 5\mathbf{f}_3\right] \ .$$

2. Describe the region of absolutely stability for this method.

3.6.4   Error Estimation

As we saw in Sect. 3.4.11, linear multistep methods can compare two numerical methods of successive orders to estimate the local error and select a suitable timestep. For the Runge-Kutta methods we have examined so far, a more suitable approach would be to estimate the error via Richardson extrapolation (see Sect. 2.2.4), and use the estimated error to select a new timestep. Suppose that we integrate an initial value problem using a timestep of size h via a method of order p, thereby producing an approximate solution $\widetilde{\mathbf{y}}(h)$ at time $t+h$. We could use the same method to take two timesteps of size $h/2$ and produce a presumably more accurate approximate solution $\widetilde{\mathbf{y}}(h)$ at time $t + h$. Then we have

$$\widetilde{\mathbf{y}}(h) - \mathbf{y}(t + h) = \mathbf{c}h^{p+1} + O\left(h^{p+2}\right) \text{ and}$$

$$\widetilde{\mathbf{y}}(h/2) - \mathbf{y}(t + h) = \mathbf{c}(h/2)^{p+1} + O\left(h^{p+2}\right) \ .$$

We can subtract the latter equation from the former to see that

$$\widetilde{\mathbf{y}}(h) - \widetilde{\mathbf{y}}(h/2) = \mathbf{c}\left(1 - 2^{-p-1}\right)h^{p+1} + O\left(h^{p+2}\right) \ .$$

We can solve for $\mathbf{c}h^{p+1}$ to see that the local error in $\widetilde{\mathbf{y}}(h)$ is

$$\widetilde{\mathbf{y}}(h) - \mathbf{y}(t + h) = \frac{\widetilde{\mathbf{y}}(h) - \widetilde{\mathbf{y}}(h/2)}{1 - 2^{-p-1}} + O\left(h^{p+2}\right) \ .$$

If we had used a timestep of size γh, we would expect the local error to be

$$\widetilde{\mathbf{y}}(\gamma h) - \mathbf{y}(t + h) = \frac{\widetilde{\mathbf{y}}(h) - \widetilde{\mathbf{y}}(h/2)}{1 - 2^{-p-1}} \gamma^{p+1} + O\left(h^{p+2}\right) .$$

If we want this local error to have norm ε, then we should use a timestep of size γh where

$$\gamma^{p+1} = \frac{\varepsilon}{\|\widetilde{\mathbf{y}}(h) - \widetilde{\mathbf{y}}(h/2)\|} \left(1 - 2^{-p-1}\right) .$$

This approach is also discussed by Hairer et al. [94, p. 165f] and by Lambert [124, p. 183].

The difficulty with Richardson extrapolation for error estimation and timestep selection is the computational cost. If our Runge-Kutta method uses s stages, then we require $3s$ function evaluations to evaluate both $\widetilde{\mathbf{y}}(h)$ and $\widetilde{\mathbf{y}}(h/2)$. As suggested by Merson [133], special Runge-Kutta methods can be designed to provide an estimate of the local truncation error at the cost of a single additional function evaluation. Such methods compute two numerical solutions of successive orders, using shared function evaluations.

Suppose that we integrate an initial value problem using a timestep of size h via a method of order p, thereby producing an approximate solution $\mathbf{y}_{n+1}(h)$ at time $t_n + h$. We could use the higher-order method to take a timestep of size h and produce an approximate solution $\widetilde{\mathbf{y}}_{n+1}(h)$ at time $t_n + h$. Then we have

$$\mathbf{y}_{n+1}(h) - \mathbf{y}(t + h) = \mathbf{c}_1 h^{p+1} + O\left(h^{p+2}\right) \text{ and}$$

$$\widetilde{\mathbf{y}}_{n+1}(h) - \mathbf{y}(t + h) = \mathbf{c}_2 h^{p+2} + O\left(h^{p+3}\right) .$$

We can subtract the latter equation from the former to see that

$$\mathbf{y}_{n+1}(h) - \widetilde{\mathbf{y}}_{n+1}(h) = \mathbf{c}_1 h^{p+1} + O\left(h^{p+2}\right) .$$

We can solve for $\mathbf{c}_1 h^{p+1}$ to see that the local error in $\widetilde{\mathbf{y}}_1(h)$ is

$$\mathbf{y}_{n+1}(h) - \mathbf{y}(t + h) = \mathbf{y}_{n+1}(h) - \widetilde{\mathbf{y}}_{n+1}(h) + O\left(h^{p+2}\right) .$$

If we had used a timestep of size γh, we would expect the local error to be

$$\mathbf{y}_{n+1}(\gamma h) - \mathbf{y}(t + \gamma h) = [\mathbf{y}_{n+1}(h) - \widetilde{\mathbf{y}}_{n+1}(h)] \gamma^{p+1} + O\left(h^{p+2}\right) .$$

If we want this local error to have norm ε, then we should use a timestep of size γh where

$$\gamma^{p+1} = \frac{\varepsilon}{\|\mathbf{y}_{n+1}(h) - \widetilde{\mathbf{y}}_{n+1}(h)\|} .$$

This approach is also discussed by Hairer et al. [94, pp. 166–174] and by Lambert [124, pp. 183–189].

Our first example of an embedded Runge-Kutta method is the **Runge-Kutta-Fehlberg 2–3 method** [75], which can be implemented as follows:

$$\mathbf{f}_1 = \mathbf{f}(\mathbf{y}_n)$$

$$\mathbf{f}_2 = \mathbf{f}(\mathbf{y}_n + h\mathbf{f}_1)$$

$$\mathbf{f}_3 = \mathbf{f}\left(\mathbf{y}_n + \frac{h}{4}[\mathbf{f}_1 + \mathbf{f}_2]\right)$$

$$\mathbf{y}_{n+1} = \mathbf{y}_n + \frac{h}{2}[\mathbf{f}_1 + \mathbf{f}_2]$$

$$\widetilde{\mathbf{y}}_{n+1} = \mathbf{y}_n + \frac{h}{6}[\mathbf{f}_1 + \mathbf{f}_2 + 4\mathbf{f}_3] .$$

Here $\mathbf{y}_{n+1}$ is second-order accurate, and $\widetilde{\mathbf{y}}_{n+1}$ is third-order accurate. The associated Butcher array is

$$
\begin{array}{c|ccc}
0 & & & \\
1 & 1 & & \\
1/2 & 1/4 & 1/4 & \\
\hline
& 1/2 & 1/2 & 0 \\
& 1/6 & 1/6 & 2/3
\end{array}
$$

GNU Scientific Library users can access this method via the C routine gsl_odeiv2_step_rk2.

England's method [72] has modified Butcher array

$$
\begin{array}{c|cccccc}
0 & & & & & & \\
1/2 & 1/2 & & & & & \\
1/2 & 1/4 & 1/4 & & & & \\
1 & 0 & -1 & 2 & & & \\
2/3 & 7/27 & 10/27 & 0 & 1/27 & & \\
1/5 & 28/625 & -1/5 & 546/625 & 54/625 & -378/625 & \\
\hline
& 1/6 & 0 & 2/3 & 1/6 & 0 & 0 \\
& 1/24 & 0 & 0 & 5/48 & 27/56 & 125/336
\end{array}
$$

In this method, $\mathbf{y}_{n+1}$ is fourth-order accurate, and $\widetilde{\mathbf{y}}_{n+1}$ is fifth-order accurate. Since England's method is explicit, the Butcher Barrier Theorem 3.6.1 proves that it was necessary to use 6 stages to obtain the fifth-order scheme. The design feature of England's method is that the fourth-order scheme does not use the last two function values. Consequently, we can obtain substantial savings by not estimating the local truncation error at each step of this method.

The most popular of the fourth-/fifth-order embedded Runge-Kutta methods is Fehlberg's **RKF45** [74, 75]. This method has modified Butcher array

$$
\begin{array}{c|cccccc}
0 \\
1/4 & 1/4 \\
3/8 & 3/32 & 9/32 \\
12/13 & 1932/2197 & -7200/2197 & 7296/2197 \\
1 & 439/216 & -8 & 3680/513 & -845/4104 \\
1/2 & -8/27 & 2 & -3544/2565 & 1859/4104 & -11/40 \\
\hline
& 25/216 & 0 & 1408/2565 & 2197/4104 & -1/5 & 0 \\
& 16/135 & 0 & 6656/12825 & 28561/56430 & -9/50 & 2/5
\end{array}
$$

RKF45 has been designed so that the error coefficient for the lower order method is minimized. A Fortran implementation of this method by Watts and Shampine is available from Netlib as rkf45. GNU Scientific Library users can access this method via the C routine gsl_odeiv2_step_rk45. MATLAB users can call command ode45.

Dormand and Prince have developed several Runge-Kutta schemes with error estimation. Their fourth- and fifth-order pair [68] has modified Butcher array

$$
\begin{array}{c|ccccccc}
0 \\
1/5 & 1/5 \\
3/10 & 3/40 & 9/40 \\
4/5 & 44/45 & -36/15 & 32/9 \\
8/9 & 19372/6561 & -25360/2187 & 64448/6561 & -212/729 \\
1 & 9017/3168 & -355/33 & 46732/5247 & 49/176 & -5103/18656 \\
1 & 35/384 & 0 & 500/1113 & 125/192 & -2187/6784 & 11/64 \\
\hline
& 5179/57600 & 0 & 7571/16695 & 393/640 & -92097/339200 & 187/2100 & 1/40 \\
& 35/384 & 0 & 5000/1113 & 125/192 & -2187/6784 & 11/64 & 0
\end{array}
$$

This scheme has been designed so that the error coefficient for the *higher order* method is minimized. The authors use the combination of the lower-order and higher-order schemes to adjust the step-length, and extrapolation of the higher-order method to estimate the local error. Unlike the other fourth-order, fifth-order Runge-Kutta pairs, the Dormand-Prince method uses seven stages. However, the function at the seventh stage is evaluated at the same point as the higher-order approximation (i.e., the seventh and ninth rows of the Butcher array are the same). Prince and Dormand [145] also provide a fifth- and sixth-order pair, and a seventh- and eighth-order pair. GNU Scientific Library users can access the latter method via the C routine gsl_odeiv2_step_rk8pd.

Another six-stage Runge-Kutta scheme with error estimation is due to Cash and Karp [33]. This method has modified Butcher array

$$
\begin{array}{c|cccccc}
0 \\
1/5 & 1/5 \\
3/10 & 3/40 & 9/40 \\
3/5 & 3/10 & -9/10 & 6/5 \\
1 & -11/54 & 5/2 & -70/27 & 35/27 \\
7/8 & 1631/55296 & 175/512 & 575/13824 & 44275/110592 & 253/4096 \\
\hline
 & 1 & 0 & 0 & 0 & 0 & 0 \\
 & -3/2 & 5/2 & 0 & 0 & 0 & 0 \\
 & 19/54 & 0 & -10/27 & 55/54 & 0 & 0 \\
 & 2825/27648 & 0 & 18575/48384 & 13525/55296 & 277/14336 & 1/4 \\
 & 37/378 & 0 & 250/621 & 125/594 & 0 & 512/1771
\end{array}
$$

The design feature of the Cash-Karp method is that it produces numerical solutions of order 1 through 5, corresponding to the last 5 lines of the modified Butcher array. If the solutions of order 1 and 2 indicate small local error, then this method computes the solution of order 3. This process can be continued up to order 5. In this sense, the Cash-Karp method has features similar to linear multistep methods, with ability to vary order and step size. GNU Scientific Library users can access this method via the C routine gsl_odeiv2_step_rkck.

There are several publicly available programs for solving initial value problems via Runge-Kutta methods, such as Brankin, Gladwell and Shampine have written the Fortran program rksuite, which provides a choice of a $(2, 3)$, a $4, 5$ or a $(7, 8)$ embedded Runge-Kutta method. Jackson, Hull, and Enright have written the Fortran program dverk, which uses a fifth- and sixth-order embedded Runge-Kutta method due to Verner [179].

Exercise 3.6.11 Select an appropriate Runge-Kutta method to solve the pendulum problem described in the first exercise of the set 3.4.16. Describe how the timestep varies with the evolution of the solution and the required accuracy.

Exercise 3.6.12 Select an appropriate Runge-Kutta method to solve the orbit equations described in the third exercise of the set 3.4.16. Describe how the timestep varies with the evolution of the solution and the required accuracy.

3.7 Stiffness

Some initial value problems have a feature that can cause serious numerical difficulties. These problems have some modes that decay very rapidly, thereby restricting the stable timesteps for methods with finite regions of absolute stability. To make matters worse, these problems also have some modes that decay much more slowly, requiring the numerical method to perform a lengthy simulation. The fast decay modes then require a large number of small timesteps to reach the final simulation time. Such problems are called **stiff**.

In this section, we will examine the nature of stiffness and describe various numerical methods to solve stiff initial value problems.

3.7.1 Problems

Consider the system of linear ordinary differential equations

$$\mathbf{y}'(t) = \mathbf{A}\mathbf{y}(t) + \mathbf{b}(t) .$$

Theorem 3.2.5 shows us that the analytical solution of this initial value problem is

$$\mathbf{y}(t) = e^{\mathbf{A}t}\mathbf{y}(0) + \int_0^t e^{\mathbf{A}(t-s)}\mathbf{b}(s)ds .$$

Note that if all of the eigenvalues of $\mathbf{A}$ have negative real part, then the term involving $\mathbf{y}(0)$ is a transient that decays as $t \to \infty$. In particular, if some eigenvalue λ_i of $\mathbf{A}$ has very negative real part, then some component of the initial data may decay very rapidly. As we have already seen, the eigenvalue λ_i with most negative real part controls the stable timestep of a numerical method for solving this problem, because we need $h\lambda_i$ to lie in the region of absolute stability. On the other hand, the eigenvalue with least negative real part controls the integration interval, since we might have to run the computation until its transient dies out.

These observations motivate the following definition.

Definition 3.7.1 The initial value problem

$$\mathbf{y}'(t) = \mathbf{f}(\mathbf{y}(t))$$

is said to be **stiff** if and only if some eigenvalues of $\frac{\partial \mathbf{f}}{\partial \mathbf{y}}$ have negative real parts of very different magnitude.

Stiffness becomes a problem when stability controls the stepsize in numerical method, rather than accuracy.

Example 3.7.1 Suppose that we want to solve

$$\frac{d}{dt}\begin{bmatrix} \mathbf{y}_1 \\ \mathbf{y}_2 \end{bmatrix} = \begin{bmatrix} 0 & 1 \\ -1000 & -1001 \end{bmatrix}\begin{bmatrix} \mathbf{y}_1 \\ \mathbf{y}_2 \end{bmatrix} , \quad \begin{bmatrix} \mathbf{y}_1 \\ \mathbf{y}_2 \end{bmatrix}(0) = \begin{bmatrix} 1 \\ -1 \end{bmatrix} .$$

Here the matrix

$$A = \begin{bmatrix} 0 & 1 \\ -1000 & -1001 \end{bmatrix}$$

has matrices of eigenvectors and eigenvalues

$$X = \begin{bmatrix} 1 & 1 \\ -1 & 1000 \end{bmatrix} \text{ and } \Lambda = \begin{bmatrix} -1 & 0 \\ 0 & -1000 \end{bmatrix},$$

respectively. Thus the solution of the initial value problem is

$$\begin{bmatrix} y_1 \\ y_2 \end{bmatrix}(t) = \exp(At) \begin{bmatrix} 1 \\ -1 \end{bmatrix}$$

$$= X \exp(\Lambda t) X^{-1} \begin{bmatrix} 1 \\ -1 \end{bmatrix} = \begin{bmatrix} 1 & 1 \\ -1 & 1000 \end{bmatrix} \begin{bmatrix} e^{-t} & 0 \\ 0 & e^{-1000t} \end{bmatrix} \begin{bmatrix} 1 \\ 0 \end{bmatrix} = \begin{bmatrix} 1 \\ -1 \end{bmatrix} e^{-t}.$$

However, if the initial data is perturbed slightly, the solution involves a rapidly decaying exponential. If we use Euler's method with varying numbers of timesteps to compute the solution at $t = 1$, we get the following results:

h	$y_1(t)$	$y_2(t)$
10^0	0.	0.
10^{-1}	3.377×10^8	-3.377×10^{11}
10^{-2}	*overflow*	*overflow*
10^{-3}	0.3677	-0.3677

The problem is that Euler's method for $y' = -\lambda y$ requires $h < 2/\lambda$ to guarantee no growth in the numerical solution; in this case one of the eigenvalues of A is -10^3, so we require $h < 2 \times 10^{-3}$. For larger values of h, small perturbations in the rapidly decaying mode are amplified by a factor of $|1 - \lambda h|$ with each step, or $(1 - \lambda T/N)^N$ for N timesteps. For the choices of h in this example, we have

$h = 1/N$	$(1 - \lambda/N)^N$
10^0	$-999 \approx 10^3$
10^{-1}	$99^{10} \approx 10^{20}$
10^{-2}	$9^{100} \approx 10^{95}$
10^{-3}	$0.9^{1000} \approx 10^{-46}$

In numerical computations, a symptom of stiffness may be wild oscillations that get worse as the timestep is reduced, until the timestep is made sufficiently small.

3.7.2 Multistep Methods

One approach to solving stiff problems is to use A-stable methods, since they include the entire left half-plane inside their absolute stability region. If we choose to use a linear multistep methods to solve a stiff initial value problem, then we must remember the second Dahlquist barrier Theorem 3.4.4, which states that A-stable linear multistep methods have order at most two and cannot be explicit. Backward differentiation formulas (BDFs), which were discussed in Sect. 3.4.7.4, have relatively large regions of absolute stability when compared to Adams family methods. Consequently, the Brown, Hindmarsh and Byrne program vode and the Cohen and Hindmarsh program cvode both suggest the use of BDFs for stiff problems.

3.7.3 Runge-Kutta Methods

Among Runge-Kutta methods, the s-stage fully implicit methods of order $2s$, which were discussed in Sect. 3.6.3, are absolutely stable. Unfortunately, these methods also require the solution of large systems of nonlinear equations for higher-order methods.

More recently, some new Runge-Kutta methods have been developed following a reconsideration of stability properties. In order to understand the stability of Runge-Kutta methods, we will present the following definition.

Definition 3.7.2 Suppose that the general Runge-Kutta scheme (3.51) is applied to the differential equation $y'(t) = \lambda y(t)$ at some step with $y_n \neq 0$. Then the **amplification factor** for the scheme is y_{n+1}/y_n.

Let us find a formula for the amplification factor. For a Runge-Kutta scheme with Butcher arrays $\mathbf{A}$ and $\mathbf{b}$, the array $\mathbf{z}$ of vectors in Eq. (3.51b) satisfies the linear system

$$\left[\mathbf{I} - h\lambda\mathbf{A}\right]\mathbf{z} = \mathbf{e}y_n .$$

Here $\mathbf{e}$ is the vector of ones. Next, note that Eq. (3.51a) gives us

$$y_{n+1} = y_n + h\lambda\mathbf{b}^\top\mathbf{z} = \left\{1 + h\lambda\mathbf{b}^\top\left[\mathbf{I} - h\lambda\mathbf{A}\right]^{-1}\mathbf{e}\right\}y_n .$$

It follows that the amplification factor for general Runge-Kutta schemes is

$$R(h\lambda) = 1 + h\lambda\mathbf{b}^\top\left[\mathbf{I} - h\lambda\mathbf{A}\right]^{-1}\mathbf{e} .$$

For the system of differential equations

$$\mathbf{y}'(t) = \mathbf{f}(\mathbf{y}(t)) ,$$

a perturbed solution

$$\widetilde{\mathbf{y}}(t) = \mathbf{y}(t) + \boldsymbol{\epsilon}(t)$$

will be such that the perturbation satisfies

$$\boldsymbol{\epsilon}'(t) \approx \frac{\partial \mathbf{f}}{\partial \mathbf{y}}(\mathbf{y}(t))\boldsymbol{\epsilon}(t) .$$

If the Jacobian of $\mathbf{f}$ is diagonalizable

$$\frac{\partial \mathbf{f}}{\partial \mathbf{y}}(\mathbf{y}(t))\mathbf{X} = \mathbf{X}\boldsymbol{\Lambda} ,$$

then the corresponding perturbations in the numerical method will satisfy

$$\boldsymbol{\epsilon}_{n+1} \approx \mathbf{X}R(\boldsymbol{\Lambda}h)\mathbf{X}^{-1}\boldsymbol{\epsilon}_n .$$

If the original initial value problem is stiff, then $\boldsymbol{\Lambda}$ has some eigenvalues with vastly different negative real parts. Assuming that the very fast decay modes are not of interest, the following definition becomes useful.

Definition 3.7.3 A Runge-Kutta scheme is **L-stable** if and only if its amplification factor $R(h\lambda)$ tends to zero as $h\lambda$ tends to minus infinity.

Practically speaking, L-stability guarantees that a Runge-Kutta method produces decaying numerical solutions for linear problems with rapid decay. An L-stable Runge Kutta method can select the timestep for accuracy considerations related to the decay modes with relatively small negative real parts. Modes with very large negative real part will have very small amplification factors for L-stable methods, and their contribution to the numerical results will be greatly reduced.

We will return to L-stability in the design of Runge-Kutta methods shortly. Our next objective is to discuss the work involved in solving systems of nonlinear equations in implicit Runge-Kutta methods.

In order to reduce the work in solving the nonlinear equations in implicit Runge-Kutta methods, Dekker and Verwer [63] have suggested the use of **singly diagonally implicit Runge-Kutta (SDIRK) methods**. For the autonomous initial value problem

$$\mathbf{y}'(t) = \mathbf{f}(\mathbf{y}(t)) \text{ for } t > t_0 \text{ with } \mathbf{y}(t_0) = \mathbf{y}_0 ,$$

the two-stage SDIRK scheme has the Butcher array

$$
\begin{array}{c|cc}
\alpha & \alpha & 0 \\
1-\alpha & 1-2\alpha & \alpha \\
\hline
& 1/2 & 1/2
\end{array}
$$

Lemma 3.6.1 shows that such a scheme has order at least two for all choices of α. The SDIRK scheme can be implemented as follows:

$$\mathbf{f}_1 = \mathbf{f}(\mathbf{y}_n + \mathbf{f}_1\alpha h) \tag{3.58a}$$

$$\mathbf{f}_2 = \mathbf{f}(\mathbf{y}_n + [1 - 2\alpha]\mathbf{f}_1 h + \alpha\mathbf{f}_2 h) \tag{3.58b}$$

$$\mathbf{y}_{n+1} = \mathbf{y}_n + (\mathbf{f}_1 + \mathbf{f}_2)\frac{h}{2} . \tag{3.58c}$$

Simplified Newton iterations for the vectors $\mathbf{f}_1$ and $\mathbf{f}_2$ might take the forms

$$\left[\mathbf{I} - \frac{\partial \mathbf{f}}{\partial \mathbf{y}}(\mathbf{y}_n)\alpha h\right] \Delta\mathbf{f}_1 = \mathbf{f}_1 - \mathbf{f}(\mathbf{y}_n + \mathbf{f}_1\alpha h) \text{ and}$$

$$\left[\mathbf{I} - \frac{\partial \mathbf{f}}{\partial \mathbf{y}}(\mathbf{y}_n)\alpha h\right] \Delta\mathbf{f}_2 = \mathbf{f}_2 - \mathbf{f}(\mathbf{y}_n + \mathbf{f}_1[1 - \alpha]h + \mathbf{f}_2\alpha h) .$$

Both of these simplified Newton iterations involve the same matrix, which could be factored once and used repeatedly. This is the design principle of the SDIRK schemes.

For the SDIRK scheme above, the amplification factor is

$$R(h\lambda) = 1 + h\lambda\frac{1}{2}\begin{bmatrix} 1 & 1 \end{bmatrix}\begin{bmatrix} 1 - h\lambda\alpha & 0 \\ -h\lambda(1 - 2\alpha) & 1 - h\lambda\alpha \end{bmatrix}^{-1}\begin{bmatrix} 1 \\ 1 \end{bmatrix}$$

$$= 1 + \frac{h\lambda}{2(1 - h\lambda\alpha)^2}\begin{bmatrix} 1 & 1 \end{bmatrix}\begin{bmatrix} 1 - h\lambda\alpha & 0 \\ h\lambda(1 - 2\alpha) & 1 - h\lambda\alpha \end{bmatrix}\begin{bmatrix} 1 \\ 1 \end{bmatrix}$$

$$= 1 + \frac{h\lambda}{2(1 - h\lambda\alpha)^2} \{2(1 - h\lambda\alpha) + h\lambda(1 - 2\alpha)\} . \tag{3.59}$$

As $h\lambda \to -\infty$, we have

$$R(h\lambda) \to 1 - \frac{1}{\alpha} + \frac{1 - 2\alpha}{2\alpha^2} = \left(1 - \frac{1 + \sqrt{1/2}}{\alpha}\right)\left(1 - \frac{1 - \sqrt{1/2}}{\alpha}\right) .$$

Thus the SDIRK scheme is L-stable whenever $\alpha = 1 \pm \sqrt{1/2}$. We will choose $\alpha = 1 - \sqrt{1/2}$.

To show that the SDIRK scheme is A-stable, we will use the following argument. If $\zeta = h\lambda$, define the linear fractional transformation

$$w(\zeta) = \frac{\zeta}{1 - \alpha\zeta} .$$

It is easy to see that w maps the left half-plane into the interior of the circle $|2\alpha w + 1| \leq 1$. If $w = (\zeta - 1)/(2\alpha)$ where $\zeta = \cos(\psi) + i\sin(\psi)$ then it is straightforward to see that $\alpha = 1 - \sqrt{1/2}$ implies that

$$|R|^2 = \frac{5}{8} + \frac{1}{2}\cos\psi - \frac{1}{8}\cos 2\psi .$$

The extreme points of this function occur at $\psi = 0, \pi$ and 2π, where $|R| = 1, 0$ and 1, respectively. This shows that $|R(\zeta)|^2 \leq 1$ for all ζ satisfying $|2\alpha w(\zeta) + 1| \leq 1$, which in turn shows that R maps the left half-plane into a subset of the unit circle.

A discussion of more general diagonally implicit Runge-Kutta schemes can be found in Alexander [4]. Also, Jawias and Ismail [114] have written a book that describes the design of diagonally implicit Runge-Kutta schemes of order four and five.

So far, we have suggested two general groups of Runge-Kutta methods for stiff problems, namely the s-stage implicit methods of order $2s$ and L-stable SDIRK schemes. Readers should also consider **Rosenbrock methods**, which use an approximation $\mathbf{J}$ to the Jacobian of the function $\mathbf{f}$ in an initial value problem $\mathbf{y}'(t) = \mathbf{f}(\mathbf{y}(t))$ to modify the basic Runge-Kutta framework. These methods typically take the form

$$\mathbf{y}_{n+1} = \mathbf{y}_n + h\sum_{i=1}^{s}\beta_i\mathbf{f}_i \text{ where}$$

$$(\mathbf{I} - \mathbf{J}\gamma_{ii}h)\,\mathbf{f}_i = \mathbf{f}\left(\mathbf{y}_n + h\sum_{j=1}^{s}\alpha_{ij}\mathbf{f}_j\right) + \mathbf{J}h\sum_{j=1}^{i-1}\gamma_{ij}\mathbf{f}_j \text{ for } 1 \leq i \leq s .$$

For example, this idea was used in the Bader-Deuflhard semi-implicit midpoint Algorithm 3.4.2. More general discussion of this approach can be found in Hairer and Wanner [93, pp. 102–117].

For more detailed discussion of numerical methods for stiff problems, see Hairer and Wanner [93] or Lambert [124].

3.7.4 Software

There are a number of publicly available programs for solving stiff initial value problems. For example, the Brown, Hindmarsh and Byrne Fortran program vode and the Cohen and Hindmarsh C program cvode both use backward differentiation formulas. GNU Scientific Library users have several choices, including the backward differential formula routine gsl_odeiv2_step_msbdf, the implicit Runge-Kutta methods gsl_odeiv2_step_rk1imp, gsl_odeiv2_step_rk2imp and gsl_odeiv2_step_rk4imp, and the Bader-Deuflhard semi-implicit midpoint extrapolation method gsl_odeiv2_step_bsimp. MATLAB users could use the trapezoidal rule command ode23t, the variable order backward differentiation command ode15s, the **TRBDF** command ode23tb of Bank et al. [13] (which is very similar to the 2-stage SDIRK scheme), or a Rosenbrock method ode23s.

   We would like to conclude our discussion of stiff initial value problems with the following example.

Example 3.7.2 Spatial discretization of the **heat equation** provides a classic example of a stiff system of ordinary differential equations. Specifically, consider the initial value problem

$$u_i'(t) = k \frac{[u_{i+1}(t) - u_i(t)] - [u_i(t) - u_{i-1}(t)]}{\Delta x^2} \text{ for } 0 < i < M \text{ and } 0 < t$$

$$u_0(t) = 0 = u_M(t) \text{ for } 0 < t$$

$$u_i(0) = \begin{cases} 1, & 1/3 < i/M < 2/3 \\ 0, & \text{otherwise} \end{cases}$$

If the problem domain has length one, then the spatial mesh width is

$$\Delta x = 1/M .$$

It is easy to see that the matrix

$$\begin{bmatrix} 2 & -1 & & \\ -1 & 2 & \ddots & \\ & \ddots & \ddots & -1 \\ & & -1 & 2 \end{bmatrix}$$

has eigenvectors and eigenvalues of the form

$$\mathbf{x} = \begin{bmatrix} \sin(\theta) \\ \sin(2\theta) \\ \vdots \\ \sin([M-1]\theta) \end{bmatrix} \text{ and } \lambda = 2 - 2\cos(\theta) = 4\sin^2(\theta/2)$$

where

$$\sin(M\theta) = 0 .$$

It follows that

$$\theta = j\pi/M$$

for $0 < j < M$. Thus the eigenvalues of the matrix in the spatially discretized heat equation are

$$\lambda_j = -\frac{k}{\triangle x^2} 4 \sin^2 \left(\frac{j\pi}{2M} \right)$$

The least negative of these occurs for $j = 1$, with

$$\lambda_1 = -\frac{4k}{\triangle x^2} \sin^2 \left(\frac{\pi}{2M} \right) \approx -\frac{4k}{\triangle x^2} \left(\frac{\pi}{2M} \right)^2 = -k\pi^2 \ .$$

The most negative eigenvalue is

$$\lambda_{M-1} = -\frac{4k}{\triangle x^2} \sin^2 \left(\frac{[M-1]\pi}{2M} \right) \approx -\frac{4k}{\triangle x^2} = -4kM^2 \ .$$

Thus the spatially discretized heat equation becomes stiffer as the spatial mesh width $\triangle x = 1/M$ is decreased.

Recall that the absolute stability region for Euler's method is the unit circle with center -1. In order for Euler's method to be absolutely stable when solving this initial value problem, we must choose $-4kh/\triangle x^2 > -2$, which implies that $h < 1/(2kM^2)$. Euler's method requires a very small timestep for the spatially discretized heat equation.

Since the region of absolute stability for the backward Euler method is the exterior of the unit circle with center 1, the backward Euler method is absolutely stable for any choice of h. The stiffness of the spatially discretized heat equation poses no problems for the backward Euler method.

The region of absolute stability for the implicit midpoint method is the left half-plane. However, the amplification factor for this method is

$$R(\lambda h) = 1 + h\lambda \left(1 - \frac{h\lambda}{2} \right)^{-1} = \frac{1 + h\lambda/2}{1 - h\lambda/2} \ .$$

Thus the implicit midpoint method is not L-stable. Unless we choose $h\lambda/2 > -1$ for all decay rates, the implicit midpoint method will have negative amplification factors for some decay rates. This could produce unphysical oscillations that must decay, possibly slowly, as more timesteps are taken.

Numerical oscillations are reduced significantly by using the SDIRK scheme, but not completely eliminated. In this case, it is useful to choose a method and timestep to preserve a discrete local maximum principle; for more discussion of this issue, please see Trangenstein [174, p. 27ff].

We suggest that readers experiment with the following JavaScript program for the heat equation.

Exercise 3.7.1 Consider the initial-value problem

$$\mathbf{y}'(t) = \begin{bmatrix} 1/y_1(t) - y_2(t)e^{t^2}/t^2 - t \\ 1/y_2(t) - e^{t^2} - 2te^{-t^2} \end{bmatrix} \equiv \mathbf{f}(\mathbf{y}(t), t) \text{ for } t > 1 \text{ with } \mathbf{y}(1) = \begin{bmatrix} 1 \\ e^{-1} \end{bmatrix}.$$

1. Show that the analytical solution is

$$\mathbf{y}(t) = \begin{bmatrix} 1/t \\ e^{-t^2} \end{bmatrix}.$$

2. Compute the Jacobian $\partial\mathbf{f}/\partial\mathbf{y}$ and show that this initial value problem becomes stiffer as t becomes large.
3. Choose a stiff algorithm with automatic timestep control and apply it to this problem. Describe how the algorithm performs for various error tolerances.

Exercise 3.7.2 Consider the system of ordinary differential equations

$$\mathbf{y}'(t) = \begin{bmatrix} 0 & 1 \\ -1 & 0 \end{bmatrix} \mathbf{y}(t) \frac{1}{1+t^2} \equiv \mathbf{f}(\mathbf{y}(t), t).$$

1. Show that this equation has analytical solution

$$\mathbf{y}(t) = \begin{bmatrix} 1 \\ -t \end{bmatrix} \frac{y_1(0)}{\sqrt{1+t^2}} + \begin{bmatrix} t \\ 1 \end{bmatrix} \frac{y_2(0)}{\sqrt{1+t^2}}.$$

2. Show that the Jacobian $\partial\mathbf{f}/\partial\mathbf{y}$ has purely imaginary eigenvalues

$$\lambda = \pm \frac{i}{\sqrt{1+t^2}}.$$

3. Which numerical method would you recommend for solving this problem, and why?

Exercise 3.7.3 Consider the system of ordinary differential equations

$$\mathbf{y}'(t) = \begin{bmatrix} -1 - 9\cos^2(6t) + 6\sin(12t) + 12\cos^2(6t) + \frac{9}{2}\sin(12t) \\ -12\sin^2(6t) + \frac{9}{2}\sin(12t) - 1 - 9\sin^2(6t) - t\sin(12t) \end{bmatrix}$$

$$\mathbf{y}(t) \equiv \mathbf{f}(\mathbf{y}(t), t) \text{ for } t > 0$$

$$\mathbf{y}(0) = \begin{bmatrix} z_1 \\ z_2 \end{bmatrix}.$$

1. Show that the Jacobian $\frac{\partial \mathbf{f}}{\partial \mathbf{y}}$ has eigenvalues -1 and -10 for all t.
2. Show that the analytical solution is

$$\mathbf{y}(t) = \begin{bmatrix} \cos(6t) + 2\sin(6t) \\ 2\cos(6t) - \sin(6t) \end{bmatrix} e^{2t}\mathbf{z}_1 + \begin{bmatrix} \sin(6t) - 2\cos(6t) \\ 2\sin(6t) + \cos(6t) \end{bmatrix} e^{-13t}\mathbf{z}_2$$

Conclude that negative real parts for all eigenvalues of the Jacobian does not necessarily imply decay of all aspects of the solution.
3. Choose an appropriate numerical method to solve this problem, and discuss the performance of the method for various choices of the error tolerance.

Exercise 3.7.4 Consider the initial value problem

$$\mathbf{y}'(t) = \begin{bmatrix} -1/2t & 2/t^3 \\ -t/2 & -1/2t \end{bmatrix} \mathbf{y}(t) \equiv \mathbf{f}(\mathbf{y}(t), t) \text{ for } t \geq 1 \text{ with } \mathbf{y}(1) = \begin{bmatrix} z_1 \\ z_2 \end{bmatrix}$$

1. Show that the analytical solution is

$$\mathbf{y}(t) = \begin{bmatrix} t^{-3/2} \\ -\frac{1}{2}t^{1/2} \end{bmatrix} \mathbf{z}_1 + \begin{bmatrix} 2t^{-3/2}\ln(t) \\ t^{1/2}(1 - \ln(t)) \end{bmatrix} \mathbf{z}_2 .$$

2. Show that the Jacobian $\partial \mathbf{f}/\partial \mathbf{y}$ has eigenvalues $(-1 \pm 2i)/(2t)$, so both eigenvalues have negative real part.
3. Show that if $z_1 = 1, z_2 = 0$, then

$$\|\mathbf{y}(t)\| = \sqrt{t^{-3} + t/4} .$$

Conclude that $\|\mathbf{y}(t)\|$ increases monotonically for $t > 12^{1/4} \approx 1.86$.
4. Choose an appropriate numerical method to solve this problem, and discuss the performance of the method for various choices of the error tolerance.

3.8 Nonlinear Stability

In Sects. 3.4.4, 3.4.8, 3.6.2, 3.6.3 and 3.7.3, we developed linear stability analyses to understand stiffness and timestep selection for both linear multistep methods and Runge-Kutta methods. However, the exercises of the set 3.7.1 demonstrate the inadequacy of linear stability theory.

Dekker and Verwer [63] describe an approach for studying the nonlinear stability of initial value problems and numerical methods. This theory applies to certain special classes of initial value problems which are described by the following definition.

Definition 3.8.1 Let t_0 and t_1 be times, $\mathbf{y}_0$ and $\widetilde{\mathbf{y}}_0$ be vectors, $\mathbf{f}$ map vectors to vectors, and $\mathbf{y}(t)$ and $\widetilde{\mathbf{y}}(t)$ solve the initial value problems

$$\mathbf{y}'(t) = \mathbf{f}(\mathbf{y}(t)) \text{ for } t \in (t_0, t_1] \text{ with } \mathbf{y}(t_0) = \mathbf{y}_0 \text{ and}$$

$$\widetilde{\mathbf{y}}'(t) = \mathbf{f}(\widetilde{\mathbf{y}}(t)) \text{ for } t \in (t_0, t_1] \text{ with } \widetilde{\mathbf{y}}(t_0) = \widetilde{\mathbf{y}}_0 .$$

Then $\mathbf{y}(t)$ and $\widetilde{\mathbf{y}}(t)$ are **contractive** if and only if for all $t_0 \le t \le \tau \le t_1$ we have

$$\|\widetilde{\mathbf{y}}(\tau) - \mathbf{y}(\tau)\| \le \|\widetilde{\mathbf{y}}(t) - \mathbf{t}(t)\| .$$

The ordinary differential equation $\mathbf{y}' = \mathbf{f}(\mathbf{y})$ satisfies a **one-sided Lipschitz condition** on the time interval $[t_0, t_1]$ if and only if there is a positive function Φ defined on $[t_0, t_1]$ and for each $t \in [t_0, t_1]$ there is a convex set $\mathcal{M}_t$ of vectors so that for all $\mathbf{y}, \widetilde{\mathbf{y}} \in \mathcal{M}_t$ we have

$$[\mathbf{f}(\widetilde{\mathbf{y}}) - \mathbf{f}(\mathbf{y})] \cdot [\widetilde{\mathbf{y}} - \mathbf{y}] \le \Phi_t \|\widetilde{\mathbf{y}} - \mathbf{y}\|_2^2 .$$

If the ordinary differential equation $\mathbf{y}' = \mathbf{f}(\mathbf{y})$ satisfies a one-sided Lipschitz condition, then it is **dissipative** if and only if for all $t \in [t_0, t_1]$ and for all $\mathbf{y}, \widetilde{\mathbf{y}} \in \mathcal{M}_t$ we have

$$[\mathbf{f}(\widetilde{\mathbf{y}}) - \mathbf{f}(\mathbf{y})] \cdot [\widetilde{\mathbf{y}} - \mathbf{y}] \le 0 .$$

For those physical problems that can be modeled by contractive ordinary differential equations, it is important to construct numerical methods that are also contractive.

However, we will pursue a different issue related to nonlinear stability in the remainder of this section. We will be interested in the stability of fixed points for initial value problems, and for the numerical methods used to solve them.

3.8.1 Fixed Points

Let us begin with an important definition.

Definition 3.8.2 Suppose that $\mathbf{y}$ is a vector, and $\mathbf{f}(\mathbf{y})$ maps vectors to vectors. Then the vector $\mathbf{y}^*$ is a **fixed point** of the ordinary differential equation

$$\mathbf{y}'(t) = \mathbf{f}(\mathbf{y}(t), t)$$

if and only if for all t

$$\lim_{t \to \infty} \mathbf{f}(\mathbf{y}^*, t) = 0 .$$

A fixed point $\mathbf{y}^*$ of this ordinary differential equation is **stable** if and only if for all $\varepsilon > 0$ there exists $\delta > 0$ such that for all initial values $\mathbf{y}_0$ satisfying

$$\|\mathbf{y}_0 - \mathbf{y}^*\| < \delta$$

the solution $\mathbf{y}(t)$ of the initial value problem

$$\mathbf{y}'(t) = \mathbf{f}(\mathbf{y}(t), t) \text{ for } t > 0 \text{ with } \mathbf{y}(0) = \mathbf{y}_0$$

satisfies

$$\|\mathbf{y}(t) - \mathbf{y}^*\| < \varepsilon .$$

Finally, if $\mathbf{y}^*$ is a stable fixed point of $\mathbf{y}'(t) = \mathbf{f}(\mathbf{y}(t), t)$, then $\mathbf{y}^*$ is **asymptotically stable** if and only if for all $\varepsilon > 0$ there exists $\delta > 0$ and $T > 0$ so that for all solutions $\mathbf{y}(t)$ of the ordinary differential equation with initial value satisfying $\|\mathbf{y}_0 - \mathbf{y}^*\| < \delta$ and for all $t > T$ we have

$$\|\mathbf{y}(t) - \mathbf{y}^*\| < \varepsilon .$$

The following two results due to Perron [142] provide necessary and sufficient conditions guaranteeing the stability of fixed points of nonlinear ordinary differential equations. The proof of these two theorems have been adapted from Coddington and Levinson [47, p. 314ff].

Theorem 3.8.1 (Stability of Fixed Point)   *Suppose that $\mathbf{f}(\mathbf{y}, t)$ is uniformly continuously differentiable in $\mathbf{y}$ and continuous in t, and let $\mathbf{y}^*$ be a fixed point of the initial-value problem $\mathbf{y}'(t) = \mathbf{f}(\mathbf{y}(t), t)$. Define*

$$\mathbf{J} = \lim_{t \to \infty} \frac{\partial \mathbf{f}}{\partial \mathbf{y}}(\mathbf{y}^*, t) .$$

If all of the eigenvalues of $\mathbf{J}$ have negative real part, then $\mathbf{y}^*$ is a stable fixed point. In addition, if there exists $T > 0$ so that for all $t > T$ we have $\mathbf{f}(\mathbf{y}^*, t) = \mathbf{0}$, then $\mathbf{y}^*$ is also asymptotically stable.

Proof Since all of the eigenvalues of $\mathbf{J}$ have negative real part, lemma 3.3.1 that there are positive scalars σ and C so that

$$\| \exp(\mathbf{J}t) \| \le C e^{-\sigma t} . \tag{3.60}$$

Pick some stability tolerance $\varepsilon < \sigma$. Since $\mathbf{f}$ is uniformly continuously differentiable, there exists a positive $\delta_{\mathbf{J}} \le \varepsilon$ so that for all $\|\boldsymbol{\eta}\| \le \delta_{\mathbf{J}}$ we have

$$\left\| \mathbf{f}(\mathbf{y}^* + \boldsymbol{\eta}, t) - \mathbf{f}(\mathbf{y}^*, t) - \frac{\partial \mathbf{f}}{\partial \mathbf{y}}(\mathbf{y}^*, t)\boldsymbol{\eta} \right\| \le \frac{\sigma}{4C} \|\boldsymbol{\eta}\| .$$

Since $\mathbf{y}^*$ is a fixed point, for any $\varepsilon_{\mathbf{f}} > 0$ there exists $T_{\mathbf{f}} > 0$ so that for all $t \geq T_{\mathbf{f}}$

$$\|\mathbf{f}(\mathbf{y}^*, t)\| \leq \varepsilon_{\mathbf{f}} \, .$$

By the definition of $\mathbf{J}$, there exists $T_{\mathbf{J}} > 0$ so that for all $t > T_{\mathbf{J}}$ we have

$$\left\| \frac{\partial \mathbf{f}}{\partial \mathbf{y}}(\mathbf{y}^*, t) - \mathbf{J} \right\| \leq \frac{\sigma}{4C} \, .$$

Let us define the scalar

$$T = \max\{T_{\mathbf{f}}, T_{\mathbf{J}}\} \, ,$$

and the function

$$\boldsymbol{\phi}(\boldsymbol{\eta}, t) = \mathbf{f}(\mathbf{y}^*, t) - \mathbf{J}\boldsymbol{\eta} \, .$$

Then for all $\boldsymbol{\eta}$ with $\|\boldsymbol{\eta}\| \leq \delta_{\mathbf{J}}$ we have

$$
\begin{aligned}
\|\boldsymbol{\phi}(\boldsymbol{\eta}, t)\| &= \left\| \left\{ \mathbf{f}(\mathbf{y}^* + \boldsymbol{\eta}, t) - \mathbf{f}(\mathbf{y}^*, t) - \frac{\partial \mathbf{f}}{\partial \mathbf{y}}(\mathbf{y}^*, t) \right\} + \left\{ \frac{\partial \mathbf{f}}{\partial \mathbf{y}}(\mathbf{y}^*, t)\boldsymbol{\eta} - \mathbf{J}\boldsymbol{\eta} \right\} \right. \\
&\quad \left. + \mathbf{f}(\mathbf{y}^*, t) \right\| \\
&\leq \frac{\sigma}{2C}\|\boldsymbol{\eta}\| + \varepsilon_{\mathbf{f}} \, .
\end{aligned}
\tag{3.61}
$$

Next, we will define the function $\boldsymbol{\eta}(t)$ to be the solution of the ordinary differential equation

$$\boldsymbol{\eta}'(t) = \mathbf{f}(\mathbf{y}^* + \boldsymbol{\eta}(t), t) \equiv \mathbf{J}\boldsymbol{\eta}(t) + \boldsymbol{\phi}(\boldsymbol{\eta}(t), t) \, ,$$

with initial condition satisfying

$$\|\boldsymbol{\eta}(T)\| \leq \min\left\{ \frac{\delta_{\mathbf{J}}}{2}, \frac{\delta_{\mathbf{J}}}{2C} \right\} \, .$$

Integration of this initial-value problem produces

$$\boldsymbol{\eta}(t) = \exp(\mathbf{J}[t - T])\boldsymbol{\eta}(T) + \int_T^t \exp(\mathbf{J}[t - s])\boldsymbol{\phi}(\boldsymbol{\eta}(s), s) \, ds \, .$$

We can take norms of both sides of this equation and multiply by $e^{\sigma t}$ to obtain

$$e^{\sigma t}\|\boldsymbol{\eta}(t)\| \leq e^{\sigma t} \|\exp(\mathbf{J}[t - T])\| \ \|\boldsymbol{\eta}(T)\| + e^{\sigma t} \int_T^t \|\exp(\mathbf{J}[t - s])\| \ \|\boldsymbol{\phi}(\boldsymbol{\eta}(s), s)\| \, ds$$

then we can use inequality (3.60) to get

$$
\leq C\mathrm{e}^{\sigma T}\|\boldsymbol{\eta}(T)\| + C\int_T^t \mathrm{e}^{\sigma s}\,\|\boldsymbol{\phi}\,(\boldsymbol{\eta}(s),s)\|\;\mathrm{ds}
$$

and we can apply inequality (3.61) to obtain

$$
\leq \mathrm{e}^{\sigma T}\frac{\delta_{\mathbf{J}}}{2} + C\int_T^t \mathrm{e}^{\sigma s}\left\{\frac{\sigma}{2C}\|\boldsymbol{\eta}(s)\| + \varepsilon_{\mathbf{f}}\right\}\;\mathrm{ds}
$$

$$
\leq \mathrm{e}^{\sigma T}\frac{\delta_{\mathbf{J}}}{2} + \frac{C\varepsilon_{\mathbf{f}}}{\sigma}\left[\mathrm{e}^{\sigma t} - \mathrm{e}^{\sigma T}\right] + \frac{\sigma}{2}\int_T^t \mathrm{e}^{\sigma s}\|\boldsymbol{\eta}(s)\|\;\mathrm{ds} \equiv \theta(t)\;.
$$

Note that

$$
\theta'(t) = C\varepsilon_{\mathbf{f}}\mathrm{e}^{\sigma t} + \frac{\sigma}{2}\mathrm{e}^{\sigma t}\|\boldsymbol{\eta}(t)\| \leq C\varepsilon_{\mathbf{f}}\mathrm{e}^{\sigma t} + \frac{\sigma}{2}\theta(t)\;.
$$

We can multiply both sides by $\mathrm{e}^{-\sigma t/2}$ and rearrange terms to get

$$
\frac{\mathrm{d}}{\mathrm{d}t}\left[\mathrm{e}^{-\sigma t/2}\theta(t)\right] \leq C\varepsilon_{\mathbf{f}}\mathrm{e}^{\sigma t/2}\;.
$$

Then we can integrate in time to get

$$
\mathrm{e}^{-\sigma t/2}\theta(t) \leq \mathrm{e}^{-\sigma T/2}\theta(T) + \frac{2C\varepsilon_{\mathbf{f}}}{\sigma}\left[\mathrm{e}^{\sigma t/2} - \mathrm{e}^{\sigma T/2}\right]\;.
$$

Thus

$$
\|\boldsymbol{\eta}(t)\| \leq \mathrm{e}^{-\sigma t}\theta(t) \leq \mathrm{e}^{-\sigma t/2}\mathrm{e}^{-\sigma T/2}\theta(T) + \frac{2C\varepsilon_{\mathbf{f}}}{\sigma}\left[1 - \mathrm{e}^{-\sigma(t-T)/2}\right]
$$

$$
= \mathrm{e}^{-\sigma(t-T)/2}\frac{\delta_{\mathbf{J}}}{2} + \frac{2C\varepsilon_{\mathbf{f}}}{\sigma}\left[1 - \mathrm{e}^{-\sigma(t-T)/2}\right] \leq \mathrm{e}^{-\sigma(t-T)/2}\frac{\delta_{\mathbf{J}}}{2} + \frac{2C\varepsilon_{\mathbf{f}}}{\sigma}\;.
$$

We can choose

$$
\varepsilon_{\mathbf{f}} \leq \frac{\sigma\delta_{\mathbf{J}}}{4C}
$$

to prove that $\|\boldsymbol{\eta}(t)\| \leq \delta_{\mathbf{J}}$ for all $t \geq T$. Under these conditions, $\boldsymbol{\eta}(t)$ exists for all time, and $\|\boldsymbol{\eta}(t)\| \leq \delta_{\mathbf{J}} \leq \varepsilon$, proving that the fixed point $\mathbf{y}^*$ is stable. Furthermore, if $\mathbf{f}(\mathbf{y}^*,t) = 0$ for all $t \geq T$, then we can take $\varepsilon_{\mathbf{f}} = 0$ and see that

$$
\|\boldsymbol{\eta}(t)\| \leq \mathrm{e}^{-\sigma(t-T)/2}\frac{\delta_{\mathbf{J}}}{2}\;.
$$

Under this additional restriction on $\mathbf{f}$, the fixed point is asymptotically stable.

On the other hand, if the Jacobian at the fixed point has an eigenvalue with positive real part, then the next theorem shows that the fixed point is not stable.

Theorem 3.8.2 *Suppose that $\mathbf{f}(\mathbf{y}, t)$ is uniformly continuously differentiable in $\mathbf{y}$ and continuous in t, and let $\mathbf{y}^*$ be a fixed point of the initial-value problem $\mathbf{y}'(t) = \mathbf{f}(\mathbf{y}(t), t)$. Define*

$$\mathbf{J} = \lim_{t \to \infty} \frac{\partial \mathbf{f}}{\partial \mathbf{y}} (\mathbf{y}^*, t) \ .$$

If there is an eigenvalue of $\mathbf{J}$ with positive real part, then $\mathbf{y}^*$ is not a stable fixed point.

Proof Since $\mathbf{J}$ has at least one eigenvalue with positive real part, define

$$\sigma = \min \{\Re e(\lambda) : \lambda \text{ is an eigenvalue of } \mathbf{J} \text{ and } \Re e(\lambda) > 0\} \ .$$

Choose

$$\varepsilon_\sigma < \sigma/4$$

and apply Theorem 1.4.7 of Chap. 1 in Volume II to find a nonsingular $\mathbf{X}$ and diagonal Λ so that

$$\mathbf{U} = \mathbf{X}^{-1} \mathbf{J} \mathbf{X} - \Lambda$$

has entries $\upsilon_{i,j}$ satisfying

$$\upsilon_{i,j} = 0 \text{ for all } j \neq i+1 \text{ and } \max_{1 \leq i < n} |\upsilon_{i,i+1}| \leq \varepsilon_\sigma \ .$$

Assume that the diagonal entries of Λ have been ordered so that we can partition

$$\Lambda = \begin{bmatrix} \Lambda_+ & \mathbf{0} \\ \mathbf{0} & \Lambda_- \end{bmatrix} \text{ and } \mathbf{U} = \begin{bmatrix} \mathbf{U}_+ & \mathbf{0} \\ \mathbf{0} & \mathbf{U}_- \end{bmatrix} ,$$

where the diagonal entries of Λ_+ all have positive real part, and the diagonal entries of Λ_- all have nonpositive real part.

The definition of $\mathbf{J}$ implies that for all $\varepsilon_\mathbf{J} > 0$ there exists $T_\mathbf{J} > 0$ so that for all $t \geq T_\mathbf{J}$

$$\left\| \frac{\partial \mathbf{f}}{\partial \mathbf{y}} (\mathbf{y}^*, t) - \mathbf{J} \right\| \leq \varepsilon_\mathbf{J} \ .$$

We will choose $\varepsilon_{\mathbf{J}}$ so that

$$16 \left\| \mathbf{X}^{-1} \right\| \ \|\mathbf{X}\| \varepsilon_{\mathbf{J}} \leq \sigma \ .$$

Since $\mathbf{f}$ is uniformly continuously differentiable, for all $\varepsilon_{\mathbf{J}} > 0$ there exists $\delta_{\mathbf{J}} > 0$ and $T_{\mathbf{J}} > 0$ so that for all $\|\boldsymbol{\eta}\| \leq \delta_{\mathbf{J}}$ and for all $t \geq T_{\mathbf{J}}$ we have

$$\left\| \mathbf{f}\left(\mathbf{y}^*, t\right) - \mathbf{f}\left(\mathbf{y}^*, t\right) - \frac{\partial \mathbf{f}}{\partial \mathbf{y}}\left(\mathbf{y}^*, t\right) \boldsymbol{\eta} \right\| \leq \varepsilon_{\mathbf{J}} \|\boldsymbol{\eta}\| \ .$$

We will choose $\varepsilon_{\mathbf{J}}$ so that

$$16 \left\| \mathbf{X}^{-1} \right\| \ \|\mathbf{X}\| \varepsilon_{\mathbf{J}} \leq \sigma \ .$$

Suppose that $\mathbf{y}^*$ is stable. Then for all $\varepsilon > 0$ there exists a positive $\delta_\varepsilon < 2\varepsilon \|\mathbf{X}\| \ \|\mathbf{X}^{-1}\|$ and there exists $T_\varepsilon > 0$ so that for all $t \geq T_\varepsilon$ the solution $\boldsymbol{\eta}(t)$ to

$$\boldsymbol{\eta}'(t) = \mathbf{f}\left(\mathbf{y}^* + \boldsymbol{\eta}(t), t\right)$$

with initial value satisfying

$$\|\boldsymbol{\eta}(T_\varepsilon)\| \leq \delta_\varepsilon$$

is such that

$$\|\boldsymbol{\eta}(t)\| \leq \varepsilon \ .$$

We will choose

$$\varepsilon \leq \delta_{\mathbf{J}} \ .$$

Eventually, we will detect a contradiction, and conclude that $\mathbf{y}^*$ cannot be stable.

Since $\mathbf{y}^*$ is a fixed point, for all $\varepsilon_{\mathbf{f}} > 0$ there exists $T_{\mathbf{f}} > 0$ so that for all $t \geq T_{\mathbf{f}}$

$$\left\| \mathbf{f}\left(\mathbf{y}^*, t\right) \right\| \leq \varepsilon_{\mathbf{f}} \ .$$

We will choose

$$\varepsilon_{\mathbf{f}} \leq \frac{\sigma \varepsilon}{4} \frac{\delta_\varepsilon}{3\varepsilon \|\mathbf{X}\| \ \|\mathbf{X}^{-1}\| - \delta_\varepsilon} \ ,$$

and

$$T \geq \max\{T_{\mathbf{J}}, T_{\mathbf{J}}, T_\varepsilon, T_{\mathbf{f}}\} \ .$$

Define

$$\boldsymbol{\psi}(\boldsymbol{\zeta}, t) = \mathbf{X}^{-1} \left[\mathbf{f}\left(\mathbf{y}^* + \mathbf{X}\boldsymbol{\zeta}, t\right) - \mathbf{J}\mathbf{X}\boldsymbol{\zeta} \right] .$$

Then for all $t \geq T$ and for all $\|\boldsymbol{\zeta}\| \leq \delta_{\mathbf{J}} / \|\mathbf{X}\|$

$$
\begin{aligned}
\|\boldsymbol{\psi}(\boldsymbol{\zeta}, t)\| &= \left\| \mathbf{X}^{-1} \left\{ \mathbf{f}\left(\mathbf{y}^* + \mathbf{X}\boldsymbol{\zeta}, t\right) - \mathbf{f}\left(\mathbf{y}^*, t\right) - \frac{\partial \mathbf{f}}{\partial \mathbf{y}} \left(\mathbf{y}^*, t\right) \mathbf{X}\boldsymbol{\zeta} \right\} \right. \\
&\quad \left. + \mathbf{X}^{-1} \left\{ \frac{\partial \mathbf{f}}{\partial \mathbf{y}} \left(\mathbf{y}^*, t\right) - \mathbf{J} \right\} \mathbf{X}\boldsymbol{\zeta} + \mathbf{X}^{-1} \mathbf{f}\left(\mathbf{y}^*, t\right) \right\| \\
&\leq \|\mathbf{X}^{-1}\| \varepsilon_{\mathbf{J}} \|\mathbf{X}\boldsymbol{\zeta}\| + \|\mathbf{X}^{-1}\| \varepsilon_{\mathbf{J}} \|\mathbf{X}\boldsymbol{\zeta}\| + \|\mathbf{X}^{-1}\| \varepsilon_{\mathbf{f}} \\
&\leq \|\mathbf{X}^{-1}\| \left\{ (\varepsilon_{\mathbf{J}} + \varepsilon_{\mathbf{J}}) \|\mathbf{X}\| \|\boldsymbol{\zeta}\| + \varepsilon_{\mathbf{f}} \right\} .
\end{aligned}
\tag{3.62}
$$

We can partition

$$\boldsymbol{\zeta} = \begin{bmatrix} \boldsymbol{\zeta}_+ \\ \boldsymbol{\zeta}_- \end{bmatrix} \text{ and } \boldsymbol{\psi}(\boldsymbol{\zeta}, t) = \begin{bmatrix} \boldsymbol{\psi}_+(\boldsymbol{\zeta}, t) \\ \boldsymbol{\psi}_-(\boldsymbol{\zeta}, t) \end{bmatrix} ,$$

corresponding to the partitioning of $\boldsymbol{\Lambda}$. We will assume that $\boldsymbol{\zeta}(t)$ solves the system of ordinary differential equations

$$\frac{d}{dt} \begin{bmatrix} \boldsymbol{\zeta}_+(t) \\ \boldsymbol{\zeta}_-(t) \end{bmatrix} = \begin{bmatrix} \boldsymbol{\Lambda}_+ + \mathbf{U}_+ & \mathbf{0} \\ \mathbf{0} & \boldsymbol{\Lambda}_- + \mathbf{U}_- \end{bmatrix} \begin{bmatrix} \boldsymbol{\zeta}_+(t) \\ \boldsymbol{\zeta}_-(t) \end{bmatrix} + \begin{bmatrix} \boldsymbol{\psi}_+(\boldsymbol{\zeta}(t), t) \\ \boldsymbol{\psi}_-(\boldsymbol{\zeta}(t), t) \end{bmatrix} .$$

We will also assume that the initial value satisfies $\|\boldsymbol{\zeta}_+(T)\| = \delta_\varepsilon / \|\mathbf{X}\|$ and $\|\boldsymbol{\zeta}_-(T)\| = 0$. Then $\boldsymbol{\eta}(t) \equiv \mathbf{X}\boldsymbol{\zeta}(t)$ solves the differential equation

$$
\begin{aligned}
\boldsymbol{\eta}'(t) = \mathbf{X}\boldsymbol{\zeta}'(t) &= \mathbf{X}\{(\boldsymbol{\Lambda} + \mathbf{U})\boldsymbol{\zeta}(t) + \boldsymbol{\psi}(\boldsymbol{\zeta}(t), t)\} \\
&= \mathbf{X}(\boldsymbol{\Lambda} + \mathbf{U})\mathbf{X}^{-1} \boldsymbol{\eta}(t) + \mathbf{f}\left(\mathbf{y}^* + \boldsymbol{\eta}(t), t\right) - \mathbf{J}\boldsymbol{\eta}(t) = \mathbf{f}\left(\mathbf{y}^* + \boldsymbol{\eta}(t), t\right) ,
\end{aligned}
$$

and its initial value satisfies

$$\|\boldsymbol{\eta}(T)\| = \|\mathbf{X}\boldsymbol{\zeta}(T)\| \leq \|\mathbf{X}\| \, \|\boldsymbol{\zeta}(T)\| = \delta .$$

Note that

$$2\|\boldsymbol{\zeta}_+(t)\| \frac{d\|\boldsymbol{\zeta}_+(t)\|}{dt} = \frac{d}{dt} \|\boldsymbol{\zeta}_+(t)\|^2 = \boldsymbol{\zeta}_+(t)^H \boldsymbol{\zeta}_+'(t) + \boldsymbol{\zeta}_+'(t)^H \boldsymbol{\zeta}_+(t)$$

then the differential equation for ζ gives us

$$= \zeta_+(t)^H \left[(\Lambda_+ + \mathbf{U}_+) + (\Lambda_+ + \mathbf{U}_+)^H \right] \zeta_+(t) + \zeta_+(t)^H \boldsymbol{\psi}_+(\zeta(t), t)$$
$$+ \boldsymbol{\psi}_+(\zeta(t), t)^H \zeta_+(t)$$
$$= 2\Re\left(\zeta_+(t)^H \Lambda_+ + \zeta_+(t) \right) + 2\Re\left(\zeta_+(t)^H \mathbf{U}_+ \zeta_+(t) \right) + 2\Re\left(\zeta_+(t)^H \boldsymbol{\psi}_+(\zeta(t), t) \right)$$

then we use the lower bound σ on the real parts of the eigenvalues of Λ_+ to get

$$\geq 2\sigma \left\| \zeta_+(t) \right\|^2 - 2\left| \zeta_+(t)^H \mathbf{U}_+ \zeta_+(t) \right| - 2\left| \zeta_+(t)^H \boldsymbol{\psi}_+(\zeta(t), t) \right|$$

then we use the Cauchy inequality (3.15) in Chap. 3 of Volume I to obtain

$$\geq 2\sigma \left\| \zeta_+(t) \right\|^2 - 2\left\| \zeta_+(t) \right\| \left\| \mathbf{U}_+ \zeta_+(t) \right\| - 2\left\| \zeta_+(t) \right\| \left\| \boldsymbol{\psi}_+(\zeta(t), t) \right\|$$

next, we use the bound on the nonzero entries of $\mathbf{U}_+$ and inequality (3.62) to obtain

$$\geq 2\sigma \left\| \zeta_+(t) \right\|^2 - 2\varepsilon_\sigma \left\| \zeta_+(t) \right\|^2$$
$$- 2\left\| \zeta_+(t) \right\| \left\| \mathbf{X}^{-1} \right\| \left\{ (\varepsilon_\mathbf{J} + \varepsilon_\mathbf{J}) \|\mathbf{X}\| \left[\left\| \zeta_+(t) \right\|^2 + \left\| \zeta_-(t) \right\|^2 \right]^{1/2} + \varepsilon_\mathbf{f} \right\}$$

then we use the inequality $\sqrt{a^2 + b^2} \leq a + b$ to get

$$\geq 2 \left\{ \sigma - \varepsilon_\sigma - (\varepsilon_\mathbf{J} + \varepsilon_\mathbf{J}) \left\| \mathbf{X}^{-1} \right\| \, \|\mathbf{X}\| \right\} \left\| \zeta_+(t) \right\|^2$$
$$- 2\left\| \zeta_+(t) \right\| \left\| \mathbf{X}^{-1} \right\| \left\{ (\varepsilon_\mathbf{J} + \varepsilon_\mathbf{J}) \|\mathbf{X}\| \left\| \zeta_-(t) \right\| + \varepsilon_\mathbf{f} \right\} .$$

It follows that

$$\frac{d\| \zeta_+(t) \|}{dt} \geq \left\{ \sigma - \varepsilon_\sigma - (\varepsilon_\mathbf{J} + \varepsilon_\mathbf{J}) \left\| \mathbf{X}^{-1} \right\| \, \|\mathbf{X}\| \right\} \left\| \zeta_+(t) \right\|$$
$$- (\varepsilon_\mathbf{J} + \varepsilon_\mathbf{J}) \left\| \mathbf{X}^{-1} \right\| \, \|\mathbf{X}\| \, \left\| \zeta_-(t) \right\| - \left\| \mathbf{X}^{-1} \right\| \varepsilon_\mathbf{f} . \qquad (3.63)$$

Similarly,

$$2\| \zeta_-(t) \| \frac{d\| \zeta_-(t) \|}{dt} = \frac{d}{dt} \| \zeta_-(t) \|^2 = \zeta_-(t)^H \zeta_-'(t) + \zeta_-'(t)^H \zeta_-(t)$$

then the differential equation for ζ gives us

$$= \zeta_-(t)^H \left[(\Lambda_- + \mathbf{U}_-) + (\Lambda_- + \mathbf{U}_-)^H \right] \zeta_-(t) + \zeta_-(t)^H \boldsymbol{\psi}_-(\zeta(t), t)$$
$$+ \boldsymbol{\psi}_-(\zeta(t), t)^H \zeta_-(t)$$
$$= 2\Re\left(\zeta_-(t)^H \Lambda_- + \zeta_-(t) \right) + 2\Re\left(\zeta_-(t)^H \mathbf{U}_- \zeta_-(t) \right) + 2\Re\left(\zeta_-(t)^H \boldsymbol{\psi}_-(\zeta(t), t) \right)$$

then we use the assumption that the real parts of the diagonal entries of Λ_- are all nonpositive to get

$$\leq 2\left|\boldsymbol{\zeta}_-(t)^H \mathbf{U}_- \boldsymbol{\zeta}_-(t)\right| + 2\left|\boldsymbol{\zeta}_-(t)^H \boldsymbol{\psi}_-(\boldsymbol{\zeta}(t),t)\right|$$

then we use the Cauchy inequality (3.15) in Chap. 3 of Volume I to obtain

$$\leq 2\left\|\boldsymbol{\zeta}_-(t)\right\| \left\|\mathbf{U}_-\boldsymbol{\zeta}_-(t)\right\| + 2\left\|\boldsymbol{\zeta}_-(t)\right\| \left\|\boldsymbol{\psi}_-(\boldsymbol{\zeta}(t),t)\right\|$$

next, we use the bound on the nonzero entries of $\mathbf{U}_-$ and inequality (3.62) to obtain

$$\leq 2\varepsilon_\sigma \left\|\boldsymbol{\zeta}_-(t)\right\|^2$$
$$+ 2\left\|\boldsymbol{\zeta}_-(t)\right\| \left\|\mathbf{X}^{-1}\right\| \left\{(\varepsilon_{\mathbf{J}} + \varepsilon_{\mathbf{J}})\|\mathbf{X}\| \left[\left\|\boldsymbol{\zeta}_+(t)\right\|^2 + \left\|\boldsymbol{\zeta}_-(t)\right\|^2\right]^{1/2} + \varepsilon_{\mathbf{f}}\right\}$$

then we use the inequality $\sqrt{a^2 + b^2} \leq a + b$ to get

$$\leq 2\left[\varepsilon_\sigma + (\varepsilon_{\mathbf{J}} + \varepsilon_{\mathbf{J}}) \left\|\mathbf{X}^{-1}\right\| \|\mathbf{X}\|\right] \left\|\boldsymbol{\zeta}_-(t)\right\|^2$$
$$+ 2\left\|\boldsymbol{\zeta}_-(t)\right\| \left\|\mathbf{X}^{-1}\right\| \left\{(\varepsilon_{\mathbf{J}} + \varepsilon_{\mathbf{J}})\|\mathbf{X}\| \left\|\boldsymbol{\zeta}_+(t)\right\| + \varepsilon_{\mathbf{f}}\right\} .$$

It follows that

$$\frac{d\left\|\boldsymbol{\zeta}_-(t)\right\|}{dt} \leq \left\{\varepsilon_\sigma + (\varepsilon_{\mathbf{J}} + \varepsilon_{\mathbf{J}}) \left\|\mathbf{X}^{-1}\right\| \|\mathbf{X}\|\right\} \left\|\boldsymbol{\zeta}_-(t)\right\|$$
$$+ (\varepsilon_{\mathbf{J}} + \varepsilon_{\mathbf{J}}) \left\|\mathbf{X}^{-1}\right\| \|\mathbf{X}\| \left\|\boldsymbol{\zeta}_+(t)\right\| + \left\|\mathbf{X}^{-1}\right\| \varepsilon_{\mathbf{f}} . \tag{3.64}$$

We can combine inequalities (3.63) and (3.64) to get

$$\frac{d}{dt} \left\{\left\|\boldsymbol{\zeta}_+(t)\right\| - \left\|\boldsymbol{\zeta}_-(t)\right\|\right\}$$
$$\geq \left[\sigma - \varepsilon_\sigma - (\varepsilon_{\mathbf{J}} + \varepsilon_{\mathbf{J}}) \left\|\mathbf{X}^{-1}\right\| \|\mathbf{X}\|\right] \left\|\boldsymbol{\zeta}_+(t)\right\|$$
$$- (\varepsilon_{\mathbf{J}} + \varepsilon_{\mathbf{J}}) \left\|\mathbf{X}^{-1}\right\| \|\mathbf{X}\| \left\|\boldsymbol{\zeta}_-(t)\right\| - \left\|\mathbf{X}^{-1}\right\| \varepsilon_{\mathbf{f}}$$
$$- \left[\varepsilon_\sigma + (\varepsilon_{\mathbf{J}} + \varepsilon_{\mathbf{J}}) \left\|\mathbf{X}^{-1}\right\| \|\mathbf{X}\|\right] \left\|\boldsymbol{\zeta}_-(t)\right\|$$
$$- (\varepsilon_{\mathbf{J}} + \varepsilon_{\mathbf{J}}) \left\|\mathbf{X}^{-1}\right\| \|\mathbf{X}\| \left\|\boldsymbol{\zeta}_+(t)\right\| - \left\|\mathbf{X}^{-1}\right\| \varepsilon_{\mathbf{f}} .$$
$$= \left\{\sigma - \varepsilon_\sigma - 2(\varepsilon_{\mathbf{J}} + \varepsilon_\sigma) \left\|\mathbf{X}^{-1}\right\| \|\mathbf{X}\|\right\} \left\|\boldsymbol{\zeta}_+(t)\right\|$$
$$- \left\{\varepsilon_\sigma + 2(\varepsilon_{\mathbf{J}} + \varepsilon_\sigma) \left\|\mathbf{X}^{-1}\right\| \|\mathbf{X}\|\right\} \left\|\boldsymbol{\zeta}_-(t)\right\| - 2\left\|\mathbf{X}^{-1}\right\| \varepsilon_{\mathbf{f}}$$
$$\geq \left\{\sigma - \frac{\sigma}{4} - \left(\frac{\sigma}{8} + \frac{\sigma}{8}\right)\right\} \left\|\boldsymbol{\zeta}_+(t)\right\|$$

$$- \left\{ \frac{\sigma}{4} + \left(\frac{\sigma}{8} + \frac{\sigma}{8} \right) \right\} \left\| \boldsymbol{\zeta}_-(t) \right\| - 2 \left\| \mathbf{X}^{-1} \right\| \varepsilon_{\mathbf{f}}$$

$$= \frac{\sigma}{2} \left\{ \left\| \boldsymbol{\zeta}_+(t) \right\| - \left\| \boldsymbol{\zeta}_-(t) \right\| \right\} - 2 \left\| \mathbf{X}^{-1} \right\| \varepsilon_{\mathbf{f}}$$

We can solve this differential inequality to get

$$\left\| \boldsymbol{\zeta}_+(t) \right\| - \left\| \boldsymbol{\zeta}_-(t) \right\| \geq e^{\sigma(t-T)/2} \left\{ \left\| \boldsymbol{\zeta}_+(T) \right\| - \left\| \boldsymbol{\zeta}_-(T) \right\| \right\}$$
$$- \frac{4}{\sigma} \left\{ e^{\sigma(t-T)/2} - 1 \right\} \left\| \mathbf{X}^{-1} \right\| \varepsilon_{\mathbf{f}} .$$

Since $\delta_\varepsilon < 2\varepsilon \|\mathbf{X}\| \, \|\mathbf{X}^{-1}\|$, we can choose $t > T$ so that

$$e^{\sigma(t-T)/2} = 3\varepsilon \|\mathbf{X}\| \, \left\| \mathbf{X}^{-1} \right\| /\delta_\varepsilon .$$

Then the initial value for ζ and the bound on $\varepsilon_{\mathbf{f}}$ give us

$$\left\| \boldsymbol{\zeta}_+(t) \right\| - \left\| \boldsymbol{\zeta}_-(t) \right\| \geq \frac{3\varepsilon \|\mathbf{X}\| \, \|\mathbf{X}^{-1}\|}{\delta_\varepsilon \|\mathbf{X}\|} \frac{\delta_\varepsilon}{\|\mathbf{X}\|}$$
$$- \frac{4}{\sigma} \left\{ \frac{3\varepsilon \|\mathbf{X}\| \, \|\mathbf{X}^{-1}\|}{\delta} - 1 \right\} \left\| \mathbf{X}^{-1} \right\| \frac{\sigma\varepsilon}{4\|\mathbf{X}^{-1}\|} \frac{\delta_\varepsilon}{3\varepsilon \|\mathbf{X}\| \, \|\mathbf{X}^{-1}\| - \delta_\varepsilon}$$
$$= 2\varepsilon \left\| \mathbf{X}^{-1} \right\| .$$

Since $\sqrt{a^2 + b^2} \geq a - b$ for nonnegative a and b, it follows that

$$\|\boldsymbol{\eta}(t)\| = \|\mathbf{X}\boldsymbol{\zeta}(t)\| \geq \frac{\|\boldsymbol{\zeta}(t)\|}{\|\mathbf{X}^{-1}\|} \geq \frac{\|\boldsymbol{\zeta}_+(t)\| - \|\boldsymbol{\zeta}_-(t)\|}{\|\mathbf{X}^{-1}\|} \geq 2\varepsilon .$$

This contradicts the stability assumption, that $\|\boldsymbol{\eta}(t)\| \leq \varepsilon$ for all $t \geq T$. We conclude that the fixed point $\mathbf{y}^*$ is not stable.

Example 3.8.1 As an example of a nonlinear ordinary differential equation, consider the **logistics equation**

$$\frac{d\mathbf{y}}{dt} = \mathbf{y}(1 - \mathbf{y}) ,$$

which arises in population dynamics (see, for example, Murray [137]). Using separation of variables and partial fractions, we can easily see that the analytical solution of this equation satisfies

$$\left| \frac{1 - \mathbf{y}(t)}{\mathbf{y}(t)} \right| = \left| \frac{1 - \mathbf{y}_0}{\mathbf{y}_0} \right| e^t$$

Depending on the initial value, the analytical solution takes one of three forms:

$$y(t) = \begin{cases} \frac{y_0}{y_0+(1-y_0)e^{-t}}, & 0 < y_0 < 1 \\ \frac{y_0}{y_0-(y_0-1)e^{-t}}, & y_0 > 1 \\ -\frac{-y_0}{(y_0-1)e^{-t}+y_0}, & y_0 < 0 \text{ and } t < \log\left(\frac{1-y_0}{-y_0}\right) \end{cases}$$

The fixed points are obviously $y^* = 0$ or $y^* = 1$. Either from these analytical solutions, or from Theorems 3.8.1 and 3.8.2, we can see that $y^* = 0$ is unstable, and $y^* = 1$ is stable.

Exercise 3.8.1 The **Landau equation**

$$y'(t) = y(t)\left[a - by(t)^2\right]$$

arises in studying steady flow of a Newtonion fluid. Determine the fixed points of the Landau equation, and find conditions under which each is stable.

Exercise 3.8.2 Find the fixed points of the system of ordinary differential equations

$$\frac{d}{dt}\begin{bmatrix} y_1(t) \\ y_2(t) \end{bmatrix} = \begin{bmatrix} -y_2(t) + y_1(t)\left(a - \|y(t)\|_2^2\right) \\ y_1(t) + y_2(t)\left(a - \|y(t)\|_2^2\right) \end{bmatrix},$$

and determine conditions under which each is stable.

3.8.2 Numerical Fixed Points

Numerical methods for initial value problems can have fixed points, too.

Definition 3.8.3 Suppose that y is a vector, h is a timestep and the function ϕ maps pairs of vectors and timesteps to vectors. Given a value for h, the vector y^* is a **period one fixed point** of the explicit nonlinear recurrence

$$y_{n+1} = \phi(y_n, h)$$

if and only if

$$y^* = \phi(y^*, h) .$$

Similarly, if $p > 1$ then y^* is a **period p fixed point** if and only if

$$y^* = \phi^p(y^*, h) \text{ but } y^* \neq \phi^k(y^*, h) \text{ for } k < p .$$

A period one fixed point of an explicit nonlinear recurrence is **stable** for some timestep h if and only if there exists $\delta > 0$ so that for all $\|\boldsymbol{\eta}\| \leq \delta$ we have

$$\|\boldsymbol{\phi}\left(\mathbf{y}^* + \boldsymbol{\eta}, h\right) - \mathbf{y}^*\| \leq \|\boldsymbol{\eta}\| .$$

It is not hard to see that $\mathbf{y}^*$ is a stable fixed point if and only if all of the eigenvalues of $\partial\boldsymbol{\phi}/\partial\mathbf{y}$ have modulus at most one.

Ideally, the fixed points of a numerical method for an initial value problem should be close to the fixed points of the corresponding differential equation. We might also hope that the fixed points of a numerical method are stable if and only if the corresponding fixed point of the differential equation is stable. These goals are not always fulfilled.

Example 3.8.2 For an autonomous ordinary differential equation, Euler's method is

$$\mathbf{y}_{n+1} = \mathbf{y}_n + h\mathbf{f}\left(\mathbf{y}_n\right) .$$

Thus fixed points $\mathbf{y}^*$ of Euler's method satisfy

$$\mathbf{f}\left(\mathbf{y}^*\right) = \mathbf{0} .$$

This is the same as the equation for fixed points of the initial value problem, so the fixed points of Euler's method are the same as the fixed points of the original differential equation. A fixed point $\mathbf{y}^*$ of Euler's method is stable if and only if for all $\boldsymbol{\eta}$ sufficiently small

$$\|\boldsymbol{\eta}\| \geq \|\{\mathbf{y}^* + \boldsymbol{\eta} + h\mathbf{f}\left(\mathbf{y}^* + \boldsymbol{\eta}\right)\} - \mathbf{y}^*\| = \|\boldsymbol{\eta} + h\mathbf{f}\left(\mathbf{y}^* + \boldsymbol{\eta}\right)\|$$

$$\approx \left\|\left[\mathbf{I} + h\frac{\partial\mathbf{f}}{\partial\mathbf{y}}\left(\mathbf{y}^*\right)\right]\boldsymbol{\eta}\right\| .$$

Thus Euler's method is stable whenever the eigenvalues of $\mathbf{I} + h\partial\mathbf{f}/\partial\mathbf{y}(\mathbf{y}^*)$ have modulus at most one. This is different from the stability condition for the ordinary differential equation.

If we apply Euler's method to the logistics equation, we obtain

$$\mathbf{y}_{n+1} = \mathbf{y}_n + h\mathbf{y}_n\left(1 - \mathbf{y}_n\right) = \mathbf{y}_n\left[1 + h\left(1 - \mathbf{y}_n\right)\right] \equiv \boldsymbol{\phi}\left(\mathbf{y}_n, h\right) .$$

A fixed point of period 2 satisfies

$$\mathbf{y}^* = \mathbf{y}_n = \mathbf{y}_{n+2} = \mathbf{y}_{n+1}\left\{1 + h - h\mathbf{y}_{n+1}\right\}$$

$$= \mathbf{y}_n\left\{1 + h - h\mathbf{y}_n\right\}\left\{1 + h - h\mathbf{y}_n\left[1 + h - h\mathbf{y}_n\right]\right\} = \boldsymbol{\phi}(\boldsymbol{\phi}(\mathbf{y}_n, h), h) .$$

One solution is $\mathbf{y}_n = 0 = \mathbf{y}_{n+1}$; another solution is $\mathbf{y}_n = 1 = \mathbf{y}_{n+1}$. Since period two fixed points are not period one fixed points, a period 2 fixed point satisfies

$$\frac{\mathbf{y}_n(1 + h - h\mathbf{y}_n)(1 + h - h\mathbf{y}_n[1 + h - h\mathbf{y}_n]) - \mathbf{y}_n}{\mathbf{y}_n(1 - \mathbf{y}_n)} = h[2 + h - 2h\mathbf{y}_n - h^2\mathbf{y}_n + h^2\mathbf{y}_n^2] .$$

This implies that

$$\mathbf{y}_n = \frac{2 + h \pm \sqrt{h^2 - 4}}{2h}$$

provided that $h \geq 2$. As we increase the timestep h past $h = 2$, the stable period one fixed point $\mathbf{y}_n = 1$ **bifurcates** into a pair of period two fixed points.

Suppose that we run many simulations with Euler's method, all with initial values y_0 between 0 and 1, and with varying values of the timestep h. We could run the simulations for several hundred steps until they reach a limiting behavior, then plot the next few step (say 50). For $h <= 2$, all of these 50 points will lie essentially on $y = 1$. For $h > 2$, we will see that the steady-state $y = 1$ becomes unstable, and the numerical solution approaches a 2-cycle. For $h \approx 2.5$ we start to see a 4-cycle. For larger h, the numerical solution becomes chaotic, but does not grow without bound for a fixed h. In other words, for nonlinear problems instability does not necessarily lead to large growth of the solution.

A program to draw the fixed points for various numerical methods applied to the logistics equation is available in GUINonlinearStability.C. This program chooses random initial values, runs the method for some given number of steps, then plots the next few iterates. The results are shown as last the few iterates versus step size. Results of this program with the forward Euler method are shown in Fig. 3.11.

Example 3.8.3   The modified midpoint method is the linear multistep method

$$\mathbf{y}_{n+1} = \mathbf{y}_{n-1} + 2h\mathbf{f}(\mathbf{y}_n) .$$

This can be rewritten as

$$\begin{bmatrix} \mathbf{y}_{n+1} \\ \mathbf{y}_n \end{bmatrix} = \begin{bmatrix} \mathbf{y}_{n-1} + 2h\mathbf{f}(\mathbf{y}_n) \\ \mathbf{y}_n \end{bmatrix} \equiv \boldsymbol{\phi}\left(\begin{bmatrix} \mathbf{y}_n \\ \mathbf{y}_{n-1} \end{bmatrix}, h\right) .$$

At a fixed point of period 1,

$$\begin{bmatrix} \mathbf{y}^* \\ \mathbf{y}^* \end{bmatrix} = \begin{bmatrix} \mathbf{y}_n \\ \mathbf{y}_{n-1} \end{bmatrix} = \begin{bmatrix} \mathbf{y}_{n+1} \\ \mathbf{y}_n \end{bmatrix} = \begin{bmatrix} \mathbf{y}_{n-1} + 2h\mathbf{f}(\mathbf{y}_n) \\ \mathbf{y}_n \end{bmatrix} = \begin{bmatrix} \mathbf{y}^* + 2h\mathbf{f}(\mathbf{y}^*) \\ \mathbf{y}^* \end{bmatrix} ,$$

which implies that

$$\mathbf{f}(\mathbf{y}^*) = \mathbf{0} .$$

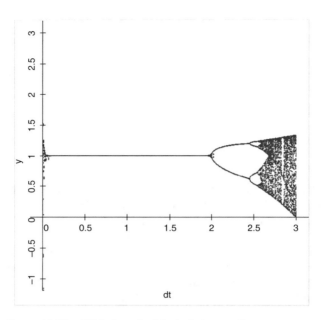

Fig. 3.11   Nonlinear stability of Euler's method for logistics equation

Thus the period 1 fixed points of the modified midpoint method are the same as the fixed points of the ordinary differential equation. Note that

$$\frac{\partial \boldsymbol{\phi}}{\partial \begin{bmatrix} \mathbf{y}_n \\ \mathbf{y}_{n-1} \end{bmatrix}} = \begin{bmatrix} 2h\frac{\partial \mathbf{f}}{\partial \mathbf{y}}(\mathbf{y}_n) & \mathbf{I} \\ \mathbf{I} & \mathbf{0} \end{bmatrix} .$$

In the special case of the logistics equation, we have

$$\frac{\partial \boldsymbol{\phi}}{\partial \begin{bmatrix} \mathbf{y}_n \\ \mathbf{y}_{n-1} \end{bmatrix}} = \begin{bmatrix} 2h\mathbf{f}'(\mathbf{y}_n) & 1 \\ 1 & 0 \end{bmatrix} ,$$

which at a fixed point $\mathbf{y}^*$ has eigenvalues λ satisfying

$$0 = -\lambda \left[2h\mathbf{f}'(\mathbf{y}^*) - \lambda \right] - 1 = \lambda^2 - 2h\mathbf{f}'(\mathbf{y}^*) \lambda - 1 .$$

Thus the eigenvalues satisfy

$$\lambda = h\mathbf{f}'(\mathbf{y}^*) \pm \sqrt{[h\mathbf{f}'(\mathbf{y}^*)]^2 + 1} .$$

For the fixed point $\mathbf{y}^* = 0$, the eigenvalues are $\lambda = h \pm \sqrt{h^2 + 1}$, so the positive root has modulus greater than 1. For the fixed point $\mathbf{y}^* = 1$, the eigenvalues are $\lambda = -h \pm \sqrt{h^2 + 1}$, so the negative root has modulus greater than 1. Thus for the logistics equation, the modified midpoint method has no stable fixed points of period 1, even though the physical problem has a stable fixed point.

Example 3.8.4 The second-order Adams-Bashforth method is the linear multistep method

$$\mathbf{y}_{n+1} = \mathbf{y}_n + \frac{h}{2} \{3\mathbf{f}(\mathbf{y}_n) - \mathbf{f}(\mathbf{y}_{n-1})\} .$$

This can be rewritten as

$$\begin{bmatrix} \mathbf{y}_{n+1} \\ \mathbf{y}_n \end{bmatrix} = \begin{bmatrix} \mathbf{y}_n + \frac{h}{2}\{3\mathbf{f}(\mathbf{y}_n) - \mathbf{f}(\mathbf{y}_{n-1})\} \\ \mathbf{y}_n \end{bmatrix} \equiv \boldsymbol{\phi}\left(\begin{bmatrix} \mathbf{y}_n \\ \mathbf{y}_{n-1} \end{bmatrix}, h \right) .$$

At a fixed point of period 1,

$$\begin{bmatrix} \mathbf{y}^* \\ \mathbf{y}^* \end{bmatrix} = \begin{bmatrix} \mathbf{y}_n \\ \mathbf{y}_{n-1} \end{bmatrix} = \begin{bmatrix} \mathbf{y}_{n+1} \\ \mathbf{y}_n \end{bmatrix} = \begin{bmatrix} \mathbf{y}_{n-1} + \frac{h}{2}\{3\mathbf{f}(\mathbf{y}_n) - \mathbf{f}(\mathbf{y}_{n-1})\} \\ \mathbf{y}_n \end{bmatrix} ,$$

which implies that

$$\mathbf{y}_{n+1} = \mathbf{y}_n = \mathbf{y}_{n-1} = \mathbf{y}^* ,$$

where

$$\mathbf{0} = 3\mathbf{f}(\mathbf{y}^*) - \mathbf{f}(\mathbf{y}^*) \Longrightarrow \mathbf{f}(\mathbf{y}^*) = \mathbf{0} .$$

Thus the fixed points of the second-order Adams-Bashforth method are the same as the fixed points of the differential equation. Now

$$\frac{\partial \boldsymbol{\phi}}{\partial \begin{bmatrix} \mathbf{y}_n \\ \mathbf{y}_{n-1} \end{bmatrix}} = \begin{bmatrix} \mathbf{I} + \frac{3h}{2}\mathbf{f}'(\mathbf{y}_n) & -\frac{h}{2}\mathbf{f}'(\mathbf{y}_{n-1}) \\ \mathbf{I} & \mathbf{0} \end{bmatrix}$$

has eigenvalues at a fixed point satisfying

$$\mathbf{0} = -\lambda \left[\mathbf{I} + \frac{3h}{2}\mathbf{f}'(\mathbf{y}^*) - \mathbf{I}\lambda \right] + \frac{h}{2}\mathbf{f}'(\mathbf{y}^*) = \mathbf{I}\lambda^2 - \left\{ \mathbf{I} + \frac{3h}{2}\mathbf{f}'(\mathbf{y}^*) \right\}\lambda + \frac{h}{2}\mathbf{f}'(\mathbf{y}^*) .$$

For the logistics equation with $\mathbf{y}^* = 0$ we have $\mathbf{f}'(0) = 1$ and

$$\lambda = \frac{1}{2} + \frac{3h}{4} \pm \sqrt{\frac{1}{2} + \frac{h}{2} + \left(\frac{3r}{4}\right)^2} \, ,$$

so the positive root has modulus greater than one, implying that this period one fixed point is unstable. For the logistics equation with $\mathbf{y}^* = 1$ we have $\mathbf{f}'(1) = -1$ and

$$\lambda = \frac{1}{2} - \frac{3h}{4} \pm \sqrt{\frac{1}{2} - \frac{h}{2} + \left(\frac{3r}{4}\right)^2} \, ,$$

For $0 < h \le \frac{2}{3}$, the dominant eigenvalue is positive and has modulus less than 1; for $\frac{2}{3} < h < 1$, the dominant eigenvalue is negative and has modulus less than 1; and for $1 \le h$ the dominant eigenvalue is negative and has modulus greater than or equal to 1. Thus for the logistics equation, the second-order Adams-Bashforth method has a single stable fixed point of period 1 for $0 < h < 1$. Results from program GUINonlinearStability.C. applied to the second-order Adams-Bashforth method are shown in Fig. 3.12.

Example 3.8.5 The modified Euler method is a two-stage Runge-Kutta method that takes the form

$$\mathbf{y}_{n+1} = \mathbf{y}_n + h\mathbf{f}\left(\mathbf{y}_n + \frac{h}{2}\mathbf{f}(\mathbf{y}_n)\right) \, .$$

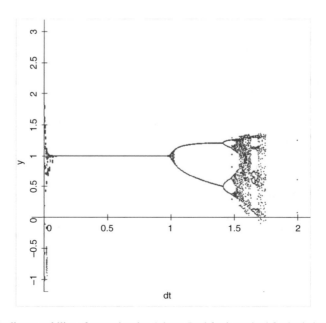

Fig. 3.12 Nonlinear stability of second-order Adams-Bashforth method for logistics equation

At a fixed point of period 1, we get

$$\mathbf{y}^* = \mathbf{y}^* + h\mathbf{f}\left(\mathbf{y}^* + \frac{h}{2}\mathbf{f}(\mathbf{y}^*)\right) ,$$

which implies that

$$\mathbf{f}\left(\mathbf{y}^* + \frac{h}{2}\mathbf{f}(\mathbf{y}^*)\right) = 0 .$$

For the logistics equation, this implies that

$$\mathbf{y}^* + \frac{h}{2}\mathbf{f}(\mathbf{y}^*) = 0 \text{ or } 1 ,$$

which in turn implies that

$$\mathbf{y}^*\left[1 + \frac{h}{2}(1 - \mathbf{y}^*)\right] = 0 \text{ or } 1 .$$

There are four solutions, namely $\mathbf{y}^* = 0, 1, 1 + 2/h$ and $2/h$. Thus the modified Euler method has more period 1 fixed points than the differential equation.

Let us analyze the stability of these period 1 fixed points. Since

$$\mathbf{F}(\mathbf{y}, h) \equiv \mathbf{y} + h\mathbf{f}\left(\mathbf{y} + \frac{h}{2}\mathbf{f}(\mathbf{y})\right) ,$$

we have

$$\frac{\partial \mathbf{F}}{\partial \mathbf{y}}(\mathbf{y}, h) = 1 + h\mathbf{f}'\left(\mathbf{y} + \frac{h}{2}\mathbf{f}(\mathbf{y})\right)\left[1 + \frac{h}{2}\mathbf{f}'(\mathbf{y})\right] .$$

For the logistics equation, we find that

$$\frac{\partial \mathbf{F}}{\partial \mathbf{y}}(0, h) = 1 + h\mathbf{f}'(0)\left[1 + \frac{h}{2}\mathbf{f}'(0)\right] = 1 + h\left[1 + \frac{h}{2}\right] > 1 \Longrightarrow \text{unstable} ,$$

$$\frac{\partial \mathbf{F}}{\partial \mathbf{y}}(1, h) = 1 + h\mathbf{f}'(1)\left[1 + \frac{h}{2}\mathbf{f}'(1)\right] = 1 - h\left[1 - \frac{h}{2}\right] = 1 - h + \frac{h^2}{2}$$

$$\Longrightarrow \text{stable for } 0 < h < 2 ,$$

$$\frac{\partial \mathbf{F}}{\partial \mathbf{y}}\left(1 + \frac{2}{h}, h\right) = 1 + h\mathbf{f}'(0)\left[-1 - \frac{h}{2}\right] = 1 - h\left[1 + \frac{h}{2}\right]$$

$$\Longrightarrow \text{stable for } 0 < h < \sqrt{5} - 1 ,$$

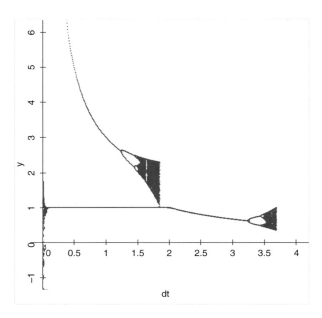

Fig. 3.13 Nonlinear stability of modified Euler method for logistics equation

$$\frac{\partial \mathbf{F}}{\partial \mathbf{y}}\left(\frac{2}{h}, h\right) = 1 + h\mathbf{f}'(1)\left[-1 + \frac{h}{2}\right] = 1 + h\left[1 - \frac{h}{2}\right]$$

$$\implies \text{stable for } 2 < h < \sqrt{5} + 1 \ .$$

Thus some of the spurious fixed points are stable for timesteps that lie within the region of absolute stability of the method. Results from program GUINonlinearStability.C applied to the modified Euler method are shown in Fig. 3.13.

Every numerical has spurious fixed points, but sometimes the spurious fixed points necessarily have period greater than one. However, it has been proved (see Yee et al. [186]) that if a linear multistep method tends to a fixed point $\mathbf{y}^* < \infty$ as $t \to \infty$, then y^* is a stable fixed point of the differential equation. Here, the theorem assumes that implicit methods solve nonlinear equations exactly. Of the methods we have examined, only Runge-Kutta and predictor-corrector methods may have spurious fixed points. It is possible for the spurious fixed points to lie between the fixed points of the differential equation.

These results suggest that significant care must be applied to Runge-Kutta and predictor-corrector methods. There are some analytical results that suggest timestep selection strategies for certain classes of Runge-Kutta methods will avoid spurious fixed points.

Exercise 3.8.3 Examine the stability of period one fixed points for the implicit midpoint rule applied to the logistics equation.

Exercise 3.8.4 Examine the stability of period one fixed points for the SDIRK method applied to the logistics equation.

Chapter 4
Boundary Value Problems

*. . . years ago many considered BVPs as some sort of a subclass
of IVPs, wherein one fiddles with the initial conditions in order
to get things right at the other end. It gradually became clear
that IVPs are actually a special and in some sense relatively
simple subclass of BVPs. The fundamental difference is that for
IVPs one has complete information about the solution at one
point (the initial point), so one may consider using a marching
algorithm which is always* local *in nature. For BVPs, on the
other hand, no complete information is available at any point,
so the end points have to be connected by the solution algorithm
in a* global *way. Only after stepping through the entire domain
can the solution at any point be determined.*

Uri M. Ascher, Robert M. M Mattheij
and Robert D. Russell *[7, p. xvii].*

*Science is a differential equation. Religion is a boundary
condition.*

Alan Turing *in epigram to Robin Gandy [104, p. 513]*

Abstract This chapter is devoted to boundary value problems for ordinary differential equations. It begins with analysis of the existence and uniqueness of solutions to these problems, and the effect of perturbations to the problem. The first numerical approach is the shooting method. This is followed by finite differences and collocation. Finite elements allow for the development of very high order methods for many boundary value problems, but their analysis typically requires sophisticated ideas from real analysis. The chapter ends with the application of deferred correction to both collocation and finite elements.

Additional Material: The details of the computer programs referred in the text are available in the Springer website (http://extras.springer.com/2018/978-3-319-69110-7) for authorized users.

J.A. Trangenstein, *Scientific Computing*, Texts in Computational
Science and Engineering 20, https://doi.org/10.1007/978-3-319-69110-7_4

4.1   Overview

In this our final chapter, we will examine another topic that will require us to apply many of the solution techniques from previous chapters. The new topic is the solution of two-point boundary value problems for ordinary differential equations. Since ordinary differential equations are involved, we should hope that some of our ideas in Chap. 3 might be useful for the new topic. Certainly numerical differentiation and integration will be important (see Chap. 2), as well as interpolation and approximation (which were discussed in Chap. 1). We will see that boundary value problems require solution techniques that act across the problem domain, often requiring the solution of large systems of linear equations (see Chaps. 3 of Volume I or 2 of Volume II) or nonlinear equations (see Chap. 3 of Volume II). Looking to the future, the discussions of initial value problems in Chap. 3 and of boundary value problems in this chapter should prepare readers for more complicated problems described by partial differential equations.

Let us introduce our new topic by examining a general application. **Hamilton's principle** in classical mechanics (see Goldstein [87, p. 30]) states that the motion of some system from a state $\mathbf{y}_0$ at some initial time t_0 to a state $\mathbf{y}_1$ at some later time t_1 is an extremum of the line integral of the Lagrangian in time. Mathematically speaking, if $\mathscr{L}(\mathbf{y}, \dot{\mathbf{y}}, t)$ is the **Lagrangian function**, then its **action function** is

$$\mathscr{H}(\mathbf{y}) \equiv \int_{t_0}^{t_1} \mathscr{L}(\mathbf{y}(t), \dot{\mathbf{y}}(t), t) \, dt \, ,$$

so the variation of $\mathscr{H}$ is

$$\delta \mathscr{H}(\mathbf{y}) = \int_{t_0}^{t_1} \{ \mathscr{L}(\mathbf{y}(t) + \delta \mathbf{y}(t), \dot{\mathbf{y}}(t) + \delta \dot{\mathbf{y}}(t), t) - \mathscr{L}(\mathbf{y}(t), \dot{\mathbf{y}}(t), t) \} \, dt$$

$$= \int_{t_0}^{t_1} \left\{ \frac{\partial \mathscr{L}}{\partial \mathbf{y}} (\mathbf{y}(t), \dot{\mathbf{y}}(t), t) \, \delta \mathbf{y}(t) + \frac{\partial \mathscr{L}}{\partial \dot{\mathbf{y}}} (\mathbf{y}(t), \dot{\mathbf{y}}(t), t) \, \delta \dot{\mathbf{y}}(t) \right\} \, dt$$

which allows us to integrate by parts and get

$$= \left[\frac{\partial \mathscr{L}}{\partial \dot{\mathbf{y}}} (\mathbf{y}(t), \dot{\mathbf{y}}(t), t) \, \delta \dot{\mathbf{y}}(t) \right]_{t_0}^{t_1}$$

$$+ \int_{t_0}^{t_1} \left\{ \frac{\partial \mathscr{L}}{\partial \mathbf{y}} (\mathbf{y}(t), \dot{\mathbf{y}}(t), t) - \frac{d}{dt} \frac{\partial \mathscr{L}}{\partial \dot{\mathbf{y}}} (\mathbf{y}(t), \dot{\mathbf{y}}(t), t) \right\} \, \delta \mathbf{y}(t) \, dt$$

but because the variation $\delta \mathbf{y}$ in the motion must be zero at the boundaries, we arrive at

$$= \int_{t_0}^{t_1} \left\{ \frac{\partial \mathscr{L}}{\partial \mathbf{y}} (\mathbf{y}(t), \dot{\mathbf{y}}(t), t) - \frac{d}{dt} \frac{\partial \mathscr{L}}{\partial \dot{\mathbf{y}}} (\mathbf{y}(t), \dot{\mathbf{y}}(t), t) \right\} \, \delta \mathbf{y}(t) \, dt \, .$$

Since the variation $\delta \mathbf{y}(t)$ is otherwise arbitrary, we deduce that the true motion $\mathbf{y}$ satisfies the **Euler-Lagrange equations**

$$\frac{\partial \mathscr{L}}{\partial \mathbf{y}}(\mathbf{y}(t), \dot{\mathbf{y}}(t), t) - \frac{d}{dt}\frac{\partial \mathscr{L}}{\partial \dot{\mathbf{y}}}(\mathbf{y}(t), \dot{\mathbf{y}}(t), t) = 0 \text{ for } t_0 < t < t_1$$

with $\mathbf{y}(t_0) = \mathbf{y}_0$ and $\mathbf{y}(t_1) = \mathbf{y}_1$. $\hspace{2cm}$ (4.1)

In classical mechanics, the **variational form** of the equations of motion is more fundamental than the derived two-point boundary value problem.

Somewhat more sophisticated discussions of static equilibrium in continuum mechanics (see Truesdell [177, p. 265ff]) develop variational principles for the deformation of a material, together with certain energy inequalities. From these, a differential equation with boundary conditions can be derived, but the energy inequalities may fall outside the standard mathematical framework for two-point boundary value problems.

Typical texts discuss the theory of two-point boundary value problems for ordinary differential equations in terms of **eigenfunction expansions** and the Sturm-Liouville theory. The most popular exposition among these is probably the book by Boyce and DiPrima [19]. We also recommend the book by Hartman [98] and the book by Kelley and Peterson [118]. The latter two books provide a more general discussion of nonlinear two-point boundary value problems.

Some numerical texts discuss two-point boundary value problems, such as Kincaid and Cheney [119] and Heath [99]. More general texts on numerical methods for partial differential equations discuss two-point boundary value problems as examples. See, for example, Iserles [113, p. 171ff], Strang and Fix [163, p. 3] and Trangenstein [174, p. 190ff].

Netlib provides a variety of programs for solving two-point boundary value problems, such as colnew or colsys by Ascher, Christiansen and Russell, MUS by Mattheij and Staarink, twpbvp, acdc and colmod by Cash and Wright, and mirkdc by Enright and Muir. MATLAB users can use the command bvp4c to solve two-point boundary value problems.

Our goals in this chapter are to examine some theory for nonlinear boundary value problems involving ordinary differential equations, and to use that theory to construct useful numerical methods. Readers should learn about the important classes of numerical methods for two-point boundary value problems, and how to select and implement an appropriate method for their applications.

Here is a brief introduction to the topics in this chapter. We will begin with a mathematical description of the two-point boundary value problem, and provide some theory regarding well-posedness of the problem. For our first numerical approach, we will use the ideas from initial value problems to develop shooting methods for boundary value problems. Then we will discuss finite difference methods and collocation methods. We will end the chapter with a discussion of finite element methods and deferred correction.

4.2   Theory

Our discussion in Sect. 1.3 of Chap. 1 in Volume I about the steps in solving a scientific computing problem emphasized the importance of examining the well-posedness of each problem. Recall that a problem is well-posed if a solution exists, is unique, and depends continuously on the data. We will begin by discussing existence and uniqueness.

4.2.1   Existence and Uniqueness

There are several kinds of two-point boundary value problems.

Definition 4.2.1 Let $t_0 < t_1$ be two real numbers, and let the function $f(y, \dot{y}, t)$ map a triple of real numbers to real numbers. Given two real numbers y_0 and y_1, the **scalar boundary value problem** for an ordinary differential equation is to find a twice differentiable function $y(t)$ mapping real numbers to real numbers so that

$$y''(t) = f(y(t), y'(t), t) \text{ for } t_0 < t < t_1 \text{ with } y(t_0) = y_0 \text{ and } y(t_1) = y_1 . \quad (4.2)$$

Alternatively, let the function $\mathbf{f}(\mathbf{y}, \dot{\mathbf{y}}, t)$ map a triple of two n-vectors and a scalar to n-vectors, and let the function $\mathbf{g}$ map a pair of n-vectors to an n-vector. Then the more general **vector boundary value problem** for an ordinary differential equation is to find a differentiable function $\mathbf{y}(t)$ mapping real numbers to n-vectors so that

$$\mathbf{y}'(t) = \mathbf{f}(\mathbf{y}(t), t) \text{ for } t_0 < t < t_1 \text{ with } \mathbf{g}(\mathbf{y}(t_0), \mathbf{y}(t_1)) = \mathbf{0} . \quad (4.3)$$

Finally, let $\mathbf{A}(t)$ map real numbers to $n \times n$ matrices, and let $\mathbf{a}(t)$ map real numbers to n-vectors. Suppose that $\mathbf{B}_0$ and $\mathbf{B}_1$ are two $n \times n$ matrices, and $\mathbf{b}$ is an n-vector. Then the general **two-point boundary value problem for a linear ordinary differential equation** is to find a differentiable function $\mathbf{y}(t)$ mapping real numbers to n-vectors so that

$$\mathbf{y}'(t) = \mathbf{A}(t)\mathbf{y}(t) + \mathbf{a}(t) \text{ for } t_0 < t < t_1 \text{ with } \mathbf{B}_0\mathbf{y}(t_0) + \mathbf{B}_1\mathbf{y}(t_1) = \mathbf{b} . \quad (4.4)$$

Here is a famous example of a scalar boundary value problem.

Example 4.2.1 Given a real number $\lambda > 0$, the **Bratu problem** is to find a function $u(t)$ so that

$$u''(t) + \lambda e^{u(t)} = 0 \text{ for } 0 < t < 1 \text{ with } u(0) = 0 = u(1) . \quad (4.5)$$

This problem arises in mathematical modeling of combustion [15]. If we look for a solution of the form

$$u(t) = -2\ln\left(\frac{\cosh\left([t - 1/2]\theta/2\right)}{\cosh(\theta/4)}\right) ,$$

we find that

$$-\frac{\theta^2}{2}\frac{1}{\cosh\left([t-1/2]\theta/2\right)} = u''(t) = -\lambda e^{u(t)} = -\lambda\frac{\cosh^2(\theta/4)}{\cosh\left([t-1/2]\theta/2\right)} .$$

This gives us the equation

$$\theta = \sqrt{2\lambda}\cosh(\theta/4) . \tag{4.6}$$

For large values of λ, the function on the right in (4.6) does not intersect the function on the left; for small values of λ there are two intersections. At the critical value of λ there is a single point of intersection where the linear function on the left is tangent to the function on the right. The equation for the slopes at the point of tangency is

$$1 = \frac{1}{4}\sqrt{2\lambda_c}\sinh\left(\frac{\theta}{4}\right) . \tag{4.7}$$

The two Eqs. (4.6) and (4.7) give us a nonlinear system with solution satisfying $\lambda_c \approx 3.51383$. For $\lambda < \lambda_c$, the Bratu problem has two solutions, corresponding to the two values of θ satisfying (4.6). For $\lambda > \lambda_c$ the Bratu problem has no solution.

We also note that the Euler-Lagrange equations (4.1) provide an example of a vector boundary value problem.

It is easiest to examine the existence and uniqueness of solutions to the linear boundary value problem.

Lemma 4.2.1 *Assume that $t_0 < t_1$ are two real numbers. Let the function $A(t)$ be a continuous mapping of real numbers to real $n \times n$ matrices, and let $a(t)$ be a continuous mapping of real numbers to real n-vectors. Suppose that B_0 and B_1 are two real $n \times n$ matrices, and that b is a real n-vector. Let $F(t)$ be a real fundamental matrix defined by*

$$F'(t) = A(t)F(t) \text{ for } t > t_0 \text{ with } F(t_0) \text{ nonsingular} .$$

Then the linear two-point boundary value problem

$$y'(t) = A(t)y(t) + a(t) \text{ for } t_0 < t < t_1 \text{ with } B_0 y(t_0) + B_1 y(t_1) = b$$

has a solution if and only if there exists an n-vector y_0 so that

$$[B_1 F(t_1) + B_0 F(t_0)]y_0 = b - B_1 F(t_1)\int_{t_0}^{t_1} F(s)^{-1}a(s) \, ds ,$$

in which case a solution of the linear two-point boundary value problem is

$$y(t) = F(t)\left\{y_0 + \int_{t_0}^{t} F(s)^{-1}a(s) \, ds\right\} .$$

Furthermore, the linear two-point boundary value problem has a unique solution if and only if $B_1 F(t_1) + B_0 F(t_0)$ is nonsingular.

Proof Theorem 3.2.4 shows that $\mathbf{F}(t)$ exists and is nonsingular for all $t \in [t_0, t_1]$. Next, Theorem 3.2.5 shows that the solution of the initial value problem

$$\mathbf{y}'(t) = \mathbf{A}(t)\mathbf{y}(t) + \mathbf{a}(t) \text{ for } t \in (t_0, t_1) \text{ with } \mathbf{y}(t_0) = \mathbf{y}_0$$

is

$$\mathbf{y}(t) = \mathbf{F}(t)\mathbf{y}_0 + \mathbf{F}(t) \int_{t_0}^{t} \mathbf{F}(s)^{-1}\mathbf{a}(s) \, ds \ .$$

The two-point boundary condition implies that we want to choose the initial value $\mathbf{y}_0$ so that

$$\mathbf{b} = \mathbf{B}_1 \left\{ \mathbf{F}(t_1)\mathbf{y}_0 + \mathbf{F}(t_1) \int_{t_0}^{t_1} \mathbf{F}(s)^{-1}\mathbf{a}(s) \, ds \right\} + \mathbf{B}_0\mathbf{y}_0$$

$$= \{\mathbf{B}_1\mathbf{F}(t_1) + \mathbf{B}_0\mathbf{F}(t_0)\} \, \mathbf{y}_0 + \mathbf{B}_1\mathbf{F}(t_1) \int_{t_0}^{t_1} \mathbf{F}(s)^{-1}\mathbf{a}(s) \, ds \ .$$

This gives us a linear system for $\mathbf{y}_0$, and the conclusions of the lemma follow easily from Lemmas 3.2.1 and 3.2.8 in Chap. 3 of Volume I.

We can present the solution of the linear boundary value problem in another interesting fashion.

Theorem 4.2.1 (Green's Function)   *Assume that $t_0 < t_1$ are two real numbers. Let the function $\mathbf{A}(t)$ be a continuous mapping of real numbers to real $n \times n$ matrices, and let $\mathbf{a}(t)$ be a continuous mapping of real numbers to real n-vectors. Suppose that $\mathbf{B}_0$ and $\mathbf{B}_1$ are two real $n \times n$ matrices, and that $\mathbf{b}$ is a real n-vector. Let $\mathbf{F}(t)$ be a fundamental matrix defined by*

$$\mathbf{F}'(t) = \mathbf{A}(t)\mathbf{F}(t) \text{ for } t > t_0 \text{ with } \mathbf{F}(t_0) \text{ nonsingular} \ ,$$

and let the matrix $\mathbf{B}_1\mathbf{F}(t_1) + \mathbf{B}_0\mathbf{F}(t_0)$ be nonsingular. Then we can choose $\mathbf{F}(t_0)$ so that

$$\mathbf{B}_0\mathbf{F}(t_0) + \mathbf{B}_1\mathbf{F}(t_1) = \mathbf{I} \ .$$

*Furthermore, there is a **Green's function** $\mathbf{G}(t, s)$ mapping pairs of real numbers to real $n \times n$ matrices so that*

$$\frac{\partial \mathbf{G}}{\partial t}(t, s) = \mathbf{A}(t)\mathbf{G}(t, s) \text{ for } t_0 < t < t_1 \text{ and for } t \neq s$$

with $\mathbf{B}_0\mathbf{G}(t_0, s) + \mathbf{B}_1\mathbf{G}(t_1, s) = \mathbf{0}$ for $t_0 < s < t_1$,

and

$$\lim_{t \downarrow s} \mathbf{G}(t, s) = \lim_{t \uparrow s} \mathbf{G}(t, s) + \mathbf{I} \ .$$

Finally, the solution of the linear boundary value problem

$$\mathbf{y}'(t) = \mathbf{A}(t)\mathbf{y}(t) + \mathbf{a}(t) \text{ for } t_0 < t < t_1 \text{ with } \mathbf{B}_0\mathbf{y}(t_0) + \mathbf{B}_1\mathbf{y}(t_1) = \mathbf{b}$$

is

$$\mathbf{y}(t) = \mathbf{F}(t)\mathbf{b} + \int_{t_0}^{t_1} \mathbf{G}(t, s)\mathbf{a}(s) \, ds . \tag{4.8}$$

Proof Theorem 3.2.4 shows that we can find a fundamental matrix $\mathbf{E}(t)$ to solve

$$\mathbf{E}'(t) = \mathbf{A}(t)\mathbf{E}(t) \text{ for } t_0 < t \text{ with } \mathbf{E}(t_0) = \mathbf{I} .$$

Then another other fundamental solution $\mathbf{F}(t)$, with nonsingular initial value $\mathbf{F}(t_0) = \mathbf{F}_0$, is of the form

$$\mathbf{F}(t) = \mathbf{E}(t)\mathbf{F}_0 .$$

This is because

$$\frac{d}{dt} \left(\mathbf{F}\mathbf{F}_0^{-1} \right)' (t) = \mathbf{A}(t) \left(\mathbf{F}(t)\mathbf{F}_0^{-1} \right) \text{ for } t_0 < t$$

satisfies the ordinary differential equation for a fundamental matrix with initial value

$$\mathbf{F}(t_0)\mathbf{F}_0^{-1} = \mathbf{I} .$$

Since

$$\mathbf{B}_0\mathbf{F}(t_0) + \mathbf{B}_1\mathbf{F}(t_1) = [\mathbf{B}_0 + \mathbf{B}_1\mathbf{E}(t_1)] \, \mathbf{F}_0$$

is nonsingular, it follows that $\mathbf{B}_0 + \mathbf{B}_1\mathbf{E}(t_1)$ is nonsingular. Thus we can choose $\mathbf{F}_0 = \mathbf{F}(t_0)$ so that

$$\mathbf{I} = [\mathbf{B}_0 + \mathbf{B}_1\mathbf{E}(t_1)] \, \mathbf{F}_0 = \mathbf{B}_0\mathbf{F}(t_0) + \mathbf{B}_1\mathbf{F}(t_1) .$$

Theorem 4.2.1 shows that the solution of the linear boundary value problem is

$$\mathbf{y}(t) = \mathbf{F}(t) \, [\mathbf{B}_0\mathbf{F}(t_0) + \mathbf{B}_1\mathbf{F}(t_1)] \left\{ \mathbf{b} - \mathbf{B}_1\mathbf{F}(t_1) \int_{t_0}^{t_1} \mathbf{F}(s)^{-1}\mathbf{a}(s) \, ds \right\}$$

$$+ \mathbf{F}(t) \int_{t_0}^{t} \mathbf{F}(s)^{-1}\mathbf{a}(s) \, ds$$

$$= \mathbf{F}(t)\mathbf{b} + \int_{t_0}^{t} \mathbf{F}(t)\mathbf{F}(s)^{-1}\mathbf{a}(s) \, ds - \int_{t_0}^{t_1} \mathbf{F}(t)\mathbf{B}_1\mathbf{F}(t_1)\mathbf{F}(s)^{-1}\mathbf{a}(s) \, ds$$

$$= \mathbf{F}(t)\mathbf{b} + \int_{t_0}^{t} \mathbf{F}(t)\left[\mathbf{B}_0\mathbf{F}(t_0) + \mathbf{B}_1\mathbf{F}(t_1)\right]\mathbf{F}(s)^{-1}\mathbf{a}(s)\,\mathrm{d}s$$

$$- \int_{t_0}^{t_1} \mathbf{F}(t)\mathbf{B}_1\mathbf{F}(t_1)\mathbf{F}(s)^{-1}\mathbf{a}(s)\,\mathrm{d}s$$

$$= \mathbf{F}(t)\mathbf{b} + \int_{t_0}^{t} \mathbf{F}(t)\mathbf{B}_0\mathbf{F}(t_0)\mathbf{F}(s)^{-1}\mathbf{a}(s)\,\mathrm{d}s - \int_{t}^{t_1} \mathbf{F}(t)\mathbf{B}_1\mathbf{F}(t_1)\mathbf{F}(s)^{-1}\mathbf{a}(s)\,\mathrm{d}s$$

$$\equiv \mathbf{F}(t)\mathbf{b} + \int_{t_0}^{t_1} \mathbf{G}(t,s)\mathbf{a}(s)\,\mathrm{d}s$$

where the Green's function $\mathbf{G}(t,s)$ is defined by

$$\mathbf{G}(t,s) = \begin{cases} \mathbf{F}(t)\mathbf{B}_0\mathbf{F}(t_0)\mathbf{F}(s)^{-1}, & t_0 < s < t < t_1 \\ -\mathbf{F}(t)\mathbf{B}_1\mathbf{F}(t_1)\mathbf{F}(s)^{-1}, & t_0 < t < s < t_1 \end{cases} .$$

This definition shows that

$$\frac{\partial \mathbf{G}}{\partial t}(t,s) = \mathbf{A}(t)\mathbf{G}(t,s) \text{ for } t_0 < t < t_1 \text{ and } t \neq s .$$

Furthermore, for $t_0 < s < t_1$ we have

$$\mathbf{B}_0\mathbf{G}(t_0,s) + \mathbf{B}_1\mathbf{G}(t_1,s) = -\mathbf{B}_0\mathbf{F}(t_0)\mathbf{B}_1\mathbf{F}(t_1)\mathbf{F}(s)^{-1} + \mathbf{B}_1\mathbf{F}(t_1)\mathbf{B}_0\mathbf{F}(t_0)\mathbf{F}(s)^{-1}$$

$$= \left[\mathbf{B}_1\mathbf{F}(t_1)\mathbf{B}_0\mathbf{F}(t_0) - \mathbf{B}_0\mathbf{F}(t_0)\mathbf{B}_1\mathbf{F}(t_1)\right]\mathbf{F}(s)^{-1}$$

$$= \left[\{\mathbf{I} - \mathbf{B}_0\mathbf{F}(t_0)\}\mathbf{B}_0\mathbf{F}(t_0) - \mathbf{B}_0\mathbf{F}(t_0)\{\mathbf{I} - \mathbf{B}_0\mathbf{F}(t_0)\}\right]\mathbf{F}(s)^{-1} = \mathbf{0} .$$

It is also easy to see that

$$\lim_{t \downarrow s} \mathbf{G}(t,s) = \mathbf{F}(s)\mathbf{B}_0\mathbf{F}(t_0)\mathbf{F}(s)^{-1}$$

and that

$$\lim_{t \uparrow s} \mathbf{G}(t,s) = -\mathbf{F}(s)\mathbf{B}_1\mathbf{F}(t_1)\mathbf{F}(s)^{-1} = -\mathbf{F}(s)\left[\mathbf{I} - \mathbf{B}_0\mathbf{F}(t_0)\right]\mathbf{F}(s)^{-1}$$

$$= -\mathbf{I} + \mathbf{F}(s)\mathbf{B}_0\mathbf{F}(t_0)\mathbf{F}(s)^{-1} .$$

Example 4.2.2 Consider the two-point boundary value problem

$$y''(t) = -y(t) \text{ for } 0 < t < \pi \text{ with } y(0) = 0 = y(1) .$$

We can write this in the form of a linear boundary value problem (4.4) as follows:

$$\begin{bmatrix} y' \\ y \end{bmatrix}'(t) = \begin{bmatrix} 0 & -1 \\ 1 & 0 \end{bmatrix} \begin{bmatrix} y' \\ y \end{bmatrix}(t) \text{ for } 0 < t < \pi \text{ with } \begin{bmatrix} 0 & 0 \\ 0 & 1 \end{bmatrix} \begin{bmatrix} y' \\ y \end{bmatrix}(\pi)$$

$$+ \begin{bmatrix} 0 & 1 \\ 0 & 0 \end{bmatrix} \begin{bmatrix} y' \\ y \end{bmatrix}(0) = \begin{bmatrix} 0 \\ 0 \end{bmatrix}.$$

The fundamental matrix is

$$\mathbf{E}(t) = \exp\left(\begin{bmatrix} 0 & -1 \\ 1 & 0 \end{bmatrix} t \right) = \begin{bmatrix} \cos t & -\sin t \\ \sin t & \cos t \end{bmatrix}.$$

Since the matrix

$$\mathbf{B}_1 \mathbf{E}(\pi) + \mathbf{B}_0 \mathbf{E}(0) = \begin{bmatrix} 0 & 0 \\ 0 & 1 \end{bmatrix} \begin{bmatrix} -1 & 0 \\ 0 & -1 \end{bmatrix} + \begin{bmatrix} 0 & 1 \\ 0 & 0 \end{bmatrix} \begin{bmatrix} 1 & 0 \\ 0 & 1 \end{bmatrix} = \begin{bmatrix} 0 & 1 \\ 0 & -1 \end{bmatrix}$$

is singular, this particular two-point boundary value problem does not have a unique solution. In fact,

$$y(t) = \alpha \sin t$$

is a solution for any real number α.

Let us consider the same ordinary differential equation with different boundary conditions:

$$y''(t) = -y(t) \text{ for } 0 < t < \pi \text{ with } y(0) = 0 \text{ and } y'(1) = -1 .$$

We can write this in the form of a linear boundary value problem (4.4) as follows:

$$\begin{bmatrix} y' \\ y \end{bmatrix}'(t) = \begin{bmatrix} 0 & -1 \\ 1 & 0 \end{bmatrix} \begin{bmatrix} y' \\ y \end{bmatrix}(t) \text{ for } 0 < t < \pi \text{ with } \begin{bmatrix} 0 & 0 \\ 1 & 0 \end{bmatrix} \begin{bmatrix} y' \\ y \end{bmatrix}(\pi)$$

$$+ \begin{bmatrix} 0 & 1 \\ 0 & 0 \end{bmatrix} \begin{bmatrix} y' \\ y \end{bmatrix}(0) = \begin{bmatrix} 0 \\ -1 \end{bmatrix}.$$

In this case, the matrix

$$\mathbf{B}_1 \mathbf{E}(\pi) + \mathbf{B}_0 \mathbf{E}(0) = \begin{bmatrix} 0 & 0 \\ 1 & 0 \end{bmatrix} \begin{bmatrix} -1 & 0 \\ 0 & -1 \end{bmatrix} + \begin{bmatrix} 0 & 1 \\ 0 & 0 \end{bmatrix} \begin{bmatrix} 1 & 0 \\ 0 & 1 \end{bmatrix} = \begin{bmatrix} 0 & 1 \\ -1 & 0 \end{bmatrix}$$

is nonsingular, so this modified boundary value problem has a unique solution, namely $y(t) = \sin t$. We leave it to the reader to verify that the Green's function for this problem is

$$
\mathbf{G}(t, s) = \begin{cases} \begin{bmatrix} \sin(t)\sin(s) & -\sin(t)\cos(s) \\ -\cos(t)\sin(s) & \cos(t)\cos(s) \end{bmatrix}, & t_0 < s < t < t_1 \\[2mm] \begin{bmatrix} -\cos(t)\cos(s) & -\cos(t)\sin(s) \\ -\sin(t)\cos(s) & \sin(t)\sin(s) \end{bmatrix}, & t_0 < t < s < t_1 \end{cases} .
$$

Next, we will study the scalar boundary value problem (4.2). First, suppose that we specify the value of the solution at both endpoints.

Theorem 4.2.2 *Let $t_0 < t_1$ be two real numbers. Suppose that the function $f(y, \dot{y}, t)$ mapping a triple of real numbers to a real number is continuous for all real y, all real $\dot{y}$ and all $t \in [t_0, t_1]$. Then the scalar boundary value problem*

$$
y''(t) = f\left(y(t), y'(t), t\right) \text{ for } t_0 < t < t_1 \text{ with } y(t_0) = y_0 \text{ and } y(t_1) = y_1
$$

has a solution for any choice of the boundary values y_0 and y_1.

Proof See, for example, Kelley and Peterson [118, p. 316].
By imposing additional continuity conditions on f and limiting the size of the problem domain $[t_0, t_1]$, we can guarantee that this scalar boundary value problem has a unique solution.

Theorem 4.2.3 *Let $t_0 < t_1$ be two real numbers. Suppose that the function $f(y, \dot{y}, t)$ mapping a triple of real numbers to a real number is continuous for all real y, all real $\dot{y}$ and all $t \in [t_0, t_1]$. Assume that there is a positive number Φ and a nonnegative number $\dot{\Phi}$ so that for all $t \in [t_0, t_1]$ and all real numbers y, $\dot{y}$, $\widetilde{y}$ and $\widetilde{\dot{y}}$ we have*

$$
\left| f\left(\widetilde{y}, \widetilde{\dot{y}}, t\right) - f(y, \dot{y}, t) \right| \le \Phi \left| \widetilde{y} - y \right| + \dot{\Phi} \left| \widetilde{\dot{y}} - \dot{y} \right| .
$$

Finally, suppose that

$$
\frac{\Phi}{8} (t_1 - t_0)^2 + \frac{\dot{\Phi}}{2} (t_1 - t_0) < 1 .
$$

Then the scalar boundary value problem

$$
y''(t) = f\left(y(t), y'(t), t\right) \text{ for } t_0 < t < t_1 \text{ with } y(t_0) = y_0 \text{ and } y(t_1) = y_1
$$

has a unique solution for any choice of the boundary values y_0 and y_1.

Proof See, for example, Hartman [98, p. 423] or Kelley and Peterson [118, p. 309].

The situation is a bit more complicated if we specify the derivative of the solution at the right-hand boundary.

Theorem 4.2.4 *Let $t_0 < t_1$ be two real numbers. Suppose that the function $f(y, \dot{y}, t)$ mapping a triple of scalars to real number is continuous for all y, all $\dot{y}$ and all $t \in [t_0, t_1]$. Assume that there are nonnegative scalars Φ and $\dot{\Phi}$ so that for all $t \in [t_0, t_1]$ and all real numbers y, $\dot{y}$, $\widetilde{y}$ and $\widetilde{\dot{y}}$ we have*

$$\left| f\left(\widetilde{y}, \widetilde{\dot{y}}, t\right) - f(y, \dot{y}, t) \right| \le \Phi \left| \widetilde{y} - y \right| + \dot{\Phi} \left| \widetilde{\dot{y}} - \dot{y} \right| .$$

Finally, suppose that

$$t_1 - t_1 < \begin{cases} 2\cos^{-1}\left(\dot{\Phi}/\left[2\sqrt{\Phi}\right]\right)/\sqrt{4\Phi - \dot{\Phi}^2}, & \dot{\Phi}^2 < 4\Phi \\ 2\cosh^{-1}\left(\dot{\Phi}/\left[2\sqrt{\Phi}\right]\right)/\sqrt{\dot{\Phi}^2 - 4\Phi}, & \dot{\Phi}^2 > 4\Phi \text{ and } \Phi > 0 \\ 2/\dot{\Phi}, & \dot{\Phi}^2 = 4\Phi \text{ and } \dot{\Phi} > 0 \\ \infty, & \dot{\Phi} \ge 0 \text{ and } \Phi = 0 \end{cases} .$$

Then the scalar boundary value problem

$$y''(t) = f\left(y(t), y'(t), t\right) \text{ for } t_0 < t < t_1 \text{ with } y(t_0) = y_0 \text{ and } y'(t_1) = \dot{y}_1$$

has a unique solution for any choice of the boundary values y_0 and $\dot{y}_1$.

Proof See, for example, Kelley and Peterson [118, p. 314].

Finally, let us discuss vector boundary value problems. Suppose that we want to solve

$$\mathbf{y}'(t) = \mathbf{f}(\mathbf{y}(t), t) \text{ for } t_0 < t < t_1 \text{ with } \mathbf{g}(\mathbf{y}(t_0), \mathbf{y}(t_1)) = \mathbf{0} .$$

Theorem 3.2.2 shows that if $\mathbf{f}$ is Lipschitz continuous in its first argument then for each vector $\mathbf{z}_0$ there is a solution $\mathbf{z}(t)$ for the initial value problem

$$\mathbf{z}'(t) = \mathbf{f}(\mathbf{z}(t), t) \text{ for } t_0 < t \text{ with } \mathbf{z}(t_0) = \mathbf{z}_0 .$$

Consider the set

$$\mathscr{L} = \{(\mathbf{z}_0, \mathbf{z}_1) : \text{ there exists } \mathbf{z}(t) \text{ so that } \mathbf{z}'(t) = \mathbf{f}(\mathbf{z}(t), t) \text{ for } t_0 < t \le t_1$$
$$\text{with } \mathbf{z}(t_0) = \mathbf{z}_0 \text{ and } \mathbf{z}_1 = \mathbf{z}(t_1)\} .$$

If $\mathscr{L}$ is nonempty, then we look for members of the set

$$\mathscr{S} = \{(\mathbf{z}_0, \mathbf{z}_1) \in \mathscr{L} : \mathbf{g}(\mathbf{z}_0, \mathbf{z}_1) = \mathbf{0}\} .$$

Each member of $\mathscr{S}$ provides a pair of boundary values for a solution of the original vector boundary value problem. This approach suggests that we can solve boundary value problems by using methods for solving initial value problem, and then employing a nonlinear equation solver to select the correct initial values. This approach is discussed in Sect. 4.3.

Exercise 4.2.1 Let $\mathbf{G}(t, s)$ be the Green's function in Theorem 4.2.1, and let $\mathbf{F}$ be the fundamental matrix in this same theorem. Show that

$$\mathbf{G}(t, s)\mathbf{F}(s) - \mathbf{G}(t, u)\mathbf{F}(u) = \begin{cases} \mathbf{F}(t), & s < t < u \\ -\mathbf{F}(t), & u < t < s \\ \mathbf{0}, & \text{otherwise} \end{cases}.$$

4.2.2   Perturbations

It is difficult to conduct a proper perturbation analysis of either nonlinear scalar boundary value problems or nonlinear vector boundary value problems. Accordingly, we will limit our discussion to linear boundary value problems, and only to perturbations in the inhomogeneities for these problems.

Suppose that $\mathbf{y}(t)$ solves

$$\mathbf{y}'(t) = \mathbf{A}(t)\mathbf{y}(t) + \mathbf{a}(t) \text{ for } t_0 < t < t_1 \text{ with } \mathbf{B}_0\mathbf{y}(t_0) + \mathbf{B}_1\mathbf{y}(t_1) = \mathbf{b}$$

and $\widetilde{\mathbf{y}}(t)$ solves the perturbed problem

$$\widetilde{\mathbf{y}}'(t) = \mathbf{A}(t)\widetilde{\mathbf{y}}(t) + \widetilde{\mathbf{a}}(t) \text{ for } t_0 < t < t_1 \text{ with } \mathbf{B}_0\widetilde{\mathbf{y}}(t_0) + \mathbf{B}_1\widetilde{\mathbf{y}}(t_1) = \widetilde{\mathbf{b}}.$$

Then we have

$$\mathbf{y}(t) = \mathbf{F}(t)\mathbf{b} + \int_{t_0}^{t_1} \mathbf{G}(t, s)\mathbf{a}(s) \, ds$$

and

$$\widetilde{\mathbf{y}}(t) = \mathbf{F}(t)\widetilde{\mathbf{b}} + \int_{t_0}^{t_1} \mathbf{G}(t, s)\widetilde{\mathbf{a}}(s) \, ds,$$

where the fundamental solution $\mathbf{F}(t)$ satisfies

$$\mathbf{F}'(t) = \mathbf{A}(t)\mathbf{F}(t) \text{ for } t_0 < t < t_1 \text{ with } \mathbf{B}_0\mathbf{F}(t_0) + \mathbf{B}_1\mathbf{F}(t_1) = \mathbf{I}$$

and the Green's function $\mathbf{G}(t, s)$ is

$$\mathbf{G}(t, s) = \begin{cases} \mathbf{F}(t)\mathbf{B}_0\mathbf{F}(t_0)\mathbf{F}(s)^{-1}, & t_0 < s < t < t_1 \\ -\mathbf{F}(t)\mathbf{B}_1\mathbf{F}(t_1)\mathbf{F}(s)^{-1}, & t_0 < t < s < t_1 \end{cases}.$$

It follows that the perturbation in the solution is

$$\widetilde{\mathbf{y}}(t) - \mathbf{y}(t) = \mathbf{F}(t)\left[\widetilde{\mathbf{b}} - \mathbf{b}\right] + \int_{t_0}^{t_1} \mathbf{G}(t,s)\left[\widetilde{\mathbf{a}}(s) - \mathbf{a}(s)\right]\, ds\ .$$

We can take vector norms of both sides to obtain

$$\|\widetilde{\mathbf{y}}(t) - \mathbf{y}(t)\| \leq \|\mathbf{F}(t)\|\ \|\widetilde{\mathbf{b}} - \mathbf{b}\| + \int_{t_0}^{t_1} \|\mathbf{G}(t,s)\|\ \|\widetilde{\mathbf{a}}(s) - \mathbf{a}(s)\|\, ds\ .$$

Finally, we can take the maximum in t to get

$$\max_{t \in [t_0, t_1]} \|\widetilde{\mathbf{y}}(t) - \mathbf{y}(t)\|$$

$$\leq \max_{t \in [t_0, t_1]} \|\mathbf{F}(t)\|\ \|\widetilde{\mathbf{b}} - \mathbf{b}\| + \max_{t \in [t_0, t_1]} \int_{t_0}^{t_1} \|\mathbf{G}(t,s)\|\ ds \max_{s \in [t_0, t_1]} \|\widetilde{\mathbf{a}}(s) - \mathbf{a}(s)\|\ .$$

Perturbations in $\mathbf{A}(t)$, $\mathbf{B}_0$ and/or $\mathbf{B}_1$ can be treated in a similar fashion. Suppose that $\mathbf{y}(t)$ solves

$$\mathbf{y}'(t) = \mathbf{A}(t)\mathbf{y}(t) + \mathbf{a}(t) \text{ for } t_0 < t < t_1 \text{ with } \mathbf{B}_0\mathbf{y}(t_0) + \mathbf{B}_1\mathbf{y}(t_1) = \mathbf{b}$$

and $\widetilde{\mathbf{y}}(t)$ solves the perturbed problem

$$\widetilde{\mathbf{y}}'(t) = \widetilde{\mathbf{A}}(t)\widetilde{\mathbf{y}}(t) + \mathbf{a}(t) \text{ for } t_0 < t < t_1 \text{ with } \widetilde{\mathbf{B}}_0\widetilde{\mathbf{y}}(t_0) + \widetilde{\mathbf{B}}_1\widetilde{\mathbf{y}}(t_1) = \mathbf{b}\ .$$

Then we have

$$\mathbf{y}(t) = \mathbf{F}(t)\mathbf{b} + \int_{t_0}^{t_1} \mathbf{G}(t,s)\mathbf{a}(s)\, ds$$

and

$$\widetilde{\mathbf{y}}(t) = \widetilde{\mathbf{F}}(t)\mathbf{b} + \int_{t_0}^{t_1} \widetilde{\mathbf{G}}(t,s)\mathbf{a}(s)\, ds$$

where the fundamental solutions $\mathbf{F}(t)$ and $\widetilde{\mathbf{F}}(t)$ satisfy

$$\mathbf{F}'(t) = \mathbf{A}(t)\mathbf{F}(t) \text{ for } t_0 < t < t_1 \text{ with } \mathbf{B}_0\mathbf{F}(t_0) + \mathbf{B}_1\mathbf{F}(t_1) = \mathbf{I}$$

and

$$\widetilde{\mathbf{F}}'(t) = \widetilde{\mathbf{A}}(t)\widetilde{\mathbf{F}}(t) \text{ for } t_0 < t < t_1 \text{ with } \mathbf{B}_0\widetilde{\mathbf{F}}(t_0) + \mathbf{B}_1\widetilde{\mathbf{F}}(t_1) = \mathbf{I}$$

and the Green's functions $\mathbf{G}(t, s)$ and $\widetilde{\mathbf{G}}(t, s)$ are

$$\mathbf{G}(t, s) = \begin{cases} \mathbf{F}(t)\mathbf{B}_0\mathbf{F}(t_0)\mathbf{F}(s)^{-1}, & t_0 < s < t < t_1 \\ -\mathbf{F}(t)\mathbf{B}_1\mathbf{F}(t_1)\mathbf{F}(s)^{-1}, & t_0 < t < s < t_1 \end{cases}.$$

and

$$\widetilde{\mathbf{G}}(t, s) = \begin{cases} \widetilde{\mathbf{F}}(t)\widetilde{\mathbf{B}}_0\widetilde{\mathbf{F}}(t_0)\widetilde{\mathbf{F}}(s)^{-1}, & t_0 < s < t < t_1 \\ -\widetilde{\mathbf{F}}(t)\widetilde{\mathbf{B}}_1\widetilde{\mathbf{F}}(t_1)\widetilde{\mathbf{F}}(s)^{-1}, & t_0 < t < s < t_1 \end{cases}.$$

It follows that the perturbation in the solution is

$$\widetilde{\mathbf{y}}(t) - \mathbf{y}(t) = \left[\widetilde{\mathbf{F}}(t) - \mathbf{F}(t) \right] \mathbf{b} + \int_{t_0}^{t_1} \left[\widetilde{\mathbf{G}}(t, s) - \mathbf{G}(t, s) \right] \mathbf{a}(s) \, \mathrm{d}s \,.$$

We can take vector norms of both sides to obtain

$$\left\| \widetilde{\mathbf{y}}(t) - \mathbf{y}(t) \right\| \leq \left\| \widetilde{\mathbf{F}}(t) - \mathbf{F}(t) \right\| \, \|\mathbf{b}\| + \int_{t_0}^{t_1} \left\| \widetilde{\mathbf{G}}(t, s) - \mathbf{G}(t, s) \right\| \, \|\mathbf{a}(s)\| \, \mathrm{d}s \,.$$

Finally, we can take the maximum in t to get

$$\max_{t \in [t_0, t_1]} \left\| \widetilde{\mathbf{y}}(t) - \mathbf{y}(t) \right\| \leq \max_{t \in [t_0, t_1]} \left\| \widetilde{\mathbf{F}}(t) - \mathbf{F}(t) \right\| \, \|\mathbf{b}\|$$

$$+ \max_{t \in [t_0, t_1]} \int_{t_0}^{t_1} \left\| \widetilde{\mathbf{G}}(t, s) - \mathbf{G}(t, s) \right\| \, \mathrm{d}s \max_{s \in [t_0, t_1]} \|\mathbf{a}(s)\| \,.$$

In practice, these bounds are overly pessimistic. Typically, fundamental solutions involve rapidly growing modes that are eliminated from solutions of boundary value problems. We should also remark that the existence and uniqueness proofs for scalar and vector boundary value problems rely on the contractive mapping theorem, and do not readily lend themselves to perturbation estimates.

Exercise 4.2.2 Analyze the sensitivity of the solution of the two-point boundary value problem

$$\varepsilon y''(t) + y'(t) = a \text{ for } 0 < t < 1 \text{ with } y(0) = 0 = y(1)$$

to perturbations in a. Assume that ε is positive but small compared to one.

Exercise 4.2.3 Analyze the sensitivity of the solution of the two-point boundary value problem

$$\varepsilon^2 y''(t) - y(t) = a \text{ for } 0 < t < 1 \text{ with } y(0) = 0 = y(1)$$

to perturbations in a. Assume that ε is small compared to one.

4.3   Shooting Methods

Our discussion at the end of Sect. 4.2.1 suggested that we could solve boundary value problems by coupling an initial value problem solver with a nonlinear equation solver. This approach would allow us to employ the sophisticated numerical methods for initial value problems which we developed in Chap. 3. The resulting techniques are called **shooting methods**. In order to avoid problems with numerical stability and the development of numerical singularities, experts have developed a variety of modifications to the basic method. For more details, we refer the reader to the book by Ascher et al. [7, Chapter 4].

4.3.1   Basic Approach

Suppose that we begin with a two-point boundary-value problem of the form

$$\mathbf{y}'(t) = \mathbf{f}(\mathbf{y}(t), t) \text{ for } t_0 < t < t_1 \text{ with } \mathbf{g}\left(\mathbf{y}(t_0), \mathbf{y}(t_1)\right) = \mathbf{0} .$$

We would like to replace this problem with a family of initial value problem of the form

$$\boldsymbol{\eta}'(t \; ; \; \mathbf{y}_0) = \mathbf{f}\left(\boldsymbol{\eta}(t \; ; \; \mathbf{y}_0), t\right) \text{ for } t_0 < t < t_1 \text{ with } \boldsymbol{\eta}(t_0 \; ; \; \mathbf{y}_0) = \mathbf{y}_0 .$$

We will use this family to choose the initial value $\mathbf{y}_0$ so that the boundary value condition

$$\mathbf{g}\left(\mathbf{y}_0, \boldsymbol{\eta}(t_1 \; ; \; \mathbf{y}_0)\right) = \mathbf{0}$$

is satisfied.

In order to choose the initial values that are correct for the original boundary value problem, we want to find a solution $\mathbf{y}_0$ of the nonlinear equation

$$\mathbf{0} = \boldsymbol{\phi}(\mathbf{y}_0) \equiv \mathbf{g}\left(\mathbf{y}_0, \boldsymbol{\eta}(t_1 \; ; \; \mathbf{y}_0)\right) .$$

Given a guess $\widetilde{\mathbf{y}}_0$, Newton's method for a solution of this equation would take the form

$$\text{solve } \frac{\partial \boldsymbol{\phi}}{\partial \mathbf{y}_0}(\widetilde{\mathbf{y}}_0) \triangle \mathbf{y}_0 = -\boldsymbol{\phi}(\widetilde{\mathbf{y}}_0)$$

$$\widetilde{\mathbf{y}}_0 = \widetilde{\mathbf{y}}_0 + \triangle \mathbf{y}_0 .$$

Note that

$$\frac{\partial \boldsymbol{\phi}}{\partial \mathbf{y}_0}(\widetilde{\mathbf{y}}_0) = \frac{\partial \mathbf{g}}{\partial \mathbf{y}_0}(\widetilde{\mathbf{y}}_0, \boldsymbol{\eta}(t_1, \widetilde{\mathbf{y}}_0)) + \frac{\partial \mathbf{g}}{\partial \mathbf{y}_1}(\widetilde{\mathbf{y}}_0, \boldsymbol{\eta}(t_1, \widetilde{\mathbf{y}}_0)) \frac{\partial \boldsymbol{\eta}}{\partial \mathbf{y}_0}(t_1 ; \widetilde{\mathbf{y}}_0) \ .$$

In order to use Newton's method to solve this problem, we need to determine the matrix

$$\mathbf{Y}(t, \widetilde{\mathbf{y}}_0) \equiv \frac{\partial \boldsymbol{\eta}}{\partial \mathbf{y}_0}(t ; \widetilde{\mathbf{y}}_0)$$

at $t = t_1$. Note that $\mathbf{Y}(t, \widetilde{\mathbf{y}}_0)$ satisfies the initial value problem

$$\mathbf{Y}'(t ; \widetilde{\mathbf{y}}_0) = \frac{\partial \mathbf{f}}{\partial \mathbf{y}}(\boldsymbol{\eta}(t \ \widetilde{\mathbf{y}}_0), t) \, \mathbf{Y}(t ; \widetilde{\mathbf{y}}_0) \ \text{ for } t_0 < t < t_1$$

$$\mathbf{Y}(t_0 ; \widetilde{\mathbf{y}}_0) = \mathbf{I} \ .$$

This initial value problem can be solved at the same time when we solve the initial value problem for $\boldsymbol{\eta}(t ; \widetilde{\mathbf{y}}_0)$. We can summarize these ideas with the following

Algorithm 4.3.1 (Shooting)

given $\mathbf{y}_0$

while not converged

solve $\begin{bmatrix} \boldsymbol{\eta} \\ \mathbf{Y} \end{bmatrix}'(t) = \begin{bmatrix} \mathbf{f}(\boldsymbol{\eta}(t), t) \\ \frac{\partial f}{\partial \mathbf{y}}(\boldsymbol{\eta}(t), t) \, \mathbf{Y}(t) \end{bmatrix}$ for $t_0 < t < t_1$ with $\begin{bmatrix} \boldsymbol{\eta} \\ \mathbf{Y} \end{bmatrix}(t_0) = \begin{bmatrix} \mathbf{y}_0 \\ \mathbf{I} \end{bmatrix}$

$\mathbf{y}_1 = \boldsymbol{\eta}(t_1)$

solve $\left[\frac{\partial \mathbf{g}}{\partial \mathbf{y}_0}(\mathbf{y}_0, \mathbf{y}_1) + \frac{\partial \mathbf{g}}{\partial \mathbf{y}_1}(\mathbf{y}_0, \mathbf{y}_1) \mathbf{Y}(t_1) \right] \triangle \mathbf{y}_0 = -\mathbf{g}(\mathbf{y}_0, \mathbf{y}_1)$

$\mathbf{y}_0 = \mathbf{y}_0 + \triangle \mathbf{y}_0$

This shooting method via Newton iteration involves significant work in order to compute the matrix $\mathbf{Y}$, even though the differential equation for $\mathbf{Y}$ is linear. The more serious difficulty is that exponential growth or decay can lead to very large or small entries in $\mathbf{Y}$ that give the Newton iteration a very small region of convergence.

Example 4.3.1 Suppose that r and f are two constants, and we want to solve the boundary-value problem

$$-y''(t) + r^2 y(t) = f \text{ for } 0 < t < 1 \text{ with } y(0) = y_0 \ , \ y(1) = y_1 \ .$$

In order to use shooting to solve this boundary value problem, we need to choose some number y_0', solve the initial value problem

$$\begin{bmatrix} \eta \\ \eta' \end{bmatrix}'(t) = \begin{bmatrix} 0 & 1 \\ r^2 & 0 \end{bmatrix} \begin{bmatrix} \eta \\ \eta' \end{bmatrix}(t) + \begin{bmatrix} 0 \\ -f \end{bmatrix} \text{ for } 0 < t \le 1 \text{ with } \begin{bmatrix} \eta \\ \eta' \end{bmatrix}(0) = \begin{bmatrix} y_0 \\ y_0' \end{bmatrix},$$

and then adjust y_0' so that $\eta(1) = y_1$.

It turns out that $\eta(t \; ; \; y_0')$ involves exponential growth in t. Theorem 3.2.5 shows that the solution of the initial value problem is

$$\begin{bmatrix} \eta \\ \eta' \end{bmatrix}(t) = \exp\left(\begin{bmatrix} 0 & 1 \\ r^2 & 0 \end{bmatrix} t\right) \begin{bmatrix} y_0 \\ y_0' \end{bmatrix} - \begin{bmatrix} 0 & 1 \\ r^2 & 0 \end{bmatrix}^{-1} \begin{bmatrix} 0 \\ -f \end{bmatrix}$$

$$= \begin{bmatrix} 1 & 1 \\ r & -r \end{bmatrix} \begin{bmatrix} e^{rt} & 0 \\ 0 & e^{-rt} \end{bmatrix} \begin{bmatrix} 1 & 1 \\ r & -r \end{bmatrix}^{-1} \begin{bmatrix} y_0 \\ y_0' \end{bmatrix} + \begin{bmatrix} f/r^2 \\ 0 \end{bmatrix}$$

$$= \begin{bmatrix} 1 & 1 \\ r & -r \end{bmatrix} \begin{bmatrix} e^{rt} & 0 \\ 0 & e^{-rt} \end{bmatrix} \begin{bmatrix} ry_0 + y_0' \\ ry_0 - y_0' \end{bmatrix} \frac{1}{2r} + \begin{bmatrix} f/r^2 \\ 0 \end{bmatrix}$$

$$= \begin{bmatrix} 1 & 1 \\ r & -r \end{bmatrix} \begin{bmatrix} e^{rt}(ry_0 + y_0') \\ e^{-rt}(ry_0 - y_0') \end{bmatrix} \frac{1}{2r} + \begin{bmatrix} f/r^2 \\ 0 \end{bmatrix}$$

$$= \begin{bmatrix} e^{rt}(ry_0 + y_0') + e^{-rt}(ry_0 - y_0') \\ e^{rt}r(ry_0 + y_0') - e^{-rt}r(ry_0 - y_0') \end{bmatrix} \frac{1}{2r} + \begin{bmatrix} f/r^2 \\ 0 \end{bmatrix}.$$

In order to obtain $\widetilde{y}(1) = y_1$, we must choose y_0' to solve

$$y_1 = e^r(ry_0 + y_0')/(2r) + e^{-r}(ry_0 - y_0')/(2r) + f/r^2.$$

This gives us

$$y_0' = \left[y_1 - y_0 \cosh(r) - f/r^2 \right] \frac{r}{\sinh(r)}.$$

Because of the hyperbolic trigonometric functions, the initial slope y_0' can become very sensitive to numerical perturbations in the initial value y_0, or in the right-hand side f.

Readers may execute a JavaScript program for shooting in the **Bratu problem.** The Bratu problem was described in Example 4.2.1. This program uses Euler's method to integrate the initial value problem for the shooting method, and employs the secant method to adjust the slope.

4.3.2   *Multiple Shooting*

We can reduce the difficulty of exponential growth in shooting by breaking the problem domain into some number of sub-problems. Then we can patch the sub-problems together by imposing continuity conditions, and finally we impose the original boundary condition on the sub-problems.

Suppose that we want to solve a two-point boundary value problem of the form

$$\mathbf{y}'(t) = \mathbf{f}(\mathbf{y}(t), t) \text{ for } t_0 < t < t_m \text{ with } \mathbf{g}\left(\mathbf{y}(t_0), \mathbf{y}(t_m)\right) = \mathbf{0} .$$

Here $m \geq 2$ is an integer. We begin by choosing a strictly increasing sequence of points $t_1, \dots, t_{m-1}$ within (t_0, t_m). Then we solve the sub-problems

$$\boldsymbol{\eta}_i'(t ; \mathbf{y}_i) = \mathbf{f}\left(\boldsymbol{\eta}_i(t ; \mathbf{y}_i), t\right) \text{ for } t_i < t \leq t_{i+1} \text{ with } \boldsymbol{\eta}_i(t_i, \mathbf{y}_i) = \mathbf{y}_i .$$

for $0 \leq i < m$. In order that the solutions on the individual intervals are continuous at the endpoints, we want to choose the initial conditions $\mathbf{y}_i$ for $0 \leq i < m$ so that

$$\mathbf{y}_{i+1} - \boldsymbol{\eta}_i\left(t_{i+1} ; \mathbf{y}_i\right) = \mathbf{0} \text{ for } 0 \leq i < m - 1 .$$

In addition, the original boundary condition requires that

$$\mathbf{g}\left(\mathbf{y}_0, \boldsymbol{\eta}_{m-1}\left(t_m ; \mathbf{y}_{m-1}\right)\right) = \mathbf{0} .$$

The continuity conditions and the original boundary condition give us m equations to solve for the m unknown initial values $\mathbf{y}_0, \dots, \mathbf{y}_{m-1}$.

Suppose that we are given guesses $\widetilde{\mathbf{y}}_i$ for the initial conditions in multiple shooting, and we want to use Newton's method to solve the nonlinear system of equations for the true initial values. As in the shooting method, for $0 \leq i < m$ we can define

$$\mathbf{Y}_i(t ; \widetilde{\mathbf{y}}_i) \equiv \frac{\partial \boldsymbol{\eta}_i}{\partial \mathbf{y}_i}(t ; \widetilde{\mathbf{y}}_i)$$

and note that this matrix satisfies the initial value problem

$$\mathbf{Y}_i'(t ; \widetilde{\mathbf{y}}_i) = \frac{\partial \mathbf{f}}{\partial \mathbf{y}}\left(\boldsymbol{\eta}_i(t ; \widetilde{\mathbf{y}}_i), t\right) \mathbf{Y}_i(t ; \widetilde{\mathbf{y}}_i) \text{ for } t_i < t \leq t_{i+1} \text{ with } \mathbf{Y}_i(t_i ; \widetilde{\mathbf{y}}_i) = \mathbf{I} .$$

Then Newton's method for the continuity conditions takes the form

$$\triangle \mathbf{y}_{i+1} - \mathbf{Y}_i(t_{i+1} ; \widetilde{\mathbf{y}}_i) \triangle \mathbf{y}_i = -\widetilde{\mathbf{y}}_{i+1} + \boldsymbol{\eta}_i(t_{i+1} ; \widetilde{\mathbf{y}}_i) \text{ for } 0 \leq i < m - 1 ,$$

and Newton's method for the original boundary condition gives us

$$\frac{\partial \mathbf{g}}{\partial \mathbf{y}_0} \left(\widetilde{\mathbf{y}}_0, \eta_{m-1} \left(t_m \; ; \; \widetilde{\mathbf{y}}_{m-1} \right) \right) \triangle \mathbf{y}_0 + \frac{\partial \mathbf{g}}{\partial \mathbf{y}_m} \left(\widetilde{\mathbf{y}}_0, \eta_{m-1} \left(t_m \; ; \; \widetilde{\mathbf{y}}_{m-1} \right) \right) \triangle \mathbf{y}_{m-1}$$

$$= -\mathbf{g} \left(\widetilde{\mathbf{y}}_0, \eta_{m-1} \left(t_m \; ; \; \widetilde{\mathbf{y}}_{m-1} \right) \right) \; .$$

Let us introduce some notation to simplify the expressions that follow. We will write

$$\mathbf{z}_{i+1} = \eta_i \left(t_{i+1} \; ; \; \widetilde{\mathbf{y}}_i \right) \text{ for } 0 \le i < m$$

and

$$\mathbf{Z}_{i+1} = \mathbf{Y}_i \left(t_{i+1} \; ; \; \widetilde{\mathbf{y}}_i \right) \text{ for } 0 \le i < m \; .$$

We also let

$$\mathbf{d}_i = \mathbf{z}_i - \widetilde{\mathbf{y}}_i \text{ for } 1 \le i < m$$

be the errors in the continuity conditions. The Newton equations for the continuity conditions lead to the linear recurrence

$$\triangle \mathbf{y}_{i+1} = \mathbf{Z}_{i+1} \triangle \mathbf{y}_i + \mathbf{d}_{i+1} \text{ for } 0 \le i < m - 1 \; , \tag{4.9}$$

which has solution

$$\triangle \mathbf{y}_{i+1} = \mathbf{Z}_{i+1} \cdot \ldots \cdot \mathbf{Z}_1 \triangle \mathbf{y}_0 + \sum_{j=1}^{i} \mathbf{Z}_{i+1} \cdot \ldots \cdot \mathbf{Z}_{j+1} \mathbf{d}_j + \mathbf{d}_{i+1} \text{ for } 0 \le i < m - 1 \; .$$

The remaining Newton equation for the original boundary condition is

$$-\mathbf{g} \left(\widetilde{\mathbf{y}}_0, \mathbf{z}_m \right) = \frac{\partial \mathbf{g}}{\partial \mathbf{y}_0} \left(\widetilde{\mathbf{y}}_0, \mathbf{z}_m \right) \triangle \mathbf{y}_0 + \frac{\partial \mathbf{g}}{\partial \mathbf{y}_m} \left(\widetilde{\mathbf{y}}_0, \mathbf{z}_m \right) \mathbf{Z}_m \triangle \mathbf{y}_{m-1}$$

$$= \frac{\partial \mathbf{g}}{\partial \mathbf{y}_0} \left(\widetilde{\mathbf{y}}_0, \mathbf{z}_m \right) \triangle \mathbf{y}_0$$

$$+ \frac{\partial \mathbf{g}}{\partial \mathbf{y}_m} \left(\widetilde{\mathbf{y}}_0, \mathbf{z}_m \right) \mathbf{Z}_m \left\{ \mathbf{Z}_{m-1} \cdot \ldots \cdot \mathbf{Z}_1 \triangle \mathbf{y}_0 + \sum_{j=1}^{m-2} \mathbf{Z}_{m-1} \cdot \ldots \cdot \mathbf{Z}_{j+1} \mathbf{d}_j + \mathbf{d}_{m-1} \right\}$$

This gives us the linear system

$$\left[\frac{\partial \mathbf{g}}{\partial \mathbf{y}_0}\left(\widetilde{\mathbf{y}}_0, \mathbf{z}_m\right) + \frac{\partial \mathbf{g}}{\partial \mathbf{y}_m}\left(\widetilde{\mathbf{y}}_0, \mathbf{z}_m\right) \mathbf{Z}_m \cdot \ldots \cdot \mathbf{Z}_1\right] \triangle \mathbf{y}_0$$

$$= -\mathbf{g}\left(\widetilde{\mathbf{y}}_0, \mathbf{z}_m\right) - \frac{\partial \mathbf{g}}{\partial \mathbf{y}_m}\left(\widetilde{\mathbf{y}}_0, \mathbf{z}_m\right) \sum_{j=1}^{m-1} \mathbf{Z}_m \cdot \ldots \cdot \mathbf{Z}_{j+1} \mathbf{d}_j \ .$$

We can solve this linear system for $\triangle \mathbf{y}_0$, and then use the linear recurrence (4.9) to compute the increments in the other initial values.

In summary, we have the following

Algorithm 4.3.2 (Multiple Shooting)

given $t_0 < t_1 < \ldots < t_m$ and $\widetilde{\mathbf{y}}_0, \ldots, \widetilde{\mathbf{y}}_{m-1}$

while not converged

$\mathbf{Z} = \mathbf{I}$

$\mathbf{d} = \mathbf{0}$

for $i = 0, \ldots, m-1$

solve $\begin{bmatrix} \boldsymbol{\eta}_i \\ \mathbf{Y}_i \end{bmatrix}'(t) = \begin{bmatrix} \mathbf{f}(\boldsymbol{\eta}_i(t), t) \\ \frac{\partial \mathbf{f}}{\partial \mathbf{y}}(\boldsymbol{\eta}_i(t), t)\mathbf{Y}_i(t) \end{bmatrix}$ for $t_i < t \le t_{i+1}$ with $\begin{bmatrix} \boldsymbol{\eta}_i \\ \mathbf{Y}_i \end{bmatrix}'(t_i) = \begin{bmatrix} \widetilde{\mathbf{y}}_i \\ \mathbf{I} \end{bmatrix}$

$\mathbf{z}_{i+1} = \boldsymbol{\eta}_i(t_{i+1})$

$\mathbf{Z}_{i+1} = \mathbf{Y}_i(t_{i+1})$

$\mathbf{d}_{i+1} = \mathbf{z}_{i+1} - \widetilde{\mathbf{y}}_{i+1}$

$\mathbf{Z} = \mathbf{Z}_{i+1}\mathbf{Z}$

$\mathbf{d} = \mathbf{Z}_{i+1}\mathbf{d} + \mathbf{d}_{i+1}$

solve $\left[\frac{\partial \mathbf{g}}{\partial \mathbf{y}_0}\left(\widetilde{\mathbf{y}}_0, \mathbf{z}_m\right) + \frac{\partial \mathbf{g}}{\partial \mathbf{y}_m}\left(\widetilde{\mathbf{y}}_0, \mathbf{z}_m\right)\mathbf{Z}\right] \triangle \mathbf{y}_0 = -\mathbf{g}(\widetilde{\mathbf{y}}_0, \mathbf{z}_m) - \frac{\partial \mathbf{g}}{\partial \mathbf{y}_m}(\widetilde{\mathbf{y}}_0, \mathbf{z}_m)\mathbf{d}$ for $\triangle \mathbf{y}_0$

for $i = 0, \ldots, m-2$

$\triangle \mathbf{y}_{i+1} = \mathbf{Z}_{i+1}\triangle \mathbf{y}_i + \mathbf{d}_{i+1}$

$\widetilde{\mathbf{y}}_{i+1} = \widetilde{\mathbf{y}}_{i+1} + \triangle \mathbf{y}_{i+1}$

Note that the initial value problems in the first loop over i can be solved separately and simultaneously on a parallel machine.

For implementations of multiple shooting, we recommend either MUS by Mattheij and Staarink, or mirkdc by Enright and Muir. For a more detailed discussion of shooting methods, we recommend Ascher et al. [7, p. 132ff].

Exercise 4.3.1 What happens if we use the shooting method to solve the Bratu problem in Example 4.2.1 with some value $\lambda > \lambda_c$? Try $\lambda = 3.6$ with 4, 8 and 16 steps with Euler's method inside shooting. Describe how the shooting method behaves as the number of timesteps used in the shooting method is increased.

Exercise 4.3.2 Consider the linear two-point boundary value problem

$$y''(t) + r^2 y(t) = 0 \text{ for } 0 < t < 1 \text{ with } y(0) = 0 , \ y(1) = 1 .$$

What happens if we use multiple shooting method to solve this problem with $r = \pi$?

4.4   Finite Differences

In this section, we will introduce a different approach for solving two-point boundary value problems. We will begin by subdividing the problem domain $[0, L]$ into mesh elements $[x_i, x_{i+1}]$ where $0 = x_0 < x_1 < \ldots < x_N = L$. Corresponding to the mesh points will be unknown values $y_i \approx y(x_i)$. We will approximate derivatives by difference quotients involving the unknowns y_i. Next, we substitute the difference quotients for the derivatives in the original ordinary differential equation, and substitute the unknowns for the function values in both the differential equation and the boundary conditions. This will give us a linear or nonlinear system for the unknowns.

Example 4.4.1 Suppose that we want to solve the **self-adjoint** boundary value problem

$$-\left[py'\right]'(t) + r(t)y(t) = f(t) , \ t_0 < t < t_m$$
$$y(t_0) = \beta_0 , \ y(t_m) = \beta_1 .$$

Here $m \geq 2$ is an integer. Let us choose mesh points

$$t_0 < t_1 < \ldots t_{m-1} < t_m .$$

The mesh points allow us to define the mesh widths

$$h_{j+1/2} = t_{j+1} - t_j \text{ for } 0 \leq j < m .$$

Let us also use the simplifying notation

$$y_j \approx y(t_j) \text{ and } p_{j+1/2} = p([t_j + t_{j+1}]/2) .$$

Then our finite difference approximation to the self-adjoint boundary value problem takes the form

$$-\left[p_{j+1/2}\frac{y_{j+1}-y_j}{h_{j+1/2}} - p_{j-1/2}\frac{y_j - y_{j-1}}{h_{j-1/2}} \right] \frac{2}{h_{j+1/2}+h_{j-1/2}} + r_j y_j = f_j \text{ for } 1 \leq j < m .$$

By introducing the notation

$$h_j = (h_{j+1/2} + h_{j-1/2})/2 ,$$

these can be scaled and expressed as the linear system

$$\begin{bmatrix} \frac{p_{1/2}}{h_{1/2}} + \frac{p_{3/2}}{h_{3/2}} + r_1 h_1 & -\frac{p_{3/2}}{h_{3/2}} & \cdots & & 0 \\ -\frac{p_{3/2}}{h_{3/2}} & \frac{p_{3/2}}{h_{3/2}} + \frac{p_{5/2}}{h_{5/2}} + r_2 h_2 & \cdots & & \vdots \\ \vdots & & \ddots & \ddots & -\frac{p_{m-3/2}}{h_{m-3/2}} \\ 0 & \cdots & & -\frac{p_{m-3/2}}{h_{m-3/2}} & \frac{p_{m-3/2}}{h_{m-3/2}} + \frac{p_{m-1/2}}{h_{m-1/2}} + r_{m-1}h_{m-1} \end{bmatrix} \begin{bmatrix} y_1 \\ y_2 \\ \vdots \\ y_{m-1} \end{bmatrix}$$

$$= \begin{bmatrix} \frac{p_{1/2}}{h_{1/2}}\beta_0 + f_1 h_1 \\ f_2 h_2 \\ \vdots \\ \frac{p_{m-1/2}}{h_{m-1/2}}\beta_1 + f_{m-1}h_{m-1} \end{bmatrix} .$$

Assuming that the numbers $p_{j+1/2}$ are positive and the numbers r_j are nonnegative, this linear system involves an irreducibly diagonally dominant symmetric tridiagonal matrix. As a result, this linear system can be solved easily by Gaussian factorization without pivoting.

In order to understand the error in this finite difference approximation, we can use Taylor expansions to see that

$$-\left[p(t_{j+1/2})\frac{y(t_{j+1})-y(t_j)}{h_{j+1/2}} - p(t_{j-1/2})\frac{y(t_j)-y(t_{j-1})}{h_{j-1/2}} \right]$$

$$\frac{2}{h_{j+1/2}+h_{j-1/2}} + r(t_j)y(t_j) - f(t_j)$$

$$\approx -\left\{ \frac{1}{3}p(t_j)y'''(t_j) + \frac{1}{2}p'(t_j)y''(t_j) + \frac{1}{4}p''(t_j)y'(t_j) \right\} [h_{j+1/2}-h_{j-1/2}]$$

$$-\left\{\frac{1}{12}p(t_j)y''''(t_j) + \frac{1}{6}p'(t_j)y'''(t_j) + \frac{1}{8}p''(t_j)y''(t_j) + \frac{1}{24}p'''(t_j)y'(t_j)\right\}$$

$$\left[h_{j+1/2}^2 - h_{j+1/2}h_{j-1/2} + h_{j-1/2}^2\right] .$$

Here we have assumed that

$$-(py')'(t_j) + r(t_j)y(t_j) - f(t_j) = 0 .$$

On a uniform mesh, for which $t_j = j/m$ for $0 \le j \le m$ and therefore

$$h_{j+1/2} = 1/m \equiv h \text{ for } 0 \le j < m ,$$

we see that the error in the finite difference approximation to the differential equation is on the order of h^2.

Example 4.4.2 Suppose that we want to solve the self-adjoint boundary value problem

$$- \left[py'\right]'(t) + r(t)y(t) = f(t) , \quad t_0 < t < t_m$$
$$y(t_0) = \beta_0 , \quad p(1)y'(t_m) = \beta_1 .$$

This problem is similar to the problem in Example 4.4.1, but the boundary condition at the right boundary involves the first derivative of the solution.

As before, we assume that $m \ge 2$ is an integer, and choose mesh points

$$t_0 < t_1 < \ldots < t_{m-1} < t_m .$$

Then our finite difference approximations to the self-adjoint boundary value problem take the form

$$-\left[p_{j+1/2}\frac{y_{j+1} - y_j}{h_{j+1/2}} - p_{j-1/2}\frac{y_j - y_{j-1}}{h_{j-1/2}}\right]\frac{2}{h_{j+1/2} + h_{j-1/2}} + r_j y_j = f_j \text{ for } 1 \le j < m$$

and

$$-\left[\beta_1 - p_{m-1/2}\frac{y_m - y_{m-1}}{h_{m-1/2}}\right]\frac{2}{h_{m-1/2}} + r_m y_m = f_m .$$

These can be expressed as the linear system

$$
\begin{bmatrix}
\frac{p_{1/2}}{h_{1/2}} + \frac{p_{3/2}}{h_{3/2}} + r_1 h_1 & -\frac{p_{3/2}}{h_{3/2}} & & \cdots & & 0 \\
-\frac{p_{3/2}}{h_{3/2}} & \ddots & & & & \\
\vdots & & \ddots & \frac{p_{m-3/2}}{h_{m-3/2}} + \frac{p_{m-1/2}}{h_{m-1/2}} + r_{m-1} h_{m-1} & -\frac{p_{m-1/2}}{h_{m-1/2}} \\
0 & & \cdots & -\frac{p_{m-1/2}}{h_{m-1/2}} & \frac{p_{m-1/2}}{h_{m-1/2}} + r_m \frac{h_{m-1/2}}{2}
\end{bmatrix}
\begin{bmatrix} y_1 \\ y_2 \\ \vdots \\ y_m \end{bmatrix}
$$

$$
= \begin{bmatrix}
\frac{p_{1/2}}{h_{1/2}} \beta_0 + f_1 h_1 \\
\vdots \\
f_{m-1} h_{m-1} \\
\beta_1 + f_m \frac{h_{m-1/2}}{2}
\end{bmatrix} .
$$

Example 4.4.3 Suppose that we want to solve the Bratu problem

$$
- y''(t) = \lambda e^{y(t)} , \ 0 < t < 1
$$
$$
y(0) = 0 = y(1) .
$$

If we define the uniform mesh width to be $h = 1/m$ and approximate $y_i \approx y(ih)$, then we obtain the finite difference equations

$$
-\frac{y_{i+1} - y_i}{h} + \frac{y_i - y_{i-1}}{h} = h\lambda e^{y_i} , \ 0 < i < m .
$$

The boundary conditions require that $y_0 = 0 = y_m$. We can write these finite difference equations as a nonlinear system

$$
\mathbf{0} = \mathbf{f}\left(\begin{bmatrix} y_1 \\ y_2 \\ \vdots \\ y_{m-2} \\ y_{m-1} \end{bmatrix} \right) \equiv
\begin{bmatrix}
2/h & -1/h & \cdots & 0 & 0 \\
-1/h & 2/h & \cdots & 0 & 0 \\
\vdots & \ddots & \ddots & \ddots & \vdots \\
0 & 0 & \ddots & 2/h & -1/h \\
0 & 0 & \cdots & -1/h & 2/h
\end{bmatrix}
\begin{bmatrix} y_1 \\ y_2 \\ \vdots \\ y_{m-2} \\ y_{m-1} \end{bmatrix}
- \begin{bmatrix} e^{y_1} \\ e^{y_2} \\ \vdots \\ e^{y_{m-2}} \\ e^{y_{m-1}} \end{bmatrix} h\lambda .
$$

The matrix of partial derivatives of this function is

$$
\frac{\partial \mathbf{f}}{\partial \mathbf{y}} =
\begin{bmatrix}
2/h - h\lambda e^{y_1} & -1/h & \cdots & 0 & 0 \\
-1/h & 2/h - h\lambda e^{y_2} & \cdots & 0 & 0 \\
\vdots & \ddots & \ddots & \ddots & \vdots \\
0 & 0 & \ddots & 2/h - h\lambda e^{y_{m-2}} & -1/h \\
0 & 0 & \cdots & -1/h & 2/h - h\lambda e^{y_{m-1}}
\end{bmatrix} .
$$

Given an initial guess $\widetilde{\mathbf{y}}$ for the vector of mesh values y_i, we can perform a Newton iteration

$$\text{solve } \frac{\partial \mathbf{f}}{\partial \mathbf{y}}(\widetilde{\mathbf{y}}) \triangle \mathbf{y} = -\mathbf{f}(\widetilde{\mathbf{y}})$$

$$\widetilde{\mathbf{y}} = \widetilde{\mathbf{y}} + \triangle \mathbf{y}$$

to improve the mesh values until convergence is achieved.

In general, suppose that we want to solve the two-point boundary-value problem

$$\mathbf{y}'(t) = \mathbf{f}(\mathbf{y}(t), t) \ , \ t_0 < t < t_m$$

$$\mathbf{g}(\mathbf{y}(t_0), \mathbf{y}(t_m)) = \mathbf{0} \ .$$

The **trapezoidal scheme** for this problem takes the form

$$\frac{\mathbf{y}_{i+1} - \mathbf{y}_i}{t_{i+1} - t_i} = \frac{1}{2}\left[\mathbf{f}(\mathbf{y}_{i+1}, t_{i+1}) + \mathbf{f}(\mathbf{y}_i, \mathbf{x}_i)\right] \text{ for } 0 \le i < m \text{ with } \mathbf{g}(\mathbf{y}_0, \mathbf{y}_m) = \mathbf{0} \ .$$

Alternatively, we could employ the **midpoint scheme**

$$\frac{\mathbf{y}_{i+1} - \mathbf{y}_i}{t_{i+1} - t_i} = \mathbf{f}\left([\mathbf{y}_{i+1} + \mathbf{y}_i]/2, [t_{i+1} + t_i]/2\right) \text{ for } 0 \le i < m \text{ with } \mathbf{g}(\mathbf{y}_0, \mathbf{y}_m) = \mathbf{0} \ .$$

$$(4.10)$$

Both the trapezoidal scheme and the midpoint scheme generate a system of nonlinear equations for the unknown vectors $\mathbf{y}_i$.

One way to develop high-order finite difference methods can be found in Fornberg [82]. Another general approach begins by writing the differential equation

$$\mathbf{y}'(t) = \mathbf{f}(\mathbf{y}(t), t)$$

in the integral form

$$\mathbf{0} = \int_{t_j}^{t_{j+1}} \mathbf{y}'(t) - \mathbf{f}(\mathbf{y}(t), t) \ dt = \mathbf{y}(t_{j+1}) - \mathbf{y}(t_j) - \int_{t_j}^{t_{j+1}} \mathbf{f}(\mathbf{y}(t), t) \ dt \ .$$

A finite difference method could be generated by approximating the integral by a quadrature rule, or by approximating the function $\boldsymbol{\phi}(t) = \mathbf{f}(\mathbf{y}(t), t)$ by an interpolating polynomial and taking its integral. Near the boundaries, the quadrature rule or interpolating polynomial would have to be centered away from the midpoint of the integration interval. For more discussion of this approach, see Ascher et al. [7, p. 208ff] or [7, p. 238ff]. Yet another way to generate high-order finite difference methods for two-point boundary value problems is to use implicit Runge-Kutta methods, particularly those that are special cases of collocation methods. We described these methods in Sect. 3.6.3. (The midpoint scheme (4.10) is the simplest example of the use of this class of implicit Runge-Kutta schemes for two-point boundary value problems.) Additional discussion of implicit Runge-Kutta schemes for two-point boundary value problems can be found in Ascher et al. [7, p. 210]. We will also see that finite difference methods can be generated by either collocation (see Sect. 4.5) or finite element methods (see Sect. 4.6). Both collocation and finite elements are easier to analyze than finite differences.

Some important theory regarding the convergence of finite difference methods for general n-th order linear two-point boundary value problems has been developed by Kreiss [121], and can also be found in Ascher et al. [7, p. 198ff]. We will not discuss the theory of finite differences in this text.

We have implemented a program to construct a finite difference approximation to the solution of the two-point boundary value problem

$$-y''(t) = \pi^2 \sin(\pi t) \text{ for } 0 < t < 1 \text{ with } y(0) = 0 = y(1) .$$

This program consists of the Fortran file finite_difference.f, which initializes the mesh and solves the linear system for the solution, the C^{++} file GUIFiniteDifference.C, which reads input parameters and runs the main program, and the makefile GNUmakefile, which compiles and assembles the program. If the user sets `ncells` to a value greater than one, then the main program will solve a boundary value problem with that number of elements. If `ncells` is equal to one, then the main program will perform a mesh refinement study, beginning with 2 elements and refining repeatedly by a factor of 2. Figure 4.1 shows some numerical results for finite difference computations with this program.

Readers may also execute a JavaScript program for finite differences with the Bratu problem, which was described in Example 4.2.1. This program uses centered differences to approximate the derivatives, and employs Newton's method to solve the nonlinear equations.

Exercise 4.4.1 Suppose that we would like to solve the boundary value problem

$$-y''(t) = \cos(t) , \ 0 < t < \pi$$
$$y(0) = 1 , \ y(\pi) = -1 .$$

The analytical solution to this problem is $y(t) = \cos(t)$. Program a finite difference method for this problem using centered differences, and perform a refinement study. Show that the maximum pointwise error in the finite difference method is second order in the mesh width.

Exercise 4.4.2 Suppose that we would like to solve the boundary value problem

$$-y''(t) = \cos(t) , \ 0 < t < \pi$$
$$y(0) = 1 , \ y'(\pi) = 0 .$$

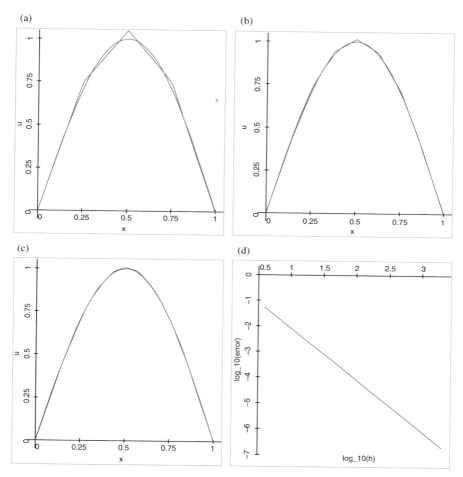

Fig. 4.1 Finite differences for $-y''(t) = \pi^2 \sin(\pi t)$ with $y(0) = 0 = y(1)$; analytical solution in red, numerical solution in blue. (**a**) 4 cells. (**b**) 8 cells. (**c**) 16 cells. (**d**) Refinement study

The analytical solution to this problem is $y(t) = \cos(t)$. Develop a finite difference approximation for this problem, including an approximation to the boundary condition at $t = \pi$. Perform a refinement study to determine the order of the method.

Exercise 4.4.3 Consider the **Bessel equation**

$$\left(ty'\right)'(t) + ty(t) = 0 \text{ for } 0 < t < 6 \text{ with } y(0) = 1 = y(6) .$$

Construct a finite difference method to solve this problem for $y(t)$, using centered finite difference on uniform meshes with $m = 2^k$ cells where $k = 2, \ldots, 10$. Use these results to perform a mesh refinement study. What power of h corresponds to the apparent rate of convergence?

4.5   Collocation

Previously in this chapter, we have developed shooting methods and finite difference methods for solving two-point boundary value problems. The former involve solving initial value problems, but require the solution of nonlinear equations to satisfy the boundary values. Often, the boundary value equations are very sensitive to the choice of initial values. On the other hand, basic low-order finite difference methods are easy to generate, but difficult to analyze.

In this section, we will introduce **collocation methods** for solving two-point boundary value problems. The basic idea is the following. Suppose that we want to solve

$$\mathbf{y}'(t) = \mathbf{f}(\mathbf{y}(t), t) \text{ for } t_0 < t < t_m \text{ with } \mathbf{g}(\mathbf{y}(t_0), \mathbf{y}(t_m)) = \mathbf{0} .$$

We choose a mesh

$$t_0 < t_1 < \ldots < t_m ,$$

a spline degree $k \geq 1$, and k scalars

$$0 < \sigma_1 < \ldots < \sigma_k < 1 .$$

Then we find a continuous vector-valued spline $\widetilde{\mathbf{y}}(t)$ so that

$$\widetilde{\mathbf{y}}'\left(t_i + \sigma_j h_{i+1/2}\right) = \mathbf{f}\left(\widetilde{\mathbf{y}}\left(t_i + \sigma_j h_{i+1/2}\right), t_i + \sigma_j h_{i+1/2}\right) \text{ for } 0 \leq i < m \text{ and } 1 \leq j \leq k$$
$$\text{with } \mathbf{g}\left(\widetilde{\mathbf{y}}(t_0), \widetilde{\mathbf{y}}(t_m)\right) = \mathbf{0} .$$

The spline has $k + 1$ unknown vector coefficients on each of m mesh intervals, which are determined by $m - 1$ vector continuity conditions, 1 boundary condition and mk vector applications of the differential equation. For differential equations involving derivatives of order n, we require the collocation approximation to have $n - 1$ continuous derivatives, and require the spline degree k to satisfy $k \geq n$.

Example 4.5.1 Suppose that we want to solve the second-order self-adjoint boundary value problem

$$-(py')'(t) + r(t)y(t) = f(t) \text{ for } t_0 < t < t_m \text{ with } y(t_0) = \beta_0 \text{ and } p(t_m)y'(t_m) = \beta_1 .$$

Given a mesh

$$t_0 < t_1 < \ldots < t_m ,$$

we will approximate $y(t)$ by a continuously differentiable piecewise quadratic function $\widetilde{y}(t)$ that satisfies the boundary conditions and the differential equation at the midpoints of the mesh intervals. Let

$$h_{i+1/2} = t_{i+1} - t_i \text{ for } 0 \le i < m$$

be the mesh widths. Our work in Sect. 1.6.2.1 shows that on the interval (t_i, t_{i+1}) we can write

$$\widetilde{y}(t) = -\widetilde{y}_i' \frac{(t_{i+1} - t)^2}{2h_{i+1/2}} + \widetilde{y}_{i+1}' \frac{(t - t_i)^2}{2h_{i+1/2}} + \widetilde{\eta}_{i+1/2} \, ,$$

where continuity of $\widetilde{y}$ at t_i requires that

$$\widetilde{\eta}_{i+1/2} = \widetilde{\eta}_{i-1/2} + \widetilde{y}_i' \frac{h_{i-1/2} + h_{i+1/2}}{2} \, .$$

The differential equation at $(t_i + t_{i+1})/2$ requires that

$$-p_{i+1/2} \frac{\widetilde{y}_{i+1}' - \widetilde{y}_i'}{h_{i+1/2}} - p_{i+1/2}' \frac{\widetilde{y}_{i+1}' + \widetilde{y}_i'}{2} + r_{i+1/2} \left[\widetilde{\eta}_{i+1/2} + \left(\widetilde{y}_{i+1}' - \widetilde{y}_i' \right) \frac{h_{i+1/2}}{8} \right] = f_{i+1/2} \, .$$

The left-hand boundary condition requires that

$$\widetilde{\eta}_{1/2} - \widetilde{y}_0' \frac{h_{1/2}}{2} = \beta_0 \, ,$$

and the right-hand boundary condition requires that

$$p_m \widetilde{y}_m' = \beta_1 \, .$$

These equations can be assembled into a linear system of the form

$$\begin{bmatrix} -h_{1/2}/2 & 1 & & & & & \\ \lambda_{1/2} & r_{1/2} & \upsilon_{1/2} & & & & \\ & -1 & \delta_1 & 1 & & & \\ & & \ddots & \ddots & \ddots & & \\ & & & -1 & \delta_{m-1} & 1 & \\ & & & & \lambda_{m-1/2} & r_{m-1/2} & \upsilon_{m-1/2} \\ & & & & & & p_m \end{bmatrix} \begin{bmatrix} \widetilde{y}_0' \\ \widetilde{\eta}_{1/2} \\ \widetilde{y}_1 \\ \vdots \\ \widetilde{y}_{m-1} \\ \widetilde{\eta}_{m-1/2} \\ \widetilde{y}_m' \end{bmatrix} = \begin{bmatrix} \beta_0 \\ f_{1/2} \\ 0 \\ \vdots \\ f_{m-3/2} \\ 0 \\ f_{m-1/2} \\ \beta_1 \end{bmatrix} .$$

Here we have used the notation

$$\delta_i = -\frac{h_{i-1/2} + h_{i+1/2}}{2} \, ,$$

$$\lambda_{i+1/2} = -\frac{p'_{i+1/2}}{2} + \frac{p_{i+1/2}}{h_{i+1/2}} - \frac{r_{i+1/2}h_{i+1/2}}{8} \quad \text{and}$$

$$\upsilon_{i+1/2} = -\frac{p'_{i+1/2}}{2} - \frac{p_{i+1/2}}{h_{i+1/2}} + \frac{r_{i+1/2}h_{i+1/2}}{8} \, ,$$

Example 4.5.2 Suppose that we want to solve the boundary value problem

$$\begin{bmatrix} y \\ \eta \end{bmatrix}'(t) = \begin{bmatrix} \eta(t)/p(t) \\ -f(t) - r(t)y(t) \end{bmatrix} \text{ for } t_0 < t < t_m \text{ with } \begin{bmatrix} y(t_0) - \beta_0 \\ \eta(t_m) - \beta_1 \end{bmatrix} = \mathbf{0} \, .$$

It is easy to see that this system is equivalent to the second-order self-adjoint boundary value problem in Example 4.5.1. Given a mesh

$$t_0 < t_1 < \ldots < t_m \, ,$$

we will approximate the vector

$$\mathbf{y}(t) = \begin{bmatrix} y(t) \\ \eta(t) \end{bmatrix}$$

by a continuous piecewise linear function $\widetilde{\mathbf{y}}(t)$ that satisfies the boundary conditions and the differential equation at the midpoints of the mesh intervals. Let

$$h_{i+1/2} = t_{i+1} - t_i \text{ for } 0 \le i < m$$

be the mesh widths. Our work in Sect. 1.6.1 shows that on the interval (t_i, t_{i+1}) we can write

$$\widetilde{\mathbf{y}}(t) = \widetilde{\mathbf{y}}_i \frac{t_{i+1} - t}{h_{i+1/2}} + \widetilde{\mathbf{y}}_{i+1} \frac{t - t_i}{h_{i+1/2}} \, .$$

The differential equation at $(t_i + t_{i+1})/2$ requires that

$$\begin{bmatrix} -p_{i+1/2}/h_{i+1/2} & -1/2 \\ r_{i+1/2}/2 & 1/h_{i+1/2} \end{bmatrix} \begin{bmatrix} \widetilde{y}_i \\ \widetilde{\eta}_i \end{bmatrix} + \begin{bmatrix} p_{i+1/2}/h_{i+1/2} & -1/2 \\ r_{i+1/2}/2 & -1/h_{i+1/2} \end{bmatrix} \begin{bmatrix} \widetilde{y}_{i+1} \\ \widetilde{\eta}_{i+1} \end{bmatrix} = \begin{bmatrix} 0 \\ f_{i+1/2} \end{bmatrix} \, .$$

The left-hand boundary condition requires that

$$\widetilde{y}_0 = \beta_0 \, ,$$

and the right-hand boundary condition requires that

$$\widetilde{\eta}_m = \beta_1 .$$

These equations can be assembled into a linear system of the form

$$
\begin{bmatrix}
[1, 0] & & & \\
\mathbf{C}_{1/2} & \mathbf{D}_{1/2} & & \\
& \ddots & \ddots & \\
& & \mathbf{C}_{m-1/2} & \mathbf{D}_{m-1/2} \\
& & & [0, 1]
\end{bmatrix}
\begin{bmatrix}
\widetilde{\mathbf{y}}_0 \\
\widetilde{\mathbf{y}}_1 \\
\vdots \\
\widetilde{\mathbf{y}}_{m-1} \\
\widetilde{\mathbf{y}}_m
\end{bmatrix}
=
\begin{bmatrix}
\beta_0 \\
\mathbf{d}_{1/2} \\
\vdots \\
\mathbf{d}_{m-1/2} \\
\beta_1
\end{bmatrix}
$$

where

$$\mathbf{C}_{i+1/2} = \begin{bmatrix} -p_{i+1/2}/h_{i+1/2} & -1/2 \\ r_{i+1/2}/2 & 1/h_{i+1/2} \end{bmatrix} ,$$

$$\mathbf{D}_{i+1/2} = \begin{bmatrix} p_{i+1/2}/h_{i+1/2} & -1/2 \\ r_{i+1/2}/2 & -1/h_{i+1/2} \end{bmatrix} \text{ and}$$

$$\mathbf{d}_{i+1/2} = \begin{bmatrix} 0 \\ f_{i+1/2} \end{bmatrix} .$$

Note that neither collocation example was able to produce a symmetric linear system for the self-adjoint boundary value problem, unlike the finite difference Example 4.4.1. We also note that yet another collocation approach will be described in Sect. 4.7; that collocation method operates on an integrated form of the differential equation.

Russell and Shampine [155] have proved that whenever a boundary value problem has a unique solution, the collocation approximation exists and is unique. Furthermore, if the solution of an n-th order boundary value problem has s bounded derivatives and the collocation method uses piecewise polynomials of degree at most k, then the error in collocation is of order at least $\min\{s, k - n\}$ in the mesh width. De Boor and Swartz [62] showed that if the points σ_j are chosen to be the Gaussian quadrature points and the solution of an n-th boundary value problem has s continuous derivatives, then the error is of order $\min\{s, k\}$ in the mesh width. These results were extended by Houstis [105].

There are several publicly available Fortran implementations of collocation methods. Readers can choose either colnew by Ascher and Bader, colsys by Ascher, Christiansen and Russell, or colmod by Wright and Cash. MATLAB users should examine the command bvp4c.

Exercise 4.5.1 Consider the fourth-order two-point boundary value problem

$$EID^4y(x) + ky(x) = q(x) \text{ for } 0 < x < L \text{ with } y(0) = 0 = y(L),$$

$$y'(0) = 0 \text{ and } y''(L) = 0.$$

Here y is the transverse deflection of a beam, $E = 3 \times 10^7$ newtons per square meter is the elastic modulus, $I = 3 \times 10^3$ newtons^4 is the second moment of area, $k = 2.604 \times 10^3$ newtons per meter is the restoring force per length in Hooke's law, and $q = 4.34 \times 10^4$ newtons is the applied body force. The boundary conditions $y(0) = 0 = y(L)$ specify zero transverse displacement at the ends of the beam. The boundary condition $y'(0) = 0$ says that the beam is clamped on the left, while the boundary condition $y''(L) = 0$ says that the beam is simply supported on the right.

1. Rewrite this two-point boundary value problem in the form

$$\mathbf{y}'(x) = \mathbf{A}\mathbf{y}(x) + \mathbf{a} \text{ for } 0 < x < L \text{ with } \mathbf{B}_0\mathbf{y}(0) + \mathbf{B}_L\mathbf{y}(L) = \mathbf{b}.$$

2. Find a fundamental matrix for this problem, and determine the analytical solution of the problem.
3. Choose a numerical method to solve this problem, and perform a convergence study to determine the effective convergence rate as a function of the mesh width.

Exercise 4.5.2 Consider the two-point boundary value problem

$$\varepsilon^2 y''(t) - y(t) = \varepsilon^2 f(t) \text{ for } -1 < t < 1 \text{ with } y(-1) = \beta_{-1} \text{ and } y(1) = \beta_1.$$

Here $f(t)$, β_{-1} and β_1 are chosen so that the solution is

$$y(t) = \cos(\pi t) + \frac{e^{-(1+t)/\varepsilon}}{1 + e^{-2/\varepsilon}} + e^{-(1-t)/\varepsilon}.$$

1. Determine $f(t)$, β_{-1} and β_1.
2. Rewrite the problem in the form

$$\mathbf{y}'(t) = \mathbf{A}\mathbf{y}(t) + \mathbf{a}(t) \text{ for } -1 < x < 1 \text{ with } \mathbf{B}_{-1}\mathbf{y}(-1) + \mathbf{B}_1\mathbf{y}(1) = \mathbf{b}.$$

3. Let $\varepsilon = 10^{-4}$. Choose a numerical method to solve this problem, and perform a convergence study.

Exercise 4.5.3 Consider the two-point boundary value problem

$$\varepsilon y''(t) - ty(t) = 0 \text{ for } -1 < t < 1 \text{ with } y(-1) = 1 \text{ and } y(1) = 1.$$

The exact solution of this problem can be expressed in terms of Airy functions (see Abramowitz and Stegun [1, p. 446]).

1. Rewrite the problem in the form

$$\mathbf{y}'(t) = \mathbf{A}(t)\mathbf{y}(t) \text{ for } -1 < x < 1 \text{ with } \mathbf{B}_{-1}\mathbf{y}(-1) + \mathbf{B}_1\mathbf{y}(1) = \mathbf{b} \ .$$

2. Let $\varepsilon = 10^{-6}$. Choose a numerical method to solve this problem, and perform a convergence study.

Warning: the solution of this problem involves dense oscillations, a **boundary layer** at the right boundary, and a turning point where the solution changes character.

Exercise 4.5.4 Consider the two-point boundary value problem

$$\varepsilon y''(t) - t y'(t) + y(t) = 0 \text{ for } -1 < t < 1 \text{ with } y(-1) = 1 \text{ and } y(1) = 2 \ .$$

1. Show that **Kummer's function**

$$M_{a,b}(z) = \sum_{n=0}^{\infty} \frac{(a)_n}{(b)_n} \frac{z^n}{n!}$$

solves **Kummer's differential equation** (see Abramowitz and Stegun [1, p. 504])

$$z u''(z) + (b - z) u'(z) - a u(z) = 0 \ .$$

Here

$$(a)_n \equiv \prod_{i=0}^{n-1} (a + i) \ .$$

2. Show that the solution of the two-point boundary value problem in this exercise is

$$y(t) = \frac{t}{2} + \frac{3}{2 M_{-1,1/2}(1/[2\varepsilon])} M_{-1,1/2}(t^2/[2\varepsilon]) \ .$$

3. Rewrite the two-point boundary value problem in the form

$$\mathbf{y}'(t) = \mathbf{A}(t)\mathbf{y}(t) \text{ for } -1 < x < 1 \text{ with } \mathbf{B}_{-1}\mathbf{y}(-1) + \mathbf{B}_1\mathbf{y}(1) = \mathbf{b} \ .$$

4. Let $\varepsilon = 1/70$. Choose a numerical method to solve this problem, and perform a convergence study.

4.6   Finite Elements

Previously in this chapter, we have developed shooting methods, finite difference methods and collocation methods. Each of these were motivated by a two-point boundary value problem written in the form of a differential equation. In this section, we will introduce a new computational approach that begins with the two-point boundary value problem formulated as a variational principle, such as Hamilton's principle in the introduction to this chapter. For example, if we begin with a Hamiltonian

$$\mathscr{H}(\mathbf{y}) \equiv \int_{t_0}^{t_1} \mathscr{L}(\mathbf{y}(t), \dot{\mathbf{y}}(t), t) \ dt \ ,$$

then its first variation is

$$\delta\mathscr{H}(\mathbf{y}) = \int_{t_0}^{t_1} \{ \mathscr{L}(\mathbf{y}(t) + \delta\mathbf{y}(t), \dot{\mathbf{y}}(t) + \delta\dot{\mathbf{y}}(t), t) - \mathscr{L}(\mathbf{y}(t), \dot{\mathbf{y}}(t), t) \} \ dt$$

$$= \int_{t_0}^{t_1} \left\{ \frac{\partial\mathscr{L}}{\partial\mathbf{y}}(\mathbf{y}(t), \dot{\mathbf{y}}(t), t) \, \delta\mathbf{y}(t) + \frac{\partial\mathscr{L}}{\partial\dot{\mathbf{y}}}(\mathbf{y}(t), \dot{\mathbf{y}}(t), t) \, \delta\dot{\mathbf{y}}(t) \right\} \ dt \ .$$

Some further integration by parts might be used to reduce the number of terms multiplied by $\delta\mathbf{y}$. Next, we would select finite-dimensional spaces for the functions $\mathbf{y}$ and $\delta\mathbf{y}$, then select an appropriate quadrature rule to approximate the integral, and thereby generate a numerical method. If the problem is originally specified in terms of a differential equation, we could develop a variational form for an ordinary differential equation, as discussed in Sect. 4.6.1.

The variational form of the two-point boundary value problem has advantages that the differential equation form cannot offer. For example, we will be able to develop approximations to differential equations with Dirac delta-function forcing.

Some authors, such as Celia and Gray [35] or Finlayson [79], call this general approach either the **method of weighted residuals** or **Petrov-Galerkin methods.** In their view, **collocation methods** are Petrov-Galerkin methods in which the **test functions** $\delta\mathbf{y}$ are chosen to be Dirac delta functions at selected points within the problem domain, and the **trial functions y** are chosen to be piecewise polynomials. **Galerkin methods** are weighted residual methods in which the test functions $\delta\mathbf{y}$ and the trial functions **y** are chosen to lie in the same finite dimensional space. **Finite element methods** are Galerkin methods in which the basis functions for the test and trial functions are nonzero over a small number of mesh intervals; usually the test and trial functions in finite element methods are piecewise polynomials.

The theory for Galerkin methods is easiest to understand when the method is applied to linear differential equations. For an analysis of Galerkin methods applied to nonlinear two-point boundary value problems, we recommend a paper by Ciarlet et al. [45]. For general texts on finite element methods, we recommend mathematical books by Babuška [10], Braess [20], Brenner and Scott [23], Chen [37], Ciarlet [43],

Strang and Fix [163], Szabó and Babuška [169] or Trangenstein [174]. We also recommend engineering-oriented books by Bathe and Wilson [14], Hughes [106] and Zienkiewicz [189].

4.6.1 Variational Form

Not all two-point boundary value problems are initially posed in variational form. Consequently, we will show how to reformulate an important class of boundary value problems in variational form.

Lemma 4.6.1 *Suppose that q, r and f are continuous functions on (t_0, t_1), p is continuously differentiable in (t_0, t_1) and y is twice continuously differentiable in (t_0, t_1). Then y solves the inhomogeneous two-point boundary value problem*

$$\mathscr{P}y(t) \equiv -(py')'(t) + 2q(t)y'(t) + r(t)y(t) = f(t) \text{ for } t_0 < t < t_1$$
$$\text{with } y(t_0) = \beta_0 \text{ and } p(t_1)y'(t_1) = \beta_1 \tag{4.11}$$

*if and only if $y(t_0) = \beta_0$ and y satisfies the following **variational form**:*

$$\mathscr{B}(\delta y, y) \equiv \int_{t_0}^{t_1} \left[\delta y'(t), \ \delta y(t) \right] \begin{bmatrix} p(t) & 0 \\ 2q(t) & r(t) \end{bmatrix} \begin{bmatrix} y'(t) \\ y(t) \end{bmatrix} dt$$
$$= \int_{t_0}^{t_1} \delta y(t) f(t) \ dt + \delta y(t_1) \beta_1 \tag{4.12}$$

for all continuously differentiable functions δy with $\delta y(t_0) = 0$.

Proof Suppose that y satisfies the boundary value problem (4.11). Let δy be continuously differentiable in (t_0, t_1) with $\delta y(t_0) = 0$. We can multiply the differential equation by $\delta y(t)$ and integrate to get

$$\int_{t_0}^{t_1} \delta y(t) f(t) \ dt = \int_{t_0}^{t_1} \delta y(t) \mathscr{P}y(t) \ dt$$
$$= \int_{t_0}^{t_1} \delta y(t) \left\{ -(py')'(t) + \delta y(t) 2q(t)y'(t) + r(t)y(t) \right\} \ dt$$

then we can integrate by parts to get

$$= -\left[\delta y(t) p(t) y'(t) \right]_{t_0}^{t_1} + \int_{t_0}^{t_1} \delta y'(t) p(t) y'(t) + \delta y(t) 2q(t) y'(t) + \delta y(t) r(t) y(t) \ dt$$
$$= \mathscr{B}(\delta y, y) - \delta y(t_1) \beta_1 \ .$$

In other words, if y satisfies the homogeneous boundary value problem (4.11). then y satisfies the variational form (4.12).

Next, suppose that y satisfies the weak form (4.12), that y is also twice continuously differentiable, and that δy is continuously differentiable. The variational form implies that

$$\int_{t_0}^{t_1} \delta y(t) f(t) \ dt + \delta y(t_1) \beta_1 = \mathscr{B}(\delta y, y) \tag{4.13}$$

$$= \int_{t_0}^{t_1} \delta y'(t) p(t) y'(t) + \delta y(t) 2q(t) y'(t) + \delta y(t) r(t) y(t) \ dt \tag{4.14}$$

then integration by parts leads to

$$= \left[\delta y p y' \right]_{t_0}^{t_1} + \int_{t_0}^{t_1} \delta y(t) \left\{ - \left[p y' \right]'(t) + 2q(t) y'(t) + r(t) y(t) \right\} \ dt \tag{4.15}$$

$$= \delta y(t_1) p(t_1) y'(t_1) + \int_{t_0}^{t_1} \delta y(t) \left\{ - \left[p y' \right]'(t) + 2q(t) y'(t) + r(t) y(t) \right\} \ dt \ . \tag{4.16}$$

We conclude that

$$\int_{t_0}^{t_1} \delta y(t) \left[\mathscr{P} y(t) - f(t) \right] \ dt = \delta y(t_1) \left[\beta_1 - p(t_1) y'(t_1) \right]$$

for all continuously differentiable functions δy satisfying $\delta y(t_0) = 0$. Because of the continuity assumptions on p, q, r, y and f, we see that $\mathscr{P} y - f$ is continuous.

We claim that $\mathscr{P} y(t) = f(t)$ for all $t \in (t_0, t_1)$. Otherwise, $\mathscr{P} y - f$ is nonzero at some point $\tau \in (t_0, t_1)$, so continuity implies that it is nonzero in some open interval I around τ. Without loss of generality, suppose that $\mathscr{P} y - f$ is positive in this interval. We can pick a continuously differentiable function δy such that δy is nonnegative on (t_0, t_1), $\delta y(\tau) > 0$ and $\delta y(t) = 0$ for $t \notin I$. Then direct evaluation of the integral shows that we must have

$$\int_{t_0}^{t_1} \delta y(t) \left[\mathscr{P} y(t) - f(t) \right] \ dt > 0 \ ,$$

which contradicts the assumption that y solves the variational form. We get a similar contradiction if $(\mathscr{P} y - f)(\tau) < 0$.

Since $\mathscr{P} y(t) - f(t) = 0$ for all t, Eq. (4.16) now shows that $p(t_1) = y'(t_1) = \beta_1$. Thus if y satisfies the variational form, then it solves the two-point boundary value problem.

Note that $\mathscr{B}$ is a linear function in each of its two arguments.

For the discussion in Lemma 4.6.3, it will be helpful for the reader to understand the concept of an inner product on functions.

Definition 4.6.1 Given a vector space $\mathscr{S}$, an **inner product** is a mapping $\mathscr{B}$: $\mathscr{S} \times \mathscr{S} \to \mathbb{R}$ satisfying the following assumptions for all scalars α and β and all members u, v and $w \in \mathscr{S}$:

1. **Linearity:** $\mathscr{B}(v, \alpha u + \beta w) = \mathscr{B}(v, u)\alpha + \mathscr{B}(v, w)\beta$;
2. **Self-Adjointness:** $\mathscr{B}(v, u) = \mathscr{B}(u, v)$;
3. **Nonnegativity:** $\mathscr{B}(v, v) \geq 0$;
4. **Definiteness:** $\mathscr{B}(v, v) = 0$ implies that $v = 0$.

Before we examine the existence and uniqueness of solutions of the linear two-point boundary value problem, we also will prove the following useful result.

Lemma 4.6.2 (Poincaré's Inequality) *Suppose that both v and v' are square integrable in (t_0, t_1). If $v(t_0) = 0$ then*

$$\int_{t_0}^{t_1} v(t)^2 \, dt \leq \frac{(t_1 - t_0)^2}{2} \int_{t_0}^{t_1} v'(t)^2 \, dt . \tag{4.17}$$

Proof First, we will assume that v is continuously differentiable and $v(t_0) = 0$. Then the Fundamental Theorem of Calculus implies that for all $t \in (t_0, t_1)$

$$v(t) = \int_{t_0}^{t} v'(s) \, ds .$$

We can square both sides and integrate to get

$$\int_{t_0}^{t_1} v(t)^2 \, dt = \int_{t_0}^{t_1} \left[\int_{t_0}^{t} v'(s) \, ds \right]^2 \, dt$$

then we can use the Cauchy-Schwarz inequality (1.58) to get

$$\leq \int_{t_0}^{t_1} \left[\int_{t_0}^{t} v'(s)^2 \, ds \int_{t_0}^{t} \, ds \right] \, dt = \int_{t_0}^{t_1} t \int_{t_0}^{t} v'(s)^2 \, ds \, dt$$

then we change the order of integration to obtain

$$= \int_{t_0}^{t_1} v'(s)^2 \int_{s}^{t_1} t \, dt \, ds = \int_{t_0}^{t_1} v'(s)^2 \frac{1}{2} (t_1^2 - s^2) \, ds \leq \frac{(t_1 - t_0)^2}{2} \int_{t_0}^{t_1} v'(s)^2 \, ds$$

The result follows by taking limits of continuously differentiable functions; see, for example, Adams [2].

It should be easy to see that the Poincaré inequality also holds if v vanishes instead at the other boundary.

The next lemma provides conditions on our two-point boundary value problem so that the associated bilinear form is an inner product.

Lemma 4.6.3 *Suppose that there is a constant* $p > 0$ *so that* $p(t) \geq \underline{p}$ *for all* $t \in (t_0, t_1)$. *Also assume that there is a constant* $\underline{r} > 0$ *so that*

$$r(t) \geq \underline{r} - \frac{2\underline{p}}{(t_1 - t_0)^2} \text{ for } t \in (t_0, t_1) .$$

Then

$$\mathscr{B}(v, u) \equiv \int_0^1 v'(t)p(t)u'(t) + v(t)r(t)u(t) \ dt$$

is an inner product on the space of functions

$$\mathscr{S} \equiv \{w : \mathscr{B}(w, w) < \infty \text{ and } w(t_0) = 0\} .$$

Proof Linearity and self-adjointness are obvious. In order to prove nonnegativity, we note that

$$\mathscr{B}(w, w) = \int_{t_0}^{t_1} p(t)w'(t)^2 + r(t)w(t)^2 \ dt \geq \int_{t_0}^{t_1} \underline{p}w'(t)^2 + \left[\underline{r} - \frac{2\underline{p}}{(t_1 - t_0)^2}\right]w(t)^2 \ dt$$

then the Poincaré inequality (4.17) gives us

$$\geq \int_{t_0}^{t_1} \underline{p}\frac{2}{(t_1 - t_0)^2}w(t)^2 + \left[\underline{r} - \frac{2\underline{p}}{(t_1 - t_0)^2}\right]w(t)^2 \ dt = \underline{r}\int_{t_0}^{t_1} w(t)^2 \ dt$$

Definiteness is a direct consequence of this inequality.

In Lemma 4.6.3, we assumed that $q(t) = 0$. The corresponding ordinary differential equation takes the form

$$-(py')'(t) + r(t)y(t) = 0 \text{ for } t \in (t_0, t_1) .$$

This is a **Sturm-Liouville problem**, and is an important topic in the study of ordinary differential equations. For more information regarding Sturm-Liouville problems, see Birkhoff and Rota [16, Chapter 10], Hartman [98, p. 337ff], or Kelley and Peterson [118, p. 197ff].

Proof of existence and uniqueness of solutions of variational forms for two-point boundary value problems typically employ ideas from functional analysis, which is beyond the scope of this book. Interested readers can obtain more detailed information on this topic from Agmon [3, p. 47ff], Babuška and Aziz [9, Part I], Braess [20, p. 34ff], Brenner and Scott [23, p. 56ff], Lions and Magenes [128, Chapter 2] or Trangenstein [174, p. 302ff]. Here is a quick summary of the ideas. The variational form of the **self-adjoint problem**

$$-(py')'(t) + r(t)y(t) = f(t) \text{ for } t \in (t_0, t_1) \text{ with } y(t_0) = \beta_0$$
$$\text{and } y(t_1) = \beta_1 \text{ or } p(t_1)y'(t_1) = \beta_1$$

has a unique solution provided that $p(t) \geq \underline{p} > 0$ and $r(t) \geq 0$ for $t_0 < t \leq t_1$. This is true for all choices of the boundary values, and even for functions f involving Dirac delta-functions. The proof of this fact uses the **Riesz representation theorem** (see Kreyszig [122, p. 188], Rudin [153, p. 130] or Yosida [187, p. 90]). More generally, the variational form of the problem

$$-(py')'(t) + q(t)y'(t) + r(t)y(t) = f(t) \text{ for } t \in (t_0, t_1)$$

$$\text{with } y(t_0) = \beta_0 \text{ and } y(t_1) = \beta_1 \text{ or } p(t_1)y'(t_1) = \beta_1$$

has a unique solution provided that $p(t) \geq \underline{p} > 0$ and $r(t) \geq \underline{r} + q(t)^2/p(t)$ for $t_0 < t \leq t_1$ and some constant $\underline{r} \geq 0$. The proof of this fact uses the **Lax-Milgram theorem** (see Ciarlet [43, p. 8] or Yosida [187, p. 92].).

Exercise 4.6.1 Find a variational formulation of the two-point boundary value problem

$$y''(t) + y(t) = 0 \text{ for } 0 < t < \pi/2 \text{ with } y(0) = 0 \text{ and } y'(\pi/2) = 1 .$$

Show that this problem does not have a solution.

Exercise 4.6.2 Find a variational formulation of the two-point boundary value problem

$$y''(t) + y(t) = 0 \text{ for } 0 < t < \pi \text{ with } y(0) = 0 = y(\pi) .$$

Show that this problem has multiple solutions.

Exercise 4.6.3 Find a variational formulation of the following two-point boundary value problem with periodic boundary conditions:

$$-(py')'(t) + r(t)y(t) = f(t) \text{ for } t \in (t_0, t_1) \text{ with } y(t_0) = y(t_1) \text{ and } y'(t_0) = y'(t_1) .$$

Exercise 4.6.4 Suppose that $p(t) > 0$ for all $t \in (t_0, t_1)$. Also assume that there is a positive constant $\underline{r}$ so that $r(t) > \underline{r} + q(t)^2/p(t)$ for $t \in (t_0, t_1)$. Show that

$$\mathcal{B}(v, u) \equiv \int_{t_0}^{t_1} v'(t)p(t)u'(t) + v(t)2q(t)u'(t) + v(t)r(t)u(t) \, dt$$

satisfies

$$\mathcal{B}(u, u) \geq \underline{r} \int_{t_0}^{t_1} u(t)^2 \, dt$$

for all u such that $\mathcal{B}(u, u) < \infty$.

4.6.2   Basic Principles

Suppose that we want to find $y(t)$ so that $y(t_0) = \beta_0$ and y solves the variational problem

$$\mathscr{B}(\delta y, y) \equiv \int_{t_0}^{t_1} \delta y'(t)p(t)y'(t) + \delta y(t)2q(t)y'(t) + \delta y(t)r(t)y(t) \, dt$$

$$= \int_{t_0}^{t_1} \delta y(t)f(t) \, dt + \delta y(t_1)\beta_1$$

for all functions δy so that $\mathscr{B}(\delta y, \delta y) < \infty$ and $\delta y(t_0) = 0$. The boundary condition $y(t_0) = \beta_0$ is called **essential**, because it must be imposed *a priori* on test functions. If this problem were written as an ordinary differential equation with boundary conditions, it would require $p(t_1)y'(t_1) = \beta_1$. The boundary condition $p(t_1)y'(t_1) = \beta_1$ is called **natural**, because it is satisfied (approximately) by continuously differentiable solutions of the variational problem.

We assume that this variational problem has a unique solution y satisfying $\mathscr{B}(y, y) < \infty$. To begin our development of a numerical method, let $\widetilde{\mathscr{S}}$ be a finite dimensional linear space of functions such that $\widetilde{s}(t_0) = 0$ for all $\widetilde{s} \in \widetilde{\mathscr{S}}$. Assume that the functions $\{\widetilde{s}_i\}_{i=1}^m$ are a basis for $\widetilde{\mathscr{S}}$. Let $b(t)$ be some appropriate function so that $b(t_0) = \beta_0$; for example, we might take $b(t) = \beta_0$. Then the **Galerkin method** approximates $y(t)$ by $\widetilde{y}(t) = b(t) + \widetilde{s}(t)$ where $\widetilde{s} \in \widetilde{\mathscr{S}}$ solves

$$\mathscr{B}(\widetilde{s}_i, b(t) + \widetilde{s}) = \int_{t_0}^{t_1} \widetilde{s}_i(t)f(t) \, dt + \widetilde{s}_i(t_1)\beta_1$$

for $1 \le i \le m$. In the **finite element method**, $\widetilde{\mathscr{S}}$ will be a linear space of splines that vanish at the left boundary, and $b(t)$ will be a spline function that satisfies the left boundary condition.

Alternatively, suppose that we want to find $y(t)$ so that $y(t_0) = \beta_0$, $y(t_1) = \beta_1$ and y solves the variational problem

$$\mathscr{B}(\delta y, y) \equiv \int_{t_0}^{t_1} \delta y'(t)p(t)y'(t) + \delta y(t)2q(t)y'(t) + \delta y(t)r(t)y(t) \, dt = \int_{t_0}^{t_1} \delta y(t)f(t) \, dt$$

for all functions δy so that $\mathscr{B}(\delta y, \delta y) < \infty$, $\delta y(t_0) = 0$ and $\delta y(t_1) = 0$. In this case, the boundary conditions at either boundary are essential. We assume that this problem has a unique solution y satisfying $\mathscr{B}(y, y) < \infty$. Let $\widetilde{\mathscr{S}}$ be a finite dimensional linear space of functions such that $\widetilde{s}(t_0) = 0 = \widetilde{s}(t_1)$ for all $\widetilde{s} \in \widetilde{\mathscr{S}}$. Assume that the functions $\{\widetilde{s}_i\}_{i=1}^{m-1}$ are a basis for $\widetilde{\mathscr{S}}$. Let $b(t)$ be some appropriate function so that

$b(t_0) = \beta_0$ and $\beta(t_1) = \beta_1$; for example, we might take

$$b(t) = \beta_0 \frac{t_1 - t}{t_1 - t_0} + \beta_1 \frac{t - t_0}{t_1 - t_0} .$$

Then the Galerkin method approximates $y(t)$ by $\widetilde{y}(t) = b(t) + \widetilde{s}(t)$ where $\widetilde{s} \in \widetilde{\mathscr{S}}$ solves

$$\mathscr{B}(\widetilde{s}_i, \beta_0 + \widetilde{s}) = \int_{t_0}^{t_1} \widetilde{s}_i(t) f(t) \, dt$$

for $1 \le i < m$. In the finite element method, $\widetilde{\mathscr{S}}$ will be a linear space of splines that vanish at both boundaries, and $b(t)$ will be a spline function that satisfies the two boundary conditions.

4.6.3   Nodal Formulation

Let us provide a simple example of a finite element approximation. Suppose that we would like to solve

$$-\left(py'\right)'(t) = f(t) \text{ for } 0 < t < 1 \text{ with } y(0) = \beta_0 \text{ and } y(1) = \beta_1 .$$

The variational form of this problem requires that

$$\int_0^1 \delta y'(t) p(t) y'(t) \, dt = \int_0^1 \delta y(t) f(t) \, dt$$

for all continuous functions δy with square-integrable derivative such that $\delta y(0) = 0 = \delta y(1)$. Let us choose some **mesh nodes**

$$0 = t_0 < t_1 < \ldots < t_m = 1 ,$$

and define the **mesh elements** to be (t_e, t_{e+1}) for $0 \le e < m$. We will also define continuous piecewise linear basis functions $\widetilde{s}_j$ for $0 \le j \le m$ by

$$\widetilde{s}_j(t) = \begin{cases} (t - t_{j-1})/(t_j - t_{j-1}), & t_{j-1} \le t \le t_j \\ (t_{j+1} - t)/(t_{j+1} - t_j), & t_j \le t \le t_{j+1} \\ 0, & \text{otherwise} \end{cases} .$$

Note that $\widetilde{s}_j(t_j) = 1$ for all $0 \le j \le m$. These continuous piecewise linear splines were previously discussed in Sect. 1.6.1.

Next, we will choose our Galerkin approximation space to be

$$\widetilde{\mathscr{S}} = \text{span}\left\{\widetilde{s}_j : 1 \le j < m\right\} .$$

Note that this linear space is finite dimensional, and that all of its members vanish at $t = 0$ and $t = 1$. We can choose our Galerkin approximation to have the form

$$\widetilde{y}(t) = \widetilde{s}_0(t)\beta_0 + \sum_{j=1}^{m-1} \widetilde{s}_j(t)\widetilde{y}_j + \widetilde{s}_m(t)\beta_1 .$$

It is easy to see that $\widetilde{y}(0) = \beta_0$, $\widetilde{y}(1) = \beta_1$ and $\widetilde{y} - \widetilde{s}_0\beta_0 - \widetilde{s}_m\beta_1 \in \widetilde{\mathscr{S}}$.

The Galerkin equations require that for all $1 \le i < m$ we have

$$\int_0^1 \widetilde{s}_i'(t)p(t) \left\{ \widetilde{s}_0'(t)\beta_0 + \sum_{j=1}^N \widetilde{s}_j'(t)\widetilde{y}_j + \widetilde{s}_{N+1}'(t)\beta_1 \right\} dt = \int_0^1 \widetilde{s}_i(t)f(t) \, dt .$$

We can decompose the integrals over the entire problem domain into a sum of integrals over elements:

$$\sum_{e=0}^{m-1} \int_{t_e}^{t_{e+1}} \widetilde{s}_i'(t)p(t) \left\{ \widetilde{s}_0'(t)\beta_0 + \sum_{j=1}^{m-1} \widetilde{s}_j'(t)\widetilde{y}_j + \widetilde{s}_m'(t)\beta_1 \right\} dt$$

$$= \sum_{e=0}^{m-1} \int_{t_e}^{t_{e+1}} \widetilde{s}_i(t)f(t) \, dt .$$

Because these particular basis functions $\widetilde{s}_i(t)$ are nonzero over at most two elements, the element sums in the Galerkin equations simplify to

$$\sum_{e=i-1}^{i} \int_{t_e}^{t_{e+1}} \widetilde{s}_i'(t)p(t) \left\{ \widetilde{s}_0'(t)\beta_0 + \sum_{j=1}^{m-1} \widetilde{s}_j'(t)\widetilde{y}_j + \widetilde{s}_m'(t)\beta_1 \right\} dt$$

$$= \sum_{e=i-1}^{i} \int_{t_e}^{t_{e+1}} \widetilde{s}_i(t)f(t) \, dt .$$

Because at most two of these particular nodal basis functions are nonzero in an element, we can simplify the sum over j in the Galerkin equations to

$$
\sum_{e=i-1}^{i} \int_{t_e}^{t_{e+1}} \widetilde{s}_i'(t)p(t) \left\{ \widetilde{s}_0'(t)\beta_0 + \sum_{j=\max\{1,e\}}^{\min\{m-1,e+1\}} \widetilde{s}_j'(t)\widetilde{y}_j + \widetilde{s}_m'(t)\beta_1 \right\} dt
$$

$$
= \sum_{e=i-1}^{i} \int_{t_e}^{t_{e+1}} \widetilde{s}_i(t)f(t)\, dt \, .
$$

For this finite element method, it is sufficient to use a single-point Gaussian quadrature rule to approximate the integrals. Note that for $t \in (t_e, t_{e+1})$,

$$
\widetilde{s}_e'(t) = -\frac{1}{t_{e+1} - t_e} \text{ and } \widetilde{s}_{e+1}'(t) = \frac{1}{t_{e+1} - t_e} \, .
$$

For $0 \le e < m$, let us use the notation

$$
h_{e+1/2} = t_{e+1} - t_e \, , \ p_{e+1/2} = p\left(\frac{t_e + t_{e+1}}{2}\right) \text{ and } f_{e+1/2} = f\left(\frac{t_e + t_{e+1}}{2}\right) \, .
$$

Then our Galerkin equations using the midpoint rule for quadrature can be written

$$
p_{i-1/2}\frac{\widetilde{y}_i - \widetilde{y}_{i-1}}{h_{i-1/2}} - p_{i+1/2}\frac{\widetilde{y}_{i+1} - \widetilde{y}_i}{h_{i+1/2}} = \frac{1}{2}f_{i-1/2}h_{i-1/2} + \frac{1}{2}f_{i+1/2}h_{i+1/2}
$$

for $1 \le i < m$. In these equations, it is understood that the boundary conditions require $\widetilde{y}_0 = \beta_0$ and $\widetilde{y}_m = \beta_1$. When written in this form, we see that the Galerkin equations have generated a finite difference approximation to the two-point boundary value problem. These equations can be organized into a linear system $\mathbf{Ay} = \mathbf{b}$, with terms involving the inhomogeneity f or the boundary conditions β_0 and β_1 contributing to $\mathbf{b}$, and the other terms contributing to $\mathbf{A}$ and $\mathbf{y}$.

4.6.4 Elemental Formulation

In the nodal formulation of the finite element method, we looped over some global ordering of the basis functions to compute the coefficients in the equation for its unknown coefficient in the Galerkin equations. In this section, we will learn how to loop over elements, compute all needed integrals on an individual element, and store those values in the appropriate locations within the matrix $\mathbf{A}$ and right-hand side $\mathbf{b}$. These computations will use mappings from a **reference element**, which we will choose to be the interval $(0, 1)$.

The user must choose a set $\{\sigma_d\}_{d=0}^{D-1}$ of **reference shape functions**, depending on the polynomial degree and smoothness of the spline functions used in the finite element method. For example, the reference shape functions for continuous piecewise linear splines are

$$\begin{bmatrix} \sigma_0(\tau) \\ \sigma_1(\tau) \end{bmatrix} = \begin{bmatrix} 1 - \tau \\ \tau \end{bmatrix} ,$$

and the reference shape functions for continuously differentiable piecewise quartics are

$$\begin{bmatrix} \sigma_0(\tau) \\ \sigma_1(\tau) \\ \sigma_2(\tau) \\ \sigma_3(\tau) \\ \sigma_4(\tau) \end{bmatrix} = \begin{bmatrix} (1 + 4\tau)(1 - 2\tau)(1 - \tau)^2 \\ \tau(1 - 2\tau)(1 - \tau)^2 \\ 4\tau^2(1 - \tau)^2 \\ \tau^2(2\tau - 1)(5 - 4\tau) \\ \tau^2(2\tau - 1)(\tau - 1) \end{bmatrix} .$$

We also need an **element-node map** that maps indices for reference shape functions to global indices for basis shape functions defined on elements in the original problem domain. For example, with linear reference shape functions the global index map is

$$v(e, d) = e + d$$

where e is the element number and d is the reference shape function number. For continuously differentiable piecewise quartic splines the global index map is

$$v(e, d) = 3e + d .$$

The obvious mapping μ_e from the reference element to an individual mesh element (t_e, t_{e+1}) is

$$\mu_e(\tau) = t_e(1 - \tau) + t_{e+1}\tau .$$

Since μ_e is linear, its derivative is constant:

$$\mu_e'(\tau) = h_{e+1/2} \equiv t_{e+1} - t_e .$$

We can also use the reference shape functions and the mappings to evaluate the original basis functions on the mesh elements as follows:

$$\widetilde{s}_{v(e,d)}(\mu_e(\tau)) = \sigma_d(\tau) \text{ for } 0 \le d < D \text{ and } \tau \in [0, 1] .$$

In order to evaluate the integrals over individual elements, we will choose a quadrature rule on the reference element:

$$\int_0^1 \phi(\tau)\, d\tau \approx \sum_{q=1}^Q \phi(\tau_q) w_q \ .$$

If we use piecewise polynomials of degree k in our finite element method, then this quadrature rule needs to use at least k quadrature points, and needs to be exact for polynomials of degree at most $2k - 1$. In other words, we can use a Gaussian quadrature rule involving k quadrature points. For more information about numerical quadrature, see Sect. 2.3. To learn more about appropriate choices of quadrature rules in finite elements methods, see Ciarlet [43, p. 178ff], Strang and Fix [163, p. 181f] or Trangenstein [174, p. 375ff].

Using the mappings to perform a change of variables in integration, the inhomogeneity integrals over the mesh elements are

$$\int_{t_e}^{t_{e+1}} \widetilde{s}_{v(e,d)}(t) f(t)\, dt = \int_0^1 \sigma_d(\tau) f(\mu_e(\tau)) \mu'(\tau)\, d\tau \approx \sum_{q=1}^Q \sigma_d(\tau_q) f(\mu_e(\tau_q)) h_{e+1/2} w_q$$

for $0 \le d < D$. Within the element (t_e, t_{e+1}), the remaining integrals needed for the Galerkin equations are

$$\int_{t_e}^{t_{e+1}} \widetilde{s}'_{v(e,d)}(t) p(t) \widetilde{s}'_{v(e,\delta)}(t)\, dt$$

$$= \int_0^1 \left[\sigma'_d(\tau) \frac{1}{\mu'_e(\tau)} \right] p(\mu_e(\tau)) \left[\sigma'_\delta(\tau) \frac{1}{\mu'_e(\tau)} \right] \mu'_e(\tau)\, d\tau$$

$$\approx \sum_{q=1}^Q \sigma'_d(\tau_q) p(\mu_e(\tau_q)) \sigma'_\delta(\tau_q) \frac{w_q}{h_{e+1/2}}$$

for $0 \le d, \delta < D$. These quadrature rule approximations suggest that we precompute the values of the reference shape functions and their derivatives at the quadrature points, before looping over the elements to assemble the linear system. For example, if we use piecewise linear shape functions and the midpoint rule for quadrature, then we precompute

$$\begin{bmatrix} \sigma'_0(1/2) \\ \sigma'_1(1/2) \end{bmatrix} = \begin{bmatrix} -1 \\ 1 \end{bmatrix} \quad \text{and} \quad \begin{bmatrix} \sigma_0(1/2) \\ \sigma_1(1/2) \end{bmatrix} = \begin{bmatrix} 1/2 \\ 1/2 \end{bmatrix} .$$

The integrals over the element (t_e, t_{e+1}) that are needed for the matrix in the linear system are the entries of the array

$$\int_{t_e}^{t_{e+1}} \begin{bmatrix} \widetilde{s}'_{v(e,0)}(t) \\ \vdots \\ \widetilde{s}'_{v(e,D-1)}(t) \end{bmatrix} p(t) \begin{bmatrix} \widetilde{s}'_{v(e,0)}(t) & \dots & \widetilde{s}'_{v(e,D-1)}(t) \end{bmatrix} dt$$

$$= \int_0^1 \begin{bmatrix} \sigma'_0(\tau) \\ \vdots \\ \sigma'_{D-1}(\tau) \end{bmatrix} \frac{1}{\mu'_e(\tau)} p(\mu_e(\tau)) \frac{1}{\mu'_e(\tau)} \begin{bmatrix} \sigma'_0(\tau) & \dots & \sigma'_{D-1}(\tau) \end{bmatrix} \mu'_e(\tau) \, d\tau$$

$$\approx \sum_{q=1}^{Q} \begin{bmatrix} \sigma'_0(\tau_q) \\ \vdots \\ \sigma'_{D-1}(\tau_q) \end{bmatrix} \frac{p(t_{e+1}\tau_q + t_e(1 - \tau_q))}{h_{e+1/2}} \begin{bmatrix} \sigma'_0(\tau_q) & \dots & \sigma'_{D-1}(\tau_q) \end{bmatrix} w_q .$$

The integrals over the element (t_e, t_{e+1}) that are needed for the right-hand side in the linear system are the entries of the vector

$$\int_{t_e}^{t_{e+1}} \begin{bmatrix} \widetilde{s}_{v(e,0)}(t) \\ \vdots \\ \widetilde{s}_{v(e,D-1)}(t) \end{bmatrix} f(t) \, dt = \int_0^1 \begin{bmatrix} \sigma_0(\tau) \\ \vdots \\ \sigma_{D-1}(\tau) \end{bmatrix} f(\mu_e(\tau))\mu'_e(\tau) \, d\tau$$

$$\approx \sum_{q=1}^{Q} \begin{bmatrix} \sigma_0(\tau_q) \\ \vdots \\ \sigma_{D-1}(\tau_q) \end{bmatrix} f(t_{e+1}\tau_q + t_e(1 - \tau_q))h_{e+1/2}w_q .$$

For continuous piecewise linear splines and single-point Gaussian quadrature, these integrals are

$$\int_{t_e}^{t_{e+1}} \begin{bmatrix} \widetilde{s}'_{v(e,0)}(t) \\ \widetilde{s}'_{v(e,1)}(t) \end{bmatrix} p(t) \begin{bmatrix} \widetilde{s}'_{v(e,0)}(t) & \widetilde{s}'_{v(e,1)}(t) \end{bmatrix} dt \approx \begin{bmatrix} -1 \\ 1 \end{bmatrix} \frac{p([t_{e+1} + t_e]/2)}{h_{e+1/2}} \begin{bmatrix} -1 & 1 \end{bmatrix}$$

and

$$\int_{t_e}^{t_{e+1}} \begin{bmatrix} \widetilde{s}_{v(e,0)}(t) \\ \widetilde{s}_{v(e,1)}(t) \end{bmatrix} f(t) \, dt \approx \begin{bmatrix} 1/2 \\ 1/2 \end{bmatrix} f([t_{e+1} + t_e]/2)h_{e+1/2} .$$

This development leads to the following algorithm to assemble the finite element linear system:

Algorithm 4.6.1 (Elemental Form of Finite Element Method)

for $1 \leq q \leq Q$ /* quadrature points */
  for $0 \leq d < D$ /* degrees of freedom (dof) */
    $\sigma_{d,q} = \sigma_d(\tau_q)$ and $\sigma'_{d,q} = \sigma'_d(\tau_q)$ /* reference shape functions and slopes */
for $0 \leq e < m$ /* mesh elements */
  for $0 \leq d < D$ /* degrees of freedom */
    determine $\nu(e,d)$ /* map to global index */
for $0 \leq e < m$ /* mesh elements */
  $\widetilde{\mathbf{A}} = \mathbf{0}$, $\widetilde{\mathbf{b}} = \mathbf{0}$
  $J_e = t_{e+1} - t_e$ /* mapping Jacobian μ'_e */
  for $1 \leq q \leq Q$ /* quadrature points */
    $t_{e,q} = \mu_e(\tau_q)$, $p_{e,q} = \frac{p(t_{e,q})}{J_e^2}$, $f_{e,q} = f(t_{e,q})$
    $w_{e,q} = J_e w_q$ /* Jacobian times quadrature weight */
    for $0 \leq d < D$ /* degrees of freedom */
      for $0 \leq \delta < D$ /* degrees of freedom */
        $\widetilde{\mathbf{A}}_{\delta,d} = \widetilde{\mathbf{A}}_{\delta,d} + \sigma'_{\delta,q} p_{e,q} \sigma'_{d,q} w_{e,q}$ /* element matrix */
      $\widetilde{\mathbf{b}}_d = \widetilde{\mathbf{b}}_d + \sigma_d(\tau_q) f_{e,q} w_{e,q}$ /* element right-hand side */
  for $0 \leq d < D$ /* degrees of freedom */
    for $0 \leq \delta < D$ /* degrees of freedom */
      $\mathbf{A}_{\nu(e,\delta),\nu(e,d)} = \mathbf{A}_{\nu(e,\delta),\nu(e,d)} + \widetilde{\mathbf{A}}_{\delta,d}$ /* global matrix */
    $\mathbf{b}_{\nu(e,d)} = \mathbf{b}_{\nu(e,d)} + \widetilde{\mathbf{b}}_d$ /* global right-hand side */
  if natural bc at left
    for $0 \leq d < D$ /* degrees of freedom */
      $\mathbf{b}_{\nu(e,d)} = \mathbf{b}_{\nu(e,d)} - \sigma_d(0) * \beta_0$
  else if natural bc at right
    for $0 \leq d < D$ /* degrees of freedom */
      $\mathbf{b}_{\nu(e,d)} = \mathbf{b}_{\nu(e,d)} + \sigma_d(1) * \beta_1$
for $0 \leq e < m$ /* mesh elements */
  if essential bc in element
    for $0 \leq d < D$ /* degrees of freedom */
      if $j = \nu(e,d)$ is global index for bc
        $\mathbf{b}_j = \beta$ /* essential boundary value */
        $\mathbf{A}_{j,j} = 1$
        for $0 \leq \delta < D$ /* degrees of freedom */
          $i = \nu(e,\delta)$
          $\mathbf{b}_i = \mathbf{b}_i - \mathbf{A}_{i,j} * \beta$ and $\mathbf{A}_{j,i} = 0 = \mathbf{A}_{i,j}$

A program to implement a Galerkin approximation to the solution of a two-point boundary value problem has been provided. This program consists of five parts. GUIFiniteElement.C contains the C^{++} main program for solving the two-point boundary value problem and plotting the results interactively. Quadrature.H and Quadrature.C contain C^{++} classes for performing various kinds of numerical quadratures. Polynomial.H and Polynomial.C contain C^{++} classes that describe various kinds of polynomials, such as Lagrange and Legendre polynomials. Point.H contains C^{++} classes to describe a point in various dimensions. GNUmakefile contains the makefile to make the executable. This executable also uses various C^{++} classes for vectors and matrices that make use of the LAPACK libraries; these C^{++} classes were described in Sects. 3.12 and 3.13 in Chap. 3 of Volume I. If the user sets `nelements` to a value greater than one, then the main program will solve a boundary value problem with that number of elements. If `nelements` is less than or equal to one, then the main program will perform a mesh refinement study, beginning with 2 elements and refining repeatedly by a factor of 2.

Numerical results with this program are presented in Fig. 4.2. The results show that for continuous piecewise linear finite element approximations, the mean square error at the mesh nodes is proportional to $\triangle x^2$, as is the maximum error at the mesh nodes.

Exercise 4.6.5 Suppose that we use continuous piecewise linear approximations in the finite element method to solve

$$-(py')'(t) = f(t) \text{ for } t_0 < t < t_m \text{ with } y(t_0) = \beta_0 \text{ and } y(t_m) = \beta_1$$

on some mesh $t_0 < t_1 < \ldots < t_m$. Also suppose that we use the trapezoidal rule to approximate integrals, instead of the midpoint rule we used in Sect. 4.6.3. Determine the finite difference form of the resulting finite element method.

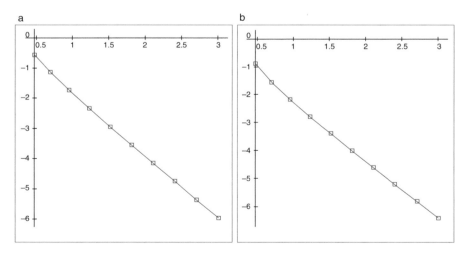

Fig. 4.2 Errors in continuous piecewise linear finite elements: log base ten of errors versus log base ten of number of basis functions. (**a**) L^2 error at Gaussian quadrature pts. $= O(\triangle x^2)$. (**b**) L^∞ error at mesh points $= O(\triangle x^2)$

Exercise 4.6.6 Suppose that p and f are constant in the previous problem. Compute the integrals in the finite element method exactly, and determine the resulting finite difference equations.

Exercise 4.6.7 If we use Simpson's rule for integration instead of the midpoint rule in the first problem, determine the finite difference form of finite element equations.

Exercise 4.6.8 Consider the two-point boundary-value problem

$$-y''(t) = \pi^2 \cos(\pi t) , \ 0 < t < 1$$
$$y(0) = 1 , \ y(1) = -1 .$$

1. Find the analytical solution of this problem.
2. Program the Galerkin method using piecewise linear approximations to y, and exact integration.
3. Plot the log of the error in the solution at the mesh points versus the log of the number of basis functions, for 2^n elements, $1 \le n \le 10$. What is the slope of this curve (i.e. the order of convergence)?
4. Plot the log of the error in the derivative of the solution at the mesh points versus the log of the number of basis functions, for 2^n elements, $1 \le n \le 10$. Note that there are two values for the derivative at each mesh point, associated with either of the two elements containing the mesh point. What is the slope of these curves (i.e. the order of convergence)?

Exercise 4.6.9 Consider the two-point boundary-value problem

$$-(py')'(t) = f(t) , \ 0 < t < 1$$
$$y(0) = 0 , \ y(1) = 0 .$$

Suppose that $f(t)$ is a Dirac delta function associated with some point $\xi \in (0, 1)$.

1. If $p(t) \equiv 1$, find the analytical solution of this problem.
2. Describe the Galerkin method for this problem, and the corresponding finite difference equations.
3. Suppose that $\xi = 1/2$, and consider uniform meshes with an even number of elements. Program the Galerkin method for this problem, and determine the log of the error in the solution at the mesh points as a function of the log of the number of basis functions.
4. Suppose that $\xi = 1/2$, and consider uniform meshes with an odd number of elements. Program the Galerkin method for this problem, and plot the log of the error in the solution at the mesh points versus the log of the number of basis functions.
5. What does the final claim of Lemma 4.6.6 below say about the error in the finite element method when ξ is located at a mesh node?

Exercise 4.6.10 Consider the two-point boundary-value problem

$$- (py')'(t) = f(t) , \ 0 < t < 1$$
$$y(0) = \beta_0 , \ p(1)y'(1) = \beta_1 .$$

Assume that p and f are continuous, and p is positive on $[0, 1]$. Let us approximate the solution of this problem by continuous piecewise cubic splines in a finite element method.

1. Which Gaussian quadrature rule should we use for this method?
2. Determine the reference shape functions for this approximation.
3. Describe the map from element number and reference degree of freedom to global index for the finite element solution.
4. Show that the matrix in the linear system for the finite element method is banded, symmetric and positive. Also find the band width of this matrix.

Exercise 4.6.11 Repeat the previous exercise, using continuously differentiable quintics.

4.6.5 Existence and Uniqueness

As we mentioned in Sect. 4.6.1, a proof of existence, uniqueness and continuous dependence on the data for solutions of variational formulations of linear-two-point boundary value problems usually involves ideas from functional analysis. Fortunately, it is relatively easy to show that the there are circumstances under which the Galerkin equations have a unique solution.

Lemma 4.6.4 *Suppose that the bilinear form $\mathscr{B}$ is positive-definite on the finite-dimensional space $\mathscr{S}$, and that we are given a function $\widetilde{b}$ so that $\mathscr{B}(\widetilde{s}, \widetilde{b}) < \infty$ for all $\widetilde{s} \in \mathscr{S}$. Then the Galerkin equations*

$$\mathscr{B}(\widetilde{s}, \widetilde{y}) = (\widetilde{s}, f) \text{ for all } \widetilde{s} \in \mathscr{S}$$

have a unique solution $\widetilde{y}$ so that $\widetilde{y} - \widetilde{b} \in \mathscr{S}$.

Proof Let $\{\widetilde{s}_1, \ldots, \widetilde{s}_N\}$ be a basis for $\mathscr{S}$. Then we can write any function $\widetilde{y}$ such that $\widetilde{y} - \widetilde{b} \in \mathscr{S}$ in the form

$$\widetilde{y}(t) = \widetilde{b} + \sum_{j=1}^{N} \widetilde{s}_j(t) \eta_j$$

for some scalars $\eta_1, \ldots, \eta_N$. This allows us to rewrite the Galerkin equation in the form of a system of linear equations:

$$\sum_{j=1}^{N} \mathscr{B}(\widetilde{s}_i, \widetilde{s}_j)\eta_j = (\widetilde{s}_i, f) - \mathscr{B}(\widetilde{s}_i, \widetilde{b}) , \ 1 \le i \le N .$$

To show that this linear system has a unique solution, it suffices to show that the only solution of the homogeneous system is zero. If

$$\sum_{j=1}^{N} \mathscr{B}(\widetilde{s}_i, \widetilde{s}_j)\eta_j = 0 \text{ for all } 1 \le i \le N ,$$

then we can multiply each equation in this system by η_i and sum to get

$$0 = \sum_{i=1}^{N} \eta_i \left[\sum_{j=1}^{N} \mathscr{B}(\widetilde{s}_i, \widetilde{s}_j)\eta_j \right] = \mathscr{B}\left(\sum_{i=1}^{N} \widetilde{s}_i \eta_i, \sum_{j=1}^{N} \widetilde{s}_j \eta_j \right) .$$

Since $\mathscr{B}$ is positive-definite, this implies that for all $t \in (0, 1)$,

$$\sum_{j=1}^{N} \widetilde{s}_j(t)\eta_j = 0 .$$

Since the functions $\widetilde{s}_1, \ldots, \widetilde{s}_N$ were assumed to form a basis for $\widetilde{\mathscr{S}}$, it follows that the η_j are all zero.

4.6.6   Energy Minimization

Certain linear two-point boundary value problems have variational formulations that are naturally related to energy functions. For these problems, it can be useful to relate the energy in the Galerkin approximation to the energy in the true solution of the problem.

Lemma 4.6.5 *Suppose that the bilinear form $\mathscr{B}$ is an inner product on the linear space $\mathscr{S}$, and that (u, v) represents the integral of the product of u and v over some problem domain. For any $s \in \mathscr{S}$, let the total energy functional be*

$$\mathscr{E}(s) = \frac{1}{2}\mathscr{B}(s, s) - (s, f) .$$

Assume that b is a function such that $\mathscr{B}(s, b)$ is finite for all $s \in \mathscr{S}$, and assume that $y \in \mathscr{S}$ solves the variational problem

$$\mathscr{B}(s, y) = (s, y) \text{ for all } s \in \mathscr{S} . \tag{4.18}$$

Then y solves the variational problem if and only if s minimizes the total energy $\mathscr{E}$ over all functions z so that $z - b \in \mathscr{S}$.

Next, Assume that $\widetilde{\mathscr{S}} \subset \mathscr{S}$ is a finite dimensional subspace. Also suppose that $\widetilde{y} - b \in \widetilde{\mathscr{S}}$, and that $\widetilde{y}$ satisfies the Galerkin equations

$$\mathscr{B}(\widetilde{s}, \widetilde{y}) = (\widetilde{s}, f) \text{ for all } \widetilde{s} \in \widetilde{\mathscr{S}} .$$

Then $\widetilde{y}$ satisfies the Galerkin equations if and only if $\widetilde{y}$ minimizes the total energy $\mathscr{E}$ over all functions $\widetilde{z}$ so that $\widetilde{z} - b \in \widetilde{\mathscr{S}}$. Furthermore,

$$\mathscr{E}(\widetilde{y}) \geq \mathscr{E}(y) .$$

Proof Suppose that $y - b \in \mathscr{S}$ and y solves the weak form (4.18). If $z - b \in \mathscr{S}$, then $z - y \in \mathscr{S}$ and

$$\mathscr{E}(z) - \mathscr{E}(y) = \frac{1}{2}\mathscr{B}(z, z) - \frac{1}{2}\mathscr{B}(y, y) - (z - y, f)$$

$$= \frac{1}{2}\mathscr{B}(z - y, z + y) - (z - y, f)$$

$$= \{\mathscr{B}(z - y, y) - (z - y, f)\} + \frac{1}{2}\mathscr{B}(z - y, z - y)$$

$$= \frac{1}{2}\mathscr{B}(z - y, z - y) \geq 0 .$$

This shows that if y solves the weak problem, then y minimizes the total energy. On the other hand, if y minimizes the total energy, then for all $s \in \mathscr{S}$ and all scalars ε we have

$$0 \leq \mathscr{E}(y + s\varepsilon) - \mathscr{E}(y) = \left\{ \frac{1}{2}\mathscr{B}(y + s\varepsilon, y + s\varepsilon) - (y + s\varepsilon, f) \right\}$$

$$- \left\{ \frac{1}{2}\mathscr{B}(y, y) - (y, f) \right\}$$

$$= \varepsilon[\mathscr{B}(s, y) - (s, f)] + \frac{1}{2}\varepsilon^2 \mathscr{B}(s, s) .$$

If $\mathscr{B}(s, y) - (s, f)$ is nonzero, we can choose ε sufficiently small with the same sign, and obtain a contradiction. We conclude that y must solve the weak problem (4.18). This completes the proof of the first claim.

If $\widetilde{s} - b \in \widetilde{\mathscr{S}}$, then a computation similar to that in the proof of the first claim shows that

$$\mathscr{E}(\widetilde{s}) - \mathscr{E}(\widetilde{y}) = [\mathscr{B}(\widetilde{s} - \widetilde{y}, \widetilde{y}) - (\widetilde{s} - \widetilde{y}, f)] + \frac{1}{2}\mathscr{B}(\widetilde{s} - \widetilde{y}, \widetilde{s} - \widetilde{y})$$

$$= \frac{1}{2}\mathscr{B}(\widetilde{s} - \widetilde{y}, \widetilde{s} - \widetilde{y}) \geq 0 \;.$$

This shows that if $\widetilde{y}$ solves the Galerkin equations, then $\widetilde{y}$ minimizes the total energy over the subspace. On the other hand, if $\widetilde{y}$ minimizes the total energy then for all $\widetilde{s} \in \widetilde{\mathscr{S}}$ and for all scalars ε we have

$$0 \leq \mathscr{E}(\widetilde{y} + \widetilde{s}\varepsilon) - \mathscr{E}(\widetilde{y})$$

$$= \left\{ \frac{1}{2}\mathscr{B}(\widetilde{y} + \widetilde{s}\varepsilon, \widetilde{y} + \widetilde{s}\varepsilon) - (\widetilde{y} + \widetilde{s}\varepsilon, f) \right\} - \left\{ \frac{1}{2}\mathscr{B}(\widetilde{y}, \widetilde{y}) - (\widetilde{y}, f) \right\}$$

$$= \varepsilon[\mathscr{B}(\widetilde{s}, \widetilde{y}) - (\widetilde{s}, f)] + \frac{1}{2}\varepsilon^2\mathscr{B}(\widetilde{s}, \widetilde{s}) \;.$$

If $\mathscr{B}(\widetilde{s}, \widetilde{y}) - (\widetilde{s}, f)$ is nonzero, we can choose ε sufficiently small with the same sign, and obtain a contradiction. We conclude that $\widetilde{s}$ must solve the Galerkin equations. This completes the proof of the second claim.

If $\widetilde{y}$ is the Galerkin approximation and y solves the variational form, then

$$\mathscr{E}(\widetilde{y}) - \mathscr{E}(y) = [\mathscr{B}(\widetilde{y} - y, y) - (\widetilde{y} - y, f)] + \frac{1}{2}\mathscr{B}(\widetilde{y} - y, \widetilde{y} - y) = \frac{1}{2}\mathscr{B}(\widetilde{y} - y, \widetilde{y} - y) \;.$$

Since $\widetilde{y} - \widetilde{y} \in \mathscr{S}$, the final equation is a consequence of the weak form for y. Since $\mathscr{B}$ is an inner product, the term on the right is nonnegative, and the third claim is proved.

4.6.7   Energy Error Estimates

Recall the Definitions 3.5.1 of Chap. 3 in Volume I of a norm, and the Definition 4.6.1 of an inner product. Theorem 1.7.1 now shows that if the bilinear form $\mathscr{B}$ satisfies the hypotheses of Lemma 4.6.3, then $\mathscr{B}$ is an inner product, and $\| \cdot \|_{\mathscr{B}}$ is a **norm**. These facts lead to some easy error estimates for Galerkin approximations.

Lemma 4.6.6 *Suppose that the bilinear form $\mathscr{B}$ is an inner product with associated* **natural norm** $\| \cdot \|_{\mathscr{B}}$, *$\mathscr{S}$ is a linear space of functions with finite natural norm and zero boundary values, and that $\widetilde{\mathscr{S}}$ is a finite-dimensional subspace of $\mathscr{S}$. Let b be a function with finite natural norm. Suppose that $y - b \in \mathscr{S}$ and y solves the weak form*

$$\mathscr{B}(s, y) = (s, f) \text{ for all } s \in \mathscr{S} \;.$$

Also assume that $\widetilde{y} - b \in \widetilde{\mathscr{S}}$ *and* $\widetilde{y}$ *solves the Galerkin equations*

$$\mathscr{B}(\widetilde{s}, \widetilde{y}) = (\widetilde{s}, f) \text{ for all } \widetilde{s} \in \widetilde{\mathscr{S}} .$$

Then the error $y - \widetilde{y}$ *is orthogonal to the subspace* $\widetilde{\mathscr{S}}$:

$$\mathscr{B}(\widetilde{s}, y - \widetilde{y}) = 0 \text{ for all } \widetilde{s} \in \widetilde{\mathscr{S}} ,$$

Furthermore, $\widetilde{y}$ *minimizes the natural norm of the error:*

$$\| y - \widetilde{y} \|_{\mathscr{B}} \leq \| y - \widetilde{s} \|_{\mathscr{B}} \text{ for all } \widetilde{w} - b \in \widetilde{\mathscr{S}} ,$$

Proof Note that if $\widetilde{s} \in \widetilde{\mathscr{S}}$, then $\widetilde{s} \in \mathscr{S}$. Suppose that y satisfies the weak form (4.12) and that $\widetilde{y}$ satisfies the Galerkin equations. Then we can subtract the latter equation from the former to get the first claim.

Since the error $y - \widetilde{y}$ is orthogonal to $\widetilde{\mathscr{S}}$, the Cauchy-Schwarz inequality (1.58) implies that for any $\widetilde{s}$ such that $\widetilde{s} - b \in \widetilde{\mathscr{S}}$ we have

$$\| y - \widetilde{y} \|_{\mathscr{B}}^2 \equiv \mathscr{B}(y - \widetilde{y}, y - \widetilde{y}) = \mathscr{B}(y - \widetilde{s}, y - \widetilde{y}) + \mathscr{B}(\widetilde{s} - \widetilde{y}, y - \widetilde{y}) = \mathscr{B}(y - \widetilde{s}, y - \widetilde{y})$$

$$\leq \| y - \widetilde{s} \|_{\mathscr{B}} \| y - \widetilde{y} \|_{\mathscr{B}}$$

If $\| y - \widetilde{y} \|_{\mathscr{B}} = 0$, the second claim is obvious. if $\| y - \widetilde{y} \|_{\mathscr{B}} > 0$, then we can cancel it on both sides of the latest inequality to prove the second claim.

Thus $\widetilde{y}$ is the best approximation to y from $\widetilde{\mathscr{S}}$, where "best" is measured in terms of the natural norm.

It is generally awkward to estimate errors for spline approximations in energy norms. In practice, the energy norms are bounded above and below by constant multiples of **Sobolev norms**. The Bramble-Hilbert lemma [21] shows that a continuous spline that reproduces all polynomials of degree k can approximate a function with one Sobolev derivative to order $O(h^k)$, where h is the maximum width of a mesh element. A duality argument can be used to prove that the mean-square error in the solution value is $O(h^{k+1})$. Proofs of these results can be found in a variety of sources, such as Babuška and Aziz [9, Part I], Babuška [10], Braess [20], Brenner and Scott [23], Chen [37], Ciarlet [43], Johnson [116], Strang and Fix [163], Szabó and Babuška [169] or Trangenstein [174].

4.6.8 Condition Numbers

For two-point boundary-value problems, finite element linear systems are banded with band width close to the polynomial degree of the spline approximations. Banded linear systems can be solved efficiently by factorization techniques described in Sect. 3.13.6 of Chap. 3 in Volume I. In practice, the errors in the band matrix factorization are small, on the order of machine precision. However, the perturbation Theorem 3.6.1 of Chap. 3 in Volume I showed that the errors dues to these factorizations can be amplified by the condition number of the matrix. Consequently, it is important to estimate the condition numbers of the matrices that arise in finite element discretizations. This issue is discussed in several texts, such as Braess [20, p. 242], Brenner and Scott [23, Section 9.6] or Trangenstein [174, Section 6.8.2]. The estimates apply to a family of meshes such that there is a constant C so that for all meshes in the family the number of elements m and the maximum mesh width h satisfy $mh \leq C$. For variational form corresponding to two-point boundary value problems involving derivatives of order at most two, if the bilinear form is bounded and positive, then the condition number of all matrices derived from finite element approximations on the meshes in such a family are on the order of h^{-2}. In general, if the differential equation involves derivatives of order at most $2n$, then the condition number would be on the order of h^{-2n}.

It is important to understand the implications of these condition number estimates. Suppose that our finite element method uses piecewise polynomials of degree k. Then the discretization error in the finite element approximation is $C_d h^{k+1}$ and the condition number of the linear system is $C_A h^{-2}$. Thus the rounding error due to solving the linear system is on the order of $C_A h^{-2} \varepsilon$, where ε represents machine precision. The **maximum attainable accuracy** occurs when the discretization error equals the rounding error, which implies that $C_d h^{k+1} = C_A h^{-2} \varepsilon$. We can solve this equation for h to get the optimal mesh width

$$h_* = \left[\frac{C_A}{C_d} \varepsilon \right]^{1/(k+3)} .$$

For this mesh width, the discretization error and the rounding error are both equal, with value

$$\delta = C_d h_*^{k+1} = C_d \left[\frac{C_A}{C_d} \varepsilon \right]^{(k+1)/(k+3)} = C_d^{2/(k+3)} C_A^{(k+1)/(k+3)} \varepsilon^{(k+1)/(k+3)} .$$

Thus we find that the maximum attainable accuracy approaches a constant times machine precision as the order k of the piecewise polynomials becomes large.

However, memory requirements per element increase with polynomial order. For continuous piecewise polynomial approximations, each element has one degree of freedom associated with each vertex and $k - 1$ degrees of freedom associated with the interior. If there are m elements, then there are $m + 1$ vertices. The total number of degrees of freedom is $m + 1 + (k-1)m = km + 1$. If for each polynomial degree k we choose the mesh width h_* that achieves the maximum attainable accuracy, then the total number of degrees of freedom will be on the order of

$$km = k/h_* = k\varepsilon^{-1/(k+3)} .$$

Suppose that $\varepsilon = 10^{-16}$. If $k = 1$, we must use on the order of 10^4 elements to achieve the maximum attainable accuracy, and we must solve a linear system involving on the order of 10^4 unknowns. On the other hand, if $k = 10$ then for maximum attainable accuracy we need only use an order of $\varepsilon^{-1/13} = 10^{16/13} \approx 17$ elements, and solve a linear system involving around 170 unknowns.

4.6.9 Static Condensation

In Sect. 4.6.8, we saw that the maximum attainable accuracy of finite element methods increases with the order of the splines. However, high-order finite element methods generate linear systems involving banded matrices with large band width. It is reasonable to examine alternatives for solving such linear systems.

Let us consider the use of continuous splines, as discussed in Sect. 1.6.1. The reference basis functions for continuous splines of degree $k > 1$ would involve 2 reference basis functions associated with interpolation at the endpoints of the reference element, and $k - 1$ interior reference basis functions. The former correspond to global basis functions that are nonzero over as many as two mesh elements, while the latter are mapped to global basis functions that are nonzero over a single mesh element. The map from element index and reference degree of freedom to global index in the linear system would be

$$v(e, d) = ke + d \text{ where } 0 \le e < m \text{ and } 0 \le d \le k \,.$$

Let y_n be the unknown coefficient of the global basis function associated with mesh node t_n for $0 \le n \le m$, and let $\mathbf{y}_{e+1/2}$ be the $(k-1)$-vector of unknown coefficients of the global basis functions associated with spline nodes interior to element e. A symmetric finite element linear system using such splines would take the form

$$
\begin{bmatrix}
\alpha_{0,0} & \mathbf{a}_{1/2,0}^{\mathsf{T}} & \alpha_{1,0} & & & & \\
\mathbf{a}_{1/2,0} & \mathbf{A}_{1/2,1/2} & \mathbf{a}_{1,1/2} & & & & \\
\alpha_{1,0} & \mathbf{a}_{1,1/2}^{\mathsf{T}} & \alpha_{1,1} & \mathbf{a}_{3/2,1}^{\mathsf{T}} & \alpha_{2,1} & & \\
& & \mathbf{a}_{3/2,1} & \mathbf{A}_{3/2,3/2} & \mathbf{a}_{2,3/2} & & \\
& & \alpha_{2,1} & \mathbf{a}_{2,3/2}^{\mathsf{T}} & \alpha_{2,2} & \cdots & \cdots \\
& & & & \vdots & \ddots &
\end{bmatrix}
\begin{bmatrix}
y_0 \\ y_{1/2} \\ y_1 \\ y_{3/2} \\ y_2 \\ \vdots
\end{bmatrix}
=
\begin{bmatrix}
\beta_0 \\ \mathbf{b}_{1/2} \\ \beta_1 \\ \mathbf{b}_{3/2} \\ \beta_2 \\ \vdots
\end{bmatrix} \,.
$$

Note that we can use the equations

$$\mathbf{a}_{e+1/2,e} y_e + \mathbf{A}_{e+1/2,e+1/2} \mathbf{y}_{e+1/2} + \mathbf{a}_{e+1,e+1/2} y_{e+1} = \mathbf{b}_{e+1/2}$$

to solve for $\mathbf{y}_{e+1/2}$ in terms of the unknowns y_e and y_{e+1}. If we factor

$$\mathbf{A}_{e+1/2,e+1/2} = \mathbf{L}_{e+1/2,e+1/2}\mathbf{L}_{e+1/2,e+1/2}{}^{\mathsf{T}}$$

as in Sect. 3.13.3.2 of Chap. 3 in Volume I, and solve

$$\mathbf{L}_{e+1/2,e+1/2}\mathbf{d}_{e+1/2,e} = \mathbf{a}_{e+1/2,e} \,,$$

$$\mathbf{L}_{e+1/2,e+1/2}\mathbf{d}_{e+1,e+1/2} = \mathbf{a}_{e+1,e+1/2} \text{ and}$$

$$\mathbf{L}_{e+1/2,e+1/2}\mathbf{c}_{e+1/2} = \mathbf{b}_{e+1/2}$$

as in Sect. 3.4.4 of Chap. 3 in Volume I, then we have

$$\mathbf{L}_{e+1/2,e+1/2}{}^{\mathsf{T}}\mathbf{y}_{e+1/2} = \mathbf{c}_{e+1/2} - \mathbf{d}_{e+1/2,e}y_e - \mathbf{d}_{e+1,e+1/2}y_{e+1} \,.$$

In this way, we can eliminate the interior unknowns; for example,

$$\alpha_{e,e}y_e + \mathbf{a}_{e+1/2,e}{}^{\mathsf{T}}\mathbf{y}_{e+1/2} + \alpha_{e+1,e+1}y_{e+1}$$

$$= \alpha_{e,e}y_e + \mathbf{d}_{e+1/2,e}{}^{\mathsf{T}}\left(\mathbf{c}_{e+1/2} - \mathbf{d}_{e+1/2,e}y_e - \mathbf{d}_{e+1,e+1/2}y_{e+1}\right) + \alpha_{e+1,e+1}y_{e+1}$$

$$= \left(\alpha_{e,e} - \mathbf{d}_{e+1/2,e}{}^{\mathsf{T}}\mathbf{d}_{e+1/2,e}\right)y_e + \left(\alpha_{e+1,e+1} - \mathbf{d}_{e+1/2,e}{}^{\mathsf{T}}\mathbf{d}_{e+1,e+1/2}\right)y_{e+1}$$

$$+ \mathbf{d}_{e+1/2,e}{}^{\mathsf{T}}\mathbf{c}_{e+1/2} \,.$$

The elimination of all interior unknowns could be accomplished by the following
Algorithm 4.6.2 (Static Condensation for Continuous Linear Splines)

> for $e = 0, \ldots, m-1$
>
> > factor $\mathbf{A}_{e+1/2,e+1/2} = \mathbf{L}_{e+1/2}\mathbf{L}_{e+1/2}{}^{\mathsf{T}}$
> >
> > solve $\mathbf{L}_{e+1/2}\mathbf{d}_{e+1/2,e} = \mathbf{a}_{e+1/2,e}$
> >
> > solve $\mathbf{L}_{e+1/2}\mathbf{d}_{e+1,e+1/2} = \mathbf{a}_{e+1,e+1/2}$
> >
> > solve $\mathbf{L}_{e+1/2}\mathbf{c}_{e+1/2} = \mathbf{b}_{e+1/2}$
> >
> > $\alpha_{e,e} = \alpha_{e,e} - \mathbf{d}_{e+1/2,e}{}^{\mathsf{T}}\mathbf{d}_{e+1/2,e}$
> >
> > $\alpha_{e+1,e} = \alpha_{e+1,e} - \mathbf{d}_{e+1/2,e}{}^{\mathsf{T}}\mathbf{d}_{e+1,e+1/2}$
> >
> > $\alpha_{e+1,e+1} = \alpha_{e+1,e+1} - \mathbf{d}_{e+1,e+1/2}{}^{\mathsf{T}}\mathbf{d}_{e+1,e+1/2}$
> >
> > $\beta_e = \beta_e - \mathbf{d}_{e+1/2,e}{}^{\mathsf{T}}\mathbf{c}_{e+1/2}$
> >
> > $\beta_{e+1} = \beta_{e+1} - \mathbf{d}_{e+1,e+1/2}{}^{\mathsf{T}}\mathbf{c}_{e+1/2}$

After static condensation has been performed (i.e., after the interior unknowns have been eliminated), we solve the tridiagonal linear system

$$
\begin{bmatrix}
\alpha_{0,0} & \alpha_{1,0} & \\
\alpha_{1,0} & \alpha_{1,1} & \cdots \\
& \vdots & \ddots
\end{bmatrix}
\begin{bmatrix}
\eta_0 \\
\eta_1 \\
\vdots
\end{bmatrix}
=
\begin{bmatrix}
\beta_0 \\
\beta_1 \\
\vdots
\end{bmatrix}.
$$

Following the solution of this tridiagonal linear system, the interior unknowns can be computed by solving the linear systems

$$
\mathbf{L}_{e+1/2,e+1/2}{}^\top \mathbf{y}_{e+1/2} = \mathbf{c}_{e+1/2} - \mathbf{d}_{e+1/2,e}\eta_e - \mathbf{d}_{e+1,e+1/2}\eta_{e+1}
$$

for $0 \le e < m$. Static condensation is also discussed by Bathe and Wilson [14, p. 259ff], Hughes [106, p. 245f] and Strang and Fix [163, p. 81].

Let us estimate the total work in static condensation. Each of the matrices $\mathbf{A}_{e+1/2,e+1/2}$ is $(k-1)\times(k-1)$ where k is the degree of the continuous spline approximation in the finite element method. Factoring each of these symmetric positive matrices requires $k(k-1)(k-2)/6$ multiplications, and solving a triangular linear system involving one of these matrix factors requires $(k-1)(k-2)/2$ multiplications. An inner product of two $(k-1)$-vectors requires $k-1$ multiplications. Factoring the condensed tridiagonal matrix requires $m-1$ multiplications, and solving the tridiagonal system for the unknowns at the mesh points requires another $2(m-1)$ multiplications. Finally, solving the tridiagonal systems for the interior unknowns would cost another $2(k-1)+(k-1)(k-2)/2$ multiplications for each. The total work would involve

$$
mk(k-1)(k-2)/6 + 4m(k-1)(k-2)/2 + 7m(k-1) + (m-1) + 2(m-1)
$$
$$
= m(k-1)(k+2)(k+8)/6 + 3(m-1)
$$

multiplications.

Next, suppose that we use a symmetric positive band solver (such as LAPACK routines dpbtrf and dpbtrs) to solve the original finite element system. The matrix would have a band width of k. The factorization would cost $m(k+1)^2 - k(k+1)(4k+5)/6$ multiplications, and the triangular system solves would cost another $2m(2k+1) - 2k(k+1)$ multiplications. If the number m of elements is large, static condensation requires more work than solving the original banded linear system for all spline degrees k.

Fortunately, in Sect. 4.7 we will discover an interesting way to "solve" the linear systems for high order finite element approximations.

Exercise 4.6.12 Suppose that we use continuously differentiable splines for our finite element approximation.

1. Show that the reference basis functions would involve 4 basis functions associated with interpolation at the endpoints of the reference interval, and $k-3$ interior basis functions.
2. Describe the block structure of the resulting finite element linear system, with the unknowns partitioned between those at mesh points, and interior unknowns.
3. Modify the static condensation algorithm for this finite element approximation.
4. Count the number of multiplications in static condensation, and compare the work to that required by a banded matrix solver for the original finite element linear system.

Exercise 4.6.13 Consider the two-point boundary value problem

$$\varepsilon^2 y''(t) - y(t) = \varepsilon^2 f(t) \text{ for } -1 < t < 1 \text{ with } y(-1) = \beta_{-1} \text{ and } y(1) = \beta_1 .$$

Here $f(t)$, β_{-1} and β_1 are chosen so that the solution is

$$y(t) = \cos(\pi t) + \frac{e^{-(1+t)/\varepsilon}}{1 + e^{-2/\varepsilon}} + e^{-(1-t)/\varepsilon} .$$

1. Determine $f(t)$, β_{-1} and β_1.
2. Find the variational form of this problem, and show that the bilinear form is *not* an inner product on continuous functions with square integrable first derivative that vanish at one of the boundaries.
3. Let $\varepsilon = 10^{-4}$. Program a finite element method using continuous piecewise linear functions for this problem, and perform a convergence study.

This problem was posed previously in Exercises 4.5.1.

Exercise 4.6.14 Consider the fourth-order two-point boundary value problem

$$EID^4 y(x) + ky(x) = q(x) \text{ for } 0 < x < L \text{ with } y(0) = 0 = y(L) , \ y'(0) = 0 \text{ and } y''(L) = 0 .$$

Here y is the transverse deflection of a beam, $E = 3 \times 10^7$ newtons per square meter is the elastic modulus, $I = 3 \times 10^3$ newtons^4 is the second moment of area, $k = 2.604 \times 10^3$ newtons per meter is the restoring force per length in Hooke's law, and $q = 4.34 \times 10^4$ newtons is the applied body force. The boundary conditions $y(0) = 0 = y(L)$ specify zero transverse displacement at the ends of the beam. The boundary condition $y'(0) = 0$ says that the beam is clamped on the left, while the boundary condition $y''(L) = 0$ says that the beam is simply supported on the right.

1. Find the variational form of this problem by multiplying the equation by $\delta y(x)$ and integrating by parts twice. Which boundary conditions are essential, and which are natural?
2. Show that the bilinear form for this problem is an inner product on continuously differentiable functions with square-integrable second derivative that are zero at both boundaries and have zero derivative at the left boundary.

3. Describe the use of continuously differentiable cubic splines in a finite element method for this problem.
4. Program the resulting finite element method, and perform a convergence study to determine the effective convergence rate as a function of the mesh width.

This problem was posed previously in Exercises 4.5.1.

4.7   Deferred Correction

In Sect. 3.5, we discussed deferred correction as an alternative method for generating high-order approximations to initial value problems. In this chapter, we will see that deferred correction is also useful as an iterative method for solving linear and nonlinear equations derived from high-order discretizations of two-point boundary value problems. These high-order approximations might arise from finite element methods, collocation or finite differences. We will discuss the application of deferred correction to both collocation methods and finite element methods. Proofs of the convergence of deferred correction methods for two-point boundary value problems can be found in Pereyra [141] and Skeel [160, 161].

4.7.1   Collocation

Suppose that we want to solve the nonlinear two-point boundary value problem

$$\mathbf{y}'(t) - \mathbf{f}(\mathbf{y}(t), t) = \mathbf{0} \text{ for } T_0 < t \leq T_M \text{ with } \mathbf{g}(\mathbf{y}(T_0), \mathbf{y}(T_M)) = \mathbf{0} .$$

Let us choose a coarse mesh

$$T_0 < T_1 < \ldots < T_{M-1} < T_M .$$

In the interior of the problem domain, the solution of the two-point boundary value problem satisfies

$$\mathbf{0} = \int_{T_I}^{T_{I+1}} \mathbf{y}'(t) - \mathbf{f}(\mathbf{y}(t), t) \ dt \frac{1}{T_{I+1} - T_I} = \frac{\mathbf{y}(T_{I+1}) - \mathbf{y}(T_I)}{T_{I+1} - T_I} - \int_{T_I}^{T_{I+1}} \mathbf{f}(\mathbf{y}(t), t) \ dt$$

for $0 \le I < M$. Let $k > 1$ be an integer and choose points

$$0 = \tau_0 < \tau_1 < \ldots < \tau_{k+1} = 1 .$$

For best practices in collocation, the points $\tau_1, \ldots, \tau_k$ should be the Gaussian quadrature points in $(0, 1)$. The mesh points for the low-order scheme are

$$t_{I(k+1)+j} = T_I + [T_{I+1} - T_I]\tau_j \text{ for } 0 \le I < M \text{ and } 0 \le j \le k$$

plus the endpoint $t_{M(k+1)} = T_M$. This suggests that we define

$$m = M(k + 1) .$$

As in the description of fully implicit Runge-Kutta methods (see Sect. 3.6.3) we define the scalars

$$\alpha_{j\ell} = \int_0^{\tau_j} \prod_{\substack{1 \le n \le k \\ n \ne \ell}} \frac{\tau - \tau_n}{\tau_\ell - \tau_n} \, d\tau \text{ for } 1 \le j, \ell \le k$$

and

$$\beta_\ell = \int_0^1 \prod_{\substack{1 \le n \le k \\ n \ne \ell}} \frac{\tau - \tau_n}{\tau_\ell - \tau_n} \, d\tau \text{ for } 1 \le \ell \le k .$$

Given a set of vectors $\{\widetilde{\mathbf{y}}_i\}_{i=0}^m$ such that

$$\mathbf{g}(\widetilde{\mathbf{y}}_0, \widetilde{\mathbf{y}}_m) = \mathbf{0} ,$$

our high-order method will define

$$\mathbf{H}\{\widetilde{\mathbf{y}}\}_{(I+1)(k+1)} \equiv \frac{\widetilde{\mathbf{y}}_{(I+1)(k+1)} - \widetilde{\mathbf{y}}_{I(k+1)}}{T_{I+1} - T_I} - \sum_{j=1}^k \mathbf{f}\left(\widetilde{\mathbf{y}}_{I(k+1)+j}, T_I + [T_{I+1} - T_I]\tau_j\right) \beta_j$$

for $0 \le I < M$ and

$$\mathbf{H}\{\widetilde{\mathbf{y}}\}_{I(k+1)+j} \equiv \frac{\widetilde{\mathbf{y}}_{I(k+1)+j} - \widetilde{\mathbf{y}}_{I(k+1)}}{T_{I+1} - T_I} - \sum_{\ell=1}^k \mathbf{f}\left(\widetilde{\mathbf{y}}_{I(k+1)+\ell}, T_I + [T_{I+1} - T_I]\tau_\ell\right) \alpha_{j\ell}$$

for $0 \leq I < M$ and $1 \leq j \leq k$. We can take the low-order scheme to be the midpoint rule on the finer mesh:

$$\mathbf{L}\{\widetilde{\mathbf{y}}\}_{i+1} \equiv \frac{\widetilde{\mathbf{y}}_{i+1} - \widetilde{\mathbf{y}}_i}{t_{i+1} - t_i} - \mathbf{f}\left([\widetilde{\mathbf{y}}_{i+1} + \widetilde{\mathbf{y}}_i]/2, [t_{i+1} + t_i]/2\right)$$

for $0 \leq i < m$. The **deferred correction** algorithm takes the form

Algorithm 4.7.1 (Deferred Correction for BVP via Collocation)

> for $i = 1, \ldots, m, \quad \mathbf{r}_i = \mathbf{0}$
>
> solve $\mathbf{L}\{\widetilde{\mathbf{y}}\}_i = \mathbf{r}_i$ for $1 \leq i \leq m$ with $\mathbf{g}(\widetilde{\mathbf{y}}_0, \widetilde{\mathbf{y}}_m) = \mathbf{0}$
>
> $n = 2$
>
> while $n < 2k$
>
> $\quad$ for $i = 1, \ldots, m, \quad \mathbf{r}_i = \mathbf{r}_i - \mathbf{H}\{\widetilde{\mathbf{y}}\}_i$
>
> $\quad$ solve $\mathbf{L}\{\widetilde{\mathbf{y}}\}_i = \mathbf{r}_i$ for $1 \leq i \leq m$ with $\mathbf{g}(\widetilde{\mathbf{y}}_0, \widetilde{\mathbf{y}}_m) = \mathbf{0}$
>
> $\quad$ $n = n + 2$

In this algorithm, n represents the current order of the approximation $\widetilde{y}$. The high-order scheme $\mathbf{H}$ has order $2k$, where k is the number of Gaussian quadrature points in each coarse mesh interval. The low-order scheme $\mathbf{L}$ has order 2. Each iteration of deferred correction improves the order of the approximation by 2, until the order of $\mathbf{H}$ is reached.

Suppose that we want to solve the equations for the high-order scheme directly. For $1 \leq I < M$, the unknown vector $\widetilde{\mathbf{y}}_{I(k+1)}$ appears in equations for $\widetilde{\mathbf{y}}_i$ with $(I - 1)(k + 1) \leq i \leq (I + 1)(k + 1)$. The linear system for a Newton iteration to solve the nonlinear system for the unknown vectors in the high-order scheme would have bandwidth $k + 1$. Factorization of the banded matrix would cost an order of $m(k + 2)^2$ multiplications, and banded triangular system solves would cost an order of $2m(2k + 3)$ multiplications. The evaluation of the functions $\mathbf{f}$ at the m fine mesh points may possibly dominate the work. The total work would be on the order of $m(k^2 + 8k + 10)$ multiplications for the linear system solve and m evaluations of $\mathbf{f}$ for each Newton iteration.

Next, let us estimate the work in deferred correction. A Newton iteration to solve the equations for the low-order midpoint method involves a tridiagonal matrix with possibly an additional entry in a corner far from the diagonal. Factoring this matrix would require an order of $2m$ multiplications, and solving the triangular linear systems would cost an order of $3m$ multiplications. Evaluation of the high-order scheme to update the residual $\mathbf{r}$ would require m evaluations of $\mathbf{f}$ for each Newton iteration. In practice, the previous value of the deferred correction solution $\widetilde{y}$ provides a very good guess for the Newton iteration in the next step of deferred correction, so typically only one Newton iteration is needed (except possibly to find the solution of the first low-order system). Deferred correction would thus require k repetitions of the previously estimated work. A true cost comparison would require *a priori* knowledge of the number of Newton iterations for each alternative approach. However, we do know that the cost to solve the linear systems in each Newton iteration within deferred correction is significantly less than the cost of the linear system solve within the original high-order system.

Cash and Wright [34] have written the Fortran program twpbvp to perform deferred correction with mono-implicit Runge-Kutta schemes for solving two-point boundary value problems. This method also performs adaptive mesh refinement. These same authors have also written the Fortran program acdc to perform deferred correction for two-point boundary value problems using Lobatto Runge-Kutta schemes and a **continuation method**. These algorithms also use deferred correction to estimate local errors and perform adaptive mesh refinement.

4.7.2 Finite Elements

Suppose that we want to solve the linear two-point boundary value problem

$$-(py')'(t) - \mathbf{f}(t) = \mathbf{0} \text{ for } T_0 < t < T_M \text{ with } y(T_0) = \beta_0 \text{ and } p(T_M)y'(T_M)) = \beta_1 .$$

Let us choose a coarse mesh

$$T_0 < T_1 < \ldots < T_{M-1} < T_M .$$

The variational form of the problem is

$$\mathbf{0} = \int_{T_0}^{T_M} \delta y'(t)p(t)y'(t) \, dt - \int_{T_0}^{T_M} \delta y(t)f(t) \, dt - \delta y(T_M)\beta_1$$

for all continuous functions δy with square-integrable first derivative that satisfy $\delta y(T_0) = 0$. Let $k > 1$ be an integer and choose points

$$0 = \tau_0 < \tau_1 < \ldots < \tau_k = 1 .$$

For example, the points $\tau_1, \ldots, \tau_{k-1}$ might be the Chebyshev interpolation points (see Sect. 1.2.6) in the unit interval $(0, 1)$. The mesh points for the low-order scheme are

$$t_{Ik+j} = T_I + [T_{I+1} - T_I]\tau_j \text{ for } 0 \le I < M \text{ and } 0 \le j \le k$$

plus the endpoint $t_{Mk} = T_M$. This suggests that we define

$$m = Mk .$$

As in the description of the elemental formulation of finite element schemes 4.6.4 we define the reference shape functions to be the Lagrange interpolation basis polynomials

$$
\sigma_j(\tau) = \prod_{\substack{0 \le \ell \le k \\ \ell \ne j}} \frac{\tau - \tau_\ell}{\tau_j - \tau_\ell}
$$

for $0 \le j$, $\ell \le k$. We also define the global index map

$$
\nu(e, d) = ek + d
$$

and the coordinate map

$$
\mu_e(\tau) = T_e(1 - \tau) + T_{e+1}\tau
$$

from the reference element $(0, 1)$ to the mesh element (T_e, T_{e+1}). Within element (T_e, T_{e+1}) for $0 \le e < M$ the shape functions are

$$
\widetilde{s}_{\nu(e,d)}(\mu_e(\tau)) = \sigma_d(\tau)
$$

for $0 \le d \le k$ and $\tau \in [0, 1]$. Given a set of scalars $\{\widetilde{y}_i\}_{i=0}^m$ such that

$$
\widetilde{y}_0 - \beta_0 = 0 ,
$$

our high-order method will define

$$
\mathbf{H}\{\widetilde{y}\}_{Ik} \equiv \int_{T_{I-1}}^{T_I} \widetilde{s}_{Ik}'(t)p(t) \left[\sum_{\ell=(I-1)k}^{Ik} \widetilde{s}_\ell'(t)\widetilde{y}_\ell \right] dt
$$

$$
+ \int_{T_I}^{T_{I+1}} \widetilde{s}_{Ik}'(t)p(t) \left[\sum_{\ell=Ik}^{(I+1)k} \widetilde{s}_\ell'(t)\widetilde{y}_\ell \right] dt - \int_{T_{I-1}}^{T_I} \widetilde{s}_{Ik} f(t) \, dt
$$

$$
- \int_{T_I}^{T_{I+1}} \widetilde{s}_{Ik} f(t) \, dt - \widetilde{s}_{Ik}(T_M)\beta_1
$$

for $0 < I \le M$ and

$$
\mathbf{H}\{\widetilde{y}\}_{Ik+j} \equiv \int_{T_I}^{T_{I+1}} \widetilde{s}_{Ik+j}'(t)p(t) \left[\sum_{\ell=Ik}^{(I+1)k} \widetilde{s}_\ell'(t)\widetilde{y}_\ell \right] dt
$$

$$
- \int_{T_I}^{T_{I+1}} \widetilde{s}_{Ik+j} f(t) \, dt - \widetilde{s}_{Ik+j}(T_M)\beta_1
$$

for $0 \le I < M$ and $1 \le j < k$. We can choose the low-order scheme to use piecewise linear shape functions on the finer mesh:

$$
\mathbf{L}\{\widetilde{\mathbf{y}}\}_i \equiv \int_{t_{i-1}}^{t_i} p(t) \frac{\widetilde{y}_i - \widetilde{y}_{i-1}}{t_i - t_{i-1}} \, dt - \int_{t_i}^{t_{i+1}} p(t) \frac{\widetilde{y}_{i+1} - \widetilde{y}_i}{t_{i+1} - t_i} \, dt
$$

$$
- \int_{t_{i-1}}^{t_i} \frac{t - t_{i-1}}{t_i - t_{i-1}} f(t) \, dt - \int_{t_i}^{t_{i+1}} \frac{t_{i+1} - t}{t_{i+1} - t_i} f(t) \, dt
$$

for $1 \le i < m$ and

$$
\mathbf{L}\{\widetilde{\mathbf{y}}\}_m \equiv \int_{t_{m-1}}^{t_m} p(t) \frac{\widetilde{y}_m - \widetilde{y}_{m-1}}{t_m - t_{m-1}} \, dt
$$

$$
- \int_{t_{m-1}}^{t_m} \frac{t - t_{m-1}}{t_m - t_{m-1}} f(t) \, dt - \beta_1 \ .
$$

The **deferred correction** algorithm then takes the form

Algorithm 4.7.2 (Deferred Correction for BVP via Finite Elements)

> for $i = 1, \ldots, m, \quad \mathbf{r}_i = \mathbf{0}$
>
> solve $\mathbf{L}\{\widetilde{\mathbf{y}}\}_i = \mathbf{r}_i$ for $1 \le i \le m$ with $\widetilde{y}_0 = \beta_0$
>
> $n = 2$
>
> while $n < k + 1$
>
> > for $i = 1, \ldots, m, \quad \mathbf{r}_i = \mathbf{r}_i - \mathbf{H}\{\widetilde{\mathbf{y}}\}_i$
> >
> > solve $\mathbf{L}\{\widetilde{\mathbf{y}}\}_i = \mathbf{r}_i$ for $1 \le i \le m$ with $\widetilde{y}_0 = \beta_0$
> >
> > $n = n + 2$

In this algorithm, n represents the current order of the approximation $\widetilde{y}$. The high-order scheme $\mathbf{H}$ has order $k + 1$, where k is the degree of the polynomials used in each coarse mesh interval. The low-order scheme $\mathbf{L}$ has order 2. Each iteration of deferred correction improves the order of the approximation by 2, until the order of $\mathbf{H}$ is reached.

Suppose that we want to solve the equations for the high-order scheme directly. Our discussion in Sect. 4.6.9 estimated that a banded matrix solver for the high-order scheme would require an order of $m(k + 1)^2 + m(4k + 2)$ multiplications. Next, let us estimate the work in deferred correction. Solution of each linear system for the low-order scheme would require an order of $3m$ multiplications, and evaluation of the high-order scheme $\mathbf{H}$ on the current deferred correction solution $\widetilde{y}$ would require an order of $M(k - 1)(k + 1) + M(2k + 1) = m(k + 2)$ multiplications. Deferred correction would thus require $(k+1)/2$ repetitions of the previously estimated work, for total work on the order of $m(k + 5)(k + 1)/2$ multiplications. For large values of k, deferred correction involves roughly half the work of solving the high-order equations directly,

Exercise 4.7.1 Program deferred correction with fully implicit Runge-Kutta methods for the Bratu problem. Compare the computational cost of deferred correction versus the cost of solving the high-order equations directly, for schemes of order 10 and 20.

Exercise 4.7.2 Suppose that you wanted to implement deferred correction for a finite element method that uses high-order continuously differentiable splines. What would you use for the low-order scheme? How would you compute the residual for the high-order scheme?

Exercise 4.7.3 Consider the two-point boundary value problem

$$\varepsilon y''(t) - ty(t) = 0 \text{ for } -1 < t < 1 \text{ with } y(-1) = 1 \text{ and } y(1) = 1 .$$

The exact solution of this problem can be expressed in terms of Airy functions (see Abramowitz and Stegun [1, p. 446]).

1. Rewrite the problem in the form

$$\mathbf{y}'(t) = \mathbf{A}(t)\mathbf{y}(t) \text{ for } -1 < x < 1 \text{ with } \mathbf{B}_{-1}\mathbf{y}(-1) + \mathbf{B}_1\mathbf{y}(1) = \mathbf{b} .$$

2. Let $\varepsilon = 10^{-6}$. Choose a numerical method that uses deferred correction to solve this problem, and perform a convergence study.

Warning: the solution of this problem involves dense oscillations, a boundary layer at the right boundary, and a turning point where the solution changes character.

Exercise 4.7.4 Consider the two-point boundary value problem

$$\varepsilon y''(t) - ty'(t) + y(t) = 0 \text{ for } -1 < t < 1 \text{ with } y(-1) = 1 \text{ and } y(1) = 2 .$$

1. Show that **Kummer's function**

$$M_{a,b}(z) = \sum_{n=0}^{\infty} \frac{(a)_n}{(b)_n} \frac{z^n}{n!}$$

solves **Kummer's differential equation** (see Abramowitz and Stegun [1, p. 504])

$$zu''(z) + (b - z)u'(z) - au(z) = 0 .$$

Here

$$(a)_n \equiv \prod_{i=0}^{n-1}(a + i) .$$

2. Show that the solution of the two-point boundary value problem in this exercise is

$$y(t) = \frac{t}{2} + \frac{3}{2M_{-1,1/2}(1/[2\varepsilon])}M_{-1,1/2}(t^2/[2\varepsilon]) .$$

3. Rewrite the two-point boundary value problem in the form

$$\mathbf{y}'(t) = \mathbf{A}(t)\mathbf{y}(t) \text{ for } -1 < x < 1 \text{ with } \mathbf{B}_{-1}\mathbf{y}(-1) + \mathbf{B}_1\mathbf{y}(1) = \mathbf{b} .$$

4. Let $\varepsilon = 1/70$. Choose a numerical method that uses deferred correction to solve this problem, and perform a convergence study.

References

1. M. Abramowitz, I.A. Stegun (eds.), *Handbook of Mathematical Functions* (Dover, New York, 1965)
2. R.A. Adams (ed.), *Sobolev Spaces* (Academic, Amsterdam, 1975)
3. S. Agmon, *Lectures on Elliptic Boundary Value Problems* (van Nostrand, Princeton, NJ, 1965)
4. R. Alexander, Diagonally implicit Runge-Kutta methods for stiff O.D.E.'s. SIAM J. Numer. Anal. **14**, 1006–1021 (1977)
5. J.R. Angelos, E.H. Kaufman Jr., M.S. Henry, T.D. Lenker, Optimal nodes for polynomial interpolation, in *Approximation Theory VI*, ed. by C.K. Chui, L.L. Schumaker, J.D. Ward (Academic, New York, 1989), pp. 17–20
6. T. Apostol, *Mathematical Analysis* (Addison-Wesley, Reading, MA, 1974)
7. U.M. Ascher, R.M.M. Mattheij, R.D. Russell (eds.), *Numerical Solution of Boundary Value Problems for Ordinary Differential Equations* (Prentice Hall, Englewood, NJ, 1988)
8. K.E. Atkinson, *An Introduction to Numerical Analysis* (Wiley, New York, 1978)
9. A.K. Aziz (ed.), *The Mathematical Foundations of the Finite Element Method with Applications to Partial Differential Equations* (Academic, New York, 1972)
10. I. Babuška, T. Strouboulis, *The Finite Element Method and Its Reliability* (Clarendon Press, Oxford, 2001)
11. G. Bader, P. Deuflhard, A semi-implicit mid-point rule for stiff systems of ordinary differential equations. Numer. Math. **41**(3), 373–398 (1983)
12. G.A. Baker Jr., P. Graves-Morris, *Padé Approximants* (Cambridge University Press, Cambridge, 1996)
13. R.E. Bank, W.M. Coughran, W. Fichtner, E.H. Grosse, D.J. Rose, R.K. Smith, Transient simulation of silicon devices and circuits. IEEE Trans. Comput. Aided Des. **4**, 436–451 (1985)
14. K.J. Bathe, E.L. Wilson, *Numerical Methods in Finite Element Analysis* (Prentice-Hall, Englewood Cliffs, 1976)
15. J. Berbernes, D. Eberly, *Mathematical Problems from Combustion Theory*. Applied Mathematical Sciences, vol. 83 (Springer, New York, 1989)
16. G. Birkhoff, G.-C. Rota, *Ordinary Differential Equations*, 3rd edn. (Wiley, New York, 1978)
17. S.M. Blinder, *Guide to Essential Math: A Review for Physics, Chemistry and Engineering Students* (Newnes, Oxford, 2013)
18. W. Böhm, G. Farin, J. Kahmann, A survey of curve and surface methods in CAGD. Comput. Aided Geom. Des. **1**, 1–60 (1984)

19. W.E. Boyce, R.C. DiPrima, *Elementary Differential Equations and Boundary Value Problems* (Wiley, Hoboken, 2012)
20. D. Braess, *Finite Elements* (Cambridge University Press, Cambridge, 2007)
21. J.H. Bramble, S. Hilbert, Bounds for a class of linear functionals with applications to hermite interpolation. Numer. Math. **16**, 362–369 (1971)
22. K.E. Brenan, S.L. Campbell, L.R. Petzold, *Numerical Solution of Initial-Value Problems in Differential-Algebraic Equations* (North-Holland, New York, 1989)
23. S.C. Brenner, L.R. Scott, *The Mathematical Theory of Finite Element Methods* (Springer, New York, 2002)
24. G. Bruügner, Rounding error analysis of interpolation procedures. Computing **33**, 83–87 (1984)
25. R.C. Buck, *Advanced Calculus* (McGraw-Hill, New York, 1965)
26. J.C. Butcher, Implic Runge-Kutta processes. Math. Comput. **18**, 50–64 (1964)
27. J.C. Butcher, On the attainable order of Runge-Kutta methods. Math. Comput. **19**(91), 408–417 (1965)
28. J.C. Butcher, The non-existence of ten stage eighth order explicit Runge-Kutta methods. BIT Numer. Math. **25**, 521–540 (1985)
29. G.D. Byrne, A.C. Hindmarsh, A polyalgorithm for the numerical solution of ordinary differential equations. ACM Trans. Math. Softw. **1**, 71–96 (1975)
30. R.E. Caflisch, Monte Carlo and quasi-Monte Carlo methods. Acta Numer. **7**, 1–49 (1998)
31. K. Calvetti, G.H. Golub, W.B. Gragg, L. Reichel, Computation of Gauss-Kronrod quadrature rules. Math. Comput. **69**(231), 1035–1052 (2000)
32. G. Casella, R.L. Berger, *Statistical Inference* (Duxbur, Pacific Grove, CA, 1990)
33. J.R. Cash, A.H. Karp, A variable order Runge-Kutta method for initial value problems with rapidly varying right-hand sides. ACM Trans. Math. Softw. **16**(3), 201–222 (1990)
34. J.R. Cash, M.H. Wright, A deferred correction method for nonlinear two-point boundary value problems: implementation and numerical evaluation. SIAM J. Sci. Stat. Comput. **12**(4), 971–989 (1991)
35. M.A. Celia, W.G. Gray, *Numerical Methods for Differential Equations* (Prentice Hall, Englewood Cliffs, NJ, 1992)
36. F. Ceschino, J. Kuntzmann, *Numerical Solutions of Initial Value Problems* (Prentice Hall, Englewood Cliffs, NJ, 1966)
37. Z. Chen, *Finite Element Methods and Their Applications* (Springer, Berlin, 1966)
38. Q. Chen, I. Babuška, Approximate optimal points for polynomial interpolation of real functions in an interval and in a triangle. Comput. Methods Appl. Mech. Eng. **128**, 405–417 (1995)
39. Q Chen, I. Babuška, The optimal symmetrical points for polynomial interpolation of real functions in the tetrahedron. Comput. Methods Appl. Mech. Eng. **137**, 89–94 (1996)
40. E.W. Cheney, *Introduction to Approximation Theory* (McGraw-Hill, New York, 1966)
41. E.M. Cherry, H.S. Greenside, C.S. Henriquez, A space-time adaptive method for simulating complex cardiac dynamics. Phys. Rev. Lett. **84**, 1343–1346 (2000)
42. C.K. Chui, *An Introduction to Wavelets* (Academic, New York, 1992)
43. P.G. Ciarlet, *The Finite Element Method for Elliptic Problems* (North-Holland, Amsterdam, 1978)
44. P.G. Ciarlet, P.-A. Raviart, General lagrange and hermite interpolation in $\mathbb{R}^n$ with applications to finite element methods. Arch. Ration. Mech. Anal. **467**, 177–199 (1972)
45. P.G. Ciarlet, M.H. Schultz, R.S. Varga, Numerical methods of high-order accuracy for nonlinear boundary value problems. Numer. Math. **13**, 51–77 (1969)
46. C.W. Clenshaw, A.R. Curtis, A method for numerical integration on an automatic computer. Numer. Math. **2**, 197 (1960)
47. E.A. Coddington, N. Levinson, *Theory of Ordinary Differential Equations* (McGraw-Hill, New York, 1955)
48. A. Cohen, *Numerical Analysis of Wavelet Methods*. Studies in Mathematics and its Applications, vol. 32 (Ellsevier, North Holland, 2003)

49. A. Cohen, I. Daubechies, J.C. Feauveau, Biorthogonal bases of compactly supported wavelets. Commun. Pure Appl. Math. **45**, 485–560 (1992)

50. J.W. Cooley, J.W. Tukey, An algorithm for the machine computation of complex Fourier series. Math. Comput. **19**, 297–301 (1965)

51. P. Costantini, R. Morandi, An algorithm for computing shape-preserving cubic spline interpolation to data. Calcolo **21**(4), 295–305 (1984)

52. G.R. Cowper, Gaussian quadrature formulas for triangles. Int. J. Numer. Methods Eng. **7**, 405–408 (1973)

53. C.W. Cryer, On the instability of high-order backward-difference multistep methods. BIT Numer. Math. **12**, 17–25 (1972)

54. C.W. Cryer, A new class of highly stable methods: a_0-stable methods. BIT Numer. Math. **13**, 153–159 (1973)

55. G. Dahlquist, Convergence and stability in the numerical integration of ordinary differential equations. Math. Scand. **4**, 33–53 (1956)

56. G. Dahlquist, A special stability problem for linear multistep methods. BIT Numer. Math. **3**, 27–43 (1963)

57. G. Dahlquist, Å. Björck, *Numerical Methods* (Prentice-Hall, Englewood Cliffs, 1974). Translated by N. Anderson

58. I. Daubechies, *Ten Lectures on Wavelets*. CBMS-NSF Regional Conference Series in Applied Mathematics (SIAM, Philadelphia, PA, 1992)

59. P.J. Davis, *Interpolation and Approximation* (Blaisdell, New York, 1965)

60. P.J. Davis, P. Rabinowitz, *Methods of Numerical Integration* (Dover, New York, 2007)

61. C. de Boor, *A Practical Guide to Splines*. Applied Mathematical Sciences, vol. 27 (Springer, New York, 1978)

62. C. de Boor, B. Swartz, Collocation at gaussian points. SIAM J. Numer. Anal. **10**(4), 582–606 (1973)

63. K. Dekker, J.G. Verwer, *Stability of Runge-Kutta Methods for Stiff Nonlinear Differential Equations* (North-Holland, Amsterdam, 1984)

64. J.W. Demmel, A numerical analyst's Jordan canonical form, Ph.D. thesis, Center for Pure and Applied Mathematics, University of California, Berkeley, 1983

65. J. Deny, J.L. Lions, Les espaces du type de Beppo levi. Ann. Inst. Fourier **5**, 305–370 (1953/1954)

66. P. Deuflhard, Order and stepsize control in extrapolation methods. Numer. Math. **41**(3), 399–422 (1983)

67. P. Deuflhard, F. Bornemann, *Scientific Computing with Ordinary Differential Equations* (Springer, New York, 2002)

68. J.R. Dormand, P.J. Prince, A family of embedded Runge-Kutta formulae. J. Comput. Appl. Math. **6**, 19–26 (1980)

69. P.G. Drazin, *Nonlinear Systems* (Cambridge University Press, Cambridge, 1992)

70. D.A. Dunavant, High degree efficient symmetrical gaussian quadrature rules for the triangle. Int. J. Numer. Methods Eng. **21**, 1129–1148 (1985)

71. A. Dutt, L. Greengard, V. Rokhlin, Spectral deferred correction methods for ordinary differential equations. BIT Numer. Math. **40**, 241–266 (2000)

72. R. England, Error estimates for Runge-Kutta type solutions to systems of ordinary differential equations. Comput. J. **12**, 166–170 (1969)

73. G. Faber, Über die interpolatorische Darstellung stetiger Funktionen. Deutsche Mathematiker-Vereinigung Jahresbericht **23**, 192–210 (1914)

74. E. Fehlberg, Classical fifth-, sixth-, seventh-, and eighth order Runge-Kutta formulas with step size control. Computing **4**, 93–106 (1969)

75. E. Fehlberg, Low-order classical Runge-Kutta formulas with step size control and their application to some heat transfer problems. Computing **6**, 61–71 (1970)

76. C. Felippa, A compendium of FEM integration formulas for symbolic work. Eng. Comput. **21**, 867–890 (2004)

77. W. Feller, *An Introduction to Probability Theory and Its Applications*. Probability and Mathematical Statistics (Wiley, New York, 1968)
78. C.T. Fike, *Computer Evaluation of Mathematical Functions* (Prentice-Hall, Englewood Cliffs, NJ, 1968)
79. B.A. Finlayson, *The Method of Weighted Residuals and Variational Principles* (Academic, New York, 1972)
80. J.D. Foley, A. van Dam, S.K. Feiner, J.F. Hughes, *Computer Graphics: Principles and Practice* (Addision-Wesley Professional, Boston, 1990)
81. G.B. Folland, *Real Analysis: Modern Techniques and Their Applications*. Pure and Applied Mathematics, 2nd edn. (Wiley, New York, 1999)
82. B. Fornberg, Generation of finite difference formulas on arbitrarily spaced grids. Math. Comput. **51**(184), 699–706 (1988)
83. C.F. Gauss, *Werke*. Gedruckt in der Dieterichschen Universitätsdruckerei, vol. 3 (W.F. Kaestner, Gottingen, 1813)
84. W. Gautschi, Questions of numerical condition related to polynomials, in *Studies in Numerical Analysis*, ed. by G.H. Golub (MAA, Washington, DC, 1984), pp. 140–177
85. C.W. Gear, The numerical integration of ordinary differential equations. Math. Comput. **21**, 146–156 (1967)
86. C.W. Gear, *Numerical Initial Value Problems in Ordinary Differential Equations* (Prentice-Hall, Englewood Cliffs, NJ, 1971)
87. H. Goldstein, *Classical Mechanics* (Addison-Wesley, Reading, MA, 1950)
88. G.H. Golub, J.H. Welsch, Calculation of Gauss quadrature rules. Math. Comput. **23**, 221–230 (1969)
89. D. Gottlieb, S.A. Orszag, *Numerical Analysis of Spectral Methods: Theory and Applications*. Regional Conference Series in Applied Mathematics, vol. 26 (SIAM, Philadelphia, PA, 1977)
90. W. Gragg, On extrapolation algorithms for ordinary initial value problems. SIAM J. Numer. Anal. **2**, 384–403 (1965)
91. A. Greenbaum, *Numerical Methods: Design, Analysis, and Computer Implementation of Algorithms* (Princeton University Press, Princeton, NJ, 2012)
92. A. Griewank, A. Walther, *Evaluating Derivatives: Principles and Techniques of Algorithmic Differentiation*. Other Titles in Applied Mathematics, vol. 105 (SIAM, Philadelphia, PA, 2008)
93. E. Hairer, G. Wanner, *Solving Ordinary Differential Equations II Stiff and Differential Algebraic Equations* (Springer, New York, 1991)
94. E. Hairer, S.P. Nørsett, G. Wanner, *Solving Ordinary Differential Equations I Nonstiff Problems* (Springer, New York, 1993)
95. P. Halmos, *Finite-Dimensional Vector Spaces*. University Series in Higher Mathematics (van Nostrand, Toronto, 1958)
96. P.C. Hammer, J.W. Hollingsworth, Trapezoidal methods of approximating solutions of differential equations. Math. Tables Other Aids Comput. **9**, 92–96 (1955)
97. A.C. Hansen, J. Strain, On the order of deferred correction. Appl. Numer. Math. **61**, 961–973 (2011)
98. P. Hartman, *Ordinary Differential Equations* (Wiley, New York, 1964)
99. M.T. Heath, *Scientific Computing: An Introductory Survey* (McGraw-Hill, New York, 2002)
100. P. Henrici, *Discrete Variable Methods in Ordinary Differential Equations* (Wiley, New York, 1962)
101. P. Henrici, *Error Propagation for Difference Methods* (Wiley, New York, 1963)
102. P. Henrici, *Elements of Numerical Analysis* (Wiley, New York, 1964)
103. F.B. Hildebrand, *Introduction to Numerical Analysis* (McGraw-Hill, New York, 1956)
104. A. Hodges, *Alan Turing: The Enigma* (Vintage, London, 1992)
105. E. Houstis, A collocation method for systems of nonlinear ordinary differential equations. J. Math. Anal. Appl. **62**(1), 24–37 (1978)
106. T.J.R. Hughes, *The Finite Element Method: Linear Static and Dynamic Finite Element Analysis* (Prentice-Hall, Englewood Cliffs, NJ, 1987)

107. T.E. Hull, W.H. Enright, B.M. Fellen, A.E. Sedgwick, Comparing numerical methods for ordinary differential equations. SIAM J. Numer. Anal. **9**(4), 603–637 (1972)
108. W. Hurewicz, *Ordinary Differential Equations* (M.I.T. Press, Cambridge, MA, 1958)
109. J.P. Imhof, On the method for numerical integration of Clenshaw and Curtis. Numer. Math. **5**, 138–141 (1963)
110. E.L. Ince, *Ordinary Differential Equations* (Dover, New York, 1958)
111. L. Infeld, *Quest: An Autobiography* (American Mathematical Society, Providence, RI, 1980)
112. E. Isaacson, H.B. Keller, *Analysis of Numerical Methods* (Dover, New York, 1994)
113. A. Iserles, *A First Course in the Numerical Analysis of Differential Equations*. Cambridge Texts in Applied Mathematics (Cambridge University Press, Cambridge, 2008)
114. N.C. Jawias, F. Ismail, *Diagonally Implicit Runge-Kutta Methods for Solving Linear Odes* (LAP Lambert Academic Publishing, Berlin, 2011)
115. A. Jensen, A. la Cour-Harbo, *Ripples in Mathematics: The Discrete Wavelet Transform* (Springer, Berlin, 2001)
116. C. Johnson, *Numerical Solution of Partial Differential Equations by the Finite Element Method* (Cambridge University Press, Cambridge, 1994)
117. P. Keast, Moderate degree tetrahedral quadrature formulas. Comput. Methods Appl. Mech. Eng. **55**, 339–348 (1986)
118. W. Kelley, A. Peterson, *The Theory of Differential Equations Classical and Qualitative* (Pearson Education, Upper Saddle River, NJ, 2004)
119. D. Kincaid, W. Cheney, *Numerical Analysis* (Brooks/Cole, Pacific Grove, CA, 1991)
120. A.S. Konrod, *Nodes and Weights of Quadrature Formulas. Sixteen-Place Tables* (Consultants Bureau, New York, 1965). (Authorized translation from the Russian)
121. H.-O. Kreiss, Difference approximations for boundary and eigenvalue problems for ordinary differential equations. Math. Comput. **21**(119), 605–624 (1972)
122. E. Kreyszig, *Introductory Functional Analysis with Applications* (Wiley, New York, 1978)
123. J.D. Lambert, *Computational Methods in Ordinary Differential Equations* (Wiley, New York, 1973)
124. J.D. Lambert, *Numerical Methods for Ordinary Differential Systems* (Wiley, New York, 1991)
125. D.P. Laurie, Anti-Gaussian quadrature formulas. Math. Comput. **65**(214), 739–747 (1996)
126. D.P. Laurie, Calculation of Gauss-Kronrod quadrature rules. Math. Comput. **66**(219), 1133–1145 (1997)
127. P.G. Lemarié, On the existence of compactly supported dual wavelets. Appl. Comput. Harmon. Anal. **3**, 117–118 (1997)
128. J.L. Lions, E. Magenes, *Non-homogeneous Boundary Value Problems and Applications* (Springer, Berlin, 1972)
129. Y. Liu, M. Vinokur, Exact integrations of polynomials and symmetric quadrature formulas over arbitrary polyhedral grids. J. Comput. Phys. **140**, 122–147 (1998)
130. C.-H. Luo, Y. Rudy, A model of the ventricular cardiac action potential: depolarization, repolarization, and their interaction. Circ. Res. **68**, 1501–1526 (1991)
131. J.N. Lyness, D. Jesperson, Moderate degree symmetric quadrature rules for the triangle. J. Inst. Math. Appl. **15**, 9–32 (1975)
132. S.G. Mallat, Multiresolution approximations and wavelet orthonormal bases of $l^2(\mathbb{R})$. Trans. Am. Math. Soc. **315**, 69–87 (1989)
133. R.H. Merson, An operational method for the study of integration processes, in *Proceedings Symposium on Data Processing*, Salisbury, South Australia, 1957. Weapons Research Establishment
134. G. Monegato, A note on extended Gaussian quadrature rules. Math. Comput. **30**(136), 812–817 (1976)
135. G. Monegato, Stieltjes polynomials and related quadrature rules. SIAM Rev. **24**(2), 137–158 (1982)
136. G. Mülbach, The general Neville-Aitken-algorithm and some applications. Numer. Math. **31**(1), 97–110 (1978)

137. J.D. Murray, *Mathematical Biology*. Lecture Notes in Biomathematics, vol. 19 (Springer, Berlin, 1989)
138. R. Neidinger, Introduction to automatic differentiation and MATLAB object-oriented programming. SIAM Rev. **52**(3), 545–563 (2010)
139. A. Nordsieck, On the numerical integration of ordinary differential equations. Math. Comput. **16**, 22–49 (1962)
140. T.N.L. Patterson, The optimum addition of points to quadrature formulae. Math. Comput. **22**(104), 847–856 (1968)
141. V. Pereyra, Iterated deferred corrections for nonlinear boundary value problems. Numer. Math. **11**, 111–125 (1968)
142. O. Perron, Über Stabilität und asymptotisches verhalten der Integrale von Differentialgleichungssystemen. Math. Z. **29**, 129–160 (1929)
143. R. Plessens, Modified Clenshaw-Curtis integration and applications to numerical computation of integral transforms, in *Numerical Integration: Recent Developments, Software and Applications*, ed. by P. Keast, G. Fairweather. NATO ASI Series (Springer, Netherlands, 1987), pp. 35–51
144. W.H. Press, S.A. Teukolsky, W.T. Vetterling, B.P. Flannery, *Numerical Recipes: The Art of Scientific Computing*, 3rd edn. (Cambridge University Press, New York, 2007)
145. P.J. Prince, J.R. Dormand, High order embedded Runge-Kutta formulae. J. Comput. Appl. Math. **7**, 19–26 (1981)
146. E.J. Putzer, Avoiding the Jordan canonical form in the discussion of linear systems with constant coefficients. Am. Math. Mon. **73**(1), 2–7 (1966)
147. D. Radunovic, *Wavelets from Math to Practice* (Springer, Berlin, 2009)
148. L.B. Rall, *Automatic Differentiation: Techniques and Applications*. Lecture Notes in Computer Science, vol. 120 (Springer, Berlin, 1981)
149. A. Ralston, P. Rabinowitz, *A First Course in Numerical Analysis* (McGraw-Hill, New York, 1978)
150. F. Riesz, B.Sz.-Nagy, *Functional Analysis* (Frederick Ungar Publishing, New York, 1965)
151. V. Rokhlin, Rapid solution of integral equations of classic potential theory. J. Comput. Phys. **60**, 187–207 (1985)
152. H.L. Royden, *Real Analysis*, 2nd edn. (Macmillan, New York, 1968)
153. W. Rudin, *Real and Complex Analysis* (McGraw-Hill, New York, 1966)
154. W. Rudin, *Principles of Mathematical Analysis*. International Series in Pure and Applied Mathematics, 3rd edn. (McGraw-Hill, New York, 1976)
155. R.D. Russell, L.F. Shampine, A collocation method for boundary value problems. Numer. Math. **19**(1), 1–28 (1972)
156. T. Sauer, Y. Xu, chapter entitled 'The Aitken-Neville scheme in several variables', in *Approximation Theory X: Abstract and Classical Analysis*, ed. by C.K. Chui, L.L. Schumaker, J. Stöckler (Vanderbilt University Press, Nashville, 2002), pp. 353–366
157. I.J. Schoenberg, *Cardinal Spline Interpolation*. CBMS-NSF Regional Conference Series in Applied Mathematics, vol. 12 (SIAM, Philadelphia, PA, 1973)
158. A. Schönhage, V. Strassen, Schnelle Multiplikation grosser Zahlen. Computing **7**, 281–292 (1971)
159. G.E. Shilov, B.L. Gurevich, *Integral, Measure and Derivative: A Unified Approach* (Prentice-Hall, Englewood Cliffs, NJ, 1966)
160. R.D. Skeel, A theoretical foundation for proving accuracy results for deferred correction. SIAM J. Numer. Anal. **19**, 171–196 (1982)
161. R.D. Skeel, The order of accuracy for deferred corrections using uncentered formulas. SIAM J. Numer. Anal. **23**, 393–402 (1986)
162. J. Stoer, *Einführung in die Numerische Mathematik*, vol. I (Springer, Berlin, 1972)
163. G. Strang, G.J. Fix, *An Analysis of the Finite Element Method* (Prentice-Hall, Englewood Cliffs, NJ, 1973)
164. G. Strang, T. Nguyen, *Wavelets and Filter Banks* (Wesley-Cambridge Press, Stockport, 1997)
165. V. Strassen, Gaussian elimination is not optimal. Numer. Math. **13**, 354–356 (1969)

166. A.H. Stroud, *Approximate Calculation of Multiple Integrals* (Prentice-Hall, Englewood Cliffs, NJ, 1971)
167. A.H. Stroud, D. Secrest, *Gaussian Quadrature Formulas* (Prentice-Hall, Englewood Cliffs, NJ, 1966)
168. G. Szegö, Uber gewisse orthogonale Polynome, die zu einer oszillerenden Belegungsfuktion gehören. Math. Ann. **110**, 501–513 (1934)
169. B. Szabó, I. Babuška, *Finite Element Analysis* (Wiley, New York, 1991)
170. M. Taylor, B. Wingate, R. Vincent, An algorithm for computing Fekete points in the triangle. SIAM J. Numer. Anal. **38**, 1707–1720 (2000)
171. G. Teschl, *Ordinary Differential Equations and Dynamical Systems* (American Mathematical Society, Providence, RI, 2012)
172. V. Thomée, *Galerkin Finite Element Methods for Parabolic Problems* (Springer, Berlin, 1997)
173. W. Tiller, Rational B-splines for curve and surface representation. IEEE Comput. Graph. Appl. **3**, 61–69 (1983)
174. J.A. Trangenstein, *Numerical Solution of Elliptic and Parabolic Partial Differential Equations* (Cambridge University Press, Cambridge, 2013)
175. L.N. Trefethen, Is Gauss quadrature better than Clenshaw-Curtis? SIAM Rev. **50**, 67–87 (2008)
176. L.N. Trefethen, *Approximation Theory and Approximation Practice* (SIAM, Philadelphia, PA, 2013)
177. C. Truesdell, *The Elements of Continuum Mechanics* (Springer, Berlin, 1966)
178. J. Tukey, The future of data analysis. Ann. Math. Stat. **33**(1), 1–67 (1962)
179. J.H. Verner, High-order explicit Runge-Kutta pairs with low stage order. Special issue celebrating the centenary of Runge-Kutta methods. Appl. Numer. Math. **22**, 345–357 (1996)
180. R. Wait, A.R. Mitchell, *The Finite Element Analysis and Applications* (Wiley, New York, 1985)
181. S. Wandzura, H. Xiao, Symmetric quadrature rules on a triangle. Comput. Math. Appl. **45**, 1829–1840 (2003)
182. O.B. Widlund, A note on unconditionally stable linear multistep methods. BIT Numer. Math. **7**, 65–70 (1967)
183. J.H. Wilkinson, *The Algebraic Eigenvalue Problem* (Oxford University Press, Oxford, 1965)
184. P. Wojtaszczyk, *A Mathematical Introduction to Wavelets* (Cambridge University Press, Cambridge, 1997)
185. P. Wynn, On a device for calculating the $e_m(s_n)$ transformations. Math. Tables Automat. Comput. **10**, 91–96 (1956)
186. H.C. Yee, K. Sweby, P, D.F. Griffiths, Dynamical approach study of spurious steady-state numerical solutions of nonlinear differential equations. 1. The dynamics of time discretization and its implications for algorithm development in computational fluid dynamics. J. Comput. Phys. **97**, 249–310 (1991)
187. K. Yosida, *Functional Analysis* (Springer, Berlin, 1974)
188. J. Yu, Symmetric gaussian quadrature formulae for tetrahedronal regions. Comput. Methods Appl. Mech. Eng. **43**, 349–353 (1984)
189. O.C. Zienkiewicz, *The Finite Element Method in Engineering Science* (McGraw-Hill, New York, 1971)
190. A. Zygmund, *Trigonometric Series* (Cambridge University Press, Cambridge, 1968)

Notation Index

$|\boldsymbol{\alpha}|$ multi-index modulus, 30
$\mathbf{f}^*$ adjoint of discrete time signal: $\mathbf{f}_n^* = \overline{\mathbf{f}_{-n}}$, 152

$\mathbf{b}$ barycentric coordinates, 34
$b_{j,n}(x)$ Bernstein polynomial, 29
$\binom{\alpha}{\beta}$ binomial coefficient for multi-indices, 30
B_k cardinal B-spline of degree $k-1$, 77

A^c complement of set A, 253
$f * g$ convolution of functions f and g, 77
$\mathbf{y} * \mathbf{x}$ convolution of two discrete time signals, 143

δ_{ij} Kronecker delta, 139
$\mathbf{D}$ vector of partial differentiation operators, 43
$Df(x)$ derivative of f at x, 225
$\frac{df}{dx}(x)$ derivative of f at x, 225
$\mathbf{D}f(\mathbf{x})$ derivative of f at $\mathbf{x}$, 226
$f'(x)$ derivative of f at x, 225
D_α dilation operator, 181
$\mathcal{X} \oplus \mathcal{Y}$ direct sum of subspaces, 219
$f[x_0, \ldots, x_\ell]$ divided difference, 11
$\downarrow_\sigma$ downsampling operator, 141

$\mathbf{e}_i$ ith axis vector, 31
$E_\Omega(f)$ expected value of random variable f on measurable set Ω, 255

$\lfloor x \rfloor$ floor : greatest integer less than or equal, 274

$\mathscr{F}\{f\}$ continuous Fourier transform, 179
$\mathscr{F}^d\{\mathbf{x}\}$ discrete Fourier transform of time signal, 146
$\mathscr{F}\{f\}$ Fourier transform, 81
$\mathbf{F}(t)$ fundamental matrix, 341

$\mathbf{G}(t, s)$ Green's function, 498

$\mathbf{Z}^H$ Hermitian = conjugate transpose of matrix, 151
h partition width, 252

$\mathbf{I}$ identity matrix, 34
$\mathbf{x} \cdot \mathbf{y}$ inner product, 92
$\mathbf{A}^{-1}$ inverse of matrix, 58

$\mathbf{J}$ Jacobian matrix, 329

$\lambda_{j,n}$ Lagrange interpolation basis polynomial, 19
$\lambda_{\boldsymbol{\alpha},n}$ Lagrange polynomial in multiple dimensions, 36
$\mathscr{L}_{d,n}$ set of lattice points, 35
$\boldsymbol{\beta} \le \boldsymbol{\alpha}$ less than or equal to inequality on multi-indices, 30
$\mathscr{L}\{f\}$ linear operator on function, 242
$\mathscr{L}_n\{f; h\}$ approximate linear operator on f of order n using mesh width h, 242

$\mathscr{M}_n$ mesh with n elements, 59

© Springer International Publishing AG, part of Springer Nature 2017
J.A. Trangenstein, *Scientific Computing*, Texts in Computational
Science and Engineering 20, https://doi.org/10.1007/978-3-319-69110-7

Author Index

© Springer International Publishing AG, part of Springer Nature 2017
J.A. Trangenstein, *Scientific Computing*, Texts in Computational
Science and Engineering 20, https://doi.org/10.1007/978-3-319-69110-7

571

Subject Index

Editorial Policy

1. Textbooks on topics in the field of computational science and engineering will be considered. They should be written for courses in CSE education. Both graduate and undergraduate textbooks will be published in TCSE. Multidisciplinary topics and multidisciplinary teams of authors are especially welcome.

2. Format: Only works in English will be considered. For evaluation purposes, manuscripts may be submitted in print or electronic form, in the latter case, preferably as pdf- or zipped ps-files. Authors are requested to use the LaTeX style files available from Springer at: http://www.springer.com/authors/book+authors/helpdesk?SGWID=0-1723113-12-971304-0 (Click on ⟶ Templates ⟶ LaTeX ⟶ monographs)
Electronic material can be included if appropriate. Please contact the publisher.

3. Those considering a book which might be suitable for the series are strongly advised to contact the publisher or the series editors at an early stage.

General Remarks

Careful preparation of manuscripts will help keep production time short and ensure a satisfactory appearance of the finished book.

The following terms and conditions hold:

Regarding free copies and royalties, the standard terms for Springer mathematics textbooks hold. Please write to martin.peters@springer.com for details.

Authors are entitled to purchase further copies of their book and other Springer books for their personal use, at a discount of 33.3% directly from Springer-Verlag.

Series Editors

Timothy J. Barth
NASA Ames Research Center
NAS Division
Moffett Field, CA 94035, USA
barth@nas.nasa.gov

Michael Griebel
Institut für Numerische Simulation
der Universität Bonn
Wegelerstr. 6
53115 Bonn, Germany
griebel@ins.uni-bonn.de

David E. Keyes
Mathematical and Computer Sciences
and Engineering
King Abdullah University of Science
and Technology
P.O. Box 55455
Jeddah 21534, Saudi Arabia
david.keyes@kaust.edu.sa

and

Department of Applied Physics
and Applied Mathematics
Columbia University
500 W. 120 th Street
New York, NY 10027, USA
kd2112@columbia.edu

Risto M. Nieminen
Department of Applied Physics
Aalto University School of Science
and Technology
00076 Aalto, Finland
risto.nieminen@tkk.fi

Dirk Roose
Department of Computer Science
Katholieke Universiteit Leuven
Celestijnenlaan 200A
3001 Leuven-Heverlee, Belgium
dirk.roose@cs.kuleuven.be

Tamar Schlick
Department of Chemistry
and Courant Institute
of Mathematical Sciences
New York University
251 Mercer Street
New York, NY 10012, USA
schlick@nyu.edu

Editor for Computational Science
and Engineering at Springer:
Martin Peters
Springer-Verlag
Mathematics Editorial IV
Tiergartenstrasse 17
69121 Heidelberg, Germany
martin.peters@springer.com

Texts in Computational Science and Engineering

For further information on these books please have a look at our mathematics catalogue at the following URL: www.springer.com/series/5151

Monographs in Computational Science and Engineering

For further information on this book, please have a look at our mathematics catalogue at the following URL: www.springer.com/series/7417

Lecture Notes in Computational Science and Engineering

24. T. Schlick, H.H. Gan (eds.), *Computational Methods for Macromolecules: Challenges and Applications.*

25. T.J. Barth, H. Deconinck (eds.), *Error Estimation and Adaptive Discretization Methods in Computational Fluid Dynamics.*

26. M. Griebel, M.A. Schweitzer (eds.), *Meshfree Methods for Partial Differential Equations.*

27. S. Müller, *Adaptive Multiscale Schemes for Conservation Laws.*

28. C. Carstensen, S. Funken, W. Hackbusch, R.H.W. Hoppe, P. Monk (eds.), *Computational Electromagnetics.*

29. M.A. Schweitzer, *A Parallel Multilevel Partition of Unity Method for Elliptic Partial Differential Equations.*

30. T. Biegler, O. Ghattas, M. Heinkenschloss, B. van Bloemen Waanders (eds.), *Large-Scale PDE-Constrained Optimization.*

31. M. Ainsworth, P. Davies, D. Duncan, P. Martin, B. Rynne (eds.), *Topics in Computational Wave Propagation.* Direct and Inverse Problems.

32. H. Emmerich, B. Nestler, M. Schreckenberg (eds.), *Interface and Transport Dynamics.* Computational Modelling.

33. H.P. Langtangen, A. Tveito (eds.), *Advanced Topics in Computational Partial Differential Equations.* Numerical Methods and Diffpack Programming.

34. V. John, *Large Eddy Simulation of Turbulent Incompressible Flows.* Analytical and Numerical Results for a Class of LES Models.

35. E. Bänsch (ed.), *Challenges in Scientific Computing - CISC 2002.*

36. B.N. Khoromskij, G. Wittum, *Numerical Solution of Elliptic Differential Equations by Reduction to the Interface.*

37. A. Iske, *Multiresolution Methods in Scattered Data Modelling.*

38. S.-I. Niculescu, K. Gu (eds.), *Advances in Time-Delay Systems.*

39. S. Attinger, P. Koumoutsakos (eds.), *Multiscale Modelling and Simulation.*

40. R. Kornhuber, R. Hoppe, J. Périaux, O. Pironneau, O. Wildlund, J. Xu (eds.), *Domain Decomposition Methods in Science and Engineering.*

41. T. Plewa, T. Linde, V.G. Weirs (eds.), *Adaptive Mesh Refinement – Theory and Applications.*

42. A. Schmidt, K.G. Siebert, *Design of Adaptive Finite Element Software.* The Finite Element Toolbox ALBERTA.

43. M. Griebel, M.A. Schweitzer (eds.), *Meshfree Methods for Partial Differential Equations II.*

44. B. Engquist, P. Lötstedt, O. Runborg (eds.), *Multiscale Methods in Science and Engineering.*

45. P. Benner, V. Mehrmann, D.C. Sorensen (eds.), *Dimension Reduction of Large-Scale Systems.*

46. D. Kressner, *Numerical Methods for General and Structured Eigenvalue Problems.*

47. A. Boriçi, A. Frommer, B. Joó, A. Kennedy, B. Pendleton (eds.), *QCD and Numerical Analysis III.*

48. F. Graziani (ed.), *Computational Methods in Transport.*

49. B. Leimkuhler, C. Chipot, R. Elber, A. Laaksonen, A. Mark, T. Schlick, C. Schütte, R. Skeel (eds.), *New Algorithms for Macromolecular Simulation.*

50. M. Bücker, G. Corliss, P. Hovland, U. Naumann, B. Norris (eds.), *Automatic Differentiation: Applications, Theory, and Implementations.*

51. A.M. Bruaset, A. Tveito (eds.), *Numerical Solution of Partial Differential Equations on Parallel Computers.*

52. K.H. Hoffmann, A. Meyer (eds.), *Parallel Algorithms and Cluster Computing.*

53. H.-J. Bungartz, M. Schäfer (eds.), *Fluid-Structure Interaction.*

54. J. Behrens, *Adaptive Atmospheric Modeling.*

55. O. Widlund, D. Keyes (eds.), *Domain Decomposition Methods in Science and Engineering XVI.*

56. S. Kassinos, C. Langer, G. Iaccarino, P. Moin (eds.), *Complex Effects in Large Eddy Simulations.*

57. M. Griebel, M.A Schweitzer (eds.), *Meshfree Methods for Partial Differential Equations III.*

58. A.N. Gorban, B. Kégl, D.C. Wunsch, A. Zinovyev (eds.), *Principal Manifolds for Data Visualization and Dimension Reduction.*

59. H. Ammari (ed.), *Modeling and Computations in Electromagnetics: A Volume Dedicated to Jean-Claude Nédélec.*

60. U. Langer, M. Discacciati, D. Keyes, O. Widlund, W. Zulehner (eds.), *Domain Decomposition Methods in Science and Engineering XVII.*

61. T. Mathew, *Domain Decomposition Methods for the Numerical Solution of Partial Differential Equations.*

62. F. Graziani (ed.), *Computational Methods in Transport: Verification and Validation.*

63. M. Bebendorf, *Hierarchical Matrices.* A Means to Efficiently Solve Elliptic Boundary Value Problems.

64. C.H. Bischof, H.M. Bücker, P. Hovland, U. Naumann, J. Utke (eds.), *Advances in Automatic Differentiation.*

65. M. Griebel, M.A. Schweitzer (eds.), *Meshfree Methods for Partial Differential Equations IV.*

66. B. Engquist, P. Lötstedt, O. Runborg (eds.), *Multiscale Modeling and Simulation in Science.*

67. I.H. Tuncer, Ü. Gülcat, D.R. Emerson, K. Matsuno (eds.), *Parallel Computational Fluid Dynamics 2007.*

68. S. Yip, T. Diaz de la Rubia (eds.), *Scientific Modeling and Simulations.*

69. A. Hegarty, N. Kopteva, E. O'Riordan, M. Stynes (eds.), *BAIL 2008 – Boundary and Interior Layers.*

70. M. Bercovier, M.J. Gander, R. Kornhuber, O. Widlund (eds.), *Domain Decomposition Methods in Science and Engineering XVIII.*

71. B. Koren, C. Vuik (eds.), *Advanced Computational Methods in Science and Engineering.*

72. M. Peters (ed.), *Computational Fluid Dynamics for Sport Simulation.*

73. H.-J. Bungartz, M. Mehl, M. Schäfer (eds.), *Fluid Structure Interaction II - Modelling, Simulation, Optimization.*

74. D. Tromeur-Dervout, G. Brenner, D.R. Emerson, J. Erhel (eds.), *Parallel Computational Fluid Dynamics 2008.*

75. A.N. Gorban, D. Roose (eds.), *Coping with Complexity: Model Reduction and Data Analysis.*

76. J.S. Hesthaven, E.M. Rønquist (eds.), *Spectral and High Order Methods for Partial Differential Equations.*

77. M. Holtz, *Sparse Grid Quadrature in High Dimensions with Applications in Finance and Insurance.*

78. Y. Huang, R. Kornhuber, O.Widlund, J. Xu (eds.), *Domain Decomposition Methods in Science and Engineering XIX.*

79. M. Griebel, M.A. Schweitzer (eds.), *Meshfree Methods for Partial Differential Equations V.*

80. P.H. Lauritzen, C. Jablonowski, M.A. Taylor, R.D. Nair (eds.), *Numerical Techniques for Global Atmospheric Models.*

81. C. Clavero, J.L. Gracia, F.J. Lisbona (eds.), *BAIL 2010 – Boundary and Interior Layers, Computational and Asymptotic Methods.*

82. B. Engquist, O. Runborg, Y.R. Tsai (eds.), *Numerical Analysis and Multiscale Computations.*

83. I.G. Graham, T.Y. Hou, O. Lakkis, R. Scheichl (eds.), *Numerical Analysis of Multiscale Problems.*

84. A. Logg, K.-A. Mardal, G. Wells (eds.), *Automated Solution of Differential Equations by the Finite Element Method.*

85. J. Blowey, M. Jensen (eds.), *Frontiers in Numerical Analysis - Durham 2010.*

86. O. Kolditz, U.-J. Gorke, H. Shao, W. Wang (eds.), *Thermo-Hydro-Mechanical-Chemical Processes in Fractured Porous Media - Benchmarks and Examples.*

87. S. Forth, P. Hovland, E. Phipps, J. Utke, A. Walther (eds.), *Recent Advances in Algorithmic Differentiation.*

88. J. Garcke, M. Griebel (eds.), *Sparse Grids and Applications.*

89. M. Griebel, M.A. Schweitzer (eds.), *Meshfree Methods for Partial Differential Equations VI.*

90. C. Pechstein, *Finite and Boundary Element Tearing and Interconnecting Solvers for Multiscale Problems.*

91. R. Bank, M. Holst, O. Widlund, J. Xu (eds.), *Domain Decomposition Methods in Science and Engineering XX.*

92. H. Bijl, D. Lucor, S. Mishra, C. Schwab (eds.), *Uncertainty Quantification in Computational Fluid Dynamics.*

93. M. Bader, H.-J. Bungartz, T. Weinzierl (eds.), *Advanced Computing.*

94. M. Ehrhardt, T. Koprucki (eds.), *Advanced Mathematical Models and Numerical Techniques for Multi-Band Effective Mass Approximations.*

95. M. Azaïez, H. El Fekih, J.S. Hesthaven (eds.), *Spectral and High Order Methods for Partial Differential Equations ICOSAHOM 2012.*

96. F. Graziani, M.P. Desjarlais, R. Redmer, S.B. Trickey (eds.), *Frontiers and Challenges in Warm Dense Matter.*

97. J. Garcke, D. Pflüger (eds.), *Sparse Grids and Applications – Munich 2012.*

98. J. Erhel, M. Gander, L. Halpern, G. Pichot, T. Sassi, O. Widlund (eds.), *Domain Decomposition Methods in Science and Engineering XXI.*

99. R. Abgrall, H. Beaugendre, P.M. Congedo, C. Dobrzynski, V. Perrier, M. Ricchiuto (eds.), *High Order Nonlinear Numerical Methods for Evolutionary PDEs - HONOM 2013.*

100. M. Griebel, M.A. Schweitzer (eds.), *Meshfree Methods for Partial Differential Equations VII.*

101. R. Hoppe (ed.), *Optimization with PDE Constraints - OPTPDE 2014.*

102. S. Dahlke, W. Dahmen, M. Griebel, W. Hackbusch, K. Ritter, R. Schneider, C. Schwab, H. Yserentant (eds.), *Extraction of Quantifiable Information from Complex Systems.*

103. A. Abdulle, S. Deparis, D. Kressner, F. Nobile, M. Picasso (eds.), *Numerical Mathematics and Advanced Applications - ENUMATH 2013.*

104. T. Dickopf, M.J. Gander, L. Halpern, R. Krause, L.F. Pavarino (eds.), *Domain Decomposition Methods in Science and Engineering XXII.*

105. M. Mehl, M. Bischoff, M. Schäfer (eds.), *Recent Trends in Computational Engineering - CE2014.* Optimization, Uncertainty, Parallel Algorithms, Coupled and Complex Problems.

106. R.M. Kirby, M. Berzins, J.S. Hesthaven (eds.), *Spectral and High Order Methods for Partial Differential Equations - ICOSAHOM'14.*

107. B. Jüttler, B. Simeon (eds.), *Isogeometric Analysis and Applications 2014.*

108. P. Knobloch (ed.), *Boundary and Interior Layers, Computational and Asymptotic Methods – BAIL 2014.*

109. J. Garcke, D. Pflüger (eds.), *Sparse Grids and Applications – Stuttgart 2014.*

110. H. P. Langtangen, *Finite Difference Computing with Exponential Decay Models.*

111. A. Tveito, G.T. Lines, *Computing Characterizations of Drugs for Ion Channels and Receptors Using Markov Models.*

112. B. Karazösen, M. Manguoğlu, M. Tezer-Sezgin, S. Göktepe, Ö. Uğur (eds.), *Numerical Mathematics and Advanced Applications - ENUMATH 2015.*

113. H.-J. Bungartz, P. Neumann, W.E. Nagel (eds.), *Software for Exascale Computing - SPPEXA 2013-2015.*

114. G.R. Barrenechea, F. Brezzi, A. Cangiani, E.H. Georgoulis (eds.), *Building Bridges: Connections and Challenges in Modern Approaches to Numerical Partial Differential Equations.*

115. M. Griebel, M.A. Schweitzer (eds.), *Meshfree Methods for Partial Differential Equations VIII.*

116. C.-O. Lee, X.-C. Cai, D.E. Keyes, H.H. Kim, A. Klawonn, E.-J. Park, O.B. Widlund (eds.), *Domain Decomposition Methods in Science and Engineering XXIII.*

117. T. Sakurai, S. Zhang, T. Imamura, Y. Yusaku, K. Yoshinobu, H. Takeo (eds.), *Eigenvalue Problems: Algorithms, Software and Applications, in Petascale Computing.* EPASA 2015, Tsukuba, Japan, September 2015.

118. T. Richter (ed.), *Fluid-structure Interactions.* Models, Analysis and Finite Elements.

119. M.L. Bittencourt, N.A. Dumont, J.S. Hesthaven (eds.), *Spectral and High Order Methods for Partial Differential Equations ICOSAHOM 2016.*

120. Z. Huang, M. Stynes, Z. Zhang (eds.), *Boundary and Interior Layers, Computational and Asymptotic Methods BAIL 2016.*

121. S.P.A. Bordas, E.N. Burman, M.G. Larson, M.A. Olshanskii (eds.), *Geometrically Unfitted Finite Element Methods and Applications.* Proceedings of the UCL Workshop 2016.

122. A. Gerisch, R. Penta, J. Lang (eds.), *Multiscale Models in Mechano and Tumor Biology.* Modeling, Homogenization, and Applications.

For further information on these books please have a look at our mathematics catalogue at the following URL: www.springer.com/series/3527

Printed in the United States
By Bookmasters